S0-AHG-413

ANNUAL REPORTS IN MEDICINAL CHEMISTRY
Volume 27

Academic Press Rapid Manuscript Reproduction

ANNUAL REPORTS IN MEDICINAL CHEMISTRY
Volume 27

Sponsored by the Division of Medicinal Chemistry of the American Chemical Society

EDITOR-IN-CHIEF:

JAMES A. BRISTOL

PARKE-DAVIS PHARMACEUTICAL RESEARCH DIVISION
WARNER-LAMBERT COMPANY
ANN ARBOR, MICHIGAN

SECTION EDITORS

*JOHN M. McCALL • WILLIAM F. MICHNE • JACOB J. PLATTNER
DAVID W. ROBERTSON • KENNETH B. SEAMON • MICHAEL C. VENUTI*

EDITORIAL ASSISTANT

LISA GREGORY

ACADEMIC PRESS, INC.
Harcourt Brace Jovanovich, Publishers
San Diego New York Boston London Sydney Tokyo Toronto

CONTENTS

I. CNS AGENTS

Section Editor: John M. McCall, The Upjohn Company, Kalamazoo, Michigan

II. CARDIOVASCULAR AND PULMONARY AGENTS

Section Editor: David W. Robertson, Ligand Pharmaceuticals, San Diego, California

VII. TRENDS AND PERSPECTIVES

Section Editor: James A. Bristol, Parke-Davis Pharmaceutical Research Division, Warner-Lambert Co., Ann Arbor, Michigan

CONTRIBUTORS

Barrett, John F.	149	Klaubert, Dieter H.	149
Benjamin, David C.	189	Koff, Wayne C.	255
Blondelle, Sylvie E.	159	Mahon, Kathleen	227
Cimini, Madeline G.	89	Masamune, Hiroko	209
Colca, Jerry R.	219	McCall, Robert B.	21
Cooper, Kelvin	209	McCall, John M.	31
Dobrusin, Ellen M.	169	Miller, Paul F.	119
Doherty, Annette M.	79	Miller, Joseph D.	11
Domagala, John M.	119	Milligan, John F.	311
Ellis, Daniel D.	321	Mirzadegan, Taraneh	291
Erickson, John W.	271	Mitchell, Mark A.	139
Fülle, Hans-Jürgen	245	Musser, John H.	301
Fesik, Stephen W.	271	Panetta, Jill A.	31
Fry, David W.	169	Perdue, Samuel S.	189
Garbers, David L.	245	Romero, Arthur G.	21
Gibson, J. Kenneth	89	Rosenberg, Steven	41
Glass, Marta J.	255	Sall, Daniel J.	99
Greenlee, William J.	59	Shea, Regan G.	311
Hagmann, William K.	199	Siegl, Peter K.S.	59
Heavner, George A.	179	Sindelar, Robert D.	199
Heffner, Thomas G.	49	Smith, Alan E.	235
Houghten, Richard A.	159	Smith, Gerald F.	99
Humblet, Christine	291	Strupczewski, June D.	321
Hutchinson, C. Richard	129	Suto, Mark J.	119
Jacobs, Robert T.	109	Tanis, Steven P.	219
Jakubowski, Joseph A.	99	Veale, Chris A.	109
Jamrich, Milan	227	Wilks, John W.	139
Joseph, Suresh K.	261	Williams, Michael	1
Katz, Leonard	129	Wise, Lawrence D.	49
Kerwin, Jr., James F.	69	Wolanin, Donald J.	109

PREFACE

Annual Reports in Medicinal Chemistry continues with its objective to provide timely updates of important areas of medicinal chemistry together with an emphasis on emerging biological science expected to provide the foundation for future therapeutic interventions.

Volume 27 retains the familiar format of previous volumes, this year with 33 chapters. Sections I - IV are related to specific medicinal agents, with annual updates on antipsychotics, antiallergy agents, vasoactive peptides and antibacterials, where the objective is to provide the reader with the most important new results in a particular field. Also included are chapters on topics accorded frequent updates, including sleep, ischemia/reperfusion, trophic factors, antithrombotic agents, angiogenesis inhibitors, antifungals, monoclonal antibodies, and diabetes. Biological mechanisms continue to be the theme of several chapters, including serotonergics, angiotensin/renin, EDRF, potassium channels, cytokines, and complement cascade. In recognition of the impact of emerging new areas of neuroscience, the initial chapter in Section I is entitled the Decade of the Brain.

Sections V and VI continue to emphasize important topics in medicinal chemistry, biology, and drug design and well as the increasingly important interfaces among these disciplines. In Section V, Topics in Biology, are included chapters on homeobox genes, guanylyl cyclases, cystic fibrosis, vaccines for AIDS, and inositol triphosphate receptors. Traditionally, areas of research reviewed in this section will appear in a chapter related to medicinal agents in a future volume, once sufficient time has passed to allow new compounds to be developed from a biological strategy. Chapters in Section VI, Topics in Drug Design and Discovery, reflect the current focus on mechanism-directed drug discovery and newer technologies These include a comprehensive treatment of macromolecular X-ray crystallography and NMR as tools for structure based drug design. Also included in this section are chapters on G-protein coupled receptors, the emerging area of glycobiology, and sequence-specific DNA binding and regulation of eucaryotic gene transcription.

Volume 27 concludes with a chapter on NCE introductions worldwide in 1991. In addition to the chapter reviews, a comprehensive set of indices has been included to enable the reader to easily locate topics in volumes 1 - 27 of this series.

Over the past year, it has been my pleasure to work with 6 highly professional section editors and 60 authors, whose critical contributions comprise this volume.

James A. Bristol
Ann Arbor, Michigan
May, 1992

SECTION I. CNS AGENTS

Editor: John M. McCall, The Upjohn Company
Kalamazoo, MI, 49001

Chapter 1. The Decade of the Brain

Michael Williams
Neuroscience Research, D - 464, Abbott Laboratories, Abbott Park, IL 60064 - 3500

Introduction - Advances in health care have dramatically increased life expectancy. Consequently, the incidence of many diseases associated with aging, including autoimmune, cardiac and metabolic diseases, as well as those directly affecting the peripheral and central nervous systems, has increased. Diseases of the elderly are associated with anxiety and depression. Improved agents for the treatment of these disorders and the various dementias as well as therapeutic agents to improve, limit or reverse the neurodegenerative process represent major challenges in CNS-related drug research. Indeed one in four people in the U.S. may anticipate progressive impairment of cognitive function and the development of Alzheimers disease (AD) (1).

The need for innovative new drugs to treat CNS disorders prompted the U.S. Congress to declare the 1990s as the 'Decade of the Brain', a major initiative (2-4) to facilitate the discovery and development of new medications not only to treat diseases of aging but also to develop new drugs for the treatment of anxiety, schizophrenia, epilepsy, pain, stroke, depression and substance abuse.

Concerns related to escalating health care costs (5-7) have led to initiatives for drug price controls despite the impact that vaccines, antibiotics, antihypertensive and psychotropic drugs have had in increasing the quality and length of life. This could jeopardize the R & D investment necessary for the effective implementation of the 'Decade' initiative. The increased need for quality health care for the aged may be seen to add to these financial problems. Health care cost issues extend beyond the pharmaceutical industry, encompassing ethical issues involving quality of life in the elderly (8,9), and require a more realistic prioritization of health care resources. Ethical pharmaceuticals may therefore be viewed as a vital part of any realistic long term solution to containing health care costs rather than their cause. The magnitude of the "Decade" challenge to both pharmaceutical reseachers and society is underscored by data from the U.S. Office of Technology Assessment indicating an annual cost in excess of $305 billion for brain related disorders with the hospitalization of more people than for any other group of diseases (10). In 1990, the costs to society for AD alone were $55 billion (11).

The Drug Discovery Process in the 1990s - The average cost of introducing a new drug is $231MM (12), reflecting increases in: the risk of addressing novel mechanistic targets, the costs of 'cutting edge' science, regulatory requirements, and the 10 -12 years of development time to market. Accordingly, there is a high premium on how R & D resources are allocated and how the process can be managed effectively without impacting creativity (13). Although major pharmaceutical companies continue to be the primary source of new drugs, venture capital interest in the industry has spread the resource base for innovative research. Many 'start up' companies represent an alternate source of 'cutting edge', high risk, biomedical technologies. Although many such ventures were founded with an exclusive focus on biotechnology, an increasing awareness of the limitations of peptides and biologics as drugs has led to a focus on the importance of 'small molecules'. The existence of these preclinical research boutiques and concurrent developments in enabling technologies and technology transfer in academia and government has irrevocably changed research strategies within major drug companies and will, for the future, lead to uncertainties in the growth and stability of 'in house' research organizations. Rather than retraining existing staff, companies can invest in 'turn key' research efforts in new technological/therapeutic areas, flexibly enhancing the innovative and competitive aspects of their discovery efforts (14).

The Congressional Initiative - The National Institutes of Neurological Disorders and Stroke (NINDS) has categorized brain diseases into four major classes (10): - a) those involving the *developing brain* that include inherited diseases like epilepsy and autism that may be amenable to treatment with gene therapy; - b) diseases involving the *injured brain* that include disorders related to trauma - head and spinal cord injury and stroke; - c) diseases involving the *failing brain* including the various dementias, multiple sclerosis, diabetic neuropathies, amyloid lateral sclerosis (ALS) and various nerve and muscle disorders that may be genetic, viral, environmental and/or autoimmune in origin and ; - d) diseases involving the *feeling brain*. This category deals with the senses and includes various mood disorders, chronic pain and substance abuse, the latter related to the anhedonic personality (15).

Drug Discovery in the Neurosciences - CNS drug discovery may be considered in three phases. The first, occuring from the early 1950s through 1975, may be termed the age of serendipity. It involved the discovery, in the clinic, of compounds originally targeted at other disease states that had unexpected CNS actions. Iproniazid, chlorpromazine and imipramine were products of this era as was L-dopa, the first treatment for Parkinson's disease based on a rational approach (16). The second phase began in the early 1970s, targeting second generation compounds with improved efficacy and reduced side effect liability as compared to those agents discovered in the first phase. The evolving mechanistic approach was further defined at this time. The clinical success of agents targeted for a defined receptor or enzyme would thus serve to validate the hypothesis related to the mechanism of action of a given class of compounds, aiding both the researcher and the clinician in the understanding of disease etiology. This approach has been extensively validated in the 5HT area, with the identification of functionally distinct receptor subtypes (17) and ligands with proven clinical use. Radioligand binding assays were developed during this phase, providing a faster, more resource efficient means to assess the receptor-ligand interaction (18). The discovery of opiate (19) and benzodiazepine (20) receptors using this technology provided evidence for 'drug' receptors and putative endogenous ligands. The application of binding assays in the form of targeted screening, the evaluation at selected molecular targets of large compound libraries and natural product sources (21) has permitted the rapid and cost effective identification of novel ligands. Antagonists for the CCK-B (MK 329; 22) and Substance P receptors (CP 96,345 (23) and RP 67580 (24)) resulted from this approach, significantly impacting conceptual approaches to peptide receptor research. Despite the increased emphasis on molecular pharmacology, drugs discovered in this period were, to a large extent, incremental in nature or, like buspirone (25), dependent on serendipitous evaluation in the clinic. The antipsychotic, clozapine (26) and the antidepressant, fluoxetine (27), are major therapeutic agents from this era.

Ongoing advances in molecular biology and molecular pharmacology have resulted in the identification of new drug targets, receptor subtypes and enzyme isoforms. These permit elucidation of receptor/enzyme function with the development of ligand/substrate SAR and compound selectivity. Such advances, fueling the third phase of CNS drug discovery, also have the potential to facilitate disease diagnosis and the understanding of disease etiology.

The emerging computer-based structural technologies (computer assisted molecular design-CAMD [28]; NMR [29]) permit the study of ligand-protein interactions in vitro provided that knowledge of the three dimensional structure of the target (receptor or enzyme), ligand or, preferably, both, is available. Unfortunately, many of CNS drug targets have yet to be crystallized, requiring approximations for structural analysis. Beyond the receptor-ligand interaction, those ligand properties that determine efficacy remain elusive and are further complicated by an ongoing revision of receptor theory involving receptor complexes, allosteric modulation (30,31), receptor-induced changes in ligand conformation (32), orphan receptors (33) and chaos theory (34).

CNS Drug Research Productivity-Cultural Influences - Despite an explosion of knowledge in the neurosciences, considerable investment from the pharmaceutical industry and over 1200 products in the worldwide R & D pipeline in 1990 (35), new drug introduction in the CNS area has not matched that seen in other therapeutic areas (36). In the past decade "fundamental advances...in the discovery of psychotropic drugs....have been remarkably limited" (37). Similarly, it has been noted that the CNS area "is one... where there has not been any recent significant ...progress towards curative therapies" (38). In contrast, major advances in antihypertensive therapy have occurred during the same period. The empirical nature of the CNS drug discovery process and the inherent complexity of clinical trials have led to a conservative attitude within the industry. This is compounded by the characterization of most CNS disorders as affecting the quality of life rather than being 'life threatening". Antibacterial and antihypertensive drugs can be sufficiently well characterized preclinically that human efficacy is all but guaranteed. Clinical trials for such agents result in positive results in a matter of weeks. For a putative antidepressant, the preclinical workup is often limited to in vitro mechanistic efficacy, CNS bioavailability and metabolism as the animal models are far from predictive ; clinical efficacy often cannot be determined for 2- 3 years and is obscured by placebo effects, refractory periods and side effects.

Resourcing of CNS research has consequently been inconsistent from a strategic viewpoint. In 1976, Squibb strategically reallocated research resources, eliminating its CNS effort. Smith, Kline and French followed suit in 1981 and, in 1990, Monsanto/Searle, in a continuing series of strategic realignments, significantly downsized CNS research. On the international front, Merck (1983; U.K.), Hoffmann - La Roche (1985; Basle), ICI (1984; U.S.A.), CIBA - Geigy (1988; Basle), Glaxo (1989; U.K.), Hoechst (1990; U.S.A.) and Rhone Poulenc Rorer (France; 1990) consolidated their CNS research efforts into single "centers of excellence". Even though the U.S. is one of the major markets for CNS drugs and a major innovator in molecular neuroscience, with the exception of ICI and Hoechst, the move in consolidation has been towards the EEC, leading to the potential conclusion that the cultural climate in the EEC may be more conducive to successful CNS drug research. Supportive of this are figures related to new market introductions. In the eight years that Annual Reports in Medicinal Chemistry has documented new drugs brought to the market (39), of the 368 compounds approved for sale in the period 1983 - 1990, some 17%, 64, were targeted for CNS indications. Of these, 8 originated in the USA with Japan producing 16 compounds and France, 12. Europe (France, Germany, Italy, the UK, Netherlands, Scandinavia) and Switzerland were thus responsible for the introduction of 40 new CNS drugs, 63% of the total marketed, providing evidence for the hypothesis that CNS-targeted drug research is more productive within the European system. The factors contributing to this apparent 'cultural' bias are not immediately obvious. Are neuroscientists in different cultures trained in a radically different manner? Is the longer term business perspective in Europe and Japan (free, to a major extent of the quarterly pressures of Wall Street [40]) a factor? Such questions are controversial, with limited hard data available for definitive conclusions.

Additional factors impacting CNS drug discovery involve: - a) a lack of knowledge related to brain function. That "neuroscience stands ... today [1991] where atomic physics was in 1919..or .. molecular biology ... in 1944" (41), suggests that it will be 40 - 50 years before CNS therapies will have a more cohesive knowledge base; - b) a lack of knowledge regarding CNS disease etiology. The descriptive nature of current psychiatric diagnosis has been compared to that existing for general medicine in the late 19th century (37); and c) an overly reductionistic, molecular approach that ignores the four principle levels of research - molecular, cellular, systems and behavior - and the necessity to objectively integrate and interpret data obtained at the various levels (42).

Neuroscience Boutiques - Venture capital financed, neuroscience based, research boutiques typically focus on high risk, often long term, albeit attractive, approaches to the major areas of unmet medical need (43). Often reflecting a single project, resourced to critical mass, with exclusive access to cutting edge technologies in the academic arena, these companies represent essential sources of enabling technology, a fact reflected by affiliations with major pharmaceutical companies (Athena/Lilly; Cephalon/Schering, etc.). In 1991, 23 neuroscience based research boutiques had agreements with 22 major pharmaceutical companies, with some companies having 3 or more separate agreements. While the outlook for the boutiques, given the roller coaster dynamics of the U.S. biotechnology industry, varies between uncertain (44) and highly optimistic (45), their approach to CNS research and the unique networks that they have established through the academic and business communities has significantly changed the nature of technology acquisition and commercialization.

Enabling technologies - In approaching the third phase of CNS drug discovery, it is imperative that emerging technologies synergize with those already existing rather than replace them. Receptor superfamily cloning and expression, and advances in structural biology and chemistry and in the development of transgenic animal models of CNS diseases, while potentially playing a major role in research strategies, have yet to be integrated into the drug discovery process. Muscarinic receptor clones (46) had little impact on the discovery of L 670,207, the reference 'super agonist' in this area (47). Similarly the identification of MK 329, MK 801, CP 96,345, RP 67580 and CI-988, occurred independently of either cloned targets or knowledge of receptor structure. Cloned and expressed human receptors and enzymes can, however, provide the chemist and biologist with relatively simple systems in which ligand SAR can be derived. In and of itself, molecular biology cannot define function since selective, efficacious and bioavailable ligands remain the driving force for drug discovery (48). Transgenic models of diseases have high potential. Of those reported for AD, involving overexpression of amyloid precursor protein, none has proven to replicate the human disease pathology in a reliable manner. Gene expression in transfected animals may also vary with the construct and the transfection site.

The use of human recombinant proteins as in vitro targets for drug design can be complicated if rodent and primate models of the disease are subsequently used. The pharmacology of the human recombinant protein must have similar properties to the target present in the brains of lower species. This is an issue that requires resolution at the earliest opportunity. Functional coupling of transfected receptors may be different from that observed in the "native" situation. Partial agonists may appear as full agonists in a transfected system while antagonists may be partial agonists. Alternative mRNA processing is an additional complication to the cloning/ expression approach (49).

Research in the area of animal behavior as related to human CNS disease states (50), both in terms of determining efficacy and side effect potential, has undergone a significant de-emphasis in the past decade as academia, government and industry have focused considerable resources on molecular aspects of neuroscience. Urgently needed predictive models of CNS disorders have not been developed so that clinical acceptance of new compounds developed on a mechanistic basis relies solely on preclinical biochemical and pharmacokinetic data. This is problematic unless the disease state has clearly defined endpoints that are directly related to compound efficacy and are amenable to ongoing measurement (37). The paucity of plasma markers for CNS-related disorders limits this type of approach. In contrast, the HMG CoA reductase inhibitors that prevent cholesterol formation had clinical endpoints based on pharmacoepidemiological studies linking elevated cholesterol levels to atherosclerosis and other vascular disorders. The determination of efficacy for the introduction of HMG CoA reductase inhibitors was thus a decrease in plasma cholesterol levels.

Biochemical parameters of CNS drug action are limited and frequently controversial. Newer animal models that extrapolate from existing empirical models predictive for a single type of agent can reduce the uncertainties involved in making the transition from preclinical studies to the clinic. The Sidman avoidance paradigm and antagonism of amphetamine-induced hyperactivity are tests used to predict antipsychotic activity based on the actions of the classical antipsychotics that act by blocking dopamine (DA) receptors. Using these models to assess the potential antipsychotic actions of σ-receptor ligands or CCK ligands, may lead to the selection of compounds with the self-same limitations of the DA antagonists. The inevitable 'Catch-22' of developing new animal models is the need for clinical feedback to validate the models.

Costs for enabling technologies have risen considerably with the commercialization of biological materials. Fees for receptor or enzyme clones can be as much as $100,000 for industry researchers. The ownership of such materials is controversial (51) making technology transfer unpredictable with high stakes for litigation between competitors. Exclusivity and cost constraints can thus limit access to 'cutting edge' technologies and have raised concerns related to the free flow of scientific knowledge.

Trained staff are essential to the implementation and integration of increasingly sophisticated enabling technologies (52). Concerns related to science education and funding in the U.S. (53,54) are similarly controversial (55). In "making the Decade of the Brain a reality" (56), the enhancement of the quality and quantity of science education as well as the adoption of more integrative, synergistic approaches to science have been highlighted (57). Education and training issues are reflected in the increasing numbers of chemists and biologists, essential to the pharmaceutical industry, who received their undergraduate training outside the U.S.

Emerging Drug Targets - Peptide neuromodulators and their receptors continue to represent an alternative approach to more conventional drug targets, a rationale based on their relative density within the CNS and the possibility that neuromodulators may be free of the side effects associated with more conventional receptor agonists and antagonists. In addition to the 'grand peptide families' (vasopressin and oxytocin, the tachykinin peptides, glucagon-related peptides, neurotensin, pancreatic-polypeptide - related peptides [NPY, PYY etc.], opioid peptides) (58), somatostatin, CCK, angiotensin II, calcitonin-gene related peptide and corticotrophin-releasing factor are potential CNS drug targets. The search for drugs acting at peptide receptors has proven difficult. Newer technologies involving novel peptide synthesis (59-61) and the discovery of CI-988, a CCK 'dipeptoid' with oral activity (62), point to a resurgence of interest in this area. From a drug discovery perspective, peptides have traditionally suffered from poor bioavailability, metabolic lability, cost of synthesis and efficacy (60). Intravenous peptides or recombinant biologics with systemic targets represent effective therapeutic agents, as evidenced by erythopoetin (EPO), provided the situation is acute and sufficiently life threatening to warrant their cost. In instances where more stable peptidomimetics (64) or small molecule mimics (fourth generation biologics; 65) become available, the usefulness of native or modified peptides is potentially decreased although some proponents of the peptide approach to drug therapy (60) have argued that the conceptual approach within the pharmaceutical industry to date has been seriously flawed. An alternative approach to identifying peptide receptor ligands involves the 'targeted screening' strategy (21) using compound libraries or natural product sources mentioned above. This approach has not proven successful for non-peptide agonists, no doubt reflecting the rigorous steric requirements for efficacy.

Nucleosides (adenosine) and nucleotides (ATP, UTP) are also potent modulators of neuronal function (acting via distinct cell surface receptors; 66) with potential CNS roles in mechanisms related to pain, schizophrenia, epilepsy and stroke (67). Second messenger systems including G proteins (68), nitric oxide synthase pathways (69), eicosanoids (70), calcium (71), phosphoinositide (PI) hydrolysis (72) and the various classes of protein kinases and phosphatases (73,74) that regulate protein phosphorylation are emerging drug targets. Selectivity is, however, a key issue in targeting compounds beyond the receptor level. Calcium antagonists, a highly sucessful class of cardiovascular drugs acting at the calcium channel level, were initially thought to be limited in use by selectivity issues that proved to be less of an issue in the clinic. The effects of lithium on PI hydrolysis appear to be selective reflecting the activity of lithium in bipolar disorders (72). The involvement of multiple second messenger pathways, and their inter-relationships, in transduction processes related to the activation of a single receptor subtype has yet to be clarified. In addition, "cross-talk" between the different second messenger products resulting from activation of heterologous receptor subtypes (D1, D2; [70,75]) appear to be important in determining the ultimate cellular response. Receptor-mediated immediate early gene expression represents a 'third messenger', effecting long term changes in neuronal function (76).

Trophic factors (NGF, BDNF, CNTF etc.[77]) modulate neuronal growth and degeneration. Their receptors and mechanisms that enhance their production are thus primary targets for neurodegenerative disease medications. Hormone responsive elements (HREs), receptors for the steroid receptor superfamily (78) present on DNA, act as transcription factors that modulate promotor activity to stimulate or inhibit gene expression and can enhance growth factor production. Several putative receptors belonging to the steroid receptor family have no known ligands and are designated as orphan receptors (33) whose activity may be modulated by phosphorylation in the absence of ligand. Interestingly, one such orphan receptor, the chick ovalbumin upstream promoter transcription factor (COUP-TF), can be activated by dopamine (79), providing an unusual example of an intracellular HRE that is responsive to a extracellular ligand, an additional example of signal transduction cross-talk. The cellular mechanisms responsible for RNA and protein processing and export represent additional targets amenable to drug targeting.

Increased knowledge of cell function has resulted in a plethora of potential drug targets beyond the receptor, restricted only by ligand specificity and access. In regard to specificity, it is important to bear in mind that disease states reflect homeostatic imbalances such that specificity may be delineated by an increased sensitivity of a hyperactive molecular target resulting from the disease.

Bioavailability and the Blood Brain Barrier (BBB) - Animal models of CNS disorders are traditionally used to gain an indirect measure of compound penetration to the CNS. If a compound is bioactive, a logical conclusion is that it must enter the brain, unless it produces its CNS actions indirectly through effects on the adrenal function (or other hormonal systems) or by vascular actions. Bioavailability models are highly empirical both in terms of endpoints and the time at which compound activity is measured. Rapidly cleared compounds or those with slow onset times due to active metabolite formation can easily be determined as 'inactive'. With increasingly stringent for compound advancement, bioavailability and pharmacokinetic parameters have become part of the lead compound selection process. For CNS drugs, in addition to assuring adequate and consistent plasma levels following oral administration, brain and/or CSF levels of parent compound and (active) metabolites must be determined to assess BBB access and brain retention.

Study of the BBB has been limited (80,81) despite its crucial role in regulating drug efficacy for CNS active agents. Instead, lipophilicity or facilitated diffusion/transport mechanisms have been used to enhance compound access to the brain. Research on BBB drug delivery systems has almost exclusively been left to academia or venture companies (Athena, Pharmatec, Alkermes). Transferrins, neural adhesion molecules and novel brain receptor systems are being studied to target drugs across the BBB. Compounds altering endothelial tight junction function have the potential to facilitate BBB access (82) via modulation of cyclic nucleotide formation.

CNS Diseases and Disorders - *Schizophrenia* is treated with DA receptor antagonists (chlorpromazine, thioridazine). Efficacy and side effect issues (extrapyramidal symptomatology, tardive dyskinesia liability) limit their use. Selective D_1 or D_2 receptor antagonists (SCH 39166, YM 09151-2), $5HT_2$ / D_2 antagonists (risperidone, sertindole), σ receptor ligands (BMY 14,802, CNS 1044, rimcazole, DuP 734; NPC 16377) and DA autoreceptor agonists (PD 118,717, SDZ 208 -912) represent approaches to 'atypical' antipsychotics, compounds with reduced side effect liability (83). The seminal atypical antipsychotic, clozapine, which has superior efficacy with reduced side effect liability is limited in use by the incidence (~ 1%) of agranulocytosis. The search for second generation 'clozapine-like' agents has proven difficult, although the affinity of clozapine for the cloned D_4 receptor (84) suggests a molecular target for this type of compound may be at hand. N-methyl-D-aspartate (NMDA) antagonists may represent another class of atypical agents (85).

Anxiety is treated by benzodiazepines (BZs; diazepam). While effective, the BZs are sedatives and muscle relaxants and potentiate the actions of alcohol producing mild dependence. Their use is limited to generalized anxiety disorders (GAD) and panic leaving obsessive compulsive disorders (OCD) and phobias largely untreated (86). $5HT_{1A}$ receptor agonists (buspirone, ipsapirone, eltoprazine etc.) and $5HT_2$ (ritanserin, ICI 170,809, RP 62203) and $5HT_3$ receptor ligands have anxiolytic activity. 5HT uptake blockers (cloimipramine, fluoxetine,etc) have proven effective in OCD. CCK-B antagonists (CI 988; 59), neurosteroids (87) and compounds selective for the various cloned $GABA_A$ receptor subtypes represent newer approaches to anxiety therapy.

Depression . Antidepressants (ADs) are typified by tricyclic monoamine uptake inhibitors, MAO (monoamine oxidase) inhibitors and 5-HT uptake inhibitors (fluoxetine; 27) that are thought to act by increasing synaptic monoamine availability potentially eliciting a functional decrease in postsynaptic responsiveness. ADs are limited by variable efficacy, tolerance, side effect liability and delayed onset (88). Newer approaches to AD therapy are: α_2-receptor antagonists (idazoxan, MK 912; napamezole, A-75200); $5HT_{1A}$ receptor agonists; $5HT_2$ antagonists; phosphodiesterase inhibitors (rolipram); adenosine antagonists (CP 66,713); ACE inhibitors; and N-methyl-D-aspartate (NMDA) antagonists (89). Carbamazepine and valproate, are efficacious in bipolar disorders (90).

Epilepsy. The mode of action of many anticonvulsant (AC) agents remains obscure reflecting the variety of seizure activities (91). Compounds that enhance GABA neurotransmission (vigabatrine) or block neuronal sodium, calcium T and chloride channels represent potentially novel ACs. GABA uptake inhibitors (tigabine), NMDA receptor antagonists (CGS 19755; MK 801), adenosine agonists and opiates are alternate approaches to anticonvulsant therapy (90).

Analgesia. The discovery of the enkephalins and endorphins represents a seminal period in the history of CNS research (92). Ascribing the various actions of opiates to receptor subtypes theoretically offered major breakthroughs in pain management based on rational drug design. Two decades later, little has altered in the management of pain. Enkephalinase inhibitors (SCH 34826) may act to potentiate the actions of endogenous opiates while 'atypical' cyclooxygenase inhibitors like ketorolac and pravadoline have analgesic activity similar to the opiates (93). NMDA (94), α_2-receptor and Substance P antagonists (93), nitric oxide synthase (NOS) inhibitors (95), bradykinin, cannabinoids, adenosine agonists (96) and calcitonin (97) are other approaches to pain control.

Neuroprotective agents. The search for agents that prevent neuronal death resulting from stroke, one of the leading causes of death and incapacitation in the U.S. (10), is intense. A diverse group of pharmacological agents appears to be beneficial in the gerbil bilateral occulsion model of ischemia which is itself somewhat controversial as a model of human stroke. Competitive NMDA antagonists (MK 801, CGS 19755, CNS 1104, LY 274614), glycine site modulators (HA 966, DCQX), polyamines, spider venom toxins and ligands for AMPA, kainate and quisqualate receptors prevent the effects of excess excitatory amino acid activity (98). Calcium entry blockers (nimodipine, flunarazine), adenosine agonists, κ-opiate agonists, PAF antagonists, gangliosides and free radical scavengers are also effective in blocking the sequence of events following interruption of cerebral blood flow (glutamate release, calcium entry, cell death; 99). Tirilazad, superoxide dismutase (SOD) and desferrioxamine are undergoing human testing. The diversity of mechanistic targets for stroke reflects the multiple steps in the process including those following ischemia-related neutrophil recruitment and reperfusion injury.

Neurodegenerative disorders and cognition: AD and related dementias (Parkinson's, Huntingtons, ALS) are major unmet medical needs. Current therapeutic approaches for AD include replacement (muscarinic - AF 102, CI - 979; SDZ ENS 163; SR 95639A; nicotinic agonists), acetylcholinesterase (AChase) inhibition (tacrine, heptylphysostigmine [L-693,487]), acetylcholine release (DuP 996) and neurorestorative (NGF, BDNF etc.) therapies (100,101). Since the disease etiology is unknown, these approaches are palliative, treating the symptoms (memory loss, cognitive and functional impairment) rather than causes. For muscarinic agonists and AChase inhibitors, side effect liability has, to date, outweighed therapeutic use. In Parkinson's Disease (PD), replacement drug therapy (L-dopa ; D2 agonists) has marked side effect liability (102); selective D1 agonists (103) and cell transplants (NeuroCRIB™) are newer approaches to PD therapy.

Neurotrophic factors support neuronal growth and plasticity (104) their removal leading to neuronal death that is reversible by NGF (105). Lack of NGF results in 'killer gene' expression (104) with the expression of proteins that lead to cell death. A potential for non-specific effects on cell growth may complicate the use of trophic factors for AD and PD.

The plaques and tangles associated with the pathophysiology of AD appear to involve abnormal cleavage products of amyloid precursor protein (APP) to form amyloid β-protein (ABP; 106). The ratio of APP 751 mRNA to APP 695 mRNA is increased in AD (107). The formation of an amyloidogenic 19 KDa product of APP (AP 19) is enhanced by phosphorylation (108). In vitro, ABP has both neurotrophic and neurotoxic action depending on concentration (109,110). Homology between ABP and the tachykinins led to studies showing peptide Substance P antagonists to have similar effects on neuronal survival as ABP (109,111) although their reported systemic actionshave yet to be confirmed. APP mutations occur in familial AD (112,113). Considerable interest has developed in a role for inflammatory mediators in amyloid plaque formation (114) inviting comparisions with atherosclerotic plaque formation. Cytokines enhance trophic factor production affecting neuronal viability (115) while NGF can modulate interleukin-1 expression (116) and the induction of the APP promotor (117). The role of nitric oxide in glutamate neurotoxicity remains controversial (118). Abnormal protein oxidation (119), angiotensin II receptor antagonists (120), and steroid modulated neuron function (121) reflect other potential targets for AD therapy .

Substance abuse represents another major target of the 'Decade' initiative reflecting the societal costs and concerns resulting from the consumption of illicit substances (cocaine, amphetamines, etc.), alcohol and nicotine. The causal factors of abuse appear to involve the anhedonic personality (15), the individual unable to experience pleasure without significant and continued reinforcment from external stimuli. Preclinical research points to a deficit in limbic DA function as well as a psychological reinforcing component as underlying this condition (122) although this remains controversial inasmuch as augmentation of brain DA levels via the use of L-dopa does not mimic the effects of cocaine. Cocaine is, however, an effective blocker of DA uptake while amphetamines, nicotine and alcohol enhance DA function (123). Drugs to treat substance abuse have had low priority in recent research agendas within the pharmaceutical industry because of issues related to the selection of appropriate targets and societal attitudes to effective approaches to the problem (124). Compounds with the same addiction liabilities would offer no clear advantage while product ownership and liability reflect additional issues similar to those plaguing the contraceptive and vaccine areas (125). A medications program sponsored by the National Institute on Drug Abuse (NIDA; 126) is working with the pharmaceutical industry via the Pharmaceutical Manufacturers Association to identify effective treatments for this major problem.

CNS active drugs have a multitude of functional effects. Anticonvulsants are active in OCD, 5HT uptake blockers function as antidepressants, as antianxiety agents for OCD and as appetite supressants. NMDA receptor antagonists may be useful in stroke therapy and as potential anticonvulsants, anxiolytics, antidepressants and antipsychotics, a wide choice of targets reflecting the importance ofglutamate in a wide variety of CNS disease states. These multiple roles for ligand classes underline the complexity of the interactions and interdependence of transmitter (and modulator) systems within the brain which are multipied several-fold (at least) when second (and third) messenger systems and RNA and protein processing systems are considered.

Neuroimmunology - The relationship between the central nervous and immune systems has led to the concept of the 'immunologic brain' (127). Immune system function is dependent on CNS viability, while many mediators of the immune system are present in the CNS and regulate neuronal function and viability. PD has been postulated to result from aberrations of the immune system (128). AIDS related dementia and dementias induced by IL-2 treatment further reinforce the concept of a pivotal interaction between the two systems in neurodegenerative disorders (129). Viral infection (130) and autoimmunity may be causal in the etiology of some CNS disorders. Anecdotal evidence suggests an inverse relationship between AD and anti-inflammatory drug use in rheumatoid arthritis patients (115). Aspirin may also have benefit in the treatment of muli-infarct dementias (131). The therapeutic potential for an anti-inflammatory drug that crosses the BBB is intriguing. The role of primary (IL-1, TNF) and secondary (IL-6, IL-8) interleukins on trophic factors are newer avenues to understanding the effect of the immune system on CNS function (132).

Prospects for the Decade - Despite the increased understanding of drug mechanisms at the molecular level, CNS drug discovery remains, in many respects, highly empirical. Significant changes in the health care and research environments, none unique to the brain, have set new guidelines for the choice of research projects within the industry and the challenges in compound advancement into and through the clinic. While molecular neurobiology offers considerable promise for the future in terms of understanding disease etiology and selecting drug targets, the brain is far more than the additive sum of its neurons, glia and receptors. Accordingly, government, academia and industry must seek to provide and focus the necessary resources to ensure that the often abstract knowledge obtained at the molecular level is integrated into a holistic perspective of the disease state. While such pragmaticism is essential to effectively implement the 'Decade of the Brain' initiative, there is an urgent need within society to address some of the conflicting agendas that impact health care policy, especially as regards the pharmaceutical industry. To ignore the cost effectiveness of pharmaceuticals while ascribing escalating health care costs to drug costs rather than an ever expanding population with increasing health care needs and expectations is untenable. Similarly, the view that U.S. pharmaceutical and biotechnology industries are major competitive assets within the global R & D marketplace is at odds with the Congressional initiative to limit the return on investment for new drug introductions by imposing price controls (7) and is potentially suicidal in the long term. The economics of health care and biomedical research are inextricably linked within the context of the 'Decade of the Brain' and its success and cannot be left to short term political 'solutions' if the quality and dignity of life in the 21st century are to be maintained (8, 9) and occupy a position beyond the merely economic (133). The practical issues facing the 'Decade' initiative are generic to the future of the drug discovery and development process, U.S. pre-eminence in biomedical research and the quality and quantity of health care for the future.

References

1. From the CDC: JAMA, 265, 313 (1990).
2. U.S. Department of Health and Human Services. "Decade of the Brain: Answers Through Scientific Research", NIH Publication 88 - 2957, 1989.
3. NINDS, Maximizing Human Potential. Decade of the Brain, 1990 - 2000, NINDS, Bethesda MD, 1991.
4. L.L.Judd, Neuropsychopharmacol. 3, 309 (1990).
5. H. Aaron and W.B. Schwartz, Science, 247, 418 (1990).
6. A. Sapienza, Technovation, in press (1992)
7. P.R.Vagelos, Science, 252, 1080 (1991).
8. D. Callahan, "Setting Limits. Medical Goals in an Aging Society", Simon and Schuster, New York, 1987.
9. D. Callahan, "What Kind of Life. The Limits of Medical Progress", Simon and Schuster, New York, 1990.
10. R.J. Porter, Presentation at NIH industrial Forum, Bethesda, MD, Nov. 8, 1990.
11. J.W. Hay, Presentation at IBC Conference "Alzheimer's Disease. Advances in Understanding and Treatment", Philadelphia, PA, December, 1991.
12. J. DiMasi, R.W. Hansen, H.G. Grabowski and L. Lasagna, J. Health Economics, 10, 107, (1991)
13. P. A. Roussel, K. N. Saad and T.J. Erickson, "Third Generation R & D", Harvard Business School Press, Boston, MA, 1991.
14. M. Williams and G.L. Neil, Prog. Drug Res. 32, 329 (1987).
15. S. Checkley, Animal Models Psychiatr. Disorders 3, 100 (1991).
16. W. Sneader, "Drug Discovery: The Evolution of Modern Medicines", Wiley, Chichester, U.K., 1985, p.186 .
17. S.J. Peroutka, A.W. Schmidt, A.J. Sleight and M.A. Harrington, Ann. N.Y. Acad. Sci., 600, 104 (1990).
18. M. Williams, Med. Res. Rev. 11, 147, (1989)
19. C.B. Pert and S.H. Snyder, Proc. Natl. Acad. Sci. U.S.A., 70, 2243 (1973).
20. R.F. Squires and C. Braestrup, Nature, 266, 732 (1977).
21. M. Williams and M.F. Jarvis, In "Drug Discovery Technologies", C.R. Clark and W.H. Moos, Eds., Ellis Horwell, Chichester, U.K. 1990, p. 129 .
22. B.E. Evans, M.G. Bock, K. E. Rittle, R.M. DiPardo, W.L. Whittier, D.F. Veber, P.S. Anderson and R.M. Freidinger, Proc. Natl. Acad. Sci. U.S.A. 83, 4918 (1986).
23. R.M. Snider, J.W. Constantine, J.A Lowe, K.P. Longo, W.S. Lebel, H.A. Woody, S.E Drozda, M.C. Desai, F.J. Vinick, R.W. Spencer and H.- J. Hess. Science, 251, 435 (1991).
24. C. Garret, A. Carruette, V. Fardin, S, Moussaoui, J.- F. Peyronel, J.- C. Blanchard and P.M. Laduron. Proc. Natl. Acad. Sci. U.S.A. 88, 10208 (1991).
25. J.S. New, Med. Res. Rev. 10, 283 (1990).
26. P.L. Bonate, Clin. Neuropharmacol. 14, 1 (1991).
27. R.W. Fuller, D.T. Wong and D. W. Robertson, Med. Res. Rev. 11, 17 (1991).
28. J.P. Snyder, Med. Res. Rev. 11, 641 (1991).

29. S. W. Fesik, J. Med. Chem. 34, 2937 (1991).
30. T.P. Kenakin, "The Pharmacologic Basis of Drug Receptor Interaction", Raven, New York, 1987.
31. M. Williams and M.A. Sills, Comp. Med. Chem. 3, 45 (1990).
32. K. Wuthrich, B. Freyberg, C. Weber, G. Wider, R. Traber, H. Widmer and W. Braun, Science 254, 953 (1991).
33. L. A. Denner, N. L. Weigel, B. L. Maxwell, W. T. Schrader and B. W. O'Malley, Science, 250, 1740 (1990).
34. R.J. Tallarida, Life Sci. 46, 1559 (1990).
35. Scrip, Review Issue 1991, p.25 (1991).
36. M. Williams, Curr. Opinion Therap. Patents. 1, 693 (1991)
37. R.J. Baldessarini, In "Advances in Behavioral Neurobiology", R.J. Birmbaum, M. Fava and J. F. Rosenbaum, Eds., Mosby-Year Book Publishers, Littelton, MA, 1992, in press.
38. K. Mansford, Scrip, 1677, 7, (1991).
39. J.D. Strupczewski, D.B. Ellis and R. C. Allen. Ann. Rep. Med. Chem. 26, 297, (1991).
40. M. T. Jacobs, "Short-Term America: The Causes and Cures of Our Business Myopia", Harvard Business School Press, Boston, MA, 1991.
41. M. Ridley, The Economist, 318, (7694), Science Survey, p. 8 (1991).
42. F.E. Bloom, Presentation, NIDA/UCLA Conference on Substance Abuse, Santa Monica, CA, Jan. 1991.
43. A. Klausner and R.J. Rodgers, Spectrum Biotech. Overview, Decision Resources, Burlington, MA, 1990.
44. C.Hall, BioVenture View, VI (6), 9 -12, (1991).
45. F - D - C Reports, 54 (2), 11 (1992).
46. T. I. Bonner, N.J. Buckley, A.C. Young and M.R. Brann, Science, 237, 527 (1987).
47. J. Saunders, M. Cassiday, S.B. Freedman, E.A. Harley, L.L. Iversen, C. Kneen, A.M. MacLeod, K.J. Merchant, R.J. Snow and R. Baker, J. Med. Chem. 22, 1128 (1990).
48. G. deStevens, J. Med. Chem. 34, 2665 (1991).
49. H.A. Lester, Science, 241 , 1058 (1988).
50. P. Willner, "Behavioral Models in Psychopharmacology", Cambridge University Press, Cambridge, U.K., 1991.
51. L. Roberts, Science, 254 , 184 (1991).
52. N. Moran, Nature, 353 , 874 (1991).
53. F.E. Bloom and M.A. Randolph, 'Funding Health Sciences Research: A Strategy to Restore Balance', National Academy Press, Washington, D.C. ,1990.
54. L.M. Lederman, "Science: The End of the Frontier", Science Supp., Jan. 1991.
55. F.E. Bloom, Neuropsychopharmacol. 3, 141 (1991).
56. F.E. Bloom, In "The Centennial of the Neuron", Fidia , Georgetown, D.C., June, 1991, Abs p. 4 (1991).
57. B.Z. Shakhashari, The Scientist, 5, (24), 11 (1991).
58. J.R. Cooper, F.E. Bloom and R.H. Roth. "The Biochemical Basis of Neuropharmacology, 6th Edn". Oxford, New York, 1991, p. 381.
59. S.P.A. Fodor, J.L. Read, M.C. Pirrung, L. Stryer, A.T. Lu, and D. Solas, Science, 251, 767 (1991).
60. K.S. Lam, S. Salamon, E.M. Hersh, V.J. Hruby, W.M. Kazmierski and R.J. Knapp, Nature, 354, 82 (1991).
61. R.A. Houghten, C. Pinilla, S.E. Blondelle, J.R. Appel, C.T. Dooley and J.H. Cuervo, Nature, 354, 84 (1991).
62. J. Hughes, P. Boden, B. Costall, A. Domeney, E. Kelly, D.C. Horwell, J.C. Hunter, R.D. Pinnock and G.N. Woodruff, Proc. Natl Acad. Sci. U.S.A., 87, 6728 (1990).
63. J. Plattner and D. Norbeck. In "Drug Discovery Technologies", C.R. Clark and W.H. Moos, Eds., Ellis Horwell, Chichester, U.K. 1990, p. 92.
64. P.S. Portoghese, J.Med. Chem. 34, 1757, (1991).
65. M. Williams, In "Drug Development, 2nd Edn", C.E. Hamner, Ed., CRC Press, Boca Raton, FL., 1990, p. 59.
66. K.A. Jacobson, P. J. M. van Galen and M. Williams, J. Med. Chem. 35 , 407 (1992).
67. M. Williams, Neurochem. Inter. 14, 249 (1989).
68. M. F.Goy, Trends Neurosci. 14, 293 (1991).
69. D.S. Bredt and S.H. Snyder , Neuron, 8, 3 (1992).
70. D. Piomelli, C. Pilon, B. Giros, P. Sokoloff, M. -P. Martres and J.-C. Schwarcz, Nature, 353 , 164 (1991).
71. N.A. Saccomano and A.H. Ganong, Ann. Rep. Med. Chem. 26, 33 (1991).
72. S.K. Fisher, A. M. Heacock and B. W. Agranoff, J. Neurochem. 58, 18 (1992).
73. J. D. Scott, Pharmac. Ther., 50, 123 (1991).
74. E.H. Fischer, H. Charbonneau and N.K. Tonks, Science, 253, 401 (1991).
75. P. G. Strange, Trends Pharmacol. Sci. 12, 48 (1991).
76. T.M. Esterle and E. Sanders - Bush, Trends Pharmacol. Sci., 12, 375 (1991).
77. D.M. Araujo, J.- G. Chabot and R. Quirion, Int. Rev. Neurobiol. 32, 141 (1990).
78. B. O'Malley, Mol. Endocrinol. 4, 363 (1990).
79. R.F. Power, J.P. Lydon, O.M. Connelly and B.W. O'Malley, Science, 252, 1546 (1991).
80. W.M. Pardridge, "Peptide Drug Delivery to The Brain", Raven, New York, 1991.
81. P. Knight, Biotechnol. 7, 1009 (1989).
82. L. Rubin, S. Porter, H. Horner and T. Yednock, Patent Pub. WO91/05038 (1991).
83. L.D.Wise and T.G. Heffner, Ann. Rep. Med. Chem. 26, 53, (1991).
84. H.H. Van Tol, J.R. Bunzow, H.C. Guan, R.K. Sunahara, P. Seeman, H.B. Niznik and O. Civelli, Nature 350, 610 (1991).

85. T. Klockgether, L. Turski, T. Honore, Z. Zhang, D.M. Gash, R. Kurlan and J.T. Greenamyre, Ann. Neurol. 30, 717 (1991).
86. M. Williams, in "Current and Future Trends In Anticonvulsant, Anxiety and Stroke Therapy", B.S. Meldrum and M. Williams, Eds., Wiley, New York ,1990, p. 131.
87. W. Sutano and E.R. de Kloet, Med. Res. Rev. 11, 617 (1991).
88. C. Holden, Science, 254 , 1450 (1991).
89. D. W. Robertson and R.W. Fuller, Ann. Rep. Med. Chem. 26, 23 (1991).
90 . R.M. Post, S. R. B. Weiss, T. Nakajima, M. Clark and A. Pert, In "Current and Future Trends In Anticonvulsant, Anxiety and Stroke Therapy" B.S. Meldrum and M. Williams, Eds., Wiley, New York, 1990, p. 45.
91. R.J. Porter, In "Current and Future Trends In Anticonvulsant, Anxiety and Stroke Therapy" , B.S. Meldrum and M. Williams, Eds., Wiley, New York ,1990, p. 1.
92. S.E. Cozzens. "Social Control and Multiple Discovery in Science: The Opiate Receptor Case", SUNY Press, Albany, NY, 1989.
93. S.M. Evans, G.R. Lenz and R.A. Lessor, Ann. Rep. Med. Chem. 25, 11 (1990).
94. A.H. Dickenson, Trends Pharmacol. Sci. 11, 307 (1990).
95. P. Klepstad, A. Maurset, E.R. Moberg and I. Oye, Eur. J. Pharmacol. 187 , 513, (1990).
96. R. Karlsten, C. Post, I. Hide and J.W. Daly, Neurosci. Letts. 121 , 267 (1991).
97. R.E. Chipkin and F. C. A. Gaeta, Ann. Rep. Med. Chem. 23 , 11 (1988).
98. G. Johnson and C.F. Bigge, Ann. Rep. Med. Chem. 26 , 11 (1990).
99. D. W. Choi and S. M. Rothman, Ann. Rev. Neurosci., 13 , 171 (1990).
100. W.H. Moos and F.M. Hershenson, Drug News Persp. 2 , 397 (1989).
101. M.R. Pavia, R.E. Davis and R.D. Schwarz, Ann. Rep. Med. Chem. 25 , 21 (1989).
102. C. G. Goetz, Neurology, 40 , (Supp. 3) 50 (1990).
103. K.J. Darney, M. H. Lewis, W.K. Brewster, D.E. Nichols and R. B. Mailman, Neuropsychopharmacol. 5 , 187 (1991).
104. L.M. Schwartz, BioEssays, 13, 389 (1991).
105. T.Shigeno, T. Mima, K. Takakura, D.I. Graham, G. Kato, Y. Hashimoto and and S. Furukawa. J. Neurosci. 11, 2914 (1991).
106. D.J. Selkoe, Neuron, 6, 487 (1991).
107. G. A. Higgins, G.A. Oyler, R.L. Neve, K.S. Chen and F.H.Gage, Proc. Natl. Acad. Sci. U.S.A. 87, 3032 (1990).
108. C. Nordstedt, S.E. Gandy, I. Alafuzoff, G.L. Caporaso, K. Iverfeldt, J. A. Grebb, B. Winblad and P. Greengard. Proc. Natl. Acad. Sci. U.S.A. 88, 8910 (1991).
109. J.S. Whitson, D.J. Selkoe and C.W. Cotman, Science, 243, 1488 (1989).
110. B. A. Yankner, L.K. Duffy and D. A. Kirschner, Science, 250, 279 (1990).
111. N.W. Kowall, M.F. Beal, J. Busciglio, L.K. Duffy and B.A. Yankner, Proc. Natl. Acad. Sci. U.S.A., 88, 7247 (1991).
112. M.- C. Chartier-Harlin, F. Crawford, H. Houlden, A. Warren, D. Hughes, L. Fidani, A. Goate, M. Rossor, P. Roques, J. Hardy and M. Mullan, Nature, 353, 844 (1991).
113. J. Murrell, M. Farlow, B. Ghetti and M.D. Benson, Science, 254, 91 (1991).
114. P. McGeer, E. McGeer, J. Rogers and J. Sibley, Lancet, 335, 1037 (1990).
115. K. Alheim, C. Andersson, S. Tingsborg, M. Ziolkowska, M. Schultzberg and T. Bartfai, Proc. Natl. Acad. Sci. U.S.A., 88, 9302 (1991).
116. X. Vige, E. Costa and B. C. Wise, Mol. Pharmacol. 40, 186 (1991).
117. R.J. Donnely, A. J. Friedhoff, B. Beer, A. J. Blume and M. P. Vitek, Cell. Mol. Neurobiol. 10, 485 (1990).
118. C. Demerle-Pallardy, M. - O., Lonchampt, P. - E., Chabrier and P. Braquet, Biochem. Biophys. Res. Comm. 181, 456 (1991).
119. C.D. Smith, J.M. Carney, P.E. Starke-Reed, C. N. Oliver, E.R. Stadtman, R. A Floyd and W. R. Markesbery, Proc. Natl. Acad. Sci. U.S.A., 88, 10540 (1991).
120. J.M. Barnes, N.M. Barnes, B. Costall, Z.P. Horovitz and R.J. Naylor. Brain. Res. 491, 136 (1989).
121. E. Roberts, Proc. Natl. Acad. Sci. U.S.A. (1992).
122. F.H. Gawin, Science, 251, 1580 (1991).
123. G.F. Koob and F.E. Bloom, Science, 242, 715 (1988).
124. A. Goldstein and H. Kalant, Science, 249, 1513 (1990).
125. C. Djerassi, Science, 245, 356 (1989).
126. D. N. Johnson and F.J. Vocci, The Pharmacologist, 33, 190 (1991).
127. W.L. Farrar, J.M. Hill, A. Harel-Bellan and M. Vinocour, Immunol. Rev. 100, 361 (1987).
128. R.S. Smith, Med. Hypothes. 34, 225 (1991).
129. L.D.Altstiel and K. Speber. Prog. Neuro-Psychopharmacol. Biol. Psychiatr., 15, 481 (1991).
130. S.B. Pruisner, Science, 252, 1515 (1991).
131. C.C. Mann and M. L. Plummer. "The Aspirin Wars", Knopf, New York, 1991, p. 333.
132. D.E. Brenneman, M. Schultzberg, T. Bartfai and I. Gozes, J. Neurochem. 58, 454 (1992).
133. P. Davies, 'The Last Election', Vintage, New York, 1987.

Chapter 2. Pharmacological Intervention in Sleep and Circadian Processes

Joseph D. Miller Ph.D.
Stanford University School of Medicine and
Department of Biological Sciences
Stanford University Stanford, CA 94305

<u>Introduction</u> - Since homeostatic mechanisms of arousal state control in mammals are strongly influenced by the circadian clock, it is essential to consider both pharmacological influences on sleep and arousal <u>per se</u>, and on the circadian timekeeping mechanism. The localization of the mammalian circadian pacemaker to the suprachiasmatic nucleus (SCN) of the hypothalamus, and subsequent <u>in vitro</u> studies of the SCN (for review, see 1), have been of great value in separating the sleep-associated and clock-associated properties of pharmacological agents. The major classes of neurotransmitter receptor ligands will be reviewed from this bipartite viewpoint. Endocrine and prostanoid sleep mediators will not be discussed, as these substances have been very adequately reviewed recently (2).

<u>GABAergic Agents</u> - Specific binding at sites on the benzodiazepine/GABA-A receptor/chloride channel complex is thought responsible for the hypnotic effects of barbiturates, alcohol, and the benzodiazepines (BZ). Oddly, a recent report has shown that the specific GABA-A receptor agonist, muscimol (<u>1</u>), lacks hypnotic effects and does not potentiate the hypnotic effects of flurazepam (<u>2</u>) in rats (3). While full dose response data for muscimol are not available, this result does imply that the hypnotic effects of the BZs are not necessarily mediated <u>in toto</u> by GABA-A receptor gating of the associated chloride channel. However, another report (4) has recently demonstrated that the GABA metabolite, γ-hydroxybutyrate (GHB; <u>3</u>), improves the quality of sleep in narcoleptic patients (and in normal subjects), measured in terms of amount of delta sleep, increases in stage 3 sleep and decreases in awakenings, and also exhibits a significant anti-cataplectic effect on the following day. Regional hypnotic sites of action for GHB and the BZs, and their relationship to GABA-A receptor distribution, remain to be evaluated.

More traditional recent studies of BZ activity include a careful analysis of sleep dynamics in humans (5) under a variety of BZs (triazolam (<u>4</u>), flunitrazepam (<u>5</u>), flurazepam). Analysis of the incidence of abortive REM (rapid eye movement) sleep episodes indicated that these drugs may impede EEG desynchronization and REM sleep signs, but do not alter the timing of REM sleep. Similarly, although the BZs depress the delta power band in SWS (slow wave sleep) and in REM sleep, the decline of delta power over consecutive episodes of SWS and the buildup of delta within an episode is preserved. Thus, homeostatic ultradian (i.e., rhythmicity with a period less than 24 hr) regulation of sleep is intact under BZs, although EEG generating mechanisms may be strongly affected.

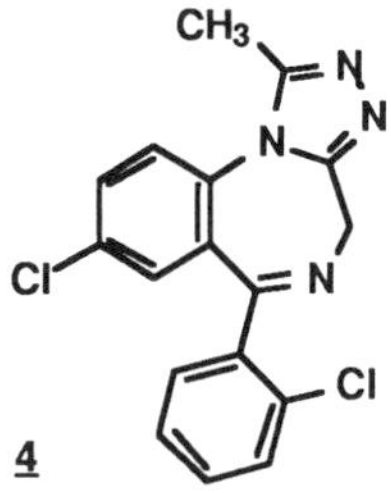

4

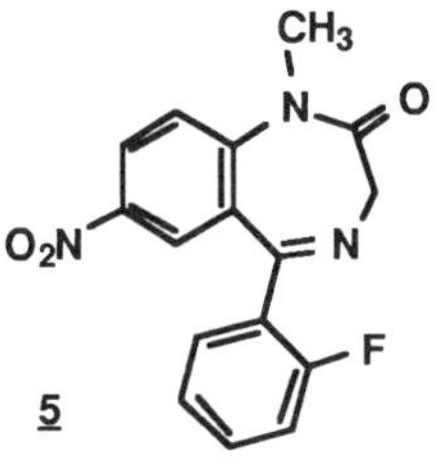

5

The non-BZ sedative-hypnotics, zolpidem (**6**) and zopiclone (**7**), are thought to bind preferentially to the ω1 BZ receptor subtype, thus avoiding the myorelaxant, anticonvulsant, and anti-anxiety effects of the BZs, presumably mediated by ω2 receptors (6). Similarly, it has been claimed that these agents are much less likely to produce anterograde amnesia than triazolam, although emesis is a fairly common side effect (7). These short acting agents reduce sleep latency and lengthen sleep duration, with no effects on sleep staging and no evidence for rebound insomnia or adverse psychomotor effects (6). While the effects of flunitrazepam on the EEG power spectrum have been related to the central distribution of the various ω receptors (8), it has also been recently claimed that the regional effects of BZ and non-BZ sedative hypnotics on CNS metabolism are essentially identical (9). Thus, it is not clear whether ω receptor binding profiles for BZs and non-BZs are of functional significance. Consideration of the central distribution of the α and γ subunit isoforms that constitute the probable BZ binding sites in the BZ/GABA-A receptor/chloride channel oligomer may give greater insight into the functional effects of BZ and non-BZ hypnotics (10).

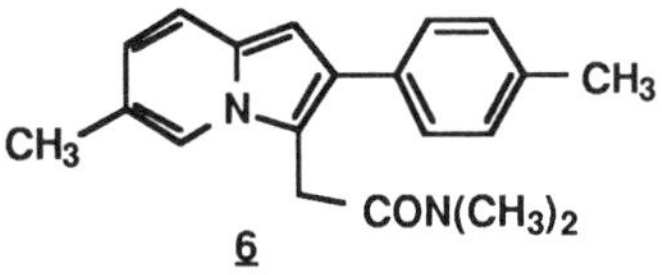

6 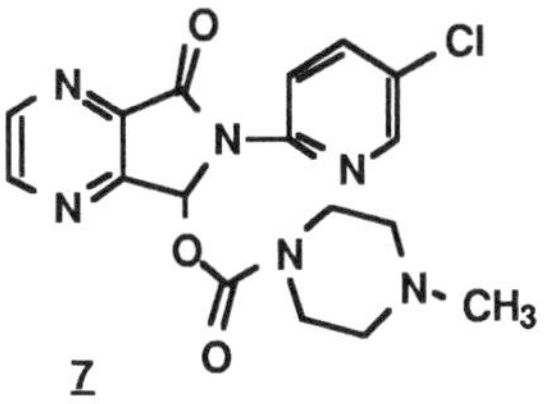   **7** 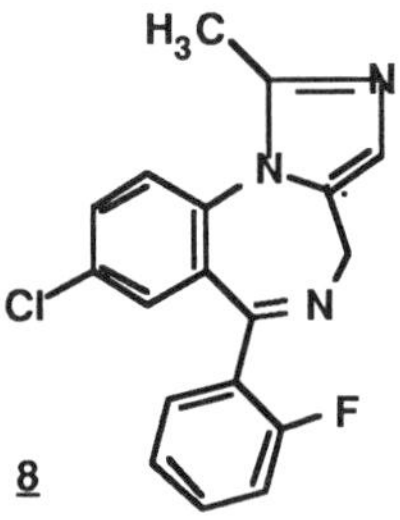   **8**

BZs, zolpidem (11) and the GABA-A receptor agonist, muscimol (12) all seem capable of resetting the circadian clock. Transient insomnia induced by an artificial 3 hr phase advance of sleep time was reversed by zolpidem (11). Triazolam, taken at bedtime by normal subjects, or midazolam (**8**), taken at the same time by 7 hr phase delayed (by a 7 time zone westward jet flight) subjects had little effect on sleep-wake architecture (13,14) or the circadian organization of hormonal secretion (14). In contrast, the sleep-disruptive effects of a seven hr phase advance were effectively reversed by midazolam at bedtime (13). The efficacy of the BZs could be predicted from the phase response curve for triazolam in the hamster (15).

The sites and modes of action of BZs in the induction of sedative or clock shifting activity are not well-understood. The sedative effects of triazolam in rats appear to depend on prior sleep debt (16). Since waking is not temporally consolidated in SCN-lesioned rats, it is not surprising that such rats are not sedated by triazolam (17). On the other hand, the phase shifting effects of triazolam, at least in the hamster, may depend on its paradoxical ability to increase locomotor activity in this species (15). The locus of action for the phase-shifting effects of triazolam is controversial. Lesion of the intergeniculate leaflet (IGL) of the lateral geniculate (which projects directly to the SCN) blocks BZ-induced phase advances in hamsters (e. g., 18). However, the IGL is nearly devoid of BZ receptors (19). The dorsal raphe nucleus projects to the IGL and the SCN and is rich in BZ receptors. However, 5,7 dihydroxytryptamine lesion of 5HT neurons in the raphe does not affect the ability of BZs to accelerate reentrainment to altered light-dark cycles (20). An important caveat here is that 80-90% of 5HT neurons must typically be destroyed before the raphe system is no longer able to compensate for cell loss by increased 5HT synthesis and postsynaptic receptor upregulation, a level apparently reached in this study. However, the median raphe nucleus also projects to the SCN and exhibits a moderate density of BZ/GABA receptors, constituting a candidate for the site of BZ action that has yet to be investigated. Direct effects of BZs at the level of the SCN may also contribute to phase shifting effects, since moderate BZ binding is present. However, although the SCN is very rich in GABA neurons, muscimol binding and immunoreactivity for the typical BZ/GABA-A receptor/chloride channel complex is very low, suggesting a novel isoform of this complex may be present in the SCN or that "naked" BZ receptors may be present there (19).

Glutamatergic Agents - In contrast to the work on GABAergic agents, relatively little is known of the role of glutamate receptors in sleep or circadian regulation. A recent paper (21) suggests that NMDA receptor blockade by DL-2-amino-5-phosphono-pentanoic acid (APV; $\underline{9}$) in the thalamus elevates deep slow wave sleep (SWS) and rapid eye movement (REM) sleep, whereas non-NMDA receptor blockade by 6-cyano-7-nitroquinoxaline-2,3-dione (CNQX, $\underline{10}$) selectively elevates light SWS. In a recent report, a sleep-promoting (both SWS and REM sleep) substance, SPS-B, isolated from the brainstem of sleep-deprived rats has been chemically identified as the oxidized form of glutathione ($\underline{11}$), one of the primary anti-oxidant free radical scavengers (22). Oxidized glutathione administered icv increases both SWS and REM sleep, in a fashion qualitatively indistinguishable from the occurrence of normal sleep. Some evidence suggests that glutathione may bind to synaptic membranes and inhibit glutamate receptor binding (22).

The primary transmitter of the retinohypothalamic tract (RHT) appears to be glutamate (23). Thus, NMDA receptor blockade prevents light-induced resetting of the circadian clock (24) or the suppressive effects of light on plasma melatonin (25). However, while a slow depolarizing potential, sensitive to NMDA receptor blockade by APV, can be detected at a relatively depolarized membrane potential in the SCN, the dominant effect of optic nerve stimulation at resting potential is a non-NMDA receptor mediated fast excitatory postsynaptic potential (26). The fact that the phase response curve (PRC) for glutamate administered directly into the SCN (27) differs radically from that of light may reflect a differential modulation of such potentials by RHT afferents that cannot be easily reproduced by bolus injection of glutamate into the SCN.

Cholinergic Agents - For some years it has been known that local cholinergic stimulation of the dorsal pontine tegmentum evokes REM sleep in cats (for brief review, 28). In humans, pilocarpine, a muscarinic agonist has the same effect (29). This year it has been shown that specific stimulation of the M-2 cholinergic receptor by cis-methyl-dioxolane ($\underline{12}$) produces a similar elevation in REM sleep and muscular atonia (30). Nicotine administration into the same region also elevates REM sleep (31). Some evidence also suggests nicotinic receptor involvement in PGO wave generation and M-1 muscarinic receptor involvement in EEG desynchronization (32).

Direct cholinergic stimulation of the SCN by carbachol in vivo can reset the biological clock (33). Nicotinic stimulation of the SCN in vitro also resets the clock in a dose-dependent fashion, but in this case, only phase advances have been observed (34). A role for muscarinic receptors in the modulation of the circadian clock has not been excluded; in fact, the co-localization of nicotinic and muscarinic receptors on a substantial population of SCN neurons suggests that muscarinic receptors may play a role in circadian function (35).

Catecholamines - A role for catecholamines in sleep is supported by the recent observation that the catecholamine neurotoxin, N-methyl-4-phenyl-1,2,3,6,-tetrahydropyridine (MPTP, $\underline{13}$), causes a selective and ultimately reversible abolition of REM sleep in cats (36). Dopamine neurons in the substantia nigra and norepinephrine neurons in the locus coeruleus accounted for most of the cell loss, although unidentified neurons were also destroyed in the nucleus basalis of the forebrain, site of a large population of cholinergic neurons. Since REM sleep recovered without REM rebound, it is possible that other cholinergic REM generator neurons in the parabrachial area were temporarily inactivated. On the other hand, narcolepsy, a disorder of REM sleep, responds therapeutically to treatment with the catecholamine precursor, L-tyrosine (37).

A role for the dopamine D-1 receptor in sleep has been suggested by the observation that the specific D-1 receptor agonist SKF38393 dose-dependently reduces REM sleep in the rat, an effect which could be antagonized by the D-1 antagonist, SCH23390 (38). The D-1 antagonist by itself elevated REM sleep. However, other authors (39) found that higher doses of the D-1 antagonist increased SWS and reduced both waking and REM sleep. The potential physiological importance of such findings is supported by the observation that sleep deprivation elevates D-1 receptors and D-1 stimulated adenylate cyclase in the limbic system (40).

The dopamine D-2 receptor has usually been associated more strongly with the waking state, and, in particular, locomotor activity in that state. However, a recent finding suggests that the D-2 receptor may be involved in canine narcolepsy (41). Autoceptor-like doses of raclopride, L-sulpiride, and (+)-3-PPP (14) suppressed cataplexy whereas D-2 agonists like quinpirole aggravated cataplexy at small doses. In contrast, both agonists and antagonists at the D-1 site suppressed cataplexy in a non-specific fashion. The D-2 results may be related to the observed elevation in dopamine metabolism in the amygdala of narcoleptic dogs (42). Pathological elevation of dopamine transmission at a D-2 heteroreceptor on noradrenergic projections in the narcoleptic dog could suppress noradrenergic transmisssion, yielding cataplexy and other narcoleptic pathology (see below). Such an interpretation could explain the therapeutic effects of D-2 blockade, but fails to explain the failure of selective dopamine reuptake blockers (e.g., amineptine) to exacerbate cataplexy (41).

In contrast, $\alpha 1$ noradrenergic agonists such as methoxamine, selective norepinephrine reuptake blockers such as nisoxetine, and $\alpha 2$ antagonists such as yohimbine all alleviate canine cataplexy, whereas $\alpha 1$ antagonists such as prazosin and $\alpha 2$ agonists such as BHT-920 (which also has considerable affinity for the dopamine autoceptor) all exacerbate cataplexy (43). Potentiation of noradrenergic transmission in normal subjects can likewise suppress REM and SWS; the atypical $\alpha 1$ agonist, modafanil (15), can maintain insomnia for four days in rhesus monkeys (44, 45) and hypnotic effects of the $\alpha 2$ agonist, dexmedetomidine (16), in humans are potently suppressed by co-administration of the $\alpha 1$ agonist, cirazoline (46). Similarly, clonidine ($\alpha 2$ agonist) is sedating in humans and idazoxan ($\alpha 2$ antagonist) increases wake activity (47).

Catecholamine effects on the biological clock are difficult to interpret. Both D-1 and D-4 dopamine receptor mRNA have been found in the SCN (48, 49), as well as $\alpha 1$ receptors (50). Acute administration of the indirect catecholamine agonist, amphetamine, does not phase shift circadian rhythms in rats (51). Chronic administration, however, will cause an apparent lengthening of the circadian period if rats have access to a running wheel (52). Interestingly, chronic amphetamine in SCN-lesioned rats can also cause the expression of infradian rhythms, with periods of 28-54 hr (53). It appears that such rhythms are best explained by the lengthening of the ultradian rest-activity cycle due to stimulant effects of the drug, rather than the unmasking of a non-SCN infradian oscillator. In contrast, chronic administration of the $\alpha 2$ agonist, clonidine, shortens the period of circadian rhythms (54).

5HT and Melatonin - An influential review has suggested that the strongest functional correlate to 5HT neurotransmission is repetitive motor activity in a behaviorally aroused state (55). The hypothesis that 5HT is a behavioral stimulant is supported by the observation that 5HT reuptake blockers like fluoxetine, clomipramine, or fluvoxamine are relatively activating or alerting agents in humans (56), suppress REM sleep in humans (57, 58) and both delta power and REM sleep in rodents (59). Since sleep deprivation with or without clomipramine results in a marked elevation in delta power in subsequent recovery sleep (59), it is apparent that clomipramine, despite its general sleep suppressant effects, does not impair the homeostatic regulation of sleep (compare to BZs above). In contrast, two recent studies have shown that $5HT_2$ receptor blockade by ritanserin increases SWS in humans (60, 61). The combination of a 5HT reuptake blocker (zimeldine) with ritanserin in rats produced a biphasic effect; an initial increase in waking, followed 4-5 hr later by a large increase in SWS (62). This result strongly suggested the involvement of other 5HT receptors, perhaps of the 5HT1 variety, in the modulation of arousal state. Indeed, the $5HT_{1A}$ agonist, buspirone (17), increases sleep latency and decreases sleep time in rats by reducing both SWS and REM sleep (63, 64)

General 5HT agonists (5HT, quipazine) induce both phase advances and phase delays, depending on circadian time of day, in the circadian rhythm in single unit activity recorded in the SCN in vitro (65, 66). 5HT-mediated phase advances seem to be mediated solely by the $5HT_{1A}$ receptor. Identical phase advances are produced by 5HT, quipazine, and the $5HT_{1A}$ agonists, (±)-8-hydroxydipropylaminotetralin (8-OHDPAT) and buspirone. Quipazine-induced phase advances are completely antagonized by the $5HT_{1A}$ antagonist, NAN-190 (18). Peculiarly, 5HT or quipazine-induced phase delays are mediated by none of the known 5HT receptors. The phase resetting properties of these drugs appear to be mediated directly by 5HT receptors on clock neurons, since the interruption of synaptic transmission by TTX or elevated magnesium treatment is without effect on the phase changes (67).

The 5HT pineal metabolite, melatonin, may have mild hypnotic effects, but these are outweighed by its effects on the circadian clock (68), apparently mediated by a dense population of melatonin receptors in the SCN (e. g., 69). Indeed, melatonin induces large phase advances in the single unit circadian rhythm in the SCN in vitro (70). Similarly, delayed sleep phase syndrome in humans can be effectively treated via the phase advancing effects of melatonin administered 5 hr before bedtime (68). In addition, two recent reports have shown that free running blind humans can entrain to oral melatonin administration, thus stabilizing sleep onset times and attaining a consolidated 24 hr sleep-wake cycle (71, 72). However, these reports disagree as to whether other rhythms besides sleep-wake (e.g., rectal temperature, urinary cortisol) are entrainable to melatonin.

Histamine - The sedative properties of antihistamines have been known for many years. The discovery of a group of histaminergic neurons in the tuberomammillary region of the hypothalamus with widely ramifying forebrain projections suggests that histamine may be a major arousal-related transmitter in the brain, in some ways similar to norepinephrine or acetylcholine (for a recent review, see 73). Neurotoxic lesion of this area, however, produces only a temporary hypersomnia, suggesting that the histaminergic projections are not necessary for arousal state control (74). H-1 receptor antagonists like diphenhydramine appear effective in the treatment of insomnia in humans (75), whereas H-2 antagonists such as zolantidine appear to have no effect on arousal state (76). The recent discovery of drugs acting at inhibitory histamine autoreceptors or H-3 receptors has provided a new means of altering histaminergic tonus in the CNS. For instance, the H-3 agonist, α-methylhistamine, significantly increases deep SWS in rats, whereas the H-3 antagonist, thioperamide (19), enhances waking in a dose-dependent fashion (77). Similarly, the H-1 blocker, mepyramine, inhibits the arousing effects of thioperamide, suggesting that the postsynaptic effects of histaminergic transmission are mediated primarily by H-1 receptors (77).

Histamine depletion by α-fluoromethylhistidine reduces SWS and REM sleep in rats in the light phase and increases it in the dark phase, resulting in an overall decrease in the circadian amplitude of the sleep-wake cycle (78). Intracerebroventricular (icv) injection of histamine in rats produced circadian time-dependent transient phase advances or permanent phase delays in locomotor or drinking rhythms (79). Histamine injection at certain circadian times can accelerate reentrainment to a shifted light-dark cycle (80). These effects may be mediated via the H-1 receptors present in the SCN.

Adenosinergic Agents - Adenosine antagonists such as caffeine and theophylline have long been known to have alerting properties. In contrast to the sedating effects of alcohol and the benzodiazepines, the alerting effects of caffeine in humans seem to be relatively independent of prior sleep debt induced by sleep restriction (81). In contrast, the ability of caffeine to reduce next day sleepiness, after nocturnal administration of triazolam or flurazepam, does seem to depend on the degree of BZ-induced prior sleepiness (82). Adenosine receptors are classified into two major functional types: the widely distributed A-1 receptor and the A-2 receptor, which has a more restricted distribution in the basal ganglia (83). A recent study suggests that the alerting effects of the non-specific adenosine antagonist, caffeine, are more similar to the alerting effects of the specific A-1 antagonist, 8-cyclopentyltheophylline (20), than to the effects of the relatively specific A-2 antagonist, alloxazine (84). Neither adenosine receptor in the brains of narcoleptic dogs appears to differ from corresponding receptors in normal dogs in terms of affinity or receptor number (83). However, the only brain regions examined were frontal cortex (A-1) and putamen (A-2).

A recent report suggests that adenosine deaminase is colocalized in histaminergic neurons of the tuberomammillary area (85), suggesting that histamine and adenosine may be cotransmitters. No evidence exists for adenosine receptors in the SCN, but physiological evidence (86) suggests that central hypnotic effects of adenosine are mediated by such receptors in the general preoptic region. If the characterized H-1 receptors in the SCN are associated with a histaminergic projection from the posterior hypothalamus, SCN neurons could be sensitive to co-released adenosine, perhaps explaining an early report of potent circadian rhythm resetting by theophylline (87). Co-release of varying amounts of transmitters with opposite effects on arousal state could sensitively modulate the functional activity of the biological clock. The fact that SCN efferents project to the same general region of the posterior hypothalamus suggests that a finely-tuned feedback system could jointly modulate arousal state and circadian timekeeping.

Neuropeptides - Agonists at mu and kappa opioid receptors are frequently sedating. For example, the kappa agonist butorphanol induces sedation indistinguishable from that induced by midazolam, with the exception that fewer disturbances of memory are reported (88). Endogenous opiate receptor ligands and other polypeptides are cleaved from the precursor, proopiomelanocortin (POMC), but most of the cleavage products lack selective effects on sleep-wake activity. An important exception is that corticotropin-like intermediate lobe peptide (CLIP; 21), administered icv, selectively elevates REM sleep in hypophysectomized rats or controls (89). Insomnia induced by the dopamine agonists, apomorphine or bromocriptine, by the $5HT_{1A}$ agonist, 8-OHDPAT, or by 5HT depletion by parachlorophenylalanine (PCPA) was in each case reversed by CLIP. CLIP selectively restored REM sleep to near control levels in PCPA treated rats (90). Another POMC cleavage product, desacetyl-α-melanocyte stimulating hormone was largely ineffective in reversing insomnia (90). Immunoreactive dynorphin and leu-enkephalin have been localized in the SCN, along with kappa, delta, and mu opioid receptors (91, 92), but their role in circadian function is completely unknown.

Arg-Pro-Val-Lys-Val-Tyr-Pro-Asn-Gly--Ala-Glu-Asp-Glu-Ser-Ala-Glu-Ala-Phe-Pro--Leu-Glu-Phe
2 1

In addition to CLIP, vasoactive intestinal polypeptide (VIP) is one of the few neuropeptides known to selectively elevate REM sleep. VIP is able to restore REM sleep in cats made insomniac by either PCPA treatment or basal forebrain lesions (93). VIP neurons constitute a large population of neurons in the ventrolateral region of the SCN. VIP levels in the SCN vary with the light-dark cycle, with highest levels reached at night (94). Administration of VIP into the SCN along with its frequent cotransmitters, peptide histidine isoleucine (PHI) and gastrin releasing peptide (GRP), generates large phase delays in the circadian rhythm in wheel running in hamsters (95).

Somatostatin is the only other neuropeptide known with the capacity to elevate REM sleep (for review, see 96). Somatostatin neurons are abundant in the SCN and somatostatin levels in rat SCN homogenates vary in a circadian fashion even in rats maintained in constant conditions (97).

Conclusion - It should be evident from this brief review that virtually all neurotransmitter-like agents that modulate arousal state, with the possible exception of the catecholamines, also have profound phase resetting effects on the circadian clock localized in the SCN (although the endogenous opiates and other POMC derivatives have yet to be tested in circadian paradigms). Furthermore, three of the four agents known to selectively augment REM sleep are found in the SCN, along with appropriate receptors (acetylcholine, VIP, somatostatin). The presence of leu-enkephalin, a potential POMC derivative, in the SCN suggests that POMC and CLIP may be present there also. This parallel between sleep-active and clock-active agents may not be accidental, but may rather imply considerable cross-talk between the circadian clock and the mechanisms responsible for the ultradian organization of arousal states. Future research will delineate the nature of this interaction and the means by which neurotransmitters modulate behavior in these radically different temporal regimes.

References

1. J. H. Meijer and W. J. Rietveld, Physiol. Rev., 69, 671 (1989).
2. A. A. Borbely and I. Tobler, Physiol. Rev. , 69, 605 (1989).
3. W. B. Mendelson and J. V. Martin, Life Sci., 47, PL-99 (1990).
4. L. Scrima, P. G. Hartman, F. H. Johnson, Jr., E. E. Thomas, and F. C. Hiller, Sleep, 13, 479 (1990).
5. A. A. Borbely and P. Achermann, Eur. J. Pharmacol., 195, 11 (1991).
6. H. D. Langtry and P. Benfield, Drugs, 40, 291 (1990).
7. S. M. Evans, F. R. Funderburk, and R. R. Griffiths, J. Pharm. Exp. Ther., 255, 1246 (1990).
8. W. Scheuler, Neuropsychobiol., 23, 213, (1990-1991).
9. M. F. Piercey, W. E. Hoffmann, and M. Cooper, Brain Res., 554, 244 (1991)
10. A. Doble and I. L. Martin, TIPS, 13, 76 (1992).
11. J. K. Walsh, P. K. Schweitzer, J. L. Sugerman, and M. J. Muehlbach, J. Clin. Psychopharmacol., 10, 184 (1990).
12. R. D. Smith, S. T. Inouye, and F. W. Turek, J. Comp. Physiol.,164, 805 (1989).
13. P. Lavie, Psychopharmacol., 101, 250 (1990).
14. G. Copinschi, A. Van Onderbergen, M. L'Hermite-Baleriaux, M. Szyper, A. Caufriez, D. Bosson, M. L' Hermite, C. Robyn, F. W. Turek, and E. Van Cauter, Sleep, 13, 232 (1990).
15. F. W. Turek, and S. Losee-Olson, Nature, 321, 167 (1986).
16. D. M. Edgar, W. F. Seidel, and W. C. Dement, Psychopharmacol., 105, 374 (1991).
17. D. M. Edgar, W. F. Seidel, C. E. Martin, P. P. Sayeski, and W. C. Dement, Neurosci. Lett., 125, 125 (1991).
18. S. M. Biello, M. E. Harrington, and R. Mason, Brain Res., 552, 47 (1991).
19. K. M. Michels, L. P. Morin, and R. Y. Moore, Brain Res., 531, 16 (1990).
20. L. Smale, K. M. Michels, R. Y. Moore, and L. P. Morin, Brain Res., 515, 9 (1990).
21. G. Juhasz, K. Kekesi, Z. Emri, I. Soltesz, and V. Crunelli, Neurosci. Lett., 114, 333 (1990).
22. Y. Komoda, H. Kazuki, and S. Inoye, Chem. Pharm. Bull., 38, 2057 (1990).
23. A. van den Pol, J. Neurosci., 11, 2087 (1991).
24. C. S. Colwell, R. G. Foster, and M. Menaker, Brain Res., 554, 105 (1991).
25. K. Ohi, M. Takashima, T. Nishikawa, and K. Takahashi, Neuroendocrinol., 53, 344 (1991).
26. Y. I. Kim and F. E. Dudek, J. Physiol., 444, 269 (1991).
27. J. H. Meijer, E. A. van der Zee, and M. Dietz, Neurosci. Lett., 86, 177 (1988).

28. J. Velazquez-Moctezuma, P. J. Shiromani , and J. C. Gillin, Prog. Brain Res., 84, 407 (1990).
29. A. Berkowitz, L. Sutton, D. S. Janowsky and J. C. Gillin, Psychiatr. Res., 33, 113 (1990).
30. J. Velazquez-Moctezuma, M. Shalauta, J. C. Gillin, and P. J. Shiromani, Brain Res., 543, 175 (1991).
31. J. Velazquez-Moctezuma, M. Shalauta, J. C. Gillin, and P. J. Shiromani, Neurosci. Lett., 115, 265 (1990).
32. J. Velazquez-Moctezuma, M. Shalauta, J.C. Gillin, and P. J. Shiromani, Psychopharmacol. Bull., 26, 349 (1990).
33. J. H. Meijer, E. A. van der Zee, and M. Dietz, J. Biol. Rhythms, 3, 333 (1988).
34. L. Trachsel, H. C. Heller, W. C. Dement, and J. D. Miller, Soc. Neurosci. Abstr., 17, 672 (1991).
35. E. A. van der Zee, C. Streefland, A. D. Strosberg, H. Schroder, and P. G. M. Luiten, Brain Res., 542, 348 (1991).
36. K. Pungor, M. Papp, K. Kekesi, and G. Juhasz, Brain Res., 525, 310 (1990).
37. J. B. Roufs, Med. Hypotheses, 33, 269 (1990).
38. M. Trampus, N. Ferri, A. Monopoli, and E. Ongini, Eur. J. Pharmacol., 194, 189 (1991).
39. J. M. Monti, M. Fernandez, and H. Jantos, Neuropsychopharmacol., 3, 153 (1990).
40. M. G. Demontis, P. Fadda, P. Devoto, M. C. Martellotta, and W. Fratta, Neurosci. Lett., 117, 224 (1990).
41. S. Nishino, J. Arrigoni, D. Valtier, J. D. Miller, C. Guilleminault, W. C. Dement, and E. Mignot, J. Neurosci., 11, 2666 (1991).
42. J. D. Miller, K. F. Faull, S. S. Bowersox, and W. C. Dement, Brain Res., 509, 169 (1990).
43. S. Nishino, L. Haak, H. Shepherd, C. Guilleminault, T. Sakai, W. C. Dement, and E. Mignot, J. Pharm. Exp. Ther., 253, 1145 (1990).
44. D. Lagarde and C. Milhaud, Sleep, 13, 441 (1990).
45. J. F. Hermant, F. A. Rambert, and J. Duteil, Psychopharmacol., 103, 28 (1991).
46. T. Z. Guo, J. Tinklenberg, R. Oliker, and M. Maze, Anesthesiol., 75, 252 (1991).
47. A. N. Nicholson and P. A. Pascoe, Neuropharmacol., 30, 367 (1991).
48. R. T. Fremeau, Jr., G. E. Duncan, M. G. Fornaretto, A. Dearry, J. A. Gingrich, G. R. Breese, and M. C. Caron, Proc. Nat. Acad. Sci., 88, 3772 (1991).
49. A. Mansour, J. Meador-Woodruff, S. Burke, J. Bunzow, H. Akil, H. H. M. Van Tol, O. Civelli, and S. J. Watson, Soc. Neurosci. Abstr., 17, 599 (1991).
50. N. G. Weiland and P. M. Wise, Endocrinol., 126, 2392 (1990).
51. P. P. Sayeski, D. M. Edgar, J. D. Miller, and W. C. Dement, Soc. Res. Biol. Rhythms Abstr., Second Annual Meeting, 51, Amelia Island, Fla. (1990).
52. S. Honma, K. I. Honma, and T. Hiroshige, Physiol. Behav., 49, 787 (1991).
53. J. F. Ruis, J. P. Buys, T. Cambras, and W. J. Rietveld, Physiol. Behav. 47, 917 (1990).
54. A. M. Rosenwasser, Pharmacol. Biochem. Behav. 35, 35 (1990).
55. B. L. Jacobs, J. Clin. Psychiat., 52, 17 (1991).
56. C. M. Beasley, Jr., B. E. Dornseif, J. A. Pultz, J. C. Bosomworth, and M. E. Sayler, J. Clin. Psychiat., 52, 294 (1991).
57. D. J. Kupfer, J. M. Perel, B. G. Pollock, R. S. Nathan, V. J. Grochocinski, M. J. Wilson, and A. B. McEachran, Biol. Psychiat., 29, 23 (1991).
58. D. J. Kupfer, B. G. Pollock, J. M. Perel, D. B. Jarrett, A. B. Mc Eachran, and J. M. Miewald, Psychiat. Res., 36, 279 (1991).
59. D. J. Dijk, A. Strijkstra, S. Daan, D. G. M. Beersma, and R. H. Van den Hoofdaker, Psychopharmacol., 103, 375 (1991).
60. A. L. Sharpley, R. A. Solomon, A. I. Fernando, J. M. da Roza Davis, and P. J. Cowen, Psychopharmacol.,101, 568 (1990).
61. C. Idzikowski, F. J. Mills, and R. J. James, Br. J. clin. Pharmacol., 31, 193 (1991).
62. B. Bjorvatn, and R. Ursin, Behav. Brain Res., 40, 239 (1990).
63. D. M. Edgar, W. F. Seidel, and W. C. Dement, Sleep Res., 19, 58 (1990).
64. W. B. Mendelson, J. V. Martin, and D. M. Rapoport, Am. Rev. Respir. Dis., 141, 1527 (1990).
65. R. A. Prosser, J. D. Miller, and H. C. Heller, Brain Res., 534, 336 (1990).
66. R. A. Prosser, J. D. Miller, and H. C. Heller, Soc. Neurosci. Abstr. 17, 672, 1991.
67. R. A. Prosser, J. D. Miller, and H. C. Heller. Brain Res., in press (1992).
68. M. Dahlitz, B. Alvarez, J. Vignau, J. English, J. Arendt, and J. D. Parkes, Lancet, 337, 1121 (1991).
69. J. A. Siuciak, J. M. Fang, and M. L. Dubocovich, Eur. J. Pharmacol., 180, 387 (1990).
70. A. J. McArthur, M. U. Gillette, and R. A. Prosser, Brain Res., 565, 158 (1991).
71. L. Palm, G. Blennow, and L. Wetterberg, Ann. Neurol., 29, 336 (1991).
72. S. Folkard, J. Arendt, M. Aldhous, and H. Kennett, Neurosci. Lett., 113, 193 (1990).
73. J. C. Schwartz, J. M. Arrang, M. Gabarg, H. Pollard, and M. Ruat, Physiol. Rev. ,71, 1 (1991).
74. M. Denoyer, M. Sallanon, C. Buda, K. Kitahama, and M. Jouvet, Brain Res., 539, 287 (1991).
75. Y. Kudo and M. Kurihara, J. Clin. Pharmacol., 30, 1041 (1990).
76. J. M. Monti, C. Orellana, M. Boussard, H. Jantos, and S. Olivera, Brain Res. Bull., 25, 229 (1990).
77. J. S. Lin, K. Sakai, G. Vanni-Mercier, J. M. Arrang, M. Gabarg, J. C. Schwartz, and M. Jouvet, Brain Res., 523, 325 (1990).
78. N. Itowi, A. Yamatodani, S. Kiyono, M. L. Hiraiwa, and H. Wada, Physiol. Behav. 49, 643 (1991).
79. N. Itowi, A. Yamatodani, K. Nagai, H. Nakagawa, and H. Wada, Physiol. Behav., 47, 549 (1990).

80. N. Itowi, A. Yamatodani, T. Mochizuki, and H. Wada, Neurosci. Lett., 123, 53 (1991).
81. L. Rosenthal, T. Roehrs, A. Zwyghuizen-Doorenbos, D. Plath, and T. Roth, Neuropsychopharmacol., 4, 103 (1991).
82. L. C. Johnson, C. L. Spinweber, and S. A. Gomez, Psychopharmacol., 101, 160 (1990).
83. M. Hawkins, S. O'Connor, M. Radulovacki, S. Bowersox, E. Mignot, and W. Dement, Pharmacol. Biochem. Behav., 38, 1 (1991).
84. R. M. Virus, S. Ticho, M. Pilditch, and M. Radulovacki, Neuropsychopharmacol., 3, 243 (1990).
85. E. P. Senba, P. E. Daddona, T. Watanabe, J. Y. Wu, and J. I. Nagy, J. Neurosci., 5, 3393 (1985).
86. S. R. Ticho, and M. Radulovacki, Pharmacol. Biochem. Behav., 40, 33 (1991).
87. D. F. Ehret, V. R. Potter, and K. W. Dobra, Science, 188, 1212 (1975).
88. M. Dershwitz, C. E. Rosow, P. M. DiBiase, and A. Zaslavsky, Anesthesiol., 74, 717 (1991).
89. N. Chastrette, R. Cespuglio, and M. Jouvet, Neuropeptides, 15, 61 (1990).
90. N. Chastrette, R. Cespuglio, Y. L. Lin, and M. Jouvet, Neuropeptides, 15, 75 (1990).
91. M. Sakanaka, S. Magari, N. Inoue, and K. Lederis, Cell and Tissue Res., 260, 549 (1990).
92. G. C. Desjardins, J. R. Brawer, and A. Beaudet, Brain Res., 536, 114 (1990).
93. M. T. Pacheco-Cano, F. Garcia-Hernandez, O. Prospero-Garcia, and R. Drucker-Colin, Sleep, 13, 297 (1990).
94. A. Morin, L. Denoroy, and M. Jouvet, Brain Res., 538, 136 (1991).
95. H. E. Albers, S. Y. Liou, E. G. Stopa, and R. T. Zoeller, J. Neurosci., 11, 846 (1991).
96. L. Vecsei and E. Widerlov, Prog. Neuro-Psychopharmacol. Biol. Psychiat., 14, 473 (1990).
97. K. Shinohara, Y. Isobe, J. Takeuchi, and S. T. Inouye, Neurosci. Lett., 129, 59 (1991).

Chapter 3. Advances in Central Serotoninergics

Arthur G. Romero, Robert B. McCall
The Upjohn Company
Kalamazoo, MI 49001

Introduction - Serotonin (5-hydroxytryptamine, 5-HT, 1) research has continued at a rapid pace. Never before has a single neurotransmitter been implicated in the etiology or treatment of so many medical problems. These include anxiety, depression, obsessive-compulsive disorder, schizophrenia, hypertension, stroke, migraine, and nausea (1, 2). The ability to treat these distinctive disease states arises from differential drug interactions at multiple 5-HT receptor subtypes (5-HT1A-D, 5-HT2, 5-HT3, 5-HT4). Great strides are being made in the elucidation of the physiological components linked to these multiple receptor subtypes. Indeed, the forefront of 5-HT research today involves the search for agents that selectively interact with one receptor subtype.

Selectivity - Despite the advances indicated above, true receptor selectivity often remains elusive (3). No agents exist with significant selectivity for 5-HT1C or 5-HT4 receptors. 5-HT1A, 5-HT1B, or 5-HT1D receptor selective antagonists are not yet available. No selective agonist yet exists that can discriminate between the 5-HT1C and 5-HT2 receptors. Agents once thought to be selective for 5-HT3 receptors are now found to have affinity for the 5-HT4 receptor.

5-HT1 RECEPTORS

The 5-HT1 receptor is a high affinity binding site and is divided into four subgroups. The 5-HT1A and 5-HT1B receptor subtypes have been elucidated on the basis of their different affinities for 8-OH-DPAT (2) and spiperone (4, 5). 5-HT1B sites, while present in mice and rats, are not found in man where it is believed that 5-HT1D sites serve a similar function (6, 7, 8). All of the 5-HT1 receptor subtypes, with the exception of the 5-HT1C site, utilize adenylate cyclase as the second messenger system (3). The 5-HT1C site should probably be moved under the 5-HT2 heading since both are coupled to a phosphatidylinositol second messenger system and both possess a high level of homology in their amino acid sequences. The 5-HT1 subtypes appear to exert different functions at both the cellular and behavioral levels and there is autoradiographic evidence for a differential distribution of 5-HT1 subtypes in the brain (9).

5-HT1A - In the CNS, the 5-HT1A receptor is broadly distributed, occurring as a somaldodendritic autoreceptor on 5-HT neurons located in raphe' nuclei, and post-synaptically in other areas such as the hippocampus (Reviews: 10, 11, 12). The 5-HT1A receptor has been widely implicated in anxiety and depression (13, 14, 15, 16).

Tetralins - 8-OH-DPAT (2) is the prototypical 5-HT1A agonist (17) and in its radiolabelled form is used to characterize the 5-HT1A receptor. Tetralin 3 exhibited nM binding at the 5-HT1A receptor and was found to lower 5HIAA levels and increase serum corticosterone levels in rats at doses between 0.1 and 1.0 mg/kg (s.c.) (18). The closely related chromane 4 also possessed high affinity binding (19). The series of fused tricyclic amines 5-8 exhibit modest binding at the 5-HT1A receptors, with 8 being the most potent (20). The cyclic ether 9 was less potent (21) than the parent phenol, 8-OH-DPAT (2). U92016A 10 was both selective and extremely potent (Ki = 0.07) at the 5-HT1A receptor and exhibited full agonist behavior in the spontaneously hypertensive rat (SHR) model (22). Substituting a pyridine for the benzene ring in 8-OH-DPAT gave tetrahydroquinoline 12 which was 15 times more potent than the dipropyl analog 11 (23). An investigation into geometrically different fused tricyclic ring systems led to 13 and 14. It was found that 13 was much less potent than 14, which had an ED_{50} for decreasing 5-HTP accumulation of

	R
11	Me
12	NHCOPh

	n	X	R
16	2	Cl	Me
17	2	Cl	allyl
18	2	OMe	allyl

5 uM/kg; the authors present a rationale for this disparity (24). A series of substituted ergoline analogs were all potent and selective at the 5-HT1A receptor. For example, LY228729 (15) was a potent and selective agonist (25, 26). Members of the tricyclic azepine series (16-18) possessed high potencies for the 5-HT1A receptor. However, they lacked good selectivity, expressing similar affinity for the alpha-2 adrenoceptor. Varying the size of the bottom ring (n = 1, 3) was not fruitful, affording analogs with reduced affinity for the 5-HT1A receptor.

Arylpiperazines - A growing body of evidence indicates that compounds in the arylpiperazine class have anxiolytic and antidepressant properties in man. Unlike the clinically efficacious buspirone and ipsapirone, which are partial agonists (27), flesinoxan (19) is a full agonist (28). Despite pronounced hypotensive effects in the SHR, 19 caused only transient and weak hypotensive effects in healthy humans (29). EMD56551 (**20**) was a selective partial agonist (30), inhibiting forskolin-stimulated adenylate cyclase levels, but also capable of antagonizing the inhibitory effects of 8-OH-DPAT. While SUN8399 (**21**) possessed a high potency for 5-HT1A binding, it also displayed some D2 and

alpha adrenergic activity *in vivo* (31). The naphthalenylpiperazine S14671 (**22**) is a potent agonist at 5-HT1A receptors with a selectivity of 30-fold over 5-HT1C and 5-HT2 receptors, where it is an antagonist (32). The potent 5HT1A agonist PAPP (LY165,163 **23**) also possesses dopaminergic activity (33). A slightly modified analog, SR-57746A (**24**) is reported to promote fecal elimination by stimulating 5-HT1A receptors, although this has not been rigorously proven (34). The closely related CM 57493 (**25**) has been reported to be neuroprotective in models of focal cerebral ischemia, with protective properties the same order of magnitude as the NMDA antagonist phencyclidine (35). This effect is thought to be mediated by an inhibitory action on neurons and not simply a result of agonist induced hypothermia (36).

Novel Structures - The unusual amidine MDL 102181 (**26**), displayed a Ki of 8 nM for 5-HT1A. The partial agonist MDL 73005EF (**27**) possessed a similar affinity and is active in animal models of anxiety such as the Vogel. Unlike the substituted pyrimidylpiperazines, it cannot be metabolically cleaved to 1-pyrimidylpiperazine (1-PP), an alpha-2 adrenoreceptor antagonist. Both **26** and **27** displayed agonist properties at dorsal raphe autoreceptors with higher potency than buspirone (37, 38). The 5-HT1A SAR has been recently reviewed (10, 39, 40).

5-HT1A Antagonists - Selective 5-HT1A antagonists remain elusive. While not selective, the most commonly used 5-HT1A antagonists are spiperone, propranolol, and pindolol (3). Propranolol and pindolol (41) bind at 5-HT1B and beta-adrenoceptors while spiperone (42) binds with high affinity at 5-HT2 and DA sites. Several compounds reported to be more selective have appeared. NAN-190 prevents the inhibitory effect of 8-OH-DPAT on forskolin stimulated adenylate cyclase activity. However, it has been shown to be a low intrinsic activity partial agonist and is a potent alpha-1 adrenoreceptor blocker (43, 44). Significantly, in the commonly used model where antagonists are used to reverse 8-OH-DPAT induced hypothermia, the mouse has been shown to be relatively insensitive to antagonists, compared with the rat (45, 46). S 10463 (**28**) inhibits the hypotensive effects of 8-OH-DPAT without any effect on beta-adrenoreceptors (42). Reported as the first selective anxiolytic to act as an antagonist in both pre- and post-synaptic receptor models,

WAY100135 (**29**) displays activity in the mouse light-dark box, suggesting that the anxiolytic potential of 5-HT1A antagonists should be further evaluated (47). The antagonist RK-153 (**30**) binds at 5-HT1A receptors with 160-fold selectivity compared to alpha-1 adrenergic receptors (48). MEP-177 (**31**) was an analog of propranolol that lacked significant affinity for 5-HT1B and beta-adrenoceptors (49). A related analog, **32**, has been found to be a potent selective 5-HT1A antag-

onist. Substituting large, sterically hindered groups on the amine renders this series more selective for 5-HT1A *vs.* beta-adrenergic receptors. The aporphine **33** represents a novel class of 5-HT1A antagonists (50). The (S)-isomer of tetralin **34** has been found to inhibit 8-OH-DPAT induced biochemical and behavioral effects in a dose dependent manner; it binds to 5-HT1A sites with modest affinity and eight-fold selectivity compared to D2 sites (51).

5-HT1B Receptors - Although the 5-HT1B receptor is found in rats and mice, its presence has not been demonstrated in humans (7), where the 5-HT1D receptor, which is found in similar brain locations in man, has been proposed to serve an analogous function. However, little correlation exists between the binding affinities of compounds at the 5-HT1B and 5-HT1D sites (52). m-Tri-fluoromethylphenylpiperazine (TFMPP) and its chloro counterpart m-CPP are frequently used as selective ligands, yet display less than three-fold selectivity for the 5-HT1B site (54). Propranolol and pindolol are used as antagonists, but they are not selective and may be partial agonists (3).

Tetrahydropyridines are the most selective 5-HT1B agents yet identified, with RU-24969 **35** serving as the prototypical ligand (53). While **35** is somewhat more selective and potent, it binds with only slightly less affinity at 5-HT1A sites (54). Indeed, in pigeons, which lack 5-HT1B receptors, **35** behaves as a 5-HT1A agonist (55). The homolog CP-96501 (**36**) possesses a similar affinity for the 5-HT1B site (IC50 = 1.5 nM), yet is more selective with a six-fold improvement for 5-HT1B over 5-HT1A receptors (56). The rotationally restricted analog CP-93129 (**38**) is much more selective for the 5-HT1B receptor, although less potent (Ki = 15 nM) than **35** (57). It is postulated that the carbonyl serves as a rotationally restricted bioisostere for the methoxy group. While this compound does not cross the blood-brain barrier, direct infusion into the brain in rodents significantly inhibits food uptake, implicating the 5-HT1B receptor in regulating feeding. CGS-12066B (**37**) was reported to be 5-HT1B selective, but this is now disputed (58, 59). Autoradiography demonstrates a preference for 5-HT1B sites for S-CM-GTNH$_2$ (**39**) (60). The pharmacology of the 5-HT1B receptor has recently been reviewed (3, 50, 61).

<u>5-HT1C Receptors</u> - Most agents that bind at the 5-HT1C receptor also bind at 5-HT2 receptors. Indeed (*vide supra*) the 5-HT1C receptor is coupled to the same second messenger system as the 5-HT2 receptor and has a high level of homology in its amino acid sequence, suggesting that the 5-HT1C receptor may belong to the 5-HT2 receptor family (62). DOI and DOB are probably 5-HT1C agonists, but are also prototypical 5-HT2 receptor ligands (63). m-CPP and quipazine have been utilized as 5-HT1C agonists, yet are far from selective (54). No 5-HT1C receptor selective ligands have been reported.

<u>5-HT1D Receptors</u> - Sumatriptan <u>40</u> has been shown to be clinically effective in the treatment of acute migraine (64). Evidencing only five fold selectivity for 5-HT1D ($Ki = 17$ nM) vs. 5-HT1A receptors and not readily crossing the blood-brain barrier (65), the precise role of 5-HT1D receptor

	R
<u>39</u>	$OCOCH_2NH$-Tyrosinamide
<u>40</u>	$CH_2SO_2NHCH_3$
<u>41</u>	$OCH_2(p\text{-}Cl\text{-}Ph)$

	R_1	R_2
<u>42</u>	H	H
<u>43</u>	OCH_3	H
<u>44</u>	H	OCH_3

activation and the locus of <u>40</u>'s effects have not been fully established, although it is thought to act on the cerebral vasculature. The 5-HT1D receptor shows a preference for indole containing structures, with CC-263 (<u>41</u>) binding at 5-HT1D receptor sites (66) with a fifteen-fold greater affinity than <u>40</u>. Many ergolines, particularly metergoline and methysergide, have high affinity for the 5-HT1D receptor (67). Naphthylpiperazines <u>42</u>-<u>44</u> show a split in affinity for the 5-HT1D receptor (62). Compounds <u>42</u> and <u>43</u> bind with very high affinity ($Ki = 14$ nM and 2 nM, respectively), whereas <u>44</u> binds with a Ki of 545 nM. No simple or substituted phenylpiperazine has been shown to be as potent as <u>42</u> and <u>43</u>. The threonine-coupled indole <u>39</u> is only one half as potent on 5-HT1D as 5-HT1B receptor sites (68). Although there are many compounds selective for 5-HT1A over 1D receptors, the reverse is not true: no 5-HT1D receptor ligands have yet been found to be truly selective over 5-HT1A. The SAR of the 5-HT1D receptor has been reviewed (3, 66, 69, 70).

<u>5-HT2 RECEPTORS</u>

<u>5-HT2 Receptor Agonists</u> - DOB and DOI are still considered the prototypical 5-HT2 receptor agonists, although they are not selective. DOI for example has equal affinity for 5-HT2 and 5-HT1C receptors (71). It has recently been shown that DOB stimulates hypothalamic corticotropin releasing factor (CRF) secretion, possibly through a 5-HT2 mechanism (72). Chronic infusion down-regulates both 5-HT2 and CRF receptors (73).

<u>5-HT2 Receptor Antagonists</u> - Although it is not selective, ketanserin is the prototypical 5-HT2 receptor antagonist. Unlike ketanserin, which is an antagonist at alpha-1 adrenoceptors and lowers blood pressure (BP), ritanserin is a more selective 5-HT2 receptor antagonist. However, ritanserin produces only modest behavioral and biochemical effects in man. These include increasing the release of DA and 5-HT in the nucleus accumbens, thus implying potential as an antipsychotic (74). A mixed D2/5-HT2 blocking effect has been suggested to contribute to the clinical efficacy of atypical antipsychotics such as clozapine (75).

There has been a longstanding interest in 5-HT2 antagonists for the prophylactic treatment of migraine. Newer 5-HT2 receptor antagonists include (+)-CV-5197 (**45**), which has good affinity for the 5-HT2 receptor and completely inhibits the 5-HT induced contraction of the isolated pig coronary artery at a concentration of 3×10^{-7} M. The (-)-isomer was completely inactive (76). Naphthosultam RP 62203 (**46**) possessed potent affinity for the 5-HT2 receptor and is orally effective and long lasting as a 5-HT2 antagonist in the mescaline - induced head twitch test in mice and rats (77). An analog combining structural elements of both ketanserin and ritanserin, **47**, has greater than twelve-fold selectivity over both, coupled with high 5-HT2 receptor affinity, giving a ratio of affinity for 5-HT2/alpha-1 adrenoceptors of 606 (78). The position of the sulfur is important for the best activity. Quinolines ICI 169,369 (**48**) and ICI 170,809 (**49**) possess good to excellent affinity for the 5-HT2 receptor, respectively, with no effect on alpha-1 adrenoceptors. These compounds have been shown to block 5-HT induced contractions in human temporal and cerebral arteries and thus may have potential for migraine therapy (79). The spiperone derivative QF-0104B (**50**) is equipotent to methysergide, a clinically effective drug used for migraine prophylaxis, in inhibiting contractions in rat aorta rings (80). Unlike the non-selective ergoline 5-HT2 antagonists methysergide, metergoline, and mianserin, LY 237733 (**51**) acts as an antagonist at the 5-HT2 receptor with no alpha-1 adrenoceptor effects (81). Ergolines often contain high 5-HT1C receptor affinity.

5-HT3 RECEPTORS

Despite the large amount of research aimed at discovering 5-HT3 receptor ligands, selective ligands have only recently become available. Thus, the majority of putative indications for 5-HT3 receptor compounds, including anxiety (82), migraine (83), schizophrenia (84), and cognition disorders (85) have yet to be validated clinically. Many 5-HT3 receptor antagonists also have affinity for the 5-HT4 receptor. Thus, some of the earlier 5-HT3 antagonists (86) which possess both anti-emetic and gastrointestinal (GI) prokinetic effects are now believed to exhibit their prokinetic effects by acting as antagonists at GI 5-HT4 receptors (87).

A large number of compounds have been prepared which bind at the 5-HT3 receptor as antagonists. Many of these compounds are clinically effective in reducing the nausea caused by radiation and chemotherapy cancer treatments (88, 89). These compounds are structurally unique compared to agents with high affinity for the other 5-HT receptor subtypes. Most analogs fall into two general structural classes, the tropine-like (bridged bicyclic amines) and the imidazole containing compounds. Two early examples, MDL-72222 (**52**) and ICS 205-930 (**53**), were both found to have anti-emetic properties (90, 91). In addition **53** possesses affinity for the 5-HT4 receptor and is currently the only available 5-HT4 receptor antagonist. Y-25130 (**54**) also acts as an anti-emetic in animal models. One focus of interest in these 5-HT3 receptor mediated models of emesis inhibition is due to interest in developing drugs as cojuncts to cancer therapy. Cancer radiation and chemotherapy may increase the level of 5-HT in the area postrema as well as activating sensory fibers in the gut, leading to nausea (92). The anti-nausea effect of these 5-HT3 antagonists is believed to be due to their antagonist activity at both peripheral and CNS 5-HT3 receptors. Peripherally, it is thought to be the result of blocking the sensory input at the sites of sensory nerve endings in the gut. The CNS effect may be due to blockade of 5-HT3 receptors in the area postrema. Evidence that some portion of the 5-HT3 antagonist anti-emetic effect is centrally mediated comes from injecting 5-HT3 antagonists directly into the area postrema in the brain, an area which contains a high density of 5-HT3 receptors, to briefly inhibit cis-platin induced emesis in ferrets (93). Oxazoles, such as in **55**, have been shown to serve as bioisosteres for the often present amide and ester functional groups (94). Zatosetron (LY 277359, **56**), a highly selective 5-HT3 antagonist, has been shown to inhibit emesis in dogs without any effect on gastric emptying (95). It was also found that **56** decreased the number of spontaneously active A10 DA cells while not changing the number of spontaneously active A9 DA cells, suggesting that it may be indicative of an atypical antipsychotic (96).

Ondansetron (GR 38032, **57**) has been shown to be clinically effective in blocking the induction of radiation and cancer chemotherapy induced nausea, an event suggested to involve

activation of 5-HT3 receptors in the area postrema (95, 97). It has also been reported that **57** has shown antipsychotic activity in at least one schizophrenic patient (98). CP-93318 (**58**) binds at the 5-HT3 receptor with a Ki of 0.42 nM, representing a forty-fold greater affinity than **57** and five-fold greater affinity than **53** as well as possessing greater efficacy in anti-emesis models than **57** (99). Both **59** and YM060 (**60**) show high affinity for the 5-HT3 receptor, demonstrating that the activity is not limited to simple imidazoles in this series (100, 101). Interestingly, the quaternary ammonium salt **61** is more active than the corresponding tertiary amine as a 5-HT3 receptor antagonist *in vivo*, suggesting that these ligands may bind in their protonated forms at the 5-HT3 site (102). Similarly, the quaternary ammonium salt of **53** displays high affinity for the 5-HT3 receptor site (103). 5-HT3 receptor antagonists have been reviewed (41, 72, 84, 104).

5-HT3 Receptor Agonists - The most widely used 5-HT3 receptor agonist is 2-methyl-5-HT (105, 106). It is lower in affinity and potency than 5-HT, but more selective for 5-HT3 receptors. Phenylbiguanide **62** is a 5-HT3 agonist or partial agonist that binds to 5-HT3 receptors with one-fifth the affinity of 5-HT (107). The chloro-substituted analog m-CPBG (**63**) binds with 10^3 greater affinity than **62** (109). The trimethyl quaternary salt of 5-HT itself displays ten-fold greater affinity with significantly greater selectivity than 5-HT (108), again demonstrating that the 5-HT3 receptor probably binds ligands in their protonated form. This may be useful in designing peripherally selective 5-HT3 agents which will not cross the BBB.

	X
62	H
63	Cl

SUMMARY

It is fascinating to consider the multitude of diverse compounds which bind to serotonin receptors, where small variations in structure often lead to greatly dissimilar affinities and selectivities for these receptor subtypes. In turn, through differential distribution in the brain and utilization of unique second messenger systems, these receptors elicit a large number of diverse biochemical and behavioral responses. The lengthy number of clinical applications currently being examined bears testimony to this. As more selective agents become available, both agonists and antagonists, a better understanding of this rapidly growing field will result.

References

1. J.R. Fozard and P.R. Saxena, Ed., "Serotonin: Molecular Biology, Receptors and Functional Effects," Birkhauser Verlag, Basel, Switzerland, 1991.
2. S.J. Peroutka in "Receptor Biochemistry and Methodology; Serotonin Receptors Subtypes - Basic and Clinical Aspects," Vol. 15, Wiley-Liss, New York, NY, 1991.
3. R.A. Glennon, M. Dukat, Pharmacol. Biochem. and Behav., $\underline{40}$, 1009 (1991).
4. N.W. Pedigo, H.I. Yamamura, D.L. Nelson, J. Neurochem., $\underline{36}$, 220 (1981).
5. D.N. Middlemiss, J. Fozard, Eur. J. Pharmacol., $\underline{90}$, 151 (1983).
6. R.E. Heuring, S.J Peroutka, J. Neurosci., $\underline{7}$, 894 (1987).
7. D. Hoyer, D.N. Middlemiss, Trends Pharmacol. Sci., $\underline{10}$, 130 (1989).
8. D. H. Hoyer, C. Weber, A. Pazos, A. Probst, J.M. Palacios, Neurosci Lett., $\underline{85}$, 357 (1988).
9. A. Pazos, J.M. Palacinos, Brain Res., $\underline{364}$, 205 (1985).
10. D.L. Nelson, Pharmacol. Biochem. and Behav., $\underline{40}$, 1041 (1991).
11. A. Frazer, S. Maayani, B.B. Wolfe, Annu. Rev. Pharmacol. Toxicol., $\underline{30}$, 307 (1990).
12. A.W. Schmidt, S.J. Peroutka, FASEB J., $\underline{3}$, 2242 (1989).
13. R.A. Glennon, Neurosci. Biobehav. Rev., $\underline{14}$, 35 (1990).
14. B.E. Suranyi-Cadotte, S.R. Bodnoff, S.A. Welner, Prog. Neuro-psychopharmacol. Biol. Psychiatry, $\underline{14}$, 633 (1990).
15. D.S. Robinson, D.R. Alms, R.C. Shrotriya, M. Messina, P. Wickramaratne, Psychopathology, $\underline{22}$ [suppl.], 27 (1989).
16. J.L. Rausch, R. Ruegg, F.G. Moeller, Psychopharmacol. Bull., $\underline{26}$, 169 (1990).
17. C.T. Dourish, S. Ahlenius, P.H. Hudson, Eds. "Brain 5-HT1A Receptors," Chichester: Ellis Horwood Ltd; 1987.
18. J. Schaus, D.L. Huser, R.D. Titus, L.S. Hoechstetter, D.T. Wong, R.D. March, R.W. Fuller, H.D. Snoddy, in "Soc. Neruosci.,", New Orleans, LA, USA, Abstr. 616.3 (1991).
19. L.-G. Larsson, R. Noreen, D. Sohn, B.E. Svensson, S.O. Thorburg, WO Patent 9,109,853 (1989).
20. A.A. Hancock, M.D. Meyer, J.F. DeBernardis, J. Recept. Res., $\underline{11}$, 177 (1991).
21. B. Hoeoek, H. Yu, T. Mezei, L. Bjoerk, B. Svensson, N.E. Anden, U. Hacksell, Eur. J. Med. Chem., $\underline{26}$, 215 (1991).
22. H.V. Wikstrom, P.A. Carlsson, B.R. Andersson, K.A. Svensson, S.T. Elebring, N.P. Stjernlof, A.G. Romero, S.R. Haadsma, C.-H. Lin, M.D. Ennis, Patent Applic. WO 9,111,435 (1990).
23. I.A. Cliffe, M.L. Mansell, A.C. White, R.S. Todd, Eur. Patent 0,395,244 (1989).
24. C. Merlin, J. Vallgaarla, D.L. Nelson, L. Bjoerk, Y. Kena, A. Hong, N.E. Anden, I. Csaeregh, L.E. Arvidsson, U. Hacksell, J. Med. Chem., $\underline{34}$, 497 (1991).
25. M.E. Flaugh, D.L. Mullen, R.W. Fuller, N.R. Mason, J. Med. Chem., $\underline{32}$, 1746 (1988).
26. J.L. Slaughter, M.A. Harrington, S.J. Peroutka, Life Sci., $\underline{47}$, 1331 (1990).
27. R.D. Clark, K.K. Weinhardt, J. Berger, L.E. Fisher, C.M. Brown, A.C. MacKinnon, A.T. Kilpatrick, M. Spedding, J. Med. Chem., $\underline{33}$, 633 (1990).
28. R. Andrade and R.A. Nicoll, Naunyn-Schmeideberg's Arch. Pharmacol., $\underline{336}$, 5 (1987).

29. W. Wouters, M.T.M. Tulp, P. Bevan, Eur. J. Pharmacol., 149, 213 (1988).
30. J.M. deVoogd and B. Uckert, in "Fourth European Meeting on Hypertension," A. Zanchetti, Ed., Milan, Italy, Abstr. (1989).
31. G.D. Bartoszyk, H. Bottcher, H. Grenier, J. Harting, C.A. Seyfried, in "Soc. Neurosci.," New Orleans, LA, USA, Abstr. 637.8 (1991).
32. I. Hirotsu, M. Harada, K. Saito, M. Shibata, K. Nomura, T. Tatsuoka, T. Ohno, T. Isihara, in "Soc. Neurosci.,", New Orleans, LA, USA, Abstr. 637.2 (1991).
33. H. Canton, J.-M. Rivet, F. Lejeune, M. Laubie, M. Brocco, K. Bervoets, G. Laville, M.J. Millan, in "Soc. Neurosci.," New Orleans, LA, USA, Abstr. 39.8 (1991).
34. M.J. Millan, K. Bervoets, M. Mavridis, Neurosci. Lett., 130, 173 (1991).
35. A. Bianchetti, T. Croci, L. Manara, Pharmacol. Res., 22 (suppl. 2), 53 (1990).
36. G.W. Bielenberg, M. Burkhardt, Stroke, 21 (suppl. IV), 16 (1990).
37. J.H.M. Prehn, C. Bakhaus, C. Karkoutly, J. Nuglisch, B. Peruche, C. Rosberg, J. Krieglstein, Eur. J. Pharmacol., 203, 213 (1991).
38. J.M. Hitchcock, T.C. McCloskey, R.A. Padich, D.R. McCarty, M.W. Dudley, J.S. Sprouse, J. Freedman, J.H. Kehne, in "Soc. Neurosci.," New Orleans, LA, USA, Abstr. 116.3 (1991).
39. M. Hilbert, P. Moser, Drugs Fut., 15, 159 (1990).
40. M. Hamon, J.-M. Cossery, U.Spampinato, H. Gozlan, Trends Pharmacol. Sci., 7, 336, (1986).
41. "5-HT1A Agonists, 5-HT3 Antagonists and Benzodiazepines," R.J. Rodgers and S.J. Cooper, Ed., Wiley-Interscience, New York, NY 1991.
42. H. Dabire, R. Baijou, K. Chaouche-Teyara, B. Fournier, G. De Nanteuil, M. Laubis, M. Satar, H. Schmitt, Eur. J. Pharmacol., 203, 323 (1991).
43. J.E. Barret, S.M. Hoffman, S.N. Olmstead, M.J. Foust, C. Harrod, B.A. Weissman, Psychopharmacology, 97, 319 (1989).
44. S. Hjorth, T. Sharp, Life Sci., 46, 955 (1990).
45. Y. Claustre, L. Rouquier, A. Serrano, J. Benavides, B. Scatton, Eur. J. Pharmacol., 204, 71 (1991).
46. P.C. Moser, Eur. J. Pharmacol., 193, 165 (1991).
47. K.M. Wozniak, M.J. Durcan, M. Linnoila, Eur. J. Pharmacol., 193, 253 (1991).
48. A. Fletcher, D.J. Bill, S.J. Bill, N.T. Brammer, I.A. Cliffe, E.A. Forster, Y. Reilly, G.K. Lloyd, "Soc. Neurosci.," New Orleans, LA, USA, Abstr. 39.3 (1991).
49. R.K. Raghupathi, L. Rydelck-Fitzgerald, M. Teiter, R.A. Glennon, "Soc. Neurosci.," New Orleans, LA, USA, Abstr. 16:1036 (1990).
50. R.A. Glennon, R.B. Westkaemper, P. Bartyzel in "Sertonin Receptor Subtypes," S.J. Peroutka, Ed., Wiley-Liss, New York, 1991, p. 19-64.
51. E.E. Beedle, D.W. Robinson, D.T. Wong, US Patent 5,013,761.
52. J.G. Cannon, H. Jackson, J.P. Long, P. Leonard, R.K. Bhatnagar, J. Med. Chem., 32, 1959 (1989).
53. S.-E. Hillver, L. Bjork, Y.-L. Li, B. Svensson, S. Ross, N.E. Anden, U. Hacksell, J. Med. Chem., 33, 1541 (1990).
54. P. Schoeffter and D. Hoyer, Naunyn-Schmeideberg's Arch. Pharmacol., 339, 675 (1989).
55. M.S. Sills, B.B. Wolfe, A. Frazer, J. Pharmacol. Exp. Ther., 231, 480 (1984).
56. D. Hoyer, P. Schoeffter, C. Waeber, J.M. Palacios in "The Neuropharmacology of Serotonin," P.M. Whitaker-Azmitia, S.J. Peroutka, Eds., The New York Academy of Sciences, New York, NY, 1990, p. 168-181.
57. J.B. Hogan, H.C. Holoway, J.E. Barrett, in "Soc. Neurosci.," St Louis, MO, USA, 16:850 (1990).
58. J.E. Macor, C.A. Burkhart, L.A. Lebel, B.K. Koe,. P.M. Chalabi, J.C. Windels, R.W. Roth, J. Labelled Comp. Radiopharm., 29, 249 (1991).
59. J.E. Macor, C.A. Burkhardt, J.H. Heym, J.L. Ives, L.A. Lebel, M.E. Newman, J.A. Nielson, K. Ryan, D.W. Schulz, L.K. Torgersen, B.K. Koe, J. Med. Chem., 23, 2087 (1990).
60. R.F. Neale, S.L. Fallon, W.C. Boyar, J.W.F. Wasley, L.M. Martin, G.A. Stone, B.S. Glaeser, C.M. Stinton, M. Williams, Eur. J. Pharmacol., 136, 1 (1987).
61. J.E. Macor, C.A. Burkhardt, J.H. Heym, J.I. Ives, L.A. Lebel, M.E. Newman, J.A. Nielsen, K. Ryan, D.W. Schulz, L.K. Torgersen, B.K. Koe, J. Med Chem., 33, 2087 (1990).
62. P. Boulenguez, J. Chauveau, L. Segu, A. Morel, J. Lanoir, M. Delange, Eur. J. Pharmacol., 194, 91 (1991).
63. D.N. Middlemiss, P. H. Hutson in "The Neuropharmacology of Serotonin," P.M. Whitaker-Azmitia and S.J. Peroutka, Eds., The New York Academy of Sciences, New York, NY, 1990, p. 132-147.
64. A.W. Schmidt and S.J. Peroutka, FASEB J., 3, 2242 (1989).
65. J.C. Garratt, E.J. Kidd, I.K. Wright, C.A. Marsden, Eur. J. Pharmacol., 199, 349 (1991).
66. P.R. Saxena and M.D. Ferrari, Trends Pharmacol. Sci., 10, 200 (1989).
67. C. Waeber, P. Schoeffter, D. Hoyer, J.M. Placios, Neurochem. Res., 15, 567 (1990).
68. R.A. Glennon, A.M. Ismaiel, C. Chaurasia, M. Titeler, Drug Dev. Res., 22, 25 (1991).
69. J.L. Slaughter, M.A. Harrington, S.J. Peroutka, Life Sci., 47, 1331 (1990).
70. P. Boulenguez, J. Chauveau, L. Segu, A. Morel, M. Delaage, J. Lanoir, J. Pharm. Exp. Therap., 259, 1360 (1991).

71.   M.A. Harrington, A.J. Sleight, J. Pitha, S.J. Peroutka, Eur. J. Pharmacol, 194 83 (1991).
72.   S.J. Peroutka, Pharm. Rev., 43, 579 (1991).
73.   D. Hoyer, Trends Pharmacol. Sci., 9, 89 (1988).
74.   M.J. Owens, D.L. Knight, J.C. Ritchie, C.B. Memeroff, J. Pharmacol. Exp. Ther., 256, 787 (1991).
75.   M.J. Owens, D.L. Knight, J.C. Ritchie, C.B. Nemeroff, J. Pharmacol. Exp. Ther., 256, 795 (1991).
76.   L.L. Devaud, E.B. Hullingsworth, Eur. J. Pharmacol., 192, 427 (1991).
77.   P.A. Janssen, C.J.E. Nimegeers, F. Wouters, K.H.L. Schellekens, A.A.H.P. Megens, T.F. Meert, J. Pharmacol. Exp. Ther., 244, 685 (1988).
78.   M. Kori, K. Kamiya, E. Kurihara, H. Sugihara, Chem. Pharm. Bull., 39, 922 (1991).
79.   J.C. Mulleron, M.T. Comte, C. Gueremy, J.F. Peyronel, A. Truchon, J.C. Blanchard, A. Doble, O. Pist, J.L. Zundel, C. Huon, B. Martin, P. Mouton, A. Viroulaud, D. Allen, J. Betschart, J. Med. Chem., 34, 2477 (1991).
80.   J.B. Press, R.K. Russell, J.J. McNally, R.A. Rampulla, R. Falotico, C. Scott, J.B. Moore, S.J. Offord, J. Tobia, Eur. J. Med. Chem., 26, 807 (1991).
81.   I. Jansen, T. Blackburn, K. Eriksen, L. Edvinsson, Pharmacol. Toxicol., 68, 8 (1991).
82.   M. Loza, I. Verde, M.E. Castro, F. Orallo, J.A. Fontenla, J.M. Calleja, E. Ravina, L. Cortizo, M.L. de Ceballos, Biorg. Med. Chem. Lett, 1, 717 (1991).
83.   P.A. McBride, J.J. Mann, E. Nimchinsky, M.L. Cohen, Life Sci., 47, 2089 (1990).
84.   M.D. Tricklebank, Trends Pharmacol. Sci., 10, 127 (1989).
85.   B. Costall, A.M. Domeney, R.J. Naylor, M.B. Tylers, Br. J. Pharmacol., 92, 881 (1987).
86.   B.J. Jones, B. Costall. A.M. Domerey, M.E. Kelly, R.J. Naylor, N.R. Oakley, M.B. Tyers, Brit. J. Pharmacol., 93, 985 (1988).
87.   J.M. Barns, N.M. Barns, B. Costall, R.J. Naylor, M.B. Tyers, Nature, 338, 762 (1989).
88.   A. Schiavone, M. Volonte, R. Micheletti, Eur. J. Pharmacol., 187, 323 (1990).
89.   D.A. Craig and D.E. Clarke, J. Pharmacol. Exp. Ther., 252, 1378 (1990).
90.   M. Marty, P. Pouillart, S. Scholl, J.P. Droz, M. Azab, N. Brion, E. Pujade-Lauraine, B. Paule, D. Paes, J. Bons, N. Eng. J. Med., 322, 816 (1990).
91.   D. Cunningham, H. Pople, H.Y. Ford, J. Hawthorn, J.C. Gazet, T. Challoner, Lancet, 1461 (1987).
92.   J.R. Fozard, M.W. Gittos, Brit. J. Pharmacol., 80, 511P (1983).
93.   B.P. Richardson, G. Engel, Donatsch, P.A. Stadler, Nature, 316, 126 (1985).
94.   T. Fukuda, M. Setuguchi, K.I. Inaba, H. Shoji, T. Tahara, Eur. J. Pharmacol., 196, 299 (1991).
95.   G.A. Higgins, G.J. Kilpatrick, K.T. Bunce, B.J. Jones, M.B. Tyers, Br. J. Pharmacol., 97, 247 (1989).
96.   C.J. Swain, R. Baker, C. Kneer, J. Moseley, J. Saunders, E.M. Seward, G. Stevenson, M. Beer, J. Stanton, K. Watling, J. Med. Chem., 34, 140 (1991).
97.   M.L. Cohen, W. Bloomquist, J.S. Gidda, W. Lacefield, J. Pharmacol. Exp. Ther., 254, 350 (1990).
98.   K.Rasmussen, M.E. Stockton, J.F. Czachura, Eur. J. Pharmacol., 205, 113 (1991).
99.   I. Monkovic and J.A. Gylys, Prog. Med. Chem., 27, 297 (1990).
100.   A. White, T.H. Corn, C. Feetham, C. Falconbridge, Lancet, 337, 1173 (1991).
101.   T. Rosen and R.A. Nagel, Drugs Fut., 16, 992 (1991).
102.   A.A. Nagel, T. Rosen, J. Rizzi, J. Daffeh, K. Guarino, J. Nowakowski, L.A. Vincent, J. Heym, S. McLean, T. Seeger, M. Connolly, A.W. Schmidt, C. Siok, J. Med. Chem., 33, 13 (1990).
103.   K. Honda, K. Miyata, T. Kamato, A. Nishida, H. Ito, H. Yuki, M. Yamano, R. Tsutsumi, FASEB J., 5, ABS.4996 (1991).
104.   M.F. Hilbert, R. Hoffmann, R.C. Miller, A.A. Carr, J. Med. Chem., 33, 1594 (1990).
105.   C.J. Swain, R. Baker, C. Kneen, J. Moseley, J. Saunders, E.M. Seward, G. Stevenson, M. Beer, J. Stanton, K. Watling, J. Med. Chem., 34, 140 (1991).
106.   M. Turconi, P. Schiantarelli, F. Borsini, C.A. Rizza, H. Ladinsky, A. Donetti, Drugs Fut., 16, 1011 (1991).
107.   G.J. Kilpatrick, K.T. Bunce, M.B. Tyers, Med. Res. Rev., 10, 441 (1990).
108.   B.P. Richardson, G. Engel, P. Donatsch, P.A. Stadler, Nature, 316, 126 (1985).
109.   G.J. Kilpatrick, A. Butler, J. Burridge, A.W. Oxford, Eur. J. Pharmacol. 182, 193 (1990).
110.   R.A. Glennon, S.J. Peroutka, M. Dukat in "The Second IUPHAR Satellite Meeting on Serotonin," Basel, Switzerland, 1990, Abstr. 42.

Chapter 4. Traumatic and Ischemia/Reperfusion Injury to the CNS

John M. McCall, The Upjohn Company, Kalamazoo, MI 49001
Jill A. Panetta, Eli Lilly and Company, Indianapolis, IN 46285

Introduction - The injury mechanisms and treatment of acute and chronic neurologic disorders are being increasingly investigated. Acute neurologic disorders include head and spinal cord injury, ischemic and hemorrhagic stroke and brain damage after cardiac arrest, while the chronic disorders include multiple sclerosis, Alzheimer's and Parkinson's Disease. This chapter will concentrate on the mechanism and treatment of acute neurologic disorders. The biologic events that lead to tissue injury in these maladies are complex (1,2). Indeed, the injury process ultimately involves glutamate release, calcium influx, metabolic dysfunction, lipid peroxidation, phospholipase activation/arachidonic acid release and metabolism, calcium activated proteases and kinases, tissue water increase, lactic acidosis and more. Blood flow often drops in the area surrounding the original injury (the ischemic penumbra). Cerebral injury can occur during this ischemic episode or after blood reperfuses the area (3). The area of CNS that surrounds the original injury is at risk. This injury mosaic implies many possible types of therapeutic interventions that include NMDA antagonists, calcium channel blockers, inhibitors of lipid peroxidation, compounds that inhibit different parts of the arachidonic acid cascade, diuretics and more.

Calcium and Sodium - Calcium influx is central to the injury cascade. Cellular calcium is regulated by several distinct mechanisms at both membrane and intracellular sites. Aberrant calcium control underlies the pathology of cell death (4). Calcium enters the cell passively because of energy failure, from glutamate activated NMDA receptors, voltage gated channels and through injured membranes. High cytoplasmic levels of calcium can swell and destroy mitochondria and activate proteinases and lipases that degrade cellular protein and lipids. Since extracellular concentrations of calcium are around 10^{-3} and intracellular are $<10^{-7}$, problems with calcium homeostasis can have devastating effects on the neuron. The role of modulation of intracellular calcium by anti-ischemic drugs has been reviewed (5,6).

Nimodipine, a calcium channel blocker with cerebral arterial dilator activity, is approved for treatment of subarachnoid hemorrhage (7). Nimodipine was effective in reversing post-traumatic ischemia and promoting electrophysiologic recovery in a rat spinal cord injury model, but only when combined with adjuvant therapy that reversed the hypotensive response of the drug (8). Nimodipine had no overall effect when treatment was begun within 48 hours in a randomized, double-blind, multi-center clinical trial in stroke comparing placebo versus nimodipine (9). A new dihydropyridine calcium channel antagonist, nilvadipine (1), reduced cerebral infarction in a middle cerebral artery occlusion (MCAO) model that produces a focal stroke in rats, at 24 hours after infarction at one-tenth the effective dose of nicardipine in the model (10).

Calcium is clearly linked to other members of the injury cascade. Lipid peroxidation of rat hepatocytes induced by $FeCl_3$ was associated with an increase of cytosolic free calcium of both mitochondrial and extramitochondrial pools. This calcium accumulation was prevented by treatment with antioxidants (vitamin E succinate) or by pretreatment with verapamil or nifedipine which blocked calcium influx while not attenuating membrane oxidation (11). Flunarizine (2) blocks both sodium and calcium channels. It is protective in various animal models of cerebral ischemia (12). Interestingly, nifedipine, flunarizine, bepridil and verapamil inhibit ascorbic acid-induced lipid peroxidation of brain microsomal membranes with IC_{50}s less than 100 µMolar. This suggests that the mechanism of action of such compounds may be complex (13).

High cytoplasmic calcium activates proteases and lipases. Inhibition of proteolysis using leupeptin (3) has been reported to protect CA1 hippocampus from ischemia-induced damage (14). Calpain inhibitor I has also been shown to be neuroprotective against the effect of hypoxia on hippocampal slices (15).

Excitatory Amino Acids - At least three subtypes of excitatory amino-acid receptor subtypes exist in the mammalian CNS: the NMDA, kainate, and AMPA (QUIS) receptors (16). Excitotoxin-induced injury to brain is triggered by presynaptic release of glutamate that induces an abnormal post-synaptic influx of calcium into cells with a high density of glutamate receptors. Increases in extracellular levels of glutamate and aspartate correlate with injury severity in brain and spinal cord (17). Glutamate acts at the NMDA receptor to increase intracellular calcium, potassium, sodium, water and chloride. Competitive and noncompetitive NMDA antagonists reportedly attenuate global ischemic damage (18), although these effects may be caused by drug-induced hypothermia (19). The use of NMDA receptor antagonists for treatment of brain ischemia has been recently reviewed (20). NMDA antagonists have been shown effective in reducing infarct size in focal ischemia (MCAO) while the AMPA receptor blocker NBQX effectively reduces infarct after global ischemia (cardiac arrest) (21). Others have shown that competitive antagonists of the AMPA receptor are more effective in complete global ischemia than NMDA antagonists (22). The coupling of glutamate release and calcium influx during ischemia has led to the testing of the NMDA antagonist dizocilpine (MK801, 4) and the calcium channel antagonist nicardipine in a gerbil model of bilateral carotid occlusion. The combination significantly protected against cell death (23). LY233053 (5), a competitive glutamate antagonist with improved bioavailability and speed of onset, is as effective as MK801 in two rabbit ischemia models, although it is less sedating than MK801 (24).

Glutamate excitation at the NMDA receptor stimulates release of nitric oxide (25). The excitotoxic effect of glutamate may be mediated by increased production of nitric oxide. Indeed, selective inhibitors of nitric oxide synthetase (NOS), such as N-nitro-L-arginine (6), have been shown to prevent glutamate-induced neurotoxicity in cortical cell cultures (26). The anion transport inhibitor L644-711 (7), which is known to reduce astrocyte swelling and excitotoxin release in primary astrocyte cultures (27), has been studied in two rabbit models of thromboembolic stroke (autologous clot embolization of one internal carotid artery in normotensive and hypotensive rabbits). Pretreatment with L644-711 (7) resulted in reduction in infarct volume in both models (28).

Reactive Oxygen - Oxygen radicals are believed important in the pathogenesis of many neurologic illnesses (29). Reactive oxygen species are involved in the expression of acute neurologic injury (stroke, head and spinal cord injury, subarachnoid hemorrhage, brain damage after cardiac arrest) and chronic neurologic injury (Parkinsonism) (30). In vivo, oxygen is reduced through superoxide anion to hydrogen peroxide and on to water. H_2O_2 is produced by dismutation of superoxide anion. Superoxide anion is released by activated phagocytic cells, by dysfunctional mitochondria and by xanthine oxidase/xanthine. Myeloperoxidase converts chloride ions and H_2O_2 to hypochlorous acid (31). Hydrogen peroxide can react with transition metals (Fe(II), Cu(I)) to yield hydroxyl radical. Reactive oxygen species that are important in vivo were reviewed (32).

Recently, the reaction product of nitric oxide (a radical) and superoxide anion has been postulated as a potentially important cytotoxin. Both nitric oxide and superoxide anion can be produced by vascular endothelium. The peroxy nitroxide anion is protonated at physiologic pH. It can release hydroxyl radical and can penetrate cellular barriers (33,34,35). Consistent with this, nitric oxide has been shown to mediate neuronal death after focal cerebral ischemia in the mouse (36). Others have argued that nitric oxide and superoxide anion may beneficially regulate vascular tone (37).

Several endogenous antioxidant mechanisms protect against oxidative injury. Vitamin E is the most important membrane bound antioxidant. Glutathione peroxidase uses H_2O_2 to oxidize reduced glutathione to oxidized glutathione. Superoxide dismutase converts superoxide anion to hydrogen peroxide. Catalase converts H_2O_2 to water. The role of endogenous antioxidants (superoxide dismutase (SOD) for conversion of superoxide anion to H_2O_2, catalase for H_2O_2 to water, vitamin E, ascorbate, adenosine, lactoferrin (iron binding protein), glutathione and carotenoids) has been reviewed (38,39,40,41, 42,43,44).

Lipid peroxidation is a chain reaction in which an allylic hydrogen of a membrane polyunsaturated fatty acid (PUFA) is removed. In the early phases of the reaction (initiation), the reaction can be efficiently terminated by antioxidants. If unchecked, the PUFA allylic radical can react with oxygen to form a lipid hydroperoxyl radical. The lipid or hydroperoxyl radical can abstract an allylic hydrogen from a neighboring PUFA. Thus begins a chain reaction that can lead to membrane destruction (see Scheme 1).

Scheme 1: Oxidation of Polyunsaturated Acids (LH)

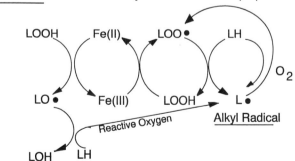

LH = polyunsaturated acid. LOH and LOOH are alcohol and peroxyl products of LH oxidation.

Peroxyl and alkoxyl radicals that are generated during lipid peroxidation can be converted to the less reactive peroxides and alcohols by vitamin E which, in turn, is oxidized to the vitamin E radical. The vitamin E radical can be converted back to parent vitamin E by ascorbic acid or glutathione, or it can oxidize to the quinone form (45). Reactive oxygen can also cause nuclear damage. Preservation of vitamin E predicts good outcome in many CNS studies. Recently, vitamin E deficient rats showed increased post-traumatic ischemia, lipid peroxidation and neuropathology relative to rats with normal vitamin E levels when subjected to spinal cord compression injury (46). Other workers have shown that vitamin E is crucial to cell survival and that its drop presages other oxidative insults to the cell. Thus, after generation of peroxyl radicals by thermal decomposition of lipophilic azo compounds (radical generation in the lipid region of the membrane), the vitamin E levels of red blood cell ghost membranes fall first to a critical level. After this initial drop of vitamin E, membrane thiols are oxidized and thiobarbituric acid-reactive substances (TBARS) increase dramatically (47). TBAR increases signify increased amounts of malonyldialdehyde, a byproduct of lipid peroxidation. In a unilateral carotid occlusion gerbil stroke model, a drop of vitamin E parallels neuronal death and a loss of calcium homeostasis in the infarct area and foretells death in these animals (48).

Steroids have a long history in the treatment of CNS injury. The recent publication of the second National Acute Spinal Cord Injury Study (NASCIS II) is a milestone in the search for therapeutic interventions that will inhibit secondary post-traumatic spinal cord degeneration. In this study, methylprednisolone hemisuccinate ($\underline{8}$) improved six month recovery of human spinal cord injury patients compared to placebo when administered at very high doses within 8 hours after injury (49). The activity of methylprednisolone was attributed to inhibition of lipid peroxidation rather than to a glucocorticoid effect (50). In a cat model of compressive injury to the spinal cord, $\underline{8}$ has a biphasic dose response curve and a limited intervention window. This pharmacology was replicated with the non-glucocorticoid U72099 ($\underline{9}$, 51). A new, antioxidant steroid that lacks glucocorticoid and mineralocorticoid activity, tirilazad mesylate (U74006F, $\underline{10}$), is undergoing clinical trial in head and spinal cord injury, subarachnoid hemorrhage and stroke. This 21-aminosteroid scavenges lipid peroxyl radicals and oxygen radicals. It spares vitamin E in several models of ischemic injury and it has membrane stabilizing effects (52). It improves outcome after experimental spinal cord trauma in the rat (53) and after concussive head injury to the mouse. It reduces vasospasm in a primate model of subarachnoid hemorrhage (54). The pharmacology of U74006F in diverse injuries (traumatic shock, tumor associated neurological dysfunction, head and spinal cord injury, myocardial ischemia, vasospasm) has been extensively reviewed (55).

Phenols are important antioxidants. U78517F (11) is a 2-methylaminochroman with the same amine side chain as U74006F. This low oxidation potential antioxidant is devoid of hypothermic or antiexcitotoxic actions, but decreases cortical neuronal necrosis in a gerbil stroke model while increasing survival (56) and enhances early neurological recovery of head-injured mice (57). Other workers have elaborated the structure of vitamin E to produce antioxidants like 12 that are among the most potent inhibitors of lipid peroxidation known (58). OPC-14117 (13) was first prepared in 1984. This compound penetrates brain and protects against cerebral ischemia in rats and mice. Recently, it was shown more effective than α-tocopherol in superoxide scavenging action in aprotic solvents (59).

A series of antioxidants that contain the backbone structure of butylated hydroxytoluene (BHT), a known antioxidant, have been evaluated in models of cerebral ischemia. LY178002 (14), a compound that inhibits both iron-dependent lipid peroxidation and key enzymes of the arachidonic acid cascade, was effective in reducing damage in hippocampal CA1 layer and in the corpus striatum in rats subjected to 30 minutes of 4-vessel occlusion (60). LY231617 (15), another di-t-butyl phenol antioxidant, has been reported to protect against CA1 hippocampal damage when administered orally or intravenously after 30 minutes of global ischemia in rats (61). Not only was LY231617 capable of reducing histological damage, it also preserved functional electro-physiological integrity of the hippocampal neuronal circuitry after the ischemic insult (62).

Spin trap agents have been evaluated in models of CNS injury. N-t-Butyl-α-phenyl-nitrone (PBN) reduced neuronal damage in gerbils subjected to forebrain ischemia when given at high dose either before or after ischemia. This suggests the involvement of radicals in the injury process (63). PBN was also effective in a closed head injury in rats. In this study, radicals were detected within 5 minutes of injury (64).

Superoxide anion production is associated with CNS injury. Xanthine oxidase has been shown to be highly concentrated in cerebral vascular endothelium and thus may be an important source of superoxide anion during ischemia (65). Allopurinol, a xanthine oxidase inhibitor, was shown to be effective in a rat spinal cord injury model where xanthine dehydrogenase and xanthine oxidase activity was shown to be high in the injured cord using a new flurometric assay (66). Superoxide dismutase (SOD) detoxifies superoxide anion. SOD and its conjugates (PEG-SOD, SOD-albumin, SOD polymer, SOD-pyran) have been reviewed (67). Human recombinant SOD inhibits transient ischemic injury to CA1 neurons in gerbils (68). The conjugates have longer circulating half lives than SOD itself. For example, conjugation of PEG monomers to SOD increases its circulatory half-life to almost 40 hours in the rat (69). A novel inhibitor of superoxide anion generation by macrophages (OPC-1561, 16) has been reported (70) along with a novel synthesis of this interesting indole (71). Liposomal delivery of SOD blunted the effect of global ischemia/reperfusion on membrane fluidity of cerebral capillary endothelial cell membranes. Membranes isolated from control animals showed, with spin label techniques, an increase in order of the mid-membrane region. This effect was blocked by liposomal SOD that was administered before reperfusion (72). Evidence for the pathologic role of superoxide anion in infarction and edema following focal cerebral ischemia comes from transgenic mice that overexpress CuZn superoxide dismutase. These animals show reduced infarct size and brain edema compared with nontransgenic mice (73).

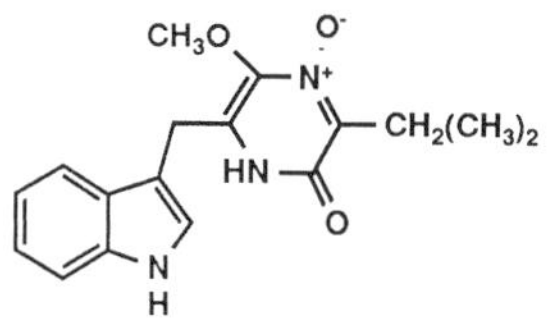

16 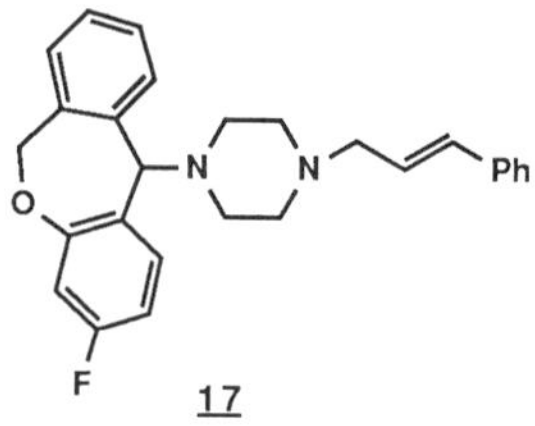   17 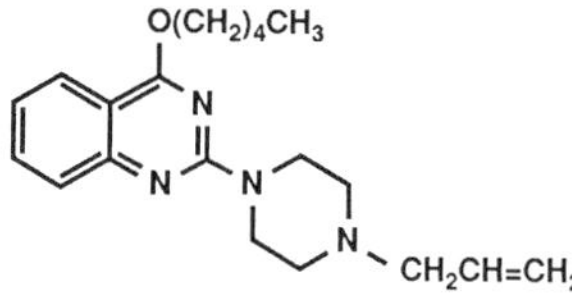   18

Many other classes of antioxidants have been evaluated. A series of dihyrodibenzoxepins was tested in models of global ischemia, hypoxia, in vitro lipid peroxidation and convulsion. AJ-3941 (17) was selected as the optimal compound (74). A novel quinazoline fumarate (KB-56666, 18) inhibits lipid peroxidation in rat brain homogenates and isolated mitochondria and, in a rat MCAO model, potently prevented brain edema and histological neuronal damage in the ischemic area (75). A novel series of 3-O-alkylascorbic acids (19) inhibits lipid peroxidation in vitro. This antioxidant activity is a function of the lipophilicity of the alkyl chain (76). This presumably relates to the way that the lipid chain anchors the antioxidant portion of the molecule in the membrane. Other workers have suggested that the interplay of antioxidant activity and lipophilicity is important in protecting endothelial from lipid radical initiated injury (77). RA-642 (20) inhibits lipid peroxidation as measured by TBARS in rat brain homogenate. If addition of the antioxidant is delayed relative to initiation of lipid peroxidation with ferrous sulfate and ascorbic acid, the RA-642 does not inhibit oxidation. Thus, it is presumably more effective in inhibiting the initiation phase of lipid peroxidation than the propagative, chain reaction phase (78). An oligomeric derivative of prostaglandin B2 and ascorbic acid (21) reduced injury in rats injured by a combination of bilateral common carotid artery occlusion and hemorrhagic hypotension. A decrease in spin adducts in the brain with PBN and a drop in TBAR formation suggest inhibition of lipid peroxidation free radical reactions (79). N-acetylcysteine, given i.p. prior to or 30 minutes after percussive brain injury in cats, preserved normal hyperventilation (80). Dihydrolipoate (DL-6,8-dithioloctanoic acid) was reported protective in a rat permanent MCAO model (81). Ebselen (22), a seleno-organic compound that exhibits glutathione peroxidase-like activity (82) and inhibits iron stimulated lipid peroxidation (83), significantly reduced both infarct size and edema progression in the MCAO model of focal stroke in rats (84).

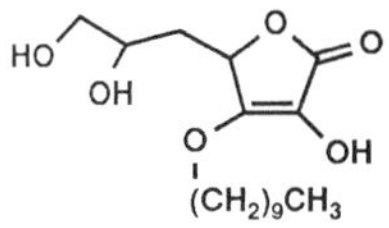

19 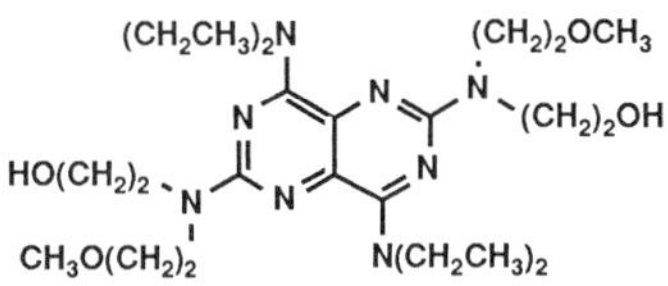   20 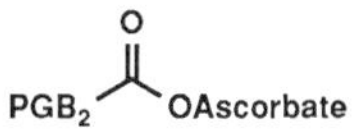   21

Platelet-activating Factor (PAF) - PAF, an endogenous phospholipid with proinflammatory and vasoactive properties, has been proposed as an important mediator of specific pathological sequelae in stroke and brain injury. The formation of PAF in the brain has been associated with other vasoactive lipid mediators (metabolites of arachidonic acid) and neurotoxins such as reactive oxygen species, cytokines and excitatory amino acids that may amplify these mediators. The cerebrovascular effects of PAF include disruption of the blood-brain barrier, edema formation and vasospasm (85). The PAF antagonist, apafant (23), a triazolobenzodiazepine that lacks affinity for central benzodiazepine receptors, has been reported to reduce infarct volume when administered before or after MCAO in rats (86). PAF antagonist BN-50739 (24) protected against progressive focal brain damage in a focal model of laser induced secondary brain damage in rats (87).

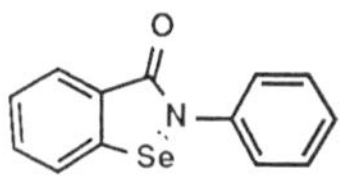

$\underline{22}$

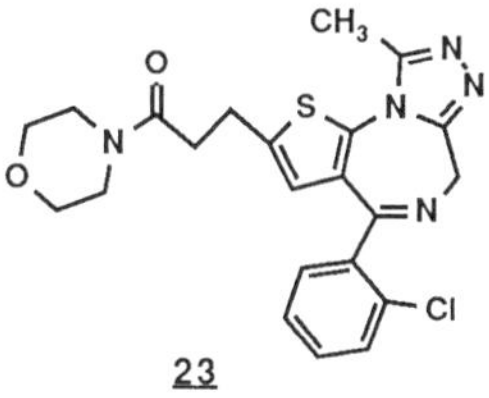

$\underline{23}$

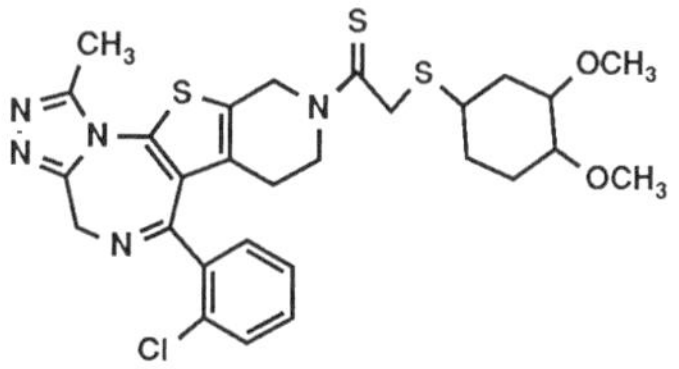

$\underline{24}$

Gangliosides - Gangliosides are sialic acid-containing glycolipids that are localized in cell membranes. They are abundant in the CNS where thy are involved in normal neuronal development and differentiation. GM1 ganglioside has been evaluated clinically in spinal cord injury where it reportedly improved neurologic outcome after spinal cord injury. Treatment was delayed until 72 hours in most patients (88). It may improve Na/K-ATPase and adenylcyclase activity (89). The efficacy of the drug has been questioned because it failed to show any improvement over placebo on the ASIA (American Spinal Injury Association) motor score after two months. The drug was negative in ALS (90). In a rat model of cortical focal ischemia where animals suffered deficits from MCAO and common carotid artery occlusion, GM1 treatment reduced edema associated sodium and calcium increases and loss of membrane ATPase activity and reduced functional motor deficits. Drug was administered immediately after injury and daily for the study's duration (1, 2, 3, 7 or 14 days) (91). GM1 failed to reduce cerebral infarction caused by permanent focal ischemia in spontaneouslly hypertensive rats.

Adenosine - This nucleoside affects both neutrophil and endothelium function and has protective effects in some reperfusion injury models.

Peptides - Lipocortin-1 is an endogenous peptide. Central administration of an active recombinant fragment (1-188 amino acids) of lipocortin-1 to rats 10 minutes after a 2 hour ischemic episode began reduced infarct size and edema at 2 hours (92). The authors concluded that lipocortin-1 is an endogenous inhibitor of cerebral ischemia. Thyrotropin-releasing factor (TRH, $\underline{25}$) was compared to naloxone in rats with spinal cord injury. TRH but not naloxone treatment starting 24 hours to 7 days after injury improved neurologic outcome in severely injured rats. The authors concluded that TRH was more effective than naloxone when treatment was 24 hours or later (93). Some stable TRH analogs (CG3509, $\underline{26}$; CG3703, $\underline{27}$; YM14673, $\underline{28}$) have shown activity in animal models of spinal cord injury. The C-terminus of TRH is essential for neuroprotection (94).

Serotonin - During recent years, the association of 5HT1a agonist activity with inhibition of neuronal excitability has been recognized. This is attributable to increased potassium conductance and a resulting hyperpolarization of cell membranes. This hyperpolarization may prevent neuronal depolarization during cerebral ischemia and thereby inhibit the

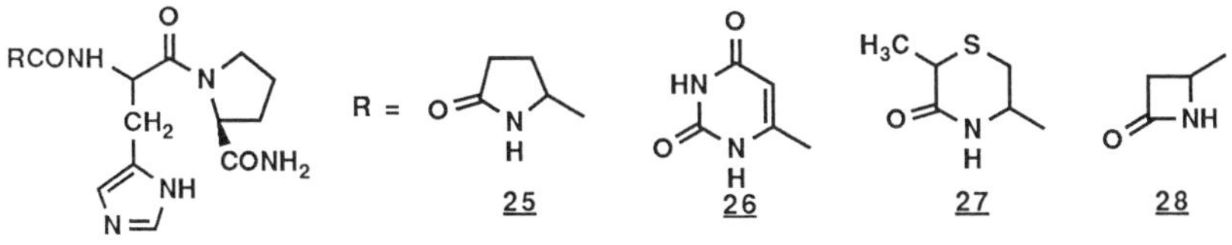

$\underline{25}$ $\underline{26}$ $\underline{27}$ $\underline{28}$

resulting rise in intracellular calcium (95). Buspirone, 8-hydroxy-DPAT, gepirone, ipsapirone and Bay R 1531 were all dosed i.p. 30 minutes before the induction of ischemia in a rat focal ischemia model (permanent occlusion of the left MCA). All of these compounds decreased cortical infarct size, although ipsapirone and Bay R 1531 were most effective (96). The serotonin antagonist mianserin improved neurological recovery after impact trauma to the thoracic region of the rat spinal cord. Ketanserin and cyproheptadine were marginally active in this model. Mianserin improved post-traumatic spinal cord blood flow in the injured cord and normalized spinal serotonin content which normally rise dramatically after trauma (97,98).

Conclusions - Effective treatments of acute neurological disorders are being developed. The animal models suggest at least the possibility of success in man. As the mosaic of acute ischemic and traumatic injury to the central nervous system is better understood, we come closer to new approaches to the treatment of chronic neurologic disorders like Parkinsonism, multiple sclerosis and even Alzheimer's. The next five years are filled with promise.

References

1.   M.D. Ginsberg, R. Busto, Stroke, 20, 1627 (1989).
2.   P. Scheinberg, Neurology, 41, 1867 (1991).
3.   R.J. Traystman, J.R. Kirsch, R.C. Koehlier, J. Appl. Physiol, 71, 1185 (1991).
4.   D.J. Triggle, DN&P, 4, 579 (1991).
5.   R. Massingham, G.W. John, P.A. VanZwieten, Drugs of Today, 27, 459 (1991).
6.   D.J. Triggle, DN&P, 4, 579 (1991).
7.   E. Martinez-Vila, F. Guillen, J.A. Villaneuva, Stroke, 21, 1023 (1990).
8.   I.B. Ross, C.H. Tator, J. Neurotrauma, 8, 229 (1991).
9.   J.P. Mohr, Stroke, 23, 3 (1992).
10.   S. Takakura, T. Susumu, H. Satoh, J. Mori, A. Shiino, J. Handa, Japan J. Pharmacol., 56, 547 (1991).
11.   E. Albano, G. Bellomo, M. Parola, R. Carini, M. Dianzani, Biochim. Biophys. Acta, 1091, 310 (1991).
12.   P.J. Pauwels, J.E. Leysen, P.A.J. Janssen, Life Sci., 48, 1881 (1991).
13.   T. Goncalves, A.P. Carvalho, C.R. Oliveira, Eur. J. Pharm., 204, 315 (1991).
14.   K.S. Lee, S. Frank, P. Vanderklish, A. Arai, G. Lynch, Proc. Natl. Acad. Sci., 88, 7233 (1991).
15.   A. Arai, M. Kessler, K. Lee, G. Lynch, Brain Res., 532, 63 (1990).
16.   P. Frey, Chimia, 45, 55 (1991).
17.   S.S. Panter, S.W. Yum, A.I. Faden, Ann. Neurol., 27, 96 (1990).
18.   J.C. Grotta, C.M. Picone, P.T. Ostrow, R.A. Strong, R.M. Earls, L.P. Yao, H.M. Rhoades, J.R. Dedman, Ann. Neurol., 27, 612 (1990).
19.   D. Corbett, S. Evans, C. Thomas, D. Wang, R.A. Jonas, Brain Res, 514, 300 (1990).
20.   B. Scatton, C. Carter, C. Dana, J. Benavides, DN&P, 4, 89 (1991).
21.   B.K. Siesjo, H. Memezawa, M.L. Smith, Fundam. Clin. Pharmacol., 5, 755 (1991).
22.   N.H. Diemer, F.F. Johansen, M.B. Jorgensen, Stroke, 21, Suppl., 39 (1990).
23.   K. Hewitt, D. Corbett, Stroke, 23, 82 (1992).
24.   K.P. Madden, W.M. Clark, A. Kochhar, J.A. Zivin, J. Neurosurg., 76, 106 (1992).
25.   J. Garthwaite, G. Garthwaite, R.M.J. Palmer, S. Moncada, Eur. J. Pharmacol., 172, 413 (1989).
26.   V.L. Dawson, T.M. Dawson, E.D. London, D.S. Bredt, S.H. Snyder, Proc. Nat. Acad. Sci. USA, 88, 6368 (1991).
27.   K.D. Barron, M.P. Dentinger, H.K. Kimelberg, L.R. Nelson, R.S. Bourke, S. Keegan, R. Mankes, E.J. Cragoe, Acta Neuropathol., 75, 295 (1988).
28.   J.J. Kohut, M.M. Bednar, H.K. Kimelberg, T.L. McAuliffe, C.E. Gross, Stroke, 23, 93 (1992).
29.   J.B. Lohr, Arch. Gen. Psychiatry, 48, 1097 (1991).
30.   J.D. Adams, I.N. Odunze, Free Radical Biol. and Med., 10, 161 (1991).
31.   S.J. Weiss, New Eng. J. Med, 320, 365 (1989)
32.   B. Halliwell, Drugs, 42, 569 (1991).
33.   J.S. Beckman, T.W. Beckman, J. Chen, P.A. Marshall, B.A. Freeman, Proc. Natl. Acad. of Sci., 87, 1620 (1990).
34.   M. Saran, C. Michel, W. Bors, Free Radical Res. Commun., 10, 221 (1990).
35.   N. Hogg, V.M. Darley-Usmar, M.T. Wilson, S. Moncada, Biochem. J., 281, 419 (1992).
36.   J.P. Nowicki, D. Duval, H. Poignet, B. Scatton, Eur. J. Pharmacol., 204, 339 (1991).
37.   B. Halliwell, Free Radical Res. Comm., 5, 315 (1989).
38.   B. Halliwell, Drugs, 42, 569 (1991).
39.   L. Packer, Am. J. Clin. Nutr., 53, 1050S (1991).
40.   J.E. Heffner, J.E. Repine, Am. Rev. Respir. Dis, 140, 531 (1989).
41.   J.M. Braughler, E.D. Hall, Free Radical Biology and Medicine, 6, 289 (1989).
42.   J.L. Farber, M.E. Kyle, J.B. Coleman, Lab. Investigation, 62, 670 (1990).
43.   P.M. Reilly, H.J. Schiller, G.B. Bulkley, Am. J. Surgery, 161, 488 (1991).

44.   R.J. Traystman, J.R. Kirsch, R.C. Koehler, J. Appl. Physiol., 71, 1185 (1991).
45.   G.W. Burton, U. Wronska, L. Stone, D.O. Foster, K.U. Ingold, Lipids, 25, 199 (1990).
46.   Y. Taoka, T. Ikata, K. Fukuzawa, J. Nutr. Sci. Vitaminol., 36, 217 (1990).
47.   Y. Takenaka, M. Miki, H. Yasuda, M. Mino, Archives Biochem. and Biophys., 285, 344 (1991).
48.   E.D. Hall, P. Yonkers, Stroke, 22, 361-365 (1991).
49.   M.B. Bracken, M.J. Shepard, W.F. Collins, New Engl. J. Med., 322, 1405 (1990).
50.   J.M. Braughler, J. Neurochem., 44, 1282 (1985).
51.   E.D. Hall, J. of Neurotrauma, 8, Supp. 1, S-31 (1991).
52.   J.M. Braughler, E.D. Hall, E.J. Jacobsen, J.M. McCall, E.D. Means, Drugs of the Future, 14, 143 (1989).
53.   A. Holtz, B. Gerdin, J. Neurotrauma, 8, 239 (1991).
54.   K. Kanamaru, B.K. Weir, I. Simpson, T. Witbeck, M. Grace, J. Neurosurg., 74, 454 (1991).
55.   Drugs of the Future, 16, 189 (1991).
56.   E.D. Hall, K.E. Pazara, J.M. Braughler, K.L. Linseman, E.J. Jacobsen, Stroke, 21, III-83 (1990).
57.   E.D. Hall, J.M. Braughler, P.A. Yonkers, K.E. Pazara, S.L. Smith, D.I. Schneider, F.J. VanDoornik, E.J. Jacobsen, J. Neurotrauma, 6, 213 (1989).
58.   J. Battioni, M. Fontecave, M. Jaouen, D. Mansuy, Biochem. and Biophys. Res. Commun., 174, 1103 (1991).
59.   J. Jinno, H. Mori, Y.Oshiro, T. Kikuchi, H. Sakurai, Free Rad. Res. Comms., 15, 223 (1991).
60.   J.A. Clemens, P.P.K. Ho, J.A. Panetta, Stroke, 22, 1048 (1991).
61.   J.A. Panetta, J.K. Shadle, E.B. Smalstig, D.R. Bennett, J.A. Clemens, J. Cereb. Blood Flow Metab., 11, S142 (1991).
62.   J.A. Clemens, B. Bhagwandin, E.B. Smalstig, D.R. Bennett, R.E. Mincy, J.K. Shadle, J.A. Panetta, J. Cereb. Blood Flow Metab., 11, S145 (1991).
63.   C. Clough-Helfman, J.W. Phillis, Free Rad. Res. Comms., 15, 177 (1991).
64.   D. Torbati, D.F. Church, M.E. Carey, W.A. Pryor, Neuroscience Abstract 3027, A891 (1991).
65.   L.S. Terada, I.R. Willingham, M.E. Rosandich, J.A. Leff, G.W. Kindt, J.E. Repine, J. Cellular Physiology, 148, 191 (1991).
66.   J. Xu, J.S. Beckman, E.L. Hogan, C.Y. Hsu, J. Neurotrauma, 8, 11 (1991).
67.   R.A. Greenwald, Free Radical Biol. and Med., 8, 201 (1990).
68.   O. Uyama, T. Matsuyama, H. Michishita, H. Nakamura, M. Sugita, Stroke, 23, 75 (1992).
69.   C.W. White, J.H. Jackson, A. Abuchowski, G.M. Kazo, R.F. Mimmack, E.M. Berger, B.A. Freeman, J.M. McCord, J.E. Repine, J. Appl. Physiol., 66, 584 (1989).
70.   Y. Nakano, T. Kawaguchi, J. Sumitomo, T. Takizawa, S. Uetsuki, M. Sugawara, M. Kido, J. Antibiotics, 44, 53 (1991).
71.   Y. Ito, H. Sato, M. Murakamai, J. Org. Chem., 56, 4864 (1991).
72.   A.M. Phelan, D.G. Lange, Bochimica et Biophysica Acta, 1 (1991).
73.   H. Kinouchi, C.J. Epstein, T. Mizui, E. Carlson, S. Chen, P. Chan, Proc. Natl. Acad. Sci. USA, 88, 11158 (1991).
74.   M. Kurokawa, F. Sato, Y. Masuda, T. Yoshida, Y. Ochi, K. Zushi, I. Fujiwara, S. Naruto, H. Uno, J. Matsumoto, Chem. Pharm. Bull., 39, 2564 (1991).
75.   H. Hara, K. Kogure, H. Kato, A. Ozaki, T. Sukamoto, Eur. J. Pharmacol., 197, 75 (1991).
76.   Y. Nihro, H. Miyataka, T. Sudo, H. Matsumoto, T. Satoh, J. Med. Chem., 34, 2152 (1991).
77.   T. Kaneko, S. Nakano, M. Matsuo, Lipids, 26, 345 (1991).
78.   I. Bellido, J.P. de la Cruz, F. S. de la Cuesta, Meth. Find. Exp. Clin. Pharmacol., 13, 371 (1991).
79.   A. Sakamoto, S. Ohnishi, T. Ohnishi, R. Ogawa, Free Radical Biology and Med., 11, 385 (1991).
80.   E.F. Ellis, L.Y. Dodson, R.J. Police, J. Neurosurg., 75, 774 (1991).
81.   J.H.M. Prehn, C. Karkoutly, J. Nuglisch, B. Peruche, J. Krieglstein, J. Cerebral Blood Flow and Metab., 12, 78 (1992).
82.   M. Hayashi, T.F. Slater, Free Rad. Res. Comms., 2, 179 (1986).
83.   A. Muller, E. Cadenas, P. Graf, H. Sies, Biochem. Pharmac., 33, 3235 (1984).
84.   T. Matsui, H.I. Johsita, T. Asano, J. Tanaka, Pharmacology of Cerebral Ischemia, J. Krieglstein, H. Oberpichler (Eds), Wissenschaftliche Verlagsgesellschaft mbtt Stuttgart, 363 (1990).
85.   P.J. Lindsberg, J.M. Hallenbeck, G. Feuerstein, Ann. Neurol., 30, 117 (1991).
86.   G.W. Bielenberg, G. Wagener, T. Beck, Stroke, 23, 98 (1992).
87.   K.U. Frerichs, P.J. Lindsberg, J.M. Hallenbeck, G. Feurerstein, J. Neurosurg., 73, 223 (1990).
88.   F.H. Geisler, F.C. Dorsey, W.P. Coleman, New Engl. J. Med, 324, 1829 (1991).
89.   C. Argentino, M.L. Sacchetti, D. Toni, Stroke, 20, 1143 (1989).
90.   P.S. Schonhofer, New Eng. J. Med., 356, 493 (1992).
91.   V.A. Bharucha, C.G. Wakade, S.P. Mahadik, S.E. Karpiak, Exptl. Neurology, 114, 136 (1991).
92.   J.K. Relton, P.J. Strjbos, C.T. O'Shaughnessy, F. Carey, R.A. Forder, F.J. Tilders, N.J. Rothwell, J. Exp. Med., 174, 305 (1991).
93.   T. Hashimoto, N. Fukuda, Eur. J. Pharmacol., 203, 25 (1991).
94.   A.I. Faden, S. Salzman, TIPS, 13, 29 (1992).

95. A. Colino, J.V. Halliwell, Nature, $\underline{328}$, 73 (1987).
96. G.W. Bielenberg, M. Burkhardt, Stroke, $\underline{21}$ (Suppl. IV), IV-161 (1990).
97. S.K. Salzman, M.A. Puniak, Z. Liu, R.P. Maitland-heriot, G.M. Freeman, C.A. Agresta, Ann. Neurol., $\underline{30}$, 533 (1991).
98. M. A. Puniak, G.M. Freeman, C.A. Agresta, L.V. Newkirk, C.A. Barone, S.K. Salzman, J. of Neurotrauma, $\underline{8}$, 193 (1991).

Chapter 5. Trophic Factors and their Receptors in the CNS

Steven Rosenberg
Chiron Corporation
4560 Horton Street, Emeryville, CA 94608

Introduction - Neurotrophic factors are molecules produced by a variety of cell types which have been functionally defined by those activities which allow the growth and maintenance of neurons in vitro and in vivo. The last review in this series focused on nerve growth factor and 3 other neurotrophic factors (1). In the intervening 3 years the tools of molecular biology have led to an explosion in the identification of novel neurotrophic factors and their receptors (2). This review will focus primarily on the neurotrophin family, of which nerve growth factor (NGF) is the progenitor. Other trophic factors, including members of the fibroblast growth factor family and ciliary neuronotrophic factor, will also be covered. The focus of this review will be on the dissection of the interactions of the receptors and factors, and the signalling pathways they subsequently activate.

NEUROTROPHINS

Identification - NGF, the prototypic neurotrophin, was first isolated from male mouse submaxillary gland as a factor which promoted the survival of neural crest derived neurons. The unique occurrence of this rich source of the protein which exists in the mouse as a complex of 3 subunits, α, β, and χ has made NGF the best characterized member of the family. The active form of NGF is a non-covalent dimer of the 13 kD, 118 amino acid, β subunits. Each subunit contains 6 cysteines arranged in 3 disulfide bonds. This subunit is synthesized as a precursor containing a relatively long pro sequence, features of which are conserved in the neurotrophin family (3).

More recently, brain derived neurotrophic factor (BDNF) was isolated from porcine brain (4,5). This very rare protein is homologous to NGF in size and location of conserved cysteines. Subsequent isolation of BDNFs from other species has been accomplished using PCR strategies, targetting common sequences between BDNF and NGF (6,7). The additional members of the neurotrophin family, as well as neurotrophins from other species have been identified as homologous proteins by similar cDNA and genomic cloning strategies using PCR (8). These strategies have been aided by the observation that the coding sequences for the neurotrophins all appear to be found on single exons. This has led to the identification of NT3 (human, mouse, rat (9,10)), NT4 (Xenopus and viper (11), and NT5 (rat and human (8)) clones from cDNA and genomic sources.

An assessment of the roles of these new neurotrophins in neuronal development and survival has depended upon the expression of the proteins using recombinant DNA strategies. In particular, all of the neurotrophins discovered to date have been expressed using mammalian cell systems. Yeast or baculovirus have been successful in a few cases (6,12). In addition, the isolation of the cDNA clones has enabled workers to locate those cells that express the various neurotrophins using in situ hybridization techniques (6,8).

Structure-Activity Relationships - The use of the PCR strategies defined above has led to the isolation and sequencing of a number of neurotrophins from a variety of species. An alignment of these sequences is presented in Figure 1.

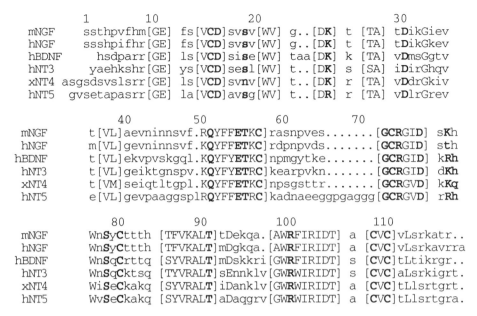

Figure 1 - Alignment of representative examples of the neurotrophin family. Conserved or identical amino acids are in upper case. The neurotrophins shown are mouse NGF, human NGF, human BDNF, human NT3, Xenopus NT4,and human NT5. Sequences in brackets are regions of 2 or more amino acid homology. The conserved cysteines and residues whose side chains are involved in tertiary interactions in mNGF (13) are in bold face.

There are several noteworthy features: 1) All of the cysteines are conserved; 2) Several regions show almost absolute conservation throughout the members of the family, especially the C terminal region from residues 99 to 110; 3) BDNF and NT3 are extremely conserved across species lines; 4) Differences between the various neurotrophins appear to cluster in a few areas, which are likely of functional significance in determining high affinity receptor specificity (see below).

These analyses have recently been given a rational structural framework from the solution of the structure of the murine NGF dimer at 2.3 Å resolution using X-ray crystallographic techniques (13). The NGF dimer consists of primarily β-structure as predicted by Raman spectroscopy (14). There are 7 β strands, 6 of which form twisted antiparallel pairs bringing together the hydrophobic core of the monomer subunit. The 3 disulfide bonds appear to be clustered at one end of the molecule and are substantially buried, contributing as well to the hydrophobic core of the molecule. The overall geometry of the monomer is quite asymmetric consisting of a long, flat molecule. The subunit interface is substantially flat and hydrophobic. The most conserved residues near the C-terminus of the molecule are located at one end. The variable regions of the various neurotrophins cluster in β-hairpin loops at the other end of the molecule (residues 29-35, 43-48, and 92-98) suggesting this region is involved in the differential high affinity receptor binding of the factors. This topology is consistent with possible models for the multiple receptor interactions involved in neurotrophin action, in which the p75 receptor binds the conserved regions and the selective *trk* tyrosine kinases bind at or near the variable loops.

A few studies have examined the role of a variety of residues in neurotrophin action by expression of mutant or chimeric proteins. In particular some conserved residues (V21, R99, R102) have been suggested as important in NGF structure and function, as judged by receptor binding and biological activity (15). However, these conclusions were drawn with crude preparations and without examination of the intact nature of the protein structure. Further studies looking at the biological activity of NGF:BDNF chimeras have suggested that no single region imparts selectivity in functional receptor binding and biological responses (16,17). Although these studies are not quantitative, they are consistent with the hypothesis that binding to the functional, high affinity receptor involves several regions of the neurotrophin molecules.

Receptors and Signal Transduction - NGF has been shown to bind to both low and high affinity sites in neuronal cells or tissues, with only the latter correlating with biological activity. The low affinity NGF receptor protein (p75) and gene were isolated previously (18), but the components of the high affinity molecule(s) remained elusive. During the past year, a variety of workers have shown that members of the trk family of tyrosine kinase receptors are key components of the high affinity receptors for NGF and the other neurotrophins (2). It had been known that tyrosine phosphorylation occurred rapidly upon NGF addition to PC12 cells, but the kinase involved was unknown (19). Trk is a member of the tyrosine kinase receptor family, first identified as an oncogenic fusion with tropomyosin (3,20). The proto-oncogene was later shown to be expressed in tissues of neural crest origin, the principal targets of NGF action. Trk is rapidly phosphorylated upon addition of NGF (21) and binds NGF with relatively high affinity (22). It is controversial whether the high affinity, functional receptor involves both trk and the 75 kD NGF receptor, or whether trk itself is sufficient for the full range of NGF biological activities (21-23). Two additional neurotrophins, NT3 and NT5, bind trk and activate tyrosine phosphorylation (8).

Further members of the trk family of receptors have been cloned and their roles in neurotrophin receptor function have been investigated (24,25). Trk B, first identified as a gene related to trk, binds BDNF, NT3, and NT5, but not NGF (24,26-28). Most recently, the trk C gene has been shown to bind NT3, but not NGF and BDNF (25). The functional consequences of binding of the neurotrophins to the trk receptors have been measured using a variety of methods, including transfection of the trk genes into 3T3 cells yielding cell lines for which the cognate neurotrophins are mitogens (29). The complex pattern of binding of the neurotrophins to more than one receptor may reflect redundancy in the role of trophic factors or temporal regulation of expression, as has been suggested for BDNF and NT3 expression (28,30).

With the identification of the role of the trk family in the high affinity neurotrophin receptors, the dissection of the subsequent signal transduction pathways becomes possible. The primary system used to investigate NGF signal transduction has been the rat phreochromocytoma cell line, PC12, which differentiates to a neuronal phenotype specifically by the addition of exogenous NGF and bFGF. It has been shown that phospholipase C-γ is phosphorylated by trk or a closely associated kinase within one minute of NGF addition in this system (31). In addition, a series of tyrosine phosphorylation events occurs, leading to modification of a cascade of protein serine/threonine kinases, known as MAP or ERK kinases (32). Work done using PC12 cells transfected with a ts v-src protein, showed a ts inducible neuritic phenotype, which is related to but distinguishable from that imposed by NGF (33). Finally, recent experiments with neutralizing antibodies suggest that both src and RAS, in that order, are necessary for PC12 cell differentiation (34).

The role of the low affinity, p75 NGF receptor in neurotrophin signalling is unresolved. All of the neurotrophins examined to date bind this molecule with moderate affinity. A series of experiments suggest that the cytoplasmic domain of this molecule is capable of playing a role in signal transduction (35); this is most dramatically seen in experiments where a hybrid receptor with the EGF receptor extracellular domain and p75 transmembrane and cytoplasmic domains, causes PC12 cells to respond to EGF by forming neurites (36). It is not clear, however, whether this represents mimicry of the normal signalling pathway, or activation of an alternative route, since the pathways of activation by NGF and EGF appear to be convergent in PC12 cells (37). It has also been postulated that the p75 receptor is involved in embryonic development of the inner ear, via a signalling pathway using glycosyl-phosphatidylinositol hydrolysis (38).

Potential Protein and Small Molecule Therapeutic Utility - A number of studies have shown that treatment of both rodents and primates with NGF prevents degeneration of basal forebrain cholinergic neurons after fimbria-fornix transection (39). Intraventricular administration of NGF in rats and primates leads to receptor-mediated retrograde transport and reduces cholinergic neuronal degeneration, suggesting that such an approach might be useful in degenerative neurological disorders such as Alzheimer's disease(40-42). This is somewhat controversial, as NGF has also been implicated in increasing the toxicity of β-amyloid (43). NGF has also been shown to prevent toxic neuropathy in mice after treatment with chemotherapeutic agents such as taxol (44). BDNF has been shown to be a neurotrophic factor for dopaminergic neurons of the substantia nigra, and blocks the toxic effects of MPTP (1-methyl-4-phenyl-1,2,3,6-tetrahydropyridine) on these cells, when administered prior to MPTP treatment (45). This suggests that BDNF may have utility in Parkinsonism.

No small molecules have been identified which act as agonists for neurotrophin activity. This is not surprising given the very recent identification of the *trk* receptors. However, two related molecules, K252a (1) and K252b (2), have been isolated which specifically block NGF action (46).

$$1 \qquad\qquad 2$$

K252a has recently been shown to directly inhibit NGF induced tyrosine phosphorylation by the *trk* receptor (47).

FIBROBLAST GROWTH FACTORS

Identification - The fibroblast growth factors, or heparin binding growth factors, are a family of seven molecules first identified as mitogens or oncogenes (48)). The first two molecules recognized in this family, acidic and basic FGF, are approximately 150 amino acid proteins with ca. 55% sequence identity. They contain no intramolecular disulfide bonds and are quite basic proteins with a characteristic ability to bind heparin and related cell surface proteoglycans (49). The other members of the group (int-2, kFGF/hst, FGF5, FGF6, and kFGF), have been identified as proto-oncogenes (int-2, hst/kFGF, FGF5), mitogens (KGF), or homologues of other members of the family (FGF6); see reference (50) for a recent review. This section will focus on aFGF and bFGF, which are the proteins whose role in the CNS has been investigated in some detail(51).

Synthesis and Structure - The synthesis of aFGF and bFGF is unusual in that the molecules are found on the cell surface but do not contain a signal sequence (52). Other members of the family including kFGF, int-2, FGF5, and KGF are secreted by a standard mechanism, via an encoded hydrophobic signal sequence (50). In addition, a number of variant forms of aFGF and bFGF are synthesized, with differing amino termini due to translational initiation with an alternate CUG codon, leading to some molecules which are retained in the nucleus (53).

Several groups recently reported the high resolution X-ray crystallographic analysis of bFGF (54-56). The structure of aFGF was also reported by one of these groups (54). The most striking feature of the structures is their similarity to each other and identical overall topology with interleukin-1-beta, the structure of which has been determined by both X-ray and NMR methods (57-59). The bFGF molecule shows overall approximate 3-fold symmetry and consists of all β structure, with 12 anti-parallel β-strands defining the core of the molecule. A peptide sequence implicated in receptor binding in bFGF is shown to exist in an irregular loop on the surface of the molecule (55,60). Two sites which bind sulfate ion in the crystal have been suggested as possible sites for binding of heparin (56).

Relatively few studies have addressed structure-function relationships within the FGF family using either synthetic peptides or site-directed mutagenesis. The most extensive work has been done showing that two of the cysteines are not required for bFGF or aFGF activity (49,61,62). Additional mutagenesis experiments have focused on trying to define the heparin binding region and to determine the nature of the receptor binding sites (63,64). These studies have yielded complex results, in part due to a lack of determination of the effects of mutations on the overall structure of the molecule. Two regions of the bFGF molecule have been implicated in biological activity using synthetic peptides. Peptides from the region 103-120 have been shown to have weak agonist activity; part of this sequence is on the surface of bFGF (60). A larger region nearer the N-terminus of the molecule (residues 24-68) has also been shown to have weak mitogenic activity. Interestingly, a mutation in this region (deletion of residues 27-32) eliminates the induction of plasminogen activator synthesis by bFGF (65).

Receptors and Signal Transduction - The FGF family has a complex set of receptors and binding molecules. Binding studies implicated two types of receptors of high (10^{-11}M) and low (10^{-9}M) affinity which have subsequently been identified as a set of tyrosine kinase receptors and heparan sulfated proteoglycans, respectively (66,67). At least 4 different gene products have been identified as tyrosine kinase receptors for the FGF family; these are currently referred to as FGFR1-4, although the nomenclature in the literature is complex (68). These molecules are characterized by extracellular domains which contain between 1 and 3 immunoglobulin-like domains, and, in some proteins, a characteristic acidic sequence in the interdomain region between regions 1 and 2 (69,70). Several variants of these species are present in many cell types, primarily due to alternate splicing, yielding molecules with 2 Ig domains, without the acidic sequence, and in some cases without a transmembrane and intracellular domain (71). The cytoplasmic domain is a "split" tyrosine kinase, as has been found for the PDGF receptors (72). There appears to be a great deal of redundancy in the receptor system; several members of the FGF family bind to most of the receptor species studied, and the mechanism by which ligand specificity is regulated in this system is just beginning to be understood (73-75). Recent work shows that splice variants in which alternate exons are expressed in the third Ig domain affect the affinity of FGFR1 for bFGF (73).

Recent studies have implicated the role of heparan sulfate proteoglycans (HSPGs) in a variety of biological processes, especially the functional binding of the FGFs to their receptors (66,76). Studies with cell lines deficient in proteoglycans showed a requirement for them for FGF activity (77). The exact mechanism by which heparan sulfate proteoglycans are involved in FGF binding to the high affinity receptor is not known; two models being considered are formation of a ternary complex of HSP:FGF:FGFR or stabilization of a receptor active conformation of FGF by the HSP (66). Soluble heparin stimulates the activity of aFGF and stabilizes it to degradation (78). Binding of FGFs to the high affinity receptors leads to receptor transphosphorylation and activation of the tyrosine kinases, even between receptor subtypes (79). This initial step is linked to a variety of secondary signalling pathways, including phosphorylation of phospholipase C-γ, phosphoinositide release and activation of protein kinase C, and activation of adenylate cyclase (51). The relative importance of these pathways in signal transduction in the CNS has not been evaluated in detail, although recent work suggests that activation of both ras and src are involved, at least in PC12 cells (34). The actions of bFGF and NGF in PC12 cells appear quite similar, although they can be distinguished by the inhibition of the latter by K252a (1), a specific inhibitor of the *trk* tyrosine kinase (47,80).

Roles in the CNS and Potential Therapeutic Utility - The FGFs implicated in the CNS are predominantly bFGF, with somewhat lesser roles for aFGF and FGF5 (51,81). bFGF is the only factor besides NGF which will induce neurite outgrowth in the PC12 cell line (82). In vivo studies have shown that bFGF and to a lesser extent aFGF bind to specific, overlapping but distinct sites in the brain, and that this binding is to high affinity receptors (83). In addition, as for NGF, bFGF is retrogradely transported to the neuronal cell body in a receptor dependent manner, presumably as a receptor:ligand complex (83). The key receptor in neurons appears to be FGFR-1, as judged by in situ hybridization (83,84). Both bFGF and aFGF can stimulate the synthesis of NGF by astrocytes, in vitro, providing a second indirect mechanism for the action of the FGFs as trophic factors (85). This also suggests a role for the FGFs in response to neuronal injury. Interestingly, bFGF is accumulated by neurons of the substantia nigra and is a trophic factor for dopaminergic neurons in culture (86). These observations suggest that bFGF might play a role in treatment of Parkinsonism.

OTHER TROPHIC FACTORS

Ciliary Neuronotrophic Factor - CNTF is a unique protein with no known homologues, unlike the FGFs and neurotrophins. CNTF is a 27 kDa protein with a single cysteine and no signal sequence; rat, rabbit, and human proteins have been identified based upon cDNA and genomic clones, and overexpressed at high levels in E.coli (87-90). Recently, a receptor for CNTF has been identified using an expression cloning strategy involving a specifically tagged ligand (91). The receptor is most closely homologous to the IL-6 receptor, but it lacks a cytoplasmic domain and is linked to the membrane via a glycosylphosphoinositol linkage. These factors suggests that a second component is required for CNTF mediated signal transduction, as is the case with the gp130 signal transducer component of the IL-6 receptor (92). Although first identified based on trophic activity for peripheral neurons, specifically, the chick ciliary ganglion, more recent results also indicate CNTF is a survival factor for hippocampal neurons (93). Moreover, in vivo studies now suggest that CNTF is a survival factor for motor neurons (94,95), and the suggestion has been made that it might be useful in treatment of ALS (Lou Gehrig's disease).

Conclusion - There has been a great deal of progress in the molecular biology of the neurotrophic factors and their receptors during the past year. A number of new factors (NT4, NT5) and new complex receptor families (the trk tyrosine kinases and p75; the HSPGs and FGFR1-4) have been discovered. Structural studies have suggested possible mechanisms of receptor activation at the molecular level, although signal transduction pathways are just beginning to be understood. Finally, the pharmacology of these receptors and their roles as targets for medicinal chemists, particularly in neurodegenerative disorders, will be elucidated as receptor subtype specific agonists and antagonists are discovered.

References

1. M. Schinstine and F.H. Gage in "Annual Reports in Medicinal Chemistry, Volume 24," J.A. Bristol, Ed.,Academic Press Inc., San Diego, CA.: 1989, p. 245.
2. M. Bothwell, Cell, 65, 915 (1991).
3. U. Suter, J.V. Heymach,Jr. and E.M. Shooter, EMBO J., 10, 2395 (1991).
4. Y. Barde, D. Edgar and H. Thoenen, EMBO J., 1, 549 (1982).
5. M.M. Hofer and Y. Barde, Nature, 331, 261 (1988).
6. A. Rosenthal, D.V. Goeddel, T. Nguyen, E. Martin, L.E. Burton, A. Shih, G.R. Laramee, F. Wurm, A. Mason, K. Nikolics and J.W. Winslow, Endocrinology, 129, 1289 (1991).
7. J. Leibrock, F. Lottspeich, A. Hohn, M. Hofer, B. Hengerer, P. Masiakowshi, H. Thoenen and Y. Barde, Nature, 341, 149 (1989).
8. L.R. Berkemeier, J.W. Winslow, D.R. Kaplan, K. Nikolics, D.V. Goeddel and A. Rosenthal, Neuron, 7, 857 (1991).
9. K.R. Jones and L.F. Reichardt, Proc.Natl.Acad.Sci.USA, 87, 8060 (1990).
10. A. Hohn, J. Leibrock, K. Bailey and Y. Barde, Nature, 344, 339 (1990).
11. F. Hallbook, C.F. Ibanez and H. Persson, Neuron, 6, 845 (1991).
12. C.F. Ibanez, F. Hallbook, S. Soderstrom, T. Ebendal and H. Persson, J.Neurochem., 57, 1033 (1991).
13. N.Q. McDonald, R. Lapatto, J. Murray-Rust, J. Gunning, A. Wlodawer and T.L. Blundell, Nature, 354, 411 (1991).
14. R. Williams, B. Gaber and J. Gunning, J.Biol.Chem., 257, 13321 (1982).
15. C.F. Ibanez, D. Hallbook, T. Ebendal and H. Persson, EMBO J., 9, 1477 (1990).
16. U. Suter, C. Angst, C.-L. Tien, C.C. Drinkwater, R.M. Lindsay and E.M. Shooter, J.Neurosci., 12, 306 (1992).
17. C.F. Ibanez, T. Ebendal and H. Persson, EMBO J., 10, 2105 (1991).
18. D. Johnson, A. Lanahan, C.R. Buck, A. Sehgal, C. Morgan, E. Mercer, M. Bothwell and M. Chao, Cell, 47, 545 (1986).
19. S.O. Meakin and E.M. Shooter, Neuron, 6, 153 (1991).
20. D. Martin-Zanca, S.H. Hughes and M. Barbacid, Nature, 319, 743 (1986).
21. D.R. Kaplan, D. Martin-Zanca and L.F. Parada, Nature, 350, 158 (1991).
22. R. Klein, S. Jing, V. Nanduri, E. O'Rourke and M. Barbacid, Cell, 65, 189 (1991).
23. G. Weskamp and L.F. Reichardt, Neuron, 6, 649 (1991).
24. R. Klein, V. Nanduri, S. Jing, F. Lamballe, P. Tapley, S. Bryant, C. Cordon-Cardo, K.R. Jones, L.F. Reichardt and M. Barbacid, Cell, 66, 395 (1991).
25. F. Lamballe, R. Klein and M. Barbacid, Cell, 66, 967 (1991).
26. R. Klein, L.F. Parada, F. Coulier and M. Barbacid, EMBO J., 8, 3701 (1989).
27. R. Klein, D. Donway, L.F. Parada and M. Barbacid, Cell, 61, 647 (1990).
28. S.P. Squinto, T.N. Stitt, T.H. Aldrich, S. Davis, S.M. Bianco, C. Radziejewski, D.J. Glass, P. Masiakowski, M.E. Furth, D.M. Valenzuela, P.S. DiStefano and G.D. Yancopoulos, Cell, 65, 885 (1991).
29. C. Cordon-Cardo, P. Tapley, S. Jing, V. Nanduri, E. O'Rourke, F. Lamballe, K. Kovary, R. Klein, K.R. Jones, L.F. Reichardt and M. Barbacid, Cell, 66, 173 (1991).
30. P.C. Maisonpierre, L. Belluscio, B. Friedman, R. Alderson, S. Weigand, M.E. Furth, R.M. Lindsay and G.D. Yancopoulos, Neuron, 5, 501 (1990).
31. M.L. Vetter, D. Martin-Zanca, L.F. Parada, J.M. Bishop and D.R. Kaplan, Proc.Natl.Acad.Sci.USA, 88, 5650 (1991).
32. N. Gómez and P. Cohen, Nature, 353, 170 (1991).
33. A. Acheson, P.A. Barker, R.F. Alderson, F.D. Miller and R.A. Murphy, Neuron, 7, 265 (1991).

34. N.E. Kremer, G. D'Arcangelo, S.M. Thomas, M. DeMarco, J.S. Brugge and S. Halegoua, J.Cell Biol., 115, 809 (1991).
35. B.L. Hempstead, N. Patil, B. Thiel and M.V. Chao, J.Biol.Chem., 265, 9595 (1990).
36. H. Yan, J. Schlessinger and M.V. Chao, Science, 252, 561 (1991).
37. M. Qiu and S.H. Grenn, Neuron, 7, 937 (1991).
38. J. Represa, M.A. Avila, C. Miner, F. Giraldez, G. Romero, R. Clemente, J.M. Mato and I. Varela-Nieto, Proc.Natl.Acad.Sci.USA, 88, 8016 (1991).
39. S.R. Whittemore, V.R. Holets, R.W. Keane, D.J. Levy and R.D.G. McKay, J.Neurosci.Res., 28, 156 (1991).
40. V.E. Koliatsos, R.E. Clatterbuck, H.J.W. Nauta, B. Knüsel, L.E. Burton, F.F. Hefti, W.C. Mobley and D.L. Price, Ann.Neurol., 30, 831 (1991).
41. M.H. Tuszynski, H.S. U, D.G. Amaral and F.H. Gage, J.Neurosci., 10, 3604 (1990).
42. I.A. Ferguson, J.B. Schweitzer, P.F. Bartlett and E.M. Johnson, J.Comp.Neurol., 313, 680 (1991).
43. B.A. Yankner, A. Caceres and L.K. Duffy, Proc.Natl.Acad.Sci.USA, 87, 9020 (1990).
44. S.C. Apfel, R.B. Lipton, J.C. Arezzo and J.A. Kessler, Ann.Neurol., 29, 87 (1991).
45. C. Hyman, M. Hofer, Y. Barde, M. Juhasz, G.D. Yancopoulos, S.P. Squinto and R.M. Lindsay, Nature, 350, 230 (1991).
46. B. Knüsel and F. Hefti, J.Neurochem., 57, 955 (1991).
47. M.M. Berg, D.W. Sternberg, L.F. Parada and M.V. Chao, J.Biol.Chem., 267, 13 (1992).
48. W.H. Burgess and T. Maciag, Ann.Rev.Biochem., 58, 576 (1989).
49. S. Ortega, M.-T. Schaeffer, D. Soderman, J. DiSalvo, D.L. Linemeyer, G. Gimenez-Gallego and K.A. Thomas, J.Biol.Chem., 266, 5842 (1991).
50. The Fibroblast Growth Factor Family " The New York Academy of Sciences, New York, N.Y.,1991,
51. J.A. Wagner, Curr.Top.Microbiol.Immunol., 165, 95 (1991).
52. J.A. Abrahan, A. Mergia, J.L. Whang, A. Tumolo, J. Friedman, A. Hjerrild, D. Gospodarowicz and J.C. Fiddes, Science, 233, 545 (1986).
53. B. Bugler, F. Amalric and H. Prats, Mol.Cell.Biol., 11, 573 (1991).
54. X. Zhu, H. Komiya, A. Chirino, S. Faham, G.M. Fox, T. Arakawa, B.T. Hsu and D.C. Rees, Science, 251, 90 (1991).
55. A.E. Eriksson, L.S. Cousens, L.H. Weaver and B.W. Matthews, Proc.Natl.Acad.Sci.USA, 88, 3441 (1991).
56. J. Zhang, L.S. Cousens, P.J. Barr and S.R. Sprang, Proc.Natl.Acad.Sci.USA, 88, 3446 (1991).
57. G.M. Clore, P.T. Wingfield and A.M. Gronenborn, Biochem, 30, 2315 (1991).
58. B.C. Finzel, L.L. Clancy, D.R. Holland, S.W. Muchmore, K.D. Watenpaugh and H.M. Einspahr, J.Mol.Biol., 209, 779 (1989).
59. J.P. Priestle, N.-P. Schar and M. Grutter, Proc.Natl.Acad.Sci.USA, 86, 9667 (1989).
60. A. Baird, D. Schubert, N. Ling and R. Guillemin, Proc.Natl.Acad.Sci.USA, 85, 2324 (1988).
61. M. Seno, R. Sasada, M. Iwane, K. Sudo, T. Kurokawa, K. Ito and K. Igarashi, Biochem.Biophys.Res.Commun., 151, 701 (1988).
62. K.A. Thomas, S. Ortega, D. Soderman, M.-T. Schaeffer, J. DiSalvo, G. Gimenez-Gallego, D. Linemeyer, L. Kelly and J. Menke in "The Fibroblast Growth Factor Family ," A. Baird and M. Klagsbrun. The New York Academy of Sciences, New York, N.Y.: 1991, p. 9.
63. J.C. Fiddes, J.A. Abraham and A. Protter, EPO Applic., 88306158.2, (1988).
64. W.F. Heath, A.S. Cantrell, N.G. Mayne and S.R. Jaskunas, Biochem, 30, 5608 (1991).
65. A. Isacchi, M. Statuto, R. Chiesa, L. Bergonzoni, M. Rusnati, P. Sarmientos, G. Ragnotti and M. Presta, Proc.Natl.Acad.Sci.USA, 88, 2628 (1991).
66. M. Klagsbrun and A. Baird, Cell, 67, 229 (1991).
67. D. Moscatelli, J.Cell.Physiol., 131, 123 (1987).
68. D.E. Johnson, J. Lu, H. Chen, S. Werner and L.T. Williams, Mol.Cell.Biol., 11, 4627 (1991).
69. A. Safran, A. Avivi, A. Orr-Urtereger, G. Neufeld, P. Lonai, D. Givol and Y. Yarden, Oncogene, 5, 635 (1990).
70. P.L. Lee, D.E. Johnson, L.S. Cousens, V.A. Fried and L.T. Williams, Science, 245, 57 (1989).
71. D.E. Johnson, P.L. Lee, J. Lu and L.T. Williams, Mol.Cell.Biol., 10, 4728 (1990).
72. A. Ullrich and J. Schlessinger, Cell, 61, 203 (1990).
73. S. Werner, D.R. Duan, C. de Bries, K.G. Peters, D.E. Johnson and L.T. Williams, Mol.Cell.Biol., 12, 82 (1992).
74. J. Partanen, T.P. Makela, E. Eerola, J. Korhonen, H. Hirvonen, L. Claesson-Welsh and K. Alitalo, EMBO J., 10, 1347 (1991).
75. T. Miki, T.P. Flembing, D.P. Bottaro, J.S. Rubin, D. Ron and S. Aaronson, Science, 251, 72 (1991).
76. E. Ruoslahti and Y. Yamaguchi, Cell, 64, 867 (1991).
77. A. Yayon, M. Klagsbrun, J.D. Esko, P. Leder and D.M. Ornitz, Cell, 64, 841 (1991).
78. J.M. Kaplow, F. Bellot, G. Crumley, C.A. Dionne and M. Jaye, Biochem.Biophys.Res.Commun., 172, 107 (1990).
79. F. Bellot, G. Crumley, J.M. Kaplow, J. Schlessinger, M. Jaye and C.A. Dionne, EMBO J., 10, 2849 (1991).

80. D.M. Loeb, J. Maragos, D. Martin-Zanca, M.V. Chao, L.F. Parada and L.A. Greene, Cell, $\underline{66}$, 961 (1991).
81. O. Haub, B. Drucker and M. Goldfarb, Proc.Natl.Acad.Sci.USA, $\underline{87}$, 8022 (1990).
82. D. Schubert, N. Ling and A. Baird, J.Cell Biol., $\underline{104}$, 635 (1987).
83. I.A. Ferguson and E.M. Johnson, J.Comp.Neurol., $\underline{313}$, 693 (1991).
84. A. Wanaka, E.M. Johnson and J. Milbrandt, Neuron, $\underline{5}$, 267 (1990).
85. K. Yoshida and F.H. Gage, Brain Res., $\underline{538}$, 118 (1991).
86. B. Knusel, P.P. Michel, J.S. Schwaber and F. Hefti, J.Neurosci., $\underline{10}$, 558 (1990).
87. P. Masiakowski, H. Liu, C. Radziejewski, F. Lottspeich, W. Oberthuer, V. Wong, R.M. Lindsay, M.E. Furth and N. Panayotatos, J.Neurochem., $\underline{57}$, 1003 (1991).
88. A. Lam, F. Fuller, J. Miller, J. Kloss, M. Manthorpe, S. Varon and B. Cordell, Gene, $\underline{102}$, 271 (1991).
89. J.R. McDonald, C. Ko, D. Mismer, D.J. Smith and F. Collins, Biochim.Biophys.Acta Gene Struct.Expression, $\underline{1090}$, 70 (1991).
90. A. Negro, E. Tolosano, S.D. Skaper, I. Martini, L. Callegaro, L. Silengo, F. Fiorini and F. Altruda, Eur.J.Biochem., $\underline{201}$, 289 (1991).
91. S. Davis, T.H. Aldrich, D.M. Valenzuela, V. Wong, M.E. Furth, S.P. Squinto and G.D. Yancopoulos, Science, $\underline{253}$, 59 (1991).
92. M. Hibi, M. Murakami, M. Saito, T. Hirano, T. Taga and T. Kishimoto, Cell, $\underline{63}$, 1149 (1991).
93. N.Y. Ip, Y. Li, I. Van de Stadt, N. Panayotatos, R.F. Alderson and R.M. Lindsay, J.Neurosci., $\underline{11}$, 3124 (1991).
94. M. Sendtner, G.W. Kreutzberg and H. Thoenen, Nature, $\underline{345}$, 440 (1990).
95. R.W. Oppenheim, D. Prevette, Q.W. Ying, F. Collins and J. MacDonald, Science, $\underline{251}$, 1616 (1991).

Chapter 6. Antipsychotics

Lawrence D. Wise and Thomas G. Heffner
Parke-Davis Pharmaceutical Research
Warner-Lambert Company
Ann Arbor, MI 48106

Introduction - Research on antipsychotic drugs has continued to focus on agents with improved efficacy and reduced neurological side effects. Recent reviews have emphasized strategies for developing new drugs (1,2,3,4,5,6) as well as problems in clinical schizophrenia research (7). Other reviews have considered the role of brain catecholamines in schizophrenia (8,9,10), the use of benzodiazepines in psychosis (11), and the neurological side effects of existing antipsychotics (12).

Dopamine (DA) D_2 Receptor Antagonists - Reports from clinical trials suggest possible improved activity profiles for several new DA D_2 antagonists. Remoxipride ($\underline{1}$) was efficacious in the treatment of schizophrenia with fewer extrapyramidal side effects (EPS, 13,14,15) and less elevation of serum prolactin (16) than with haloperidol. Studies in monkeys indicated that $\underline{1}$ produces extrapyramidal dysfunction, but at higher doses than those necessary for other CNS effects (17). Others suggested that the clinical profile of $\underline{1}$ may stem from inhibition of limbic DA release (18) or from DA receptor effects within striatal subregions (19).

	$\underline{R^1}$	$\underline{R^2}$	$\underline{R^3}$
$\underline{1}$	OMe	H	Br
$\underline{2}$	H	SO_2NH_2	H

The DA antagonist properties of novel salicylamides and benzamides was reviewed (20). While the (S)-enantiomers of [(N-ethyl-2-pyrrolidinyl)methyl]benzamides and salicylamides are more active than their (R)-enantiomers, the (R)-enantiomers are most active among the N-benzyl analogues (21). Similar effects were obtained after replacing the N-ethyl group of sulpiride ($\underline{2}$) or dihydrobenzofuran $\underline{3}$ with longer alkyl chains (22). The influence of electronic factors and lipophilicity on DA D_2 receptor binding was quantitated for a large series of benzamides and salicylamides (23,24).

The tetracyclic DA D_2 antagonist savoxepine ($\underline{4}$) was developed based on the theory that a drug capable of inducing a preferential blockade of hippocampal versus striatal DA receptors might cause reduced EPS (25). However, $\underline{4}$ produced EPS in a small clinical trial with schizophrenic patients (26). Aminomethyltetralones $\underline{5}$ and $\underline{6}$ were identified as the most potent analogues from a series of rigid butyrophenone DA antagonists (27). A pharmacophore for DA D_2 antagonist activity was developed based on the conformational properties of semi-rigid antipsychotic drugs (28). A theoretical and NMR analysis of the solution conformations of the rigid DA antagonist butaclamol was also reported (29,30).

DA D_1 Receptor Antagonists - SCH-39166 (<u>7</u>), a relatively selective DA D_1 antagonist undergoing clinical evaluation, was reported to have lesser affinity for serotonin (5HT) receptors than the prototypal D_1 antagonist, SCH-23390 (<u>8</u>, 31). In monkeys treated with the DA neurotoxin, MPTP, <u>8</u> was found to cause Parkinsonian-like signs suggesting that D_1 antagonists may have liability for EPS clinically (32). Although <u>7</u> produced neurological dysfunction in haloperidol-sensitized monkeys, an established model of antipsychotic-induced EPS, abnormal movements diminished during repeated administration (33). New evidence suggested that the behavioral effects of D_1 antagonists are mediated by D_1 receptors in primates (34).

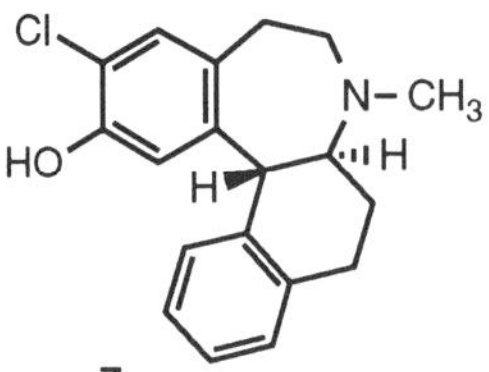

<u>7</u> 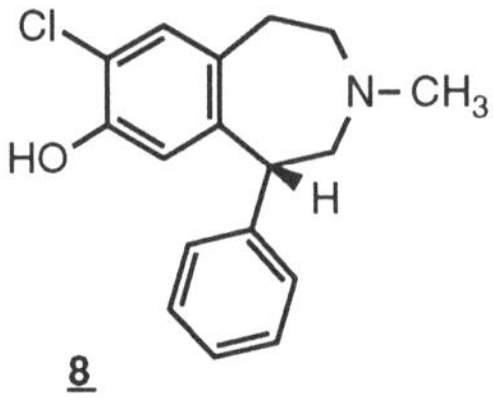   <u>8</u> 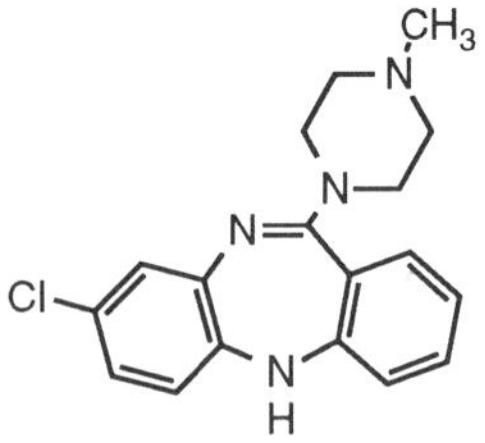   <u>9</u>

Clozapine - In the past year numerous review articles and clinical study summaries provide further documentation of the improved clinical profile of clozapine (<u>9</u>, 35,36,37,38,39,40). The biological mechanisms underlying the effects of <u>9</u> continued to be an important area of research (41). Positron emission tomography (PET) studies with the DA D_2 ligand [^{18}F]methylspiperone (MSP) demonstrated that MSP binding was decreased in the striatum of schizophrenics treated with haloperidol but not in two patients treated with <u>9</u> (42). Preclinical studies suggested that <u>9</u> may owe its unique profile to blockade of D_1 receptors (43), M_1 muscarinic antagonist activity (44,45), effects on $5HT_{1C}$ receptors (46) or inhibition of mesolimbic DA release (47,48).

Serotonin (5HT) Antagonists - Involvement of brain 5HT systems in the etiology and drug therapy of schizophrenia was reviewed (49,50,51,52). Zotepine (<u>10</u>), a potent 5HT and DA antagonist, was efficacious in clinical trials with paranoid, resistant and negative schizophrenia (53,54,55). While ritanserin (<u>11</u>) is a specific $5HT_2$ antagonist and risperidone (<u>12</u>) has higher affinity for $5HT_2$ than DA D_2 receptors, ocaperidone (R-79598, <u>13</u>) was described as an equipotent DA D_2 and $5HT_2$ antagonist (56,57). When used in combination with classical antipsychotic drugs, <u>11</u> was reported to be effective in the treatment of schizophrenia (56). In early phase II studies, <u>12</u> appeared to be effective towards both the positive and negative symptoms of schizophrenia and decreased existing EPS (58). The pharmacological profiles of sertindole (<u>14</u>) and SM- 9018 (<u>15</u>), compounds with high affinity for $5HT_2$ and DA D_2 receptors, were reviewed (59,60). The gamma-carboline WY-47,791 (<u>16</u>) had modest affinity for both $5HT_2$ and DA D_2 receptors and was active in preclinical tests predictive of antipsychotic efficacy (61).

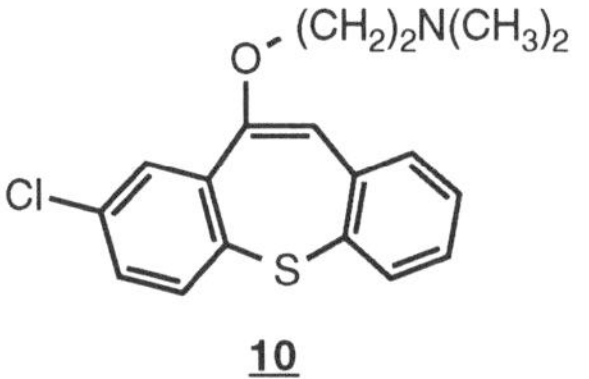

<u>10</u> 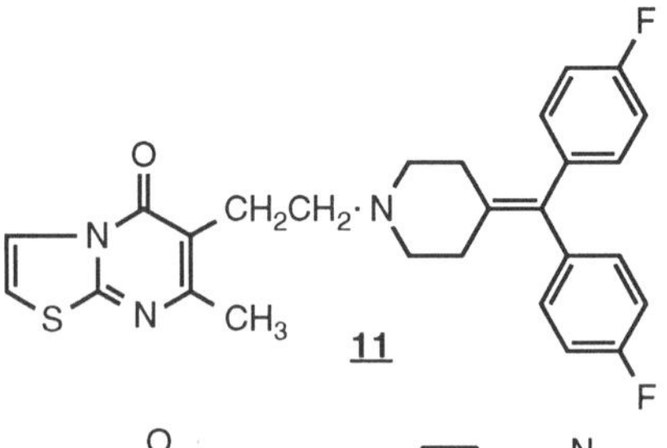   <u>11</u>

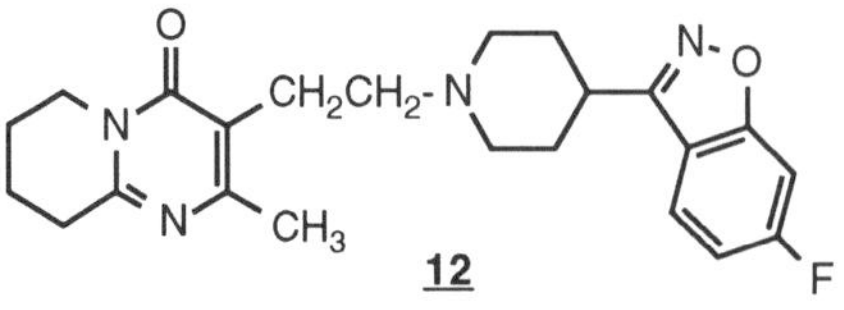

<u>12</u> 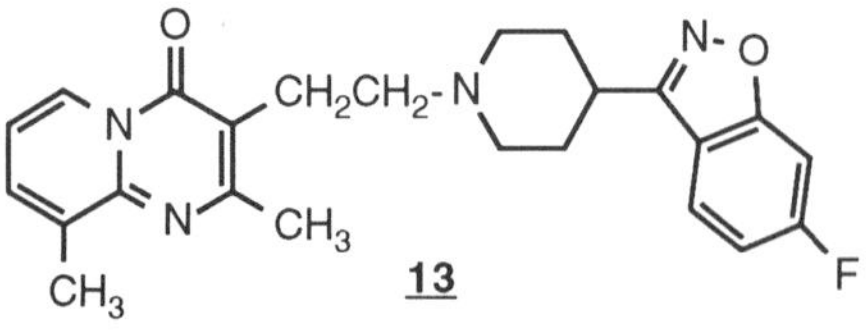   <u>13</u>

14   **15**   **16**   **17**

Umespirone (**17**), a compound with high affinity for 5HT1A, DA D2 and alpha1 adrenergic receptors, was found to be active in preclinical tests for antipsychotic and anxiolytic activity (62,63). By replacing the catechol moiety of DA with a series of heterocyclic surrogates, the naphthalenylpiperazine **18** was identified as a potential antipsychotic with affinities for 5HT1A, 5HT2, DA and adrenergic receptors (64). While both enantiomers of octoclothepin (**19**) were found to bind to 5HT2 and DA D2 receptors, the (**S**)-enantiomer was more potent (65). Two piperazine analogs, P-936 (**20**) and the N-alkoxyimide **21**, had high affinity for DA D2 as well as 5HT1A and 5HT2 receptors and were active in behavioral tests predictive of antipsychotic activity (66,67). Schizophrenics receiving the 5HT1A agonist buspirone (**22**) in combination with standard antipsychotic drugs showed improvement in an open trial (68).

18   **19**   **20**   **21**   **22**

The 5HT3 antagonist ondansetron (**23**) was reported to reduce mesolimibic DA activity in rats and marmosets (69), lending support to the proposed antipsychotic potential for such compounds. However, other preclinical studies reported more variable effects with zatosetron (**24**), granisetron (**25**), and WAY-100,289 (**26**) (70,71,72,73,74). Unlike other 5HT3 antagonists, MDL 72222 (**27**) antagonized conditioned avoidance and self-stimulation in rodents, effects seen with DA antagonist antipsychotics (75). Clozapine shared with 5HT3 antagonists the ability to block the suppression of neuron firing caused by the 5HT3 agonist, 2-methylserotonin (76).

23   **24**   **25**

26   **27**

<u>DA Autoreceptor Agonists</u> - Results from early clincial trials on DA autoreceptor agonists in schizo-
phrenia have provided mixed results. For example, calming and antipsychotic effects were reported
in a four patient trial with the partial DA agonist, (-)-3-PPP (**28** , 77). Talipexole (B-HT 920, **29**), an
agonist with high intrinsic activity, was reported to produce psychomotor agitation in an eleven
patient study (78). The partial DA agonist roxindol (EMD 49980, **30**) showed only weak
antipsychotic efficacy in acute schizophrenia in a twenty patient open label study (79). While the
partial agonists terguride (**31**) and pramipexole (**32**) had limited efficacy against positive symptoms
of schizophrenia in small trials, they were said to be more effective against negative schizophrenic
symptoms (78,80). A battery of preclinical tests indicated that SDZ 208-911 (**33**) has DA D_2 partial
agonist properties comparable to those of terguride while SDZ 208-912 (HDC-912, **34**) has lower
intrinsic activity at DA receptors (81). A clinical report stated that **34** had antipsychotic efficacy as
well as an EPS profile comparable to that haloperidol (82).

28   **29**   **30**

31 $R_1 = H$, $R_2 = N(C_2H_5)_2$
33 $R_1 = CH_3$, $R_2 = tBu$
34 $R_1 = Cl$, $R_2 = tBu$

32

In preclinical studies, U-86170F (R-**35**) was found to bind with high affinity to DA D_2 receptors and
to possess good oral activity (83). A study on rigid tricyclic aminothiazoles reported that the
(R)-(+)-enantiomer of **36** had a DA autoreceptor agonist profile and was active in rodent and monkey
tests predictive of antipsychotic efficacy (84). The (R)-(+)-enantiomer of PD 128483, **37**, was found
to be a full DA agonist while the (S)-(-)-enantiomer appeared to be a weak partial DA agonist (85). SAR
studies on benzopyranones indicated that compounds with an unsubstituted aryl group were
DA agonists while those with aryl substitutions were DA antagonists. One compound, PD 119819
(**38**), showed a DA autoreceptor agonist profile and produced antipsychotic-like effects in rodent and
monkey tests (86). PD 135222 (**39**) and PD 138276 (**40**), were identified as selective DA
autoreceptor agonists (87,88). Biochemical evidence was provided for the existence of DA D_1
autoreceptors in young but not adult rats (89).

35   **36** R = CH_3 **38**
                          37 R = C_3H_5

39   **40**

$\underline{41}$ $\underline{42}$

$\underline{43}$ $\underline{44}$ X = O, trans
                         $\underline{45}$ X = H$_2$, cis

<u>Sigma Ligands</u> - While sigma ligands have been proposed as potential antipsychotics, the functional role of brain sigma binding sites remained an enigma (90). New studies suggested the existence of two sigma binding sites in brain (91,92,93,94,95,96). Some reports suggested that the sigma site is associated with cytochrome P-450 (97). Conflicting evidence was reported regarding changes in sigma binding sites in postmortem brain tissue from schizophrenics (98,99). BMY 14802 ($\underline{41}$), a compound with potent affinity for sigma sites and weak affinity for 5HT1A receptors was well tolerated in Phase I clinical trials (100). DuP 734 ($\underline{42}$), a compound with potent affinity for sigma and 5HT2 receptors, antagonized mescaline induced effects, blocked isolation induced aggression, had little activity in conditioned avoidance tests and increased DA turnover in the rat frontal cortex (101,102,103).

The aminoalkoxychromone, NPC 16377 ($\underline{43}$), is a sigma ligand that was active in rodent behavioral and neurophysiological tests (104,105,106) but did not affect DA metabolism in brain (107). From a systematic SAR study of compounds related to the kappa agonist, U50,488, (1S,2S-(-)- $\underline{44}$), (1R,2S-(-)-$\underline{45}$) was identified as a compound with high affinity for sigma sites and little or no affinity for D2 receptors (108). SAR studies on a series of compounds derived from PCP (e.g., PRE-084 ($\underline{46}$)) indicated that alkyl or ester insertions between the amino and cycloalkyl groups increased potencies for sigma and decreased potencies for PCP receptors (109). A series of papers proposed a N-substituted 2-phenylaminoethane group as the primary sigma pharmacophore (110,111,112). Studies were also reported on JO1784 ($\underline{47}$) and L-687384 ($\underline{48}$) (113,114).

$\underline{46}$ $\underline{47}$ $\underline{48}$

<u>NEW BIOLOGICAL FINDINGS</u>

<u>Receptors</u> - The molecular cloning of DA D4 and D5, two new DA receptors, was reported. The DA D4 receptor was described as a DA D2 -like site that is localized within the limbic forebrain and that shows a particularly high affinity for clozapine (115). The DA D5 receptor was described as a DA D1-like site that has higher affinity for DA than does the D1 receptor and that is also localized within the limbic forebrain (116). DA D1 and D2 receptors were found to be distributed differentially in the cerebral cortex of monkeys suggesting that these receptor subtypes may mediate different aspects of cortical DA function (117). The density of striatal and mesolimbic DA D1 and D2 receptors was found to be unaffected by impoverished or enriched rearing conditions (118). The distribution of DA D2 receptor mRNA was described in primate brain (119) and the distribution of DA D3 mRNA was described in rat brain (120). A review compared the distribution and regulation of DA D2 mRNA and receptors (121). Site directed mutagenesis studies suggested different agonist and antagonist

binding domains for DA D_1 (122) and D_2 (123,124,125) receptors. A molecular model was constructed for the DA D_2 receptor (126). Haloperidol was found to initially decrease and then increase rat brain DA D_2 mRNA and receptor levels during chronic administration (127). Multiple genes were identified for the human DA D_1 (128) and D_5 receptor (129,130).

Peptides - Reduced levels of neurotensin were found in the cerebrospinal fluid from a subgroup of psychotic women who showed a delayed response to antipsychotic therapy (131). A study on the interaction between antipsychotic drugs and brain neurotensin (NT) in rats revealed that when given chronically, haloperidol increased NT levels in the caudate and nucleus accumbens, rimcazole and sulpiride increased NT in the caudate only and remoxipride failed to alter NT in either region (132). NT was found to reduce the affinity of agonists at brain DA D_2 receptors in vitro (133). Reviews summarized the neurophysiological effects of NT on central neurons, emphasizing interactions with brain DA that may be involved in the mediation of antipsychotic-like effects (134,135).

The levels of CCK in the cerebrospinal fluid were found to be reduced in a small group of schizophrenic patients (136). Evidence was presented that CCK-A receptors may contribute to the reduction in number of active brain DA neurons during chronic haloperidol or clozapine treatment (137). Repeated administration of a CCK-A antagonist produced an increase in the number of active A10, but not A9, DA neurons (138). A CCK-B antagonist was reported to decrease the number of active brain DA neurons in both A9 and A10 (139). A review appeared on the interactions between CCK and brain DA with implications for a role of CCK in psychiatric disorders (140).

Etiology of Schizophrenia - Studies of the brain DA D_2 receptor from schizophrenics revealed no structural changes (141). A positron emission tomography study failed to reveal a change in the density of DA D_2 receptors in the striatum of unmedicated schizophrenics (142). A genetic linkage study failed to find an alteration in the DA D_2 gene locus in schizophrenia (143). A reduction in the level of mRNA for a non-NMDA glutamate receptor was found in hippocampal tissue from a small group of schizophrenic patients (144). A study reported reduced release of glutamate and GABA from brain synaptosomes obtained from schizophrenics (145). Apparent decreases in the density of serotonin $5HT_2$ (146) and histamine H_1 receptors (147) in frontal cortex tissue from schizophrenics were reported. A review considered the hypothesis that an imbalance between GABA and glutamate contributes to psychosis (148). Another review proposed that schizophrenia might result from alterations in the regulation of phasic and tonic components of brain DA release (149).

Animal Models - Further studies on prepulse inhibition of acoustic startle supported the selectivity of this procedure for identifying antipsychotic drugs (150,151); however, the weak activity of clozapine raised questions as to the sensitivity of this test for atypical agents. Haloperidol and sulpiride were claimed to enhance latent inhibition in a potential model of schizophrenic attentional disorder (152). Amphetamine-induced release of DA from brain slices was proposed as a biochemical model for psychosis (153).

References

1.   H.Y. Meltzer, Schizophr. Bull., 17, 263 (1991).
2.   A.Y. Deutch, B. Moghaddam, R.B. Innis, J.H. Krystal, G.K. Aghajanian, B.S. Bunney, D.S. Charney, Schizophr. Res., 4, 121 (1991).
3.   D.M. Coward in "Advances in Neuropsychiatry and Psychopharmacology," Vol. 1, C.A. Tamminga, S.C. Schulz, Ed., Raven Press, New York, N.Y., 1991, p. 297.
4.   D. Holland, M.D. Watanabe, R. Sharma, Psychiatr. Med., 9, 5 (1991).
5.   J. Gerlach, Schizophr. Bull., 17, 289 (1991).
6.   P.H. Andersen, E.B. Nielsen, Drug News & Perspectives, 4, 150 (1991).
7.   J.M. Kane, Schizophr. Bull., 17, 353 (1991).
8.   K.L. Davis, R.S. Kahn, G. Ko, M. Davidson, Am. J. Psychiat., 48, 1474 (1991).
9.   D.P. van Kammen, M. Kelley, Schizophr. Res., 4, 173 (1991).
10.   D. Healy, Br. J. Psychiat., 159, 319 (1991).
11.   O.M. Wolkowitz, D. Pickar, Am. J. Psychiat., 148, 714 (1991).
12.   D.E. Casey, Schizophr. Res., 4, 109 (1991).
13.   M. Blomqvist, B. Gustafsson, T. Lewander, Psychopharmacol., 103, B9 (1991).
14.   T. Lewander, M. Blomqvist, B. Gustafsson, Psychopharmacol., 103, B9 (1991).
15.   G.F. Hebenstreit, G. Laux, H. Schubert, H. Beckmann, J. Amman, J. Bunse, G. Eikmeier, C.

Geretsegger, R.D. Kanitz, W.T. Kanzow, P. Konig, M. Mair, H. Mayr, T. Platz, H. Rittmannsberger, H.W. Schony, M. Struck, Z. Viski-Hanka, M. Wibmer, R. Zochling, Pharmacopsychiat., 24, 153 (1991).
16.   A.G. Awad, Y.D. Lapierre, K.G. Jostell, Prog. Neuropsychopharmacol. Biol. Psychiat., 14, 769 (1990).
17.   D.E. Casey, Psychopharmacol. Bull., 27, 47 (1991).
18.   J. Ichikawa, H.Y. Meltzer, Soc. Neurosci. Abst., 17, 687 (1991)
19.   K. Fuxe, S.O. Ogren, Acta Physiol. Scand., 141, 577 (1991).
20.   T. Hogberg, Drugs of the Future, 16, 333 (1991).
21.   T. Hogberg, P. Strom, T. de Paulis, B. Stensland, I. Csoregh, K. Lundin, H. Hall, S.O. Ogren, J. Med. Chem., 34, 948 (1991).
22.   S. Murakami, N. Marubayashi, T. Fukuda, S. Takehara, T. Tahara, J. Med. Chem., 34, 261 (1991).
23.   T. de Paulis, N. E. Tayar, P.-A. Carrupt, B. Testa, H. van de Waterbeemd, Helv. Chem. Acta, 74, 241 (1991).
24.   U. Norinder, T. Hogberg, Quant. Struct.-Act. Relat., 10, 1 (1991).
25.   S. Bischoff, J. Krauss, A. Vassout, A. Bruinink, Schizophr. Res., 4, 313 (1991).
26.   H. Wetzel, K. Wiedemann, F. Holsboer, O. Benkert, Psychopharmacol., 103, 280 (1991).
27.   L. Cortizo, L. Santana, E. Ravina, R. Orallo, J.A. Fontenla, E. Castro, J.M. Calleja, M.L. de Ceballos, J. Med. Chem., 34, 2242 (1991).
28.   M. Froimowitz, S. Ramsby, J. Med. Chem., 34, 1707 (1991).
29.   M.G. Casarotto, D.J. Craik, E.J. Lloyd, J. Med. Chem., 34, 2043 (1991).
30.   M.G. Casarotto, D.J. Craik, E.J. Lloyd, A.C. Partridge, J. Med. Chem., 34, 2036 (1991).
31.   R.D. McQuade, C.E. Tedford, R.A. Duffy, L.A. Taylor, M.A. Hunt, J.K. Wamsley, A. Barnett, Soc. Neurosci. Abst., 17, 1348 (1991).
32.   M.S. Lawrence, D.E. Redmond, Jr., J.D. Elsworth, J.R. Taylor, R.H. Roth, Life Sci., 49, PL-229 (1991).
33.   D. McHugh, V. Coffin, Eur. J. Pharmacol., 202, 133 (1991).
34.   J. Bergman, B.K. Madras, R.D. Spealman, J. Pharmacol. Exp.Ther., 258, 910 (1991).
35.   M.W. Jann, Pharmacotherapy, 11, 179 (1991).
36.   S.L. McElroy, E.C. Dessain, H.G. Pope, J.O. Cole, P.E. Keck, F.R. Frankenberg, Y.H. Aizle, S. Obrien, J. Clin. Psychiat., 52, 411 (1991).
37.   A. Brieir, R.W. Buchanan, B. Kirpatrick, W.T. Carpenter, O. Davis, L.A. Moricle, D. Irish, Schizophr. Res., 4, 315 (1991).
38.   D. Naber, H. Hippius, Schizophr. Res., 4, 322 (1991).
39.   A. Safferman, J.A. Lieberman, J.M. Kane, S. Szymanski, B. Kinon, Schizophr. Bull., 17, 247 (1991).
40.   R.J. Baldessarini, F.R. Frankenburg, N. Engl. J. Med., 324, 746 (1991).
41.   B. Moghaddam, R.B. Innis, J.H. Krystal, G.K. Aghajanian, B.S. Bunney, D.S. Charney, Schizophr. Res., 4, 121 (1991).
42.   H. Karbe, K. Wienhard, K. Hamacher, M. Huber, K. Herholz, H.H. Coenen, G. Stocklin, A. Lovenich, W.D. Heiss, J. Neural Transm. [GenSect], 86, 163 (1991).
43.   B.A. Ellenbroek, M.T. Artz, A.R. Cools, Eur. J. Pharmacol., 196, 103 (1991).
44.   R. Neeper, E. Richelson, A. Nelson, Neuropharmacol., 30, 527 (1991).
45.   C. Bolden, B.Cusack, E. Richelson, Eur. J. Pharmacol., 192, 205 (1991).
46.   H. Canton, L. Verriele, F.C. Coppaert, Eur. J. Pharmacol., 191, 93 (1991).
47.   J. Chen, W. Paredes, E.L. Gardner, Neurosci. Lett., 122, 127 (1991).
48.   J. Ichikawa, H.Y. Meltzer, J. Pharmacol. Exp. Ther., 256, 348 (1991).
49.   L.J. Siever, R.S. Kahn, B.A. Lawlor, R.L. Trestman, T.L. Lawrence, E.F. Coccaro, Pharmacol. Rev., 43, 509 (1991).
50.   S.J. Peroutka, Pharmacol. Rev., 43, 579 (1991).
51.   H.Y. Meltzer, J.F. Nash, Pharmacol. Rev., 43, 587 (1991).
52.   A. Bleich, S.-L. Brown, H.M. Van Pragg in "Clinical and Experimental Psychaitry," No. 4, S.-L. Brown, H.M. Van Pragg, Ed., Brunner/Mazel, New York, N.Y., 1991, p. 183.
53.   H. Wetzel, V. Bardeleben, F. Holsboer, O. Benkert, Fortschr. Neurol. Psychiatr., 59,(Suppl.1), 23 (1991).
54.   T. Harada, S. Otsuke, Y. Fujiwara, Fortschr. Neurol. Psychiat., 59, (Suppl.1), 41 (1991).
55.   F. Muller-Spahn, D. Dieterle, M. Ackenheil, Fortschr. Neurol. Psychiatr., 59, (Suppl.1), 30 (1991).
56.   A.A.H.P. Megens, F.H.L. Awouters, T.F. Meert, K.H.L. Schellenkens, C.J.E. Niemegeers, P.A.J. Janssen, J. Pharmacol. Exp. Ther., 260, 146 (1992).
57.   A.A.H.P. Megens, C.J.E. Niemegeers, F.H.L. Awouters, Pharmacol. Exp. Ther., 260, 160 (1992).
58.   M. Sloth Nielson, S.L.E. Heylen, Y.G. Gelders, Psychopharmacol., 103, B9 (1991).
59.   C. Sanchez, J. Arnt, N. Dragsted, J. Hyttel, H.L. Lembol, E. Meier, J. Perregaard, T. Skarsfeldt, Drug Dev. Res., 22, 239 (1991).
60.   Drugs of the Future, 16, 122 (1991).
61.   M. Abou-Gharbia, K. Marquis, T. Andree, Drugs of the Future, 16, 1008 (1991).
62.   F. Krijzer, H. Krahling, Drugs of the Future, 16, 437 (1991).
63.   N.M. Barnes, B. Costall, A.M. Domeney, P.A. Gerrard, M.E. Kelly, H. Krahling, R.J. Naylor, D.M. Tomkins, T.J. Williams, Pharmacol. Biochem. Behav., 40, 89 (1991).

64.   J.A. Lowe,III, T.F. Seeger, A.A. Nagel, H.R. Howard, P.A. Seymour, J.H. Heym, F.E.Ewing, M.E. Newman, A.W. Schmidt, J.S. Furman, L.A. Vincent, P.R. Maloney, G.L. Robinson, L.S. Reynolds, F.J. Vinick, J. Med. Chem., 34, 1860 (1991).
65.   K.P. Bogeso, T. Liljefors, J. Arnt, J. Hyttel, H. Pedersen, J. Med. Chem, 34, 2023 (1991).
66.   N.J. Hrib, J. G. Jurcak, F.P. Huger, C. L. Errico, R. W. Dunn, J. Med. Chem., 34, 1068 (1991).
67.   N.J. Hrib, J.G. Jurcak, D.E. Bregna, R.W. Dunn, H.M. Geyer,III, H.B. Hartman, J.E. Roehr, K.L. Rogers, D.K. Rush, A.M. Szczepanik, M.R. Szewczak, C.A. Wilmot, P.G. Conway, Abstr. Am. Chem. Soc. (202mtg), 4th Chem. Cong. N.A. MEDI:85 (1991).
68.   D.C. Goff, K.K. Midha, A.W. Brotman, S. McCormick, M. Waites, E.T. Amico, J. Clin. Psychopharmacol, 11, 193 (1991).
69.   J.M. Bell, M.B. Tyers, J. Clin. Pharmacol., 31, 268 (1991).
70.   K. Rasmussen, M.E. Stockton, J.F. Czachura, Eur. J. Pharmacol., 205, 113 (1991).
71.   R.Y. Wang, Y. Minabe and C.R. Ashby, Jr., Soc. Neurosci. Abst., 17 , 600 (1991).
72.   Y. Minabe, C.R. Ashby, Jr., J.E. Schwartz, R.Y. Wang, Eur. J. Pharmacol., 209, 143 (1991).
73.   Y. Minabe, C.R. Ashby, Jr., R. Y. Wang, Eur. J. Pharmacol., 209, 151 (1991).
74.   C.W. Uzzle, J. T. Haskins, Soc. Neurosci. Abst., 17, 600 (1991).
75.   R.W. Dunn, W.A. Carlezon, R. Corbett, Drug Dev. Res., 23, 289 (1991).
76.   C.R. Ashby, Y. Minabe, E. Edwards, R.Y. Wang, Synapse, 8, 155 (1991).
77.   C.A. Tamminga, N. Cascella, R. Lahti, A. Carlsson, Soc. Neurosci. Abst., 17, 690 (1991).
78.   M. Abou-Gharbia, E.A. Muth, Drug News & Perspectives, 4, 647 (1991).
79.   A. Klimke, E. Klieser, Pharmacopsychiat., 24, 107 (1991).
80.   R. Olbrich, H. Schanz, J. Neural Transm., 84, 233 (1991).
81.   K. Svensson, A. Ekman, M.F. Piercey, W.E. Hoffmann, J.T. Lum, A. Carlsson, Naunyn-Schmiedeberg's Arc. Pharmacol., 344, 263 (1991).
82.   D. Naber, Schizophr. Res., 4, 323 (1991).
83.   R.A. Lahti, D.L. Evans, L.M. Figur, R.M. Huff, H.W. Moon, Naunyn-Schmiedeberg's Arc. Pharmacol., 344, 509 (1991).
84.   B.W. Caprathe, J.C. Jaen, L.D. Wise, T.G. Heffner, T.A. Pugsley, L.T. Meltzer, M. Parvez, J. Med. Chem., 34, 2736 (1991).
85.   J.C. Jaen, B.W. Caprathe, L.D. Wise, T.A. Pugsley, L.T. Meltzer, T.G. Heffner, Bioorg. Med. Chem. Lett., 1, 539 (1991).
86.   J.C. Jaen, L.D. Wise, T.G. Heffner, T.A. Pugsley, L.T. Meltzer, J. Med. Chem., 34, 248 (1991).
87.   L.M. Ball, L.W. Cooke, A.E. Corbin, T.A. Pugsley, F.W. Ninteman, L.T. Meltzer, T.G. Heffner, Soc. Neurosci. Abst., 17, 689 (1991).
88.   L.D. Wise, J.C. Jaen, B.W. Caprathe, S.J.Smith, T.A. Pugsley, T.G. Heffner, Soc. Neurosci. Abst., 17, 689 (1991).
89.   M.H. Teicher, A.L. Gallitano, H.A. Gelbard, H.K. Evans, E.R. Marsh, R.G. Booth, R.J. Baldessarini, Dev. Brain Res., 63, 229 (1991).
90.   C.D. Ferris, D.J. Hirsch, B.P. Brooks, S.H. Snyder, J. Neurochem., 57, 729 (1991).
91.   X.Z. Wu, J.A. Bell, C.E. Spivak, E.D. London, T.P. Su, J. Pharmacol. Exp. Ther., 257, 351 (1991).
92.   A.R. Knight, J. Gillard, E.H.F. Wong, D.N. Middlemiss, Neurosci. Lett., 131, 233 (1991).
93.   G.-Z. Zhou, J.M. Musacchio, Eur. J. Pharmacol. Mol. Pharmacol., 206, 261 (1991).
94.   A. Georg, A. Friedl, J. Pharm. Exp. Ther., 259, 479 (1991).
95.   R.B. Rothman, A. Reid, A. Mahboubi, C.-H. Kim, B.R. De Costa, A.E. Jacobson, K.C. Rice, Mol. Pharmacol., 39, 222 (1991).
96.   C.D. Ferris, D.J. Hirsch, B.P. Brooks, A.M. Snowman, S.H. Snyder, Mol. Pharmacol., 39, 199 (1991).
97.   J. Lehmann, Drug News & Prospectives, 4, 208 (1991).
98.   G.P. Reynolds, J.E. Brown, D.N. Middlemiss, Eur. J. Pharm., 194, 235 (1991).
99.   A.D. Weissman, M.F. Casanova, J.E. Kleinman, E.D. London, E.B. De Souza, Biol. Psychiat., 29, 41 (1991).
100. G.R. Gewirtz, J. Volavka, J.M. Gorman, R. Pyke, P.Q. Owen, C.A. Kaufmann, R.L. Borison, Schizophr. Res., 4, 317 (1991).
101. S.W. Tam, G.F. Steinfels, P.J. Gilligan, J.F. McElroy, V.J. DeNoble, A.L. Johnson, L. Cook, Soc. Neurosci. Abst., 17, 333 (1991).
102. C. Rominger, S.W. Tam, Soc. Neurosci. Abst., 17, 333 (1991).
103. L. Cook, S.W. Tam, W.K. Schmidtand, K.W. Rohrbach, Soc. Neurosci. Abst., 17, 333 (1991).
104. W. Karbon, M. Bailey, S. Borosky, M. Abreu, L. Martin, R. Erickson, K. Natalie, M. Pontecorvo, J. Ferkany, Soc. Neurosci. Abst., 17, 334 (1991).
105. D.B. Clissold, T. Hartman, M.E. Abreu, R. Erickson, E.W. Karbon, M.J. Pontecorvo, J.W. Ferkany, Soc. Neurosci. Abst., 17, 334 (1991).
106. P.D. Shepard, H. Romeyn, Soc. Neurosci. Abst., 17, 334 (1991).
107. M.E. Abreu, L.A. Martin, R. Erickson, E.W. Karbon, M.J. Pontecorvo, J.W. Ferkany, Soc. Neurosci. Abst., 17, 334 (1991).
108. L. Radesca, W.D. Bowen, L. Di Paolo, B.R. de Costa, J. Med. Chem., 34, 3058 (1991).

109. T.-P. Su, X.-Z. Wu, E.J. Cone, K. Shukla, T.M. Gund, A.L. Dodge, D.W. Parish, J. Pharm. Exp. Ther., 259, 543 (1991).
110. R.A. Glennon, A.M. Ismaiel, J.D. Smith, M. Yousif, M. El-Ashmawy, J.L. Herndon, J.B. Fischer, K.J. Burke Howie, A.C. Server, J. Med. Chem., 34, 1855 (1991).
111. R.A. Glennon, J.D. Smith, A.M. Ismaiel, M. El-Ashmawy, G. Battaglia, J.B. Fischer, J. Med. Chem., 34, 1094 (1991).
112. R.A. Glennon, M.Y. Yousif, A.M. Ismaiel, M.B. El- Ashmawy, J.L. Herndon, J.B. Fischer, A.C. Server, K.J. Burke Howie, J. Med. Chem., 34, 3360 (1991).
113. J.L. Junien, F.J. Roman, G. Brunelle, X. Pascaud, Eur. J. Pharmacol., 200, 343 (1991).
114. D.N. Middlemiss, D. Billington, M. Chambers, P.H. Hutson, A. Knight, M. Russell, L. Thorn, M.D. Tricklebank, E.H.F. Wong, Brit. J. Pharmacol., 102, (Suppl), 153P (1991).
115. H.H.M. Van Tol, J.R. Bunzow, H.-C. Guan, R.K. Sunahara, P. Seeman, H.B. Niznik, O. Civelli, Nature, 350, 610 (1991).
116. R.K. Sunahara, H.-C. Guan, B.F. O'Dowd, P. Seeman, L.G. Lauier, G. Ng, S.R. George, J. Torchia, H.H.M. Van Tol, H.B. Niznik, Nature, 350, 614 (1991).
117. M.S. Lidow, P.S. Goldman-Rakic, D.W. Gallager, P. Rakic, Neurosci., 40, 657 (1991).
118. M.T. Bardo, R.P. Hammer, Neurosci., 45, 281 (1991).
119. J.H. Meador-Woodruff, A. Mansour, O. Civelli, S.J. Watson, Prog. Neuro-Psychopharmacol. Bio. Psychiat., 15, 885 (1991).
120. M.L. Bouthenet, E. Souil, M.P. Martres, P. Sokoloff, B. Giros, J.C. Schwartz, Brain Res., 564, 203 (1991).
121. J.H. Meador-Woodruff, A. Mansour, Bio. Psychiat., 30, 985 (1991).
122. N.J. Pollock, A.M. Manelli, C.W. Hutchins, R.G. MacKenzie, D.E. Frail, Soc. Neurosci. Abst., 17, 85 (1991).
123. F. Meng, A. Mansour, J. Meador-Woodruff, L.P. Taylor, H. Akil, Soc. Neurosci. Abst., 17, 599 (1991).
124. K.A. Neve, B.A. Tester, R.A. Henningsen, G.C. Hughes, A. Spanoyannis, R.L. Neve, Soc. Neurosci. Abst., 17, 815 (1991).
125. K.A. Neve, B.A. Cox, R. A. Henningsen, A. Spanoyannis, R.L. Neve, Mol. Pharmacol., 39, 733 (1991).
126. S.G. Dahl, O. Edvardsen, I. Sylte, Proc. Natl. Acac. Sci. USA, 88, 8111 (1991).
127. J.A. Angulo, H. Coirini, M. Ledoux, M. Schumacher, Mol. Brain Res., 11, 161 (1991).
128. R.L. Weinshank, N. Adham, M. Macchi, M.A. Olsen, T.A. Branchek, P.R. Hartig, J. Bio. Chem., 266, 22427 (1991).
129. D.K. Grandy, Y.A. Zhang, C. Bouvier, Q.Y. Zhou, R.A. Johnson, L. Allen, K. Buck, J.R. Bunzow, J. Salon, O. Civelli, Proc. Natl. Acad. Sci. USA, 88, 9175 (1991).
130. T. Nguyen, R. Sunahara, A. Marchese, H.H.M. Vantol, P. Seeman, B.F. Odowd, Biochem. Biophysical. Res. Com., 181, 16 (1991).
131. D.L. Garver, G. Bissette, J.K. Yao, C.B. Nemeroff, Am. J. Psychiat., 148, 484 (1991).
132. B. Levant, G. Bissette, E. Widerlov, C.B. Nemeroff, Regul. Pept., 32, 193 (1991).
133. G. Von Euler, Brain Res., 561, 93 (1991).
134. Z.N. Stowe, C.B. Nemeroff, Life Sci., 49, 987 (1991).
135. Z.N. Stowe, G. Bissette, C.B. Nemeroff, Yakubutsu Seishin Kodo, 11, 49 (1991).
136. M.C. Beinfeld, D.L. Garver, Prog. Neuro-Psychopharmacol. Biol. Psychiat., 15, 601 (1991).
137. Y. Minabe, C.R. Ashby, Jr., R.Y. Wang, Brain Res., 549, 151 (1991).
138. J. Zhang, L.A. Chiodo, A.S. Freeman, Peptides, 12, 339 (1991).
139. K. Rasmussen, M.E. Stockton, J.F. Czachura, J.J. Howbert, Eur. J. Pharmacol., 209, 135 (1991).
140. J.N. Crawley, TiPS, 12, 232 (1991).
141. G. Sarkar, S. Kapelner, D.K. Grandy, M. Marchionni, O. Civelli, J. Sobell, O.N. Heston, S.S. Sommer, Genomics, 11, 8 (1991).
142. J.L. Martinot, M.L. Pailleremartinot, C. Loch, P. Hardy, M.F. Pouirier, B. Mazoyer, B. Beaufils, B. Maziere, J.F. Allilaire, A. Syrota, Br. J. Psychiat., 158, 346 (1991).
143. H.W. Moises, J. Gelemter, L.A. Giuffra, V. Zarcone, L. Wetterberg, O. Civelli, K.K. Kidd, L. L. Cavalli-Sforza, D.K. Grandy, J.L. Kennedy, S. Vinogradov, J. Mauer, M. Litt, B. Sjögren, Arch. Gen. Psychiat., 48, 643 (1991).
144. P.J. Harrison, D. McLaughlin, R.W. Kerwin, Lancet, 337, 450 (1991).
145. A.D. Sherman, A.T. Davidson, S. Baruah, T.S. Hegwood, R. Waziri, Neurosci. Lett., 121, 77 (1991).
146. R.C. Arora, H.Y. Meltzer, J. Neural. Trans. (Gen.Sec.), 85, 19 (1991).
147. T. Nakai, N. Kitamura, T. Hashimoto, Y. Kajimoto, N. Nishino, T. Mita, C. Tanaka, Biol. Psychiat., 30, 349 (1991).
148. R.F. Squires, E. Saederup, Neurochem. Res., 16, 1099 (1991).
149. A.A. Grace, Neurosci., 41, 1 (1991).
150. G.C. Rigdon, K. Viik, Drug Dev. Res., 23, 91 (1991).
151. N.R. Swerdlow, V.A. Keith, D.L. Braff, M.A. Geyer, J.Pharmacol. Exp. Ther., 256, 530 (1991).
152. J. Feldon, I. Weiner, Biol. Psychiat., 29, 635 (1991).
153. S. M. Lillrank, S.S. Oja, P. Saransaari, T. Seppala, Int. J. Neurosci., 60, 1 (1991).

SECTION II. CARDIOVASCULAR AND PULMONARY AGENTS

Editor: David W. Robertson
Ligand Pharmaceuticals
San Diego, California, 92121

Chapter 7. Angiotensin / Renin Modulators

William J. Greenlee and Peter K. S. Siegl
Merck Sharp and Dohme Research Laboratories
Rahway, NJ 07065

Introduction - Inhibition of the renin-angiotensin system continues to be an important target for drug development (1). Angiotensin-converting enzyme (ACE) inhibitors have led the way with impressive efficacy for the treatment of hypertension and heart failure. Currently, the most active focus of drug discovery in the renin-angiotensin system area is the development of non-peptidic angiotensin II (AII) receptor antagonists. Since AII antagonists are potentially more selective than ACE inhibitors for blocking the renin-angiotensin system, it is of great interest to see how their clinical efficacy and safety compare with those of ACE inhibitors.

ACE INHIBITORS

During 1991, four new ACE inhibitors (ramipril, fosinopril, benzazepril, quinapril) were approved in the U.S., and others await approval. The antihypertensive efficacy of ACE inhibitors (e.g. captopril, enalapril and lisinopril) is well established, and new data on their therapeutic benefit in heart failure has been impressive. Survival of patients with severe heart failure was improved when enalapril was added to standard therapy in the CONSENSUS I study (2). During 1991, reports from two large trials, SOLVD (3) and VHeFT II (4), found ACE inhibitors also had a favorable effect on survival in patients with mild to moderate heart failure. Furthermore, results from VHeFT II suggest that enalapril is superior to nonspecific vasodilators, perhaps due to lack of sympathetic nervous system activation and better regression of ventricular hypertrophy after treatment with the ACE inhibitor. Prophylactic therapy with ACE inhibitors is presently under study in SAVE (5) and the prevention arm of SOLVD, where the benefits of ACE inhibitors in patients with heart disease but not overt heart failure are being explored. Additional utilities of ACE inhibitors (e.g. restenosis) seem likely (6,7). The success of currently-available ACE inhibitors provides a significant challenge for development of superior second-generation ACE inhibitors or other renin-angiotensin system blockers.

RENIN INHIBITORS

Clinical Pharmacology - Renin inhibitors have been reviewed recently (8, 9), and an overview of recent clinical results has appeared (10). There are new clinical data on several renin inhibitors. Pharmacokinetics of enalkirin, and its acute blood pressure (BP) effects compared to those of enalapril have been reported (11, 12). In vivo pharmacology and clinical studies of the potent inhibitor ES-8891 (1) have been reviewed (13). This inhibitor suppressed renin gene expression and blocked renin secretion (14). Hemodynamic effects of terlakirin (CP80,794; 2) have been reported (15). How best to measure the biological activity of renin inhibitors continues to be discussed. Discrepancies between measurements of plasma renin activity (PRA) and AII levels after treatment with RO-42,5892 (3) were resolved by use of an antibody trapping assay (16). This inhibitor appears to act mainly by inhibition of renin in an extra plasma compartment (17). BP responses to PD-132002 (4) in sodium-depleted dogs correlated well to PRA measured by the antibody-trapping method, but not to that determined by a conventional PRA assay (18). The importance of plasma active renin levels in determining plasma AI and AII levels has been discussed (19) in a study of RO-42,5892 (3), as has the importance of displacement of inhibitor from plasma proteins by the angiotensinase inhibitors used in plasma renin assays (16).

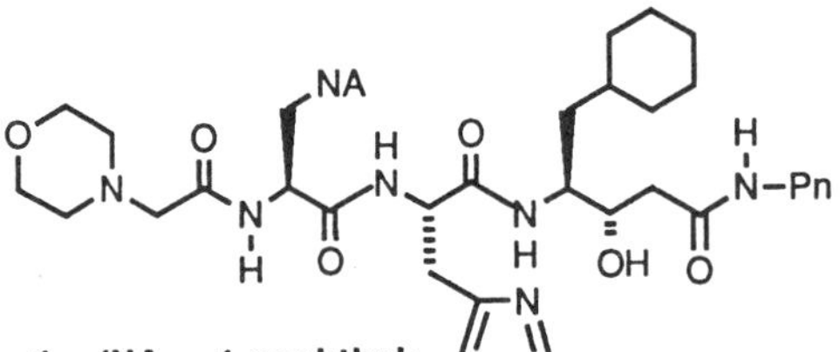

1 (NA = 1-naphthyl; Pn - n-pentyl)

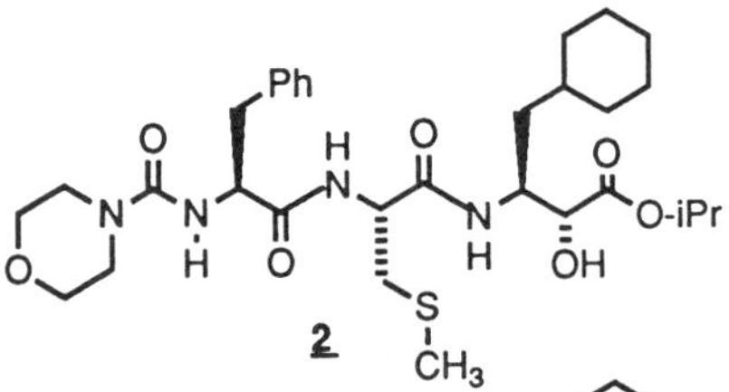

2

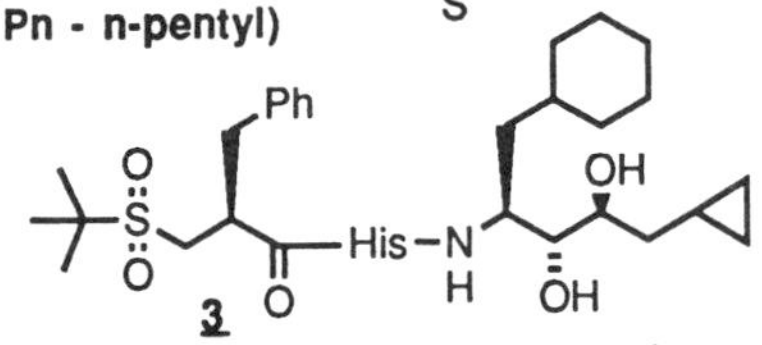

3

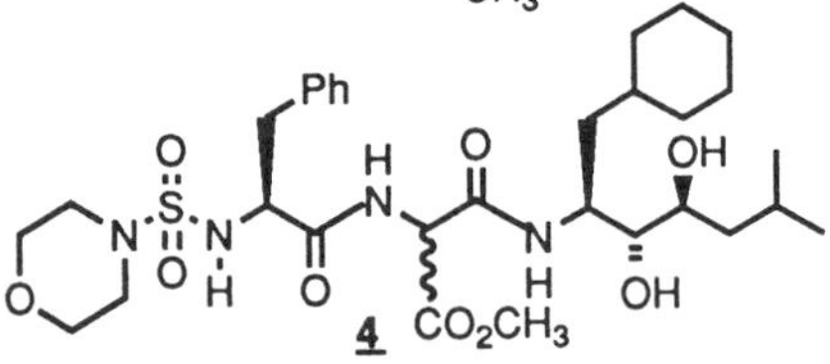

4

New Inhibitor Designs - New work continues to focus on improving oral bioavailability of renin inhibitors by modifying their physical properties and reducing their peptide character. Introduction of heterocycles (21) or hydrophilic amino (22) or phosphonate (23) groups at the carboxy-terminus has yielded potent inhibitors. Modifications at the amino-terminus of renin inhibitors include a malonate replacement for the P_2 His element in PD-132002 (18,24) and replacement of the P_3-P_2 bond by a sulfone (25,26) or a beta-amino acid (26). Replacement of the P_2-P_3 amide bond by a thiazole ring yields inhibitors with modest potency (27). The potent inhibitors A-74273 (5) and PD-134672 (6), which show improved oral bioavailability, incorporate stable hydrophilic groups at the amino-terminus (28-32). The inhibitor FK-744 (7), reported to show 50% oral bioavailability in cynomolgus monkeys, has entered clinical trials (33). Incorporation of an N-linked saccharide element at the amino terminus of a series of peptide inhibitors reduced their rate of biliary clearance (34).

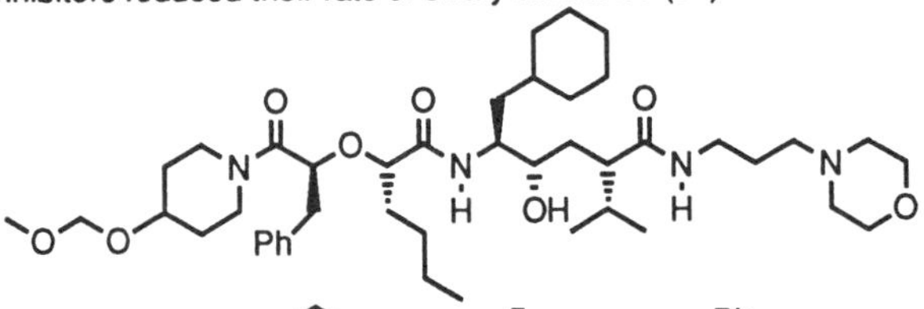

5

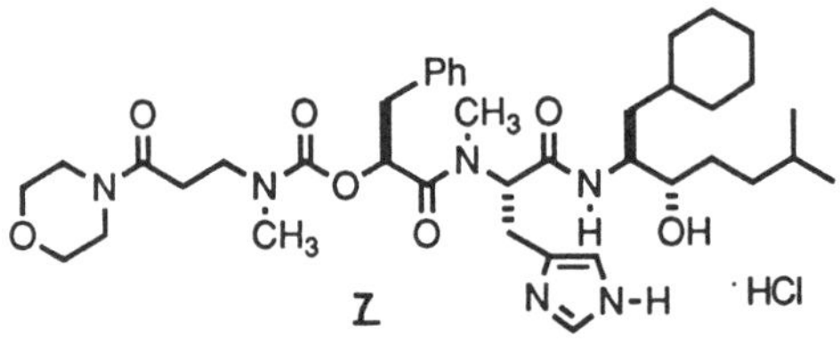

7

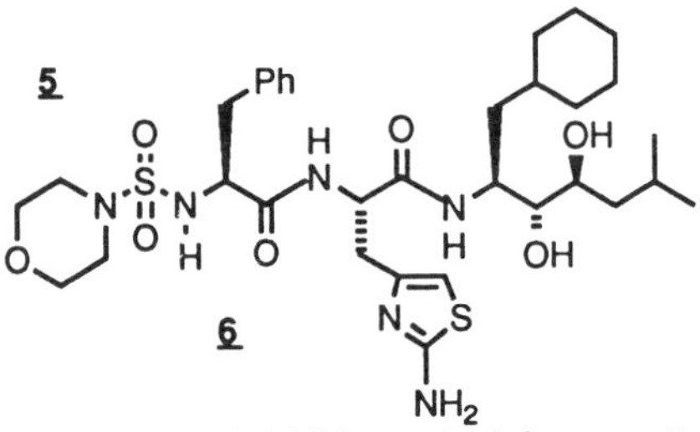

6

Conformationally-constrained inhibitors continue to be reported. Inhibitors which incorporate carboxy-terminal lactam (35) and oxazolidinone (36) constraints are highly potent. Potent macrocyclic inhibitors which link P_2 and P_4 elements (37) or P_1 and P_3 elements (38) have been reported, as have macrocycles linking P_2 and P_1' with amide (39,40), ester (41,42) or sulfide-containing (39,43) elements. An extensive review of the use of molecular modeling in renin-inhibitor design has appeared (44). An X-ray crystallographic study of a recombinant human renin-inhibitor complex has been published (45). The complex natural product cyclothiazomycin is a renin inhibitor with micromolar potency (20).

ANGIOTENSIN II RECEPTOR ANTAGONISTS

Peptidic Antagonists - Peptide analogs of AII continue to provide important insights into how AII interacts with its two binding sites, the AT_1 and AT_2 receptors (see below). The (pentafluoro)Phe[8] analog of AII (Asp-Arg-Val-Tyr-Ile-His-Pro-Phe) retains agonist activity, but is also a slowly reversible antagonist (46). Incorporation of cyclohexylalanine (Cha) at position 8 of [Sar[1]]-AII and of sarmesin (giving Sar-Arg-Val-(OMe)Tyr-Ile-His-Pro-Cha) yields potent antagonists (47); while the former retains significant partial agonist activity, the latter does not. Deletion of Sar^1 from the antagonist [Sar^1, Ile^8]AII or replacement of Arg^2 of AII by neutral amino acids reduces antagonist potency and eliminates the slow reversibility (48). Dimerization of analogs of AII at the carboxy-terminus (CGP-37534; **8**) led to a substantial loss of binding potency, confirming the importance of the C-terminal carboxylic acid (49). Dimerization at the amino-terminus yielded the potent agonist CGP-39026 (**9**) and the antagonist CGP-37346 (**10**). The pentapeptide **11** has nanomolar potency for the AT_2 receptor (see below), although it retains only micromolar potency at the AT_1 receptor. The potent antagonist **12** incorporates an amino acid with a polar side-chain in place of Ile^5 (50). An analog of [Sar^1, Ile^8]AII which incorporates a Val-His isostere (**13**) was 700-fold less active (51). Work aimed at introduction of conformational restraints into AII peptides has continued. Introduction of a conformationally-constrained amino acid in place of Ile^5, yields **14**, a potent agonist (52). The potent cyclic antagonist **15**, derived by connecting the side-chains of homocysteine (Hcy) residues at positions 3 and 5 via a disulfide bridge (53), is similar to cyclic antagonists reported earlier (1). This antagonist has been shown to bind to both AT_1 and AT_2 receptors, suggesting a common bound conformation at the two receptors. Further conformational restraint has been introduced by incorporation of 4-mercapto-proline in place of one homocysteine residue, giving a potent cyclized ligand **16** which binds to both receptors. Replacement of Pro^7 with azetidinecarboxylic acid or pipecolic acid results in analogs (agonists or antagonists) which retain high potency (54).

8   Sar-Arg-Val-Tyr-Val-His-Pro-Ile-NH-CH₂⎤
        Sar-Arg-Val-Tyr-Val-His-Pro-Ile-NH-CH₂⎦

9   ⎡CH₂-CO-Asn-Arg-Val-Ile-His-Pro-Phe-OH
        ⎣CH₂-CO-Asn-Arg-Val-Ile-His-Pro-Phe-OH

10 ⎡CH₂NHCH₂CO-Asn-Arg-Val-Ile-His-Pro-Phe-OH
        ⎣CH₂NHCH₂CO-Asn-Arg-Val-Ile-His-Pro-Phe-OH

11 Ac-Tyr-Val-His-Pro-Ile

12 Aib-Arg-Val-Tyr-Thr(OMe)-His-Pro-Phe

13 Sar-Arg-Val-Tyr-Valψ[CH(CONH₂)NH]His-Pro-Ile

14 Sar-Arg-Val-Tyr—N——CO-His-Pro-Phe (with S-containing ring)

15 Sar-Arg-Hcy-Tyr-Hcy-His-Pro-Phe (S——S disulfide bridge)

16 Sar-Arg-...Tyr-N——CO-His-Pro-Phe (S——S bridge)

Non-Peptidic Antagonists - The development of potent, orally-active non-peptidic AII receptor antagonists from a series of benzyl-substituted imidazoles (55) such as S8308 (**17**) has been reviewed (56-59). High potency and oral activity were achieved with losartan (DuP753; MK-954; **18**), a potent and selective AII antagonist in clinical trials (see below). Structure-activity relationships of related imidazole-based antagonists have been reported (60). An unusual tetrazole N2-glucuronide is a metabolite of losartan, a mode of metabolism which other tetrazole-containing antagonists may share (61). The potent imidazoleacrylate antagonist SK&F-108566 (**19**), also derived from the benzylimidazole lead (**17**), incorporates elements proposed to mimic the carboxy-terminus of AII (62).

The availability of non-peptidic antagonists has permitted the study of AII receptor subtypes in numerous tissues. The non-peptidic antagonists cited above have high selectivity for a receptor subtype found in several tissues, including aorta, liver and adrenal, which is now designated AT_1 (63). Other antagonists with high selectivity for a second receptor subtype (designated AT_2) found in adrenal, uterus and brain have been reported (1, 64). These include PD-121981 ($\underline{20}$), PD-123177 ($\underline{21}$, EXP655) and PD-123319 ($\underline{22}$). Selectivity for the AT_2 receptor has been reported for several peptides (1), including CGP-37065 ($\underline{11}$). A coupling mechanism and pharmacological roles for the AT_2 receptor have not yet been determined (see below).

Losartan and its more-potent metabolite EXP3174 ($\underline{23}$) have served as leads for the development of potent AT_1-selective antagonists in the imidazole and other heterocyclic series (65). The imidazolecarboxylic acid DuP532 ($\underline{24}$) has higher potency than losartan, and (like EXP-3174) is an "insurmountable" antagonist of AII (66,67). Both benzimidazoles (1,68) and imidazoles fused with heterocyclic rings have been reported. The latter include imidazopyridines (1, 69-71), among which is L-158,809 ($\underline{25}$), a potent antagonist (72-75). Particularly interesting are antagonists which (like EXP-3174) incorporate a carboxylic acid as a heterocyclic substitutent (76,77), such as L-158,978 ($\underline{26}$) and the benzimidazole $\underline{27}$. The fused imidazocycloheptadienone $\underline{28}$ has moderate antagonist potency (78). Potent xanthine (79), pyrazolopyrimidine (80) and imidazopyridazine (81) antagonists have also been disclosed.

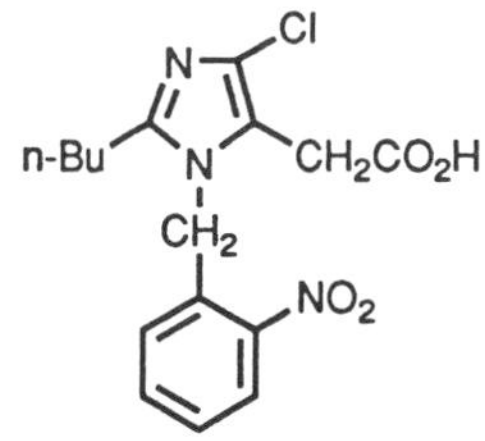

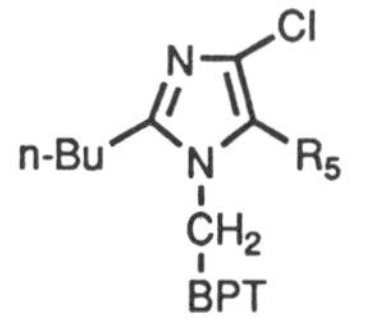

18 R_5 = -CH_2OH

23 R_5 = -CO_2H

17

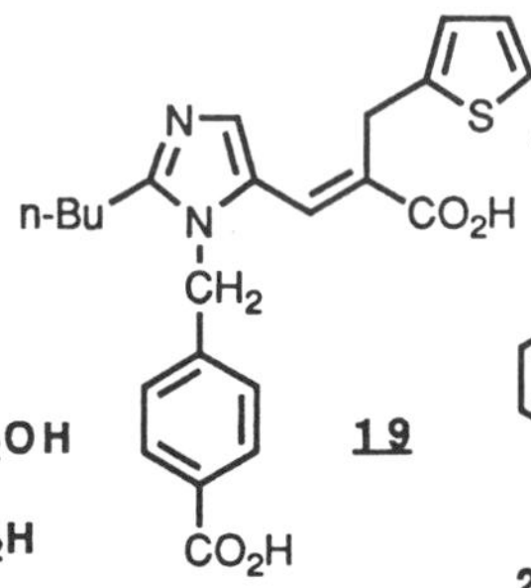

19

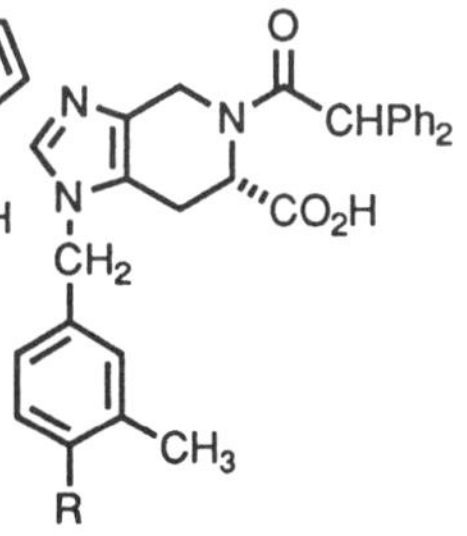

20 R = - OCH_3

21 R = -NH_2

22 R = -$N(CH_3)_2$

BPT =

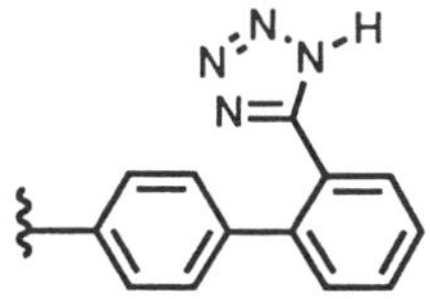

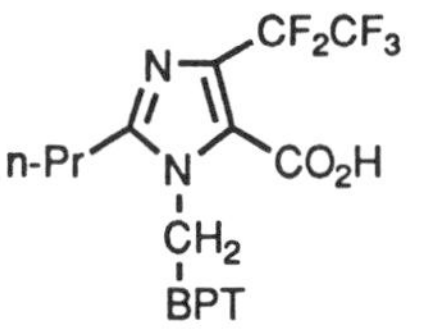

24

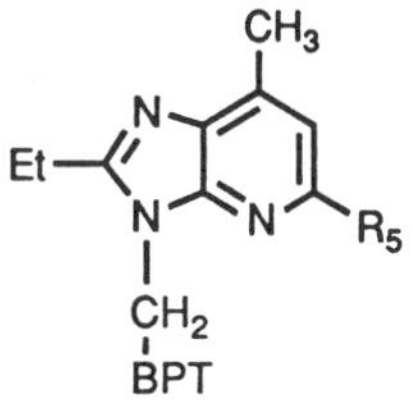

25 R_5 = -CH_3

27

28

26 R_5 = -CO_2H

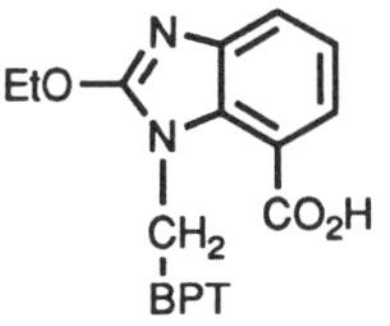

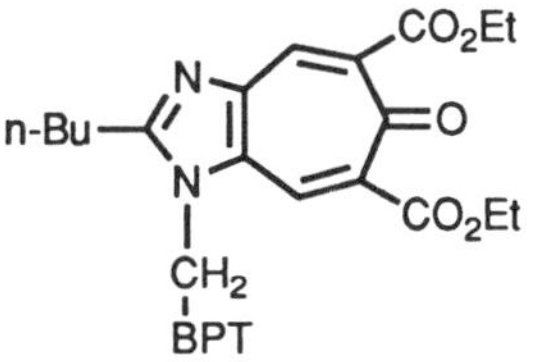

The potent antagonist SR-47436 ($\underline{29}$) incorporates an imidazolinone ring in place of imidazole (82-85). Other 5-membered ring antagonists disclosed include 1-substituted (86) or 4-substituted (87) triazoles and triazolinones (1,88), and pyrazoles in which the substituted biphenyl moiety is attached to the heterocycle at a carbon atom (89-91). Antagonists with a nitrogen- or carbon-linked six-membered ring heterocycle include pyrimidines (92,93), pyrimidinones (1,94,95), pyrimidinediones (96), and quinazolinones (1, 95). Potent quinolines such as D-8731 ($\underline{30}$) have been reported (97,98), as have related naphthyridines (99) and pyridines (100,101); the link to the biphenyl element is through an oxymethylene group. Thienopyridinone antagonists have been disclosed in which a substituted thiophene ring appears to take the place of the 2-(butyl)imidazole substituent of losartan (102).

Antagonists which incorporate variations of the biphenyltetrazole element have also been reported, including GR-117289 ($\underline{31}$), a potent benzoxazole antagonist (103-107). Antagonists in which the tetrazole-bearing phenyl ring is replaced by various carboxylic acid-containing elements have been disclosed (108-110). Introduction of methoxy groups on the upper ring of the biphenyl results in an analog **32** with reduced antagonist potency (111), possibly due to conformational effects. Other conformationally-restrained structures include the tetrahydronaphthyl antagonist **33,** which retains significant antagonist potency (112). Other novel designs (113,114) are exemplified by the non-heterocyclic antagonist **34** and the substituted imidazole **35**. Potent benzimidazoles which also have substantial ($< \mu M$) AT_2 potency have been reported (115), including BIBS39 (**36**). The natural product arrivicin A has modest AT_1 antagonist potency (116).

In Vitro Pharmacology - Exciting progress has been made in the molecular characterization of AII receptors. The AT_1 receptor has been cloned and sequenced from rat aorta (117), bovine adrenal (118), rat kidney (119) and from human lymphocytes (120). The rat, bovine and human receptors all contain seven transmembrane-spanning domains and have a high degree of homology. Both rat and bovine receptors are reported to bind losartan, but not the AT_2-selective antagonist PD-123,177 (**21**). Characterization and cloning of additional AT_1 receptor subtypes and of the AT_2 receptor is awaited with interest. Other proteins which bind angiotensin have been reported. A novel gene product "mrg" encodes a receptor (121) with 35% homology to the mas oncogene (1), reported earlier to encode a putative angiotensin receptor. The mRNA's for both proteins increase the electrophysiological response to angiotensin peptides after injection into Xenopus oocytes (122), but neither binds AII. More recently, transfection of Cos-1 cells with mas has resulted in increased responsiveness of the endogenous AII signalling system (121). Binding sites for AII have been reported which are not blocked by non-peptide antagonists of AT_1 and AT_2 receptors, including sites on the microorganism Mycoplasma hyorhinis (123), on mouse neuroblastoma cells (124) and on amphibian cells (125). Evidence suggesting non AT_1 / AT_2 binding sites in rat hypothalamus has also been reported (126). Purification and characterization of a soluble AII binding protein from liver cells has been described (127).

The occurrence of AII receptor subtypes and their functions have been reviewed (128). The AT_1 receptor is coupled to the phospho-inositide (PI) turnover transduction system, but a second messenger system for the AT_2 receptor has remained elusive (129). Evidence for a link between the AT_2 receptor and a particulate guanylate cyclase, through the activity of a protein tyrosine phosphatase (130) has been reported. Possible functions of the AT_2 receptor have been the object of much study. Astrocyte cultures from neonatal rat brains contain mostly AT_1 receptors coupled to stimulation of PI hydrolysis, while neuronal cultures contain mostly AT_2 receptors that are associated with a reduction in basal cGMP levels (131). AT_2 receptors predominate in rat anterior cerebral arteries where they are proposed to be linked to cerebral circulation (132). Both receptor subtypes are present in renal cortex of rats and rhesus monkeys, but only AT_1 antagonists stimulate renin release (133). The AT_2 receptors are reported to mediate prostaglandin synthesis in human astrocytes (134), and to mediate neurophysiological responses of paraventricular neurons to the angiotensin peptide A(1-7) (135). The activation by AII of vasopressin release from the rat hypothalamoneurohyophysial system is blocked by AT_2 antagonists (but not by antagonists of AT_1) (136). The AT_2-selective antagonist PD-123177 (__21__) has been reported to produce diuresis in rats (137), but the involvement of AT_2 receptors has not been confirmed. Both AT_1 and AT_2 receptors have been implicated in growth. Thymidine incorporation into human neuroblastoma cells (138) is blocked by an ACE inhibitor and by either losartan or PD-123,177 (__21__). AII inhibits an ATP-sensitive potassium channel in porcine coronary artery smooth muscle cells (139), but a link to AII receptors has not been established.

In Vivo Pharmacology - *In vivo* pharmacology of losartan has been reviewed (140, 141). Other antagonists which have shown efficacy in blocking the pressor response to AII or in models of hypertension include DuP532 (__24__), SK&F-108566 (__19__), L-158,809 (__25__), GR-117289 (__31__), D-8731 (__27__), SR-47436 (__29__) and BIBS39 (__36__). Prolonged (7-day) treatment of spontaneously hypertensive rats with losartan gave a sustained drop in blood pressure (equal to that resulting from treatment with the ACE inhibitor benzazeprilat), despite a 10-fold increase in AII levels (142). The blood pressure in the treated rats was not reduced to levels as low as those in normotensive (WKY) rats (whose blood pressure was also reduced to some extent by losartan). Losartan blocked AII pressor response in normotensive humans (143-145) and was as effective as enalapril in reducing blood pressure in mild hypertensives (146).

Losartan crosses the blood-brain barrier in rats, as judged by the resulting blockade of brain AT_1 receptors (147). The AT_1-selective losartan analog EXP3880, but not PD-123177 (__21__), reversed a renin-induced deficit of passive avoidance in rats (148). Losartan reverses apomorphine-induced stereotypy in rats (149), and enhances cognition (150). In a rat model of heart failure, losartan and captopril gave similar hemodynamic changes (151). Both losartan (152, 153) and the ACE inhibitor benzazeprilat (152) reduced intimal lesion size after balloon catheterization in rats.

New Directions - AII induces the synthesis and secretion of endothelin (ET) by cultured rat vascular smooth muscle cells, suggesting a role for AII in mediating the action of ET (154). These two vasoactive peptides show a chronic synergistic effect on BP when administered to rats at subpressor intravenous doses (155). Enzymes capable of the specific conversion of prorenin to active renin (1) continue to be reported, including renal cathepsin B (156). A synthetic peptide substrate for putative prorenin-converting enzymes is cleaved by trypsin at the authentic Arg-Leu processing site (157). Smaller angiotensin peptides continue to be of interest. A review of the occurrence and actions of A(1-7) has appeared (158). New potential utilities for blockers of the renin-angiotensin system continue to be explored. The involvement of sperm ACE in fertilization has been reported (159). Evidence suggesting the possible use of captopril in treatment-resistant depression has been published (160). The utility of renin inhibitors in treatment of psoriasis has been claimed (161).

References

1. W. J. Greenlee, P. K. S. Siegl, Annu. Rep. Med. Chem., __26__, 63 (1991).
2. CONSENSUS Trial Study Group, N. Engl. J. Med., __316__, 1429 (1987).
3. SOLVD Investigators, N. Engl. J. Med., __325__, 293 (1991).
4. J. N. Cohn, N. Engl. J. Med., __325__, 303 (1991).
5. E. Braunwald, N. Engl. J. Med., __325__, 351 (1991).
6. J. S. Powell, J. P. Clozel, R. K. M. Muller, H. Kuhn, F. Hefti, M. Hosang, H. R. Baumgartner, Science, __245__, 186 (1989).
7. R. D. Farhy, K.-L. Ho, O. A. Carretero, A. G. Scicli, Biochem. Biophys. Res. Commun., __182__, 283 (1992).
8. H. D. Kleinert, W. R. Baker, H. H. Stein, Adv. Pharmacol., __22__, 207 (1991).
9. T. D. Ocain, M. Abou-Gharbia, Drugs of the Future, __16__, 37 (1991).
10. R. R. Luther, H. N. Glassman, R. S. Boger, Clin. Nephrol., __36__, 181 (1991).
11. M. Karol, M. Weber, W. Hsueh, R. Luthur, R. Boger, R. Granneman, Am. J. Hypertens., __4__, 31A, Abstract 43 (1991).
12. J. M. Neutel, R. R. Luthur, R. S. Boger, M. A. Weber, Am. Heart J., __122__, 1094 (1991).
13. T. Kokubu, K. Hiwada, E. Murakami, S. Muneta, Y. Kitami, K. Oizumi, H. Takahagi, H. Koike, Cardiovasc. Drug. Rev., __9__, 49 (1991).
14. Y. Kitami, K. Hiwada, E. Murakami, T. Iwata, S. Muneta, T. Kokubu, Clin. Sci., __81__, 387 (1991).
15. W. R. Murphy, R. T. Wester, R. L. Rosati, D. J. Hoover, I. M. Purcell, J. T. MacAndrew, T. M. Schelhorn, D. E. Wilder, A. H. Smith, W. F. Holt, Amino Acids: Chem. Biol. Med., (Int. Cong. Amino Acid Res.) G. A. Lubec, G. Rosenthal, Eds., Leiden, The Netherlands, 1990, p. 676.

16. F. H. M. Derkx, A. H. van der Meiracker, W. Fischli, P. J. J. Admiraal, A. J. Man in't Veld, P. van Brummelen, M. A. D. H. Schalekamp, Am. J. Hypertens., 4, 602 (1991).
17. W. Fischli, J.-P. Clozel, K. E. Amrani, W. Wostl, W. Neidhart, H. Stadler, Q. Branca, Hypertension, 18, 22 (1991).
18. J. T. Repine, R. J. Himmelsbach, J. C. Hodges, J. S. Kaltenbronn, I. Sircar, R. W. Skeean, S. T. Brennan, T. R. Hurley, E. Lunney, C. C. Humblet, R. E. Weishaar, S. Rapundalo, M. J. Ryan, D. G. Taylor, Jr., S. C. Olson, B. M. Michniewicz, B. E. Kornberg, D. T. Belmont, M. D. Taylor, J. Med. Chem., 34, 1935 (1991).
19. E. Camenzind, J. Nussberger, L. Juillerat, A. Munafo, W. Fischli, P. Coassolo, P. van Brummelen, C. H. Kleinbloesem, B. Waeber, H. R. Brunner, J. Cardiovasc. Pharmacol., 18, 299 (1991).
20. M. Shibasaki, M. Asano, Y. Fukunaga, T. Usui, M. Ichihara, Y. Marakami, K. Nakano, T. Fujikura, Am. J. Hypertens., 4, 932 (1991).
21. C. J. Gardner, D. J. Twissell, P. A. Charlton, P. C. Cherry, Eur. J. Pharmacol., 192, 329 (1991).
22. M. A. Poss, EPA # 452,587 (1991).
23. M. J. Ryan, S. T. Rapundalo, R. K. Palmer, B. L. Batley, D. G. Taylor, Am. J. Hypertens., 4, 31A, Abstract 42 (1991).
24. R. A. Rivero, W. J. Greenlee, Tetrahedron Lett., 39, 5263 (1991).
25. P. Raddatz, A. Jonczyk, K.-O. Minck, C. J. Schmitges, J. Sombroek, J. Med. Chem., 34, 3267 (1991).
26. S. G. dePaolis, W. G. Earley, J. A. Gainor, T. D. Gordon, B. A. Morgan, E. D. Pagani, J. Singh, Peptides 1990, Proc. 21st Eur. Pept. Symp., E. Girald, D. Andreu, Eds., ESCOM Science Publishers, B. V., 1991, p. 380.
27. H. Stein, A. Fung, J. Cohen, W. Baker, S. Rosenberg, S. Boyd, B. Dayton, Y.-L. Armiger, S. Condon, R. Mantei, H. Kleinert, FASEB J., 5, A1766, Abstract 8102 (1991).
28. H. D. Kleinert, H. H. Stein, W. R. Baker, K. M. Verburg, G. Young, A. Fung, S. Boyd, S. Rosenberg, J. Polakowski, J. Barlow, R. Mantei, V. Klinghofer, J. Cohen, FASEB J., 5, A1766, Abstract 8104 (1991).
29. H. D. Kleinert, S. Boyd, A. Fung, H. H. Stein, W. R. Baker, K. M. Verburg, J. Polakowski, J. Barlow, J. Cohen, V. Klinghofer, R. Mantei, S. Cepa, S. Rosenberg, J. F. Denissen, Hypertension, 18, 407, Abstract P39 (1991).
30 M. J. Ryan, B. L. Batley, G. W. Hicks, C. A. Painchaud, S. T. Rapundalo, D. G. Taylor, FASEB J., 5, A1575, Abstract 6991 (1991).
31. W. C. Patt, H. W. Hamilton, M. J. Ryan, C. A. Painchaud, M. D. Taylor, S. T. Rapundalo, B. L. Batley, C. J. C. Connolly, D. G. Taylor, Jr., Abstracts, A. C. S. 201st Natl. Meeting, Atlanta, GA, April 14-19, 1991, MEDI No 65 (1991).
32. T. F. Andoh, K. Sogabe, I. Nagatomi, Y. Shimazaki, H. Horiai, J. Mori, M. Kohsaka, Am. J. Hypertens., 4, 31A, Abstract 44 (1991).
33. J. F. Fisher, A. W. Harrison, G. L. Bundy, K. F. Wilkinson, B. D. Rush, M. J. Ruwart, J. Med. Chem., 34, 3140 (1991).
34. M. Aoki, T. Ohtsuka, Y. Itezono, K. Yokose, K. Furihata, H. Seto, Tetrahedron Lett., 32, 217 (1991).
35. P. D. Williams, D. S. Perlow, L. S. Payne, M. K. Holloway, P. K. S. Siegl, T. W. Schorn, R. J. Lynch, J. J. Doyle, J. F. Strouse, G. P. Vlasuk, K., Hoogsteen, J. P. Springer, B. L. Bush, T. A. Halgren, A. D. Richards, J. Kay, D. F. Veber, J. Med. Chem., 34, 887 (1991).
36. S. Rosenberg, H. D. Kleinert, H. H. Stein, D. L. Martin, M. A. Chekal, J. Cohen, D. A. Egan, K. A. Tricarico, W. R. Baker, J. Med. Chem., 34, 469 (1991).
37. S. Thaisrivongs, J. R. Blinn, D. T. Pals, S. R. Turner, J. Med. Chem., 34, 1276 (1991).
38. C. E. Brotherton-Pleiss, S. R. Newman, L. D. Waterbury, M. S. Schwartzberg, Peptides, Chemistry and Biology, Proceedings of the Twelfth American Peptide Symposium, J. A. Smith, J. E. Rivier, Eds., ESCOM, Leiden, 1992, p. 816.
39. A. E. Weber, M. G. Steiner, L. Yang, D. Dhanoa, J. R. Tata, T. A. Halgren, P. K. S. Siegl, W. H. Parsons, W. J. Greenlee, A. A. Patchett, Peptides, Chemistry and Biology, Proceedings of the Twelfth American Peptide Symposium, J. A. Smith, J. E. Rivier, Eds., ESCOM, Leiden, 1992, p. 749.
40. R. A. Rivero, W. J. Greenlee, Tetrahedron Lett., 32, 2453 (1991).
41. A. E. Weber, T. A. Halgren, J. J. Doyle, R. J. Lynch, P. K. S. Siegl, W. H. Parsons, W. J. Greenlee, A. A. Patchett, J. Med. Chem., 34, 2692 (1991).
42. S. J. Wittenberger, W. R. Baker, B. Donner, C. W. Hutchins, Tetrahedron Lett., 32, 7655 (1991).
43. L. Yang, A. E. Weber, J. J. Doyle, T. W. Schorn, P. K. S. Siegl, T. A. Halgren, W. J. Greenlee, A. A. Patchett, Abstracts, A. C. S. 202nd National Meeting, New York, N. Y., 1991, MEDI No. 122 (1991).
44. C. Hutchins, J. Greer, Crit. Rev. Biochem. Mol. Biol., 26, 77 (1991).
45. J. Rahuel, J. P. Priestle, M. G. Grutter, J. Struct. Biol., 107, 227 (1991).
46. P. R. Bovy, D. P. Getman, J. M. Matsoukas, G. J. Moore, Biochim. Biophys. Acta, 1079, 23 (1991).
47. J. Hondrelis, J. Matsoukas, P. Cordopatis, R. C. Ganter, K. J. Franklin, G. J. Moore, Int. J. Pep. Prot. Res., 37, 21 (1991).
48. G. J. Moore, R. C. Ganter, M. H. Goghari, K. J. Franklin, Int. J. Pep. Prot. Res., 38, 1 (1991).
49. M. de Gasparo, S. Whitebread, B. Kamber, L. Criscione, H. Thomann, B. Riniker, R. Andreatta, J. Receptor Research, 11, 247 (1991).
50. P. Cordopatis, E. Manessi-Zoupa, D. Theodoropoulos, R. Bosse, S. Gagnon, E. Escher, Peptides 1990, Proc. 21st Eur. Pept. Symp., E. Girald, D. Andreu, Eds., ESCOM Science Publishers, B. V., 1991, p. 657.

51.   R. Mohan, Y.-L. Chou, R. Bihovsky, W. C. Lumma, Jr., P. W. Erhardt, K. J. Shaw, J. Med. Chem., 34, 2402 (1991).
52.   J. Samanen, T. Cash, D. Narindray, E. Brandeis, W. Adams, Jr., H. Weideman, T. Yellin, J. Med. Chem., 34, 3036 (1991).
53.   G. R. Marshall, K. Kaczmarek, T. Kataoka, K. Plucinska, R. Skeean, B. Lunney, C. Taylor, D. Dooley, G. Lu, R. Panek, C. Humblet, Peptides 1990, Proc. 21st Eur. Pept. Symp., E. Girald, D. Andreu, Eds., ESCOM Science Publishers, B. V., 1991, p. 594.
54.   J. Matsoukas, J. Hondrelis, G. Agelis, R. Yamdagni, R. C. Ganter, G. J. Moore, Peptides 1990, Proc. 21st Eur. Pept. Symp., E. Girald, D. Andreu, Eds., ESCOM Science Publishers, B. V., 1991, p. 659.
55.   Y. Furukawa, S. Kishimoto and S. Nishikawa, U. S. Pat. 4,340,598; 4,355,040 (1982).
56.   P. B. M. W. M. Timmermans, P. C. Wong, A. T. Chiu, W. F. Herblin, Trends Pharm Sci., 12, 55 (1991).
57.   Nonpeptide Angiotensin II Receptor Antagonists: A New Concept for Pharmacological Control of the Renin System, Supplement Issue, J. H. Laragh, R. D. Smith, Eds., Am. J. Hypertens., 4, 271S (1991).
58.   J. V. Duncia, D. J. Carini, A. T. Chiu, A. L. Johnson, W. A. Price, P. C. Wong, R. R. Wexler, P. B. M. W. M. Timmermans, Med. Res. Rev., 12, 149 (1992).
59.   J. V. Duncia, D. J. Carini, A. T. Chiu, M. E. Pierce, W. A. Price, R. D. Smith, G. J. Wells, P. C. Wong, R. R. Wexler, A. L. Johnson, P. B. M. W. M. Timmermans, Drugs of the Future, 16, 305 (1991).
60.   D. J. Carini, J. V. Duncia, P. E. Aldrich, A. T. Chiu, A. L. Johnson, M. E. Pierce, W. A. Price, J. B. Santella III, G. J. Wells, R. R. Wexler, P. C. Wong, S.-E. Yoo, P. B. M. W. M. Timmermans, J. Med. Chem., 34, 2525 (1991).
61.   R. A. Stearns, G. A. Doss, R. R. Miller, S.-H. L. Chiu, Drug Metab. Disp., 19, 1160 (1991).
62.   J. Weinstock, R. M. Keenan, J. Samanen, J. Hempel, J. A. Finkelstein, R. G. Franz, D. E. Gaitanopoulos, G. R. Girard, J. G. Gleason, D. T. Hill, T. M. Morgan, C. E. Peishoff, N. Aiyar, D. P. Brooks, T. A. Fredrickson, E. H. Ohlstein, R. R. Ruffolo, Jr., E. J. Stack, A. C. Sulpizio, E. F. Weidley, R. M. Edwards, J. Med. Chem., 34, 1514 (1991).
63.   F. M. Bumpus, K. J. Catt, A. T. Chiu, M. DeGasparo, T. Goodfriend, A. Husain, M. J. Peach, D. G. Taylor, Jr., P. B. M. W. M. Timmermans, Hypertension, 17, A724 (1990).
64.   C. J. Blankley, J. C. Hodges, S. R. Klutchko, R. J. Himmelsbach, A. Chucholowski, C. J. Connolly, S. J. Neergaard, M. S. Van Nieuwenhze, A. Sebastian, J. Quin, III, A. D. Essenburg, D. M. Cohen, J. Med. Chem., 34, 3248 (1991).
65.   P. C. Wong, P. B. M. W. M. Timmermans, J. Pharmacol. Exp. Therapeut., 258, 49 (1991).
66.   A. T. Chiu, D. J. Carini, J. V. Duncia, K. L. Leung, D. E. McCall, W. A. Price, Jr., P. C. Wong, R. D. Smith, R. R. Wexler, P. B. M. W. M. Timmermans, Biochem. Biophys. Res. Commun., 177, 209 (1991).
67.   P. C. Wong, S. D. Hart, A. T. Chiu, W. F. Herblin, D. J. Carini, R. D. Smith, R. R. Wexler, P. B. M. W. M. Timmermans, J. Pharmacol. Exp. Ther., 259, 861 (1991).
68.   M. Fortin, D. Frechet, G. Hamon, S. Jouquey, J. -P. Vevert, EPA # 461,039 (1991).
69.   K. Miyake, M. Matuskura, N. Yoneda, O. Hiroshima, N. Mori, H. Ishihara, T. Musha, T. Matsuoka, S. Hamano, N. Minami, EPA # 420,237 (1991).
70.   T. Oku, H. Setoi, H. Kayakiri, T. Inoue, A. Kuroda, A. Katayama, M. Hashimoto, S. Satoh, H. Tanaka, EPA # 426,021 (1991).
71.   M. Fortin, D. Frechet, G. Hamon, S. Jouquey, J. -P. Vevert, EPA # 461,040 (1991).
72.   N. B. Mantlo, P. K. Chakravarty, D. L. Ondeyka, P. K. S. Siegl, R. S. Chang, V. J. Lotti, K. A. Faust, T.-B. Chen, T. W. Schorn, C. S. Sweet, S. E. Emmert, A. A. Patchett, W. J. Greenlee, J. Med. Chem., 34, 2919 (1991).
73.   R. S. L. Chang, P. K. S. Siegl, B. V. Clineschmidt, N. B. Mantlo, P. K. Chakravarty, W. J. Greenlee, A. A. Patchett, V. J. Lotti, FASEB J., 5, A1575, Abstract 6993 (1991).
74.   P. K. S. Siegl, R. S. L. Chang, W. J. Greenlee, V. J. Lotti, N. B. Mantlo, A. A. Patchett, C. S. Sweet, FASEB J., 5, A1576, Abstract 6995 (1991).
75.   S. D. Kivlighn, R. A. Gabel, P. K. S. Siegl, FASEB J., 5, A1576, Abstract 6994 (1991).
76.   N. B. Mantlo, D. Ondeyka, R. S. L. Chang, V. J. Lotti, S. D. Kivlighn, P. K. S. Siegl, A. A. Patchett, W. J. Greenlee, Abstracts, American Chemical Society, Fourth Congress of North America, New York, N. Y., 1991, MEDI No. 103 (1991).
77.   T. Naka, K. Nishikawa, T. Kato, EPA # 459,136 (1991).
78.   P. R. Bovy, J. O'Neal, J. T. Collins, G. M. Olins, V. M. Corpus, S. K. Burrows, E. G. McMahon, M. Palomo, K. Koehler, Med. Chem. Res. 1, 86 (1991).
79.   A. Morimoto, K. Nishikawa, EPA # 430,300 (1991).
80.   E. E. Allen, S. X. Huang, R. S. L. Chang, V. J. Lotti, P. K. S. Siegl, A. A. Patchett, W. J. Greenlee, Abstracts, A. C. S. 202nd National Meeting, New York, N. Y., 1991, MEDI No. 104 (1991).
81.   P. R. Bovy, T. S. Chamberlain, J. T. Collins, PCT # WO91/19715 (1991).
82.   D. Nisato, C. Cazaubon, C. Lacour, J. Gougat, P. Guiradou, C. Bernhart, P. Perreaut, J. C. Breliere, G. Le fur, Proceedings of the British Pharmacological Society, University College, London / Sandoz Institute for Medical Research, December 17-19, 1991, Brit. J. Pharmacol. (Proceedings Suppl.), 105, Abstract 84P (1992).
83.   C. Bernhart, J.-C. Brelier, J. Clement, D. Nisato, P. Perreaut, EPA 454,511 (1991).
84.   C. Bernhart, J.-C. Brelier, J. Clement, D. Nisato, P. Perreaut, PCT # 91/14679 (1991).
85.   B. Cristophe, R. Libon, P. Chatelain, A. S. Manning, Proceedings of the British Pharmacological Society, University College, London / Sandoz Institute for Medical Research, December 17-19, 1991, Brit. J. Pharmacol. (Proceedings Suppl.), 105, Abstract 259P (1992).

86.　D. B. Reitz, PCT # WO91/17148 (1991).
87.　L. L. Chang, R. A. Strelitz, M. MacCoss, R. S. L. Chang, V. J. Lotti, S. D. Kivlighn, P. K. S. Siegl, Abstracts, A. C. S. 202nd National Meeting, New York, N. Y., 1991, MEDI No. 105 (1991).
88.　R. E. Manning, D. B. Reitz, H.-C. Huang, PCT # WO91/18888 (1991).
89.　B. C. Ross, D. Middlemiss, C. D. Eldred, J. G. Montana, P. Shah, EPA # 446,062 (1991).
90.　N. Bru-Magniez, E. Nocohai, J.-M. Teulon, EPA # 449,699 (1991).
91.　E. E. Allen, W. J. Greenlee, M. MacCoss, W. T. Ashton, PCT # WO 91/15479 (1991).
92.　P. Herold, P. Buhlmayer, EPA # 424,317 (1991).
93.　E. E. Allen, W. J. Greenlee, M. MacCoss, A. A. Patchett, PCT # WO 91/15209.
94.　P. Herold, P. Buhlmayer, EPA 435, 827 (1991).
95.　A. Morimoto, K. Nishikawa, EPA # 445,811 (1991).
96.　T. Naka, K. Nishikawa, EPA # 442,473 (1991).
97.　A. A. Oldham, C. P. Allott, J. S. Major, R. J. Pearce, D. A. Roberts, S. T. Russel, Proceedings of the British Pharmacological Society, University College, London / Sandoz Institute for Medical Research, December 17-19, 1991, Brit. J. Pharmacol. (Proceedings Suppl.), 105, Abstract 83P (1992)..
98.　R. H. Bradbury, M. P. Edwards, R. W. A. Luke, J. S. Major, A. A. Oldham, R. J. Pearce, D. A. Roberts, Abstracts, American Chemical Society, Fourth Congress of North America, New York, N. Y., 1991, MEDI No. 102 (1991).
99.　D. A. Roberts, R. J. Pearce, R. H. Bradbury, PCT # WO91/07407 (1991).
100.　D. A. Roberts, R. H. Bradbury, A. H. Ratcliffe, EPA # 453,210 (1991).
101.　K. Katano, H. Ogino, E. Shitara, H. Watanabe, J. Nagura, N. Osada, Y. Ichimaru, F. Konno, T. Machinami, T. Tsuroka, PCT # WO91/19697.
102.　A. Morimoto, K. Nishikawa, T. Naka, # EPA 443,568 (1991).
103.　D. Middlemiss, G. M. Drew, B. C. Ross, M. J. Robertson, D. I. C. Scopes, M. D. Dowle, J. Akers, K. Cardwell, K. L. Clark, S. Coote, C. D. Eldred, J. Hamblett, A. Hilditch, G. C. Hirst, T. Jack, J. Montana, T. A. Panchal, J. M. S. Paton, P. Shah, G. Stuart, A. Travers, Bioorg. Med. Chem. Lett, 1, 711 (1991).
104.　F. H. Marshall, J. C. Barnes, J. D. Brown, A. D. Michel, M. B. Tyers, Abstracts of the British Pharmacological Society Meeting, University of Southampton, England, September 18-20, 1991, Brit. J. Pharmacol., 104 , Abstract 425P (1991).
105.　M. J. Robertson, D. Middlemiss, B. C. Ross, G. M. Drew, D. I. C. Scopes, M. D. Dowle, Abstracts of the British Pharmacological Society Meeting, University of Southampton, England, September 18-20, 1991, Brit. J. Pharmacol., 104, Abstract 300P (1991).
106.　A. Hilditch J. S. Akers, A. Travers, A. A. E. Hunt, M. J. Robertson, G. M. Drew, D. Middlemiss, B. C. Ross, Abstracts of the British Pharmacological Society Meeting, University of Southampton, England, September 18-20, 1991, Brit. J. Pharmacol., 104 , Abstract 423P (1991).
107.　A. A. E. Hunt, A. Hilditch, M. J. Robertson, G. M. Drew, C. J. Gardner, D. J. Twissel, D. Middlemiss, B. C. Ross, Abstracts of the British Pharmacological Society Meeting, University of Southampton, England, September 18-20, 1991, Brit. J. Pharmacol., 104 , Abstract 424P (1991).
108.　W. J. Greenlee, A. A. Patchett, D. G. Hangauer, W. Ashton, T. Walsh, K. J. Fitch, R. A. Rivero, D. Dhanoa, PCT # WO91/11909.
109.　W. J. Greenlee, A. A. Patchett, D. G. Hangauer, T. Walsh, K. J. Fitch, R. A. Rivero, D. Dhanoa, PCT # WO91/11999; 12001; 12002 (1991).
110.　J. C. Hodges, I. Sircar, U. S. Pat. 5,045,540 and 5,041,552 (1991).
111.　P. R. Bovy, J. T. Collins, G. M. Olins, E. G. McMahon, W. C. Hutton, J. Med. Chem., 34, 2410 (1991).
112.　P. Buhlmayer, L. Criscione, W. Fuhrer, P. Furet, M. de Gasparo, S. Stutz, S. Whitebread, J. Med. Chem., 34, 3105 (1991).
113.　P. Buhlmayer, F. Ostermayer, T. Schmidlin, EPA 443,983 (1991).
114.　S. L. Lifer, W. S. Marshall, F. Mohamadi, J. K. Reel, R. L. Simon, M. I. Steinberg, C. A. Whitesitt, EPA # 438,869 (1991).
115.　J. S. Zhang, P. A. van Zwieten, Proceedings of the British Pharmacological Society, University College, London / Sandoz Institute for Medical Research, December 17-19, 1991, Brit. J. Pharmacol. (Proceedings Suppl.), 105 , Abstract 85P (1992).
116.　Y. Chen, M. F. Bean, C. Chambers, T. Francis, M. J. Huddleston, P. Offen, J. W. Westley, B. K. Carte, Tetrahedron Lett., 47, 4869 (1991).]
117.　T. J. Murphy, R. W. Alexander, K. K. Griendling, M. S. Runge, K. E. Bernstein, Nature, 351, 233 (1991).
118.　K. Sasaki, Y. Yamano, S. Bardhan, N. Iwai, J. J. Murray, M. Hasegawa, Y. Matsuda, T. Inagami, Nature, 351 , 230 (1991).
119.　N. Iwai, Y. Yamano, S. Chaki, F. Konishi, S. Bardhan, C. Tibbetts, K. Sasaki, M. Hasegawa, Y. Matsuda, T. Inagami, Biochem. Biophys. Res. Commun., 177, 299 (1991).
120.　H. Furuta, D.-F. Guo, T. Inagami, Biochem. Biophys. Res. Commun., 183, 8 (1992).
121.　C. Monnot, V. Weber, J. Stinnakre, C. Bihoreau, B. Teutsch, P. Corvol, E. Clauser, Mol. Endocrinol., 5, 1477 (1991).
122.　C. Ambroz, A. J. L. Clark, K. J. Catt, Biochem. Biophys. Acta, 1133, 107 (1991).
123.　C. Bergwitz, S. Madoff, A.-B. Abou-Samra, H. Juppner, Biochem. Biophys. Res. Commun., 179, 1391 (1991).
124.　S. Chaki and T. Inagami, Biophys. Res. Commun., 182, 388 (1992).
125.　K. Sandberg, H. Ji, M. A. Millan, K. J. Catt, FEBS Lett., 284, 281 (1991).

126.   N. Obermuller, T. Unger, J. Culman, P. Gohlke, M. deGasparo, S. P. Bottari, Neuroscience Letters, 132, 11 (1991).

127.   R. L. Soffer, M. A. Ravi Kiron, A. Mitra, S. J. Fluharty, Methods in Neurosciences, Vol. 5, P. M. Conn, Ed., Academic Press, San Diego, 1991, p. 192.

128.   P. C. Wong, A. T. Chiu, J. V. Duncia, W. F. Herblin, P. B. M. W. M. Timmermans, Trends Endocrinol. and Metab., in press, 1992.

129.   D. T. Dudley, S. E. Hubbell, R. M. Summerfelt, Mol. Pharmacol., 40, 360 (1991).

130.   S. P. Bottari, I. N. King, S. Reichlin, I. Dahlstroem, N. Lydon, M. de Gasparo, Biochem. Biophys. Res. Commun., 183, 206 (1992).

131.   C. Sumners, W. Tang, B. Zelezna, M. H. Raizada, Proc. Natl. Acad. Sci. USA, 88, 7567 (1991).

132.   K. Tsutsumi, J. M. Saavedra, Am. J. Physiol., 261, H667, 1991.

133.   R. E. Gibson, H. H. Thorpe, M. E. Cartwright, J. D. Frank, T. W. Schorn, P. B. Bunting, P. K. S. Siegl, Am. J. Physiol., 261 (Renal Fluid Electrolyte Physiol. 30), F512 (1991).

134.   N. Jaiswal, E. A. Tallant, D. I. Diz, M. C. Khosla, C. M. Ferrario, Hypertension, 17, 1115 (1991).

135.   D. Felix, M. C. Khosla, K. L. Barnes, H. Imboden, B. Montani, C. M. Ferrario, Hypertension, 17, 1111 (1991).

136.   M. T. Schiavone, K. B. Brosnihan, M. C. Khosla, C. M. Ferrario, Hypertension, 17, 425, Abstact 74 (1991).

137.   M. G. Cogan, F.-Y. Liu, P. C. Wong, P. B. M. W. M. Timmermans, J. Pharmacol. Exp. Therapeut., 259, 687 (1991).

138.   L. Chen, R. N. Re, Am. J. Hypertens., 4, 82A, Abstract 235 (1991).

139.   Y. Miyoshi, Y. Nakaya, Biochem. Biophys. Res. Commun., 181, 700 (1991).

140.   R. D. Smith, A. T. Chiu, P. C. Wong, W. F. Herblin, P. B. M. W. M. Timmermans, Ann. Rev. Pharmacol., 32, 135 (1992).

141.   P. C. Wong, T. B. Barnes, A. T. Chiu, D. D. Christ, J. V. Duncia, W. F. Herblin, P. B. M. W. Timmermans, Cardiovascular Drug Reviews, in press, 1992.

142   B. Bunkenburg, C. Schnell, H.-P. Baum, F. Cumin, J. M. Wood, Hypertension, 18, 278 (1991).

143.   Y. Christen, B. Waeber, J. Nussberger, M. Porchet, R. M. Borland, R. J. Lee, K. Maggon, L. Shum, P. B. M. W. M. Timmermans, H. R. Brunner, Circulation, 83, 1333 (1991).

144.   Y. Cristen, B. Waeber, J. Nussberger, R. J. Lee, P. B. M. W. M. Timmermans, H. R. Brunner, Am. J. Hypertens., 4, 350S (1991).

145.   H. R. Brunner, J. Nussberger, B. Waeber, Curr. Opinion in Cardiology, 6, 724 (1991).

146.   E. Nelson, D. Merrill, C. Sweet, T. Bradstreet, D. Panebianco, R. Byyoy, T. Herman, K. Lassiter, B. Levy, G. Lewis, F. G. McMahon, R. Reeves, D. Ruff, A. Shepherd, D. Weidler, J. Irvin, Abstracts, European Society of Hypertension, Fifth European Meeting on Hypertension, Milan, Italy, June 7-10, 1991, Abstract 512 (1991).

147.   K. Song, J. Zhuo, F. A. O. Mendelsohn, Br. J. Pharmacol., 104, 771 (1991).

148.   V. J. DeNoble, K. F. DeNoble, K. R. Spencer, A. T. Chiu, P. C. Wong, P. B. M. W. M. Timmermans, Brain Res., 561, 230 (1991).

149.   R. J. A. Banks, C. T. Dourish, Abstracts of the British Pharmacological Society Meeting, University of Southampton, England, September 18-20, 1991, Br. J. Pharmacol., 104 , Abstract 63P (1991).

150.   R. P. Dennes, J. C. Barnes, A. D. Michel, M. B. Tyers, Proceedings of the British Pharmacological Society, University College, London / Sandoz Institute for Medical Research, December 17-19, 1991, Brit. J. Pharmacol. (Proceedings Suppl.), 105, Abstract 88P (1992).

151.   T. E. Raya, S. J. Fonken, R. W. Lee, S. Daugherty, S. Goldman, P. C. Wong, P. B. M. W. M. Timmermans, Am. J. Hypertens., 4, 334S (1991).

152.   M. F. Prescott, R. L. Webb, M. A. Reidy, Am. J. Pathology, 139, 1291 (1991).

153.   R. F. Kauffman, J. S. Bean, K. M. Zimmerman, R. F. Brown, M. I. Steinberg, Life Sci., 49, PL-223 (1991).

154.   T. Scott-Burden, T. R. Resink, A. W. A. Hahn, P. M. Vanhoutte, J. Cardiovasc. Pharmacol., 17 (Suppl. 17), S96 (1991).

155.   K. Yoshida, M. Yasujima, M. Kohzuki, K. Tsunoda, K. Kudo, M. Kanazawa, T. Yabe, K. Abe, K. Yoshinaga, J. Cardiovasc. Pharmacol., 17 (Suppl. 7), S514 (1991).

156.   P. H. Wang, Y. S. Do, L. Macauley, T. Shinagawa, P. W. Anderson, J. D. Baxter, W. A. Hsueh, J. Biol. Chem., 266, 12633 (1991).

157.   D. Dubin, R. E. Pratt, K. Y. Hui, V. J. Dzau, J. Hypertens., 9, 483 (1991).

158.   C. M. Ferrario, K. B. Brosnihan, D. I. Diz, N. Jaiswal, M. C. Khosla, A. Milsted, E. A. Tallant, Hypertension, 18 (Suppl. III), III-126 (1991).

159.   C. Foresta, R. Mioni, M. Rossato, A. Varotto, M. Zorzi, Int. J. Androl., 14, 333 (1991).

160.   A. Vuckovic, B. M. Cohen, G. S. Zubenko, J. Clin. Psychopharmacol., 11, 395 (1991).

161.   R. S. Boger, S. R. Crowley, PCT # WO91/16031 (1991).

Chapter 8. EDRF, an Emerging Target for Drug Design

James F. Kerwin, Jr.
Pharmaceutical Products Division
Abbott Laboratories, Abbott Park, IL 60064

<u>Introduction</u> - In 1980 Furchgott and Zawadzki (1) demonstrated an endothelial-derived diffusible factor that was responsible for the vasodilation elicited in rabbit arteries by acetylcholine. The endothelial cell layer was thus important and necessary for normal vascular function. The factor was termed endothelium-derived relaxing factor (EDRF) and has been an important component in the rapidly developing field of vascular biology (2, 3). During the ensuing decade, research has demonstrated EDRF to be a ubiquitous intra- and intercellular messenger (4) with important roles in vascular homeostasis (5-7), neuronal function (8), and immunological function (9). While early workers had sought to equate EDRF with arachidonic acid metabolites or oxygen free radical species, attention was soon focused on nitric oxide (NO) as a likely candidate for EDRF (10, 11). The common properties of EDRF and NO are: *i*) lability, with approximately similar half lives in physiological solution (3-50 sec); *ii*) relaxation of vascular smooth muscle; *iii*) activation of cytosolic guanylate cyclase; *iv*) inhibition by hemoglobin and methylene blue; and *v*) potentiation of pharmacolgical effect by superoxide dismutase (SOD), a scavenger of superoxide radical anion. Pharmacological comparison of NO and EDRF utilizing a cascade bioassay, as well as direct chemiluminescent detection of NO, gave equivalent results (12). More recently a number of nitrosothiols have been suggested as possible EDRFs including S-nitrosocysteine and S-nitrosoproteins (13-17). Although some reports suggest that EDRF may not be directly equivalent with NO, the majority of evidence equates EDRF with NO or a ready source of NO (eg. nitrosothiols), and therefore EDRF has been termed the "endogenous nitrovasodilator" (12, 18).

<u>The Arginine - Nitric Oxide Pathway</u> - Initial evidence suggested that the biochemical conversion of L-arginine to NO or an NO-like molecule provided EDRF-like activity in vascular endothelial cells (19, 20). More definitive evidence arose from mass spectrometry experiments with [$^{15}N^G$]-L-arginine demonstrating the incorporation of [^{15}N] into NO and the NO derived products, nitrate and nitrite. L-Arginine but not D-arginine, was a substrate for the enzyme while N^G-methyl-L-arginine (L-NMA, <u>1</u>) stereoselectively inhibited EDRF formation (21). L-NMA had also been utilized to inhibit the formation of L-citrulline and nitrogen oxides from L-arginine in macrophages, and this key finding helped explain the source of urinary nitrates in clinical cases of infection and inflammation (22, 23). In macrophages, the NO and L-citrulline forming activity is induced <u>via</u> the action of cytokines or endotoxin, which release cytokines. It therefore appeared that there were constitutive as well as inducible EDRF-like activities. EDRF-like activity was also identified in the CNS (24). It now appears that the L-arginine - NO pathway is a ubiquitous signal transduction mechanism and a pathway for L-arginine metabolism. The elucidation of the biochemical mechanism(s) by which L-arginine is converted to NO (EDRF), the enzyme(s) responsible for the oxidation of L-arginine to NO, and the role of NO in pharmacology and physiology of various systems represent ongoing research activities.

<u>Nitric Oxide Synthases (NOS)</u> - Many research groups have pursued the purification of NO synthase (NOS) from various cell and tissue sources. The primary means by which the enzymatic activity can be monitored are *i*) <u>via</u> the synthesis of [^3H]-L-citrulline from [^3H]-L-arginine (25); *ii*) <u>via</u> the indirect stimulation of soluble guanylate cyclase by the produced EDRF (26); *iii*) <u>via</u> the synthesis of nitrogen oxides measured as either nitrites and/or nitrates; *iv*) <u>via</u> the chemiluminescent reaction of NO with ozone (27); and *v*) <u>via</u> the trapping of NO with hemoglobin (28, 29). NADPH dependent NOS derived from brain or endothelial sources was calcium dependent and calcium dependency suggested a role for calmodulin (30-32). Tetrahydrobiopterin was later found to be necessary for activity by some researchers although recent reports have cast doubt on its role as a cofactor (33, 34). The first NOS isoform to be purified was from rat brain utilizing a DEAE column, and a 2', 5'- adenosine diphosphate (ADP)-linked agarose column which takes advantage of NOS' affinity for the cofactor, NADPH (35). The DEAE separation removed the necessary cofactor, calmodulin, which was added back to monitor activity. The reported K_m for L-arginine was 1.5 µM, and the purified NOS produced a single 155 kDa band on SDS-PAGE and an

apparent molecular mass of 200 kDa on Superose-6 chromatography. As shown in Table 1, this isoform has been named NOS Type I (36). Other researchers have also isolated the Type I NOS in a similar manner, some utilizing calmodulin affinity chromatography, and reported consistent findings (37). In 1985 it was reported that macrophages could be induced with lipopolysaccharide (LPS) to produce both nitrate and nitrite (38). Further work utilizing cytokine-induced macrophages identified the enzyme involved in this nitrate formation as an NOS localized to both cytosolic and particulate fractions (39). This cytosolic NOS isoform (Type II) required NADPH, 5,6,7,8-tetrahydrobiopterin, flavin-adenine-dinucleotide (FAD), flavin-mononucleotide (FMN), and thiols for full activity. Three groups have reported the purification of this NOS isoform. Kawai and coworkers isolated it from induced rat peritoneal macrophages and report a molecular mass of 150 kDa on SDS-PAGE and a molecular mass of 300 kDa for the native protein (40). Stuehr and coworkers utilized the RAW 264.7 murine macrophage cell line and reported a tight triplet of bands of molecular mass between 125-135 kDa on SDS-PAGE and a molecular mass of 250 kDa for the native protein (41). Marletta and coworkers reported similar results with one band of 130 kDa on SDS-PAGE and 260 kDa for the native dimer for the NOS purified from murine macrophage (42). These discrepancies have yet to be resolved. The endothelial-derived NOS has been designated Type III and appears to be predominantly particulate (43). Type III NOS isoform requires NADPH, Ca^{2+}, calmodulin, and 5,6,7,8-tetrahydrobiopterin for optimal activity and has a molecular weight of 135 kDa (44). A particulate NOS activity present in macrophages has been tenatively classed as isoform Type IV and has yet to be purified (36).

Table 1. Isoenzymes of Nitric Oxide Synthase

NOS Isoform	Cosubstrates Cofactors	Regulated by	M_r of Denatured Protein	Location
I	NADPH, FAD FMN, tetrahydro-biopterin	Ca^{2+}, calmodulin	155 kDa native dimer	brain, cerebellum
II	NADPH, FAD, FMN tetrahydrobiopterin	induced by cytokines and endotoxin, Ca $^{2+}$ independent	125-35 kDa native dimer 150 kDa	RAW macrophages peritoneal macrophage
III	NADPH, FAD FMN, tetrahydro-biopterin	Ca^{2+}, calmodulin	135 kDa	endothelial cells
IV	NADPH, FAD, FMN tetrahydrobiopterin	induced by cytokines and endotoxin, Ca $^{2+}$ independent	?	macrophages

Substrates and Inhibitors of NOS - Beside L-NMA a number of other potent inhibitors of NOS have been identified. These include N^G-nitro-L-arginine (L-NNA, **2**), N^G-nitro-L-arginine methyl ester (L-NAME, **3**), N^G-amino-L-arginine (**4**), and N^5-ethylimino-L-ornithine (NIO, **5**). Most of these agents have been utilized as pharmacological probes and approximate inhibitory potencies in vitro (Table 2) have been reported (45-48). Interest in the mechanism of the biological oxidation of L-arginine to L-citrulline and NO has sparked further studies aimed at elucidating this 5 electron oxidation process (49). Earlier work had demonstrated the source of the NO nitrogen. More recent work with [^{18}O] water and O_2 demonstrated that the source of the ureido oxygen in L-citrulline is molecular oxygen (50). Likewise the NO oxygen appears to be derived from molecular oxygen (51). Other workers have been able to demonstrate the intermediacy of N^G-hydroxy-L-arginine (**6**); the oxidation of the hydroxylated guanidino-nitrogen of **6** to NO and the requirement of less NADPH for this conversion (52, 53). Besides L-arginine other substrates have been limited to L-arginine esters, L-homoarginine (54) or L-arginine containing peptides (55). One report has presented some evidence for the presence of a non-arginine substrate pool for the inducible NOS isoforms (22).

Molecular Biology of NOS Isoforms - One report has appeared on the cloning and expression of NOS, the Type I isoform (56). Using a rat brain cDNA library three clones were derived encoding a protein of 1,429 amino acids (molecular mass ~160K); this protein incorporated peptide sequences derived from trypsin digests of the native NOS protein. Comparison of the amino acid sequence of NOS with that from cytochrome P_{450} reductase revealed a high degree of homology (58%) within the C-terminal domain, while the N-terminal region demonstrated no homologies to any known protein. In addition, examination of the amino acid sequence revealed consensus sites for calmodulin, NADPH, FAD and FMN binding domains. Several groups are currently working on the cloning, expression, and sequencing of the type II and type III NOS isoforms.

Table 2. Inhibitors and Substrates of Nitric Oxide Synthases

				IC_{50} (μM)			
		R'	R	CNS	Macrophage	Endo.	Ref.
H_2N CO_2R'	**1**	H	$NHCH_3$	9.8	11.8	2.9	(57,58)
	2	H	$NHNO_2$	0.05	212		(57)
	3	CH_3	$NHNO_2$			3.1	(58)
	4	H	$NHNH_2$	3.0	7.4		(57)
NH	**5**	H	CH_3			0.5	(58)
HN R	**6**	H	NHOH				

A polyclonal antibody raised against rat Type I NOS was used for the immunohistochemical localization of NOS within the rat (59). NOS is concentrated in the cerebellar molecular and granule layers, in the olfactory bulb, the dentate gyrus of the hippocampus, the bed nucleus of the stria terminalis, the posterior pituitary, the autonomic nerve fibers in the retina, the nerve fibers of the myenteric plexus of the intestinal tract, the adrenal medulla and in the nerve fibers surrounding the adventitia in cerebral blood vessels. Staining of endothelial layers of larger blood vessels was also observed. Monoclonal antibodies to Type III NOS were developed recently and one of these showed no reactivity to Type I NOS (60). The ability to differentiate these two constitutive NOS isoforms is important to the elucidation of NOS-related physiological events. The structural homology of NOS with cytochrome P_{450} reductase and other oxidative enzymes requiring NADPH, FAD and FMN has been highlighted. Therefore it was not surprising to find that NOS has diaphorase activity similar to that of cytochrome P_{450} reductase and that NOS activity within the CNS could account for most of the diaphorase activity localized by immunohistochemical staining (61-63).

A number of possible phosphorylation sites were discovered via the sequencing of NOS. Protein kinase C and calcium calmodulin-dependent protein kinase II have been reported to increase NOS activity in brain derived NOS (64). There is one report of inhibited NOS activity by protein kinase C activation in NOS-transfected kidney cells (65). Protein kinase A has been shown to phosphorylate brain-derived NOS stoichiometrically without apparent change in activity (66).

Physiological and Pharmacological Actions of EDRF/NO - EDRF/NO is a labile humoral agent released from the vascular endothelium by the action of several endothelium-dependent vasodilators such as acetylcholine, bradykinin, substance P, oxytocin, histamine, thrombin, serotonin, α_2-adrenergic agonists, and ADP. The L-arginine - NO pathway transduces the signal from the endothelial cell to the vascular smooth muscle via the action of readily diffusible NO on cytosolic guanylate cyclase contained within smooth muscle. Guanylate cyclase activation produces cyclic GMP which then mediates further signal transduction in the smooth muscle cell and leads ultimately to vasorelaxation. EDRF/NO pathways have been shown to exist in many parts of the vascular system including pulmonary, coronary, systemic, mesenteric, and cerebro- vasculature. NOS activity is responsible for the maintenance of a basal dilatory vascular tone which vasconstrictors such as endothelin then modify. The basal release of EDRF/NO has been demonstrated via the use of NOS inhibitors and the loss of endothelial dependent vasodilation (67). In addition to receptor mediated activation, mechanical stimuli such as shear stress on vascular endothelium are sufficient to evoke EDRF/NO synthesis and release (68, 69). NOS activity may act to minimize cardiac load by optimally dilating local systemic vasculature (70, 71). Thus the NOS pathway is not only under local receptor-mediated pharmacological regulation but also under mechanical regulation in response to the vascular load itself.

Hypoxia is another pathophysiological regulatory mechanism of EDRF/NO. Systemic administration of L-NNA attenuated the fall in arterial pressure induced by systemic hypoxia in sino-denervated rats (72). These effects were reversible by L-arginine and independent of the basal arterial pressure. The hypoxic depressor response is mainly mediated by the NO pathway. The initial hypoxic depressor response is followed by a delayed increase in arterial pressure which is neurogenically mediated. The short half life and gaseous stability of NO have been utilized in experiments on hypoxic pulmonary vasoconstriction in dogs. Small amounts of NO (40 and 80 ppm) administered via a blended ventilation system reversed hypoxic vasoconstriction (73, 74). No sign of inhaled NO in the systemic vascular system was apparent by nitrite and methemoglobin monitoring. A regulatory role for NO has been proposed in the normal oxygenation of blood in the pulmonary circulation via ventilation-perfusion matching (75). Changes in NO levels of expired air versus oxygen in inspired air support this idea (76, 77).

The NO pathway contributes to the prevention of abnormal platelet aggregation, the prevention of vasoconstriction induced by aggregating platelets and possible ensuing vasospasms (78). Platelet soluble guanylate cyclase is activated by NO leading to a decrease in intracellular calcium concentrations which suppress adhesion and aggregation. NO and PGI_2 appear to act together in their anti-platelet aggregation effects since both are released by endothelium and reduce platelet intracellular calcium (78). NOS and the NO pathway is of prime importance in the regulation of vascular tone and vascular homeostasis and therefore has led to the study of NOS and EDRF/NO pathways in several vascular disorders (79, 80) related to hypercholesterolaemia, atherosclerosis (81, 82), angina (83, 84), hypertension (85, 86), arterial transplants (87), and angioplasty (88). In hypercholesterolemic models the function of the NO pathway generally remains intact, L-arginine treatment having no effect or some benefit of increased vasodilation, depending on the vasculature and species examined (89, 90). In hypercholesterolemic humans treated with L-arginine, a correction of endothelial dysfunction of coronary microcirculation has been reported (91). Changes in SOD levels or activity has been suggested to account for endothelial dysfunction of the EDRF/NO system leading to atherogenesis and ultimately atherosclerosis (92, 93). Oxidized low-density lipoproteins (LDL) or native LDL were also shown to attenuate NOS linked vasodilator function although there was no effect on NO synthesis and NOS activity (94, 95). The oxidation of lipo-proteins may involve NO-derived species. NO reacts with superoxide anion to form peroxynitrite which decomposes rapidly to provide oxidative species such as the hydroxyl radical (96, 97). Another source of NO that could lead to damaging oxidative pathways are neutrophils (98). Overproduction of NO in cholesterol-rich neutrophils could result in processes that contribute to atherogenesis (99).

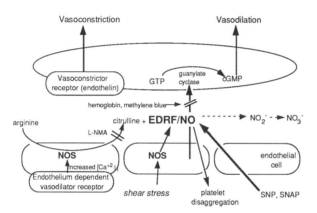

Fig. 1. Signal Transduction Pathway of Nitric Oxide Synthase in Vascular Tissue. Endothelium dependent vasodilators activate NOS leading to the production of endogenous NO and citrulline from L-arginine. Nitrovasodilators such as SNP or SNAP provide exogenous NO which diffuses into vascular smooth muscle activating guanylate cyclase and ensuing vasodilation.

The balance between the vasodilation provided by EDRF/NO and vasoconstrictors such as endothelin has been studied in several labs. Endothelin receptors are coupled to the phosphoinositol second messenger system, and activation thereby increases internal calcium concentrations (see Chapter 9 in this volume). Concomitant with the rise in intracellular calcium is the activation of NOS and NO formation (100). Thus endothelin, a potent vasoconstrictor, possesses a bifurcated pharmacology. While the vasoconstrictor response of endothelin is functionally antagonized by this activation of the NO pathway, other actions such as the antinatriuretic response and decreases in the glomerular filtration rate are not inhibited (101, 102). The separation of these pathways via transduction mechanisms or receptor subtypes is an ongoing area of research. Loss or absence of the NO pathway leads to the overexpression of vasoconstrictor responses. In spontaneously hypertensive rats the basal release of NO as measured via NOS inhibition is lower than in normotensive rats (103). A similar finding in human forearm vasculature was found by studying the response of hypertensive and normotensive patients to acetylcholine challenges. Sodium nitroprusside (SNP) was used as a control (86). These findings suggest that some vasoconstrictive diseases such as hypertension and congestive heart failure which are accompanied by profound vasoconstriction, may be due to functional abnormalities of the NO system in resistance vessels. While the actual physiological and biochemical changes that accompany vascular disease remain unknown, the current focus is clearly on endothelial function and in particular the loss of endothelium-dependent vasodilation as mediated by the NO pathway.

Pathophysiological Implications of EDRF/NO - The vascular endothelial NOS (Type III) and the neuronal NOS (Type I) are constitutive NOS isoforms which produce picomolar levels of NO of which only a small fraction survives to elicit physiological responses. The cytokine-induced NOS isoforms (Types II and IV) produce nanomolar levels of NO which act as a primary defense mechanism of the immune system (104, 105). NO in macrophages activates soluble guanylate cyclase to produce cGMP which may further regulate the cytotoxicity of the macrophage. The primary cytokines studied in vitro have been interferon (IFN-γ), interleukins, and tumor necrosis factor (TNF). LPS has also been utilized to induce NOS through the release of TNF causing induction of NOS activity. A number of cytokines or cytokine stimulators have been shown to synergize in promoting NOS activity (106). NO or NOS activity is necessary for the cytotoxic and microbiostatic effects of macrophages and examples exist with the pathogens Toxoplasma gondii (107), Cryptococcus neoformans (108), Corynebacterium parvum (109), Leishmania major (110, 111), Plasmodium berghei (112), Trypanosoma musculi (113), Mycobacterium leprae (114), as well as tumors (115). NO cytotoxicity in immune function is not fully understood. However a number of mechanisms have been suggested including the inhibition of ferrous-containing proteins and metabolic enzymes (28, 116), inhibition of electron transport and mitochondrial respiration in general (117), and nitrosylation of DNA (118). The inhibition of DNA replication and/or translation by NO fits with the altered metabolism of L-arginine which normally produces necessary polyamines for protein synthesis and cell growth (119-121). The observation that NOS activity is generally inversely related to cell growth (122) can be rationalized in terms of the partitioning of L-arginine metabolic pathways.

Induction of NOS activity can be blocked by treatment with transforming growth factor-beta (123, 124), glucocorticoids (125), and in some cell types by specific interleukins. The anti-inflammatory effect of glucocorticoids may in part be mediated by this inhibition of NOS activity. NOS induction requires protein synthesis and transcription and is accompanied by the induction of tetrahydrobiopterin synthesis (126). A consequence of the stimulation of the NOS pathway by pathogen-derived toxins may be inappropriate production of NO leading to pathophysiological disease states (117, 127). The inappropriate induction of NOS activity has been advanced as a cause for type I diabetes (128-131). In addition to cytokine induction in monocytes and macrophages, NOS activity has been induced in neutrophils, Kupffer cells, hepatocytes (132), endothelial cells, fibroblasts (133) and vascular smooth muscle cells. Induction of NOS activity in Kupffer cells and/or hepatocytes appears to suppress protein synthesis. Induction of NOS (Type II) activity in endothelial and vascular smooth muscle cells appears to be associated with pathophysiological hypotensive episodes resulting from the over-modulation of the vasodilatory NO pathway. In particular, the inducible NOS in vascular smooth muscle appears to have robust activity and is clearly capable of over-stimulation of the soluble guanylate cyclase in these cells, leading to profound hypotension. Administration of LPS provides a model of sepsis in which NOS in vascular smooth muscle and macrophages has been induced. Nitrosylated hemoglobin has been detected in these LPS-treated animals (134) and L-NMA has reversed the decrease in mean arterial blood pressure (135).

EDRF/NO in the Nervous System - NOS exists in many cells and tissues where its function is primarily that of an inter- or intracellular messenger and/or autacoid. In the CNS, NOS (Type I) is located within neurons and is important in neurotransmission, the central control of the vascular system, and possibly learning and memory. The identification of an EDRF-like activity in the CNS which was activated by NMDA receptor stimulation has spurred much interest in the role of NOS in the brain (24). Receptor-mediated intracellular increases in calcium levels leads to the activation of NOS. The NO formed then activates soluble guanylate cyclase and thereby increases cyclic GMP levels (136-138). The guanylate cyclase target may be intracellular, a neighboring neuron, or even neighboring glial cells (139). Investigation of the role of NOS in mediating glutamate-induced neurotoxicity has demonstrated some neuronal protection in vitro with NOS inhibitors (140) although there have been conflicting reports (141). NO itself in high enough concentrations is cytotoxic and one report of prevention of focal ischemia with an NOS inhibitor has appeared (142). The coadministration of NMDA and L-arginine (icv) produced convulsive effects that were blocked by L-NAME (143). The role of NO in glutaminergic pharmacology suggests that optimal levels are necessary for normal neuronal function while either reduced production or over production (as in glutaminergic excitotoxicity) may lead to CNS dysfunction (144). In ischemic and hypoxic insults the reperfusion injury associated with the formation of peroxynitrite and associated oxidants may contribute significantly to radical-mediated neuronal damage (145).

Evidence for the role of NOS in long term potentiation (LTP) is developing. NO is a retrograde messenger capable of diffusing from postsynaptic neurons to presynaptic neurons and potentiating their responses (146). More interestingly EDRF/NO potentiates neuronal responses in the apparent NO diffusion locus that are not being activated by the prime excitatory input (147). Research aimed at elucidating a possible feedback loop between NO and the NMDA receptor has led to equivocal results. NO donors have been shown to inhibit NMDA receptor function via a redox event (148). However the use of SNP as an NO precursor clouds the issue since the ferrous ion itself inhibits NMDA receptor activation (149, 150). Another role for NO in synaptic plasticity is in long term depression (LTD) in the cerebellum. LTD in cerebellum slices initiated by conjunctive stimulation of climbing fibers and parallel fibers was blocked by L-NMA and hemoglobin (151). NOS activity has also been reported in astrocytes and microglial cells though a specific role for NOS in these cells remains to be determined (152). There also appears to be a role for NOS in the neurogenic control of systemic blood pressure. Injection of S-nitroso-L-cysteine (25 pmol) into the rat nucleus tractus solitarius (NTS) lowered arterial pressure and heart rate (153). This NO-mediated baroreceptor response has been linked to glutaminergic receptors via microinjection of glutamate into the NTS (154). SNP had similar effects when injected into the rostral ventrolateral medulla (RVLM) of the cat and these effects could be reversed via injection of SNP into the caudal ventrolateral medulla (CVLM) (155). Sympathetic neuronal activity has also been indicated in the regulation of vascular NOS function (156). In addition to the role that the NO pathway plays in LTP, LTD and the control of cerebral blood flow and cardiovascular function, a possible role in the development of the nervous system has been hypothesized. The shape and function of the developing CNS has been shown to depend on the pattern of correlated neuronal activity. A diffusible labile signal as provided by NO could serve this function (157).

NOS in the peripheral nervous system appears to be linked to nonadrenergic noncholinergic (NANC) neuronal pathways and has been demonstrated to be involved with gastrointestinal function. Gastric mucosal blood flow induced by secretagogues such as pentagastrin in rat is inhibited by L-NMA (12.5 and 50 mg/kg, iv). Pentagastrin-stimulated gastric acid secretion remained unaffected by L-NMA treatment. Protection afforded the gastric mucosa from acid and other factors is derived at least in part by the NO-mediated gastric mucosal blood flow (158, 159). NO also plays a role in pressure-induced adaptive relaxation of the fundus in the guinea pig. The fundus relaxation occurs via two pathways both mediated by NO, a reflex relaxation and a ganglionic nicotinic-activated relaxation (160). Other NANC tissues containing NO pathways are the canine ileocolonic junction, rat anococcygeus, esophageal smooth muscle, and the bovine retractor penis muscle (161, 162). The penile corpus cavernosum smooth muscle in rabbits and humans contains a NANC inhibitory pathway. Corpus cavernosum smooth muscle relaxation to acetylcholine, bradykinin, substance P, or electrical field stimulation is endothelium dependent, antagonized by NOS inhibitors and mimicked by S-nitroso-N-acetyl-penicillamine (SNAP) or SNP. Some forms of impotence may result from defects in the NO-mediated smooth muscle relaxation in corpus cavernosum (163, 164).

Therapeutic Directions for NOS - Among the first therapeutic utilities suggested for NOS inhibitors are for sepsis and septic shock, and the hypotension associated with interleukin therapies (135). Inhibition of the induction of NOS in vascular tissues presents a clear and measurable biochemical target. Early studies have demonstrated that non-selective NOS inhibition may be detrimental in septic shock therapy since it results in a decreased peripheral organ perfusion and lower cardiac output (165). A conjunctive treatment of an NOS inhibitor and an NO donor has been evaluated in a sepsis model with limited success and a suggestion has been advanced that selective induced-NOS (Type II) inhibition may be desirable (166). Recent uncontrolled clinical studies have demonstrated a positive benefit of NOS inhibition by L-NMA and L-NAME on cardiovascular function for sepsis patients who had failed to respond to traditional treatments (167, 168). Anaphylactic shock and hemorrhagic shock have also been ameliorated by NOS inhibition in animal models (169, 170). In atherogenesis and atherosclerosis the NOS enzyme appears to be functioning normally, however, the NO appears to be shunted away from its normal target, guanylate cyclase, and vasorelaxation. If NO is involved in atherogenesis by the indirect oxidation of lipoproteins, then NOS inhibition might be a viable target. The immune system utilizes NO as a cytotoxic agent but very little is understood about deleterious NO formation in macrophages and lymphocytes. The role of NOS in the etiology of diabetes remains to be proven but may involve inappropriate activation of NOS by the immune system. Even chronic stimulus and release of NO from the immune system at appropriate levels may lead to disease pathologies. Since NO is capableof damaging DNA, the chronic exposure of tissues to NO could lead to abnormalities. For example, chronic stimulation of microglia in the CNS would result in selective neuronal loss and neurodegenerative disease. Radical-induced neurodegeneration (171) remains a reasonable rationale for the etiology of some CNS dysfunctions and NO may be involved. Evidence suggests that some hypertensive patients have impaired NOS transduction pathways. An understanding of the nature of EDRF/NO and its transduction pathways may lead to better nitrovasodilators.

A role for NOS inhibitors in CNS ischemia therapy is emerging. The role of glutaminergic pathways in neuronal degeneration (172) suggests that inhibition of the NO signal transduction pathway may be a viable means of blocking glutamate-mediated excitotoxicity. However the timing of effective NOS inhibitor treatment for excitotoxic glutaminergic insults has yet to be addressed. A major issue for NOS in cerebral ischemia is whether NO produced under hypoxic conditions is more or less detrimental than the NO under reperfusion conditions. In cerebral ischemia, the need for a selective NOS inhibitor appears clear since cerebral vasoconstriction could be detrimental to reversing the ischemic insult. A related area of interest is the role of NOS in cerebrovascular disorders such as migraine (173, 174).

Future Directions - The initial decade of EDRF/NO research has focused attention on the importance of this regulatory pathway in vascular, immunological, and CNS physiology. Much more research will be needed to understand the contribution of the NO pathway to the etiology of vascular and neuronal disease states, and in particular, the regulation of inducible NOS isoforms which may contribute to a number of degenerative disease states. Although a number of NOS isoforms have been discovered there may be others yet to be defined and the mechanisms by which these NOS isoenzymes are regulated and modulated could provide new targets for drug discovery. Present NOS inhibitors are not particularly potent nor selective and the advent of such agents will help to clarify several issues in the EDRF/NO field. The selectivity of any therapeutically interesting NOS inhibitor will be an issue since the NOS - NO pathway is ubiquitous and important in the regulation of a large number of cells. As NOS isoforms become more readily available, biophysical and structural studies may aid the medicinal chemist in inhibitor design. An appreciation of the importance of the NO pathway coupled with new NOS inhibitors will hopefully allow further advances in the understanding of the vascular, neuronal, and immune systems.

References

1.   R.F. Furchgott and J.F. Zawadzki, Nature, $\underline{288}$, 373 (1980).
2.   P.A. Marsden, M.S. Goligorsky, and B.M. Brenner, J. Am. Soc. Nephrol., $\underline{1}$, 931 (1991).
3.   R.C. Becker, Cardiology, $\underline{78}$, 13 (1991).
4.   F. Murad, K. Ishii, U. Forstermann, L. Gorsky, J.F. Kerwin, Jr., J. Pollock, and M. Heller, in "Advances in Second Messenger and Phosphoprotein Research", Vol. 24, Y. Nishizuka, M. Endo and C. Tanaka Eds. New York: Raven Press (1990) p. 441.
5.   J.A. Angus and T.M. Cocks, Pharmacol. Ther., $\underline{41}$, 303 (1989).
6.   T.F. Luscher, H.A. Bock, Z.H. Yang, D. Diederich, Kidney Int., $\underline{39}$, 575 (1991).
7.   S. Moncada, R.M. Palmer and E.A. Higgs, Pharmacol. Rev., $\underline{43}$, 109 (1991).
8.   J. Garthwaite, Trends Neurosci., $\underline{14}$, 60 (1991).
9.   D.J. Stuehr and C.F. Nathan, J. Exp. Med., $\underline{169}$,1543 (1989).
10.   L.J. Ignarro, R.E. Byrnes and K.S. Wood, in "Mechanisms of Vasodilatation", Raven Press: New York. (1988) p. 427.
11.   R.M. Furchgott, in "Mechanisms in Vasodilatation", Raven Press: New York. (1988) p. 401.
12.   S. Moncada, R.M.J. Palmer and E.A. Higgs, Hypertension, $\underline{12}$, 365 (1988).
13.   G.M. Rubanyi, A. Johns, D. Wilcox, F.N. Bates and D. Harrison, J. Cardiovasc. Pharmacol., $\underline{17}$, S41 (1991).
14.   P.R. Myers, R.L. Minor, Jr., R. Guerra, Jr., J.N. Bates and D.G. Harrison, Nature, $\underline{345}$, 161 (1990).
15.   E.A. Kowaluk and H.L. Fung, J. Pharmacol. Exp. Ther., $\underline{255}$, 1256 (1990).
16.   K.D. Thornbury, S.M. Ward, H.H. Dalziel, A. Carl, D.P. Westfall and K.M. Sanders, Am. J. Physiol., $\underline{261}$, G553 (1991).
17.   J.S. Stamler, D.I. Simon, J.A. Osbourne, M.E. Mullins, O.T. Jaraki, D. Michel, J. Singel and J. Loscalzo, Proc. Natl. Acad. Sci. USA, $\underline{89}$, 444 (1992).
18.   T.F. Luscher, Eur. Heart J., $\underline{12}$, 2 (1991).
19.   R.M.J. Palmer, D.S. Ashton and S. Moncada, Nature, $\underline{333}$, 664 (1988).
20.   I. Sakuma, D.J. Stuehr, S.S. Gross, C. Nathan and R. Levi, Proc. Natl. Acad. Sci. USA, $\underline{85}$, 8664 (1988).
21.   R.M.J. Palmer, D.D. Rees, D.S. Ashton and S. Moncada, Biochem. Biophys. Res. Commun., $\underline{153}$, 1251 (1988).
22.   D.L. Granger, J.B.J. Hibbs, and L.M. Broadnax, J. Immunol., $\underline{146}$, 1294 (1991).

23. J.M. Langrehr, R.A. Hoffman, T.R. Billiar, K.K. Lee, W.H. Schraut and R.L. Simmons, Surgery, 110, 335 (1991).
24. J. Garthwaite, S.L. Charles and R. Chess-Williams, Nature, 336, 385 (1988).
25. R.M.J. Palmer and S. Moncada, Biochem. Biophys. Res. Commun., 158, 348 (1989).
26. K. Ishii, H. Sheng, T. D. Warner, U. Forstermann and F. Murad, Am. J. Physiol., 261, H598 (1991).
27. J.F. Brien, B.E. McLaughlin, K. Nakatsu and G.S. Marks, J. Pharmacol. Methods, 25, 19 (1991).
28. J.R.J. Lancaster and J.B.J. Hibbs, Proc. Natl. Acad. Sci. USA, 87, 1223 (1990).
29. M. Ogata, K. Ishii, N. Ioku and T. Meguro, Physiol. Chem. Phys. Med. NMR, 22, 125 (1990).
30. R. Busse and A. Mulsch, FEBS Lett., 265, 133 (1990).
31. U. Forstermann, J.S. Pollock, H.H. Schmidt, M. Heller and F. Murad, Proc. Natl. Acad. Sci. USA, 88, 1788 (1991).
32. B. Mayer, M. John and E. Bohme, J. Cardiovasc. Pharmacol., 17, S46 (1991).
33. J. Giovanelli, K.L. Campos and S. Kaufman, Proc. Natl. Acad. Sci. USA, 88, 7091 (1991).
34. B. Mayer, M. John, B. Heinzel, E.R. Werner, G. Wachter, H. Schultz, and E. Bohme, FEBS Lett., 288, 187 (1991).
35. D.S. Bredt and S.H. Snyder, Proc. Natl. Acad. Sci. USA, 87, 682 (1990).
36. U. Forstermann, H.H.H.W. Schmidt, J.S. Pollock, H. Sheng, J.A. Mitchell, T.D. Warner, M. Nakane and F. Murad, Biochem. Pharmacol., 42, 1849 (1991).
37. H.H. Schmidt, J.S. Pollock, M. Nakane, L.D. Gorsky, U. Forstermann and F. Murad, Proc. Natl. Acad. Sci. USA, 88, 365 (1991).
38. D.J. Stuehr and M.A. Marletta, Proc. Natl. Acad. Sci. USA, 82, 7738 (1985).
39. T.R. Billiar, R.D. Curran, B.G. Harbrecht, D.J. Stuehr, A.J. Demetris and R.L. Simmons, J. Leukoc. Biol., 48, 565 (1990).
40. Y. Yui, R. Hattori, K. Kosuga, H. Eizawa, K. Hiki and C. Kawai, J. Biol. Chem., 266, 12544 (1991).
41. D.J. Stuehr, H.J. Cho, N.S. Kwon, M.F. Weise and C.F. Nathan, Proc. Natl. Acad. Sci. USA, 88, 7773 (1991).
42. J.M. Hevel, K.A. White and M.A. Marletta, J. Biol. Chem., 266, 22789 (1991).
43. J.A. Mitchell, U. Forstermann, T.D. Warner, J.S. Pollock, H.H. Schmidt, M. Heller and F. Murad, Biochem. Biophys. Res. Commun., 176, 1417 (1991).
44. J.S. Pollock, U. Forstermann, J.A. Mitchell, T.D. Warner, H.H.H.W. Schmidt, M. Nakane and F. Murad, Proc. Natl. Acad. Sci. USA, 88, 10480 (1991).
45. G. Thomas and P.W. Ramwell, Biochem. Biophys. Res. Commun., 179, 1677 (1991).
46. P.K. Moore, A.O. Oluyomi, R.C. Babbedge, P. Wallace and S.L. Hart, Br. J. Pharmacol., 102, 198 (1991).
47. H.M. Vargas, J.M. Cuevas, L.J. Ignarro and G. Chaudhuri, J. Pharmacol. Exp. Ther., 257, 1208 (1991).
48. J.M. Fukuto, K.S. Wood, R.E. Byrns and L.J. Ignarro, Biochem. Biophys. Res. Commun., 168, 458 (1990).
49. M.A. Marletta, Trends Biochem. Sci., 14, 488 (1989).
50. N.S. Kwon, C.F. Nathan, C. Gilker, O.W. Griffith, D.E. Matthews and D.J. Stuehr, J. Biol. Chem., 265, 13442 (1990).
51. A.M. Leone, R.M.J. Palmer, R.G. Knowles, P.L. Francis, D.S. Ashton and S. Moncada, J. Biol. Chem., 266, 23790 (1991).
52. D.J. Stuehr, N.S. Kwon, C.F. Nathan, O.W. Griffith, P.L. Feldman and J. Wiseman, J. Biol. Chem., 266, 6259 (1991).
53. G.C. Wallace and J.M. Fukuto, J. Med. Chem., 34, 1746 (1991).
54. M.E. Gold, K.S. Wood, R.E. Byrns, G.M. Buga and L.J. Ignarro, Am. J. Physiol., 259, H1813 (1990).
55. M. Hecker, D.T. Walsh, and J.R. Vane, FEBS Lett., 294, 221 (1991).
56. D.S. Bredt, P.M. Hwang, C.E. Glatt, C. Lowenstein, R.R. Reed, S.H. Snyder, Nature, 351, 714 (1991).
57. L.E. Lambert, J.P. Whitten, B.M. Baron, H.C. Cheng, N.S. Doherty and I. A. McDonald, Life Sci., 48, 69 (1991).
58. D.D. Rees, R.M. Palmer, R. Schulz, H.F. Hodson and S. Moncada, Br. J. Pharmacol., 101, 746 (1990).
59. D.S. Bredt, P.M. Hwang and S.H. Snyder, Nature, 347, 768 (1990).
60. J.S. Pollock, M. Nakane, U. Forstermann and F. Murad, FASEB J., 6, 123A (1992).
61. T.M. Dawson, D.S. Bredt, M. Fotuhi, P.M. Hwang and S.H. Snyder, Proc. Natl. Acad. Sci. USA, 88, 7797 (1991).
62. D.S. Bredt, C.E. Glatt, P.M. Hwang, M. Fotuhi, T.M. Dawson and S.H. Snyder, Neuron, 7, 615 (1991).
63. S.R. Vincent and H. Kimura, Neuroscience, 46, 755 (1992).
64. M. Nakane, J. Mitchell, U. Forstermann and F. Murad, Biochem. Biophys. Res. Commun., 180, 1396 (1991).
65. D.S. Bredt and S.H. Snyder, Neuron, 8, 3 (1992).
66. B. Brune and E.G. Lapetina, Biochem. Biophys. Res. Commun., 181, 921 (1991).
67. F.M. Faraci, Am. J. Physiol., 261, H1038 (1991).
68. G.M. Buga, M.E. Gold, J.M. Fukuto and L.J. Ignarro, Hypertension, 17, 187 (1991).
69. R.F. Furchgott and P.M. Vanhoutte, FASEB J., 3, 2007 (1989).
70. T.M. Griffith and D.H. Edwards, J. Theor. Biol., 146, 545 (1990).

71. G.M. Rubanyi, A.D. Freay, K. Kauser, A. Johns and D.R. Harder, Blood Vessels, 27, 246 (1990).
72. M. Sun and D.J. Reis, Life Sci., 50, 555 (1992).
73. R.N. Channick, S.P. Bradley, F.W. Johnson, J.W. Newhart, R.G. Konopka and K.M. Moser, Clin. Res., 40, 38A (1992).
74. E. Weitzberg, Acta Physiol. Scand., 143, 451 (1991).
75. M.G. Persson, L.E. Gustafsson, N.P. Wiklund, S. Moncada and P. Hedqvist, Acta Physiol. Scand., 140, 449 (1990).
76. L.E. Gustafsson, A.M. Leone, M.G. Persson, N.P. Wiklund and S. Moncada, Biochem. Biophys. Res. Commun., 181, 852 (1991).
77. G. Cremona, A.T. Dinh Xuan and T.W. Higenbottom, Lung, 169, 185 (1991).
78. E. Bassenge, Z. Kardiol., 80, 17 (1991).
79. T.F. Luscher, Eur. Heart J., 10, 847 (1989).
80. J.P. Tolins, P.J. Shultz and L. Raij, Hypertension, 17, 909 (1991).
81. M.A. Creager, J.P. Cooke, M.E. Mendelsohn, S.J. Gallagher, S.M. Coleman, J. Loscalzo and V.J. Dzau, J. Clin. Invest., 86, 228 (1990).
82. C. Bossaller and E. Fleck, Z. Kardiol., 78, 59 (1989).
83. C.J.M. Vrints, H.B. Bult, E. Hitter, A.G. Herman and J.P. Snoeck, J. Am. Coll. Cardiol., 19, 21 (1992).
84. J.T. Shepherd, Z.S. Katusic, Y. Vedernikov and P.M. Vanhoutte, J. Mol. Cell. Cardiol., 23, 125 (1991).
85. Y. Dohi and T.F. Luscher, Br. J. Pharmacol., 100, 889 (1990).
86. J.A. Panza, N. Engl. J. Med., 323, 22 (1990).
87. T.F. Luscher, Eur. Heart J., 10, 847 (1989).
88. J. Van de Voorde, Int. J. Artif. Organs, 13, 647 (1990).
89. A. Mugge and D.G. Harrison, Blood Vessels, 28, 354 (1991).
90. J.P. Cooke, N.A. Andon, X.J. Girerd, A.T. Hirsch and M.A. Creager, Circulation, 83, 1057 (1991).
91. H. Drexler, A.M. Zeiher, K. Meinzer and H. Just, Lancet, 338, 1546 (1991).
92. A. Mugge, J.H. Elwell, T.E. Peterson and D.G. Harrison, Am. J. Physiol., 260, C219 (1991).
93. Z.S. Katusic and P.M. Vanhoutte, Semin. Perinatol., 15, 30 (1991).
94. K. Kugiyama, S.A. Kerns, J.D. Morrisett, R. Roberts and P.D. Henry, Nature, 344, 160 (1990).
95. J. Galle, A. Mulsch, R. Busse and E. Bassenge, Arterioscler. Thromb., 11, 198 (1991).
96. J.S. Beckman, T.W. Beckman, J. Chen, P.A. Marshall and B.A. Freeman, Proc. Natl. Acad. Sci. USA, 87, 1620 (1990).
97. R. Radi, J.S. Beckman, K.M. Bush and B.A. Freeman, Arch. Biochem. Biophys., 288, 481 (1991).
98. F.A. Nicolini and J.L. Mehta, Biochem. Pharmacol., 40, 2265 (1990).
99. J.L. Mehta, D.L. Lawson, F.A. Nicolini, M.H. Ross and D.W. Player, Am. J. Physiol., 261, H327 (1991).
100. H. Aoki, S. Kobayashi, J. Nishimura, H. Yamamoto and H. Kanaide, Biochem. Biophys. Res. Commun., 181, 1352 (1991).
101. Y. Yamashita, T. Yukimura, K. Miura, M. Okumura, S. Yamanaka and K. Yamamoto, J. Cardiovasc. Pharmacol., 17, S332 (1991).
102. Y. Hirata, H. Matsuoka, K. Kimura, T. Sugimoto, H. Hayakawa, E. Suzuki and T. Sugimoto, J. Cardiovasc. Pharmacol., 17, S169 (1991).
103. V.B. Schini, N.D. Kim and P.M. Vanhoutte, J. Cardiovasc. Pharmacol., 17, S267 (1991).
104. M.A. Marletta, M.A. Tayeh, and J.M. Hevel, Biofactors, 2, 219 (1990).
105. C.F. Nathan and J.B.J. Hibbs, Curr. Opin. Immunol., 3, 65 (1991).
106. F.Y. Liew, Y. Li and S. Millott, J. Immunol., 145, 4306 (1990).
107. L.B. Adams, J.B. Hibbs, Jr., R.R. Taintor and J.L. Krahenbuhl, J. Immunol., 144, 2725 (1990).
108. J.A Alspaugh. and D.L. Granger, Infect. Immun., 59, 2291 (1991).
109. I.E. Flesch and S.H. Kaufmann, Infect. Immun., 59, 3213 (1991).
110. S.J. Green, R.M. Crawford, J.T. Hockmeyer, M.S. Meltzer and C.A. Nacy, J. Immunol., 145, 4290 (1990).
111. F.Y. Liew, S.Millott, C. Parkinson, R.M. Palmer and S. Moncada, J. Immunol., 144, 4794 (1990).
112. S. Mellouk, S.J. Green, C.A. Nacy and S.L. Hoffman, J. Immunol., 146, 3971 (1991).
113. P. Vincendeau and S. Daulouede, J. Immunol., 146, 4338 (1991).
114. L.B. Adams, S.G. Franzblau, Z. Vavrin, J.B. Hibbs, Jr. and J.L. Krahenbuhl, J. Immunol., 147, 1642 (1991).
115. L.L. Thomsen, L.M. Ching, L. Zhuang, J.B. Gavin and B.C. Baguley, Cancer Res., 51, 77 (1991).
116. D.W. Reif and R.D. Simmons, Arch. Biochem. Biophys., 283, 537 (1990).
117. R. Dijkmans and A. Billiau, Eur. J. Biochem., 202, 151 (1991).
118. D.A. Wink, K.S. Kasprazak, C.M. Maragos, R.K. Elespuru, M. Misra, T.M. Dunams, T.A. Cebula, W.H. Koch, A.W. Andrews, J.S. Allen and L.K. Keefer, Science, 254, 1001 (1991).
119. M. Lepoivre, B. Chenais, A. Yapo, G. Lemaire, L. Thelander and J.P. Tenu, J. Biol. Chem., 265, 14143 (1990).
120. T. Nakaki, M. Nakayama and R. Kato, Eur. J. Pharmacol., 189, 347 (1990).
121. R.A. Hoffman, J.M. Langrehr, T.R. Billiar, R.D. Curran and R.L. Simmons, J. Immunol., 145, 2220 (1990).
122. J.M. Langrehr, A.R. Muller, P.M. Markus, R.L. Simmons and R.A. Hoffman, Transplant. Proc., 23, 3260

(1991).
123. B.J. Nelson, P. Ralph, S.J. Green and C.A. Nacy, J. Immunol., 146, 1849 (1991).
124. J. Pfeilschifter and K. Vosbeck, Biochem. Biophys. Res. Commun., 175, 372 (1991).
125. M.W. Radomski, R.M. Palmer and S. Moncada, Proc. Natl. Acad. Sci. USA, 87, 10043 (1990).
126. E.R. Werner, G. Werner-Felmayer, D. Fuchs, A. Hausen, R. Reibnegger, J.J. Yim and H. Wachter, Pathobiology, 59, 276 (1991).
127. D.J. Fast, B.J. Shannon, M.J. Herriott, M.J. Kennedy, J.A. Rummage and R.W. Leu, Infect. Immun., 59, 2987 (1991).
128. K.D. Kroncke, V. Kolb-Bachofen, B. Berschick, V. Burkart and H. Kolb, Biochem. Biophys. Res. Commun., 175, 752 (1991).
129. M.L. Lukic, S. Stosic-Grujicic, N. Ostojic, W.L. Chan and F.Y. Liew, Biochem. Biophys. Res. Commun., 178, 913 (1991).
130. H.H.H.W. Schmidt, T.D. Warner, K. Ishii, H. Sheng and F. Murad, Science, 255, 721 (1992).
131. N. Welsh, D.L. Eizirik, K. Bendtzen and S. Sasndler, Endocrinol., 129, 3167 (1991).
132. R.D. Curran, T.R. Billiar, D.J. Stuehr, J.B. Ochoa, B.G. Harbrecht, S.G. Flint and R.L. Simmons, Ann. Surg., 212, 462 (1990).
133. G. Werner-Felmayer, E.R. Werner, D. Fuchs, A. Hausen, G. Reibnegger and H. Wachter, J. Exp. Med., 172, 1599 (1990).
134. Q.Z. Wang, J. Jacobs, J. DeLeo, H. Kruszyna, R. Kruszyna, R. Smith and D. Wilcox, Life Sci., 49, 60 (1991).
135. R.G. Kilbourn, A. Jubran, S.S. Gross, O.W. Griffith, R. Levi, J. Adams and R.F. Lodato, Biochem. Biophys. Res. Commun., 172, 1132 (1990).
136. C.A. Ross, D. Bredt and S.H. Snyder, Trends Neurosci., 13, 216 (1990).
137. Y. Ishizaki, L.J. Ma, I. Morita and S. Murota, Neurosci. Lett., 125, 29 (1991).
138. S.J. East and J. Garthwaite, Neurosci. Lett., 123, 17 (1991).
139. L. Kiedrowski, E. Costa and J.T. Wroblewski, Neurosci. Lett., 135, 59 (1992).
140. V.L. Dawson, T.M. Dawson, E.D. London, D.S. Bredt and S.H. Snyder, Proc. Natl. Acad. Sci. USA, 88, 6368 (1991).
141. C. Demerle-Pallardy, M.-O. Lonchampt, P.-E. Chabrier and P. Braquet, Biochem. Biophys. Res. Commun., 181, 456 (1991).
142. J.P. Nowicki, D. Duval, H. Poignet and B. Scatton, Eur. J. Pharmacol., 204, 339 (1991).
143. V. Mollace, G. Bagetta and G. Nistico, NeuroReport, 2, 269 (1991).
144. A. Rengasamy and R.A. Johns, J. Pharmacol. Exp. Ther., 259, 310 (1991).
145. J.S. Beckman, J. Develop. Physiol., 15, 53 (1991).
146. T.J. O'Dell, R. D. Hawkins, E. R. Kandel and O. Arancio, Proc. Natl. Acad. Sci. USA, 88, 11285 (1991).
147. E.M. Schuman and D.V. Madison, Science, 254, 1503 (1991).
148. N.J. Sucher and S.A. Lipton, J. Neurosci. Res., 30, 582 (1991).
149. S.J. East, A.M. Batchelor and J. Garthwaite, Eur. J. Pharmacol., 209, 119 (1991).
150. L. Kiedrowski, H. Manev, E. Costa and J.T. Wroblewski, Neuropharmacol., 30, 1241 (1991).
151. K. Shibuki and D. Okada, Nature, 349, 326 (1991).
152. S. Murphy, R.L. Minor, Jr., G. Welk and D.G. Harrison, J. Neurochem., 55, 349 (1990).
153. S.J. Lewis, H. Ohta, B. Machado, J.N. Bates and W.T. Talman, Eur. J. Pharmacol., 202, 135 (1991).
154. E.D. Di Paola, M.J. Vidal and G. Nistico, J. Cardiovasc. Pharmacol., 17, S269 (1991).
155. L.N. Shapoval, V.F. Sagach, and L.S. Pobegailo, Neurosci. Lett., 132, 47 (1991).
156. P.J. Lacolley, S.J. Lewis and M.J. Brody, Hypertension, 17, 881 (1991).
157. J.A. Gally, P. R. Montague, G.N. Reeke, Jr. and G.M. Edelman, Proc. Natl. Acad. Sci. USA, 87, 3547 (1990).
158. J.M. Pique, J.V. Esplugues and B.J. Whittle, Gastroenterol., 102, 168 (1992).
159. B.L. Tepperman and B.J. Whittle, Br. J. Pharmacol., 105, 171 (1992).
160. K.M. Desai, W.C. Sessa and J.R. Vane, Nature, 351, 477 (1991).
161. J.A. Mitchell, H. Sheng, U. Forstermann and F. Murad, Br. J. Pharmacol., 104, 289 (1991).
162. C. Du, J. Murray, J.N. Bates and J.L. Conklin, Am. J. Physiol., 261, G1012 (1991).
163. K.M. Azadzoi, N. Kim, M.L. Brown, I. Goldstein, R.A. Cohen and I. Saenz de Tejada, J. Urology, 147, 220 (1992).
164. J. Rajfer, W.J. Aronson, P.A. Bush, F.J. Dorey and L.J. Ignarro, N. Engl. J. Med., 326, 90 (1992).
165. R.E. Klabunde and R.C. Ritger, Biochem. Biophys. Res. Commun., 178, 1135 (1991).
166. C.E. Wright, D.D. Rees and S. Moncada, Cardiovasc. Res., 26, 48 (1992).
167. E. Nava, R.M.J. Palmer and S. Moncada, Lancet, 338, 1555 (1991).
168. A. Petros, D. Bennett and P. Vallance, Lancet, 338, 1557 (1991).
169. S. Amir and A.M. English, Eur. J. Pharmacol., 203, 125 (1991).
170. W. Lieberthal, A. E. McGarry, J. Shiels and C.R. Valeri, Am. J. Physiol., 261, F868 (1991).
171. L. Volicier and P.B. Crino, Neurobiol Aging, 11, 567 (1990).
172. D.W. Choi, Neuron, 1, 623 (1991).
173. E.J. Mylecharane, J. Neurol., 238, S45 (1991).
174. O. Appenzeller, Med. Clin. North Am., 75, 763 (1991).

Chapter 9. Endogenous Vasoactive Peptides

Annette M. Doherty
Parke-Davis Pharmaceutical Research Division
Warner-Lambert Company, Ann Arbor, Michigan 48105.

<u>Introduction</u> - Maintenance of circulatory homeostasis depends on the combined effects of vasoconstrictor and vasodilator mechanisms (1-3). Vasoactive peptides may be classified into three general groups: those that are produced by endocrine glands and act as circulating hormones; those produced in the central and peripheral nervous systems and termed neuropeptides; and those peptides produced by enterochromaffin cells, modulating intestinal function. Vasoactive peptides modulate vascular tone either by the direct activation of vascular smooth muscle or by the release of other endogenous factors. The discovery of compounds designed to modulate the effects of endogenous vasoactive peptides will clarify the role(s) and interactions between vasodilator and vasoconstrictor mechanisms.

ENDOTHELIN

Pharmacological studies of the vasoconstrictor family of endothelin (ET) peptides have been reviewed recently (4-8). The role of ET in cardiovascular and other diseases awaits the study of specific receptor antagonists and processing inhibitors in relevant disease models (9).

<u>Endothelin Processing</u> - Human and porcine ET (termed ET-1) are derived from a 203 amino acid precursor known as preproendothelin, which is cleaved by an unknown dibasic endopeptidase to produce a 38 (human) or 39 (porcine) amino acid peptide, known as proendothelin or big ET-1. An unknown endothelin converting enzyme (ECE) cleaves big ET-1 to the biologically active, 21-amino-acid peptide, ET-1. Most reports indicate ECE to be a membrane-bound metalloproteinase, although recently a thiol protease, isolated from a soluble extract of porcine aortic endothelial cells has been suggested as a potential ECE (10-14). ECE activity has been detected in both the membranous and cytosolic fractions of cultured bovine and porcine endothelial cells (10-12). Phosphoramidon $\underline{1}$, a non-specific metalloproteinase inhibitor, antagonized the intracellular conversion of big ET-1 to ET-1 in cultured endothelial cells at micromolar concentrations, while the aspartic proteinase inhibitor pepstatin did not affect endothelin processing (13,15). ECE isolated from renal epithelial cells showed similar biochemical properties to ECE derived from endothelial cells with an IC_{50} of 4.5 µM for phosphoramidon inhibition (16). Phosphoramidon caused an increase in the concentration of both big ET-1 and ET-1 in the culture medium of a permanent vascular endothelial cell line, known to express ET-1 mRNA and to secrete both peptides (17). Phosphoramidon inhibited the vasoconstrictor effects evoked by big ET-1 after intravenous administration to pigs, although arterial plasma levels of ET-1 were not affected (18). In contrast, intracisternally administered big ET-1 to cerebral arteries of dogs caused a marked increase in immunoreactive ET levels in the cerebrospinal fluid that was inhibited by pretreatment with $\underline{1}$ (19). Phosphoramidon potently inhibited delayed cerebral vasospasm after subarachnoid hemorrhage in the two-hemorrhage canine model. ET-1 concentrations in the cerebrospinal fluid (CSF) of the phosphoramidon-treated group did not appear to correlate directly with clinical improvements (20). Infusion of big ET-1 to anesthetized euvolemic rats produced a significant increase in mean arterial blood pressure and a decrease in effective renal plasma flow. Simultaneous infusion of $\underline{1}$ completely abolished these effects while the neutral endopeptidase inhibitor (NEP), thiorphan, had no effect (21). Studies with substrate analogs of big ET-1 have shown the importance of residues 32-37 for recognition by the phosphoramidon-sensitive ECE (22). There have been no reports of specific ECE inhibitors.

Structural Studies - Structural studies of ET-1 by NMR, molecular dynamics / energy minimization, fluorescence and circular dichroism (CD) have been reported (23-25). CD studies show that ET-1 is about 35% helical, the helicity existing between Lys^9 and Cys^{15} (23,25). There have been conflicting reports of the conformation of the biologically important C-terminal hexapeptide, but most studies are unable to define this flexible region. A NMR study of big ET-1 indicated that [Leu^{17}-Ser^{38}], reported to be important for cleavage by ECE, did not adopt a defined conformation (22,26).

ET Receptors - Distribution of two cloned receptor subtypes, ET_A and ET_B, has been studied extensively (27,28). The ET_A, or vascular smooth muscle receptor, is widely distributed in cardiovascular tissues and certain regions of the brain (27,29). The ET_B receptor, originally cloned from rat lung, has been found in rat cerebellum and endothelial cells, although it is not known if the ET_B receptors are the same from these sources. The human ET receptor subtypes have been cloned and expressed (30,31). The ET_A receptor clearly mediates vasoconstriction, and there have been a few reports implicating ET_B receptors in the initial vasodilatory response to ET (32). However recent data have shown that the ET_B receptor mediates vasoconstriction in some tissues (33,34).

Structure-Activity Relationships - Structure-activity studies of the ET's and closely related sarafotoxin (SRTX) peptides have been reviewed (7,8). ET-3 is the weakest vasoconstrictor in the ET family, acting as a vasodilator in some vascular beds (8). The Ser^2 to Thr^2 substitution in ET-3 is shared by two weak constrictor peptides, SRTX-c and SRTX-d, suggesting its possible importance for receptor affinity (35). However structure-activity studies of substituted SRTX's have indicated that the Lys^9 to Glu^9 substitution causes a much larger loss of biological activity than either the Ser^2 to Thr^2, Lys^4 to Asn^4, or Tyr^{13} to Asn^{13} substitutions (35). Further reports describing receptor binding and functional activities for C-terminal hexapeptides and ET monocyclic derivatives have appeared (36,37). D-amino acid substitutions at His^{16} and Ile^{20} increased binding affinity in rabbit aorta, pulmonary artery and rat heart compared with the corresponding L-amino-acid analogues (36).

ET Receptor Agonists - As shown in Table 1, the receptor affinities of the ET's and SRTX's in rat aorta and atria (ET_A) or cerebellum and hippocampus (ET_B), indicate that SRTX-c is a selective ET_B ligand (38). SRTX-c causes only vasodilation in rat aortic rings, possibly through the release of endothelium-derived relaxing factor (32). As shown in Table 2, the structural requirements for binding to the ET_B receptor do not require the presence of the disulfides (39,40). Selective ET_B agonists, compounds **2**, **3** and **4** caused vasorelaxation in isolated, endothelium-intact porcine pulmonary arteries (40). However, some of these analogs, **2** and **3**, are potent vasoconstrictors in the rabbit pulmonary artery, a tissue that possesses an ET_B-like receptor (33).

Table 1. Receptor Subtype Binding of ET Receptor Agonists in a Variety of Rat Tissues.

Ligand	Aorta	Atrium	Cerebellum	Hippocampus
		K_i , nM		
ET-1	0.11	0.034	0.015	0.010
ET-3	2.50	1.60	0.021	0.036
SRTX-6b	0.24	0.087	0.013	0.012
SRTX-6c	>5000	4200	0.016	0.023

Table 2. Receptor Subtype Binding of ET Receptor Agonists in a Variety of Porcine Tissues.

Agonist		Aorta	Cerebellum
		IC_{50} , nM	
2	ET[$Ala^{1,3,11,15}$]	570	0.33
3	ET[8-21, $Ala^{11,15}$]	1400	1.2
4	N-acetyl-ET[10-21]	9900	12.0

ET Receptor Antagonists - Initial reports of various ET receptor antagonists have appeared. An ET analog, **5**, has been reported as an ET antagonist of unknown specificity (41). A non-peptide series of ET antagonists of unknown specificity, discovered by random screening, has been disclosed (42). The most potent compound, **6** (FR901367) with a binding affinity of $IC_{50} = 0.67$ μM in porcine aorta (ET_A), inhibited the contractile response of ET-1 in rabbit thoracic aorta at high concentrations (42). The bombesin antagonist [D-Arg[1], D-Phe, D-Trp[7,9], Leu[11]]substance P (SP), previously reported to inhibit [^{125}I]ET-1 to rat cardiac membranes, is not a competitive antagonist, due to the large increase in non-specific binding observed at higher concentrations (43).

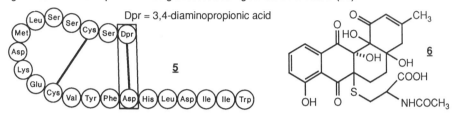

Dpr = 3,4-diaminopropionic acid

Structure-activity studies of a cyclic pentapeptide ET_A receptor antagonist, **7**, (BE-18257B), discovered by random screening of *Streptomyces misakiensis*, have led to more potent analogs, **8** and **9** [BQ-123 and BQ-153, respectively] (44-48). Compound **10**, BQ-162, is considerably less potent (Table 3). Compounds **8** and **9** antagonize ET-1 induced contraction in isolated porcine coronary arteries (49). The first rationally designed non-selective ET_A/ET_B antagonist **11**, (PD 142893), has been reported (50). Compound **11** antagonizes ET-induced contraction in rabbit renal, femoral and pulmonary arteries. Tripeptidic compounds **12-17** are ET_A antagonists of unknown specificity (51). Compounds **12** and **13** inhibit ET-1 induced contraction in isolated rabbit aorta and thoracic arteries. *N*-pentanoyltryptophan has been claimed as a functional ET antagonist in rat pulmonary artery (52).

Table 3. ET Antagonists

Compound			ET_A[a]	ET_B[b]
				IC_{50} ,nM
7	cyclo[*D*-Trp-*D*-Glu-Ala-allo-*D*-Ile-Leu]		1400	>10[5]
8	cyclo[*D*-Trp-*D*-X-Pro-*D*-Val-Leu]	X=Asp	7.3	18,000
9		X= Glu	8.6	54,000
10		X=sulfoalanine	230	>10[5]
11	Ac-*D*-HNC(CHPh$_2$)CO-Leu-Asp-Ile-Ile-Trp		15.0[c]	150
12	Cycloheptyl-NCO-Leu-*D*-Trp(Me)-*D*-Pya-OH (FR 139317)[d]		2.5[e]	4600[f]
13	Cyclohexyl-HNCO-Leu-*D*-Trp(Me)-*D*-Pya-OH.HCl		2.3	g
14	Cycloheptyl-NCO-Leu-*D*-Trp(Me)-*D*-Pya-Sar-ONa		7.6	g
15	Cycloheptyl-NCO-Leu-*D*-Trp(Me)-*D*-Pya-Phe-ONa		21.0	g
16	Cycloheptyl-NCOCH$_2$CH(*S*)(CH$_2$CH(Me)$_2$)CO-*D*-Trp(Me)-*D*-Pya-NHMe		7.6	g
17	[(2-Chlorophenyl)acetyl]Leu-*D*-Trp(Me)-*D*-Leu-OH		32.0	g

[a]Binding affinity in porcine aortic membranes or vascular smooth muscle (VSM) cells
[b] rat cerebellum   [c] rabbit renal artery VSM cells   [d]*D*-Pya = *D*-(2-pyridyl)alanine
[e] human aorta   [f] porcine brain   [g] not reported

There are major tissue and species differences in the biological activities of ET analogs and fragments, and it will be important to define localization of receptor subtypes in human tissues in order to develop specific non-peptidic receptor antagonists. The physiological roles of the two known receptor subtypes are not clear at present and whether it will eventually prove beneficial to block both ET_A and ET_B receptors, in certain pathological situations, is not known.

TACHYKININS

Substance P - Substance P (SP) is one of the endogenous tachykinin peptides implicated in pain and neurogenic inflammation, and may be critically involved in asthma, rheumatoid arthritis, cognitive disorders and inflammatory bowel disease (53-57). It has a wide spectrum of biological actions including smooth muscle contraction, powerful vasodilation in certain tissue beds and potent secretory activity (58). Microinjections of SP into the anterior and ventromedial hypothalmus produce a marked increase in mean arterial pressure and heart rate (59). The release of SP has been reported to increase in response to shear stress, cytokines, and AII (60-62). Classification of the tachykinin receptor types (NK1, 2 and 3) and signalling mechanisms has been reviewed (57,63).

Peptide Receptor Antagonists - Structure-activity studies of a tripeptide SP (NK1) antagonist, FR106506 [Boc-Gln-D-Trp(CHO)-Phe-OBzl], have been reported. The most potent analog from this series, FR113680 [Ac-Thr-D-Trp(CHO)-Phe-NHCH$_2$Ph], is a competitive antagonist with an IC$_{50}$ of 5.8 nM in the guinea pig trachea (64,65). This compound inhibited the contraction of guinea-pig tracheal strips by SP and suppressed SP-induced airway constriction and edema (64). Further pharmacological studies of the SP antagonists, spantide I and II have been described (66-68). Intrathecal (i.t.) infusion of the SP antagonist, (D-Pro^2,D-Trp^{7,9})SP, lowered blood pressure (BP) profoundly with no significant change in heart rate (69). [Pro^9]-SP was reported as a selective NK1 ligand (57,70). The peptide antibiotic, dactinomycin, structurally similar to the known NK2 receptor antagonist L-659877 (cyclo[Leu-Met-Gln-Trp-Phe-Gly]), is an NK2 receptor antagonist (71).

Non-Peptide Receptor Antagonists - The selective and competitive NK1 receptor antagonist, $\underline{18}$ (RP 67,580), was active in two classical analgesic tests in mice (72). Compound $\underline{18}$ competitively inhibited contractions induced by SP and septide in guinea pig ileum (72). Intravenous administration (i.v.) of $\underline{18}$ to rats totally inhibited plasma extravasation induced by SP in the urinary bladder and the antidromic electrical stimulation of the sphenous nerve in the hind paw skin (72).

Further synthetic and pharmacological studies of potent, selective nonpeptide antagonists of the SP receptor, $\underline{19}$ (CP-96,345), and its chloro derivative $\underline{20}$ have been reported (73-80). Administration of racemic $\underline{19}$, via intratracheal (i.t.)or intraperitoneal (i.p.) routes, selectively blocked the adverse behaviour induced by i.t. SP-administration (78). In addition, the compound produced thermal analgesia in the hot plate test (78). Compound $\underline{19}$ has also been shown to dose-dependently block SP-induced hypotension in anaesthetized dogs (79). Compound $\underline{21}$ has recently been reported as a competitive antagonist of SP, with an IC$_{50}$ of 50 nM (81). For furthur reading on this subject, see chapter 12 of this volume.

ZINC-METALLOPROTEINASE INHIBITORS

NEP/ACE Inhibition - The zinc-containing peptidases, neutral endopeptidase EC 3.4.24.11 (NEP) and angiotensin converting enzyme (ACE), are involved in the regulation of two important hormonal systems, the atrial natriuretic peptide (ANP) and renin-angiotensin systems, respectively (82,83). Compounds inhibiting NEP and ACE simultaneously include glyoprilat (**22**) and alatrioprilat (**23**) (and their corresponding ester prodrugs, **24** and **25**), SQ 28,133 (**26**) and MDL 100,173 (**27**). In rodent models these compounds exert typical actions of ACE and NEP inhibitors by prevention of AII induced hypertension and ANP degradation (84-86). Whether these compounds will provide advantages over the single agents or combination of both agents is not known.

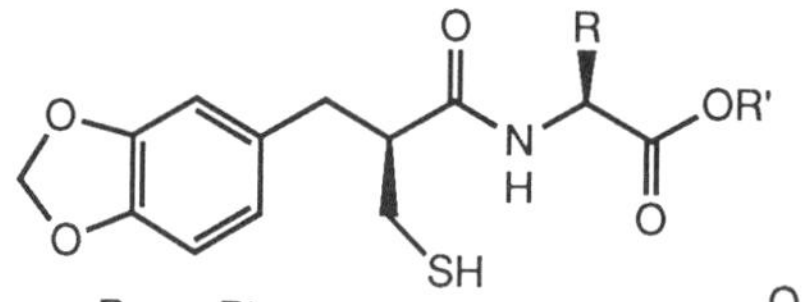

	R	R'
22	H	H
23	CH₃	H
24	H	CH₂Ph
25	CH₃	CH₂Ph

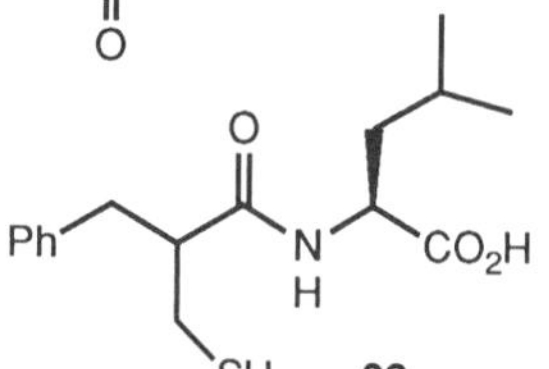

26

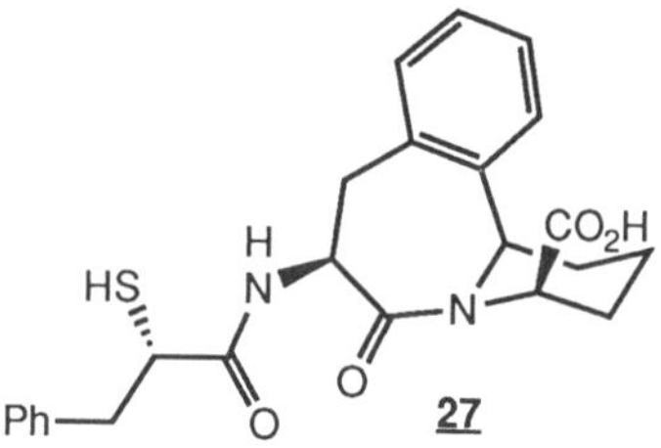

27

NEP Inhibition - Further *in vivo* studies with various NEP inhibitors have appeared. Compound **28**, SCH 34826, had no effect on BP or heart rate in 8 healthy volunteers maintained on a high sodium intake for 5 days (87). However, **28** promoted significant increases in excretion of urinary sodium, phosphate and calcium (87). Repeated administration of **28** to rats reduced dopamine receptor density, suggesting potential antidepressant activity (88). Treatment of **28** to SHR for one month caused a regression in cardiac hypertrophy (89). SCH 39370 (**29**), when administered to rats with heart failure, caused elevation of ANP and cGMP levels (90). Candoxatrilet, **30**, has been shown to reduce ANP clearance in intact and nephrectomized rats (91). Candoxatril, **31**, administered to human subjects caused a rise in basal ANF levels at 10-200 mg doses. Natriuresis was only observed at the highest dose and no change in BP was recorded after acute dosing (92). Sinorphan caused a transient rise in sodium urinary excretion in cirrhotic patients with ascites (93).

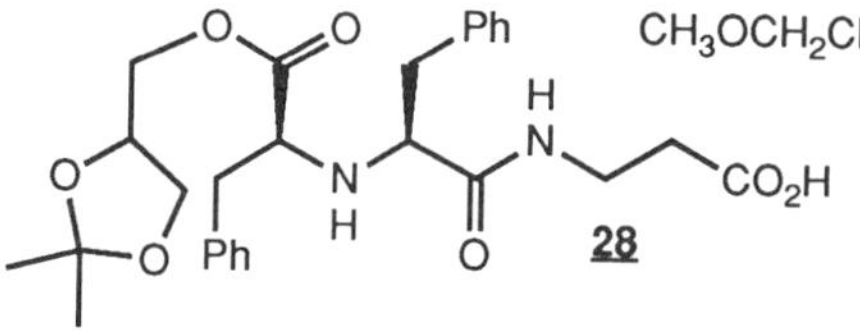

28

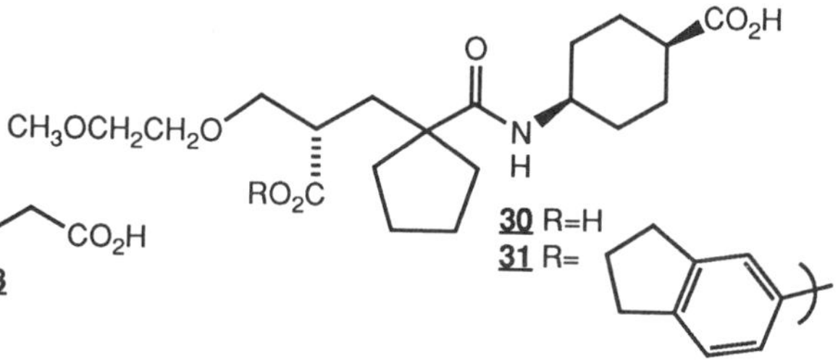

30 R=H
31 R=

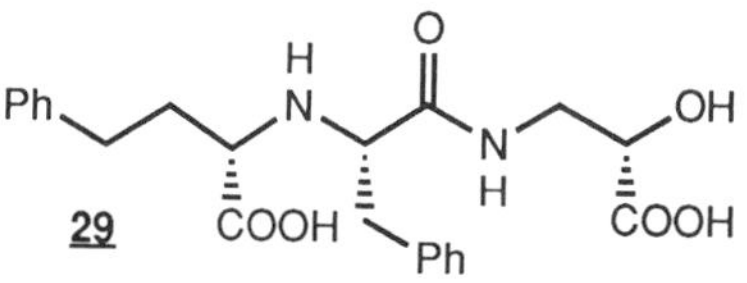

29

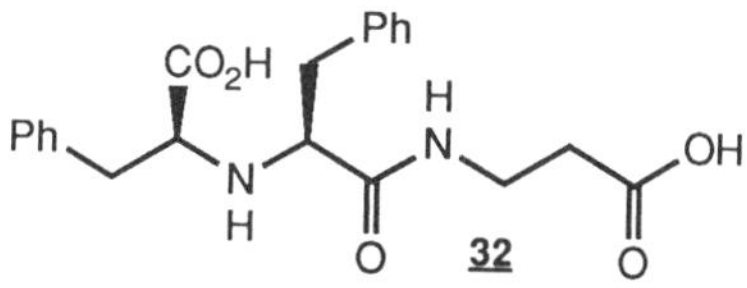

32

Inhibitors of SP degradation may be useful therapeutic agents. It is known that ACE and NEP cause SP degradation, in addition to an unidentified highly specific membrane-bound metalloendoproteinase (94). Plasma levels of SP and neuropeptide Y are increased in patients with moderate and severe congestive heart failure (94,95). CSF-SP levels were increased after i.v. administration of the NEP inhibitor **32**, SCH32615, to halothane-anesthetized and -ventilated cynomologus macaque monkeys (96). Analogue [pGlu^6, AHPA^8]SP6-11 (AHPA=(R,S)-α-hydroxy, (S)-β-amino-4-phenylbutyric acid) has been reported as a specific inhibitor of SP degradation (97).

EP-24.15 Inhibition - Metalloendopeptidase-24.15, EC 3.4.24.15 (EP-24.15), known also as thimet oligopeptidase, cleaves bradykinin (BK), AI and AII, neurotensin, luteinizing hormone releasing hormone and several enkephalin-containing peptides, and may be involved in BP control (98,99). Intravenous administration of the specific EP-24.15 inhibitor **33** caused a marked, but transient, fall in BP in phenobarbitol- anesthetized, normotensive rats (98,99). This effect was almost abolished by a kinin receptor antagonist, supporting a role for kinins in the regulation of normal BP. It is not clear whether the effects were due to specific EP-24.15 inhibition.

33

NEUROPEPTIDE Y

Neuropeptide Y (NPY) is a 36 amino acid peptide present in the brain, adrenal medulla and peripheral sympathetic nerves and may play an important role in BP control, atherosclerosis and myocardial ischemia (100-102). Y1 and Y3 receptors have recently been cloned and expressed (103-105). NPY[2-36] was 5-fold less potent than NPY in lowering body temperature, while this analogue was more potent than NPY in increasing food intake in animals (106). In the anesthetized rat, NPY, NPY[13-36] (Y2 receptor agonist), and a stabilized 13-36 analogue of NPY exerted pressor actions and attenuated vagal action at the heart (107,108). In contrast, [Pro^{34}]NPY (Y1 receptor agonist) caused increases in arterial BP but did not attenuate vagal action in rats or dogs (107,108).

The recently described NPY functional antagonist, D-myo-inositol-1.2.6-trisphosphate (PP56), completely antagonized NPY-induced potentiation of the noradrenaline concentration-response curve in rabbit femoral arteries (109,110). It has been shown to act as a locomoter stimulant upon i.c.v. administration. However, PP56 does not displace [^{125}I]NPY binding at either Y1 or Y2 receptors, and its mechanism of action is unknown (111). Peptidic fragment analogs of NPY, Ac-[3-(2,6-dichlorobenzyl)Tyr^{27}, D-Thr^{32}]NPY(27-36) amide and Ac-[3-(2,6-dichlorobenzyl)Tyr^{27,36} D-Thr^{32}]NPY(27-36) amide, bind weakly to the Y2 receptor (110,112). Affinity labelling of iodinated NPY to bovine hippocampal binding proteins showed that the α2-receptor antagonist, benextramine, irreversibly blocked specific NPY binding to a Y2 binding protein (113). A series of imidazoylalkyl guanidine derivatives, including the previously reported non-specific and weakly potent antagonist He-90481 have been disclosed (110,114). No other non-peptide NPY antagonists have been reported.

CALCITONIN-GENE RELATED PEPTIDE

The potent vasodilator calcitonin-gene related peptide (CGRP), acting directly or through the renin-angiotensin-aldosterone system, may play a role in regulating peripheral vascular tone and/or vascular volume. The biology and possible clinical relevance of the CGRP family has been reviewed extensively (115). Infusion of CGRP to conscious rats significantly reduced the pressor effects of norepinephrine or AII (116). N-truncated human CGRP fragments, hCGRP[8-37] and hCGRP[12-37], behave as competitive antagonists in certain bioassays. Functional and binding data demonstrate the importance of residues Thr^9 and His^{10} for potent antagonistic properties (117). hCGRP[8-37] antagonizes amylin inhibition of insulin-stimulated glucose uptake into isolated rat soleus muscle and inhibits amylin-evoked elevation of plasma lactate and glucose in fasted anaesthetized rats (118).

VASOPRESSIN

Attempts to develop a vasopressin antagonist for the treatment of various cardiovascular disorders have encountered problems of marked species differences and inconsistencies between *in vitro* and *in vivo* assay systems (119-121). Indeed some of the early competitive antagonists in animal models were found to be full agonists in man (121,122). The first orally active nonpeptide vasopressin V1 receptor antagonist, OPC-21268 (**34**), blocked the pressor response to arginine vasopressin (AVP), after oral administration to pithed and conscious rats (123). Molecular cloning and expression of the rat VIa AVP receptor has been reported (124).

OXYTOCIN

Oxytocin (OT) is a nonapeptide, primarily synthesized in the hypothalmus and closely related to AVP. It may be physiologically important for its milk-ejecting and uterine-contracting activities in mammals. In addition OT may play a role in renal development and function, and in cancer (125-127). The structure and expression of the human oxytocin receptor complementary DNA, isolated by expression cloning, has recently been described (128).

Several novel cyclic hexapeptides, L-366,682 [cyclo-(*L*-Pro-*D*-Trp-*L*-Ile-*D*-pipecolic acid-*L*-pipecolic acid-*D*-His)] and L-366,948 (*D*-2-naphthyl-alanine instead of *D*-Trp) have been characterized as potent, long-acting and selective OT antagonists with no agonist activity *in vitro* and *in vivo* (129,130). In conscious, freely moving, pregnant rhesus monkeys, L-366,948 or L-366,682 given i.v. or s.c. were effective antagonists of OT-induced uterine contractions (129,130). *N*-sulfonylindoline oxytocin antagonists such as compound **35** are claimed as vasopressor agents for the treatment of hypertension and ischemia (131). Spirocyclic compounds such as compounds **36** and **37** are claimed to be useful in the treatment of preterm labour and dysmenorrhea (132).

References

1.　　　D. Ganten, M. Paul, and R.E. Lang, Cardiovasc. Drugs Ther., 5, 119 (1991)
2.　　　J. Pernow, A. Hemsen, A.Hallen, and J.M. Lundberg, Acta Physiol. Scand., 140, 311 (1990).
3.　　　M.R. Moore and P. McL. Black, Neurosurg. Rev., 14, 97 (1991).
4.　　　M.D. Randall, Pharmac. Ther., 50, 73 (1991).
5.　　　W. G. Naylor in "The Endothelins" Springer-Verlag, Berlin, Heidelberg, New York,1990.
6.　　　V.B. Schini and P.M. Vanhoutte, Pharmacol. Toxicol., 69, 303 (1991).
7.　　　A.M. Doherty, Annu. Rep. Med. Chem., 26, 83 (1991).
8.　　　A.M. Doherty and R. E. Weishaar, Annu. Rep. Med. Chem., 25, 89 (1990).
9.　　　A.M. Doherty, J. Med. Chem., 35, 1493 (1992).
10.　　J. Takada, K. Okada, T. Ikenaga, K. Matsuyama, and M. Yano, Biochem. Biophys. Res. Commun., 176, 860 (1991).

11. Y. Matsumura, R. Ikegawa, Y. Tsukahara, M. Takaoka, and S. Morimoto, Biochem. Biophys. Res. Commun., $\underline{178}$, 899 (1991).

12. Y. Matsumura, R. Ikeyawa, Y. Tsukahara, M. Takaoka, and S. Morimoto, Biochem. Biophys. Res. Commun., $\underline{178}$, 531 (1991).

13. T. Sawamura, Y. Kasuya, Y. Matsushita, N. Suzuki, O. Shinmi, N. Kishi, Y. Sugita, M. Yanagisawa, K. Goto, T. Masaki, and S. Kimura, Biochem. Biophys. Res. Commun., $\underline{174}$, 779 (1991).

14. Y. Deng, P. Savage, S. S. Shetty, L.L. Martin, and A.Y. Jeng, J. Biochem., $\underline{111}$, 346 (1992).

15. P.P Shields, T.A. Gonzales, D. Charles, J.P. Gilligan, amd W. Stern, Biochem. Biophys. Res. Commun., $\underline{177}$, 1006 (1991).

16. J. Takada, M. Hata, K. Okada, K. Matsuyama, and M. Yano, Biochem. Biophys. Res. Commun., $\underline{182}$, 1383 (1992).

17. O. Saijonmaa, T. Nyman, U. Hohenthal, and F. Fyhrquist, Biochem. Biophys. Res. Commun., $\underline{181}$, 529 (1991).

18. A. Modin, J. Pernow, and J.M. Lundberg, Life Sci., $\underline{49}$, 1619 (1991).

19. H. Shinyama, T. Uchida, H. Kido, K. Hayashi, M. Watanabe, Y. Matsumura, R. Ikegawa, M. Takaoka, and S. Morimoto, Biochem. Biophys. Res. Commun., $\underline{178}$, 24 (1991).

20. Y. Matsumura, R. Ikegawa, Y. Suzuki, M. Takaoka, T. Uchida, H. Kido, H. Shinyama, K. Hayashi, M. Watanabe, and S. Morimoto, Life Sci., $\underline{49}$, 841 (1991).

21. D.M. Pollock, and T.J. Opgennorth, Am. J. Physiol., $\underline{261}$, R257 (1991).

22. K. Okada, J. Takada, Y. Arai, K. Matsuyama, and M. Yano, Biochem. Biophys. Res. Commun., $\underline{180}$, 1019 (1991).

23. M.D. Reily and J.B. Dunbar, Biochem. Biophys. Res. Commun., $\underline{178}$, 570 (1991).

24. J.T. Pelton, Neurochem. Int., $\underline{18}$, 485 (1991).

25. V. Saudek, J. Hoflack, and J.T. Pelton, Int. J. Pept. Protein Res., $\underline{37}$, 174 (1991).

26. H. Inooka, S. Endo, T. Kikuchi, M. Wakimasu, E. Mizuta, and M. Fujino, "Peptide Chem." 1990, Y. Shimonishi (Ed.), Protein Research Foundation, Osaka 1991.

27. H. Arai, S. Hori, I. Aramori, H. Ohkubo, and S. Nakanishi, Nature (Lond.), $\underline{348}$, 730 (1990).

28. T. Sakurai, M. Yanagisawa, Y. Takuwa, H. Miyazaki, S. Kimura, K. Goto, and T. Masaki, Nature (Lond.), $\underline{348}$, 732 (1990).

29. H.Y. Lin, E.H. Kaji, G.K. Winkel, H.E. Ives, and H.F. Lodish, Proc. Natl. Acad. Sci. USA, $\underline{88}$, 3185 (1991).

30. A. Sakamoto, M. Yanagisawa, T. Sakurai, Y. Takuwa, H. Yanagisawa, and T. Masaki, Biochem. Biophys. Res. Chem., $\underline{178}$, 656 (1991).

31. K. Hosoda, K. Nakao, H. Arai, S. Suga, Y. Ogawa, M. Mukoyama, G. Shirakami, Y. Saito, S. Nakanishi, and H. Imura, FEBS Lett., $\underline{287}$, 23 (1991).

32. R. Takayanagi, K. Kitazumi, C. Takasaki, K. Ohnaka, S. Aimoto, K. Tasaka, M. Ohashi, and H. Nawata, FEBS Lett., $\underline{282}$, 103 (1991).

33. R.L. Panek, T.C. Major, G.P. Hingorani, A.M. Doherty, D.G. Taylor, and S.T. Rapundalo, Biochem. Biophys. Res. Commun., $\underline{183}$, 566 (1992).

34. M. Yoshinaga, Y. Chijiiwa, T. Misawa, N. Harada, and H. Nawata, Am. J. Physiol., $\underline{262}$, G308 (1992).

35. C. Takasaki, S. Aimoto, K. Kitazumi, K. Tasaka, T. Shiba, K. Nishiki, Y. Furukawa, R. Takayanagi, K. Ohnaka, and H. Nawata, Eur. J. Pharmacol., $\underline{198}$, 165 (1991).

36. A.M. Doherty, W.L. Cody, N.L. Leitz, P.L. DePue, M.D. Taylor, S.T. Rapundalo, G.P. Hingorani, T.C. Major, R.L. Panek, and D.G. Taylor, J. Cardiovasc. Pharmacol., 17 (Suppl. 7), S59 (1991).

37. W.L. Cody, A.M. Doherty, X. He, S. Rapundalo, S.T. Rapundalo, G.P. Hingorani, R.L. Panek, and T.C. Major, J. Cardiovas. Pharmacol., 17 (Suppl. 7), S62 (1991).

38. D.L. Williams, K.L. Jones, D.J. Pettibone, E.V. Lis, and V. Clineschmidt, Biochem. Biophys. Res. Commun., $\underline{175}$, 556 (1991).

39. C.R. Hiley, C.R. Jones, J.T. Pelton, and R.C. Miller, Br. J. Pharmacol., $\underline{101}$, 319 (1990).

40. T. Saeki, M. Ihara, T. Fukuroda, M. Yamagiwa, and M. Yano, Biochem. Biophys. Res. Commun., $\underline{179}$, 286 (1991).

41. M.J. Spinella, A.B. Malik, J. Everitt, and T.T. Anderson, Proc. Natl. Acad. Sci. USA, $\underline{88}$, 7443 (1991).

42. N. Oohata, M. Nishikawa, S. Kiyoto, S. Takase, K. Hemmi, H. Murai, and M. Okuhara, EP 0 405 421 A2 (1990).

43. X.H. Gu, D.J. Casley, and W. G. Naylor, Biochem. Biophys. Res. Commun., $\underline{179}$, 130 (1991).

44. M. Ihara, T. Fukuroda, T. Saeki, M. Nishikibe, K. Kojiri, H. Suda, and M. Yano, Biochem. Biophys. Res. Commun., $\underline{178}$, 132 (1991).

45. K. Kojiri, M. Ihara, S. Nakajima, K. Kawamura, K. Funaishi, M. Yano, and H. Suda, J. Antibiot., $\underline{44}$, 1342 (1991).

46. M. Hashimoto, M. Nishikawa, M. Esaki, S. Kiyoto, M. Okuhara, S. Takase, K. Henmi, M. Neya, N. Fukami, and M. Hashimoto, JO 3130-299-A (1990).

47. T. Fukami, T. Hayama, K. Niiyama, T. Nagase, T. Mase, K. Fujita, U. Kumagai, Y. Urakawa, M. Ihara, S. Kimura, and M. Yano, Twelfth American Peptide Symposium, Cambridge, Massachusetts, June 16-21, 1991, P-506.

48. I. Kiyofumi, F. Takehiro, H. Takashi, N. Kenji, N. Toshio, M. Toshiaki, F. Kagari, N. Masaru, I. Masaki, and M. Yano, EPA 0 436 189 A1 (1990).

49. M. Ihara, K. Noguchi, T. Saeki, T. Fukuroda, S. Tsuchida, S. Kimura, T. Fukami, K. Ishikawa, M. Nishikibe, and M. Yano, Life Sci., 50, 247 (1992).

50. A.M. Doherty, W.L. Cody, X. He, P.L. DePue, D.M. Leonard, D.T. Dudley, S.T. Rapundalo, G.P. Hingorani, R.L. Panek, T.C. Major, K.E. Hill, M.A. Flynn, and E.E. Reynolds, ACS, San Francisco, April 1992, MEDI 174.

51. K. Hemmi, M. Neya, N. Fukami, M. Hashimoto, H. Tanaka, and N. Kayakiri, EPA 91107554.7 (1991)

52. H. Keiji, N. Masahiro, F. Naoki, H. Masashi, H. Tanaka, and N. Kayakiri, EPA 0457 195 A2 (1991).

53. S.L.Ollerenshaw, D. Jarvis, C.E. Sullivan, and A.J. Woolcock, Eur. Respir. J., 4 , 673 (1991).

54. R. Barker, Neuropeptides, 20, 73 (1991).

55. J. Marx, Science, 253, 857 (1991).

56. N.W. Kowall, M.F. Beal, J. Busciglio, L.K. Duffy, and B.A. Yankner, Proc. Natl. Acad. Sci. USA, 88, 7247 (1991).

57. M.E. Logan, R. Goswami, B.E. Tomczuk, and B. R. Venepalli, Annu. Rep. Med. Chem., 26, 43 (1991).

58. Y. Nakamura, R. Parent, and M. Lavallee, Circulation, 84, 300 (1991).

59. K. Itoi, N. Jost, E. Badoar, C. Tschope, J. Culman, and T. Unger, Brain Res., 558, 123 (1991).

60. P. Milner, K.A. Kirkpatrick, V. Ralevic, V. Toothill, J. Pearson, and G. Burnstock, Proc. R. Soc. Lond. [Biol], 241, 245 (1990).

61. M. Freidin and J.A. Kessler, Proc. Natl. Acad. Sci. USA, 88, 3200 (1991).

62. K.L. Barnes, D.I. Diz, and C.M. Ferrario, Hypertension, 17, 1121 (1991).

63. S.Guard and S.P. Watson, Neurochem. Int., 18, 149 (1991).

64. D. Hagiwara, H. Miyake, H. Morimoto, M. Murai, T. Fujii, and M. Matsuo, J. Pharmacobiodyn., 14, S104 (1991).

65. D. Jukic, N. Rouissi, R. Laprise, M. Boussougou, and D. Regoli, Life Sci., 49, 1463 (1991).

66. C.A. Maggi, R. Patacchini, D.M. Feng, and K. Folkers, Eur. J. Pharmacol., 199, 127 (1991).

67. Z. Wiesenfeld-Hallin, X.J. Xu, R. Hakanson, D.M. Feng, and K. Folkers, Brain Res., 526, 284 (1990).

68. R. Hakanson, S. Leander, N. Asano, D.M. Feng, and K. Folkers, Regul. Pept., 31, 75 (1990).

69. Q. Lin, C.D. Li, and P. Li, Brain Res., 530, 317 (1990).

70. F. Petitet, J.C. Beaujouan, M. Saffroy, Y. Torrens, G. Chassaing, S. Lavielle, J. Besseyre, C. Garret, A. Carruette, and J. Glowinski, J. Neurochem., 56, 879 (1991).

71. P. Delay-Goyet and J.M. Lundberg, Biochem. Biophys. Res. Commun., 180, 1342 (1991).

72. C. Garret, A. Carruette, V. Fardin, S. Moussaoui, J.- F. Peyronel, J.- C. Blanchard, and P.M. Laduron, Proc. Natl. Acad. Sci., USA, 88, 10208 (1991).

73. J.A. Lowe, S.E. Drozda, R.M. Snider, K. P. Longo, and J. Bordner, Bioorg. Med. Chem. Lett., 1, 129 (1991).

74. R.M. Snider, J.W. Constantine, J.A. Lowe, K.P. Longo, W.S. Lebel, H.A. Woody, S.E. Drozda, M.C. Desai, F.J. Vinick, R.W. Spencer, and H.- J. Hess, Science, 251, 435 (1991).

75. S. McLean, A.H. Ganong, T.F. Seeger, D.K. Bryce, K.G. Pratt, L.S. Reynolds, C.J. Siok, and J.A. Lowe, and J. Heym, Science, 251, 437 (1991).

76. R. Hakanson, Z. Wang, and K. Folkers, Biochem. Biophys. Res. Commun., 178, 297 (1991).

77. N. Rouissi, B.D.Gitter, D.C. Waters, J.J. Howbert, J.A. Nixon, and D. Regoli, Biochem. Biophys. Res. Commun., 176, 894 (1991).

78. A. Lecci, S. Giuliani, R. Patacchini, G. Viti, and C.A. Maggi, Neuroscience Lett., 129, 299 (1991).

79. J.W. Constantine, W.S. Lebel, and H.A. Woody, Naurnyn-Schmeidelberg's Arch. Pharmacol., 344, 471 (1991).

80. R.M. Snider, K.P. Longo, S.E. Drozda, J.A. Lowe III, and S.E. Leeman, Proc. Natl. Acad. Sci. USA, 88, 10042 (1991).

81. B.R. Venepalli, L.D. Aimone, K.C. Appell, M.R. Bell, J.A. Dority, R. Goswani, P.L. Hall, V. Kumar, K.B. Lawrence, M.E. Logan, P.M. Scensny, J.A. Seelye, B.E. Tomczuk, and J.M. Yanni, J. Med. Chem., 35, 374 (1992).

82. F.M. Gutierrez Marcos and A. Fernandez-Cruz, Rev. Clin. Esp., 188, 418 (1991).

83. V.J. Dzau, Circulation, 77(Suppl I), I-4 (1988).

84. C. Gros, N. Noel, A. Souque, J.-C. Schwartz, D. Danvy, J.-C. Plaquevent, L. Duhamel, P. Duhamel, J.-M. Lecomte, and J. Bralet, Proc. Natl. Acad. Sci. USA, 88, 4210 (1991).

85. A.A. Seymour, J.N. Swerdel, and B. Abboa-Offei, J. Cardiovasc. Pharmacol. 17, 456 (1991).

86. G.A. Flynn, A.M. Warshawsky, T.P. Burkholder, S. Mehdi, E.L. Girox, and R.C. Dage, ACS San Francisco, April 1992, MEDI 180.

87. M. Burnier, M. Ganslmayer, F. Perret, M. Porchet, T. Kosoglou, A. Gould, J. Nussberger, B. Waeber, and H.R. Brunner, Clin. Pharmacol. Ther., 50, 181 (1991).

88.   M. Trampus, E. Ongini, K. Varani, and P.A. Borea, Eur. J. Pharmacol., 194, 17 (1991).
89.   A. Monopoli, A. Forlani, and E. Ongini, J. Hypertension, 9, S246 (1991).
90.   K. Helin, I. Tikkanen, T. Tikkanen, O. Saijonmaa, E.J. Sybertz, S. Vemulapalli, H. Sariola, and F. Fyhrquist, Eur. J. Pharmacol., 198, 23, (1991).
91.   P.L.Barclay, J.A. Bennett, G.M.R. Samuels, and N.B. Shepperson, Biochem. Pharmacol., 41, 841 (1991).
92.   J.E. O'Connell, A.G. Jardine, G. Davidson, and J.M.C. Connell, J. Hypertension, 10, 271 (1992).
93.   J.C. Dussaule, J.D. Grange, J.P. Wolf, J.M. Lecomte, C. Gros, J.C. Schwartz, F. Bodin, and R. Ardaillou, J. Clin. Endocrinol. Metab., 72, 653 (1991).
94.   L. Edvinsson, R. Ekman, P. Hedner, and S. Valdemarsson, Eur. J. Clin. Invest., 20, 85 (1990).
95.   S. Valdemarsson, L. Edvinsson, R. Ekman, P. Hedner, and S. Sjoholm, J. Int. Med., 230, 325 (1991).
96.   T.L. Yaksh, M.B. Sabbe, D. Lucas, E. Mjanger, and R.E. Chipkin, J. Pharmacol. Exp. Ther., 256, 1033 (1991).
97.   A. Ewenson, R. Laufer, M. Chorev, Z. Selinger, and C. Gilon, Eur. J. Med. Chem., 26, 435 (1991).
98.   E. M. Genden and C.J. Molineaux, Hypertension, 18, 360 (1991).
99.   O.A. Carretero and A. G. Scicli, Hypertension, 68, 366 (1991).
100.   P. Walker, E. Grouzmann, M. Burnier, and B. Waeber, Trends Pharmacol.,12, 111 (1991).
101.   B. Waeber, M. Burnier, J. Nussberger, and H.R. Brunner, Horm. Res., 34, 161 (1990).
102.   K. Kawamura, T.L. Smith, Q. Zhou, and F.A. Kummerow, Biochem. Biophys. Res. Commun., 179, 309 (1991).
103.   C. Eva, K. Keinanen, H. Monyer, P.Seeburg, and R. Sprengel, FEBS Lett., 271, 81,1990.
104.   J. Rimland, W. Xin, P. Sweetnam, K. Saijoh, E. J. Nestler, and P.S. Duman, Mol. Pharmacol., 40, 869 (1991).
105.   M.C. Michel, Trends Pharmacol. Sci. 12, 389 (1991).
106.   F.B. Jolicoeur, J.N. Michaud, D. Menard, and A. Fournier, Brain Res. Bull., 26, 309 (1991).
107.   D.I. McCloskey and E.K. Potter, Clin. Exp. Pharmacol. Physiol., 18, 47 (1991).
108.   E.K. Potter, J. Fuhlendorff, and T. Schwartz, Eur. J. Pharmacol., 193, 15 (1991).
109.   M. Adamsson and L. Edvinsson, Neuropeptides, 19, 13 (1991).
110.   M. C. Michel and A. Buschauer, Drugs of the Future, 17, 39 (1992).
111.   M. Heilig, L. Edvinsson, and C. Wahlestedt, Eur. J. Pharmacol., 209, 27 (1991).
112.   K. Tatemoto, M.J. Mann, and M. Shimizu, Proc. Natl. Acad. Sci. USA, 89, 1174 (1992).
113.   W. Li, R.G. MacDonald, and T.D. Hexum, Eur. J. Pharmacol., 207, 89 (1991).
114.   M.C. Michel, J.P. Moersdorf, H. Engler, H. Schickaneder, K.-H. Ahrens, EP 448-765-A, (1990).
115.   M. Zaidi, B.S. Moonga, P.J.R. Bevis, Z.A. Bascal, and L.H. Breimer, Crit. Rev. Clin. Lab. Sci., 28, 109 (1990).
116.   S. Fujioka, O. Sasakawa, H. Kishimoto, K. Tsumura, and H. Morii, J. Hypertens., 9, 175 (1991).
117.   M. Mimeault, A. Fournier, Y. Dumont, S. St-Pierre, and R. Quirion, J. Pharmacol. Exp. Ther., 258, 1084 (1991).
118.   M.W. Wang, A.A. Young, T.J. Rink, and G.J. Cooper, FEBS Lett., 291, 195 (1991).
119.   M. Boscaro, E. Mani, M. Pistorello, and A. Rampazzo, Minerva Endocrinol., 15, 95 (1990).
120.   J.K. Kim, J.B. Michel, F. Soubrier, J. Durr, P. Corvol, and R.W. Schrier, Kidney Int., 38, 818 (1990).
121.   Vasopressin in Heart Failure, L.F. Arnolda, A. Katapothis, P.A. Philips, and C. I. Johnston, Colloq. INSERM, 208, 565 (1991).
122.   H. Gavras, Clin. Chem., 37/10B, 1828 (1991).
123.   Y. Yamamura, H. Ogawa, T. Chihara, K. Kondo, T. Onogawa, S. Nakamura, T. Mori, M. Tominaga, and Y. Yabuuchi, Science, 252, 572 (1991).
124.   A. Morel, A.-M. O'Carroll, M.J. Brownstein, and S.J. Lolait, Nature, 356, 523 (1992).
125.   A. Schmidt, S. Jard, J.J. Dreifuss, and E. Tribollet, Am. J. Physiol., 259, F872 (1990).
126.   M.B. Turla, M.M. Thompson, M.H. Corjay, and G.K. Owens, Circ. Res., 68, 288 (1991).
127.   A.H. Taylor, V.T. Ang, J.S. Jenkins, J.J. Silverlight, R.C. Coombes, and Y.A. Luqmani, Cancer Res., 50, 7882 (1990).
128.   T. Kimura, O. Tanizawa, K. Mori, M.J. Brownstein, and H. Okayama, Nature, 356, 526 (1992).
129.   D.J. Pettibone, B.V. Clineschmidt, E.V. Lis, D.R. Reiss, J.A. Totaro, C.J. Woyden, M.G. Bock, R.M. Freidinger, R.D. Tung, D.F. Veber, P.D. Williams, and R.I. Lowensohn, J. Pharmacol. Exp. Ther., 256, 304 (1991).
130.   B.V. Clineschmidt, D.J. Pettibone, D.R. Reiss, E.V. Lis, G.J. Haluska, M.J. Novy, M.J. Cook, M.A. Cukierski, M.J. Kaufman, M.G. Bock, R.M. Freidinger, D.F. Veber, and P.D. Williams, J. Pharmacol. Exp. Ther., 256, 827 (1991).
131.   J. Wagnon, P. de Cointet, D. Nisato, C. Plouzane, and C. Serradeil-Legal, EP 0469984 A2 (1991).
132.   B.E. Evans, R.M. Freidinger, and D.J. Pettibone, EP 0444945 A2 (1991).

Chapter 10. Recent Advances in Antiarrhythmic Therapy: Potassium Channel Antagonists

Madeline G. Cimini and J. Kenneth Gibson
Upjohn Laboratories
Kalamazoo, MI 49001

<u>Introduction</u> - Although we have seen tremendous gains in our understanding of cardiac electrophysiology, sudden cardiac (arrhythmic) death remains the leading cause of mortality in the United States (1). The Cardiac Arrhythmia Suppression Trial (CAST) studies, however, demonstrated the need to re-evaluate therapeutic/pharmacologic strategies. In view of this study, increasing emphasis has been placed on class III antiarrhythmic agents, which have potassium channels as their putative targets. This report reflects that emphasis.

Numerous studies have suggested that class III antiarrhythmic agents exert their electrophysiological effects by blocking potassium channels in the heart. This action delays the repolarization of the cardiac action potential and increases cardiac refractoriness. A few agents such as ibutilide (1) appear to increase the duration of the cardiac action potential by increasing the inward sodium current (2). Cardiac electrophysiologic actions have also been observed with a series of vasodilators that open potassium channels in the heart and enhance the repolarization of the cardiac action potential, shortening the duration of the action potential (3,4). Some reports have indicated that these agents can exert antiarrhythmic actions (5,6) while others have suggested they have proarrhythmic actions (7,8). It is not clear, however, that the antiarrhythmic actions of these potassium channel openers result from their electrophysiologic actions. These agents have also been described as "cardioprotective", that is they ameliorate cardiac cell death and dysfunction that occurs following myocardial ischemia and reperfusion and their antiarrhythmic actions may result from this ability to maintain cell viability (5,9). In light of the controversy regarding potassium channel openers, and their review in this series last year (10) we have focused this review on those agents that block potassium channels.

1

<u>Classification and Mechanisms of Action of Antiarrhythmic Agents</u> - In 1970 Vaughan Williams (11) proposed a system to classify antiarrhythmic agents based upon the drug's ability to exhibit clearly definable pharmacological actions. Four groups were originally described: Class I drugs had direct membrane activity; class II drugs had sympatholytic activity; class III compounds prolonged the action potential; and class IV drugs were centrally acting. At the molecular level, class I compounds (e.g., quinidine, lidocaine and encainide) act on Na^+ currents; class II compounds (e.g., propranolol) possess β-adrenoceptor blocking activity; class III drugs (e.g., amiodarone, sotalol and bretylium) block K^+ channels; and class IV drugs (e.g., verapamil and diltiazam) target Ca^+ channels. Eventually, a fifth class was added, drugs (e.g., alinidine) that target Cl^- channels. Later, class I drugs were subdivided into class IA, IB and IC based on the kinetics of blockade (12) and it became clear that almost all drugs possess multiple actions, (e.g., α-adrenoceptor blockade, muscarinic blockade, ganglionic blockade, and multi-channel effects). A weakness of the system is that it was based on the drug's action on normal tissue, which could (and does) respond differently than tissue stressed by conditions such as hypoxia, ischemia or disease (13). Equally

important shortcomings are the lack of information on a compound's efficacy or actions dependent on concentration, rate and tissue type (14). *In vitro* electrophysiological activity does not always translate into *in vivo* therapeutic activity. That the system has been widely used in the past decade is confirmation of its utility, but, as fundamental understanding of electrophysiologic mechanisms grew, it became evident that the system did not totally meet clinical or pharmacological needs.

Recently (1991) the Task Force of the Working Group on Arrhythmias of the European Society of Cardiology published "The Sicilian Gambit - A New Approach to the Classification of Antiarrhythmic Drugs Based on Their Actions on Arrhythmogenic Mechanisms" (15). Their premise was that a "more sophisticated framework for understanding drug actions and their relationship to clinical drug efficacy is needed". The proposed system is a flexible catalog with drug and major activity (channel, receptor, and/or pump) and intensity/potency of activity (low, moderate, high) included in spreadsheet format. The major advantage of this approach is the ability to assimilate new information into the system as it becomes available; however, that flexibility may also be viewed as a liability. The inclusion of qualitative and quantitative data, along with the ever increasing number of compounds discovered and developed may make recollection difficult (amiodarone, for example has five actions listed and within those actions are listed two intensity levels). Only time will tell if this system will prove too cumbersome for general acceptance, or be readily accepted.

The CAST and CAST II Study - The CAST study was reviewed previously in this series; therefore this review will not provide extensive detail on the design, interim analysis, and subsequent removal of encainide and flecainide from the CAST study (16). Ventricular premature depolarizations and left ventricular dysfunction are independent predictors for cardiac death after myocardial infarction. Specifically, several studies have shown an independent association between the presence of ventricular premature complexes and increased risk of arrhythmic death following myocardial infarction (17,18). The CAST study was designed to test the hypothesis that pharmacologic suppression of asymptomatic or mildly symptomatic ventricular premature beats in survivors of acute myocardial infarction would reduce arrhythmic death (19). Based upon an earlier Cardiac Arrhythmia Pilot Study (CAPS), the agents used included encainide, flecainide and moricizine.

Interim analysis of the CAST study after 10 months of follow up, demonstrated increased deaths from arrhythmia in patients treated with encainide or flecainide. While much has been speculated, the causes of this excess mortality in patients receiving encainide or flecainide treatment remain unknown. Potential causes of this enhanced mortality may include a proarrhythmic action of these agents enhanced in the presence of myocardial ischemia, negative inotropic effects of these agents resulting in a reduced cardiac oxygen supply in the face of increased oxygen demand and/or a low placebo mortality (20,21).

The investigators of the CAST trial concluded that despite an increased risk of death, asymptomatic or mildly symptomatic patients with ventricular premature beats or non-sustained ventricular tachycardia after a myocardial infarction, may not benefit from the use of encainide or flecainide to treat these arrhythmias. The FDA Cardiorenal Advisory Committee subsequently recommended that CAST conclusions be applied to all Class 1C drugs such as encainide or flecainide.

Despite the disappointing results from the interim analysis of CAST, significant changes in design were made and the CAST II trial was continued with moricizine. The left ventricular ejection fraction for acceptance into the trial was lowered, and the entry window was narrowed to 90 days after infarction. These changes permitted enrollment of patients with more severe disease and a higher event rate. In August 1991, however, CAST II was stopped after recommendations to discontinue the study were made by the CAST II data and safety monitoring board and the CAST II executive committee (22). At the interim analysis in April 1989, only 4 patients receiving moricizine had died compared to 11 receiving placebo, but this trend was reversed as CAST II continued. A review of the mortality data showed that of the 1,325 CAST II patients, 68 patients in the placebo group had died versus 98 patients in the moricizine treated group. Although not statistically significant, a trend of increased mortality was observed in the first 14 days of treatment (23).

CAST and CAST II have had a major impact on many groups including patients, physicians, the pharmaceutical industry and regulatory agencies (24). It has been reported that patients now refuse treatment with antiarrhythmic agents. The medical community may have become reluctant to use any antiarrhythmic agent, even when there are legitimate medical reasons for its use to prevent serious supraventricular or ventricular arrhythmias. Industry will now reevaluate new drug development in this area in the face of a shrinking market, possibly producing a substantial reduction in the number of new drugs being developed for what remains an unmet medical need. Finally, in the face of the uncertainties observed in these groups, regulatory agencies now have a major responsibility to provide new guidelines for the approval of antiarrhythmic agents and restore confidence that this process is based upon the best possible scientific evaluation. This is particularly true if the newer agents presented in this review are to be carefully studied in the clinic.

<u>CLASS III ANTIARRHYTHMIC AGENTS</u>

The methylsulfonamide moiety is the most consistent functional group in class III antiarrhythmic compounds, and, as such, it facilitates categorization. Therefore, we have grouped these compounds into those that contain this functional group, and those that do not.

<u>Compounds containing a methylsulfonamide group</u> - Racemic sotalol (<u>2</u>) remains a prototype antiarrhythmic agent whose pharmacologic activity stems from depression of the delayed rectifier potassium current (25). Clinical studies continue with the β-adrenoceptor antagonist racemate, although interest is increasing for the d-isomer which is devoid of β-adrenoceptor blocking activity (26,27). Racemic sotalol (at 320 and 640 mg/day) is efficacious against chronic ventricular premature complexes (28). A dose of 1.5 mg/kg converted paroxysmal supraventricular tachycardias of mixed etiologies to normal sinus rhythm in 83% of patients, vs. 16% for placebo (29). Racemic sotalol is orally effective at preventing primary paroxysmal atrial fibrillation (30), and drug refractory ventricular arrhythmias (31). The "reverse-use dependence" seen with other class III antiarrhythmic agents has also been demonstrated with sotalol; QT interval prolongation was more pronounced at slow (RR interval > 800 ms) heart rates and was not apparent during exercise-induced tachycardia in normal human volunteers (32). Proarrhythmia was seen in 6% (25 of 433) of patients undergoing long-term sotalol therapy for ventricular arrhythmias (33).

Sematilide (CK-1752, $\underline{3}$) was produced by replacement of the N-acetyl group in N-acetylprocainamide with a methanesulphonyl group. *In vitro* studies on a variety of isolated tissues, showed sematilide to be a selective class III agent (34). Microelectrode voltage clamp studies demonstrated that sematilide increases action potential duration by inhibiting the delayed rectifier current (35). In an effort to combine the class III actions of sematilide and the class II actions of propranolol, CK-3579 ($\underline{4}$) was synthesized (36). CK-3579 has selective β-1 receptor blocking activity and class III actions (37). While the class III activity is not enantioselective, the S-enantiomer, CK-4000 has the highest affinity for the cardiac β-receptor. Evaluation of sematilide and CK-3579 in a canine model of sudden cardiac death demonstrated that only sematilide could prevent the induction of ventricular tachyarrhythmias via programmed stimulation, but both agents prevented sudden death following arterial thrombosis in the presence of a previous myocardial infarction (38). In 14 patients with chronic, non-sustained arrhythmias, sematilide exerted a class III action and prolonged QT intervals in a dose-related and plasma concentration-related fashion (39). Increases in QT interval of 25% were observed at plasma concentrations of 2 ug/ml. Plasma concentrations > 0.8 ug/ml suppressed arrhythmias in 5 patients while aggravating them in 3 others. One patient required cardioversion for *torsades de pointes* at a plasma concentration of 2.7 ug/ml (39).

UK-68,798 ($\underline{5}$) prolonged action potential duration (APD) and effective refractory period (ERP) but did not change resting membrane potential, upstroke velocity or action potential amplitude recorded from isolated guinea pig papillary muscle, canine Purkinje fibers or ventricular muscle at concentrations greater than 10^{-8} M. Concentrations as high as 10^{-5} M had no effect, however, on rat papillary muscle or atria (40,41). Studies on voltage-clamped myocytes demonstrated that this class III action resulted from blockade of a time-dependent potassium current with little or no effect on the time-independent potassium current (35). Binding studies with ^{3}H-UK-68798 in guinea pig ventricular muscle indicated that d-sotalol, E-4031 or sematilide displace $\underline{5}$ from a single population of non-interacting sites (42). In a canine model of sudden cardiac death, $\underline{5}$ suppressed programmed stimulation of ventricular tachycardia in dogs following occlusion and reperfusion of the left anterior descending coronary artery (43). Subsequent induction of occlusion thrombus in the circumflex coronary artery produced sudden death in 83% of the vehicle treated animals and 33% of the animals treated with $\underline{5}$. In patients with coronary artery disease, $\underline{5}$ significantly prolonged the QT interval in a dose-dependent manner without affecting other ECG intervals, heart rate or blood pressure (44). In patients with atrial fibrillation (AF) or atrial flutter (AFL), $\underline{5}$ converted AF to sinus rhythm in 62% of the patients and AFL was converted to sinus rhythm in 100% of the patients (45). *Torsades de pointes* has been seen after high doses of $\underline{5}$ in man and laboratory animals (46,47).

A series of phenethanolamines and phenoxyalkanolamines with diminished β-adrenoceptor blocking activity (accomplished by transforming the nitrogen atom to a tertiary or quartenary species) was produced (48). Substitution at R with quinolin-2-yl or 6-[methylsulfonyl)-amino]quinolin-2-yl yielded $\underline{6}$ (WAY123,223) and $\underline{7}$ (WAY125,971). Compound $\underline{7}$ blocked the delayed rectifier potassium current more potently than $\underline{6}$ in voltage-clamped isolated feline myocytes. Compound $\underline{7}$ was also more potent than $\underline{6}$ in open-chest dogs where atrial and ventricular effective refractory periods were prolonged by 53% and 26%, respectively (48).

Incorporation of a heteroaromatic ring into a risotilide (see below) template produced WAY-123,398 ($\underline{8}$), a specific inhibitor of the delayed rectifier potassium current in isolated cat myocytes (49). In canine Purkinje fibers, $\underline{8}$ increased action potential duration > 30% at cycle lengths of 300 ms, and > 50% at cycle lengths of 1000 ms. In open-chest anesthetized dogs atrial and ventricular effective refractory periods were prolonged by > 40% and > 20%, respectively, but atrial and ventricular conduction times were unchanged. Compound $\underline{7}$ produced a 3-fold increase in ventricular fibrillation threshold, with reversion to normal sinus rhythm without countershock in 2 of 6 dogs. Oral (intragastric) bioavailability was good, and the compound was without apparent behavioral side-effects in conscious dogs (49).

The sotalol derivative E-4031 ($\underline{9}$), caused dose-dependent increases in developed tension and its maximal rate of rise, and increased effective refractory periods in isolated rabbit papillary muscles (50). Intravenous administration of $\underline{9}$ in open chest dogs produced positive inotropic and negative lusitropic effects, both seemingly related to prolongation of action potential duration (50). In a canine model of atrial flutter, **9** terminated atrial flutter by abolishing the excitable gap through a greater prolongation of refractoriness relative to a lesser (compared to flecainide, propafenone and disopyramide) slowing of conduction (51). In a canine model of post-infarct myocardial ischemia, **9** protected against ventricular fibrillation (sudden coronary death) in 8 of 10 animals, and increased the relative and effective refractory periods of the non-infarcted and infarct-related regions of the left ventricle (52). Pretreatment with the β-adrenergic-specific agonist isoproterenol prevented refractory period increases in isolated guinea pig papillary muscles and also antagonized prolongation of action potential duration in isolated guinea pig ventricular myocytes (53).

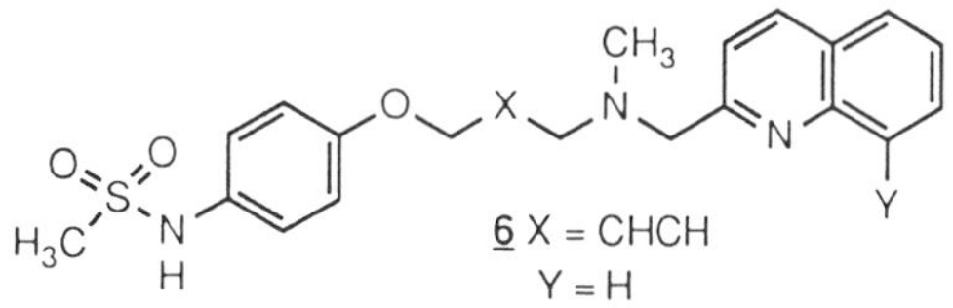

6 X = CHCH   Y = H

7 X = CH₂   Y = SO₂CH₃

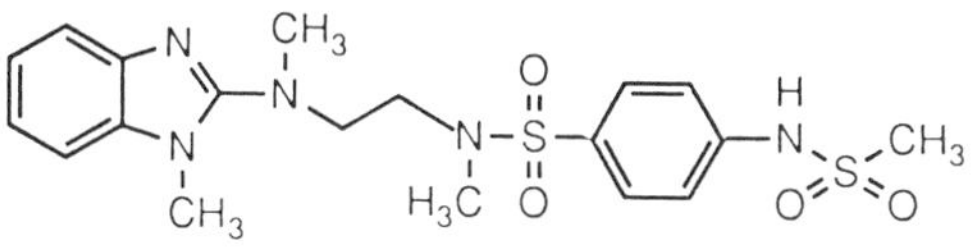

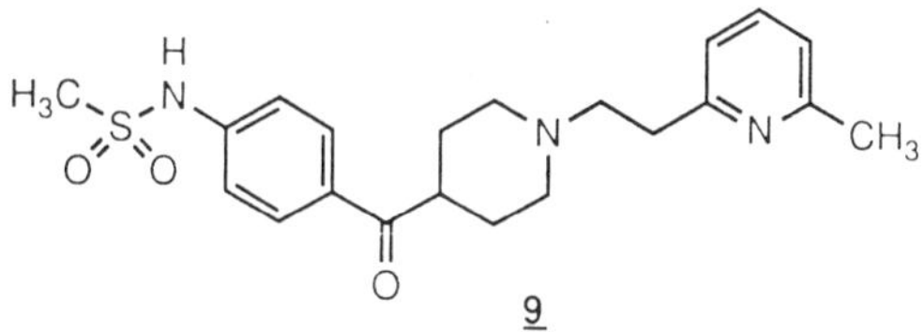

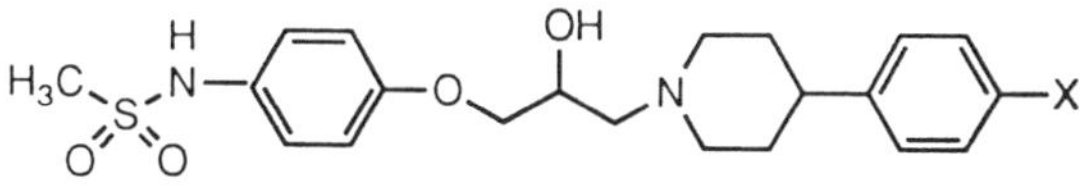

The (methylsulfonyl)amino moiety of sotalol was the starting point for a series of oxypropanolamine derivatives. Incorporating the nitrogen atom into a ring and introducing hydrophobic chlorine (**10**) or trifluoromethyl (**11**) groups resulted in two compounds that prolonged action potential duration > 40% at nM concentrations in isolated guinea pig ventricular myocytes (54). It appears that these bulky and hydrophobic tertiary amino groups confer selectivity for the delayed rectifier potassium current because these compounds had no effect on atrial conduction velocity (54).

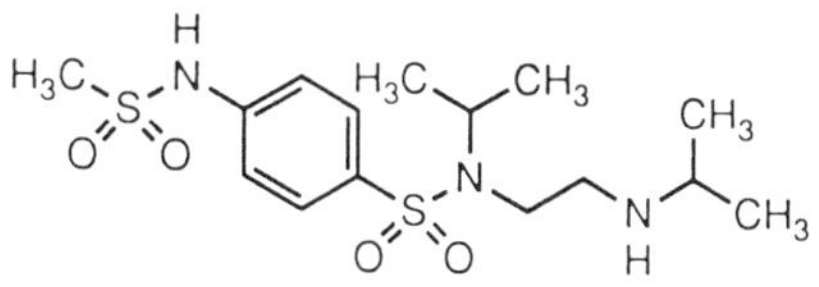

The benzenesulfonamide risotilide (**12**), an inhibitor of the voltage-dependent potassium channel is devoid of β-adrenoceptor binding activity (55). Voltage-independent potassium channel mean open times in rat hippocampal neurons were decreased 38% by 5 μM risotilide; however, in this system risotilide was less potent than **5** or tedisamil (56). Risotilide (30 mg/kg, i.v.), completely abolished occlusion-induced ventricular fibrillation in rats (57). In a feline model of acute ischemia, pretreatment with $\underline{12}$ prevented post-occlusion changes in monophasic action potential duration, dispersion of refractoriness and ventricular fibrillation threshold (58). Narrowed dispersion of recovery of excitability and protection against ventricular fibrillation were produced by 5 mg/kg, i.v. risotilide in a feline model of left ventricular hypertrophy. In this study the compound prolonged refractoriness and repolarization time in both the normal and hypertrophied ventricle (59).

In a canine model of myocardial infarction, UK-66,914 ($\underline{13}$) prolonged ventricular effective refractory periods but did not alter conduction velocity (either in regions exhibiting normal or slowed conduction). In chronically instrumented conscious dogs $\underline{13}$ caused only a modest increase in stroke volume (60); however, in chloralose-anesthetized dogs, $\underline{13}$ increased myocardial contractility in a concentration dependent manner (61). In this study, $\underline{13}$ was more potent than d-sotalol at prolonging ventricular refractoriness (as were $\underline{5}$ and $\underline{9}$).

Linkage of sematilide with a 1-arylpiperazine yielded a series of compounds with both class III antiarrhythmic activity and β-adrenoceptor blocking activity (62). In canine Purkinje fibers 0.2-3.9 μM of $\underline{14}$ prolonged action potential duration > 20% from control, but did not affect the fast inward sodium current. Compound $\underline{14}$ was as effective as sotalol in canine models of programmed stimulation and epinephrine-induced arrhythmia; $\underline{14}$ was less depressant and better tolerated than sotalol at equivalent doses, perhaps due to its β_1 selectivity (62).

Compound L-706,000 ($\underline{15}$) increased effective refractory periods in isolated ferret papillary muscles and in chloralose-anesthetized dogs. In animal models $\underline{15}$ was less efficacious but possessed a longer duration of action than previous leads, all of which were chosen from a series of spiro-benzopyran-2,4'-piperidines with class III antiarrhythmic activity (63).

Compounds without a methylsulfonamide group - Almokalant (H 234/09, $\underline{16}$), was developed from β-adrenoceptor blocking structures with p-substitution in the benzene ring. In anesthetized guinea pigs, an intravenous dose of 0.07 ± 0.06 μmol/kg prolonged the duration of the monophasic action potential (MAP) by 20%, while in isolated perfused guinea pig hearts the EC_{50} for MAP prolongation was 0.11μM (64). In open chest dogs $\underline{16}$ prolonged atrial and ventricular refractory periods but did not affect atrioventricular conduction (65). In addition, the drug showed positive inotropic activity (20% of maximal response to isoprenaline) in isolated feline papillary muscles (66) and during acute ischemic left ventricular failure in dogs (67). It appears that in rabbit ventricular cells almokalant lengthens the action potential duration via a selective blockade of the delayed rectifier potassium current in a time- and voltage-dependent manner 68, and is devoid of β-adrenoceptor blocking activity. A 4.5 mg intravenous infusion of the compound significantly increased the QT interval in healthy human volunteers in a double-blind, crossover, placebo-controlled Phase I clinical study (69). Phase II clinical studies are currently in progress (59).

Tedisamil (KC 8857, **17**), a heterocyclic hydrochloride based on sparteine inhibits the transient outward K^+ current in isolated rat ventricular myocytes (70). In a model of occlusion induced arrhythmia using conscious rats tedisamil (4 mg/kg) completely suppressed ventricular fibrillation. Compound **17** prolonged epicardial intracellular action potential durations in anesthetized rats at every dose (0.5 to 8 mg/kg) tested (71). In these studies the higher doses of **17** slightly depressed the maximum rise rate of phase 0 of the action potential, consistent with earlier findings that tedisamil also blocks sodium channels. Mean residence time at a Ca^{++} and voltage-dependent potassium channel was 1-2 ms in guinea pig smooth muscle cells, indicating that tedisamil is an "intermediate" (as opposed to a "fast" or "slow") K^+ channel blocker (72).

At concentrations of 0.1 to 10µM, MS-551 (**18**) prolonged action potential duration but did not affect the fast inward sodium current in rabbit papillary muscles, while in isolated rabbit ventricular cells **18** inhibited the delayed rectifier potassium current (73). Prolongation of the action potential duration in isolated canine Purkinje fibers was more prominent at slow driving rates, but was maintained during high potassium (8-12 mM) conditions (74). In a canine model of atrial flutter, doses of 0.03-0.3 mg/kg, i.v. converted atrial flutter to sinus rhythm in 7 of 8 dogs (75). Atrioventricular conduction time appeared to be unaffected (74,76) as were the hemodynamic parameters, mean blood pressure, aortic flow and myocardial contractility in anesthetized dogs (77).

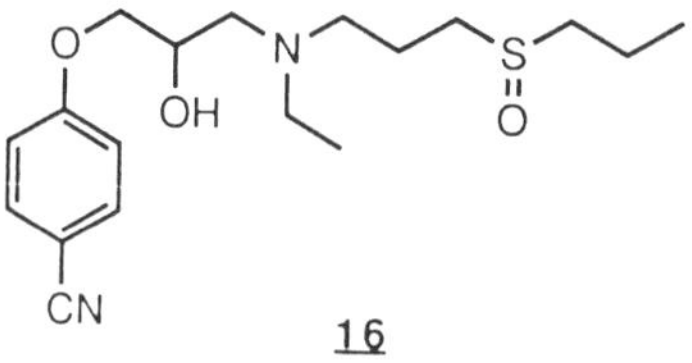

16

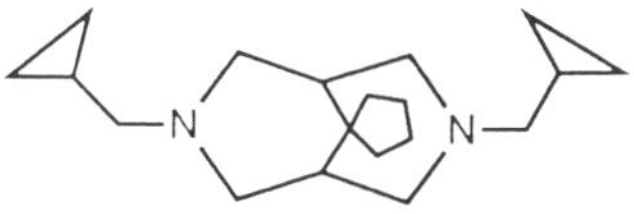

17

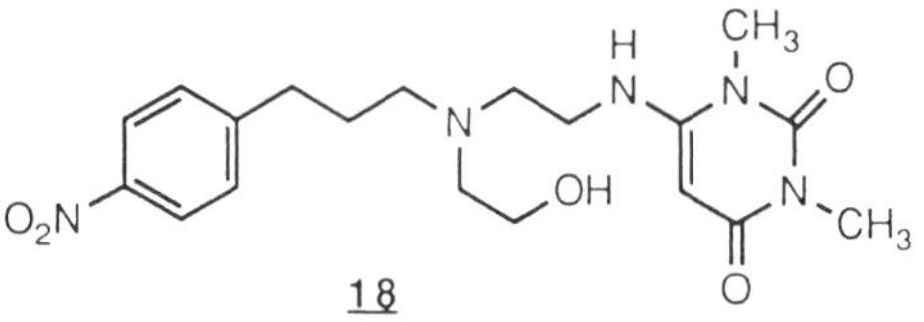

18

In isolated guinea pig ventricular myocytes RP-58866 (**19**), a benzopyran derivative, preferentially blocks the inward rectifier potassium current (78). Intravenous doses of 0.3 mg/kg (5 minutes prior to coronary artery occlusion) protected dogs and micropigs against reperfusion-induced ventricular fibrillation and ST segment changes; however, the drug caused a marked reduction in heart rate and an increase in QT time interval (79). In isolated guinea pig left ventricular papillary muscles, 0.2 and 2.0 µM of **19** significantly improved both contractility and relaxation (80).

In a liquid nitrogen produced anisotropic ring of perfused rabbit hearts RP-62719, the active enantiomer of **19** terminated ventricular tachycardia. In this study cycle length increased, ventricular refractory period was prolonged and the excitable gap was decreased, while conduction velocity was slowed only at high concentrations (81). Double wave reentry produced by programmed electrical stimulation was not possible after infusion of 0.03 µmol/l RP-62719, again in Langendorff perfused anisotropic rabbit epicardium (82). In cultured CA1 hippocampal neurons, introduction of 5 µM of RP-62719 resulted in the mean open time of the K(Ca) channel to be diminished by 85%

at 24°C, but amplitudes of the unitary currents were unchanged (83). These results are consistent with simple open channel, voltage-independent block of the K(Ca) channel.

$$\underline{19}$$

Acecainide (N-acetylprocainamide) a major, active metabolite of the class I agent procainamide, possesses class III activity. In normal canine Purkinje fibers, doses of 0.01-0.24 mM did not significantly affect resting membrane potential or the fast inward sodium current (84). In voltage-clamped guinea pig ventricular myocytes, 500 μM of acecainide blocked an early component of the outward potassium current (85). In intact neonatal canine hearts, administration of 40 mg/kg, i.v. acecainide after lidocaine resulted in a significant amplification of lidocaine-induced conduction delay in the His-Purkinje system (86). Action potential duration was prolonged by 50% in the atria and by 60% in the ventricles (87), indicating that neonatal myocardium may be exceptionally sensitive to class III antiarrhythmic agents.

Ambasalide (LU-47710, $\underline{20}$) caused a dose-dependent (0.01 to 10 μM) inhibition of the delayed rectifying potassium current in voltage clamped isolated guinea pig myocytes (88). At 10^{-5} M ambasalide prolonged action potential duration in canine ventricular muscle and Purkinje fibers paced at 0.5 Hz; however, in Purkinje fibers the effect diminished with higher (2.0 Hz) stimulation (89). In these studies a frequency dependent depression of sodium channel activity at 10^{-5} M in both ventricular muscle and Purkinje fibers indicated class I activity.

$$\underline{20}$$

Conclusions - The medical need for safe and effective antiarrhythmic compounds continues, and population demographics demonstrate that this need will only increase (90). Whereas atrial arrhythmias may have been viewed as clinically less important than ventricular arrhythmias, the presence of atrial fibrillation confers a significant increase in mortality (91), and attention is focusing on therapeutic interventions that treat supraventricular and ventricular arrhythmias. In the past decade increasing interest has focused on the general therapeutic potential of potassium channel modulators (92). These compounds (especially potassium channel antagonists, that prolong the cardiac action potential duration) have the potential to be of major therapeutic importance for the treatment of cardiac dysrhythmias.

References

1. Centers for Disease Statistics, Feb. 1992.
2. E.W. Lee, M.C. McKay and K.S. Lee, J.Mol.Cell.Cardiol., $\underline{22}$, Suppl. 1, S.15 (1990).
3. T.J. Colatsky and C.H. Follmer, Cardiovasc.Drug Rev., $\underline{7}$, 3, 199(1989).
4. G. Edwards and A.H. Weston, TIPS, $\underline{11}$, 417 (1990).
5. L. Carlsson, C. Abrahamsson, L. Drews and G. Duker, Circulation, $\underline{85}$, 1491 (1992).
6. G.J. Grover, P.G. Sleph and S. Dwonczyk, J.Cardiovasc.Pharmacol., $\underline{16}$, 853 (1990).
7. A.C.G. U'Prichard, L.Chi, E.M. Driscoll and B.R. Lucchesi, J.Mol.Cell.Cardiol., $\underline{21}$, S.13 (1989).
8. C.D. Wolleben, M.C. Sanguinetti and P.K. Siegl, J.Mol.Cell.Cardiol., $\underline{21}$, 783 (1989).
9. G.J. Grover, J.R.McCullough, D.E. Henry, M.L. Conder and P.G. Sleph, J.Pharm.Exp.Ther., $\underline{251}$, 1, 98 (1989).

10.   J.M. Evans and S.D. Longman, Ann.Rep.Med.Chem., 26, 73 (1991).
11.   E.M. Vaughan Williams in "Symposium on Cardiac Arrhythmias", E. Sandoe, E. Flensted-Jensen, and K.H. Olsen, Eds., Astra, Södertälje, 449 (1970).
12.   D.C. Harrison. Am.J.Cardiol., 56, 185 (1985).
13.   S.M. Cobbe. Cardiovasc.Res., 22, 847 (1988).
14.   S. Nattel. Drugs, 41, 672 (1991).
15.   Task Force of the Working Group on Arrhythmias of the European Society of Cardiology, Circulation, 84, 1831 (1991).
16.   J.E. Arrowsmith and P.E. Cross, Ann.Rep.Med.Chem., 25, 79 (1989).
17.   J.T. Bigger, Jr., J.L. Fleiss, R. Kleiger, J.P. Miller and L.M. Rolnitzky. Multicenter Post-Infarction Research Group, Circulation, 69, 250 (1984).
18.   C.J. Mukharji, R.E. Rude, W.K. Poole, et al., Am.J.Cardiol., 54, 31 (1984).
19.   D.S. Echt, P.R. Liebson, L.B. Mitchell, R.W. Peters, D. Obias-Manno, A.H. Barker, D. Arensberg, A. Baker, L. Friedman, H.L. Greene, M.L. Huther, D.W. Richardson and the CAST Investigators, New Eng.J.Med., 324, 12, 781 (1991).
20.   A.E. Epstein, J.T. Bigger, Jr., D.G. Wyse, D.W. Romhilt, R.A. Reynolds-Haertle, A.P. Hallstrom and the CAST Investigators, J.Am.Coll.Cardiol., 18, 14 (1991).
21.   C.M. Pratt and L.A. Moye, Am.J.Cardiol., 65, 20B (1990).
22.   FDC Pink Sheet. August 19, 1991.
23.   CAST Investigators, J.Am.Col.Cardiol., 19, 3, 237A (1992).
24.   P.R. Kowey, R.A. Marinchak and S.J. Rials, J.Cardiovasc.Electrophysiol., 1, 457 (1990).
25.   A. Varro, P.P. Nanassi and D.A. Lathrop, Life-Sci., 49, 4, 7 (1991).
26.   B. Avitall, J.W. Hare, G.L. Zander, C.S. Wilhelm, M.R. Jazayeri and P.J. Tchou, J.Am.Coll.Cardiol., 17, 2, Suppl. A, 39A (1991).
27.   H.M. Hoffmeister, S. Muller and L. Seipel, J.Cardiovasc.Pharmacol., 17, 4, 581, (1991)
28.   M.I. Anastasiou-Nana, E.M. Gilbert, R.H. Miller, S. Singh, R.A. Freedman, D.L. Keefe, S. Saksena, D.J. MacNeil and J.L. Anderson, Am.J.Cardiol., 67, 6, 511 (1991).
29.   L. Jordaens, A. Gorgels, R. Stroobandt and J. Temmerman, Am.J.Cardiol., 68, 1, 35 (1991).
30.   A. Nobile, A. Sampinato, G. Ansalone, O. Sacchetti, R. Polzella and P. Zecchi, Eur.Heart J., 12, Abstr. Suppl., 337 (1991).
31.   A.C. Rankin, L. Hamilton, D. Du, J. Newell, H. Garan and J.N. Ruskin, Br.Heart J.,66, 1, 108 (1991).
32.   C. Funck-Bretano, Y. Kibleur, F. Le-Coz, J.M. Poirier, A. Mallet and P. Jaillon, Circulation,83, 2, 536 (1991).
33.   P. Deedwania, Eur.Heart J., 12, (Suppl.) 223 (1991).
34.   T.M. Argentieri, M.S. Carroll and M.E. Sullivan, J.Cardiovasc.Pharmacol., 18, 1, 67, (1991).
35.   T.M. Argentieri and M.S. Carroll, J.Mol.Cell.Cardiol., 22, III, S.81 (1990).
36.   R. Lis, T.K. Morgan, Jr., A.J. Marisca, R.P. Gomez, J.M. Lind, D.D. Davey, G.B. Phillips and M.E. Sullivan, J.Med.Chem., 33, 10, 2883 (1990).
37.   T.M. Argentieri, M.E. Sullivan, H. Troy, M. Carroll, D. Hausamann, J. Creasy, E. Cantor, G. Phillips, J. Lampe and H.J. Reiser, FASEB J., 4, 3, A454 (1990).
38.   L. Chi, D-X. Mu, E.M. Driscoll and B.R. Lucchesi, J.Cardiovasc.Pharmacol., 16, 312 (1990).
39.   W. Wong, H.N. Pavlou, U.M. Birgersdotter, D.E. Hilleman, S.M. Mohiuddin and D.M. Roden, Am.J.Cardiol., 69, 206 (1992).
40.   P.M. Tande, H. Bjornstad, T. Yang and H. Refsum, J.Cardiovasc.Pharmacol., 16, 3, 401 (1990).
41.   M. Gwilt, J.E. Arrowsmith, K.J. Blackburn, R.A. Burges, P.E. Cross, H.W. Dalrymple and A.J. Higgins, J.Pharmacol.Exp.Ther., 256, 1, 318 (1991).
42.   P.M. Greengrass, F.L. Sanders and M.G. Wyllie, Fundam.Clin.Pharmacol., 5, 5, 408 (1991).
43.   S.C. Blackburn, L. Chi, D-X. Mu and B.R. Lucchesi, J.Pharmacol.Exp.Ther., 258, 2, 416 (1991).
44.   M. Sedgwick, H.S. Rasmussen, D. Walker and S.M. Cobbe, Br.J.Clin.Pharmacol., 31, 515 (1991).
45.   M.J. Suttorp, P.E. Polak, A. van't Hof, H.S. Rasmussen, P. Lacante and P.H. Dunselman, Eur.Heart J., 12, Suppl., 368 (1991).
46.   T.C.K. Tham, B.A. MacLennan, D.W.G. Haron, P.E. Coates, D. Walker and H.S. Rasmussen, Br.J.Pharmacol., 31, 2, 243 (1991).
47.   L. Carlsson, O. Almgren and G. Duker, J.Cardiovasc.Pharmacol., 16, 2, 276 (1991).
48.   J.A. Butera, W. Spinelli, V. Anantharaman, N. Marcopulos, R.W. Parsons, I.F. Moubarak, C.Cullinan and J.F.Bagli, J.Med.Chem., 34, 3212 (1991).
49.   J.W. Ellingboe, W. Spinelli, M.W. Winkley, T.T. Nguyen, R.W. Parsons, I.F. Moubarak, J.M. Kitzen, D. VonEngen and J.F. Bagli, J.Med.Chem., 35, 705 (1992).
50.   H.E. Cingolani, R.T. Wiedmann. J.J. Lynch, E.P. Baskin and R.B. Stein, J.Cardiovasc.Pharmacol., 17, 83 (1991).
51.   H. Inoue, T. Yamashita, A. Nozaki and T. Sugimoto, J.Am.Coll.Cardiol., 18, 1098 (1991).
52.   L. Chi, D-X. Mu and B.R. Lucchesi, J.Cardiovasc.Pharmacol., 17, 2, 285 (1991).
53.   M.Sanguinetti, N.K. Jurkiewicz, A. Scott and P.K.S. Siegl, Circ.Research, 68, 77 (1991).

54.   S.P. Connors, P.D. Dennis, E.W. Gill and D.A. Terrar, J.Med.Chem., 34, 5, 1570 (1991).
55.   S.F. Pong, C.M. Kinney and T.J. Moorehead, FASEB J., 5, A1215 (1991).
56.   J.G. McLarnon and X-P. Wang, Mol.Pharmacol., 39, 540 (1991).
57.   J.A. Kauffman, J.J. Salata, R. Falotico and A.J. Tobia, FASEB J., 5, A1215 (1991).
58.   S. Rials, J. Sewter, Y. Wu, R. Marinchak and P. Kowey, FASEB J., 5, A1216 (1991).
59.   P.R. Kowey, T.D. Friehling, J. Sewter, Y. Wu, A. Sokil, J. Paul and J. Nocella, Circulation, 83, 2067 (1991).
60.   J.R. Onufer, H.W. Dalrymple and P.B. Corr, J.Cardiovasc.Electrophysiol., 2, 117 (1991).
61.   A.A. Wallace, R.F. Stupienski, L.M. Brookes, H.G. Selnick, D.A. Claremon and J.J. Lynch, Jr., J.Cardiovasc.Pharmacol., 18, 687 (1991).
62.   G.B. Phillips, T.K. Morgan, Jr., W.C. Lumma,Jr., R.P. Gomez, J.L. Lind, R. Lis, T. Argentieri and M.E. Sullivan, J.Med.Chem., 35, 743 (1992).
63.   J. M. Elliot, J.J. Baldwin, J.W. Butcher, D.A. Claremon, J.J. Lynch, G.S. Ponticello, D.C. Remy and H.G. Selnick, 203rd ACS Meeting (Division of Medicinal Chemistry) Abstracts, 157 (1992).
64.   G.D. Duker and O.S. Almgren, J.Mol.Cell.Cardiol., 22, Suppl. 3 S.82 (1991).
65.   L. Carlsson, O. Almgren and G. Duker, Eur.Heart J., 11, Suppl., 441 (1990).
66.   O. Almgren, R. Bergstrand, C. Bäärnhielm and G. Duker, Cardiovasc.Drugs Ther., 5, Suppl. 3 401 (1991).
67.   E. Mortensen, T. Yang and H. Refsum, J.Mol.Cell.Cardiol., 22, Suppl. 3, S.87 (1991).
68.   E. Carmeliet, J.Mol.Cell.Cardiol., 23, Suppl. 3, S.79 (1991).
69.   B. Darpö, O. Almgren, R. Bergstrand, C. Gottfridsson, B. Sanstedt and N. Edvardsson, Cardiovasc.Drugs Ther., 5, (Suppl. 3) 364 (1991).
70.   H.D. Lux, I.D. Dukes and M. Morad, Biophys.J., 59, Suppl. 2, Part 2, 267a (1991).
71.   G.N. Beatch, S. Abraham, B.A. MacLeod, N.R. Yoshida and M.J.A. Walker, Br.J.Pharmacol., 102, 13 (1991).
72.   D. Pfründer and V.A.W. Kreye, Pflügers Arch., 418, 308 (1991).
73.   H. Nakaya, Y. Takeda and M. Kanno, Jpn.J.Pharmacol., 55, Suppl. 1, 398P (1991).
74.   K. Hashimoto, M. Ishii and J. Kamiya, J.Mol.Cell Cardiol., 22, Suppl. III, S.83 (1990).
75.   J. Kamiya, M. Hirayama, M. Ishii, T. Yokoyama and T. Katakami, J.Mol.Cell Cardiol., 22, Suppl. III, S.70, (1990).
76.   T. Yamagishi, S. Motomura and K. Hashimoto, Japn.J.Pharmacol., 55, Suppl. 1, 399P (1990).
77.   J. Kamiya, H. Banno, K. Yoshihara, M. Ishii and T. Katakami, Eur.J.Pharmacol., 183, 4, 1776 (1990).
78.   D. Escande, S. LeGuern, M. Laville and J. Courteix, Eur.J.Pharmacol., 183, 2, 270 (1990).
79.   M. Mestre, D. Escande and I. Cavero, Eur.J.Pharmacol., 183, 4, 1239 (1990).
80.   J.P. Beregi, D. Escande, N. Coudray, P. Méry, D. Chemla, M. Mestre and Y. Lecarpentier, J.Mol.Cell.Cardiol., 23, Suppl. IV, S.7 (1991).
81.   J. Brugada, L. Boersma, D. Escande, C. Kirchof and M. Allessie, J.Am.Coll.Cardiol., 17, 2, 42A (1991).
82.   J. Brugada, L. Boersma, C. Kirchof, P. Brugada, M. Havenith, H.J. Wellens and M. Allessie, Circulation, 81, 163 (1990).
83.   J.G. McLarnon and X-P. Wang, Biophys.J., 60, 1278 (1991).
84.   D.D. Coyle , C.A. Carnes, S.F. Schaal and W.W. Muir, Clin.Pharmacol.Ther., 49, 2, 188 (1991).
85.   J. Turgeon, P.B. Bennet and D.M. Roden, Circulation, 82, 4, Suppl. III, III-528 (1990).
86.   T.L. Dise, A. Stolfi, C.W. Clarkson and A.S. Pickoff, J.Pharmacol.Exp.Ther., 259, 2, 535 (1991).
87.   T.L. Dise, A. Stolfi, S. Yamasaki and A.S. Pickoff, J.Cardiovasc.Pharmacol., 17, 96 (1991).
88.   Z-H. Zhang, J.S.M. Sarma, F. Chen and B.N. Singh, Circulation, 82, No.4, Suppl. 527 (1990).
89.   C. Takanaka, J.S.M. Sarma and B.N. Singh, J.Cardiovasc.Pharmacol., 19, 2, 290 (1992).
90.   B. Surawicz, J.Am.Coll.Cardiol., 14, 1401 (1989).
91.   L.J. Repique, S.N. Shah and G.E. Marais, Chest, 104, 4 (1992).
92.   D.W. Robertson and M.I. Steinberg, J.Med.Chem., 33, 6 (1990).

Chapter 11. Future Antithrombotic Therapy

Joseph A. Jakubowski, Gerald F. Smith and Daniel J. Sall
Lilly Research Laboratories, Eli Lilly and Company
Indianapolis, IN 46285

<u>Introduction</u> - Over the past 50 years substantial progress has been made in the treatment of thromboembolic diseases (1). Nevertheless, such cardiovascular diseases remain the leading cause of morbidity and mortality in developed societies. The last 5 years have provided fresh insight into the mechanisms and pathways that lead to occlusion of blood vessels. Exploitation of this knowledge will ultimately result in future generations of antithrombotic agents which, in contrast to today's established agents (*i.e.*, aspirin, heparin and warfarin), will provide more specific, mechanism-based therapies. While this more rational approach promises more potent pharmaceutical agents, it remains to be established whether these agents will ultimately improve the overall quality of life of patients. This chapter describes recent advances in the synthesis and characterization of select agents that hold the promise of being future antithrombotic therapeutics. Emerging drug discovery targets in thrombolysis have recently been discussed in this series (2).

<u>Hemostasis and Thrombosis</u> - Hemostasis is the natural physiological response of blood to damage to the vasculature, ultimately leading to cessation of blood loss (3). Blood loss is halted by the formation of platelet aggregates intermeshed with and consolidated by an insoluble fibrin matrix. Thrombosis, which occurs in both arteries and veins, is the virtual or total occlusion of a blood vessel by elements of the hemostatic system (platelet aggregates and fibrin). Arterial thrombi typically form in arteries that have preexisting atherosclerotic lesions (plaque). Plaque rupture with resulting platelet adhesion and aggregation followed by fibrin formation is a hallmark of coronary thrombosis (4,5). Venous thrombi contrast with those found in arteries in that they are fibrin- and erythrocyte-rich with relatively little platelet content. The major factors implicated in venous thrombosis are blood stasis and/or activation. The most common manifestations of venous thrombosis are deep vein thrombosis and pulmonary embolus. The pathogenesis of venous thromboembolism has recently been reviewed (6). Since the pathogenesis of arterial thrombosis clearly differs from that of venous thrombosis, it is not surprising that agents that inhibit platelet activity have most benefit in the treatment of arterial thrombosis and, until recently, agents that inhibit the coagulation pathways (anticoagulants) were used primarily for venous disease. However, since thrombin also plays a role in the activation of platelets in arterial thrombosis, anticoagulants additionally have several arterial indications. For the sake of clarity this chapter, dealing with recent advances in the design of antithrombotics, will be divided into sections dealing with antiplatelet and anticoagulant agents.

ANTIPLATELET AGENTS

Antiplatelet agents are considered those agents that interact directly with platelets and ultimately inhibit their aggregation. While many agents are labelled platelet aggregation inhibitors, this effect typically is secondary to the agent's effect on an enzyme or agonist system that modulates platelet function. While a multitude of agents have been reported to have antiplatelet activity, this section will deal with a limited selection that holds relatively near term promise.

<u>Thromboxane Synthase Inhibitor/Thromboxane Receptor Antagonist (TSI/TRA)</u> - The widespread clinical study of aspirin in arterial thrombosis has resulted in the general acceptance of aspirin as a useful antithrombotic with a variety of arterial indications. Aspirin inhibits platelet aggregation by inhibiting generation of thromboxane A_2 (TXA_2) by activated platelets *via* acetylation of the enzyme cyclooxygenase (for review see 7). The utility of agents modulating TXA_2 generation and/or activity in pulmonary disease has recently been reviewed (8). TXA_2 promotes platelet aggregation and vascular constriction and its production has been shown to be elevated in patients with coronary artery disease (9). With the delineation of aspirin's efficacy and mode of action has come the search for more efficacious agents that are without aspirin's limitations, *i.e,* aspirin's non specific block of vascular prostacyclin and platelet prostaglandin (PG) D_2 production and aspirin's inability to block platelet aggregation induced by higher concentrations of various agonists. Two approaches have

been taken to address non-specificity. These have been the development of specific TRAs and specific TSIs (reviewed in 10 and 11). Many different classes of TRAs have been identified, however, it is only more recently that compounds with appropriate potency, duration of action and freedom from partial agonism have been identified. Since a recent comprehensive review of this subject appeared (11), a novel class of extremely potent 7-oxabicycloheptane oxazole TRAs, including 1-4, has been described (12). Amide substitution with a range of lipophilic moieties (1-3) was well tolerated (IC_{50} for inhibition of U46619-induced platelet aggregation was 5 nM for 1 and 25 nM for 3). N-methyl substitution (4) reduced activity ($IC_{50} = 81$ nM). The compound with an optimal profile of activities against human platelets, SQ 33,961 (2), demonstrated a remarkably high affinity for the platelet receptor with a K_d of 100 pM. More recently BMS-180291 (5), in which the cyclohexyl group of SQ 33,961 was replaced with a methyl group, demonstrated resistance to first-pass metabolism while maintaining potency ($K_d = 4$ nM), with prolonged *in vivo* activity (13). Whether an agent with this superior potency and duration of action will be clinically more effective than previous TRAs remains to be determined.

An alternative means of controlling TXA_2-dependent aspects of platelet aggregation is by specific inhibition of thromboxane synthase. The majority of TSIs contain 1-imidazole or 3-pyridyl rings (10), the nitrogen of the heterocycle interacting with the heme group of the enzyme. The optimal distance between the nitrogen and the requisite carboxylic acid function has been determined to be 8.5-10 Å (14,15). A major limitation of this class of agent in antithrombotic therapy is the ability of the precursor of TXA_2, PGH_2, to activate the thromboxane receptor (16). As recently reviewed, agents that combine TSI and TRA activities may provide superior antithrombotic effects than either agent alone (16). The TSI aspects of such a dual agent promotes local prostacyclin (PGI_2) production by mediating an accumulation of PGH_2 that is utilized by PGI_2 synthase in the vessel wall. In addition, diversion of PGH_2 within the platelet increases the synthesis of PGD_2 (17). Both PGI_2 and PGD_2 provide particularly effective inhibition of platelet activity by raising platelet cyclic AMP levels. Since the generation of PGI_2 and PGD_2 is localized, the side effects (*i.e.*, hypotension) commonly noted with the systemic presence of agents that have similar biochemical effects, *e.g.*, prostacyclin mimetics and phosphodiesterase inhibitors, are avoided. The TRA moiety of dual TSI/TRA agents prevent PGH_2 from activating the platelet TXA_2/PGH_2 receptor. CV4151 (6 ; isbogrel) and the related R 68 070 (7; ridogrel) represent the first meaningful generation of such dual TSI/TRA compounds. Both compounds are long acting TSIs, inhibiting the generation of platelet-derived TXA_2 in serum with IC_{50} values of 20-50 nM. At the platelet receptor level the compounds are relatively impotent, exhibiting K_d values of approximately 1 µM. Data from several studies of combination dosing support the conceptual advantages of this approach (16). However, while the Ridogrel versus Aspirin Coronary Patency Trial (RAPT) is ongoing (18), to date no large clinical trial testing the antithrombotic efficacy of this approach versus aspirin, or either agent alone, has been reported in full.

Molecular modelling studies have been utilized to aid the design of a novel class of dual TSI/TRA (15). The family of compounds synthesized was based on 9 which has the desired 9-10 Å C_1-nitrogen distance. Of the compounds synthesized, 8-12 had the better mix of enzyme inhibition and receptor antagonism in purified platelet test systems, with 11 and 12 having better receptor antagonism, and 8-10 having better synthase inhibitory activity. When tested in a plasma-based assay system $\underline{8}$-$\underline{12}$ offered little advantage over the comparator $\underline{7}$. Notably, the receptor antagonist activity of the sulfonamide-containing $\underline{12}$ dropped ~ 350-fold when tested in plasma (15). The same group described an acyclic sulfonamide, CGS 22652, $\underline{13}$, with a more optimal mix of activities (19). Compound $\underline{13}$ inhibited the synthesis of TXA_2 by human platelet microsomes with an IC_{50} of ~ 2 nM and appeared to be relatively specific (IC_{50} for cyclooxygenase > 100 µM). As a platelet TRA, $\underline{13}$ demonstrated an IC_{50} value of approximately 19 nM and provided superior benefit to several comparators in a canine model of thrombolysis (20).

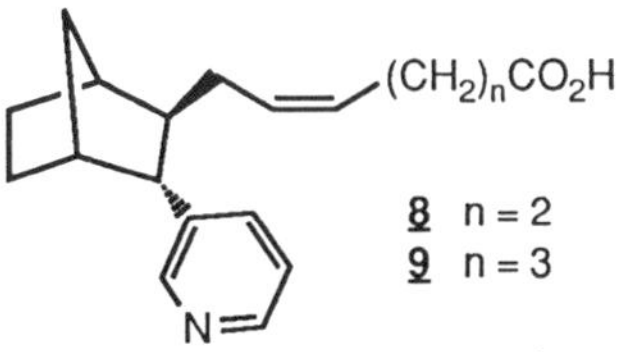

$\underline{8}$ n = 2
$\underline{9}$ n = 3

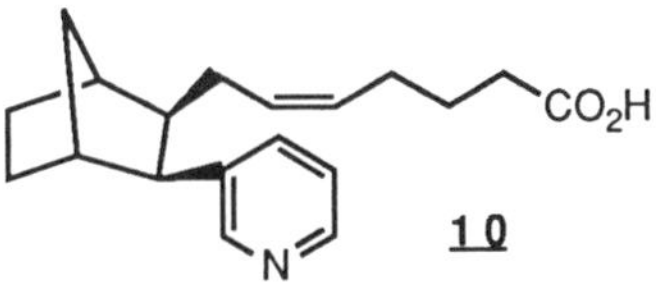

$\underline{10}$

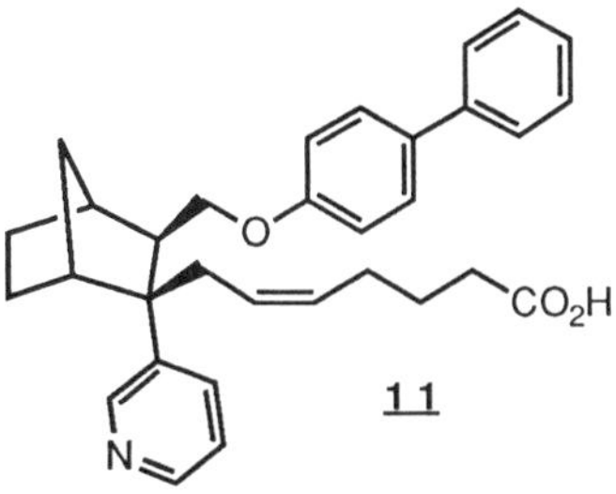

$\underline{11}$

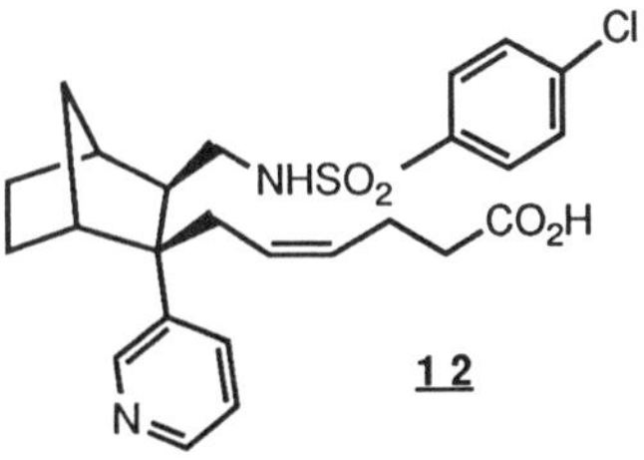

$\underline{12}$

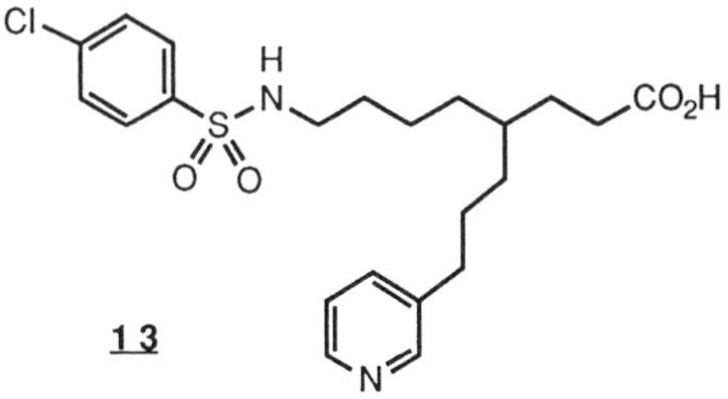

$\underline{13}$

The amino-prostanoid, GR 70067, $\underline{14}$, is a potent, orally active and long acting TRA (pA_2 ~ 8 at both platelet and aortic receptors) and is devoid of partial agonism (21). Incorporation of substituted 3-pyridyl moieties into this compound (22) generated a family of compounds, several of which had both TSI and TRA activity $\underline{15}$-$\underline{19}$. Substitution in the pyridyl ring increased receptor antagonism (platelet pA_2 = 7.9 for $\underline{15}$, ≥ 8.2 for $\underline{16}$-$\underline{19}$). TSI activity of these compounds was variable; while the IC_{50} for enzyme inhibition of the 4-n-butyl derivative ($\underline{17}$) was 0.1 µM, the 4-t-butyl analog ($\underline{18}$) had an IC_{50} of > 10 µM, and $\underline{19}$ had intermediate enzyme inhibitory activity. The compound chosen for further biological evaluation in this series $\underline{16}$ (GR 83783), exhibited a pA_2 of 8.6 at the platelet receptor and specifically inhibited TXA_2 synthase with an IC_{50} of 100 nM. However, while exhibiting prolonged antagonism of the receptor following $p.o.$ administration, synthase inhibition was inhibited for just 2 hours (22).

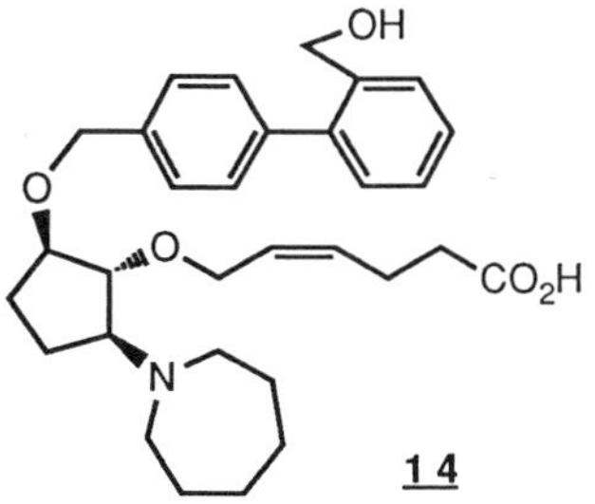

$\underline{14}$

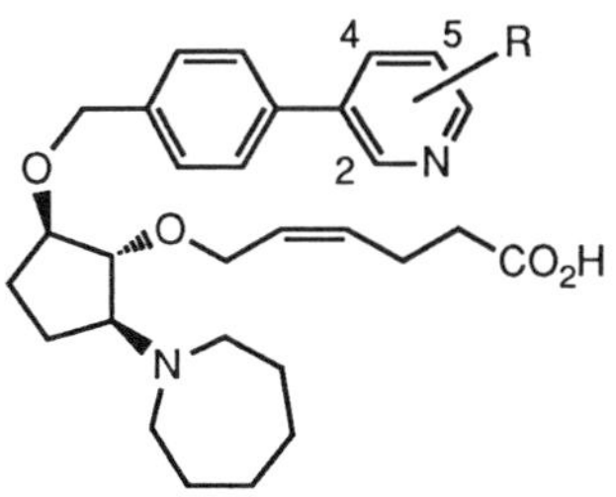

$\underline{15}$ R = H
$\underline{16}$ R = 4-n-Pr
$\underline{17}$ R = 4-n-Bu
$\underline{18}$ R = 4-t-Bu
$\underline{19}$ R = 4-CH_2OH

Several groups have recently described strategies to increase the TRA activity of **6** and **7** by incorporation of aryl sulfonamide groups (23,24), a widely used TRA pharmacophore (10,11). An example of this approach is illustrated in **20** which represents the compound with optimal mix of TSI/TRA activities derived from a large family of sulfonamide derivatives of **6**. Biological testing identified **20** (GR 85305) as the compound with optimal mix of activities associated with two methylene groups in the sulfonamide linker in combination and a C_5 pyridyl-acid tether (24). This compound retained the potent TSI activity of **6** (CV4151) and increased TRA activity by several orders of magnitude at both platelet and arterial TXA_2 receptors (24). Introduction of a gem-dimethyl group into the sulfonamide linker of **20** resulted in **21** (GR 108774), a compound demonstrating increased receptor antagonism with maintenance of effective enzyme inhibitory activity. *In vivo* testing of **21** documented a prolonged duration of activity following oral dosing suggesting that this compound represents an attractive clinical candidate. Similar modifications to the oxime-containing R 68 070 (**7**) identified a compound (GR 103237) which exhibited excellent TRA activity compared to **7** but suffered a substantial loss (> 10-fold) in enzyme inhibitory activity (24).

Thienopyridines - Ticlopidine (**22**) is a thienopyridine derivative that has undergone extensive clinical testing and is a moderately active inhibitor of platelet aggregation. While its mode of action is not entirely clear, it appears to interrupt ADP-dependent pathways of platelet aggregation (for comprehensive review of ticlopidine see ref. 25). More recently, an analog of ticlopidine, PCR-4009 (**23**) and its *d*-enantiomer (SR25990C; clopidogrel), have been described. These agents are more rapidly acting, and 5-10 times more active than the parent molecule and are currently undergoing clinical evaluation (26-27). If these analogs are devoid of the neutropenia associated with ticlopidine, they may prove to be more useful in the chronic therapy of arterial thrombotic disorders.

Glycoprotein Receptor Antagonists - Given the numerous receptor-mediated pathways *via* which platelets may be activated, antagonism of single receptor types (*e.g.*, by TXA_2, 5-HT_2, PAF or epinephrine receptor antagonists) may not effectively control platelet aggregation. Accordingly, agents that block the glycoprotein (GP) receptors mediating initial platelet adhesion (GPIb-IX) or the final aggregation event (GPIIb-IIIa) are under development. The GPIb antagonist RG 12986 is a 33-kDa fragment (Ser 445-Val 733) of von Willebrand factor - the adhesive protein that mediates platelet adhesion (28). This recombinant fragment has demonstrated efficacy in a primate model of thrombus formation and is undergoing clinical evaluation (28). Antagonism of the "final common pathway" of platelet aggregation, namely the binding of fibrinogen by platelet GPIIb-IIIa and subsequent cross-linking of platelets, has recently gained much attention as an interventional target. The advantage of this approach is the potential to control platelet aggregation irrespective of the activator. The first successful agent that has proceeded to phase III clinical testing is the monoclonal antibody 7-E3 directed against GPIIb-IIIa (29). More recently, realization that the tripeptide sequence arginine-glycine-aspartic acid (Arg-Gly-Asp; R-G-D) on fibrinogen is a major locus of fibrinogen binding to GPIIb-IIIa has led to alternative sequences that are potent antagonists of the fibrinogen receptor GPIIb-IIIa (2). While the linear peptide R-G-D inhibits fibrinogen binding and platelet aggregation, affinity for GPIIb-IIIa is low. Potency has been dramatically improved by cyclization of the R-G-D sequence, most often by flanking R-G-D with sulfur-containing residues and subsequent cyclization *via* a disulfide bridge. This approach is typified in a series of peptides **24-28** recently described in detail (30). The cyclic peptide **25** was approximately 5.6-fold more effective as a platelet aggregation inhibitor (IC_{50} =16.2 μM) than the linear **24**. Activity was further improved

Ac-Arg-Gly-Asp-Ser-NH₂   Ac-Cys-Arg-Gly-Asp-Cys-NH₂   Ac-Cys-Arg-Gly-Asp-Pen-NH₂

24   **25**   **26**

Ac-Cys-D-Arg-Gly-Asp-Pen-NH₂   **27**   Ac-Cys-(N-Me)Arg-Gly-Asp-Pen-NH₂   **28**

by cyclization *via* a carboxy-terminal penicillamine (**26**, IC_{50} = 4.1 µM) and by N-methylating the Arg (**28**, IC_{50} = 0.36 µM). Preclinical evaluation demonstrated that **28** (SK & F 106760) was an efficacious intravenous antithrombotic. In combination with streptokinase, **28** reduced by > 50% the time to reperfusion in thrombosed canine coronary arteries and reduced the incidence of subsequent reocclusion by > 80% (31). Specificity of this series of compounds for GPIIb-IIIa over other integrin receptors has not been fully documented. A recently described series of novel cyclic peptides are relatively specific and potent platelet receptor antagonists (32). Representatives of this series, **29** and **30**, containing Dtc (5,5-dimethylthiazolidine-4-carboxylic acid), inhibit platelet aggregation at sub-micromolar concentrations. While **29** had similar inhibitory effects on other integrin receptors, replacement of Arg with Amf (p-aminomethylphenylalanine) yielded **30** (L-367,073), which had > 3000-fold selectivity for the platelet fibrinogen receptor over other vascular integrin receptors (32). Clinical evaluation of **30** has necessitated the large scale synthesis of this compound, as recently described (33).

Ac-Cys-Asn-Dtc-Arg-Gly-Asp-Cys-OH   **29**   Ac-Cys-Asn-Dtc-Amf-Gly-Asp-Cys-OH   **30**

Novel peptidomimetic GPIIb-IIIa antagonists **31** and **32** have also been described (34). These benzamidino acid derivatives block GPIIb-IIIa with IC_{50} values in the low nanomolar range, effectively inhibit platelet aggregation and have little effect on the vitronectin receptor (35). Efficacy in animal models of unstable angina has also been reported (35,36). If protease inhibition by the benzamidine moiety on these compounds is inconsequential, this series may contain attractive clinical candidates.

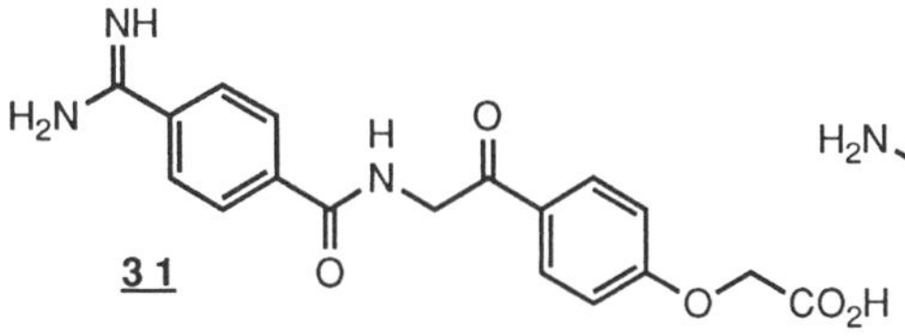

31 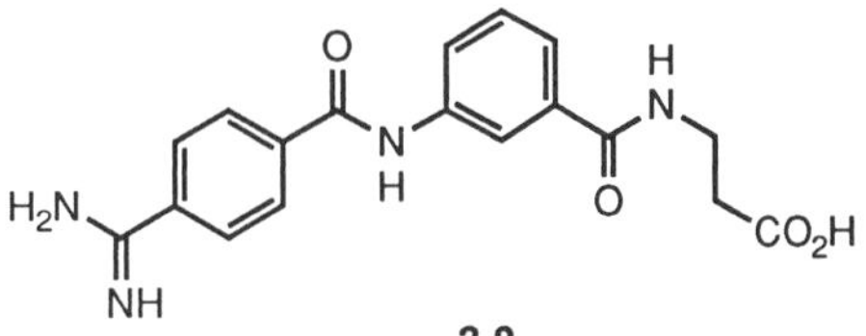   **32**

<u>ANTICOAGULANTS</u>

<u>Factor Xa Inhibitors</u> - Factor Xa is an arginine directed endopeptidase responsible for the conversion of prothrombin to thrombin. Inhibition of this enzyme is a potential means to control aberrant coagulation and resulting thrombosis. The pharmacology of the leech-derived polypeptide antistasin (ATS) and the tick anticoagulant peptide (TAP) has been reviewed (2). Recent studies have found recombinant antistasin to be efficacious as an anticoagulant (37) and as an adjunct to thrombolytic therapy (38). An 18-kDa polypeptide inhibitor of factor Xa has also been isolated from the black fly salivary glands and its primary sequence determined (39). The peptide displayed tight binding inhibition of factor Xa with no detectable affinity for thrombin. The antithrombotic activities of synthetic low MW weight inhibitors of factor Xa, **33** and **34**, have been compared with the antithrombotic effects achieved with low MW thrombin inhibitors (40). At comparable anticoagulant doses, superior antithrombotic effects were achieved using synthetic thrombin inhibitors.

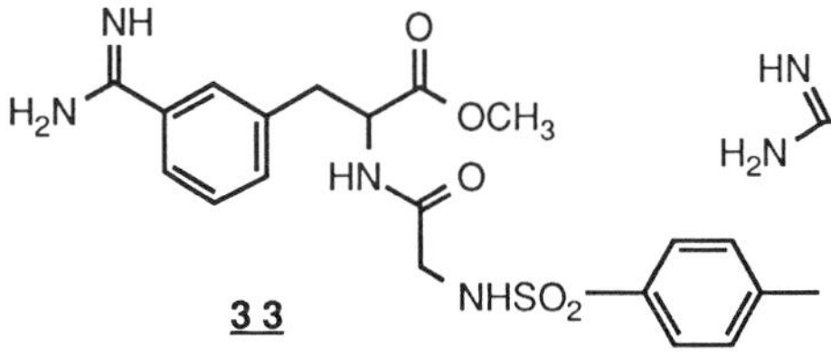

33 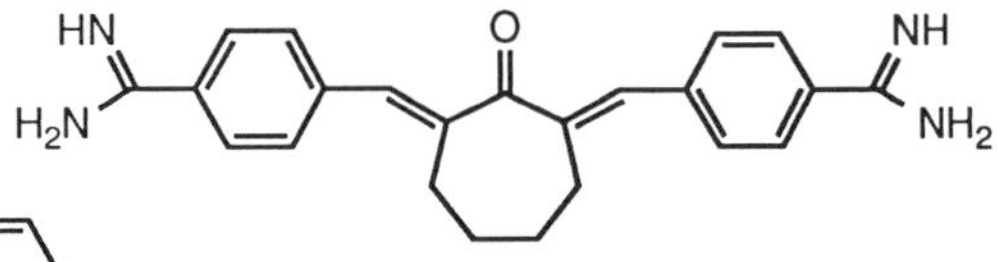   **34**

<u>Thrombin Inhibitors</u> - Thrombin, a trypsin-like serine protease, cleaves Arg-Gly bonds in both the Aα and Bβ chains of fibrinogen. Classical routes to the design and synthesis of active-site directed inhibitors have employed a highly basic guanidine or amidine as an arginine surrogate which imparts specificity for trypsin-like proteases over other serine proteases (41). In addition, recent advances in x-ray crystallography continue to provide valuable information about active-site inhibitor-thrombin interactions and may direct the design of more selective agents relative to other arginine- or lysine-directed endopeptidase coagulation factors and fibrinolytic enzymes (42-45). The fibrinogen binding site on thrombin is also being explored as a potential site of action of therapeutic agents. In this section we will describe three classes of thrombin inhibitors: reversible and irreversible active-site directed inhibitors of thrombin, and hirudin-based antithrombins.

A number of transition state analogues have been prepared that are potent, reversible, active-site inhibitors of thrombin by replacing the C-terminal carboxyl in the fibrinogen-like sequence D-phenylalanine-L-proline-L-arginine (D-Phe-Pro-Arg; <u>35</u>) with a suitable electrophile (46,47). Following inhibitor binding, the electrophilic tail reacts with the active-site serine to form a tetrahedral intermediate which mimics the high energy tetrahedral intermediate formed during normal substrate hydrolysis. A series of arginal thrombin inhibitors has been developed in which the C-terminal acid of <u>35</u> has been replaced by an aldehyde. The compound D-Me-Phe-Pro-Arg-H, <u>36</u>, was prepared as a stable, potent inhibitor of thrombin which effectively prolongs the thrombin time (TT) and activated partial thromboplastin time (APTT) without substantially interfering with fibrinolysis induced by plasmin, urokinase or t-PA (48-50). Oral antithrombotic efficacy in the rabbit has been reported at 10 mg/kg (48). An extensive SAR surrounding the N-terminal residue has led to a number of more potent and selective agents (51). D-Me-Phg-Pro-Arg-H (Phg = phenylglycine) has been extensively studied as an antithrombotic agent and as an adjunct to t-PA-induced thrombolysis (51,52). Trigonal boron possesses an empty 2p-orbital which is capable of reacting with nucleophiles such

<u>35</u> R = H; X = NHC(NH)NH$_2$; Y = CO$_2$H
<u>36</u> R = Me; X = NHC(NH)NH$_2$; Y = CHO
<u>37</u> R = Ac; X = NHC(NH)NH$_2$; Y = B(OH)$_2$
<u>38</u> R = H; X = NHC(NH)NH$_2$; Y = COCF$_3$
<u>39</u> R = H; X = NHC(NH)NH$_2$; Y = CN

as alkoxides. Equipping the tripeptide <u>35</u> with a C-terminal boronic acid resulted in a series of boroarginines which displayed potent, slow, tight-binding inhibition of thrombin (53). One of these compounds, DUP-714, <u>37</u>, effectively prolonged the APTT in rabbits following both i.v. and s.c. administration (53). It was shown to be efficacious in venous and arterial models of thrombosis and as an adjunct to thrombolytic therapy using streptokinase (54,55). It proved to be orally effective in the rat at 3 mg/kg (56) and in the dog at 5 mg/kg (57). Additional modifications to the carboxy terminus of <u>35</u> have included electrophilic ketones, phosphonate esters and nitriles. Arginine fluoroalkyl ketones react with the active-site serine in trypsin-like proteases (58) and incorporation of a trifluoromethyl ketone into the D-Phe-Pro-Arg sequence resulted in a potent, tight-binding thrombin inhibitor (<u>38</u>) which prolonged the APTT and TT in a dose-dependent manner (59). Nitrile <u>39</u> proved to be an effective inhibitor both *in vitro* and *in vivo* (60,61). However, the hemodynamic and pharmacokinetic profiles of <u>39</u> limits its potential clinical use (62). The arginine sidechain has been successfully replaced by a number of hydrophobic groups in the C-terminal boronic (63,64) and diphenyl phosphonate (65,66) ester analogs, resulting in peptides <u>40</u> and <u>41</u> which are potent and selective inhibitors of thrombin *in vitro* (63-66). Incorporation of the more hydrophobic diphenylalanine at the N-terminus, (<u>41</u>), provided greater thrombin affinity and has been studied in the aforementioned series of aldehydes (64), as well as the C-terminal piperidides (64) such as <u>42</u>.

<u>40</u> R$_1$ = H; R$_2$ = Cbz; Y = *S*-Pinanediol Boronate   <u>42</u>
<u>41</u> R$_1$ = C$_6$H$_5$; R$_2$ = H; Y = P(O)(OC$_6$H$_5$)$_2$

Peptidyl thrombin inhibitors containing arginine have been isolated from the marine sponge *Theonella sp.*. Both cyclotheonamide A (**43**), and nazumamide A (**44**), are potent and selective inhibitors of thrombin *in vitro* (67, 68).

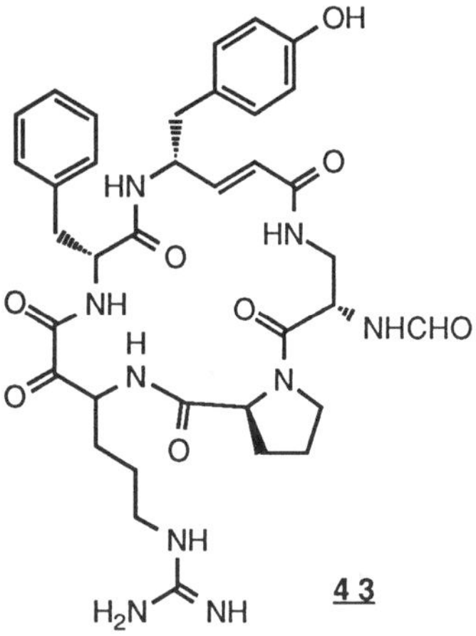

43

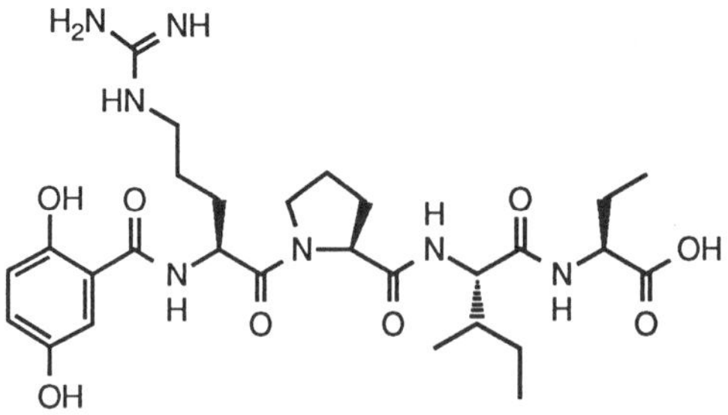

44

Irreversible active-site inhibitors of thrombin, which target the serine in the catalytic site of thrombin, have also been developed. Incorporation of a basic guanidinium ion into a β-lactam nucleus led to a series of 3-guanidinoalkyl-2-azetidinones which are slow-binding, irreversible inhibitors (69). Monobactam **45** proved to be efficacious at micromolar concentrations in the thrombin clotting assay and displayed high selectivity for thrombin over trypsin. The isocoumarin nucleus has also been equipped with a basic guanidine as in compound **46**, resulting in an effective plasma anticoagulant (70). Thrombin inhibition was time-dependent and irreversible. Although **46** displayed significant antithrombin activity in the thrombin time assay, its instability in plasma ($t_{1/2} \sim 8$ min) limits its use *in vivo*. Despite the lack of a positive charge, the neutral coumarin **47** was shown to be a potent inhibitor which displayed high thrombin specificity relative to both plasmin and t-PA (71). The potency of **47** suggests that in the absence of a basic group, tight binding to thrombin can be achieved by interactions between secondary recognition elements of both the inhibitor and enzyme. Isatoic anhydrides such as **48**, also irreversibly inhibit thrombin *in vitro* (72). The N-benzyl substituent imparts thrombin selectivity over trypsin, chymotrypsin and plasmin.

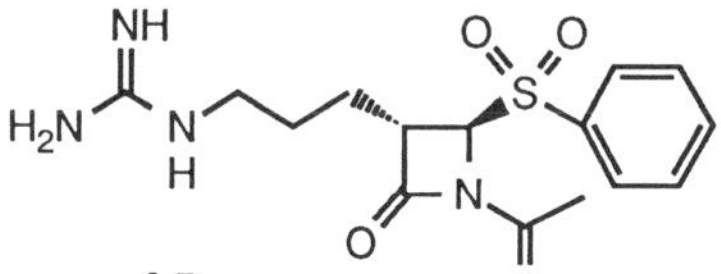

45

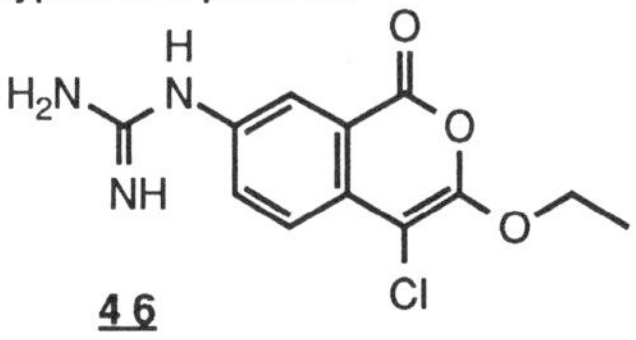

46

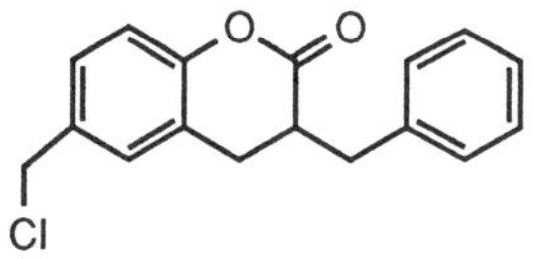

47

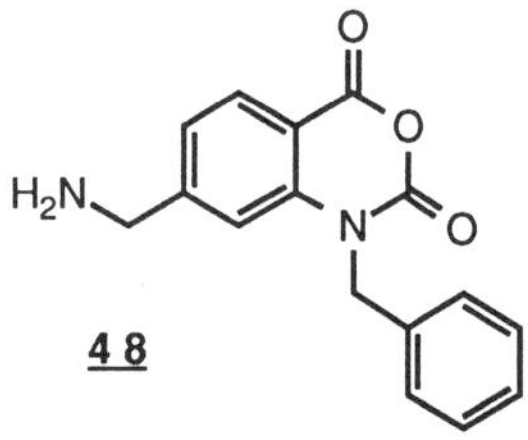

48

Hirudin-based antithrombins are modelled after the natural 65-66 amino acid protein found in the medicinal leech (*Hirudo medicinalis*). Hirudin is the most potent reversible thrombin inhibitor yet discovered (73). Recombinant desulfato hirudins, generated in yeast and *E. coli,* were efficacious in animal models of thrombosis (74) and have performed well in early testing in man (75). A synthetic, hirudin carboxy-terminal (Hir_{53-64}), Tyr-sulfated, dodecapeptide (hirugen, **49**) binds to thrombin's anion-binding exosite and displayed moderate anticoagulant and antithrombotic activity (76-78). By incorporation of an active-site directed D-Phe-Pro-Arg moiety into the dodecapeptide together with the appropriate spacer, a thrombin inhibitor, hirulog-1 (50), rivalling hirudin in potency (hirulog K_i = 2.3 nM) has been constructed (79). Preclinical and clinical testing of hirulog are encouraging (80,81). Further modifications to exosite directed peptides are also underway (82-86).

Ac-Asn-Gly-Asp-Phe-Glu-Glu-Ile-Pro-Glu-Glu-(SO₃-Tyr)-Leu   **49**

D-Phe-Pro-Arg-Pro-(Gly)₄-Asn-Gly-Asp-Phe-Glu-Glu-Ile-Pro-Glu-Glu-(SO₃-Tyr)-Leu   **50**

Factor XIII Inhibitors - Factor XIIIa is the transglutaminase that catalyzes interchain cross-linking reactions and serves to stabilize newly formed fibrin clot. The observation that the factor XIIIa inhibitor, L-722,151 (51), accelerated thrombolysis in both the rabbit and dog suggest that inhibitors of this enzyme may have potential in thrombolytic regimens (87,88).

51   **52**   **53**

A series of 2-thiol imidazoles (52) and imidazolium salts (53) have been studied as factor XIIIa inhibitors (89,90). Both series inhibit the Factor XIIIa catalyzed incorporation of [^{14}C]putrescine into casein, however the imidazolium salts are more effective. In an effort to develop more selective, active-site directed agents, a substrate-based inhibitor program has been initiated based on the finding that the N-terminal dodecapeptide of α-2-antiplasmin, 54, is a good substrate for factor XIIIa (91). Dissection of the dodecapeptide has led to a pentapeptide substrate, 55, which displays binding kinetics similar to native 54.

H-Asn-Gln-Glu-Gln-Val-Ser-Pro-Leu-Thr-Gly-Leu-Lys-OH   **54**

H-Asn-Gln-Glu-Gln-Val-OH   **55**

Conclusions -These are exciting days for those studying hemostasis and thrombosis. Based on the opportunities afforded by advances in molecular biology and by elucidation of the molecular basis of receptor/ligand and enzyme/substrate interactions, great strides have been made in our understanding of hemostasis and thrombosis. Supported by studies utilizing potent inhibitors found naturally - *e.g.*, the potent antithrombin activities of hirudin and antiplatelet activities found in a variety of snake venoms - a large effort has been launched to develop more powerful, mechanism-based antithrombotic agents. With these novel agents now moving towards large scale clinical testing there will no doubt be successes and failures. It remains to be determined whether the large potential benefit afforded by more effective agents outweighs the inherent bleeding liability.

References

1. S. Sherry, J. Int. Med., 229,, 113 (1991).
2. R.J. Shebuski, Annu. Rep. Med. Chem., 26, 93 (1991).
3. R.W.Colman, V.J. Marder, E.W. Salzman, and J. Hirsh, In: "Hemostasis and Thrombosis," 2nd edition, R.W. Colman, J. Hirsh, V.J. Marder and E.W. Salzman (Eds), J.B. Lippincott, New York, NY, 1987, p 3.
4. V. Fuster, L. Badimon, J. Badimon and J.H. Cheesebro, N. Engl. J. Med., 326, 242 (1992).
5. V. Fuster, L. Badimon, J. Badimon and J.H. Cheesebro, N. Engl. J. Med., 326, 310 (1992).
6. J. Hirsch and E.W. Salzman In: "Hemostasis and Thrombosis," 2nd edition, R.W. Colman, J. Hirsh, V.J. Marder and E.W. Salzman (Eds), J.B. Lippincott, New York, NY, 1987, p. 1199.
7. C.H. Hennekens, J.E. Buring, P. Sandercock, R. Collins and R. Peto, Circulation, 80, 749 (1989).
8. C.A. Veale, R.T. Jacobs and D.J. Wolanin, Annu. Rep. Med. Chem., 27, chapter 12 (1992).
9. C. Patrono, G. Davi and G. Ciabattoni, Trends Cardiovasc. Med., 2, 15 (1992).
10. E.W. Collington and H. Finch, Ann. Rep. Med. Chem., 25, 99 (1990).
11. S.E. Hall, Med. Res. Revs., 11, 503 (1990).
12. R.N. Misra, B.R. Brown, P.M. Sher, M.M. Patel, H.J. Goldenberg, I.M. Michel and D.N. Harris, Bioorg. Med. Chem. Lett., 1, 461 (1991).

13. R.N. Misra, B.R. Brown, P.M. Sher, M.M. Patel, S.E. Hall, W-C Han, J.C. Barrish, D.M. Floyd, P.W. Sprague, R.A. Morrison, R.E. Ridgewell, R.E. White, G.C. DiDonato, D.N. Harris, A. Hedberg, W.A. Schumacher, M.L. Webb and M.L. Ogletree, Bioorg. Med. Chem. Lett., 2, 73 (1992).

14. K. Kato, S. Ohkawa, S. Terao, Z. Terashita , K. Nishikawa, J. Med. Chem., 28, 287(1985).

15. S.S. Bhagwat, C. Gude, D.S. Cohen, W. Lee, P. Furness and F.H. Clarke, J. Med. Chem., 34, 1790 (1991).

16. P. Gresle, H. Deckmyn, G.G. Nenci and J. Vermylen, Trends Pharmacol. Sci., 12, 158 (1991).

17. I.S. Watts, K.A. Wharton, B.P. White and P.Lumley, Br. J. Pharmacol., 102, 497 (1991).

18. J.A. Hsia and A.M. Ross, Cor. Artery Dis., 3, 103 (1992).

19. D.S. Cohen, S. Bhagwat, R. Dotson, J. Mathis, P. Furness, M. Louzan, W. Lee, J. Peppard and C. Gude, Thomb. Haemo., 65, 1077 (1991).

20. R.L. Webb, R.W. Olson R. Dotson, J. Mathis and D.S. Cohen, Thomb. Haemost., 65, 1177 (1991).

21. I.B. Campbell, E.W. Collinton, H. Finch, P. Hallett, R. Hayes, P. Lumley, K. Mills, C.J. Wallis and B.P. White, Bioorg. Med. Chem. Letts., 1, 689 (1991).

22. I.B. Campbell, E.W. Collinton, H. Finch, R. Hayes, P. Lumley, K. Mills, G.M. Robertson, K. Wharton and I.S. Watts, Bioorg. Med. Chem. Lett., 1, 695 (1991).

23. A. Heckel, J. Nickl, R. Soyka, W. Eisert, T. Muller, J. Weisenberger, C. Meade and G. Muacevic, European Patent 397044A2 (1990).

24. I.B. Campbell, E.W. Collinton, H. Finch, R. Hayes, P. Lumley, K. Mills, N.B. Pike, G.M. Robertson and I.S. Watts, Bioorg. Med. Chem. Lett., 1, 699 (1991).

25. D. McTavish, D. Faulds and K.L. Goa, Drugs, 40, 238 (1990).

26. H. Caplain, G. Kieffer, J.F. Thiercelin, and J. Thebault, Thromb. Haemo., 62, 410 (1989).

27. G.D.O. Lowe, A. Rumley, P. Griffiths, B.G. Shaw, K.H. McGregor and C. Bouloux, Thromb. Haemo., 62, 414 (1989).

28. C.J. Kasiewski, Cook J.J., M.E. Hrinda, J. Newman, M.H. Perrone, and J.A. Barrett, Circulation, 84 (suppl II), II-247 (1991).

29. B.S. Coller, L.E. Scudder, J. Beer, H.K. Gold, J.D. Folts, J. Cavagnaro, R. Jourdan, C. Wagner, J. Luliucci, D. Knight, J. Ghrayeb, C. Smith, H.F. Weisman and H. Berger, Ann. New York Acad. Sci., 614, 193 (1990).

30. J. Samanen, F. Ali, T. Romoff, R. Calvo, E. Sorenson, J. Vasko, D. Berry, D. Bennett, M. Strohsacker, D. Powers, J. Stadel, and A. Nichols, J. Med. Chem., 34, 3114 (1991).

31. A. Nichols, J. Vasko, P. Koster, J. Smith, F. Barone, A. Nelson, J. Stadel, D. Powers, G. Rhodes, Miller-Stein C. V. Boppana, D. Bennett, D. Berry, T. Romoff, R. Calvo, F. Ali, E. Sorenson and J. Samanen, Eur. J. Pharmacol., 183, 2019 (1990).

32. R.F. Nutt, S.F. Brady, C.D. Colton, J.T. Sisko, T.M. Ciccarone, M.R. Levy, M.E. Duggan, I.S. Imagire, R.J. Gould, P.S. Anderson and D.F. Veber, in "Peptides: Chemistry and Biology", Proceedings of the 12th American Peptide Symposium, J.A. Smith and J.E. Riveier Eds., ESCOM Science Publishers B.V., Leiden, The Netherlands, 1992, p. 914.

33. S.F. Brady, J.T. Sisko, T.M. Ciccarone, C.D. Colton, M.R. Levy, K.M. Witherup, M.E. Duggan, J.F. Paycheck, O.E. Moreno, M.S. Egbertson, G.D. Hartman, W. Halczenko, W.L. Laswell, T-J. Lee, W.J. Holtz, W. Hoffman, G.E. Stokker, R.L. Smith, D.F. Veber and R.F. Nutt, in "Peptides: Chemistry and Biology", Proceedings of the 12th American Peptide Symposium, J.A. Smith and J.E. Riveier Eds., ESCOM Science Publishers B.V., Leiden, The Netherlands, 1992, p. 657.

34. L. Alig, A. Edenhofer, M. Müller, A. Trzeciak and Weller T, U.S. Patent 5,039,805 (1991).

35. B. Steiner, P. Hadváry, S. Roux and T. Weller, Thomb. Haemo., 65, 748 (1991).

36. S. Roux, B. Steiner, P. Hadváry, and T. Weller, Thomb. Haemo., 65, 812 (1991).

37. L.W. Schaffer, J.T. Davidson, C.T. Dunwiddie, G.P. Vlasuk, E.M. Nutt and P.K.S. Siegl, Circulation, 84 (suppl. II), A522 (1991).

38. C. Dunwiddie, M. Holahan, G. Vlasuk, J. Lynch Jr. and M. Mellott, Circulation, 84 (suppl. II), A522 (1991).

39. J.W. Jacobs, E.W. Cupp, M. Sardana and P.A. Friedman, Thromb. Haemos., 64, 235 (1990).

40. J. Hauptmann, B. Kaiser, G. Nowak, J. Stürzebecher and F. Markwardt, Thromb. Haemo., 63, 220 (1990).

41. For an extensive review on thrombin inhibitors prior to 1984 see J. Stürzebecher in The Thrombin; R. Machovich; CRC Press, Inc.: Boca Raton, FL, 1984; Vol 1, Chapter 7.

42. W. Bode and D. Turk, Thromb. Haemo., 65, 774 (1991).

43. W. Bode, D. Turk and J. Stürzebecher, Eur. J .Biochem., 193, 175 (1990).

44. D.W. Banner and P. Hadváry, J. Biol. Chem., 266, 20085 (1991).

45. D. Turk, J. Stürzebecher and W. Bode, FEBS Letters, 287, 133 (1991).

46. S. Bajusz, E. Barabás, E. Széll and D. Bagdy, in "Peptides: Chemistry, Structure and Biology", Proceedings of the Fourth American Peptide Symposium; R. Walter, and J. Meienhofer Eds.; Ann Arbor, Science Publishers Inc.: Ann Arbor, MI 1975; p 603.

47. F. Ni and H.A. Scheraga, Biochemistry, 27, 4481 (1988).

48. S. Bajusz, E. Szell, D. Bagdy, E. Barabas, G. Horvath, M. Dioszegi, Z. Fittler, G. Szabo, A. Juhasz, E. Tomori and G. Szilagyi. J. Med. Chem., 33, 1729 (1990).

49. D. Bagdy, E. Barabás, S. Bajusz and E. Széll, Thromb. Haemo., 67, 325 (1992).

50. D. Bagdy, E. Barabás, G. Szabó, S. Bajusz and E. Széll, Thromb. Haemo., 67, 357 (1992).

51. R.T. Shuman, R.B. Rothenberger, C.S. Campbell, G.F. Smith, D.S. Gifford-Moore and P.D. Gesellchen, in "Peptides: Chemistry and Biology", Proceedings of the12th American Peptide Symposium; J.A. Smith and J.E. Rivier Eds., ESCOM Science Publishers B.V.: Leiden, The Netherlands, 1992, p 801.

52. G.F. Smith, R.T. Shuman, P.D. Gesellchen, T.J. Craft, D. Gifford, K.D. Kurz, C.V. Jackson, G.E. Sandusky and P.D. Williams, Circulation, 84 (suppl. II), II-579 (1991).

53. C. Kettner, L. Mersinger and R. Knabb, J. Biol. Chem., 265, 18289 (1990).

54. R.M. Knabb, C.A. Kettner, P.B.M.W.M. Timmermans and T.M. Reilly, Thromb. Haemo., 67, 56 (1992).

55. R.M. Knabb, C.A. Kettner and T.M. Reilly, Circulation, 84 (suppl. II), II-467 (1991).

56. M.A. Hussain, R. Knabb, B.J. Aungst and C. Kettner, Peptides, 12, 1153 (1991).

57. R.M. Knabb, J.H. Shaw, J.M. Luettgen, C.A. Kettner, P.B. Timmermans and T.M. Reilly, Circulation, 84 (suppl. II), II-580 (1991).

58. T. Ueda, C.-M. Kam and J.C. Powers, Biochem. J., 265, 539 (1990).

59. B. Neises and T. Tarnus, Thrombo. Haemo., 65, 1290 (1991).

60. W. Stüber, US Patent 4,927,809 (1990).

61. B. Kaiser, M. Richter, J. Hauptmann and F. Markwardt, Pharmazie, 46, 128 (1991).

62. B. Kaiser, J. Hauptmann and F. Markwardt, Pharmazie, 46, 131 (1991).

63. G. Claeson, M. Philipp, R. Metternich, E. Agner, T. DeSoyza, M.F. Scully and V.V. Kakker, Thromb. Haemo., 65, 1289 (1991).

64. G. Claeson, L. Cheng, N. Chino, J. Deadman, S. Elgendy, V.V. Kakkar, in "Peptides: Chemistry and Biology, Proceedings of the 12th American Peptide Symposium; J.A. Smith and J.E. Rivier Eds.; ESCOM Science Publishers B.V., The Netherlands, 1992, p 824.

65. L. Cheng, C.A. Goodwin, M.F. Scully, V.V. Kakkar and G. Claeson, Tetrahedron Lett., 32, 7333 (1991).

66. L. Cheng, C. Goodwin, M.F. Scully, V.V. Kakkar and G. Claeson, In Peptides: Biology, Proceedings of the Twelth American Peptide Symposium; J.A. Smith and J.E. Rivier Eds.; ESCOM Science Publishers B.V.: The Netherlands, 1992; p 822.

67. N. Fusetani and S Matsunaga, J. Am. Chem. Soc., 112, 7053 (1990).

68. N. Fusetani, Y. Nakao and S. Matsunaga, Tetrahedron Lett., 32, 7073 (1991).

69. W.T. Han, US Patent 5,037,819, (1991).

70. C.-M. Kam, K. Fujikawa and J.C. Powers, Biochemistry, 27, 2547 (1988).

71. A. Mor, J. Maillard, C. Favreau and M. Reboud-Ravaux, Biochim. Biophys. Acta, 1038, 119 (1990).

72. M.H. Gelb and R.H. Abeles, J. Med. Chem., 29, 585 (1986).

73. P.H. Johnson, P. Sze, R. Winant, P.W. Payne and J.B. Lazar, Semin. Thromb. Haemo., 15, 302 (1989).

74. L. Badimon, J.J. Badimon, R. Lassila, M. Heras, J.H. Cheesebro, and V. Fuster, Blood, 78, 423 (1991).

75. F. Markwardt, G. Nowak and J. Sturzebecher, Haemostasis, 21 (suppl.1), 133 (1991).

76. J.M. Maraganore, B. Chao, J. Jablonski, and K.L. Ramachandran, J. Biol. Chem., 264, 8692 (1989).

77. J.A. Jakubowski and J.M. Maraganore, Blood, 75, 399 (1990).

78. Y. Cadroy, J.M. Maraganore, S.R. Hanson and L.A. Harker, Proc. Natl. Acad. Sci. USA, 88, 1177 (1991).

79. J.M. Maraganore, P. Bourdon, J. Jablonski, K.L. Ramachandran and J.W. Fenton, Biochemistry, 29, 7095 (1990).

80. A. Dawson, P. Loynds, K. Findlen, E. Levin, T. Mant, J. Maraganore, D. Hanson, J. Wagner and I. Fox, Thromb. Haemo., 65, 830 (1991).

81. A.B. Kelly, J.M. Maraganore, P. Bourden, S.R. Hanson and L.A. Harker, Proc. Natl. Acad. Sci. USA, in press (1992).

82. J.L. Krstenansky, M.H. Payne, T.J. Owen, M.I. Yates and S.J.T. Mao, Thromb. Res., 54, 319 (1989).

83. C.F. Church, J.E. Phillips and J.L. Woods, J. Biol. Chem., 266, 11975 (1991).

84. T. Kline, C. Hammond, P. Bourdon and J.M. Maraganore, Biochem. Biophys. Res. Comm., 177, 1049 (1991).

85. P. Bourdon, J-A. Jablonski, B.H. Chao and J.M. Maraganore, FEBS Letts., 294, 163 (1991).

86. J. DiMaio, F. Ni, B. Gibbs and Y. Konishi, FEBS Letts., 282, 47 (1991).

87. E.M. Leidy, A.M. Stern, P.A. Friedman and L.R. Bush, Thromb. Res., 59, 15, (1990).

88. R.J. Shebuski, G.R. Sitko, D.A. Claremon, J.J. Baldwin, D.C. Remy and A.M. Stern, Blood, 75, 1455 (1990).

89. J.J. Baldwin, D.C. Remy, and D.A. Claremon, US Patent 4,968,713, (1990).

90. D.A. Claremon, J.J. Baldwin, and D.C. Remy, US Patent 5,019,572, (1991).

91. P.M. Doyle, C.J. Harris, K.R. Carter, D.S.A. Simpkin, P. Bailey-Smith, D. Stone, L. Russell and G.J. Blackwell, Biochem. Soc. Trans., 18, 1318 (1990).

Chapter 12. Pulmonary and Anti-Allergy Agents

Robert T. Jacobs, Chris A. Veale and Donald J. Wolanin
ICI Americas, Inc.
Wilmington, Delaware 19897

<u>Introduction</u> - Asthma continues to be the major focus of pulmonary research (1-4). Further substantiation of the involvement of peptidoleukotrienes in allergen, exercise and aspirin-induced asthma has been obtained from clinical evaluation of potent leukotriene antagonists (5-7). In contrast, trials with 5-lipoxygenase inhibitors (8) and platelet-activating factor antagonists (9) have been less definitive. Investigations of neuropeptide antagonists as an approach to the treatment of asthma have gained momentum wtih the discovery of non-peptidic tachykinin antagonists (10-13). The literature debate over the role of β-adrenergic agonists in the treatment of asthma has intensified. Past studies have linked increased usage of β-agonists to increased morbidity and mortality in asthmatics; critics have cautioned that the design of these studies do not allow extension of their conclusions to all asthmatics and to all β-agonists (14,15). However, results from a new study - commonly referred to as the Saskatchewan study - suggest that inhaled β-agonists are indeed associated with higher mortality and that this is an effect common to all β-agonists (16,17). An hypothesis has been advanced which suggests that β-agonist inhibition of mast cell degranulation, and thus suppression of a natural anti-inflammatory mechanism, may be responsible for increased asthmatic morbidity and mortality (18).

REGULATORS OF LIPID MEDIATORS

<u>Leukotriene D_4/E_4 Antagonists</u> - The current status of research on leukotriene antagonists and their role in asthma have been reviewed (19,20). Clinical studies showed that ICI 204,219 (<u>1</u>), given orally, could reduce antigen-induced bronchoconstriction in asthmatic patients (5,21-24). One study concluded that <u>1</u> also decreased late phase hyperreactivity to histamine (5), although this conclusion has been questioned (25). Evaluation of MK-571 (<u>2</u>) as an inhibitor of LTD_4-induced bronchoconstriction following intravenous administration to healthy volunteers and asthmatic patients has been described (6, 26). Development of the (R)-isomer of <u>2</u>, MK-679 (<u>3</u>), which was selected for clinical evaluation based on preclinical in vitro (pK_B = 7.9-9.0 on human trachea) and in vivo (blockade of antigen-induced bronchoconstriction in rats at 0.03 mg/kg, po) pharmacology, has been terminated due to poor tolerance (27-29). Compound RG 12525 (<u>4</u>), when dosed orally to mild asthmatics, was found to cause a 7.5-fold shift of the dose-response to inhaled LTD_4 (30). Examination of ONO-1078 (<u>5</u>), orally dosed to normal and asthmatic subjects, showed inhibition of both LTD_4 and antigen-induced bronchoconstriction (31). Oral administration of SR 2640 (<u>6</u>) to asthmatic patients was found to significantly reduce LTD_4-induced bronchoconstriction (32). Aerosol administration of SKF 104353 (<u>7</u>) was effective in reducing bronchoconstriction induced by LTC_4 or LTE_4 (33), aspirin (7) or platelet-activating factor (34) in man. Compound CGP 45715A (<u>8</u>) has high potency versus LTD_4 in isolated guinea pig lung ileum (pA_2 = 10.1) and conscious guinea pigs (IC_{50} = 0.3 mg/kg, po) with a long duration of action (35). A new class of LT antagonist was exemplified by Ro 24-5913 (<u>9</u>), which was active both in isolated human bronchi (pK_B = 9.3) and conscious guinea pigs (ID_{50} = 0.12 mg/kg, po) (36, 37).

5- Lipoxygenase (5-LO) Inhibitors - Zileuton (A-64,077, **10**), dosed orally to atopic asthmatics, only partially inhibited allergen-induced 5-LO activity *in vivo*, despite almost complete *ex vivo* 5-LO inhibition (8). An interesting correlation between pseudoperoxidase activity and 5-LO inhibitory potency has been described for a series of N-hydroxyureas (**11**), suggesting that redox properties of this class contribute to their 5-LO inhibition (38). A series of quinoline-containing inhibitors of LT biosynthesis, exemplified by L-674,573 (**12**), have been found to bind to the 5-lipoxygenase activating protein (FLAP). Their ability to interact with FLAP correlated well with inhibition of LT biosynthesis. Other quinoline-based compounds such as REV-5,901 (**13**) and WY-49,232 (**14**) also bind to FLAP, suggesting that these compounds act *via* inhibition of 5-LO translocation (39). Compound A-69,412 (**15**) was reported to be a potent, selective inhibitor of 5-LO with a long duration of action. Oral administration of **15** to rats effectively inhibited *ex vivo* ionophore-stimulated LTB_4 biosynthesis (40). ICI-211,965 (**16**) emerged from a structure-activity relationship (SAR) study of a (methoxyalkyl)thiazole series designed to inhibit 5-LO (41). Mechanistically, the inhibitory activity of this series was not dependent on redox properties or iron chelation. Compound **16** was orally active (ED_{50} = 10 mg/kg, *ex vivo*) and showed no inhibition of cyclooxygenase.

Thromboxane Receptor Antagonists/Thromboxane Synthetase Inhibitors - The development of thromboxane receptor antagonists (TXRAs) has been reviewed (42). Oral administration of the

TXRA AA-2414 (<u>17</u>) to asthmatic subjects favorably attenuated their response to methacholine challenge (43). Pharmacodynamic studies on the TXRA ICI 192,605 (<u>18</u>) demonstrated significant inhibition of U46619-induced platelet aggregation up to 12 hours after a single oral dose (44). The same dose, however, had no significant effect on bronchial response to antigen challenge in asthmatic subjects (45). In contrast, oral administration of the thromboxane synthetase inhibitor CGS 13080 (<u>19</u>) to asthmatic subjects inhibited fall in forced expiratory volume (FEV_1) following antigen challenge (46). TXRA S-1452 (<u>20</u>), the (+)-isomer of S-145, had an ED_{50} of 0.026 mg/kg (po) against antigen-induced bronchoconstriction in sensitized guinea pigs (47). The TXRA U-3405 (<u>21</u>) exhibited a $pA_2 = 8.8$ against U46619-induced contraction of human bronchus (48). Intravenous administration of <u>21</u> to sheep protected them from increases in pulmonary resistance induced by U46619 (49). Further modification of a 7-oxabicyclo[2.2.1]heptane series of TXRA by incorporation of a phenylene spacer in the α-chain afforded SQ 35,091 (<u>22</u>), which exhibited an IC_{50} of 3 nM versus arachidonic acid induced platelet aggregation and a long duration of action (T_{50}=16h) versus U46619-induced sudden death in mice (50). The oxazole-containing analog SQ-33,961 (<u>23</u>) had a similar pharmacological profile, with less potential toxicity (51).

<u>PAF Antagonists</u> - The 1988-1990 literature on the possible clinical implications of platelet-activating factor (PAF) has been reviewed (52,53). The potential effects of PAF on pulmonary function through activation of human eosinophils, neutrophils and macrophages have been examined (54,55). The hetrazepine PAF antagonist apafant (WEB-2086, <u>24</u>) has been utilized in various clinical trials (56). A study with healthy volunteers demonstrated that <u>24</u> was rapidly absorbed following oral administration and produced no significant adverse effects (57). The potential of PAF antagonists as therapeutic agents in asthma was equivocated by a study in which short term treatment (100 mg, po, tid for 1 week) with <u>24</u> did not affect allergen-induced asthmatic responses in human subjects (9). In a separate trial, oral dosing of <u>24</u> in man provided potent inhibition of *in vivo* cutaneous and *ex vivo* platelet responses to PAF (58). Clinical results have also appeared on SCH-37370 (<u>25</u>), a representative from a series of compounds which display dual antagonism of PAF and histamine (59). *Ex vivo* studies with plasma from subjects who received a single oral dose of <u>25</u> demonstrated that antihistamine activity persisted longer than anti-PAF activity (60). Oral dosing of <u>25</u> attenuated histamine-induced airway hyperreactivity in a study with stable asthmatics (61). The hetrazepine class of PAF antagonists has been extensively studied; WEB-2347 (<u>26</u>), E-6123 (<u>27</u>) and <u>28</u> have emerged as orally active derivatives with profiles superior to that of <u>24</u> when measured by inhibition of PAF-induced bronchoconstriction in guinea pigs (62-65). Structurally novel PAF antagonist SR-27417 (<u>29</u>) displayed potent and long-lasting antagonism of PAF-induced responses in a variety of biological models (66). The synthesis and PAF antagonist activity of furan derivative MK-287 (L-680,573, <u>30</u>) have been summarized (67). Nicotinamide derivative TCV-309 (<u>31</u>) demonstrated potent inhibition of PAF-induced rabbit platelet aggregation *in vitro* and PAF-induced hypotension in conscious rats following either

intravenous or oral dosing (68). A short communication summarized new work on a PAF receptor model which may allow *de novo* design of PAF antagonists (69).

ANTIALLERGICS

The basic pharmacology and clinical results from several important second-generation histamine antagonists have been summarized (70,71). Recent studies on some of these agents have shown that they may possess characteristics in addition to their ability to antagonize histamine receptors. Cetirizine can block the release of some mediators and also reduces eosinophil migration (72). Loratadine suppresses histamine release (73), while azelastine is capable of preventing platelet aggregation and degranulation induced by stimulated polymorphonuclear neutrophils (74). Because several different assay systems were utilized for characterization of the second-generation antihistamines, quantitative comparisons of these agents are lacking. However, a recent paper describes a comparative study of *in vitro* and *in vivo* binding at both peripheral and central histamine receptors by a new drug, noberastine (R-72075, **32**), and several non-sedating antihistamines (75). This paper also reported results which refute current dogma about blood-brain barrier penetration, including the notion that penetration by antihistamines is directly related to drug lipophilicity. For example, while **32**, astemizole and terfenadine represent a broad range of lipophilicities, none of these antihistamines occupy brain receptors at doses which give maximal occupation of lung receptors. KW-4994 (**33**), a novel zwitterionic antiallergic, was orally active in the rat passive cutaneous anaphylaxis (PCA) test (ED_{50} = 0.92 mg/kg) and displayed anti-histaminic properties with no significant CNS effects (76,77). Compound **34** was an effective inhibitor of histamine release from mast cells (IC_{50} = 4.7 nM) and was active in the rat PCA model following intravenous or oral dosing (78). Asobamast (Z-1819, **35**) was orally active (ED_{50} = 4.7 mg/kg) against IgE-mediated passive pulmonary anaphylaxis in rats and showed a potent inhibition of antigen-induced mediator release from sensitized guinea pig lung fragments (79).

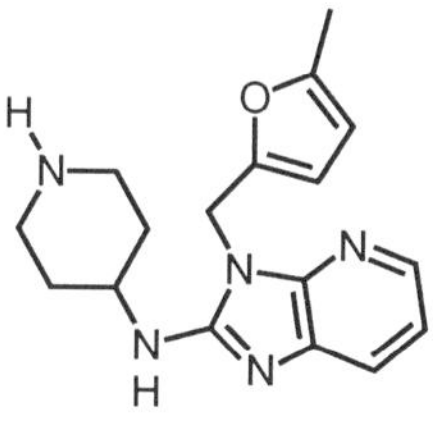

32 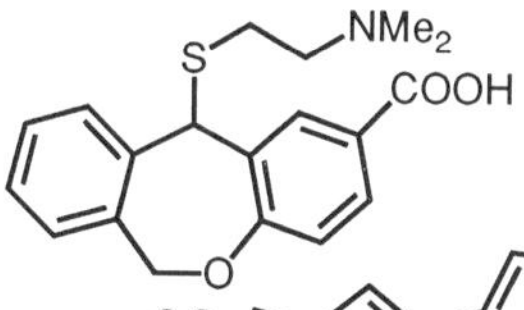   33 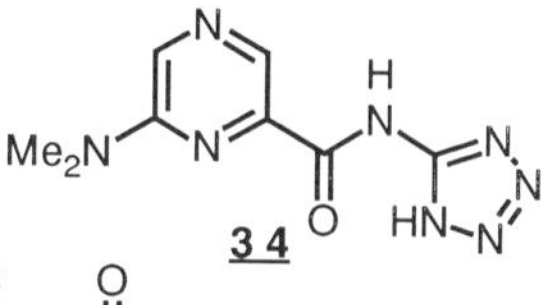   34 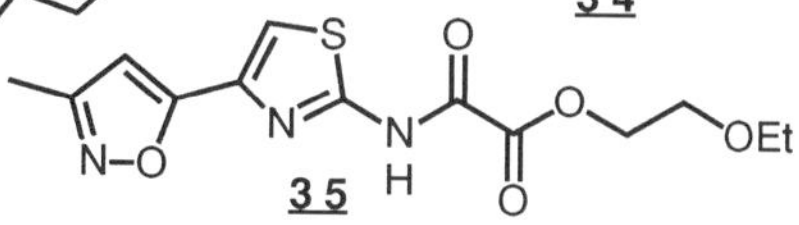   35

OTHER TOPICS

Potassium Channel Openers - The search for applications of potassium channel openers (PCO's) in asthma has continued (80). The (3S,4R)-isomer of cromakalim (lemakalim, **36**) inhibited cigarette smoke-induced plasma exudation and goblet cell secretion in guinea pig trachea, suggesting a role for PCO's in control of mucus secretion (81). An analog of **36**, BRL 55834 (**37**), has been reported to exhibit better selectivity for airway smooth muscle relative to vascular smooth muscle (82). Ring contraction of the pyran ring of **36** afforded a series of indanols exemplified by **38** , reported to have good activity versus histamine-induced dyspnea in conscious guinea pigs (83). The design (84) and synthesis (85) of Ro 31-6930 (**39**) have been reported; **39** inhibited agonist and allergen-induced bronchoconstriction in guinea pigs and cats (86).

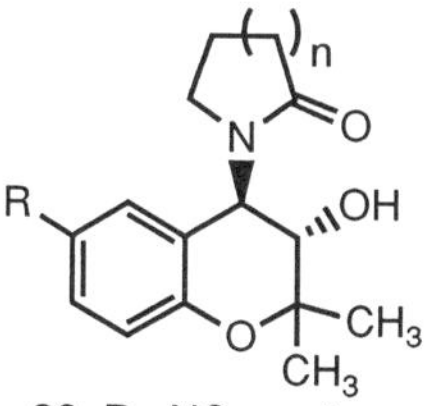

36, R= NC-; n=1

37, R= F$_3$CF$_2$C-; n=2

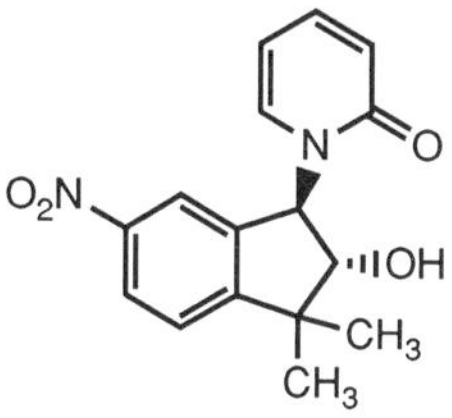

38 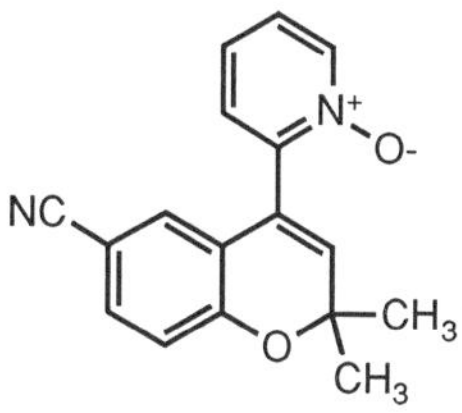   **39**

Bradykinin Antagonists - Bradykinin (BK, **40**) and bradykinin antagonists received increased attention and have been the subject of several recent reviews (87-90). The B$_2$ receptor from rat uterus was cloned and expressed (91). The primary sequence of this receptor suggests that it is a member of the G-protein-coupled receptor superfamily. In vitro autoradiography has been used to show that B$_2$ receptors are widely distributed in human lung tissue. The highest density of receptors were found on bronchial and pulmonary blood vessels, suggestive of BK playing a role in the regulation of airway and pulmonary blood flow, as well as airway epithelial function (92). The results of a clinical trial in volunteers failed to show a statistical difference between the effects of the B$_2$ receptor antagonist NPC-567 (**41**) and placebo in blocking the response to BK via intranasal administration (93). A study using aerosolized **41** to assess the effects of BK on allergen-induced early and late phase bronchial responses in allergic sheep showed a significant reduction in the late phase airway response and a corresponding decrease in associated late phase mediators (94). Subcutaneous administration of the potent BK antagonist HOE-140 (**42**) to anaesthetized guinea-pigs produced dose dependent antagonism of BK-induced bronchoconstriction with a t$_{1/2}$ of at least 4 hours (95). A study probing the topography of the β-turn accepting portion of the BK receptor has shown that the combination of a substituted L-4-hydroxyproline and an L-Oic residue in positions 7 and 8 of **42** greatly enhanced the probability of β-turn formation, leading to very potent peptide antagonists (96). NPC-17,730 (**43**) was reported to be 600-800 times more potent than NPC-567 and pharmacologically effective for up to 4 hours (97).

$\underline{40}$ BK = Arg^1-Pro^2-Pro^3-Gly^4-Phe^5-Ser^6-Pro^7-Phe^8-Arg^9

$\underline{41}$ D-Arg^0[Hyp^3, D-Phe^7]BK

$\underline{42}$ D-Arg^0[Hyp^3, Thi^5, D-Tic^7-Oic^8]BK

$\underline{43}$ D-Arg-Arg-Pro-Hyp-Gly-Thi-Ser—N ... Oic-Arg

Tachykinins - The role of tachykinins in pulmonary diseases has been reviewed (98-102). Molecular cloning, characterization and functional expression of both the NK-1 (103) and NK-2 (104) receptors was reported. Photoaffinity labeling of the NK-1 receptor by a photoreactive Substance P (SP) analog was described (105). The existence of two subtypes of NK-2 receptors, denoted as NK-2A and NK-2B, has been suggested on the basis of differential affinities of peptide antagonists (106, 107). Current evidence suggests that the NK-2A subtype is found in the guinea pig trachea, rabbit pulmonary artery and human bronchus (HB), whereas the NK-2B subtype is found in hamster urinary bladder muscle and hamster trachea preparations. Evidence for mediation of macromolecule secretion in airways by SP has been reported from studies on ferret trachea (108). Examination of contractions of isolated HB induced by neuropeptide γ ($\underline{44}$), a selective NK-2 receptor agonist, and antagonism of these contractions by NK-2 antagonists MDL 29913 ($\underline{45}$) and L-659,877 ($\underline{46}$), have suggested that the receptor found in this tissue is of the NK-2A subtype (109). A non-peptidic NK-1 receptor antagonist, CP-96,345 ($\underline{47}$) has been reported (10, 11); the ($2S,3S$) isomer has an IC_{50} of 3.4 nM versus [^{3}H]-SP in bovine brain, with no significant activity at NK-2 or NK-3 receptors. In addition, the ($2R,3R$) isomer of $\underline{47}$ was inactive. Another non-peptide antagonist of SP, the perhydroisoindole RP 67580 ($\underline{48}$), had an IC_{50} of 4.16 nM in rat brain (12). Evaluation of $\underline{48}$ versus SP-induced contraction in GP ileum provided further evidence of competitive antagonism (pA_2 = 7.16). No effect in binding assays or tissue preparations for NK-2 or NK-3 receptors was observed. Actinomycin D has been found to be a competitive antagonist at NK-2 receptors (110-112). An extremely potent non-peptide NK-2 antagonist, SR 48968 ($\underline{49}$), with a pA_2 = 9.2 on isolated HB has been revealed (13).

$\underline{44}$ Asp-Ala-Gly-His-Gly-Gln-Ile-Ser-His-Lys-Arg-His-Lys-Thr-Asp-Ser-Phe-Val-Gly-Leu-Met-NH_2

$\underline{45}$ cyclo-[Gln-Trp-Phe-Gly-Leu-CH_2NCH_3-Leu]

$\underline{46}$ cyclo-[Gln-Trp-Phe-Gly-Leu-Met]

Emphysema - A recent conference outlined the advances and present challenges in the treatment of pulmonary emphysema (113, 114). Reviews have appeared which detail the strategies and potential problems in designing inhibitors of human neutrophil elastase (HNE), the enzyme whose unrestrained action may result in the lung damage underlying the development of emphysema (115,116). Research towards gene therapy for persons hereditarily deficient in the production of α1-antitrypsin (α1AT, the natural inhibitor of HNE) advanced with the demonstration that recombinant human α1AT gene, transferred into epithelial cells in the respiratory tract of rats by an adenovirus, caused secretion of human α1AT (117). L-680,833 (50) , a potent and selective inhibitor of HNE, inhibited elastase-induced lung hemorrhage in hamsters (ED_{50} =1.5 mg/kg) when dosed orally 5 hours prior to challenge with enzyme (118). ONO-5046 (51) is a competitive inhibitor of HNE which is reported to be active in animal models by intratracheal or intravenous administration (119). Intratracheal administration of ICI 200,880 (52) to hamsters prevented the progression of lesions induced by pre-treatment with HNE (120). Compound L-658,758 (53) emerged as a clinical candidate from a study of mechanism-based inhibitors (121).

Cystic Fibrosis (CF)- A number of developments have increased the likelihood for improved therapy in CF. In vivo transfer of the human cystic fibrosis transmembrane conductance regulatory gene into rat airway epithelium was accomplished and expression demonstrated (122). Clinical trials using aerosolized recombinant DNase showed significant improvements in sputum volume, vital capacity, and FEV_1 compared to placebo (123,124). Chronic amiloride therapy increased both the sodium and chloride content of sputum in cystic fibrotics (125). Extracellular nucleotides (UTP and ATP) were found to be effective in vivo chloride secretagogues in the nasal epithelia of CF patients (126). The role of mast cell and neutrophil proteases in airway mucus hypersecretion was further elucidated; neutrophil elastase, cathepsin G and mast cell chymase were found to be potent mucus secretagogues(127). HNE inhibitor ICI-200,355 (54) decreased CF sputum-induced mucus secretion from bovine tracheal gland serous cells in a dose-dependent manner (128). This suggests that elastase inhibitors offer a potential strategy for therapeutic intervention in CF.

References

1.   "Advances in the Understanding and Treatment of Asthma", P.J. Piper and R.D. Krell, Eds., NY Acad. Sci., New York, NY, 1991.
2.   A.B. Kay, J. Allergy Clin. Immunol., 87, 893 (1991).
3.   P.J. Barnes, Trends in Biol. Sci., 16, 365 (1991).
4.   P.G. Bardin and S.T. Holgate, Drugs Today, 27, 107 (1991).
5.   I.K. Taylor, K.M. O'Shaughnessy, R.W. Fuller and C.T. Dollery, Lancet, 337, 690 (1991).
6.   J.C. Kips, G.F. Joos, I. De Lepeleire, D.J. Margolskee, A. Buntinx, R.A. Pauwels and M.E. van der Straeten, Am. Rev. Respir. Dis., 144, 617 (1991).
7.   P.E. Christie, C.M. Smith and T.H. Lee, Am. Rev. Respir. Dis., 144, 957 (1991).
8.   K.P. Hui, I.K. Taylor, G.W. Taylor, P. Rubin, J. Kesterson, N.C. Barnes, and P.J. Barnes, Thorax, 46, 184 (1991).
9.   A. Freitag, R.M. Watson, G. Hatsos, C. Eastwood and P.M. O'Byrne, Am. Rev. Respir. Dis., 143 , (4 Part 2), A157 (1991).
10. R.M. Snider, J. W. Constantine, J.A. Lowe III, K.P. Longo, W.S. Lebel, H.A. Woody, S.E. Drozda, M.C. Desai, F.J. Vinick, R.W. Spencer and H.-J. Hess, Science, 251, 435 (1991).
11. J.A. Lowe, III, S.E. Drozda, R.M. Snider, K.P. Longo and J. Bordner, Bioorg. Med. Chem. Lett., 1, 129 (1991).
12. C. Garret, A. Carruette, V. Fardin, S. Moussaoui, J.-F. Peyronel, J.-C. Blanchard and P.M. Laduron, Proc. Natl. Acad. Sci. USA, 88, 10208 (1991).
13. X. Emonds-Alt, P. Vilain, P. Goulaouic, V. Proietto, D. Van Broeck, C. Advenier, E. Naline, G. Neliat, G. Le Fur and J.C. Breliere, Life Sci., 50, PL101 (1992).

14. H.S. Nelson, S.J. Szefler and R.J. Martin, Am. Rev. Respir. Dis., 144, 249 (1991).
15. N. Pearce, J. Crane, C. Burgess, R. Jackson and R. Beasley, Clin. Exper. Allergy, 21, 401 (1991).
16. R.I. Horwitz, W. Spitzer, S. Buist, D. Cockcroft, P. Ernst, B. Habbick, B. Hemmelgarn, M. McNutt, A. Rebuck and S. Suissa, Chest, 100, 1586 (1991).
17. W.O. Spitzer, S. Suissa, P. Ernst, R.I. Horwitz, B. Habbick, D. Cockcroft, J.-F. Boivin, M. McNutt, A.S. Buist and A.S. Rebuck, N. Engl. J. Med., 326, 501 (1992).
18. C.P. Page, Lancet, 337, 717 (1991).
19. A. Shaw and R.D. Krell, J. Med. Chem., 34, 1235 (1991).
20. W. W. Busse and J. N. Gaddy, Am. Rev. Respir. Dis., 143, S103 (1991).
21. S.E. Dahlen, B. Dahlen, E. Eliasson, H. Johansson, T. Bjorck, M. Kumlin, K. Boo, J. Whitney, S. Binks, B. King, R. Stark and O. Zetterstrom, Adv. Prostaglandin, Thromboxane and Leukotriene Res., 21A, 461 (1991).
22. K.P. Hui and N.C. Barnes, Lancet, 337, 1062 (1991).
23. M. Glass, Ann. NY Acad. Sci., 629, 143 (1991).
24. R.A. Nathan, W.W. Storms, S.F. Bodman, M.C. Minkwitz and M. Glass, J. Allergy Clin. Immunol., 87, 256 (1991).
25. R. Aalbers and J.G.R. deMonchy, Lancet, 338, 445 (1991).
26. D.J. Margolskee, Ann. NY Acad. Sci., 629, 148 (1991).
27. T.R. Jones, R. Zamboni, M. Belley, E. Champion, L. Charette, A.W. Ford-Hutchinson, J.-Y. Gauthier, S. Leger, A. Lord, P. Masson, C.S. McFarlane, K. Metters, H. Piechuta, S. Pong, M. Therien and R.N. Young, Am. Rev. Respir. Dis., 143, A643 (1991).
28. Scrip, 1612, 23 (1991).
29. J-W. Lammers, P. vanDaele, C. vanHerwaarden, M. Decramer, I. De Lepeleire and B.S. Friedman, Am. Rev. Respir. Dis., 143, A643 (1991).
30. I. Wahenda, A.S. Wisniewski and A.E. Tattersfield, Br. J. Clin. Pharmacol., 32, 512 (1991).
31. T. Nakagawa, Y. Mizushima, A. Ishii, F. Nambu, M. Motoishi, Y. Yui, T. Shida and T. Miyamoto, Adv. Prostaglandin, Thromboxane and Leukotriene Res., 21A, 465 (1991).
32. L. Frolund, F. Madsen and J. Nielsen, Allergy, 46, 355 (1991).
33. P.E. Christie, B.W. Spur and T.H. Lee, J. Allergy Clin. Immunol., 88, 193 (1991).
34. D.A. Spencer, J.M. Evans, S.E. Green, P.J. Piper and J.F. Costello, Thorax, 46, 441 (1991).
35. M.A. Bray, W.H. Anderson, N. Subramanian, U. Niederhauser, M. Kuhn, M. Erard and A. von Sprecher, Adv. Prostaglandin, Thromboxane and Leukotriene Res., 21A, 503 (1991).
36. M. O'Donnell, H.J. Crowley, B. Yaremko, N. O'Neill and A.F. Welton, J. Pharm. Exp. Ther., 259, 751 (1991).
37. M. O'Donnell, Ann. NY Acad. Sci., 629, 413 (1991).
38. D. Riendeau, J.P. Falgueyret, J. Guay, N. Udea, and S. Yamamoto, Biochem. J. 274, 287 (1991).
39. J.F. Evans, C. Leveille, J.A. Mancini, P. Prasit, M. Therien, R. Zamboni, J.Y. Gauthier, R. Fortin, P. Charleson, D.E. Macintyre, S. Luell, T.J. Bach, R. Meurer, J. Guay, P.J. Vickers, C.A. Rouzer, J.W. Gillard, and D.K. Miller, Molecular Pharmacology, 40, 22 (1991).
40. P.R. Young, R.L. Bell, J. Bouska, C. Lanni, J.B. Summers, D.W. Brooks and G.W. Carter, FASEB J., 5, Abstract 5235 (1991).
41. T.G. Bird, P. Bruneau, G.C. Crawley, M.P. Edwards, S.J. Foster, J.M. Girodeau, J.F. Kingston, and R. M. McMillan, J. Med. Chem., 34, 2176 (1991).
42. S.E. Hall, Med. Res. Rev., 11, 503 (1991).
43. M. Fujimura, S. Sakamoto, M. Saito, Y. Miyake and T. Matsuda, J. Allergy Clin. Immunol., 87, 23 (1991).
44. S.B. French, R. Prasad, P. Leese, C. Yeh, and P.T. Thyrum, Clin. Pharm. Ther., 49, 141 (1991).
45. N. Eiser, Am. Rev. Respir. Dis., 143, A645 (1991).
46. P.J. Manning, W.H. Stevens, D.W. Cockroft and P.M. O'Byrne, Eur. Respir. J., 4, 667 (1991).
47. A. Arimura, F. Asanuma, A. Kurosawa, M. Harada, H. Nagai and A.Koda, J. Allergy Clin. Immunol., 87, 254 (1991)
48. M.G. McKenniff, P. Norman, N.J. Cuthbert and P.J. Gardiner, Br. J. Pharmacol., 104, 585 (1991).
49. N.A. Fitch, B.A. Briggs, D.L. Francis and Y. Naik, Am. Rev. Respir. Dis., 143, A644 (1991).
50. R.N. Misra, B.R. Brown, W.-C. Han, D.N. Harris, A. Hedberg, M.L. Webb and S.E. Hall, J. Med. Chem., 34, 2882 (1991).
51. R.N. Misra, B.R. Brown , P.M. Sher, M.M. Patel, H.J. Goldenberg, I.M. Michel and D.N. Harris, Bioorg. Med. Chem. Lett., 1, 461 (1991).
52. M. Koltai, D. Hosford, P. Guinot, A. Esanu and P. Braquet, Drugs, 42(1), 9 (1991).
53. M. Koltai, D. Hosford, P. Guinot, A. Esanu and P. Braquet, Drugs, 42(2), 174 (1991).
54. E.M. Zoratti, J.B. Sedgwick, R.R. Vrtis and W.W. Busse, J. Allergy Clin. Immunol., 88, 749 (1991).
55. T. Schaberg, H. Haller and H. Lode, Biochem. Biophys. Res. Commun., 177, 704 (1991).
56. Drugs Future, 16(3), 252 (1991).

57. H.M. Brecht, W.S. Adamus, H.O. Heuer, F.W. Birke and E.R. Kempe, Arzneim.-Forsch./Drug Res., 41(I), 51 (1991).
58. J.P. Hayes, S.M. Ridge, S. Griffith, P.J. Barnes and K.F. Chung, J. Allergy Clin. Immunol., 88, 83 (1991).
59. J.J. Piwinski, J.K. Wong, M.J. Green, A.K. Ganguly, M.M. Billah, R.E. West, Jr. and W. Kreutner, J. Med. Chem., 34, 457 (1991).
60. M.M. Billah, H.G. Gilchrest, S. Eckel, C. Granzow, E. Radwanski, M. Affrime, D. Pharm and M.I. Siegel, J. Allergy Clin. Immunol., 87, 310 (1991).
61. R. Berkowitz, J. Zora, D. Tinkelman, F. Cuss, M. Danzig, J. Fourre and J. Mooney, J. Allergy Clin. Immunol., 87, 167 (1991).
62. H.O. Heuer, J. Lipid Mediators, 4, 39 (1991).
63. Y. Sakuma, H. Tsunoda, M. Shirato, S. Katayama, I. Yamatsu and K. Katayama, Prostaglandins, 42(5), 463 (1991).
64. H. Tsunoda, Y. Sakuma, M. Shirato, H. Obaishi, K. Harada, K. Yamada, N. Shimomura, Y. Machida, I. Yamatsu and K. Katayama, Arzneim.-Forsch./Drug Res., 41(I), 224 (1991).
65. A. Walser, T. Flynn, C. Mason, H. Crowley, C. Maresca, B. Yaremko and M. O'Donnell, J. Med. Chem., 34, 1209 (1991).
66. J.M. Herbert, A. Bernat, G. Valette, V. Gigo, A. Lale, M.C. Laplace, L. Lespy, P. Savi, J.P. Maffrand and G. Le Fur, J. Pharmacol. Exp. Ther., 259, 44 (1991).
67. S.P. Sahoo, D.W. Graham, J. Acton, T. Biftu, R.L. Bugianesi, N.N. Girotra, C.-H. Kuo, M.M. Ponpipom, T.W. Doebber, M.S. Wu, S.-B. Hwang, M.-H. Lam, D.E. MacIntyre, T.J. Bach, S. Luell, R. Meurer, P. Davies, A.W. Alberts and J.C. Chabala, Bioorg. Med. Chem. Lett., 1, 327 (1991).
68. M. Takatani, N. Maezaki, Y. Imura, Z. Terashita, K. Nishikawa and S. Tsushima, Adv. Prostaglandin, Thromboxane and Leukotriene Res., 21B, 943 (1991).
69. J.P. Batt, A. Lamouri, F. Tavet, F. Heymans, G. Dive and J.-J. Godfroid, J. Lipid Mediators, 4, 343 (1991).
70. F. Estelle, R. Simons and K.J. Simons, Ann. Allergy, 66, 5 (1991).
71. F. Estelle, R. Simons and K.J. Simons, Clin. Pharmacokinet., 21, 372 (1991).
72. R.M. Naclerio, Allergy Proc., 12, 187 (1991).
73. M. Andersson, H. Nolte, C. Baumgarten and U. Pipkorn, 46, 540 (1991).
74. P. Renesto, V. Balloy, B.B. Vargaftig and M. Chignard, Br. J. Pharmacol., 103, 1435 (1991).
75. J.E. Leysen, W. Gommeren, P.F.M. Janssen and P.A.J. Janssen, Drug. Dev. Res., 22, 165 (1991).
76. E. Ohshima, T. Kumazawa, H. Takizawa, H. Harakawa, H. Sato, H. Obase, Y. Oiji, A. Ishii, H. Ishii and K. Oshmori, Chem. Pharm. Bull., 39, 2724 (1991).
77. T. Kumazawa, E. Ohshima, H. Harakawa, H. Sato, H. Obase, Y. Oiji, A. Ishii, H. Ishii and K. Ohmori, Chem. Pharm. Bull., 39, 2729 (1991).
78. E. Makino, N. Iwasaki, N. Yagi, T. Ohashi, H. Kato, Y. Ito and H. Azuma, Chem. Pharm. Bull., 38, 201 (1991).
79. D. Chiarino, G. Grancini, V. Frigeni, I. Biasini and A. Carenzi, J. Med. Chem., 34, 600 (1991).
80. H. Nagai, K. Kitagaki, S. Goto, H. Suda and A. Koda, Japan J. Pharamcol., 56, 13 (1991).
81. D.F. Rogers, Y.Lei, H.-P. Kuo, J.A.L. Rohde and P.J. Barnes, Am. Rev. Respir. Dis., 143, A754 (1991).
82. N. E. Bowring, D.R. Buckle, J.F. Taylor and J.R.S. Arch, Br. J. Pharmacol, in press (1992).
83. D.R. Buckle, J.R.S. Arch, C. Edge, K.A. Foster, C.S.V. Houge-Frydrych, I.L. Pinto, D.G. Smith, J.F. Taylor, S.G. Taylor, J.M. Tedder and R.A.B. Webster, J. Med. Chem., 34, 919 (1991).
84. M.R. Attwood, P.S. Jones, P.B. Kay, P.M. Paciorek and S. Redshaw, Life Sciences, 48, 803 (1991).
85. M.R. Attwood, I. Churcher, R.M. Dunsdon, D.N. Hurst and P.S. Jones, Tetrahedron Lett., 32, 811 (1991).
86. P.M. Paciorek, D.T. Burden, P.R. Gater, Y.M. Hawthorne, A.M. Spence, J.C. Taylor and J.F. Waterfall, Pulm. Pharmacol., 4, 225 (1991).
87. J.M. Bathon and D. Proud, Annu. Rev. Pharmacol. Toxicol., 31, 129 (1991).
88. S.G. Farmer, Immunopharmacology, 22, 1 (1991).
89. S.G. Farmer and R.M. Burch, Ann. NY Acad. Sci., 629, 237 (1991).
90. "Bradykinin Antagonists", R.M. Burch, Ed., Marcel Dekker, Inc. New York, N.Y. 1991.
91. A.E. McEachern, E.R. Shelton, S. Bhakta, R. Obernolte, C. Bach, P. Zuppan, J. Fujisaki, R. Aldrich, and K. Jarnagin, Proc. Natl. Acad. Sci. USA, 88, 7724 (1991).
92. J.C.W. Mak and P.J. Barnes, European J. Pharmacology, 194, 37 (1991).
93. J.A. Pongracic, R.M. Naclerio, C.J. Reynolds, and D. Proud, Br. J. Clin. Pharmacol., 31, 287 (1991).
94. W.M. Abraham, R.M. Bruch, S.G. Farmer, M.W. Sielczak, A. Ahmed, and A. Cortes, Am. Rev. Respir. Dis., 143, 787 (1991).
95. K. Wirth, F.J. Hock, U. Albus, W. Linz, H.G. Alpermann, H. Anagnostopoulos, S. Henke, G. Breipohl, W. Konig, J. Knolle and B.Z.Scholkens, Br. J. Pharmacol., 102, 774 (1991).

96. D.J. Kyle, J.A. Martin, R.M. Bruch, J.P. Carter, S. Lu, S. Meeker, J.C. Prosser, J.P. Sullivan, J. Togo, L. Noronha-Blob, J.A. Sinsko, R.F. Walters, L.W. Whaley, and R.N. Hiner, J. Med. Chem., 34, 2649 (1991).
97. Scrip, 1645, 6 (1991).
98. N. Frossard and C. Advenier, Life Sci., 49, 1941 (1991).
99. P.J. Barnes, Int. Arch. Allergy Appl. Immunol., 94, 303 (1991).
100. P.J. Barnes, J.N. Baraniuk and M.G. Belvisi, Am. Rev. Respir. Dis., 144, 1187 (1991).
101. N. Frossard and P.J. Barnes, Neuropeptides, 19, 157 (1991).
102. T. Matsuse, R.J. Thomson, X.-R. Chen, H. Salari and R.R. Schellenberg, Am. Rev. Respir. Dis., 144, 368 (1991).
103. Y.Takeda, K.B. Chou, J. Takeda, B.S. Sachais and J.E. Krause, Biochem. Biophys. Res. Commun., 179, 1232 (1991).
104. A. Graham, B. Hopkins, S.J. Powell, P. Danks and I. Briggs, Biochem. Biophys. Res. Commun., 177, 8 (1991).
105. N.D. Boyd, C.F. White, R. Cerpa, E.T. Kaiser and S.E. Leeman, Biochemistry, 30, 336 (1991).
106. C.A. Maggi, S. Giuliani, L. Ballati, A. Lecci, S. Manzini, R. Patacchini, A.R. Renzetti, P. Rovero, L. Quartara and A. Giachetti, J. Pharm. Exp. Ther., 257, 1172 (1991).
107. S.J. Ireland, F. Bailey, A. Cook, R.M. Hagan, C.C. Jordan and M.L. Stephens-Smith, Br. J. Pharmacol., 103, 1463 (1991).
108. S.E. Gentry, Life Sci., 48, 1609 (1991).
109. E. Burcher, L.A. Alouan, P.R.A. Johnson and J.L. Black, Neuropeptides, 20, 79 (1991).
110. R. Patacchini, M. Astolfi, M.C.S. Brown and C.A. Maggi, Neuropeptides, 20, 109 (1991).
111. P. Delay-Goyet and J.M. Lundberg, Biochem. Biophys. Res. Commun., 180, 1342 (1991).
112. T. Fujii, M. Murai, H. Morimoto, M. Nishikawa and S. Kiyotoh, Eur. J. Pharmacol., 194, 183 (1991).
113. "Pulmonary Emphysema: The Rationale for Therapeutic Intervention", G. Weinbaum, R.E. Giles, and R.D. Krell, Eds., NY Acad. Sci., New York, NY, 1991.
114. "Pulmonary Emphysema: The Rationale for Therapeutic Intervention, Proceedings of a Follow-Up Workshop on Treating the Underlying Causes of Emphysema", A.B. Cohen, Ed., NY Acad. Sci., New York, NY, 1991.
115. S. Eriksson, Eur. Respir. J., 4, 1041 (1991).
116. J. Travis, H. Fritz, Am. Rev. Respir. Dis., 143, 1412 (1991).
117. M.A. Rosenfeld, W. Siegfried, K. Yoshimura, K. Yoneyama, M. Fukayama, L. Stier, P.K. Paakko, P. Gilardi, L.D. Stratford-Perricaudet, M. Perricaudet, S. Jallat, A. Pavirani, J.P. Lecocq, R.G. Crystal, Science, 252, 431 (1991).
118. R.A. Mumford, J.B. Doherty, S.K. Shah, P.E. Finke, C.P. Dorn, W.K. Hagmann, J.J. Hale, A.L. Kissinger, K.R. Thompson, K. Brause, G.O. Chandler, W.B. Knight, A.L. Maycock, B.M. Ashe, H. Weston, P. Gale, D.S. Fletcher, P.S. Dellea, K.M. Hand, D.G. Osinga, R.J. Bonney, L.B. Peterson, J.M. Metzger, D.T. Williams, O.F. Anderson, H.R. Williams, D. Underwood, J.L. Humes, S. Pacholok, W. Hanlon, D. Frankenfield, T. Nolan, and P. Davies, abstracts of KININN '91 MUNICH, Munich, Germany, Sept. 8-14, (1991), abs. PW-2.1.
119. K. Kawabata, M. Suzuki, M. Sugitani, K. Imaki, M. Toda, and T. Miyamoto, Biochem. Biophys. Res. Commun., 177, 814 (1991).
120. J.C. Williams, R.C. Falcone, C. Knee, R.L. Stein, A.M. Strimpler, B. Reaves, R.E. Giles, and R.D. Krell, Am. Rev. Respir. Dis., 144, 875 (1991).
121. W.K. Hagmann, S.K. Shah, C.P. Dorn, L.A. O'Grady, J.J. Hale, P.E. Finke, K.R. Thompson, K.A. Brause, B.M. Ashe, H. Weston, M.E. Dahlgren, A.L. Maycock, P.S. Dellea, K.M. Hand, D.G. Osinga, R.J. Bonney, P. Davies, D.S. Fletcher, J.B. Doherty, Bioorg. Med. Chem. Lett., 1, 545 (1991).
122. M.A. Rosenfeld, K. Yoshimura, B.C. Trapnell, K. Yoneyama, E.R. Rosenthal, W. Dalemans, M. Fukayama, J. Bargon, L.E. Stier, L. Stratford-Perricaudet, M. Perricaudet, W.B. GugginoA. Pavirani, J.P. Lecocq, and R.G. Crystal, Cell, 68, 143 (1992).
123. R.C. Hubbard, N. Mcelvaney, P. Birrer, S. Shak, A.B. Montgomery, J. Healy, W. Robinson, C. Jolley, M. Chernick, and R.G. Crystal, Am. Rev. Respir. Dis., 143, (4 part 2), A605 (1991).
124. M.L. Aitken, W. Burke, G. McDonald, M. Villalon, S. Shak, A.B. Montgomery, and A. Smith, Am. Rev. Respir. Dis., 143, (4 part 2), A298 (1991).
125. R.P. Tomkiewicz, E.M. App, R.C. Boucher, M.R. Knowles, and M. King, Am. Rev. Respir.Dis., 143, (4 part 2), A296 (1991).
126. M.R. Knowles, L.L. Clarke, and R.C. Boucher, N. Engl. J. Med., 325, 533, (1991).
127. J.A. Nadel, Am. Rev. Respir. Dis., 144, S48 (1991).
128. A. Schuster, I. Ueki, J.A. Nadel, abstracts of the Fifth Annual North American CF Conference, Dallas, Texas, Oct 2-5, (1991).

SECTION III. CHEMOTHERAPEUTIC AGENTS

Editor: Jacob J. Plattner
Abbott Laboratories, Abbott Park, IL 60064

Chapter 13. Antibacterial Agents, Targets and Approaches

Mark J. Suto, John M. Domagala and Paul F. Miller
Parke-Davis Pharmaceutical Research Division
Warner-Lambert Company
Ann Arbor, Michigan 48105

<u>Introduction</u> - The development of resistance to current antibacterial therapy continues to drive the search for more effective agents. Several reviews have appeared illustrating the problems encountered by today's infectious disease clinicians (1-3). Studies are underway investigating the use of combination strategies to overcome both resistant *S. aureus* (4) and *P. aeruginosa* infections (5). Reports on the ability of animal chemotherapy models to effectively predict clinical success (6) and the use of tissue concentrations in man to accurately assess an antibiotic's therapeutic range have also appeared (7,8). Molecular modifications of existing classes of antibacterial agents have centered on improving activity against resistant organisms. Recently, a better understanding of resistance has also led to the development of new mechanistic approaches to antibacterial research. To address these new emerging technologies, this chapter has been structured to review the relevant developments of the past year from a mechanistic viewpoint. The topics considered include inhibition of DNA and protein synthesis, the interference of cell-wall formation, and miscellaneous items not falling into any of the above categories. In each area, along with a discussion of more traditional classes of compounds, new developments offering potential for the discovery of new agents will be included.

<u>DNA Synthesis Inhibitors</u> - DNA gyrase, an essential type II bacterial topoisomerase, is known to be the target of quinolone antibiotics in gram-negative bacteria (9). Recent studies in *E. coli* suggested that the N-terminal portion of the GyrA subunit was critical for quinolone action (10). If the concentration of a quinolone, such as nalidixic acid, exceeded the optimum bactericidal concentration (OBC), the bactericidal activity of the compound decreased (11). This decrease was a result of the gyrase induced relaxation of chromosomal DNA that occurred at high concentrations of quinolones, which prevented transcription into RNA, a step required for the bactericidal activity of these compounds. Some newer fluoroquinolones, at concentrations above their OBC's, did not inhibit their own bactericidal activity to the same extent as nalidixic acid, suggesting a secondary mechanism of action for these compounds (11). DNA gyrase was recently shown to be the target of quinolones in gram-positive organisms as well. Mutations in the genes encoding DNA gyrase were associated with the development of clinical resistance of *S. aureus* to fluoroquinolones. The IC_{50}'s against isolated *S. aureus* (FDA 209) gyrase and the MIC's against the organism itself were closely correlated, indicating that the site of action was DNA gyrase (12). In *S. aureus* (MS 16405), resistance to the fluoroquinolones was shown to be due mainly to mutations in GyrA, but reduced drug uptake was also observed (13). Ciprofloxacin-resistant *Staphylococcus spp.* was shown to have alterations in the Ser 84 residue of the GyrA subunit, indicating that this residue was critical for the interaction of quinolones with gyrase (14). Many of the resistant strains studied showed an associated decrease in quinolone uptake, related to changes in outer membranes of gram-negative isolates (15), but changes in uptake alone did not account for the differences seen between norfloxacin-resistant and -sensitive *S. aureus* (16). The role of outer membranes in quinolone uptake and possible mechanisms involved in this process were reviewed (17). The similarity of quinolones to lysine and imipenem, with respect to the

spatial orientation of their charges, may be critical for permeation through cells via protein D_2 and therefore could be used to improve the cell permeation of quinolones (18).

A number of reports illustrated that quinolones, in addition to inhibiting bacterial topoisomerase II (DNA gyrase), also inhibit mammalian topoisomerase II. This inhibition caused chromosomal aberrations i.e., induction of micronuclei and mammalian cell cytotoxicity (19). Structure-activity studies have revealed no clear correlation between mammalian cell cytotoxicity and antibacterial activity, but trends suggested that certain structural features could reduce mammalian cell cytotoxicity (20). Based on those studies, PD 140248 (1) and PD 138312 (2) were synthesized and showed reduced genotoxic risk and potent gram-positive activity (21,22). Overall, the quinolones used clinically have proven to be valuable antimicrobial agents with generally low toxicity (23). Their use in a wide variety of bacterial infections was reviewed, as were their pharmacokinetic properties, adverse effects, and drug interactions (23-26). A comprehensive review covering aspects of quinolone chemistry, structure-activity relationships, mechanisms of actions and biology also appeared (27). Many of the new compounds were designed and synthesized to investigate the effect of the N-1 substituent on antibacterial (particularly anti-gram-positive organisms) activity.

A detailed reevaluation of the optimum van der Waals volumes required at the N-1 position of quinolones was performed (28). These studies defined a receptor model of N-1 with two distinct acceptable regions: the first corresponding to the cyclopropyl group existing above the plane of the quinolone, and the other corresponding to the region occupied by the fluorine and hydroxy group at the para-position of the N-1 phenyl group. This work clarified earlier calculations, which failed to account for the activity of the N-1 phenyl group. The relationship between the N-1 substituent (ethyl, cyclopropyl, difluorophenyl) and substituents at C-7 (piperazines, pyrrolidines), C-8 (hydrogen, fluorine, chlorine), and particularly C-5 (hydrogen, amino, hydroxy, methyl), was also defined (29,30). The results indicated that the activity seen with various C-5, C-7, and C-8 substituents was determined primarily by the substituent at N-1.

A number of novel N-1 substituents appeared which, in general, resulted in compounds with increased gram-positive activity. Addition of a fluorine atom to the N-1 cyclopropyl group, DU-6859 (3), or the 1-t-butyl substituent (4), resulted in compounds with overall improved activity (31,32). Even though 1 induced chromosomal aberrations, its clastogenic potential was 1/10 that of the non-fluorinated analogue (33); adding a second fluorine atom resulted in a loss of activity (30). Replacement of the N-1 cyclopropyl with oxetane provided WQ 1101 (5), which had increased gram-positive activity relative to ciprofloxacin, but decreased gram-negative activity (34). In general, the oxetane-containing compounds approached the activity of the corresponding cyclopropyl derivatives. Use of a bridged azetidine, U-87947 (6), provided an alternative to the cyclopropyl group, but offered no advantage (35). Potent gram-positive activity was retained when the N-1 substituent was 1,1-dimethylpropargyl (36). U-9139E (7) was 2-3 times more active than ciprofloxacin versus gram-positive organisms, but its gram-negative activity was reduced. Connecting the hydroxy group at C-8 to a dimethylamino group at N-1 provided Ro 09-1168 (8), an ofloxacin type derivative (37). This compound not only displayed superior in $vitro$ and in $vivo$ antibacterial activity compared to ofloxacin and ciprofloxacin, but also had good bioavailabilty (94%) and a $t_{1/2}$ of 4.6 hours. This compound is under development for veterinary use.

A 7-endo-(3-amino-8-azabicyclo[3.2.1]oct-8-yl) derivative (9) emerged from the study of a series of bicyclopiperazine analogues (38). This compound was active against both gram-positive and gram-negative organisms, inhibited DNA gyrase, and was active in $vivo$ against $E.$ $coli$ and $S.$ $pneumonia$. The reason for the increased efficacy of this compound relative to other members of the series is unknown at this time. A highly soluble dipeptide prodrug of tosufloxacin (A-70826, norval-norvalyl) was shown to sequentially break down to tosufloxacin in $vivo$, yielding a Cmax of tosufloxacin that was 3 times higher than that obtained when tosufloxacin itself was administered (39). Polyethylene glycol ethers, attached to the carboxylic acid moiety of oxolinic acid, also served as effective prodrugs of the parent compound (40).

Two series of benzonaphthyridines were examined for antibacterial activity. In the first series, benzo[b][1,6]naphthyridine, AT 5755 (10), containing a new replacement for the 3-carboxylic acid moiety of quinolones, retained both in $vitro$ and in $vivo$ activity (41). In the second series, Ru 67829 (11) was more active against gram-positive organisms than ciprofloxacin, but departed from the normal quinolone SAR since it contained a N-1 methyl group and a 3-phenyl piperazine as the side-chain (42).

Recent work has indicated that DNA gyrase may also be the target of several naturally occurring antibiotics. The bactericidal peptide microcin B17 (MccB17), a known inhibitor of DNA replication, was found to trap gyrase in an enzyme-DNA cleavable complex (43,44). Using a combined genetic and biochemical approach, MccB17 resistant mutants of *E. coli* were mapped to the GyrB region of the chromosome. Subsequent cloning and sequencing showed that the specific genetic change in GyrB was different from the mutations that are normally associated with quinolone resitance, indicating that MccB17 may provide an additional tool for the study of gyrase function. This was the first report of a peptide inhibitor of DNA gyrase and underscores the accessibility of this essential enzyme to antibiotic attack. Cinodine, a glycocinnamoylspermidine antibiotic, was also shown to inhibit bacterial DNA gyrase *in vitro* (45). Previously, it had been established that cinodine interacted directly with DNA and rapidly inhibited its synthesis (46). However, results from the current work indicated that DNA gyrase activity was inhibited at low drug concentrations where no *in vitro* effect was observed with two other DNA binding enzymes, topoisomerase I and the restriction enzyme BamHI (45). While this observation suggested a possible specific effect on gyrase, the lack of success in isolating cinodine resistant mutants might be an indication that other DNA enzymatic functions were also affected at bactericidal concentrations.

Inhibitors of Cell-Wall Synthesis - The most widely used compounds currently available clinically are the cell-wall inhibitors such as the ß-lactams and glycopeptides. The success of these inhibitors underscores the essential nature and attractiveness of the cell wall as an antibacterial target. Many of the enzymatic steps prior to the terminal crosslinking reactions that are inhibited by ß-lactams are unique to bacteria and represent a number of potential targets (47); consequently, new examples of both targets and chemical classes continue to be described.

As the use of ß-lactams increases, the number of organisms resistant to ß-lactams continues to increase. Recent reviews covering the mechanisms of resistance to ß-lactams and the emergence of extended spectrum ß-lactamases have appeared (48,49). Mechanistic studies provide insight into ß-lactam resistance. For example, in methicillin-resistant *Staphylococci* (MRS), a unique ß-lactam-insensitive penicillin-binding protein (PBP2A) has been found. A truncated form of the mecA gene from a MRS strain, which encodes PBP2A, has been reported (50) The construct directs the expression of a soluble form of PBP2A in *E. coli* that retains its weak penicillin-binding characteristics. Purification of this protein and subsequent crystallographic studies should give insights into structural differences between this protein and the normal ß-lactam-sensitive penicillin-binding proteins, which may facilitate the design of new agents.

In addition to targeting resistance in *Staphylococcus*, ß-lactam research has also focused on improving potency against *Pseudomonas* and *Enterobacter*. Chemical modifications have focused on three areas: the side chain R, the ring substituent X and the Z-group (compounds 12-20). Increasing the polarity of the Z-substituent with a thiazoliomethyl group resulted in CS 461 (12), which not only had improved *in vitro* and *in vivo* acitivity against both gram-positive and gram-negative bacteria, but was associated with less toxicity in mice (51). Activity against *Pseudomonas* was improved with the dithiane derivative 13, but not without a loss in activity against *Staphylococcus* and *Enterobacter* (52). Both FK 312 (14) and FK 037 (15) demonstrated excellent activty against *Staphylococcus*, including methicillin- and oximinoacid-resistant strains (53,54). In a variety of *Staphylococcal* infection models, 14 was one of the most active cephalosporins thus far reported and its activity approached that of imipenem in overall efficacy. The Z-substituent also influences oral activity. The cyanomethylthio-containing compound (16) had improved activity against *Staphylococcus* while maintaining good pharmacokinetics (55). When the more traditional vinyl group was replaced by a 2-propenyl group, both antimicrobial activity and the pharmacokinetic profile were adversely affected (56). In the carbapenem series, the basicity of a series of pyridines as the Z-substituent was shown to have a dramatic effect on the stability to renal dehydropeptidase 1 (DHP-1), as well as potency (57). A structure-activity study indicated that certain substituted phenyl groups attached directly to the penem as the Z-substituent resulted in compounds with broad spectrum activity with the 3- and 4- carbamoyl substituents preferred (58). The methylene spacer in the Z-group of 17 produced a three-fold increase in DHP-1 stability, but also led to a concomitant loss in *Pseudomonas* activity (59). The 2-(1,3,4-thiadiazinyl)thiomethyl group of 18 conferred *in vitro* potency superior to imipenem, but without the corresponding *in vivo* efficacy (60).

A further extension into substituents at Z involved the so-called dual action cephalosporins (DACS). Several papers appeared that examined the nature of the linkage between the 3'-position of a cephalosporin and a quinolone (61-63). Attaching the quinolone through a basic nitrogen of the side chain via a urethane, tertiary amine or quarternary amine linkages all produced DACS with far greater stability than those which linked the quinolone via an ester bond. In each case, broad spectrum activity and *in vivo* efficacy was demonstrated. Extending this concept to the penems and carbapenems resulted in antibacterials equipotent to imipenem *in vitro*, and in the case of **19**, with decreased susceptibility to DHP-1 (64,65).

New synthetic methodology was developed to provide a variety of substituted alkyls at the X-position of carbapenems (66). One of the better compounds was the chemically stable amidino

Compound Number	n	R^a	X	Z
12	1	$2-AT-\overset{NOCH_3}{\underset{\|\|}{C}}-CONH-$	S	(thiazolium side chain with $-CH_2-\overset{+}{N}$ ring, $-OH$)
13	1	$2-AT-\overset{\|\|}{C}-CONH-$ (with $N-O-C(Me)(Et)-CO_2H$)	S	$-CH$ (dithiole with CO_2H, $CONH_2$)
14	1	$2-AT-\overset{NOH}{\underset{\|\|}{C}}-CONH-$	S	$-CH_2-$ (thiadiazole)
15	1	$2-AT-\overset{NOCH_3}{\underset{\|\|}{C}}-CONH-$	S	$-CH_2-\overset{+}{N}$ (pyrazolium, $-NH_2$, $-OH$)
16	1	$2-AT-\overset{NOCH_2CO_2H}{\underset{\|\|}{C}}-CONH-$	S	$-SCH_2CN$
17	0	(hydroxyethyl, OH)	$\overset{CH_3}{\underset{}{CH-}}$	$-S-$ (pyrrolidine, CH_2CONH_2, NH)
18	0	(hydroxyethyl, OH)	$\overset{CH_3}{\underset{}{CH-}}$	$-S-$ (thiadiazole, N-N, S)
19	0	(hydroxyethyl, OH)	$\overset{CH_3}{\underset{}{CH-}}$	$-CH_2OC(O)-N$ (piperazinyl quinolone: F, O, CO_2H, cyclopropyl)
20	0	(hydroxyethyl, OH)	$\overset{CH_2CH_2\overset{NH}{\overset{\|\|}{C}}-NH_2}{\underset{}{CH-}}$	$-SCH_2CH_2CN$

derivative **20**, which had gram-negative activity superior to imipenem, but diminished potency versus gram-positive organisms (67). The stability of these analogues could be enhanced further by acylation with amino acids, providing compounds which acted as prodrugs *in vivo* (68). Changing the R-substituent had a variety of effects depending upon the ß-lactam nucleus. For example, in the cephems, it was shown that the pKa's of various 2-aminobenzimidazoles and imidazolines correlated not only with bacterial penetration, but also with penicillin-binding protein affinity (69). In the cephalosporins and monobactams, catechols attached via the oxime moiety enhanced gram-negative potency, presumably through utilization of the bacterial iron uptake system (70-72). In a related example, natural spermidine-based hydroxamate sideropheres were directly attached via an amide bond to an acyl amino R-substituent (73). The compounds displayed modest antibacterial activity and showed some evidence for a dual mechanism of action. The stability of a series of carbapenems to DHP-1, as well as their water solubility, was increased with a 1-fluoroethyl group as the R-substituent; however, the stability to ß-lactamases was decreased (74).

In the penicillins, the catechol-containing compound **21** had outstanding activity against *Pseudomonas*, and stability to gram-negative ß-lactamases, but not to the lactamases of *S. aureus* (75). Stability to gram-positive ß-lactamases could be achieved through the use of a (benzimidazol-2-yl)hydroxy methyl group at the R-position. Compound **22** demonstrated activity versus penicillin-resistant *Staphylococcus*, *Streptococcus* and *Haemophilus*. The compound was itself a ß-lactamase inhibitor and displayed synergy with ampicillin (76).

Recent attention has also focused on the glycopeptides, which represent an important therapy for difficult gram-positive bacteria. Vancomycin resistance in *E. faecium* BM 4147 has been shown to be associated with the production of Van A and Van H proteins (77). Purified Van A displayed D-Ala-D-Ala ligase activity, while purified Van H had some sequence identity to D-specific α-keto acid reductases. Further studies using various model dipeptides and depsipeptides have indicated that the peptidoglycan of resistant strains had a novel chemical structure where the D-Ala-D-Ala portion had been replaced by D-Ala-D-HBut (hydroxybutyric acid). This modified peptidoglycan can undergo normal cross-linking, but because of the ester linkage, lacks the hydrogen bonding capability critical for productive binding to vancomycin. Studies are in progress to determine the connection between this proposed mechanism of resistance in *E. faecium* and the mechanism of intrinsic glyocopeptide resistance, particularly in *Enterococcus* strains. The results from synergy experiments with ß-lactams suggested that the intrinsic resistance of *E. gallinarum* to vancomycin may be due to the high carboxypeptidase activity of PBP6 in this organism, which resulted in fewer unmodified D-Ala-D-Ala targets available for recognition by the antibiotic (78).

Structure-activity relationships of the glycopeptides focusing on the six amino acids and some semi-synthetic N-aryl and N-alkyl derivatives were defined against *S. aureus* and *Streptococcus* with reference to vancomycin (79). The data obtained were consistent with the proposed mode of action. Two glycopeptides produced by *Amycolatopsis sp.*, MM 55266 and MM 55268, possessed lipophilic side chains similar to other compounds in this class (80). These glycopeptides differed from the more traditional agents in that the acyl-bearing sugar was attached to amino acid six rather than four, the compounds were highly chlorinated, and they contained three attached sugars. Overall, MM 55268 was slightly less active than vancomycin versus a spectrum of gram-positive organisms. The helvecardins (A and B) were isolated by D-alanyl-D-alanine affinity chromatography (81). Their structures consisted of the same pseudoaglycones as found in ß-apovarin and contained the same basic complement of neutral sugar, amino acid and amino sugar. However, these compounds contained a 2'-O-methyl rhamnose sugar and helvecardin B contained no mannose (82). The MIC's against aerobic gram-positive bacteria approached that of vancomycin, but against aneorobes, they were more active (83). *In vivo*, they were generally non-toxic and protected mice from *S. aureus* infections.

Ramoplanin, a lipoglycopeptide structurally unrelated to vancomycin and teicoplanin, inhibited the transfer of N-acetylglucosamine to lipid-bound N-acetylmuramyl pentapeptide (84), while the lipopeptide daptomycin appeared to act at an earlier stage in precursor formation (85). Thus, in addition to the traditional enzymes in cell-wall biosynthesis, several possible targets in peptidoglycan assembly have been identified for further exploitation.

Inhibitors of Protein Synthesis - The use of agents which affect protein synthesis continues to play an important role in the chemotherapy of infectious diseases. A review appeared outlining the role of macrolides and potential structural modifications which could be investigated to improve this class of compounds (86). Azithromycin recently received a considerable amount of study due to its unusual pharmacokinetic properties (87). The rational design and structure-activity relationships of a variety of oxime ethers of erythromycin, which led to the discovery of roxithromycin, was presented (88). Resistance to macrolides and related antibiotics can develop by a variety of mechanisms; acquired target-modified resistance was the most prevalent and confered cross-resistance to MLS antibiotics (macrolides, lincosamide, streptogramin) (89). Active efflux and antibiotic modification also occurred, but these mechanisms were more class- and organism-specific (90).

Development of resistance and nephrotoxicity have limited the use of aminoglycosides, even though they are still effective against certain gram-negative infections (91). Recent efforts to decrease the toxicity have focused on two areas. The first attempt involved increasing the basicity of the compound, in an effort to improve antibacterial activty without a concomitant increase in toxicity. Replacing the amino group at C-1 of kanamycin A with N-(4-amidino-2-hydroxybutyl) provided **23**, which not only displayed good antibacterial activity, but had renal levels that were 53% less than amikacin at equal doses (92). Structure-activity relationships of the 6" position of kanamycin B were also investigated (93). It was found that small substituents were acceptable, and such groups had little effect on inhibition of lysosomal phospholipase A_1, a potential predictor of nephrotoxicity (94). It was also found that C-1 hydroxymethyl derivatives such as **24** were not superior to kanamycin B, even when they contained favorable C-6" substituents (95).

New information regarding the interaction of the aminoglycosides with their ribosomal targets has renewed interest in this drug class. The recent isolation of resistant mutants that map to the ribosomal RNA, as well as the demonstration that certain aminoglycosides interact with specific nucleotides on the ribosomal RNA, suggested that it was the ribosomal RNA itself that was the primary target and not the ribosomal proteins (96-98). Using chemical methods to probe the aminoglycoside-ribosome complexes *in vitro*, it was established that the sites of drug interaction with the ribosomal RNA were frequently located in regions that had been implicated in specific functions in protein synthesis (99). Importantly, the specific step in translation that was affected by a given aminoglycoside (e.g. initiation, translocation) correlated well with the functional site on the ribosomal RNA where the interaction was detected (97). These results suggested that a more detailed understanding of these interactions could stimulate new efforts in drug design.

Additional support for the RNA activity of the aminoglycosides came from studies of their effects on the self-splicing of group I introns (100). Many of the aminoglycosides known to affect the fidelity of protein synthesis were also found to block pre-mRNA splicing *in vitro*, at concentrations similar to those required to inhibit bacterial protein synthesis. These observations suggested that the self-splicing group I RNAs may share some structural similarity with essential regions of the ribosome and may assist in the study of aminoglycoside/RNA interactions (100,101).

A novel peptide antiobiotic, GE2270 A (**25**), was isolated from the fermentation broth of *Planobispora rosea* (102). This thiazolyl peptide antibiotic quickly blocked protein synthesis by increasing the stability of the elongation factor Tu/GTP complex and inhibited ternary complex formation by blocking subsequent binding of aminoacylated tRNA (103). In addition, __25__ had good activity against all strains of gram-positive bacteria, particularly the anaerobe *P. acnes*. It was also active against *Mycobacterium tuberculosis*, and *B. fragilis* and protected mice against *S. aureus* infection. These data substantiated the importance of elongation factor TU (EF-Tu) as a critical target for inhibition of bacterial cell growth.

Miscellaneous - Activity against anaerobic organisms was reported for a series of 2-substituted-5-nitroimidazoles and a series of tricylic N-oxides. The 5-nitroimidazole __26__ was unique in that it contained a lactam connected via an ethylene bridge to the 2-position of a 5-nitroimidazole. This compound was found to be more potent both *in vitro* and *in vivo* than metronidazole (104). The N-oxide, SC-44942A (__27__), was two to four times more potent than metronidazole over a broad range of anaerobes (105). This compound was active in the Ames test, but did not show any apparent toxicities in subsequent testing. Due to the nature of its redox potential, it was thought to elicit its antibacterial activity via the same mechanism as metronidazole. The proton pump inhibitor, lansoprazole (AG 1749, __28__) and a related analog __29__, were shown to be active against *H. pylori* (106). The data suggested that the compound acted locally, as well as systemically, in the gastric mucosa of patients. Of particular interest was the fact that the three major metabolites were also active against the organism. These compounds were inactive against bacteria other than *H. pylori* and therefore would not induce the acquisition of antibiotic resistance, an important consideration of such a compound.

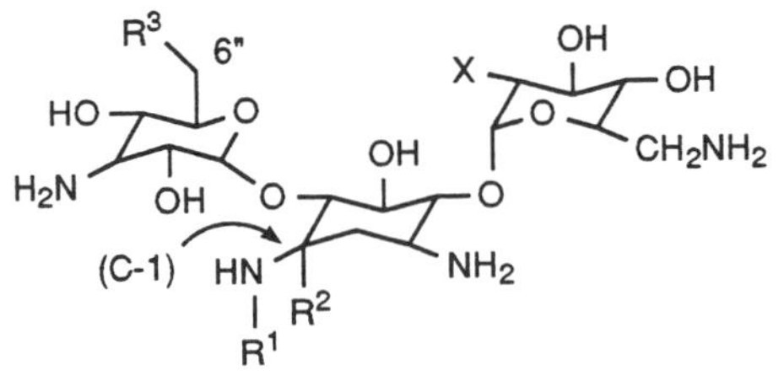

__23__ R^1 = COCH(OH)CH$_2$CH$_2$NHCH=NH, R^2 = H, R^3 = X = OH
__24__ R^1 = H, R^2 = CH$_2$OH, R^3 = Cl, N$_3$, NHAc, X = NH$_2$

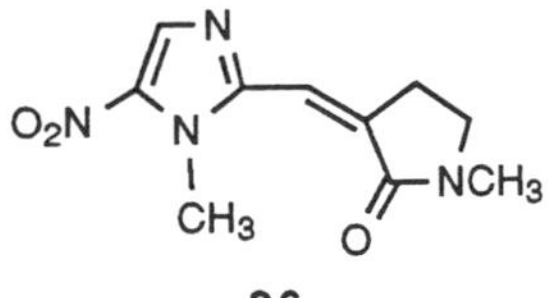

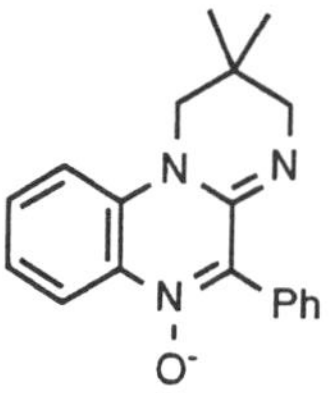

__27__

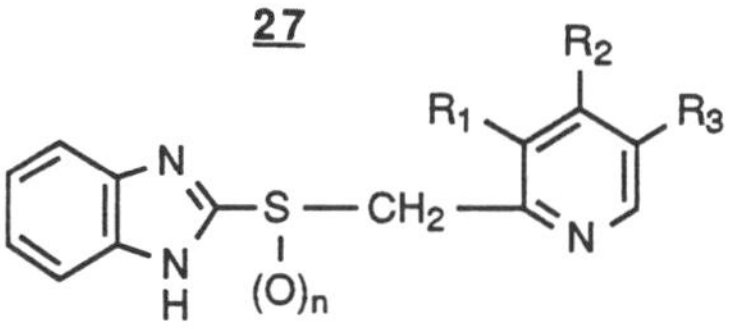

__28__ n = 1, R_1 = CH$_3$, R_3 = H, R_2 = OCH$_2$CF$_3$
__29__ n = 0, R_1 = R_3 = H, R_2 = OCH$_2$CF$_2$CF$_3$

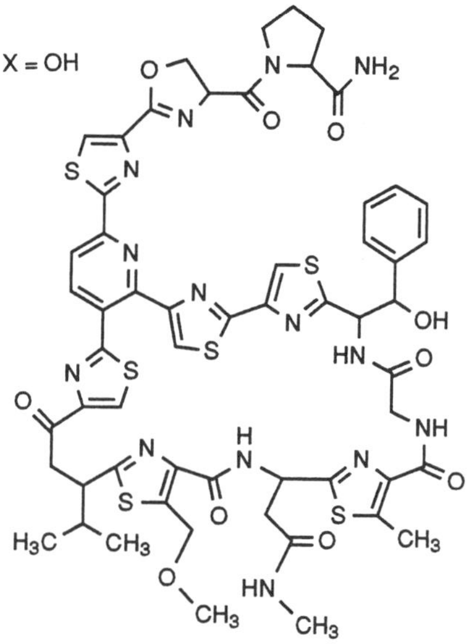

__26__

__25__

Many pathogenic bacteria produce a protective polysaccharide capsule (107). While the actual structure of the capsule varies significantly from organism to organism, recent evidence suggests that a common regulatory pathway is utilized by a variety of bacteria to control capsule production (108). In addition, their is evidence that several of the genes whose products are involved in the transport of capsular polysaccharides from the cytoplasm to the cell surface are evolutionarily conserved among gram-negative bacteria (109). These results indicate that capsule biosynthesis might be a process that is accessible to intervention.

References

1.   G.A. Jacoby and G.A. Archer, New Eng. J. Med., 324, 601 (1991).
2.   L.O. Gentry, Orthop. Clinics N. Amer., 22, 379 (1991).
3.   R.P. Wenzel, M.D. Nettleman, R.N. Jones, and M.A. Pfaller, Am. J. Med., 91, 2215 (1991).
4.   C. Chuard, M. Herrmann, P. Vaudaux, F.A. Waldvogel and D.P. Lew, Antimicrob. Agents Chemother., 35, 2611 (1991).
5.   J.A. Korvick and V.L. Yu, Antimicrob. Agents Chemother., 35, 2167 (1991).
6.   O. Zak, and T. O'Reilly, Antimicrob. Agents Chemother., 35, 1527 (1991).
7.   D.E. Nix, S.D. Goodwin, C.A. Peloquin. D.L. Rotella and J.J. Schentag, Antimicrob. Agents Chemother., 35, 1947 (1991).
8.   D.E. Nix, S.D. Goodwin, C.A. Peloquin. D.L. Rotella and J.J. Schentag, Antimicrob. Agents Chemother., 35, 1953 (1991).
9.   R.J. Reece and A. Maxwell, Critical Rev. Biochem. Mol. Biol., 26, 335 (1991).
10.   D.C. Hooper and J.S. Wolfson, Eur. J. Clin. Microbiol. Infect. Dis., 10, 223 (1991).
11.   C.S. Lewin, I. Morrissey and J.T. Smith, Eur. J. Clin. Microbiol. Infect. Dis., 10, 240 (1991).
12.   M. Tanaka, K. Sato, Y. Kimura, I. Hayakawa, Y. Osada and T. Nishino, Antimicrob. Agents Chemother., 35, 1489 (1991).
13.   N. Nakanishi, S. Yoshida, H. Wakebe, M. Inoue, T. Yamaguchi and S. Mitsuhashi, Antimicrob. Agents Chemother., 35, 2562 (1991).
14.   S. Sreedharan, L.R. Peterson and L.M. Fisher, Antimicrob. Agents Chemother., 35, 2151 (1991).
15.   J.M. Diver, T. Schollaardt, H.R. Rabin, C. Thorson and L.E. Bryan, Antimicrob. Agents Chemother., 35, 1538 (1991).
16.   K.V. Cundy, C.E. Fasching, K.E. Willard and L.R. Peterson, J. Antimicrob. Chemother., 28, 491 (1991).
17.   L.J.V. Piddock, J. Antimicrob. Chemother., 27, 399 (1991).
18.   M. Michea-Hamzehpour, Y.X. Furet and J. Pechere, Antimicrob. Agents Chemother.,35, 2091 (1991).
19.   S.J. Gracheck, M. Mychajlonka, L. Gambino, M.A. Cohen, G. Roland, V. Ciarvino, D. Worth, J. C. Theiss and C. L. Heifetz, 31st ICAAC, 1488 (1991).
20.   M.J. Suto, J.M. Domagala, D. Worth, G. Roland amd M.A. Cohen, 31st ICAAC, 1489 (1991).
21.   S. Hagen, J. Domagala, C. Hiefetz, J. Sanchez, J. Sesnie, M. Stier, M. Suto and D. Szotek, 31st ICAAC, 1438 (1991).
22.   J.M. Domagala, M. Cohen, J. Kiely, S. Hagen, E. Laborde, M. Schroeder, J. Sesnie and M.J. Suto, 31st ICAAC, 1439 (1991).
23.   J.H. Paton and D.S. Reeves, Drug Safety, 6, 8 (1991).
24.   D.C. Hooper and J.S. Wolfson, N. Eng. J. Med., 324, 384 (1991).
25.   J.S. Wolfson and D.C. Hooper, Eur. J. Clin. Microbiol. Infect. Dis., 10, 267 (1991).
26.   S.R. Norrby, Eur. J. Clin. Microbiol. Infect. Dis., 10, 378 (1991).
27.   D.T.W. Chu and P.B. Fernandse in "Advances in Drug Research," Vol. 21, B. Testa, Ed., Academic Press, New York, N.Y., 1991, p. 39.
28.   M. Ohita and H. Koga, J. Med. Chem., 34, 131 (1991).
29.   J.M. Domagala, A.J. Bridges, T.P. Culbertson, L. Gambino, S.E.Hagan, G. Karrick. K. Porter, J.P. Sanchez, J.A. Sesnie, F.G. Spense, D. Szotek and J. Wemple, J. Med. Chem., 34, 1142 (1991).
30.   S.E. Hagan, J.M. Domagala, C.L. Heifetz and J. Johnson, J. Med. Chem., 34, 1155 (1991).
31.   I. Hayakawa, S. Atarashi, Y. Kimura, K. Kawakami, T. Saito, T. Yafune, K. Sato, T. Une and M. Sato, 31st ICAAC, 1504 (1991).
32.   P. Remuzon, D. Bouzard, P. Di Cesare, M. Essiz, J.P. Jacquet, J.R. Kiechel, B. Ledoussal, R.E. Kessler and J. Fung-Tomc, J. Med. Chem., 34, 29 (1991).
33.   H. Shimada and S. Itoh, 31st ICAAC, 1507 (1991).
34.   T. Nishino, M. Otsuki, N. Hayashi and T. Yatsunami, 31st ICAAC, 1448 (1991).
35.   M.R. Barbachyn, D.K. Hutchinson, R.J. Reid and D.S. Toops, 31st ICAAC, 1456 (1991).
36.   E.L. Fritzen, K.S. Kim, D.R. White, K.C. Grega, M.R. Barbachyn, D.R. Hutchinson, R.J. Reid and L.C. Davenport, 31st ICAAC, 1455 (1991).
37.   M. Arisawa, T. Hirata, Y. Aoki, Y. Fujioka, Y. Sekine, H. Tahara and A. Kuwahara, 31st ICAAC, 1454 (1991).
38.   J.S. Kiely, M.P. Hutt, T.P. Culbertson, R.A. Bucsh, D.F. Worth, L.E. Lesheski, R.D. Gogliotti, J.C. Sesnie, M. Solomon and T.F. Mich, J. Med. Chem., 34, 656, (1991).
39.   D.T.W. Chu, K.C. Marsh, R. Hallas, N. Shipkowitz, J.J. Plattner and A.G. Pernet, 31st ICAAC, 1440 (1991).
40.   B. Loubinoux, J.L. Colin and V. Thomas, Eur. J. Med. Chem., 26, 461 (1991).
41.   J. Nakano, T. Hirose, K. Yamamoto, K. Shibamori, M. Fujita, M. Kataoka, H. Okada, Y. Nishimura, K. Chiba, A. Minamida, T. Hirose, T. Miyamoto and J. Matsumoto, 31st ICAAC, 1494 (1991).
42.   M. Barreau, M. Antoine, J.L. Benichon, J.F. Desconclois, P. Girard, M. Robin, S. Wentzler, G. Picaut, J.F. Desnottes, O. Rolin and D.H. Bouanchaud, 31st ICAAC, 369 (1991).
43.   J.L. Vizan, C. Hernandez-Chico, I. del Castillo and F. Moreao, EMBO J. 10, 467 (1991).
44.   F. Moreao and F. Baquero, 31st ICAAC, 424 (1991).
45.   M.S. Osburne, W.M. Maiese and M. Greenstein, Antimicrob. Agents Chemother., 34, 1450 (1990).
46.   M. Greenstein, J.L. Speth and W.M. Maiese, Antimicrob. Agents Chemother., 20, 425 (1981).
47.   N.H. Georgopapadakoa, in" Antibiotic Inhibition of the Bacterial Cell Surface Assembly and Function", P. Actor et al eds. American Society for Microbiology, Washington, D.C. 1988, P. 505.
48.   D.M. Livermore, Scand. J. Infect. Dis., Suppl., 78, 7 (1991).
49.   G.A. Jacoby and A.A. Medeiros, Antimicrob. Agents and Chemother., 35, 1697 (1991).
50.   C.Y.E. Wu, J. Hoskins, S. Unal, D. Preston, L. Blaszczak and P. Skatrud, 31st ICAAC, 493 (1991).
51.   E. Nakayama, K. Watanabe, M. Miyauchi, K. Fujimoto, S. Muramatsu, H. Yasuda, M. Fukami and J. Ide, J. Antibiot., 44, 864 (1991).

52. N. Naganu, K. Nakano, T. Shibanuma and R. Hara, J. Antibiot., 44, 415 (1991).
53. Y. Inamoto, J. Goto, K. Sakane, T. Kamimura and T. Takaya, J. Antibiot., 44, 507 (1991).
54. 31st ICAAC, Session 74; 849-857 (1991).
55. C. Yokoo, M. Goi, A. Onodera, M. Murata, T. Nagate and Y. Watanabe, J. Antibiot., 44, 498 (1991).
56. W. Kim, K.Ko, H. Kim and J. Oh, J. Antibiot., 44, 1073 (1991).
57. J.M. Balkovec, M.J. Szymonifka, J.V. Heck and R.W. Ratcliffe, J. Antibiot., 44, 1172 (1991).
58. S. Connolly, K.W. Moore, M.D. Cooke, J.G. Walmsley and P.H. Bentley, J. Antibiot., 44, 1169 (1991).
59. M. Sunagawa, H. Matsumura and T. Inoue, J. Antibiot., 44, 459 (1991).
60. M. Imuta, H. Itani, H. Ona, Y. Hamada, S. Uyeo and T. Yoshida, Chem. Pharm. Bull., 39, 663 (1991).
61. H.A. Albrecht, G. Beskid, J.G. Christenson, N.H. Georgopapadakou, D.D. Keith, F.M. Konzelmann, D.L. Pruess, P.L. Rossman and C. Wei, J. Med. Chem., 34, 2857 (1991).
62. H.A. Albrecht, G. Beskid, J.G. Christenson, J.W. Durkin, V. Fallat, N.H. Georgopapadakou, D.D. Keith, F. M. Konzelmann, E.R. Lipschitz, D.H. McGarry, J. Siebelist, C. Wei, M. Weigele and R. Yang, J. Med. Chem., 34, 669 (1991).
63. H.A. Albrecht, G. Beskid, J.G. Christenson, K.H. Deitcher, N.H. Georgopapadakou, D.D. Keith, F.M. Konzelmann, D.L. Pruess and C. Wei, 31st ICAAC, 262 (1991).
64. E. Perrone, F. Zarini, G. Visentin, M. Alpegiani, D. Jabes, R. Rossi and C. Dellabruna, 31st ICAAC, 825 (1991).
65. D.D. Keith, D.L. Pruess, P.L. Rossman, J. Unowski and C. Wei, 31st ICAAC, 826 (1991).
66. 31st ICAAC, Session 73, 827-830 (1991).
67. J. Banville, C. Bachand, J. Corbeil, J. Desiderio, J.F. Tomc, P. Lapointe, A. Martel, V.S. Rao, R. Remillard, M. Menard, R.E. Kessler and R.A. Partyka, 31st ICAAC, 828 (1991).
68. V.S. Rao, C. Bachand, J. Banville, G. Bouthillier, J. Desiderio, J.F. Tomc, P. Lapointe, A. Martel, H. Mastalerz, R. Remillard, A. Michel, M. Menard, R.E. Kessler and R.A. Partyka, 31st ICAAC, 830 (1991).
69. F. Jung, C. Delvare, D. Boucherot and A. Hamon, J. Med. Chem., 34, 1110 (1991).
70. K. Sakagami, K. Iwamatsu, K. Atsumi and M. Hatanaka, Chem. Pharm. Bull., 38, 3476 (1990).
71. S. Jendrzejewski, J.E. Sundeen and R. Zahler, 31st ICAAC, 834 (1991).
72. I.A. Critchley, M.J. Baker, R.A. Edmondson and S.J. Knott, J. Antimicrob. Chemother., 28, 377 (1991).
73. E.K. Dolence, A.A. Minnick, C.E. Lin and M.J. Miller, J. Med. Chem., 34, 968 (1991).
74. A. Watanabe, M. Sakamoto, Y. Fukagawa, and T. Yoshioka, Chem. Pharm. Bull., 39, 335 (1991).
75. D. Bartkovitz, A. Corraz, K. Deitcher, N. Georgopapadakou, D. Keith, R. Mook, D. Pruess and C.C. Wei, 31st ICAAC, 824 (1991).
76. Y.L. Chen, K. Hedberg and K. Guarino, J. Antibiot., 44, 870 (1991).
77. T.D.H. Bugg, G.D. Wright, S. Dutka-Malen, M. Arthur, P. Courvalin and C.T. Walsh, Biochem., 30, 10408 (1991).
78. S. Vincent and D.M. Shlaes, 31st ICAAC, 990 (1991).
79. R. Nagarajan, Antimicrob. Agents Chemother., 35, 605 (1991).
80. S.J. Box, N.J. Coates, C.J. Davis, M.L. Gilpin, C.S.V. Houge-Frydrych and P.H. Milner, J. Antibiot., 44, 807 (1991).
81. M. Takeuchi, R. Enokita, T. Okazaki, Y. Kagasaki and M. Inukai, J. Antibiot., 44, 263 (1991).
82. M. Takeuchi, S. Takahashi, M. Inukai, M. Nakamura and T. Kinoshita, J. Antibiot., 44, 271 (1991).
83. M. Takeuchi, T. Katayama and M. Inukai, J. Antibiot., 44, 278 (1991).
84. E.A. Somner and P.E. Reynolds, Antimicrob. Agents Chemother., 34, 413 (1990).
85. N.E. Allen, J.N. Hobbs, Jr. and W.E. Alborn, Jr., Antimicrob. Agents Chemother., 31, 1093 (1991).
86. N. Woodford, A.P. Johnson and R.C. George, J. Antimicrob. Chemother., 28, 483 (1991).
87. E. Rubinstein, Eur. J. Clin, Microb. Inf. Dis., Volume 10, (1991).
88. J. Gasc, S.G. d'Ambrieres, A. Lutz and J. Chantot, J. Antibiotics, 44, 313 (1991).
89. R. Leclercq and P. Courvalin, Antimicrob. Agents Chemother., 35, 1267 (1991).
90. R. Leclercq and P. Courvalin, Antimicrob. Agents Chemother., 35, 1273 (1991).
91. R.S. Edson and C.L. Terrell, Mayo Clin. Proc., 66, 1158 (1991).
92. T. Yamaski, Y. Narita, H. Hoshi, S. Aburaki, H. Kamei, T. Naito and H. Kawaguchi, J. Antibiot., 44, 646 (1991).
93. A. Van Schepdael, J. Delcourt, M. Mulier, R. Busson, L. Verbist, H.J. Vanderhaeghe, M.P. Mingeot-Leclercq, P.M. Tulkens and P.J. Claes, J. Med. Chem., 34, 1468 (1991).
94. M.P. Mingeot-Leclercq, A. Van Schepdael, R. Brasseur, R. Busson, H.J. Vanderhaeghe, P.J. Claes, and P.M. Tulkens, J. Med. Chem., 34, 1476 (1991).
95. A. Van Schepdael, R. Busson, H.J. Vanderhaeghe, P.J. Claes, L. Verbist, M.P. Mingeot-Leclercq, R. Brasseur and P.M. Tulkens, J. Med. Chem., 34, 1483 (1991).
96. E.A. De Stasio, D. Moazed, H.F. Noller and A.E. Daahlberg, EMBO J., 8, 1213 (1991).
97. D. Moazed and H.F. Noller, Nature, 327, 389 (1991).
98. J. Woodcock, D. Moazed, M. Cannon, J. Davies and H.F. Noller, EMBO J., 10, 3099 (1991).
99. D. Moazed and H.F. Noller, J. Mol. Biol., 211 135 (1991).
100. U. von Ahsen, J. Davies and R. Schroeder, Nature, 353, 368 (1991).
101. H.F. Noeller, Nature, 353, 302 (1991).
102. E. Selva, G. Beretta, N, Montanini, G.S. Saddler, L. Gastaldo, P. Ferrari, R. Lorenzetti, P. Landini, F. Ripamonti, B.P. Goldstein, M. Berti, L. Montanaro and M. Denaro, J. Antibiot., 44, 693 (1991).
103. P.H. Anborgh and A. Parmeggiani, EMBO J. 10, 779 (1991).
104. O. Jentzer, P. Vanelle, M.P. Crozet, J. Maldonado and M. Barreau, Eur. J. Med. Chem., 26, 687 (1991).
105. P.A. Mueller, J. Oppermann, S. Nugent, C. Piper, R. Guzzie, R. Harkin, R. Partis, Y. Oshiro, R. Boreus, S. Baldcock, G. Ellames, R. Upton, P. Buckle and A. Pope, 31st ICAAC, 373 (1991).
106. T. Iwahi, H. Satoh, M. Nakao, T. Iwasaki, T. Yamazaki, K. Kubo, T. Tamura and A. Imada, Antimicrob. Agents Chemother., 35, 490 (1991).
107. K. Jann and B. Jann, Curr. Top. Microbiol., 150, 19 (1990).
108. S. Gottesman and V. Stout, Mol. Microbiol., 5, 1599 (1991).
109. M. Frosch, U. Edwards, K. Bousset, B. Kraube and C. Weisberger, Mol. Microbiol., 5, 1251 (1991).

Chapter 14. Genetic Engineering of Antibiotic Producing Organisms

Leonard Katz[a] and C. Richard Hutchinson[b]
[a]Abbott Laboratories, Abbott Park, IL 60064

[b]School of Pharmacy and Department of Bacteriology,
University of Wisconsin, Madison, WI 53706

<u>Introduction</u> - The application of genetic engineering technologies to antibiotic biosynthesis was made possible by the discoveries, beginning in the late 1970's, of methods which permitted the stable introduction of DNA into actinomycetes (1-3), and by the rapid development of vectors used to perform gene cloning experiments in these hosts (4,5). With this technology came the promise that manipulation of antibiotic biosynthesis genes would result in the overproduction of antibiotics and lead to the synthesis of novel products.

The creation of novel structures by gene manipulation is based on the observation that the enzymes involved in antibiotic pathways do not have absolute specificity for their substrates but can utilize compounds resembling natural substrates in analogous reactions (6). This has been demonstrated for many antibiotic pathways by conversion of exogenously fed false precursors into the corresponding altered structures. An example is the conversion of sodium phenoxyacetate into penicillin V by *Penicillium chrysogenum* that normally produces penicillin G (7). In "mutasynthesis", substrate analogs fed to mutants blocked in the pathway prior to the accumulation of the normal substrate are converted by the remaining enzymes of the pathway into compounds resembling the antibiotic (8). In a co-synthesis variation, a compound accumulated by a mutant blocked in one pathway at a given step is converted, by a second organism, into a novel structure when the two organisms are co-fermented. Both approaches to biotransformation have been demonstrated for a number of antibiotics including aminoglycosides (9), macrolides (10) and β-lactams (11). The logical extension of biotransformation is to have a single organism produce the substrate analog and convert it to the final product. This could require, in an elaborate example, the assembly of two sets of genes from different pathways in a single host. Though endeavors of this kind are still in their infancy, significant progress has been made and is the subject of this report. In addition, genetic manipulations resulting in the overproduction of antibiotics have achieved some measure of success and are also reviewed here.

<u>Organization of Antibiotic Biosynthesis Genes</u> - It has been demonstrated, from studies on more than a dozen pathways in bacteria, that antibiotic biosynthesis genes are clustered in a single, usually chromosomal, segment. Genes determining self-resistance to the antibiotic are normally found in the cluster as well, although some exceptions to this rule have been noted. Since antibiotics are produced through multi-step pathways, DNA segments range usually from 20 to >60 kilobases (kb) and encompass upwards of twenty genes. The organization of genes currently identified for the biosynthesis of the macrolide antibiotics erythromycins A, B, C and D (<u>1</u> - <u>4</u>, respectively) in *Saccharopolyspora erythraea,* comprising about 45 kb is shown in Fig. 1. Erythromycin is composed of a 14-membered macrocyclic ring to which are attached the amino sugar desosamine at C-5 and the neutral sugar mycarose (or its *O*-methylated derivative, cladinose) at C-3. The *ery* cluster contains: *ermE*, the erythromycin resistance gene (12); three *eryA* genes, involved in the synthesis of 6-deoxyerythronolide B (6dEB) (13-16); *eryF*, involved in hydroxylation of 6dEB at C-6 (17); two *eryB* genes, involved in synthesis and/or attachment of mycarose (14,18); two *eryC* genes, involved in synthesis and/or attachment of desosamine (14,19,20); and *eryG*, for the methylation of mycarose (21,22). Additional genes are presumably involved in the synthesis of the sugars. The scheme for erythromycin synthesis is shown in Fig. 2.

<u>Methodologies for Gene Manipulation in Actinomycetes</u> - Production of novel structures requires either the inactivation of one or more genes of a pathway or the introduction of heterologous genes into an antibiotic-producing host. Gene disruption, resulting from insertion of DNA into a coding sequence, is technically the simplest method to inactivate genes and has been used for a number of antibiotics including erythromycin (14,16,17,19), spiramycin (23) and methylenomycin (24). A single, reciprocal recombination event between a vector (plasmid or phage), which carries a

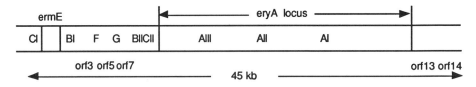

Figure 1. Genetic organization of the *ery* cluster. The 45 kb region indicated has been sequenced. The *ery* genes are designated by the functions described in the text. Functions have not yet been assigned to open reading frames (orf) 3, 5, 7, 13 and 14. *ermE* is the resistance gene.

segment of the gene to be disrupted, and the homologous sequence in the chromosome will cause integration of the construct into the chromosome and disruption of the gene. In practice, the vector carries a selectable marker and is incapable of episomal maintenance, thus enabling isolation of a colony carrying the integrated segment. The marker can also be cloned between two segments of the targeted gene on an appropriate vector. Independent recombination of the two segments and their chromosomal homologs will drive the marker into the chromosome, causing the gene to be disrupted, but the remaining sequences on the vector will not be integrated. Where disruption of an open reading frame is not desired (for example, where the gene requiring inactivation is part of an operon or may encode a multi-functional protein, of which only a single function is to be inactivated), the two-step recombination process can be used to replace the wild type version of the gene with an allele that carries a point mutation or in-frame deletion. Introduction of genes into heterologous hosts can be as straightforward as sub-cloning the gene onto a plasmid and introducing the plasmid into the desired host. If the gene has not been cloned previously, 'shot-gun' cloning (creating a library of genomic sequences in a population of plasmids, transforming the population into the host and screening for the transformant of choice) can be used to find genes. In cases where the plasmids (or cloned genes) are unstable, plasmids or phages are available that site-specifically integrate into the chromosome at locations removed from antibiotic biosynthesis genes (5). Insertion of DNA sequences into the biosynthesis regions can be accomplished by the gene replacement methods outlined above.

NOVEL STRUCTURES

Macrolides - Genetic manipulation of *Sac. erythraea*, has resulted in the production of three novel compounds and the overproduction of an important pathway intermediate. Weber and Losick (25) found that the plasmid pIJ702 (26), capable of replication in a variety of *Streptomyces* species, was not replication-proficient but could be used as a vector for gene disruption in *Sac. erythraea*. Employing uncharacterized segments of chromosomal DNA adjacent to *ermE* in pIJ702, and examining the fermentation broths of strains in which the plasmid had integrated, Weber *et al.* found a strain which produced the compound 6-deoxyerythromycin A (**5**), along with its B - D congeners (17). The disruption was located in a gene designated *eryF*, whose deduced polypeptide sequence shares significant homology with a family of cytochrome P-450 monooxygenases (27) and which encodes an activity that hydroxylates 6dEB at C-6 (28). These findings showed that a step in the biosynthesis of **1** could be bypassed and that the 6-deoxy analogs could be converted to **5** along the normal pathway, by its cognate enzymes, as shown in Fig. 2. Because **1** is readily inactivated under acidic conditions, through formation of 8,9-anhydroerythromycin-6,9-hemiketal, acid-stable compounds that have replaced or lost the C-6 OH, but which retain full potency, have long been desired. 6-Deoxyerythromycin A has the same anti-bacterial spectrum and nearly equivalent *in vitro* potency as **1** but has superior *in vivo* potency due, presumably, to its enhanced acid-stability. One attractive feature of **5** is that it can be produced as a natural product from the genetically engineered strain. The acid-stable compound 6-*O*-methylerythromycin (clarithromycin) has excellent properties but must be produced through synthetic derivatization of **1** (29).

	R_1	R_2	R_3	R_4
1	OH	Me	OH	Me
2	H	Me	OH	Me
3	OH	H	OH	Me
4	H	H	OH	Me
5	OH	Me	H	Me
8	OH	Me	OH	H

The gene disruption technique employing pIJ702 was used originally to identify *eryG* (21). Disruption of *eryG* resulted in accumulation of $\underline{3}$ and $\underline{4}$, which had previously been difficult to obtain in high quantitities. *eryG* encodes an *O*-methyltransferase that converts $\underline{3}$ to $\underline{1}$ and $\underline{4}$ to $\underline{2}$ (18,21). An *eryG* mutant was later uncovered after mutagenesis of *Sac. erythraea* and the screening of more than 10,000 fermentation broths by TLC for production of $\underline{4}$ (22). A mutant lacking the C-12 hydroxylase activity and which accumulates $\underline{2}$ (and, possibly, $\underline{4}$) has not yet been uncovered. If the corresponding gene lies within the *ery* cluster, it should be possible to use the gene disruption technique to identify the correct mutant.

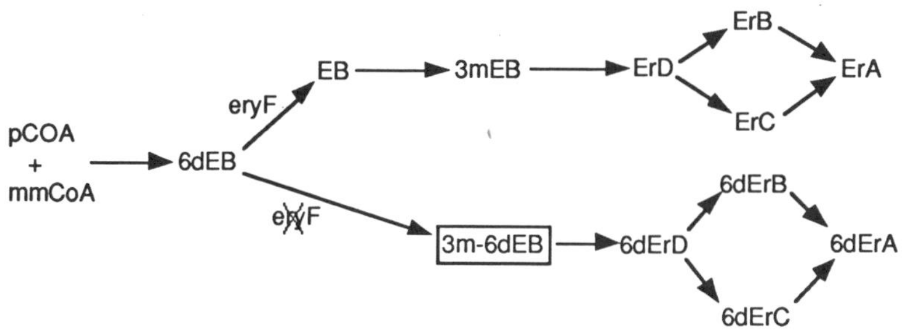

Figure 2. Scheme for the synthesis of erythromycin and 6-deoxyerythromycin. All compounds have been identified except for the one boxed. Abbreviations: 6dEB, 6-deoxyerythronolide B; 6dErA-D, 6-deoxyerythromycin A-D; EB, erythronolide B; ErA-D, erythromycin A-D; 3mEB, 3-α-mycarosylerythronolide B, 3m-6dEB, 3-α-mycarosyl-6-deoxyerythronolide B; mmCoA, methylmalonyl Coenzyme A; pCoA, propionyl Coenzyme A.

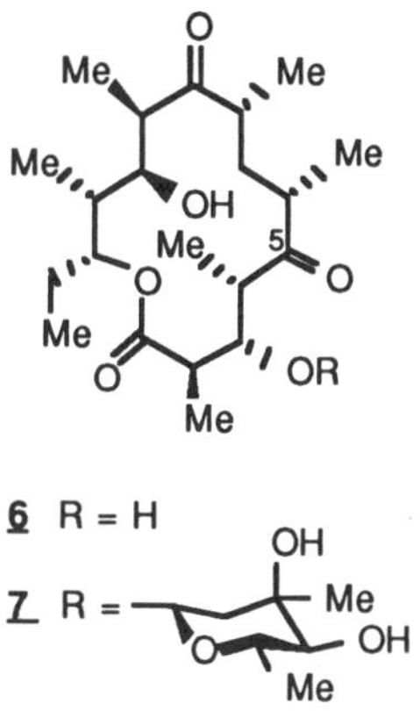

$\underline{6}$ R = H

$\underline{7}$ R =

Genetic manipulation of *eryAIII*, involved in the synthesis of the polyketide portion of $\underline{1}$, 6dEB, has resulted in the production of specific derivatives, 5,6-dideoxy-5-oxoerythronolide B ($\underline{6}$) and 3-α-mycarosyl-5,6-dideoxy-5-oxoerythronolide B ($\underline{7}$) (16). These structures were predicted from the knowledge of the mechanism of synthesis of 6dEB, obtained from determining the nucleotide sequence of the *eryA* genes. As shown in Fig. 3, synthesis of 6dEB is conducted by three multi-functional polypeptides containing a total of 28 enzymatic domains that are organized into six "synthase units". The DNA sequence encoding each synthase unit has been given the designation "module". The "modular hypothesis" proposed that each synthase unit is involved in one condensation step in the growth of the polyketide chain towards its ultimate lactonization into 6dEB and that, within each synthase unit, each enzymatic domain is responsible for one unique step (16). This hypothesis was tested by the creation, through gene replacement, of an 813 base pair in-frame deletion at a site corresponding to a domain containing a β-ketoreductase which was predicted to be involved in reducing the carbonyl group that would ultimately appear at C-5 of 6dEB (see Fig. 3). If this step could be bypassed without interfering with the otherwise normal completion of acyl chain growth, elimination of the KR function from the polyketide synthase was predicted to yield 5-deoxy-5-oxoerythronolide B. A small amount of 5-deoxy-5-oxoerythronolide B was, in fact, observed but the great majority of the compounds produced lacked the C-6 OH group, reflecting a partial bypass of the C-6 hydroxylation step that takes place after completion of synthesis of the macrolactone ring.

2-Norerythromycin A ($\underline{8}$) and its B-D congeners were detected in a mutant of *Sac. erythraea* carrying an in-frame deletion of the AT domain (Fig. 3) of the sixth *eryA* module (16), which had been transformed with an uncharacterized segment of genomic DNA from *Streptomyces antibioticus*, the producer of the 14-membered macrolide oleandomycin (30). Labelling experiments indicated that $\underline{8}$ resulted from the incorporation of malonate in place of methylmalonate at the sixth condensation step, suggesting that the AT function determines the extender unit incorporated during synthesis of the polyketide portion of the macrolide.

The only true "hybrid" macrolides produced by genetic engineering to date, are 4"-iso-valerylspiramycins I, II and III ($\underline{9}$ - $\underline{11}$), composites of spiramycins I - III ($\underline{12}$ - $\underline{14}$) and carbomycin ($\underline{15}$).

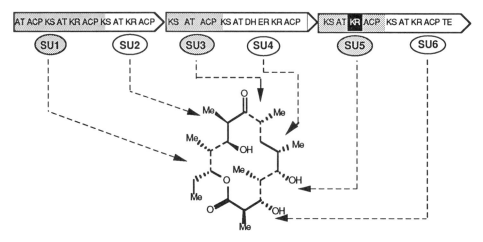

Figure 3. Scheme of 6dEB synthesis (16). The bars show the polypeptides, from left to right, encoded by *eryAI*, *eryAII* and *eryAIII*, respectively (Fig. 2), that contain the synthase units (SU). Enzymatic activities are displayed with shading to discriminate between SUs. The structure shown is employed to illustrate the role of each SU in the synthesis of 6dEB, where the carbons in the ring introduced by the odd and even numbered SUs are represented by dashed and solid lines, respectively. The KR of the fifth SU, highlighted in black, was deleted in the host that produced **6** and **7**. Abbreviations: ACP, acyl carrier protein; AT acyltransferase; DH, dehydratase; ER, enoylreductase; KR, β-ketoreductase; KS, β-ketoacyl:ACP synthase; TE, thioesterase.

The finding that a mutant of *Streptomyces thermotolerans*, blocked in the synthesis of the polyketide portion of **15**, could convert exogenously fed **12** to **9** indicated that the enzyme that acylates the 4"-position of the mycarose residue can utilize spiramycin as a substrate (31). Epp *et al.* inserted various segments of chromosomal DNA adjacent to the carbomycin resistance gene *carB* (32) into pIJ702 and introduced the plasmids into *Streptomyces ambofaciens*, the producer of **12** - **14** (33). One of the transformants produced **9** - **11**. When the gene which determined the acylating activity, designated *carE*, was isolated from the transformant and introduced into *Streptomyces lividans,* the *S. lividans* (*carE*) strain converted exogenously fed **12** to **9**. In this instance, therefore, the production of a novel structure was used as a means of identifying an antibiotic biosynthesis gene.

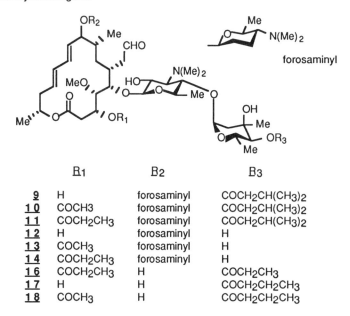

	R₁	R₂	R₃
9	H	forosaminyl	COCH₂CH(CH₃)₂
10	COCH₃	forosaminyl	COCH₂CH(CH₃)₂
11	COCH₂CH₃	forosaminyl	COCH₂CH(CH₃)₂
12	H	forosaminyl	H
13	COCH₃	forosaminyl	H
14	COCH₂CH₃	forosaminyl	H
16	COCH₂CH₃	H	COCH₂CH₃
17	H	H	COCH₂CH₂CH₃
18	COCH₃	H	COCH₂CH₂CH₃

15

In an analogous experiment, a *carE*-hybridizing segment of chromosomal DNA from *Streptomyces mycarofaciens*, producer of the 16-membered macrolide midecamycin (**16**), when introduced via a derivative of pIJ702 into *S. lividans*, was found to confer on the host the ability to convert exogenously fed **12** to **13** and **14** and leucomycin A5 (**17**) to leucomycin A4 (**18**) (34). The cloned gene, therefore, encoded a C-3-acetylation/propionylation activity rather than the C-4" *O*-acylation activity encoded by *carE*.

Aromatic Polyketides - The aromatic polyketides comprise a large family of antibiotics and antitumor agents. The most commercially important types are the tetracyclines and the anthracycline doxorubicin (adriamycin). Included in this family are the isochromanequinones dihydrogranaticin (**19**) and actinorhodin (**20**), whose biosynthesis genes are the best characterized for all antibiotics. Genes for the synthesis of **19** and **20** were employed in the first examples of the biosynthesis of hybrid antibiotics (35,36). When the *actVa* locus from *Streptomyces coelicolor*, recently shown to contain six ORFs, of which four are involved in ring hydroxylation (37), was cloned in a low copy vector and introduced into *Streptomyces sp.* AM-7161, the producer of medermycin (**21**), the transformant produced two novel, hybrid antibiotics designated 'mederrhodins' A (**22**) and B (35,36). Since the *actVa* locus normally acts at a late step in the synthesis of **20**, after the rings have closed, it is likely that, in the cell producing the hybrid, the *actVa*-encoded hydroxylase acts on the medermycin polyketide at the analogous step in the sequence of reactions. It is interesting to note, therefore, that introduction of the 32 kb segment containing the genes for the complete synthesis of **20** (38) into *Streptomyces sp.* AM-7161 resulted in the production of only **20** and **21**, but not **22**. In a parallel experiment, a plasmid containing the 32 kb *act* segment, when introduced into *Streptomyces violaceoruber* Tü22, producer of **19**, yielded the novel compound named 'dihydrogranatirhodin' (**23**) (35,36). Though not demonstrated at the time, from the knowledge obtained later of the roles of the various *act* genes in the synthesis of **20**, it is likely that the *actVI* locus interacted with the *gra* genes to produce the hybrid (39).

Actinorhodin (**19**) is made in *S. coelicolor* from the condensation of one acetate and seven malonates, governed by the products of the *actI*, *actIII* and *actVII* loci which compose the polyketide synthase complex (40,41). Since *actI* mutants are blocked in the initial steps of the pathway and do not accumulate identifiable intermediates, the role played by the PKS components in determining the structure of the polyketide has been difficult to assess. The anthracyclines aklavinone (**24**) and 2-hydroxyaklavinone (**25**) are produced in *Streptomyces galilaeus* ATCC31133 and *S. galilaeus* ATCC31671, respectively, from the condensation of a propionate starter and nine malonates. Introduction of the segment containing *actI* and *actVII* into the two *S. galilaeus* strains resulted in production of the compounds aloesaparonin II (**26**) or its 2-hydroxy derivative, desoxyerythrolaccin (**27**), respectively (42). When the *actIII* locus was included in the segment introduced into *S. galilaeus* ATCC31671, only **26** was produced. This indicated that the structure of **26** was determined, at least in part, by both the *actI* locus from *S. coelicolor* and the *actIII* counterpart (*dauB*) from *S. galilaeus* (42). These data and the finding that a shunt product with a different ring structure is produced in an *actVII*-deficient *S. coelicolor* strain, suggest that reduction of the fourth keto group incorporated during the synthesis of actinorhodin may be important in the folding of the polyketide to allow cyclization into the correct structure (39,42).

In the examples described above, the novel compounds made were the result of the addition of a single, new enzymatic step to an existing pathway, or the bypassing of one or more steps through inactivation of one of the enzymatic components. In the case to be described next, a novel, multi-step pathway was engineered into a fungus by the introduction of a number of genes. The gene cloning employed an *E. coli* plasmid that contained a marker selectable in the fungal host.

In fungi, stable transformation generally results from the integration of incoming DNA into the chromosome, often at multiple sites.

β-Lactams - The compound 7-aminocephalosporanic acid (7ACA) (31), the substrate for chemical conversion into a number of clinically important third-generation cephalosporins, has been usually produced by a complex chemical and enzymatic deacylation process of cephalosporin C (28). It was found that the cephalosporin acylase, cloned from *Pseudomonas diminuta*, was 20-fold more active on the derivative GL-7ACA (30) than on 28 (43). Thus it was predicted that a three-step metabolic pathway (Fig. 4) would provide efficient *in vivo* conversion of 28 to 31. The pathway was engineered into a strain of the fungus *Acremonium chrysogenum* in the following way (44). Segments containing the complementary DNA sequence from *Fusarium solani* encoding the enzyme D-amino acid oxidase (DAO), and a genomic DNA sequence from *P. diminuta* encoding the α and β subunits of the cephalosporin acylase, were placed under the control of the promoter of an alkaline phosphatase gene from *A. chrysogenum* and inserted into a plasmid containing the selectable hygromycin resistance marker. The construct was used to transform *A. chrysogenum* protoplasts. Southern hybridization blots of a number of transformants indicated that the DAO and acylase genes had integrated at multiple sites in the genome of the host. Two transformants produced 31 at levels up to 150 μg/ml, approximately 3% of that of 28. Other transformants produced less of 31 but up to 150 μg/ml of the intermediate 30. Thus, although the levels of 31 produced through fermentation in these experiments were too low to be considered commercially acceptable, this work should be regarded as a major achievement in that it demonstrates the feasibility of engineering a new biochemical pathway.

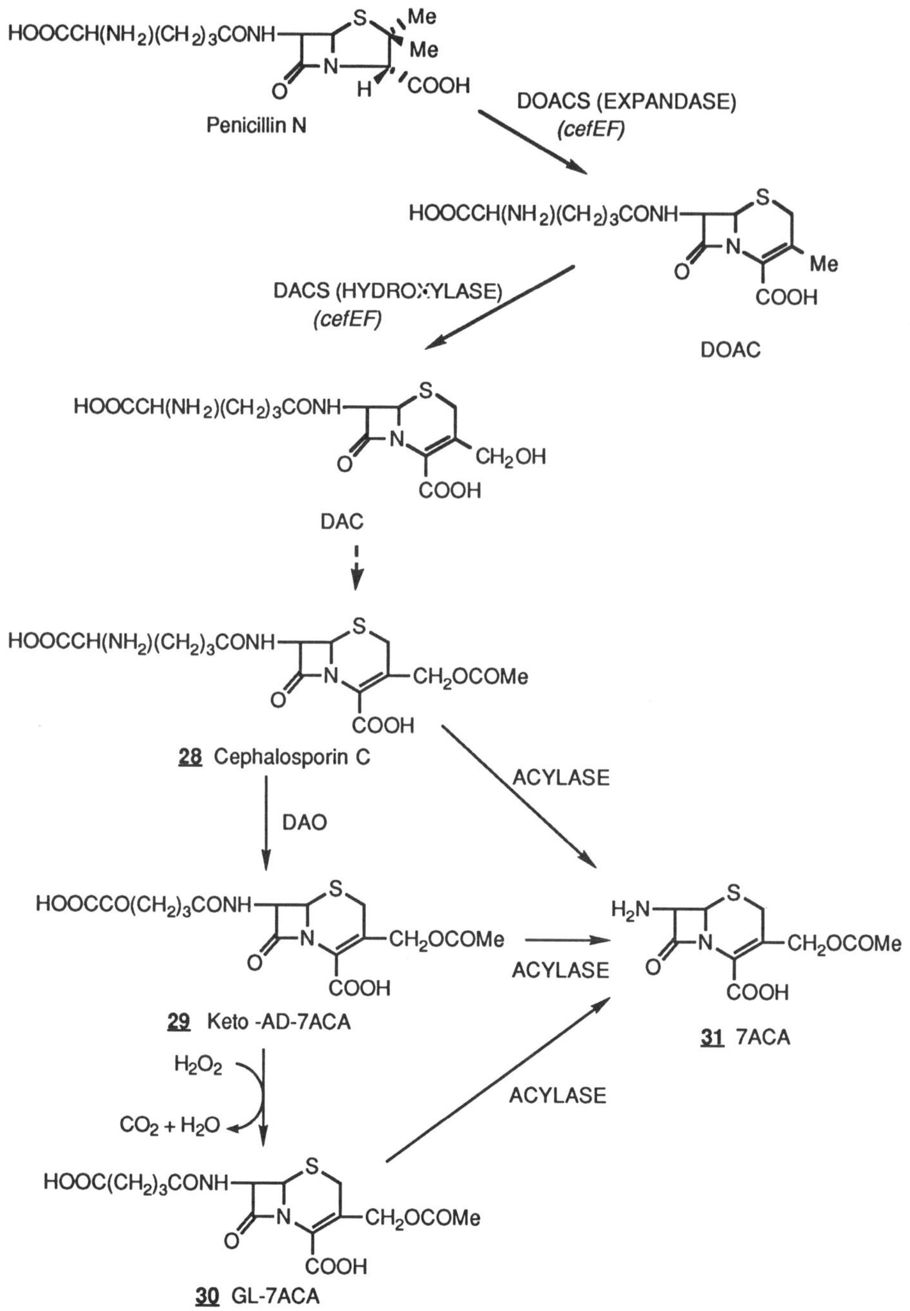

Figure 4. Biosynthesis of 7ACA from cephalosporin C in *A. chrysogenum* (43). Conversion of **29** to **30** is non-enzymatic. The cloning of genes for DAO and ACYLASE is described in the text. Abbreviations: 7ACA, 7-aminocephalosporanic acid; DAC, deacetylcephalosporin C; DACS, deacetylcephalosporin C synthase; DAO, D-amino acid oxidase; DAOC, deacetoxycephalosporin C; DAOCS, deacetoxycephalosporin C synthase; GL-7ACA, 7-β-(4-carboxybutanamido)-cephalosporanic acid; Keto - AD-7ACA, 7-β-(5-carboxy-5-oxopentanamido)-cephalosporanic acid; ACYLASE, cephalosporin acylase.

INCREASED TITERS

Increased Dosage of Structural Genes - A rate limiting step in the biosynthesis of cephalosporin C in the industrial strain 394-4 of the filamentous fungus *Cephalosporium acremonium* is the conversion of penicillin N to deacetoxycephalosporin C (DAOC), which is determined by the gene *cefEF* (Fig. 4). Penicillin N is produced in 394-4 to a level of ca. 16 units/L, about one-third the molar amount of cephalosporin C (45). *cefEF* encodes a bifunctional protein that carries out two sequential steps in the cephalosporin C pathway: expansion of the five-membered ring (penicillin N) to a six-membered one (DAOC) and hydroxylation of DAOC to form deacetylcephalosporin C (DAC) (Fig. 4). The *cefEF* gene, placed under the control of the promoter of the *C. acremonium* gene determining isopenicillin N synthetase (IPNS), was inserted in a plasmid containing a hygromycin resistance gene and then introduced into protoplasts of strain 394-4 (45). The transformants produced varying levels of cephalosporin C, ranging from 4 to 50% above the level in the untransformed host. The transformant with the highest level also showed reduced titers of penicillin N, to ca. 2 units/L in pilot plant fermentors. Southern blots showed the presence of one additional copy of the *cefEF* gene in this transformant. This work demonstrated for the first time that a rate-limiting step in an antibiotic biosynthesis pathway could be overcome by increasing the dosage of the corresponding gene. This work also showed that improvements in antibiotic titers can be accomplished by rational genetic manipulation of industrial strains, in addition to the more commonly employed laboratory strains which normally produce low levels of antibiotics.

A notable increase in the production of tetracenomycin C biosynthetic intermediates has also been achieved by a combination of increased gene dosage and enhanced gene expression. In this case, three of the type II PKS genes for the earliest step of the tetracenomycin C pathway, formation of the 20 carbon decapolyketide (39), were cloned under the control of the strong, constitutive ermE* promoter in a high copy number plasmid (46). Introduction of a plasmid carrying the two β-ketoacyl:ACP synthase genes and the ACP gene into wild-type or blocked mutants of *Streptomyces glaucescens* ETH22794 resulted in a 10 to 40-fold increase in the production of tetracenomycin D3 (**32**) but not tetracenomycin C (**33**), the final product of the pathway. A similar effect was obtained with just the ACP gene, suggesting that the titer of this protein is rate-limiting. The fact that **33** was not also overproduced either means that a later step becomes rate-limiting under these conditions or, less likely, that the endogenous level of resistance to **33** limits the amount that can be made.

32   **33**

Regulatory Genes - The observation that the *S. coelicolor actII* mutants appeared to be deficient in all of the steps of actinorhodin biosynthesis, even though they still contained the entire *act* cluster, has recently led to the demonstration that *actII* is a key regulator of antibiotic production (40,47). Many other clusters of antibiotic production genes contain such regulatory elements (48). In the case of *Streptomyces peucetius*, the *dnrlJ* genes and the *dnrR$_2$* DNA segment have been used to achieve an 8 to 10-fold increase in the titer of daunorubicin (the immediate precursor of doxorubicin) in the wild-type strain (49). [Introduction of the *actII* gene into *S. coelicolor* has a similar, but larger effect on actinorhodin production (47).] *afsR* and *redD* are other putative regulatory genes of *S. coelicolor* that have been demonstrated to enhance the output of secondary metabolism dramatically when introduced into the wild-type *S. coelicolor* strain or its close relative *S. lividans* (50,51). Such findings point to the high potential of overexpression of regulatory genes for yield enhancement as a complementary approach to the augmentation of the copy number of structural genes.

The exact mechanism by which genes like *actII*, *redD* and *dnrlJ* exert their effects that result in enhancement of antibiotic synthesis is not currently known. They may stimulate transcription of structural genes, resulting in increased levels of the pathway enzymes or the proteins that confer self-resistance. Since some of the self-resistance proteins are involved in antibiotic export, the increased expression of the corresponding gene through the action of these key regulators could also lead to enhanced antibiotic titers (49).

Bacterial Hemoglobin - The bacterium *Vitreoscilla* employs hemoglobin to grow under microaerophilic conditions (52). The gene encoding this hemoglobin (VHb) was found to have a dramatic effect on growth and actinorhodin production when introduced, on a high copy plasmid, into *S. coelicolor* (53). The strain produced VHb and grew to ca. 45% higher cell densities under various conditions of oxygen limitation, paralleling the effects seen when VHb was expressed in *E. coli* (52). In addition, under reduced aeration conditions, actinorhodin production was ten-fold higher in the strain containing the VHb gene. No difference in actinorhodin production was seen when oxygen was not limiting in the fermentation, however. It is not known whether the effect of VHb on production of the antibiotic is direct or whether it is a secondary effect of increased oxygen consumption or increased cell yield.

Future Prospects - The limited examples presented here illustrate that the application of genetic engineering technology to antibiotics can result in the production of novel structures and improved titers. Both applications depend on detailed knowledge of the biochemistry and genetics of the pathways concerned. Approaches to novel structures require determination of the nucleotide sequences of the structural genes for the pathway. Since only a small number of genes have been sequenced [in fact, very little information has been obtained on aminoglycosides with the exception of streptomycin (54) and fortimicin A (55)], the search for novel structures will be slowed until more sequencing has been performed. In the macrolide area, predicted changes can be made to the lactone ring. It should also be possible to change the sugars once the corresponding genes for their syntheses have been characterized. Modification of the rings by genetic approaches may become important in generating improved versions of the antifungal macrolides FK506 and rapamycin that are under intense investigation for their potent immunosupressive effects. Though the aromatic polyketides are the best studied at the genetic level of all antibiotics, the current lack of understanding of how the aromatic rings are determined limits rational approaches to changing structures. More work, using the actinorhodin and tetracenomycin C model systems, will be required before predicted novel structures can be engineered (38,39,56).

As shown for the synthesis of cephalosporin C and the tetracenomycins, increasing the gene dosage of a structual gene can lead to an increase in yield. Understanding the nature of regulation of the structural genes is essential to adopting a genetic approach to improving titers. Since antibiotics are produced as secondary metabolites, studies on the nature of the regulation of secondary metabolism will also have an impact on antibiotic titers.

References

1. M.J. Bibb, J.M. Ward and D.A. Hopwood, Nature, **274**, 398 (1978).
2. M.J. Bibb, J.L. Schottel and S.N. Cohen, Nature, **284**, 526 (1980).
3. J.E. Suarez and K.F. Chater, Nature, **286**, 527 (1980).
4. D.A. Hopwood, T. Kieser, D.J. Lydiate and M.J. Bibb, in "The Bacteria, Vol, 9," S.W. Queener and L.E. Day, Ed., Academic Press, New York, 1986, p.159.
5. T. Kieser and D.A. Hopwood, Methods Enzymol., **204**, 431 (1991).
6. J.A. Robinson, Chem. Soc. Rev., **17**, 383 (1988).
7. O.K. Behrens, J. Corse, J.P. Edwards, L. Garrison, R.G. Jones, Q.F. Soper, F.R. van Abeele and C.W. Whitehead, J. Biol. Chem., **175**, 793 (1948).
8. K.L. Rinehart, Jr., Pure and Appl. Chem., **49**, 361 (1977).
9. K.L. Rinehart, Jr., J.-R. Fang, W.-Z. Jin, C.J. Pearce, K.-I. Todano and T. Toyakuni, Devel. Industrial Microbiol., **26**, 117 (1984).
10. P.F. Wiley, Devel. Industrial Microbiol., **26**, 97 (1984).
11. D.A. Lowe, Devel. Industrial Microbiol., **26**, 143 (1984).
12. C.J. Thompson, T. Kieser, J.M. Ward and D.A. Hopwood, Gene, **20**, 51 (1982).
13. J.S. Tuan, J.M. Weber, M.J. Staver, J.O. Leung, S. Donadio and L. Katz, Gene **90**, 21, (1990).
14. J.M. Weber, J.O. Leung, G.T. Maine, R.H.B. Potenz, T.J. Paulus and J.P. Dewitt, J. Bacteriol., **172**, 2373 (1990).
15. J. Cortes, S.F. Haydock, G.A. Roberts, D.J. Bevitt and P.F. Leadlay, Nature, **348**, 176 (1990).
16. S. Donadio, M.J. Staver, J.B. McAlpine, S.J. Swanson and L. Katz, Science, **252**, 675 (1991).
17. J.M. Weber, J.O. Leung, S.J. Swanson, K.B. Idler and J.B. McAlpine, Science, **252**, 114 (1991).
18. S.F. Haydock, J.A. Dowson, N. Dhillon, G.A. Roberts, J. Cortes and P.F. Leadlay, Mol. Gen. Genet., **230**, 120 (1991).
19. N. Dhillon, R.S. Hale, J. Cortes and P.F. Leadlay, Mol. Microbiol., **3**, 1405 (1989).
20. J. Vara, M. Lewandowska-Skarbek, Y.-G. Wang, S. Donadio and C.R. Hutchinson, J. Bacteriol., **171**, 5872 (1989).
21. J.M. Weber, B.E. Schoner and R. Losick, Gene, **75**, 235 (1989).
22. T.J. Paulus, J.S. Tuan, V.E. Luebke, G.T. Maine, J.P. Dewitt and L. Katz, J. Bacteriol., **172**, 2541 (1990).

23. M.A. Richardson, S. Kuhstoss, M.L.B. Huber, L. Ford, O. Godfrey, J.R. Turner and R.N. Rao, J. Bacteriol., 172, 3790 (1990).
24. K.F. Chater and C.J. Bruton, Gene, 26, 67 (1983).
25. J.M. Weber and R. Losick, Gene, 68, 173, (1988).
26. E. Katz, C.J. Thompson and D.A. Hopwood, J. Gen. Microbiol., 129, 2703 (1983).
27. D.P. O'Keefe and P.A. Harder, Mol. Microbiol., 5, 2099 (1991).
28. J.F. Andersen and C.R. Hutchinson, J. Bacteriol., 174, 725 (1992).
29. P.B. Fernandes, R. Bailer, R. Swanson, C.W. Hansen, E. McDonald, N. Ramer, D. Hardy, N. Shipkowitz, R.R. Bower and E. Gade, Antimicrob. Agents Chemother., 30, 856 (1986).
30. J.B. McAlpine, J.S. Tuan, D.P. Brown, K.B. Grebner, D. Whittern, A. Buko and L. Katz, J. Antibiotics, 40, 1115 (1987).
31. H.A. Kirst, M. Debono, K.E. Willard, B.A. Trudell, T.E. Toth, J.R. Turner, D.R. Berry, B.B. Briggs, D.S. Fukuda, V.M. Daupert, A.M. Felty-Duckworth, J.L. Ott and F.T. Counter, J. Antibiotics, 39, 1724 (1986).
32. J.K. Epp, S.G. Burgett and B.E. Schoner, Gene, 53, 73 (1987).
33. J.K. Epp, M.L.B. Huber, T. Goodson and B.E. Schoner, Gene, 85, 293 (1989).
34. O. Hara, K. Nagaoka and C.R. Hutchinson, Japan Patent Application No. 07616901 (1991).
35. D.A. Hopwood, F. Malpartida, H.M. Kieser, H. Ikeda, J. Duncan, I. Fujii, B.A.M. Rudd, H.G. Floss and S. Omura, Nature, 314, 642 (1985).
36. S. Omura, H. Ikeda, F. Malpartida, H.M. Kieser and D.A. Hopwood, Antimicrob. Agents Chemother., 29, 13 (1986).
37. J.L. Caballero, E. Martinez, F. Malpartida and D.A. Hopwood, Mol. Gen. Genet., 230, 401 (1991).
38. F. Malpartida and D.A. Hopwood, Nature, 309, 462 (1984).
39. D.H. Sherman and D.A. Hopwood, Annu. Rev. Genet., 24, 37 (1990).
40. B.A.M. Rudd and D.A. Hopwood, J. Gen. Microbiol., 114, 35 (1979).
41. F. Malpartida and D.A. Hopwood, Mol. Gen. Genet., 205, 66 (1986).
42. P.L. Bartel, C.-B. Zhu, J.S. Lampel, D.C. Dosch, N.C. Conners, W. R. Strohl, J.M. Beale and H. G. Floss, J. Bacteriol., 172, 4816 (1990).
43. A. Matsuda, K. Matsuyama, K. Yamamoto, K. Ichikawa and S. Komatsu, J. Bacteriol., 169, 5815 (1987).
44. T. Isogai, M. Fukagawa, I. Aramori, M. Iwami, H. Kojo, T. Ono, Y. Ueda, M. Kohsaka and H. Imanaka, Bio/Technology, 9, 188 (1991). [see also T. Isogai et al., Actinomycetol., 5, 102 (1991).]
45. P.L. Skatrud, A.J. Tietz. T.D. Ingolia, C.A. Cantwell, D.L. Fisher, J.L. Chapman, and S.W. Queener, Bio/Technology, 7, 477 (1989).
46. H. Gramajo, J. White, C.R. Hutchinson and M.J. Bibb, J. Bacteriol., 173, 6475 (1991).
47. M.A. Fernandez-Moreno, J.L. Caballero, D.A. Hopwood and F. Malpartida, Cell, 66, 769 (1991).
48. K.F. Chater, Bio/Technology, 8, 115 (1990).
49. K.J. Stutzman-Engwall, S.L. Otten and C.R. Hutchinson, J. Bacteriol., 174, 144 (1992).
50. S. Horinouchi, M. Kito, N. Nishiyama, K. Furuya, S.-K. Hong, K. Miyake and T. Beppu, Gene, 95, 49 (1990).
51. K.E. Narva and J.S. Feitelson, J. Bacteriol., 172, 326 (1990).
52. C. Khosala and J.E. Bailey, Nature, 331, 633 (1988).
53. S.K. Magnolo, D.L. Leenutaphong, J.A. DeModena, J.E. Curtis, J.E. Bailey, J.L. Galazzo and D.E. Hughes, Bio/Technology 9, 473 (1991).
54. K. Mansouri and W. Piepersberg, Mol. Gen. Genet., 228, 459 (1991).
55. M. Hasegawa, Actinomycetol. 5, 126 (1991).
56. R.G. Summers, E. Wendt-Pienkowski, H. Motamedi and C.R. Hutchinson, J. Bacteriol., 174, 1810 (1992).

Chapter 15. Inhibitors of Angiogenesis

Mark A. Mitchell and John W. Wilks
The Upjohn Company, Kalamazoo, MI 49001

<u>Introduction</u> - Angiogenesis, the formation of new blood vessels, is a rare event in the healthy individual. It is an essential, but tightly regulated, component of normal growth and wound healing. Uncontrolled angiogenesis is a driving force in the growth of solid tumors, and a prominent feature of arthritis, psoriasis, diabetic retinopathy, and a variety of ophthalmic diseases. There is an unmet medical need for effective anti-angiogenic (angiostatic) agents to intervene in these destructive diseases of neovascularization. Angiogenesis is a complex, morphogenic event requiring the orchestrated temporal and spatial interaction of multiple cell types to produce a functional new blood vessel. The recent progress in identifying the key cellular and molecular features of angiogenesis is notable, but many details of the angiogenic cascade remain poorly defined (1-4). There are likely to be multiple sites for pharmacologic intervention. Since this is the first review of this subject for Annual Reports in Medicinal Chemistry, we have included a historical perspective. Many of the agents which modulate neovascularization are commonly known compounds, and their structures are not illustrated here. The recent, novel compounds, or those which have potential clinical utility, are illustrated.

ANGIOGENESIS MODELS AND ASSAYS

The unavailability of simple, quantitative, and reproducible assays is the greatest impediment for discovering anti-angiogenic therapies, and the ability of any assay to select clinically active agents has not been established. Despite these obstacles, inhibitors of angiogenesis are being chosen for clinical evaluation. A synopsis of the important considerations in choosing an angiogenesis model and interpreting the results is provided here; a detailed analysis was recently published (5).

<u>In Vivo</u> Assays - In vivo assays possess the greatest similarities to human disease, but they are labor intensive and difficult to quantitate. Four in vivo assays enjoy common use: the cornea assay, sponge implant assays, the chicken chorioallantoic membrane (CAM) assay, and the intradermal assay. Induction of angiogenesis in the mouse, rat or rabbit cornea is widely favored because the cornea is an avascular tissue and any blood vessels which develop are unequivocally new vessels. Inhibitors may be applied to the eye topically, or implanted near the angiogenic stimulus in a slow release pellet. The cornea, like other avascular tissues may contain endogenous inhibitors of neovascularization, and this must be considered when evaluating results (6,7). Subcutaneous implantation of polyurethane or polyvinyl alcohol sponges in rodents induces an inflammatory response and subsequent penetration of the sponge by a neovasculature. Any vessels occupying the sponge are clearly new vessels, but they are hard to visualize except by histology. The CAM assay and intradermal assays involve angiogenic responses on a background of existing vasculature, and it is not easy to distinguish the new blood vessels. The CAM assay is the most frequently used model because chicken embryos are inexpensive, and they can be grown in shell-free culture for easy and repeated observation. Inhibitors may be tested for their ability to produce avascular zones in the rapidly growing CAM, or to inhibit neovascularization stimulated by the application of a growth factor or physical stimulus (8). Quantitation of the results is often subjective, although the use of image analysis to evaluate responses is gaining acceptance. The intradermal assay involves inoculation of an angiogenic stimulus into the skin of mice, followed by examination of the response at autopsy 2-3 days later; responses are quantitated by counting vessels surrounding the inoculation site.

<u>In Vitro</u> Assays - In vitro assays are easy to perform and quantitate, but are limited in scope because they simulate only a few aspects of the angiogenic process. Many in vitro assays are

restricted to endothelial cells, and ignore the contributions of other cell types in neovascularization. Endothelial cell proliferation is important in angiogenesis, although plasticity in the vascular system permits limited development of new blood vessels without mitogenesis (9,10). An advantage of *in vitro* endothelial cell proliferation assays for drug discovery is that human cells from umbilical veins are easily obtained, and the specificity of antiproliferative agents for endothelial cells can be investigated. Angiogenic growth factors stimulate endothelial cell migration, and therefore the ability of compounds to prevent motility is often examined in chemokinesis and/or chemotaxis assays (1). However, not all *in vivo* inhibitors of angiogenesis affect endothelial *in vitro* cell migration. Endothelial cells form tubular structures reminiscent of capillaries when cultured under the appropriate conditions in a suitable extracellular matrix. These capillary "sprouting" or tubule formation assays are intriguing because they permit the complex cell-cell interactions necessary to form a blood vessel. Versions of the sprouting assay which utilize tissue specimens, such as fragments of aortas from rodents, may more closely approximate the *in vivo* condition (11).

NATURALLY-OCCURRING AND LARGE MOLECULAR WEIGHT INHIBITORS

<u>Cytokines</u> - Several cytokines stimulate neovascularization when administered *in vivo* by recruiting and activating macrophages or lymphocytes (12). These same peptide factors inhibit many aspects of endothelial cell function *in vitro* (1). This paradox is exemplified by transforming growth factor $\beta1$ (TGF$\beta1$) which inhibits endothelial cell proliferation and migration (13), and diminishes the proteolysis essential for capillary sprouting *in vitro* (14), yet TGF$\beta1$ elicits a strong angiogenic response *in vivo* when injected subcutaneously in rats (15), implanted into rodent corneas (12), or applied to the surface of the CAM (16). Recent work to reconcile these disparate observations has emphasized the need to carefully define the timing and intensity of the *in vivo* TGF$\beta1$ dose, as well as the responsiveness of non-endothelial host cells at the site of angiogenesis (12,16,17). Thus, TGF$\beta1$ not only plays a role in initiating neovascularization during wound healing and inflammation, but may also bring about completion of the tightly regulated angiogenic process in these situations (18). Tumor necrosis factor (TNFα), lymphotoxin, and interferon-γ also inhibit the *in vitro* proliferation of endothelial cells; interferon-γ also suppresses capillary tubule formation *in vitro* (19-22). TNFα *in vivo* elicits neovascularization in the cornea and CAM assays (21,23).

Interferon-α, deserves special mention since it is the first clinically successful inhibitor of angiogenesis. Early work demonstrated that interferon inhibits the *in vitro* motility of capillary endothelial cells, and reduces the *in vivo* angiogenesis resulting from intradermal injection of tumor cells in mice (24,25). These observations prompted limited clinical studies using prolonged administration of recombinant interferon-α in patients with life-threatening hemangiomas (26,27). The successful resolution of disease in these patients prompted a larger series of studies in infants with hemangiomas that had failed corticosteroid therapy; regression of hemangiomas occurred in 11/13 patients with little or no toxicity (28). These encouraging results will undoubtedly foster evaluations of interferon-α in other diseases of neovascularization.

<u>Tissue Inhibitor of Matrix Metalloproteinases (TIMP)</u> - The observation that certain tissues are avascular (e.g., cartilage) led investigators to postulate that these tissues contain endogenous inhibitors of angiogenesis. Studies initiated in the 1970s to identify these inhibitors led to the recent isolation of a 27 kDa protein with strong anti-angiogenic activity; this protein was named cartilage-derived inhibitor (CDI) (6,7,29). CDI inhibits endothelial cell proliferation and migration *in vitro*, and blocks angiogenesis in the CAM assay. The amino acid sequence at the NH_2-terminus of CDI is highly homologous with the sequences of TIMP-1 and TIMP-2 (30). TIMP-1 and TIMP-2 are specific inhibitors of the matrix metalloproteinases (MMPs) which degrade components of the extracellular matrix during embryogenesis, wound healing, and diseases such as cancer (31). The purification of CDI highlights the important role of MMPs in angiogenesis, and this concept is strengthened by the observation that both TIMP-1 and TIMP-2 inhibit angiogenesis in a chicken embryo assay; TIMP-1 also inhibits fibroblast growth factor (FGF) stimulated neovascularization in the rat cornea (32,33). CDI reduces the neovascularization induced by V2 carcinoma implanted in the rabbit cornea, and rat tumor cells stably transfected to express TIMP-2 grow more slowly than control cells when implanted in nude mice (30,34). These observations are encouraging investigators to search for small molecular weight inhibitors of MMPs.

Extracellular Matrix Proteins - Acquisition of the angiogenic phenotype is a consequence of the genetic changes that occur as a tumor cell progresses toward the malignant state (35). A tumor suppressor gene in hamster cells controls the secretion of a 140 kDa glycoprotein, which is a monomer of thrombospondin (TSP) without the heparin binding domain (36,37). TSP is a component of the extracellular matrix. Authentic human TSP, and the 140 kDa glycoprotein, inhibit the neovascularization of the rat cornea elicited by FGF (37). TSP inhibits the proliferation of endothelial cells, but not other cell types, in culture; TSP also suppresses endothelial cell migration (37-39). The mechanism of action of TSP is unknown, but recent studies have focused on the ability of this protein to alter the shape of endothelial cells and consequently cell behavior (40). The importance of extracellular matrix molecules in controlling capillary blood vessel morphogenesis has recently been reviewed (41).

Laminin, a major component of basement membranes, is a 800 kDa glycoprotein with multiple domains. Two synthetic peptides based on amino acid sequences found in active domains of laminin blocked the formation of capillary sprouts *in vitro* (42). One of these peptides contained a YIGSR sequence important in cell-cell adhesion, and the other contained a RGD sequence important for cell-matrix adhesion. A laminin-derived peptide (CDPGYIGSR-NH$_2$) inhibits angiogenesis in the CAM assay, and also inhibits the growth of a murine sarcoma when grown as a solid tumor but not in the ascites form (43). Other RGD peptides inhibit capillary sprout formation *in vitro*, and tumor-mediated angiogenesis in a mouse intradermal assay (44,45). Fragments of fibronectin, another extracellular matrix protein, also inhibit endothelial cell proliferation in culture (4).

Ribonuclease Inhibitor - Angiogenin is a potent angiogenic factor of 123 amino acids which has 35% sequence homology with pancreatic ribonuclease (1). Angiogenin is a unique inducer of neovascularization because it is not a mitogen for endothelial cells, yet it does activate phospholipases A$_2$ and C (46,47). Although the ribonuclease activity of angiogenin is dramatically less than that of RNases, human placental ribonuclease inhibitor (PRI) binds to angiogenin with high avidity (48). PRI blocks the angiogenic activity of angiogenin in the CAM assay suggesting an *in vivo* role for PRI and related inhibitors in the regulation of angiogenesis (48). A synthetic peptide from the C-terminal region of angiogenin (angiogenin 108-121) blocked the neovascularization elicited by angiogenin in the CAM assay (49). Based on the crucial roles for His-12 and His-119 in the function of RNase A, site-directed mutagenesis was used to replace the His-13 or His-114 in angiogenin with alanine; both of these proteins inhibited the angiogenic activity of angiogenin in the CAM assay (50).

Platelet Factor 4 - Platelet factor 4 (PF4) is a 70 amino acid polypeptide originally isolated from the α granule of platelets. PF4 was studied as an angiogenesis inhibitor because it binds to heparin. Another heparin-binding protein, protamine, was the first identified inhibitor of angiogenesis (51). Early work demonstrated that PF4 produces avascular zones in the chorioallantoic membrane of the chicken embryo (51). Recombinant PF4 also inhibits angiogenesis in the CAM assay, reduces the *in vitro* migration of endothelial cells, and specifically blocks the proliferation of human endothelial cells, but not fibroblasts or keratinocytes (52,53). Synthesis of peptide fragments of PF4 show that the angiostatic activity resides in the carboxy-terminal, heparin-binding region (52). Recombinant human PF4 inhibits the growth of murine melanoma and human colon carcinoma cell lines when injected directly into the tumor inoculation site of mice, but importantly has no effect on the growth of these tumor cells in culture (53). A recombinant analog of PF4, in which two lysine pairs of the carboxy-terminus essential for heparin binding were modified, yielded a protein lacking affinity for heparin. This PF4 analog retains all the anti-angiogenic and anti-tumor activity of PF4 indicating that heparin binding is not essential for angiostatic activity (54).

Heparinoids - Early studies documented increased concentrations of mast cells at sites of angiogenesis (55). Although mast cells alone do not induce angiogenesis in the CAM assay, cell lysates stimulate the migration of endothelial cells *in vitro*; the active agent in these lysates is heparin (55). Heparin also enhances the angiogenic activity present in tumor extracts. Many angiogenic growth factors contain heparin binding domains, and binding to heparin protects these factors from inactivation (56). Sequestration of FGF to heparan sulfate on the cell surface and in the extracellular matrix provides an important biological reservoir for the growth factor, and is essential for the recognition of FGF by its receptor (56,57). Therefore, the discovery that heparin would dramatically potentiate the anti-angiogenic activity of small molecules was unexpected (55). Significant inhibitions of angiogenesis in CAM assays, the cornea, or other *in vivo* models are achieved when heparin is co-administered with angiostatic steroids or collagen biosynthesis inhibitors (58,59). Combination treatments of heparin and corticosteroids significantly reduce the growth of tumors in mice (60). Heparin preparations are highly heterogeneous and not all of them will facilitate the angiostatic or anti-tumor activity of angiogenesis inhibitors; this has resulted in confusion in the literature. Some preparations of heparin alone give small inhibitions of angiogenesis in the CAM assay (8). Heparin fragments containing at least six saccharide residues augment the angiostatic activity of steroids in the chicken embryo, but smaller fragments are ineffective (60). The mechanism of action for heparin-mediated inhibition of angiogenesis is not understood, but may in part involve an inhibition of basement membrane synthesis and a reduction in endothelial cell proliferation (58,61,62).

The heterogeneity of heparin preparations fostered a search for anti-angiogenic heparin substitutes. Synthetic chemistry efforts will be reviewed below. A sulfated polysaccharide-peptidoglycan complex (SP-PG) of bacterial origin not only inhibits angiogenesis in the CAM assay when given alone, but significantly enhances the activity of angiostatic steroids (63). SP-PG also inhibits the growth of murine tumors when administered with angiostatic steroids or alone (63). Curiously, the anti-angiogenic and antitumor properties of SP-PG are increased by co-administration with the anti-estrogen, tamoxifen (64). Pentosan polysulfate, a semi-synthetic polysaccharide, is reported to both stimulate and inhibit endothelial cell proliferation *in vitro* (65,66). Nevertheless, pentosan polysulfate reduces the growth and vascularity of a human tumor inoculated into nude mice (67). A sulfated derivative of chitin inhibits endothelial cell migration *in vitro*, and reduces tumor angiogenesis in mice when injected intradermally with a murine melanoma (68). High molecular weight hyaluronic acid also inhibits capillary blood vessel formation when applied to granulation tissue (69).

Miscellaneous - A naturally occurring, 16 kDa fragment of the pituitary hormone prolactin inhibits the *in vitro*, FGF-stimulated proliferation of capillary endothelial cells but not baby hamster kidney cells (70). Maltose tetrapalmitate is reported to inhibit tumor vascularization and growth in mice, and to enhance the antitumor activity of cortisone (71,72).

SYNTHETIC AND SMALL MOLECULAR WEIGHT INHIBITORS

Fumagillin and its Analogs - The identification of a naturally secreted antibiotic of *Aspergillus fumigatus*, fumagillin (1), as an angiogenesis inhibitor is providing the impetus for discovery of novel small molecular weight inhibitors of neovascularization. Fumagillin inhibits endothelial cell proliferation *in vitro*, angiogenesis in the CAM assay, and tumor angiogenesis in the mouse (73).

Synthetic efforts have focused on modification of the core, fumagillol (2), structure. Modifications of fumagillol have included the synthesis of 6-O-acylated, alkylated, carbamoylated, and sulfonylated analogs, and its side chain hydrogenated, oxidized or aminated derivatives (74). The 6-amino-6-desoxy, 6-epi, and products of nucleophilic ring opening of the spiro oxirane system of fumagillol derivatives have been reported (75-78).

One analog of fumagillol, AGM-1470 (3), is 50-times more potent than fumagillin in inhibiting the proliferation of human umbilical vein endothelial cells in culture (73). AGM-1470 inhibits angiogenesis in the CAM assay, and in several rat models including the cornea assay, a sponge implantation assay, and formation of capillary-like tubules in a thoracic vein organ culture assay (79). Subcutaneous administration of AGM-1470 reduces the growth rate of Lewis Lung carcinoma and B16 melanoma in mice, and significantly increases the survival of mice inoculated with M5076 reticulum cell sarcoma (73). Intraperitoneally inoculated P388 leukemia, a tumor which grows in ascites form independent of angiogenesis, is not inhibited by AGM-1470. Compound 4 inhibited corneal angiogenesis greater than 70% in all rats; and compound 5 inhibited tumor growth 91% in mice bearing subcutaneous M5076 reticulum cell sarcoma (76). Both fumagillin and fumagillol were administered to cancer patients in the 1950's with no therapeutic benefit, so anticipated clinical trials with the more potent AGM-1470, or another analog, will be watched with great interest (80).

Synthetic Heparin Substitutes - As discussed above, the variable anti-angiogenic activity of heparin preparations, when combined with angiostatic steroids, prompted a search for chemically-defined heparin substitutes. A synthetic pentasaccharide inhibits angiogenesis in the CAM assay when administered with hydrocortisone (81). Unsubstituted α- and β-cyclodextrins are not effective heparin substitutes, but sulfation of hydroxyl groups on α-, β-, or γ-cyclodextrin yields compounds which inhibit angiogenesis in the CAM assay when co-administered with hydrocortisone (82). Increased levels of sulfation are directly correlated with anti-angiogenic activity; for example β-cyclodextrin tetradecasulfate (6), is superior to β-cyclodextrin heptasulfate. β-cyclodextrin tetradecasulfate is 100-1000 times more potent than heparin, and is active over a wider dose range in the CAM assay (82). β-cyclodextrin tetradecasulfate blocks inflammation-induced angiogenesis in the rabbit cornea when co-administered with hydrocortisone, or steroids devoid of corticosteroid properties (83). β-cyclodextrin tetradecasulfate has a high binding affinity for the angiogenic growth factor, FGF (84).

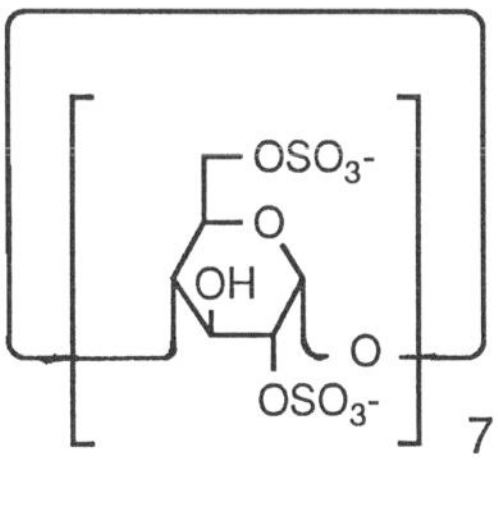

6

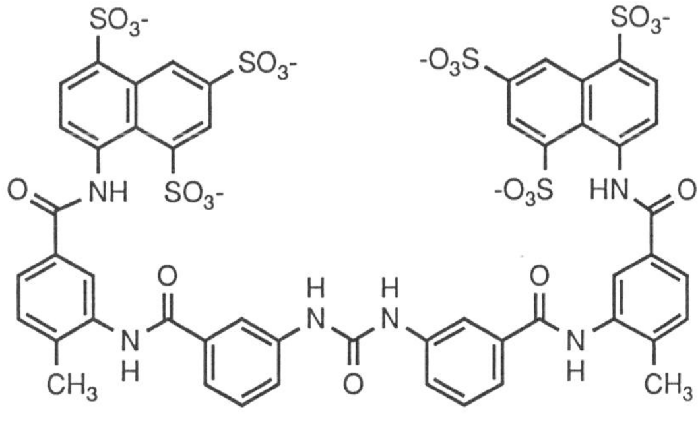

7

Another anionic compound, suramin (7), slightly inhibits angiogenesis in the CAM assay when given alone, but is most effective when given in combination with angiostatic steroids (8). The inhibition of angiogenesis seen with steroids plus suramin is similar to that observed for steroids plus heparin. The anti-angiogenic mechanism of suramin is not known, but may involve inhibition of growth factor binding to their receptors, inhibition of cell proliferation at other sites, or blockade of cell migration (85,86). Suramin plus an angiostatic steroid inhibits the growth of B16 melanoma in mice, and suramin has antitumor activity in humans (85,87).

Angiostatic Steroids - Early studies documented that 6α-methylprednisolone and dexamethasone block tumor-induced angiogenesis in the hamster and rabbit, probably because of the antiproliferative action of corticosteroids (88,89). Interest in steroids as small molecular weight inhibitors of angiogenesis increased dramatically with the demonstration that co-administration of corticosteroids with heparin would potentiate their anti-angiogenic properties, and could cause complete regression of some--but not all--murine tumors (60). Two obstacles limited the clinical application of these discoveries. The first was the heterogeneity of heparin preparations as discussed above. The second was the side-effects associated with the high dose corticosteroid therapy.

The angiostatic steroids were originally defined as inhibitors of angiogenesis without glucocorticoid and mineralocorticoid activity, although in the literature the term angiostatic steroid has been used to describe all anti-angiogenic steroids (81). 11α-Hydrocortisone retains the anti-angiogenic properties of hydrocortisone ($\underline{8}$), in the CAM assay, but lacks glucocorticoid and mineralocorticoid activity (81). The 4,5 double bond in the A-ring, and the 11-hydroxyl on the C-ring of hydrocortisone are not essential for anti-angiogenic activity; however the 17-hydroxyl and C-20 and -21 side chain of hydrocortisone contribute significantly to its anti-angiogenic activity (81). The tetrahydro derivative of cortisone ($\underline{9}$), previously considered biologically inactive, has twice the anti-angiogenic activity of hydrocortisone when applied on the CAM with heparin (81). A synthetic steroid without glucocorticoid activity, 6α-fluoro-7,21-dihydroxy-16β-methyl-pregna-4,9,(11)-diene-3,20-dione (U-42129, $\underline{10}$), and its 16α-methyl epimer (U-24067), have strong angiostatic activity when administered in the CAM assay with heparin or suramin (8,81). Although inhibition of tumor growth is observed in mice given combination treatments of heparin or suramin with true angiostatic steroids, the complete regression experienced with some corticosteroids is not observed (63,87). However, the angiostatic steroids are very effective when applied to the cornea in combination with β-cyclodextrin tetradecasulfate, and may have application in ophthalmic diseases of neovascularization (83). The anti-angiogenic mechanism of action of heparin-steroid combinations remains ill-defined, but these treatments do block the synthesis of collagen in the extracellular matrix (89-91).

Another steroid, medroxyprogesterone acetate (MPA, $\underline{11}$), inhibits tumor vascularization and tumor growth in the rabbit cornea (89). MPA also reduces the synthesis of angiogenic factors by a variety of human gynecologic tumors in culture, and inhibits the endothelial cell plasminogen activator production important for capillary sprouting (92,93). The anti-angiogenic properties of MPA may contribute to its anti-cancer properties in humans.

Collagen Biosynthesis Inhibitors - Inhibitors of collagen biosynthesis and processing are inhibitors of angiogenesis (59). Proline analogs, including L-azetidine-2-carboxylic acid (LACA), cis-hydroxyproline, D,L-3,4-dehydroproline, and thioproline, disrupt collagen biosynthesis and inhibit angiogenesis in the CAM assay (59). The anti-angiogenic activity of LACA is significantly potentiated by co-administration with heparin, and LACA administered with angiostatic steroids and heparin produces avascular zones in 100% of CAMs (59). Co-administration of L-proline with the proline analogs reverses their anti-angiogenic effect confirming an action through perturbation of collagen metabolism. *Cis*-hydroxyproline also inhibits capillary sprouting *in vitro* (94). LACA, *cis*-hydroxyproline and thioproline all limit the growth of a mammary tumor in rats, but L-3,4-dehydroproline actually increased tumor growth in mice inoculated with B16 melanoma (95,96).

β-aminopropionitrile (BAPN) irreversibly inhibits lysyl oxidase, the enzyme which crosslinks collagen fibers through oxidative deamination of lysine to aminoadipic semialdehydes. BAPN inhibits angiogenesis in the CAM assay, and augments the activity of heparin plus steroids (59). D-penicillamine reduces collagen biosynthesis by inhibiting lysyl oxidase, preventing the condensation of lysyl-derived aldehydes, and chelating copper. Copper is a cofactor for lysyl oxidase, and is essential for angiogenesis (97). D-penicillamine inhibits endothelial cell proliferation *in vitro*, and blocks neovascularization in the rabbit cornea (98). An inhibitor of prolyl and lysyl oxidase, α,α-dipyridyl, produces avascular zones in the CAM assay (59). GPA-1734 (12), prevents the hydroxylation of proline and lysine by interaction with Fe^{++}, resulting in inhibition of basement membrane synthesis (99,100). GPA-1734 inhibits collagen synthesis and angiogenesis in the CAM assay, and reduces the growth of Walker 256 carcinoma in rats (101).

Miscellaneous - Numerous small molecular weight compounds have been reported to inhibit angiogenesis in CAM or cornea assays. These inhibitors include: a) 2-hydroxy-5-nitro toluenesulfonate, an arylsulfatase inhibitor and thus a potentiator of heparin (102), and xylosides which modulate glycosaminoglycan synthesis (103); b) minocycline (13), a matrix metalloproteinase inhibitor (104); c) the antibiotics herbimycin A (105,106), 15-deoxyspergualine (107), and eponemycin (108); d) vitamin D_3 analogs (109,110) and retinoids (59,111,112); e) the opiods β-endorphin and morphine (113); f) the cancer chemotherapeutic drugs bleomycin, methotrexate, mitoxantrone, bisantrene, and razoxane (114-117); g) inhibitors of arachidonic acid metabolism (118); and h) miscellaneous compounds including α-difluoromethylornithine, tranexamic acid, and gold sodium thiomalate (119-122).

Conclusions and State of the Art - The inhibitors of angiogenesis comprise a lengthy and diverse group of biochemical and chemical entities acting at many points in the angiogenic cascade. While many of these inhibitors will not be useful clinically, they have provided insights into the biochemical pathways relevant for neovascularization. Our current level of knowledge provides a foundation for the future selection of endothelial-cell-selective pharmacologic targets, including anti-proliferative agents, inhibitors of cell migration, and modulators of cell-matrix interactions. More detailed molecular definition of the angiogenic processes will permit the design of carefully controlled, quantitative, and rapid cell-free or cell-based assays. New assays are necessary for the medicinal chemist to design novel inhibitors based on carefully determined structure-activity relationships. Discovery of the efficacy of interferon-α in some diseases of neovascularization is a stimulus to continue the search for angiogenesis inhibitors, and portends the discovery of important new agents to control the diseases of neovascularization.

References

1.   J. Folkman and M. Klagsbrun, Science, 235, 442 (1987).
2.   C.H. Blood and B.R. Zetter, Biochim. Biophys. Acta, 1032, 89 (1990).
3.   R.S. Kerbel, BioEssays, 13, 31 (1991).

4.   R. Eisenstein, Pharmac. Ther., 49, 1 (1991).
5.   R. Auerbach, W. Auerbach, and I. Polakowski, Pharmac. Ther., 51, 1 (1991).
6.   R. Eisenstein, K.E. Kuettner, C. Neapolitan, L.W. Soble, and N. Sorgente, Am. J. Pathol., 81, 337 (1975).
7.   R. Langer, H. Conn, J. Vacanti, C. Haudenschild, and J. Folkman, Proc. Natl. Acad. Sci. USA, 77, 4331 (1980).
8.   J.W. Wilks, P.S. Scott, L.K. Vrba, and J.M. Cocuzza, Int. J. Radiat. Biol., 60, 73 (1991).
9.   J. Denekamp and B. Hobson, Br. J. Cancer, 46, 711 (1982).
10. M.M. Sholley, G.P. Ferguson, H.R. Seibel, J.L. Montour, and J.D. Wilson, Lab. Invest., 51, 624 (1984).
11. R.F. Nicosia and A. Ottinetti, Lab. Invest., 63, 115 (1990).
12. P.J. Polverini, Cytokines, 1, 54 (1989).
13. G. Muller, J. Behrens, U. Nussbaumer, P. Bohlen, and W. Birchmeier, Proc. Natl. Acad. Sci. USA, 84, 5600 (1987).
14. M.S. Pepper, D. Belin, R. Montesano, L. Orci, and J.-D. Vassalli, J. Cell Biol., 111, 743 (1990).
15. A.B. Roberts, M.B. Sporn, R.K. Assoian, J.M. Smith, N.S. Roche, L.M. Wakefield, U.I. Heine, L.A. Liotta, V. Falanga, J.H. Kehrl, and A.S. Fauci, Proc. Natl. Acad. Sci. USA, 83, 4167 (1986).
16. E.Y. Yang and H.L. Moses, J. Cell Biol., 111, 731 (1990).
17. A.B. Sutton, A.E. Canfield, S.L. Schor, M.E. Grant, and A.M. Schor, J. Cell Sci., 99, 777 (1991).
18. S. Tsunawaki, M. Sporn, A. Ding and C. Nathan, Nature, 334, 260 (1988).
19. R. Friesel, A. Komoriya, and T. Maciag, J. Cell Biol., 104, 689 (1987).
20. N. Tsuruoka, M. Sugiyama, Y. Tawaragi, M. Tsujimoto, T. Nishihara, T. Goto, and N. Sato, Biochem. Biophys. Res. Commun., 155, 429 (1988).
21. M. Frater-Schroder, W. Risau, R. Hallmann, P. Gautschi, and P. Bohlen, Proc. Natl. Acad. Sci. USA, 84, 5277 (1987).
22. R.K. Maheshwari, V. Srikantan, D. Bhartiya, H.K. Kleinman, and D.S. Grant, J. Cell. Physiol., 146, 164 (1991).
23. S.J. Leibovich, P.J. Polverini, H.M. Shepard, D.M. Wiseman, V. Shively, and N. Nuseir, Nature, 329, 630 (1987).
24. D. Brouty-Boye and B.R. Zetter, Science, 208, 516, (1980).
25. Y.A. Sidky and E.C. Borden, Cancer Res., 47, 5155 (1987).
26. C.W. White, H.M. Sondheimer, E.C. Crouch, H. Wilson, and L.L. Fan, N. Engl. J. Med., 320, 1197 (1989).
27. P.J. Orchard, C.M. Smith III, W.G. Woods, D.L. Day, L.P. Dehner, and R. Shapiro, Lancet, ii, 565 (1989).
28. A. Ezekowitz, J. Mulliken, and J. Folkman, Brit. J. Haematol., 79 Suppl. 1, 67 (1991).
29. M.A. Moses, J. Sudhalter, and R. Langer, Science, 248, 1408 (1990).
30. M.A. Moses and R. Langer, J. Cell. Biochem., 47, 230 (1991).
31. J.F. Woessner, Jr., FASEB J., 5, 2145 (1991).
32. M. Takigawa, Y. Nishida, F. Suzuki, J. Kishi, K. Yamashita, and T. Hayakawa, Biochem. Biophys. Res. Commun., 171, 1264 (1990).
33. M.D. Johnson, H.R. Choi-Kim, L. Chesler, G.S. Tsao-Wu, N. Bouck, and P. Polverini, Proc. Am. Assoc. Cancer Res., 32, 81 (1991).
34. Y.A. DeClerck, N. Perez, H. Shimada, T.C. Boone, K.E. Langley, and S. Taylor, Cancer Res., 52, 701 (1991).
35. N. Weidner, J.P. Semple, W.R. Welch, and J. Folkman, New Engl. J. Med., 324, 1 (1991).
36. F. Rastinejad, P.J. Polverini, and N.P. Bouck, Cell, 56, 345 (1989).
37. D.J. Good, P.J. Polverini, F. Rastinejad, M.M. LeBeau, R.S. Lemons, W.A. Frazier, and N.P. Bouck, Proc. Natl. Acad. Sci. USA, 87, 6624 (1990).
38. P. Bagavandoss and J.W. Wilks, Biochem. Biophys. Res. Commun., 170, 867 (1990).
39. G. Taraboletti, D. Roberts, L.A. Liotta, R. Giavazzi, J. Cell Biol., 111, 765 (1990).
40. E.H. Sage and P. Bornstein, J. Biol. Chem., 266, 14831, (1991).
41. D. Ingber, J. Cell. Biochem., 47, 236 (1991).
42. D.S. Grant, H.K. Kleinman, and G.R. Martin, Ann. N. Y. Acad. Sci., 588, 61 (1990).
43. N. Sakamoto, M. Iwahana, N.G. Tanaka, Y. Osada, Cancer Res., 51, 903 (1991).
44. R.F. Nicosia and E. Bonanno, Am. J. Pathol., 138, 829 (1991).
45. I. Saiki, J. Murata, T. Makabe, N. Nishi, S. Tokura, and I. Azuma, Jpn. J. Cancer Res., 81, 668 (1990).
46. R. Bicknell and B.L. Vallee, Proc. Natl. Acad. Sci. USA, 85, 5961 (1988).
47. R. Bicknell and B.L. Vallee, Proc. Natl. Acad. Sci. USA, 86, 1573 (1989).
48. R. Shapiro and B.L. Vallee, Proc. Natl. Acad. Sci. USA, 84, 2238 (1987).
49. S.M Rybak, D.S. Auld, D.K. St. Clair, Q.-Z. Yao, and J.W. Fett, Biochem. Biophys. Res. Commun., 162, 535 (1989).
50. R. Shapiro and B.L. Vallee, Biochemistry, 28, 7401 (1989).
51. S. Taylor and J. Folkman, Nature, 297, 307 (1982).

52.   T.E. Maione, G.S. Gray, J. Petro, A.J. Hunt, A.L. Donner, S.I. Bauer, H.F. Carson, and R.J. Sharpe, Science, 247, 77 (1990).
53.   R.J. Sharpe, H.R. Byers, C.F. Scott, S.I. Bauer, and T.E. Maione, J. Natl. Cancer Inst., 82, 848 (1990).
54.   T.E. Maione, G.S. Gray, A.J. Hunt, and R.J. Sharpe, Cancer Res., 51, 2077 (1991).
55.   J. Folkman, Biochem. Pharmacol., 34, 905, (1985).
56.   I. Vlodavsky, R. Bar-Shavit, R. Ishai-Michaeli, P. Bashkin, and Z. Fuks, Trends Biochem. Sci., 16, 268 (1991).
57.   A.C. Rapraeger, A. Krufka, and B.B. Olwin, Science, 252, 1705 (1991).
58.   J. Folkman and D.E. Ingber, Ann. Surg., 206, 374 (1987).
59.   D. Ingber and J. Folkman, Lab. Invest., 59, 44 (1988).
60.   J. Folkman, R. Langer, R.J. Linhardt, C. Haudenschild, and S. Taylor, Science, 221, 719 (1983).
61.   N. Sakamoto, N.G. Tanaka, A. Tohgo, and H. Ogawa, Cancer J., 1, 55 (1986).
62.   S.J. Busch, G.A. Martin, R.L. Barnhart, M. Mano, A.D. Cardin, and R.L. Jackson, J. Cell Biol., 116, 31 (1992).
63.   N.G. Tanaka, N. Sakamoto, K. Inoue, H, Korenaga, S. Kadoya, H. Ogawa, and Y. Osada, Cancer Res., 49, 6727 (1989).
64.   N.G. Tanaka, N. Sakamoto, H, Korenaga, K. Inoue, H. Ogawa, and Y. Osada, Int. J. Radiat. Biol., 60, 79 (1991).
65.   J.M. Herbert, M. Cottineau, F. Driot, J.M. Pereillo, and J.P. Maffrand, Biochem. Pharmacol., 37, 4281 (1988).
66.   C. Klein-Soyer, A. Beretz, J.-P. Cazenave, E. Wittendorp-Rechenmann, J.-L. Vonesch, R.V. Rechenmann, F. Driot, and J.-P. Maffrand, Arteriosclerosis, 9, 147 (1989).
67.   A. Wellstein, G. Zugmaier, J.A. Califano III, F. Kern, S. Paik, and M.E. Lippman, J. Natl. Cancer Inst., 83, 716 (1991).
68.   J. Murata, I. Saiki, T. Makabe, Y. Tsuta, S. Tokura, and I. Azuma, Cancer Res., 51, 22 (1991).
69.   D.C. West and S. Kumar, Ciba Foundat. Sympos., 143, 187 (1989).
70.   N. Ferrara, C. Clapp, and R. Weiner, Endocrinology, 129, 896 (1991).
71.   O. Benrezzak, E. Bissonnette, P. Madarnas, and V.N. Nigam, Anticancer Res., 9, 1815 (1989).
72.   O. Benrezzak, P. Madarnas, R. Pageau, V.N. Nigam, and M.M. Elhilali, Anticancer Res., 9, 1883 (1989).
73.   D. Ingber, T. Fujita, S. Kishimoto, K. Sudo, T. Kanamaru, H. Brem, and J. Folkman, Nature, 348, 555 (1990).
74.   S. Kishimoto, T. Fujita, T. Kanamaru, M.J. Folkman, and D. Ingber, EP 357061 (1990).
75.   S. Kishimoto, S. Marui, and T. Fujita, EP 386667 (1990).
76.   S. Kishimoto, S. Marui, and T. Fujita, EP 415294 (1991).
77.   T. Oku, C. Kasahara, T. Ohkawa, M. Hashimoto, EP 354767 (1990).
78.   T. Oku, C. Kasahara, T. Ohkawa, M. Hashimoto, EP 354787 (1990).
79.   M. Kusaka, K. Sudo, T. Fujita, S. Marui, F. Itoh, D. Ingber, and J. Folkman, Biochem. Biophys. Res. Comm., 174, 1070 (1991).
80.   J.A. DiPaolo, D.S. Tarbell, and G.E. Moore in "Antibiotics Annual 1958-1959," H. Welch and F. Marti-Ibanez, Eds., Medical Encyclopedia, Inc. New York, N.Y., 1959, p. 541.
81.   R. Crum, S. Szabo and J. Folkman, Science, 230, 1375 (1985).
82.   J. Folkman, P.B. Weisz, M.M. Joullie, W.W.Li, and W.R.Ewing, Science, 243, 1490 (1989).
83.   W.W. Li, R. Casey, E.M. Gonzalez, and J. Folkman, Invest. Ophthalmol. Vis. Sci., 32, 2898 (1991).
84.   Y. Shing, J. Folkman, P.B. Weisz, M.M. Joullie, and W.R. Ewing, Anal. Biochem., 185, 108 (1990).
85.   C.A. Stein, R. LaRocca, and C. Myers, PPO Updates, 4, 1 (1990).
86.   M. Nakajima, A. DeChavigny, C.E. Johnson, J. Hamada, C.A. Stein, and G.L. Nicolson, J. Biol. Chem., 266, 9661 (1991).
87.   J.W. Wilks, T.F. DeKoning, J.M. Cocuzza, A.Tomilo, P.S. Scott, and L.K. Vrba, Proc. Am. Assoc. Cancer Res. (Washington, D.C.), 31, 60 (1990).
88.   P. Shubik, R. Feldman, H. Garcia, and B.A. Warren, J. Natl. Cancer Inst., 57, 769 (1976).
89.   J. Gross, R.G. Azizkhan, C. Biswas, R.R. Bruns, D.S.T. Hsieh, and J. Folkman, Proc. Natl. Acad. Sci. USA, 78, 1176 (1981).
90.   D.E. Ingber, J.A. Madri, and J. Folkman, Endocrinology, 119, 1768 (1986).
91.   M.E. Maragoudakis, M. Sarmonika, and M. Panoutsacopoulou, J. Pharmacol. Exp. Ther., 251, 679 (1989).
92.   J. Fujimoto, S. Hosoda, H. Fujita, and H. Okada, Invasion Metastasis, 9, 269 (1989).
93.   H. Ashino-Fuse, Y. Takano, T. Oikawa, M. Shimamura, and T. Iwaguchi, Int. J. Cancer, 44, 859 (1989).
94.   R.F. Nicosia, P. Belser, E. Bonanno, and J. Diven, In Vitro Cell. Dev. Biol., 27A, 961 (1991).
95.   W.R. Kidwell, M. Bano, and S.J. Taylor, Biol. Responses Cancer, 4, 47 (1985).
96.   W.B. Ershler, R.L. Gamelli, A.L. Moore, M.P. Hacker, and A.J. Blow, 19, 367 (1984).
97.   M. Ziche, J. Jones, and P.M. Gullino, J. Natl. Can. Inst., 69, 475 (1982).

98.   T. Matsubara, R. Saura, K. Hirohata, and M. Ziff, J. Clin. Invest., 83, 158 (1989).

99.   M.E. Maragoudakis, M. Sarmonika, and M. Panoutsacopoulou, J. Pharmacol. Exp. Ther., 244, 729 (1988).

100.   M.E. Maragoudakis, H.J. Kalinsky, and J. Wasvary, J.Pharmacol. Exp. Ther., 204, 372 (1978).

101.   E. Missirlis, G. Karakiulakis, and M.E. Maragoudakis, Invest. New Drugs, 8, 145 (1990).

102.   N.T. Chen, E.J. Corey, and J. Folkman, Lab. Invest. 59, 453 (1988).

103.   A.M. Schor and S.L. Schor, In Vitro Cell. Dev. Biol., 24, 659 (1988).

104.   R.J. Tamargo, R.A. Bok and H. Brem, Cancer Res. 51, 672 (1991).

105.   T. Yamashita, M. Sakai, Y. Kawai, M. Aono, and K. Takahashi, J. Antibiot., 42, 1015 (1989).

106.   T. Oikawa, K. Hirotani, M. Shimamura, H. Ashino-Fuse, and T. Iwaguchi, J. Antibiot., 42, 1202 (1989).

107.   T. Oikawa, M. Shimamura, H. Ashino-Fuse, T. Iwaguchi, M. Ishizuka, and T. Takeuchi, J. Antibiot. 44, 1033 (1991).

108.   T. Oikawa, M. Hasegawa, M. Shimamura, H. Ashino, S. Murota, and I. Morita, Biochem. Biophys. Res. Comm., 181, 1070 (1991).

109.   T. Oikawa, K. Hirotani, H. Ogasawara, T. Katayama, O. Nakamura, T. Iwaguchi, and A. Hiragun, Eur. J. Pharmacol., 178, 247 (1990).

110.   T. Oikawa, Y. Yoshida, M. Shimamura, H. Ashino-Fuse, T. Iwaguchi, and T. Tominaga, Anti-Cancer Drugs, 2, 475 (1991).

111.   T. Oikawa, K. Hirotani, O. Nakamura, K. Shudo, A. Hiragun, and T. Iwaguchi, Cancer Lett., 48, 157 (1989).

112.   T. Matsubara and M. Ziff, J. Clin. Invest., 79, 1440 (1987).

113.   A. Pasi, B. Qu, R. Steiner, H.-J. Senn, W. Bar, and F.S. Messiha, Gen. Pharmac., 22, 1077 (1991).

114.   T. Oikawa, K. Hirotani, H. Ogasawara, T. Katayama, H. Ashino-Fuse, M. Shimamura, T. Iwaguchi, and O. Nakamura, Chem. Pharm. Bull, 38, 1790 (1990).

115.   S. Hirata, T. Matsubara, R. Saura, H. Tateishi, and R.R. Anderson, J. Am. Acad. Dermatol, 20, 1123 (1989).

116.   K. Hellmann, Abstracts, International Symposium on Angiogenesis, St. Gallen, Switzerland, 1991, p. 133.

117.   P.J. Polverini and R.F. Novak, Biochem. Biophys. Res. Commun., 140, 901 (1986).

118.   W.L. Haynes, A.D. Proia, and G.K. Klintworth, Invest. Ophthalmol. Vis. Sci., 30, 1588 (1989).

119.   M. Takigawa, M. Enomoto, Y. Nishida, H.-O. Pan, A. Kinoshita, and F. Suzuki, Cancer Res., 50, 4131 (1990).

120.   A. Sundbeck, R. Myrhage, H.-I. Peterson, Anticancer Res., 1, 295 (1981).

121.   A.E. Koch, J.C. Burrows, P.J. Polverini, M. Cho, and S.J. Leibovich, Agents Actions, 34, 350 (1991).

122.   L. Rudnicka, M. Marczak, A. Szmurlo, B. Makiela, A. Skiendzielewska, M. Skopinska, S. Majewski, and S. Jablonska, Skin Pharmacol., 4, 150 (1991).

Chapter 16. Recent Advances in Antifungal Agents

John F. Barrett and Dieter H. Klaubert

The R.W. Johnson Pharmaceutical Research Institute, Raritan, NJ

Introduction- Primary and opportunistic fungal infections continue to increase rapidly because of the increased number of immunocompromised patients (AIDS, cancer and transplants) (1). The last two years have been marked by new azole approvals (2), attempts at reformulating existing drugs, reports of many new natural product antifungals, and a resurgence in studies attempting to define and understand fungal pathogenicity. Fluconazole, a well absorbed, oral antifungal with good tissue penetration, was launched in the United States in 1990, and has been well received (3). Few breakthroughs have been made with the polyenes, the main therapeutics against systemic mycoses (4).

Polyenes- Amphotericin B (AmB), a polyene given parenterally, has been reformulated in an attempt to maintain serum level and reduce toxicity (5, 6). Fungizone, a desoxycholate formulation of AmB (7), has been used successfully to treat systemic $Candida$ infections and $Aspergillosis$ infections but still has limited use in therapy (8, 9). Dimyristoylglycerol and liposomes prepared from dimyristoylphosphatidylcholine have been investigated as delivery systems (7, 10, 11). $In\ vitro$ and $in\ vivo$ testing of these formulations using different models have given variable results; further experimentation is required before a clearly preferred formulation emerges (7, 10-14). Hamycin, a heptaene member of the polyene family of antifungals with clinical activity against $C.\ albicans$, $Histoplasma\ capsulatum$, $Cryptococcus\ neoformans$, $Blastomyces\ dermatitidis$, and $Aspergillus$ $niger$, has also been reformulated with liposomes in an attempt to develop a less toxic product for parenteral use (15).

Improvements in combination therapy have been a primary focus of physicians in attempts to reduce resistance emergence, improve dosage regimen, and decrease toxicity. Numerous reports of the interactions of the two most widely used systemic antifungal classes of drugs have been reported. There have been conflicting reports about polyene-imidazole interactions, varying from antagonism (16-19) to synergy (20, 21), but more recent studies indicate that polyene antifungals were antagonized by imidazoles, whereas imidazole activity was increased in the presence of a polyene (22, 23). The importance of the growth stage and the location of a $Candida$ $albicans$ infection site in therapy was made apparent with the observation that the amphotericin fungicidal effect in macrophages is less than that for extracellular $C.\ albicans$, but germ tube formation by $C.\ albicans$ is completely blocked both extracellularly and in macrophages. Amphotericin may facilitate host defenses by inhibiting an important virulence factor which normally shields $C.\ albicans$ from phagocytosis (24).

Nystatin (NYS), used as a systemic antifungal (25), continues to be studied in different formulations such as γ-cyclodextrin complexes against $C.\ albicans$ (26), and as a tool for studying polyene effects on membranes. In membrane reconstitution studies (27), the channel-forming ability of NYS has been shown to be dependent on ergosterol (consistent with resistance emergence in $Aspergillus\ niger$ (28)), whereas the facilitated loss of K^+ from $Saccharomyces$ $cerevisiae$ cultured aerobically or anaerobically (varying lipid composition) is not correlated with cell death (29). Trichomycin, an antifungal antibiotic similar to AmB but used to treat vaginal infections, has had several structural components (of more than seventeen) identified (30). Trichomycin A (1) and its 8-fold weaker sister trichomycin B (2) have good anti-candidal activity when compared to miconazole $in\ vitro$, but are weaker than miconazole against most other fungi (31, 32). The structure was published without stereochemical assignments (33, 34) but the spectral characteristics were so close to amphotericin B that it would be reasonable to assume that both compounds have the same stereochemistry.

Another approach to improved polyene antifungal agents is the hydrazine derivatization (3) of clinically useful polyenes such as AmB, candidin, NYS and aerofacin (35). The diaspartate salts improve water solubility and decrease toxicity to host membranes. Polyene-resistant strains of

S. cerevisiae have been used to screen for non-polyene antifungal agents from natural product isolations (36).

$\underline{1}$ R$_1$=OH, R$_2$=H

$\underline{2}$ R$_1$=H, R$_2$=OH

In the continued evaluation of toxicity of AmB and other antifungals, more sophisticated studies on the effects of antifungal agents on human neutrophils have been conducted (37, 38). Polymorpho-nuclear leukocytes (PMNs) were examined *in vitro* at therapeutic levels of AmB with five other antifungal agents. AmB, fluconazole, flucytosine, and cilofungin had no effect on PMN function (chemotaxis, phagocytosis, and intracellular killing) in contrast to previous reports (37). The use of pentoxifylline to modulate AmB effects against PMNs *in vitro* has been suggested (38). *In vitro* susceptibility studies demonstrated synergy between the antineoplastic drugs and AmB (below MIC levels) as a possible improvement in treatment of cancer patients (39).

$\underline{3}$ R= polyene

$\underline{4}$ R= C(NH$_2$)=NH$_2^+$ OTf$^-$

A different strategy of inhibiting/disrupting lipid in fungal membranes is to target a step unique to fungi in the sterol biosynthesis pathway. An effort to identify novel antifungal agents that inhibit S-adenosyl-L-methionine:Δ^{24}-sterol methyltransferase (EC 2.1.1.41) led to a series of side chain modified analogs of cholesterol and lanosterol such as $\underline{4}$, which lack selectivity with respect to the fungal enzyme (40,41).

Lipopeptides- 1,3-β–D-glucan synthesis inhibition is an important target for providing protection against candidiasis normally not adequately provided by existing antifungal agents. A review of glucan biosynthesis has appeared (42). Cilofungin (LY 121019) is the 4-n-octylbenzoyl analog of echinocandin B (43, 44); it shows excellent activity against *Candida* but has little activity against other species of yeasts (45-47). It is a noncompetitive inhibitor of 1,3-β-D-glucan synthase with a K$_{iapp}$ =2.5 μM but has no effect on chitin synthesis (48, 49). Cilofungin's effect on cell wall sterol content (where a 55-60% decrease in ergosterol content and a concomitant 4-13% decrease in lanosterol content accompanies a 77-79% decrease in glucan synthesis in *C. albicans*) is not understood (50). SAR studies have implicated the essential nature of the lipophilic side chain for glucan synthesis inhibition (51). Reports of variable blood serum levels from species to species (52) indicate that additional formulation work needs to be done to provide clinical efficacy. In fact, some reports have questioned the fungicidal activity *in vivo* (53) while other studies have shown efficacy against *Aspergillus* spp. *in vivo* (in the absence of *in vitro* activity (54)). Pharmacokinetic studies have indicated a complex, non-linear plasma pharmacokinetic pattern in rabbits (55, 56) which may explain the efficacy data. The high degree of target specificity will probably limit usage to treating *C. albicans*, *C. tropicalis*, and *Malassezia pachydermatis*, with reduced utility for clinically relevant pathogenic fungi such as *C. pseudotropicalis*, *C. parapsilosis*, *C. krusei*, *B. dermitidis*, *Aspergillus* species, *C. neoformans*, *Torulopsis glabrata*, *M. furfur*, and *Paracoccidioides brasiliensis* (57, 58). Much interest has been shown in combination therapy of cilofungin with other cell wall inhibitors such as anticapsin (59) as well as with AmB (54). The latter combination was shown to be additive or synergistic *in vivo* in disseminated candidiasis in mice (60, 61), whereas cilofungin alone has been shown to be ineffective in the murine candidiasis model (62, 63). Still others have shown a mix of antagonism and synergy with combinations of cilofungin and flucytosine or AMB against fifty strains of *Candida* (64). The direct correlation of *in vitro* and *in vivo* effects of antifungal agents is still not clear. Cilofungin can inhibit *Coccidioides immitis* mycelial growth, but has much less effect on

C. immitis spherule development *in vitro*, indicating its specificity for a particular growth stage (65). Cilofungin may actually decrease the killing effect of human neutrophils since cilofungin interferes with the opsonization of most *C. albicans* strains (66).

L688,786 (<u>5</u>) and its water soluble prodrug, L693,989 (<u>6</u>), are also under development as an anti-candidal agents (67-71). A related member of the echinocandin family, L671,329 (<u>7</u>), is active in a rat model of *P. carinii* pneumonia (72 , 73), indicating potential use in AIDS patients where *P. carinii* pneumonia is a major cause of death. Compound <u>5</u> was compared to other anti-candidal agents *in vitro* and *in vivo* and was shown to be superior to ketoconazole and fluconazole, cilofungin, and amphotericin B against *C. albicans in vivo* even though the prodrug <u>6</u> was less potent *in vitro*. The IC$_{50}$ inhibiting values of 1,3-β-D-glucan synthetase correlated well with the antifungal activities *in vitro*, although the killing ratios were lower than AmB. Derivative <u>6</u> was also shown to be less active than cilofungin against *C. parapsilosis* and *C. pseudotropicalis in vitro*.

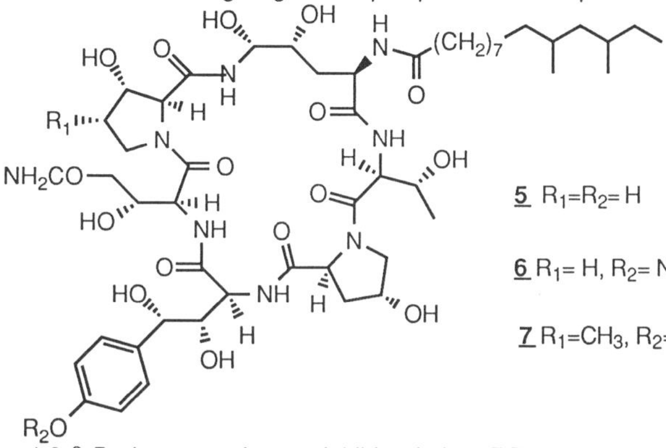

<u>5</u> R$_1$=R$_2$= H

<u>6</u> R$_1$= H, R$_2$= NaHPO$_3$

<u>7</u> R$_1$=CH$_3$, R$_2$=H

Another 1,3-β-D-glucan synthetase inhibitor is L-687,781 (<u>8</u>), produced by the cultures of *Dictochacta simplex* ATCC 20960. It is a member of the papulacandin family of antibiotics (74, 75), but in addition to inhibiting *C. albicans*, <u>8</u> also inhibits *C. parapsilosis in vitro*, and inhibits *P. carinii in vivo* as measured by the reduction of cysts in rats (76).

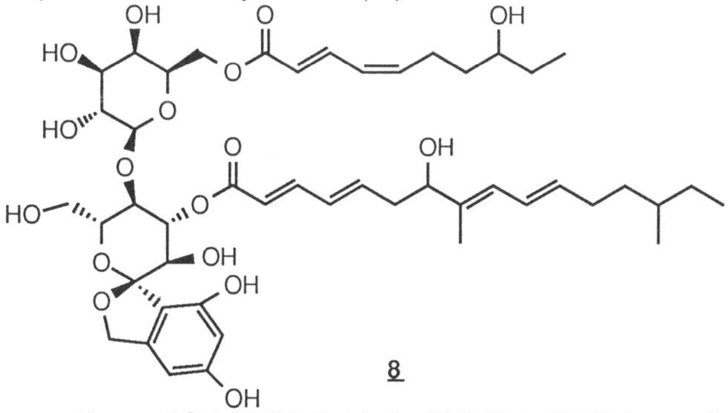

<u>8</u>

<u>Azoles</u>- Only one enantiomer of Sch 39304, that is the (R,R-)Sch 42427, has antifungal activity but further development has been stopped (23) because of heptacellular adenomas and carcinomas in rodents after prolonged (18 to 24 months) treatment (77, 78). Fluconazole has little effect on the immune system while itraconazole has a strong immunosuppressive effect; ketoconazole and miconazole are intermediate (79-81). This effect may be related to their inhibition of 5-lipoxygenase in human PMNLs (82). Saperconazole is another azole compound under development with broad spectrum fungicidal activity especially against disseminated candidosis and cryptococcosis (83, 84). This lipophilic azole may be developed as an i.v. formulation using β-cyclodextrin as the carrier (85-87). All of these azoles inhibit C-14-lanosterol demethylase but have different effects on various other enzymes. Fenticonazole (and 5-fluorocytosine) appears to inhibit the synthesis of the acid proteinase secreted by *C. albicans* (88) while ICI 195,739 inhibits *T. cruzi* proliferation via an additional, rapid, and as yet unidentified mechanism (89, 90). A series of triazoles (<u>9</u>) designed as inhibitors of fungal cytochrome P-450 14-α–demethylase have been reported as having good potency in a systemic *C. albicans* infection model (oral or parenteral administration) (91). Modeling of the C(4)-C(5)(4R,5R) bond of the active enantiomer and the C(13)-C(14) bond of lanosterol led to

the conclusion that the acyclic analogs (**10**) would better fit the enzyme active site (92). These compounds show excellent oral anticandidal activity in the mouse. A series of vinylimidazoles (**11,12**) showed good anti-dermatophyte and anti-yeast activity *in vitro* as compared to clotrimazole and miconazole (93).

9   **10**   **1 1** R= C_5H_{11}
                              1 2 R= C_6H_{13}

Chitin Synthase Inhibitors- Polyoxins and nikkomycins (fermentation products and chemically synthesized) competitively inhibit the chitin synthase of fungi with *C. albicans* as the primary target organism (31, 94-96). These relatively non-toxic agents inhibit chitin synthase of many fungi at micromolar levels, but require high concentrations to inhibit growth (95-97). A thorough SAR analysis of the nikkomycins has highlighted three important structural factors that determine activity against the whole cell: 1) affinity for chitin synthase; 2) susceptibility to peptidase degradation; and 3) efficiency of peptide transport (95). In addition, definitive assignment of which chitin synthase is required for cell viability (and therefore should be targeted for inhibition) has been shown with *S. cerevisiae* (97, 98) and *C. albicans* (99). A significant improvement in nikkomycins must be made before the members of this family can be considered as bona fide anti-candidal agents (95). Despite the lack of single agent anti-candidal activity, nikkomycin X/Z has been shown to act synergistically *in vitro* with fluconazole, ketoconazole and tioconazole against *C. albicans* at sub-MIC levels of azoles. This is believed to arise via the phenomenon of azole-induced uncoordinated chitin synthesis and deposition when combined with nikkomycins (100). Such an effect can best be explained by the azole-mediated inhibition of membrane ergosterol, thereby affecting membrane fluidity, and ultimately chitin synthase activity. A definitive explanation will eventually arise from genetic and biochemical experiments that dissect the 1,3-β-D-glucan synthesis pathway (101-103). Most promising for the nikkomycins X and Z is the demonstration of *in vitro* and *in vivo* anti-coccidioidomycosis, histoplasmosis and blastomycosis activity (104). Nikkomycin Z was found to be more effective than nikkomycin X, and more potent than most azoles in inhibiting highly chitinous, dimorphic fungal pathogens such as *Coccidioides immitis* and *Blastomyces dermatitidis* (104).

Miscellaneous Agents- Nitroxoline (**13**), a synthetic quinoline useful as a urinary tract antiseptic, has been shown to have some antifungal activity *in vitro* (105). As an anti-candidal agent, it inhibited 186 representative clinical isolates of *Candida* with MICs of 0.25 to 2 µg/ml; this is approximately the same activity as amphotericin B, superior to the imidazoles tested (miconazole, ketoconazole), but less active than flucytosine (105). Phosphonates used as antifungal agents in agriculture for the reduction of fungal plant pathology continue to be viewed solely as phyto-fungicides (106) without application to human fungal infections. A series of fused oxazoles has been studied to try to understand the QSAR of this series. The 2-substituted oxazolo(4,5-b)pyridine derivatives (**14**) had the best inhibitory activity against *C. albicans* (107, 108). Flavones and isoflavones (such as genistein) as well as their dihydro derivatives have been reported as causing *in vitro* inhibition of Aspergillus spp. (109, 110).

A review of naftifine discusses its placement as a potent anti-dermatophyte agent (*Trichophyton* spp., *Epidermophyton floccosum*, and *Microsporum* spp.) (111). Its complex mechanism of action, including squalene epoxidase inhibition, is not completely understood but its once-daily application and absence of systemic side effects makes it a useful addition to the physician's repertoire. Despite poor *in vitro* activity, naftifine has good activity *in vivo* against cutaneous *Candida* spp. (112).

Natural Products- The sampangines have been isolated from the bark of *Cleistopholis patens*, and subsequently synthesized (113, 114). These compounds, as illustrated by 3-methoxysampangine (**15**), are potent cytotoxic agents, with antiviral activity against *Herpes simplex* (115), and DNA intercalators, so selectivity for non-mammalian cells is an issue. Meridine (**16**) inhibits growth of *C. albicans* and *C. neoformans* by interacting with nucleic acid biosynthesis (116).

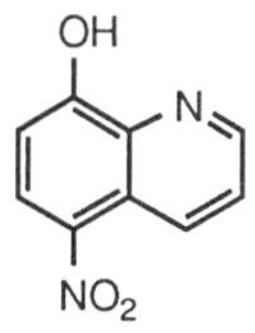

13 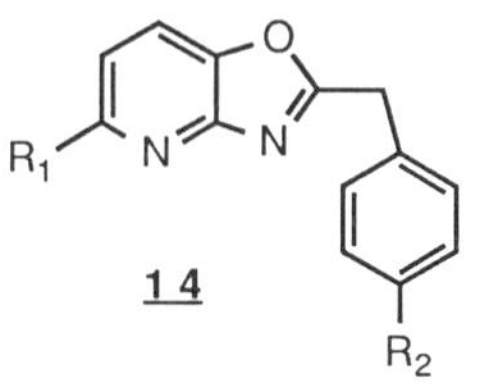　　　　**14** 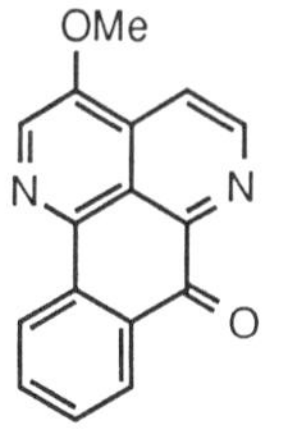　　　　**15** 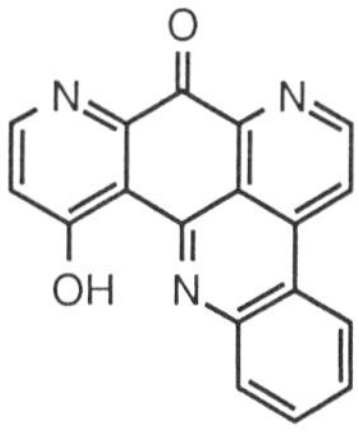　　　　**16**

The pradimicins isolated from *Actinomadura hibisca* are a family of closely related analogs, with pradimicin A (**17**) being very effective in systemic infections of *C. albicans* in normal and immunocompromised mice after i.v. injection. In this case, cytotoxicity to cultured mammalian cells occured at 10 to 100 times higher concentrations than the antifungal MICs (75, 117). A water soluble (>40 mg/ml at pH 7.2) semi-synthetic analog, BMY 28864 (**18**) has similar activity (74, 118). Benanomicin B, isolated from *Actinomycete,* is stated to be identical to pradimicin C (**19**) (119).

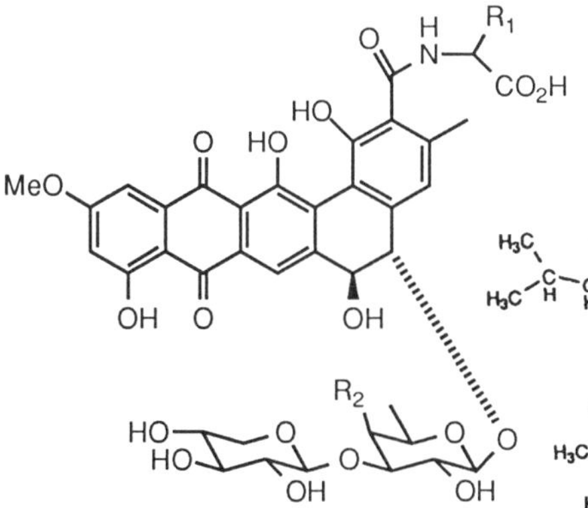

17 $R_1 = CH_3$, $R_2 = NHMe$

18 $R_1 = CH_2OH$, $R_2 = NMe_2$

19 $R_1 = CH_3$, $R_2 = H$

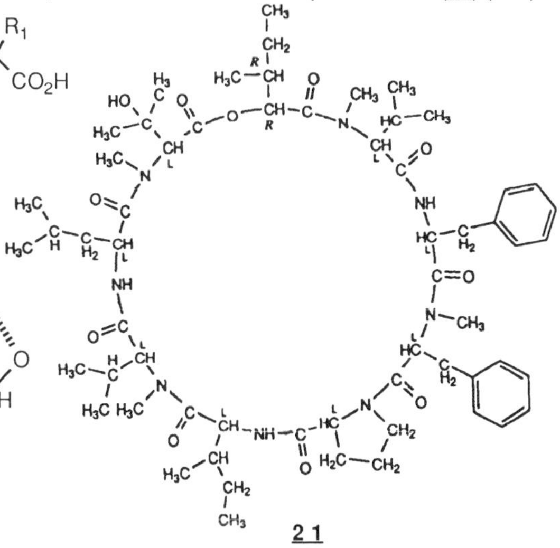

21

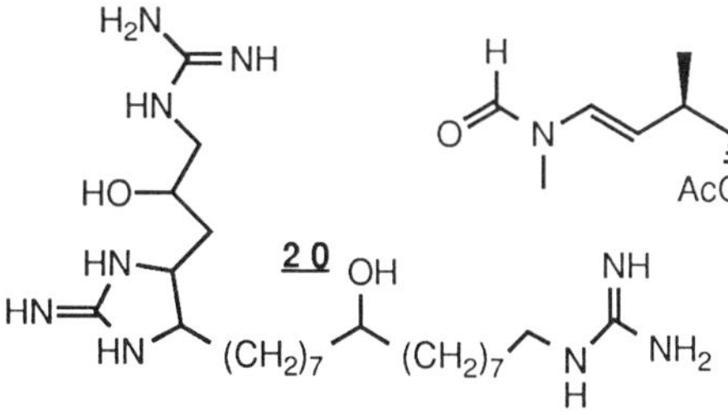

20

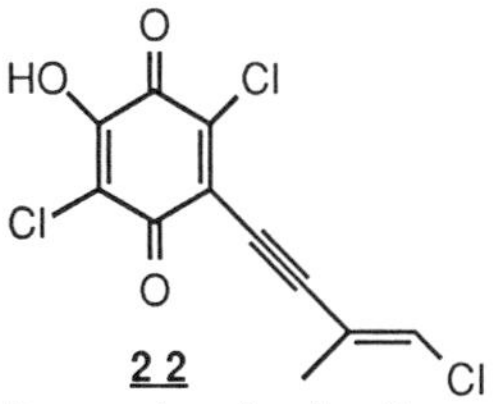

22

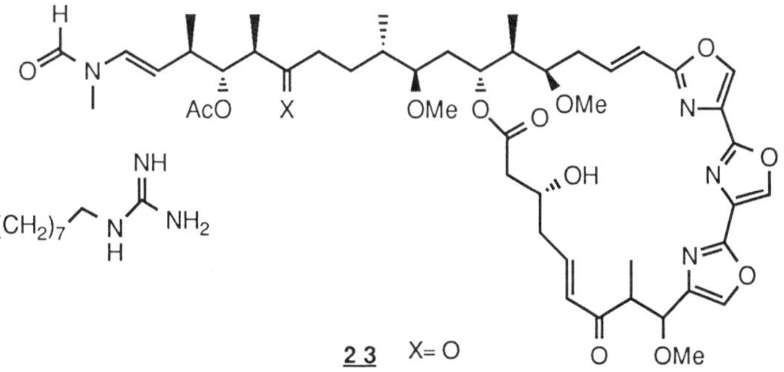

23　X= O

24　X= $CH_2CH_2COCH(OMe)CH_2OMe$

25　X= $CH_2CH_2COCH(OMe)CH_3$

Actinomadura is also the source for Sch 40873 (**20**), a potent inhibitor of the mycelial phase of *C. albicans* (120). Another macrocyclic family of antifungal antibiotics, aureobasidin A-R, was isolated from the fermentation broth of a strain of black yeast, *Aurebasidium pullulans*. Aureobisidin A (**21**) shows excellent *in vitro* anticandidal activity, and moderate *C. neoformans* activity (121). Mycenon ($\underline{22}$) is an isocitrate lyase inhibitor showing moderate activity against *S. cerevisiae* but not against *C. albicans* (122).

Isolation of marine natural products has produced a variety of macrocycles such as the myacolides A-C ($\underline{23}$-$\underline{25}$) the patellazoles A-C ($\underline{26}$-$\underline{28}$) and the ulithiacyclamides ($\underline{29}$) (123-126). All of these compounds are general cytotoxins and antifungal agents and would need to be modified to gain selectivity. An approach to the synthesis of the ulapualide triazole ring system has been published (127).

Complex macrocycles are not the only natural products isolated as antifungal agents. Additional members of the allosamidine family ($\underline{30}$-$\underline{32}$) showed *in vitro* activity against *C. albicans*, *S. cerevisiae* and *Trichoderma* sp. chitinase and interestingly, the hydrolysis product ($\underline{33}$) showed activity only against *C. albicans* chitinase (128).

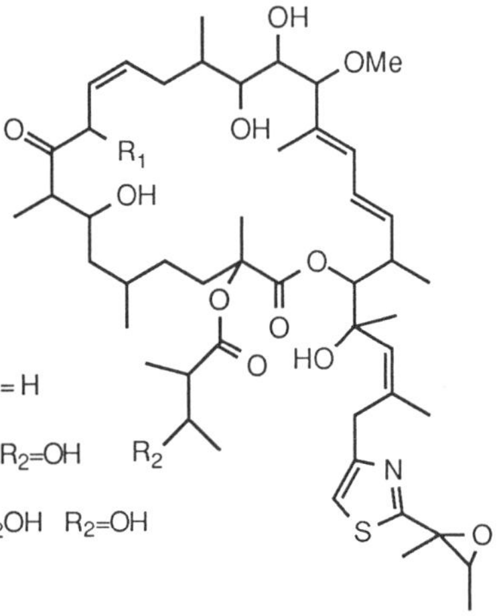

$\underline{26}$ $R_1=R_2=H$

$\underline{27}$ $R_1=H$ $R_2=OH$

$\underline{28}$ $R_1=CH_2OH$ $R_2=OH$

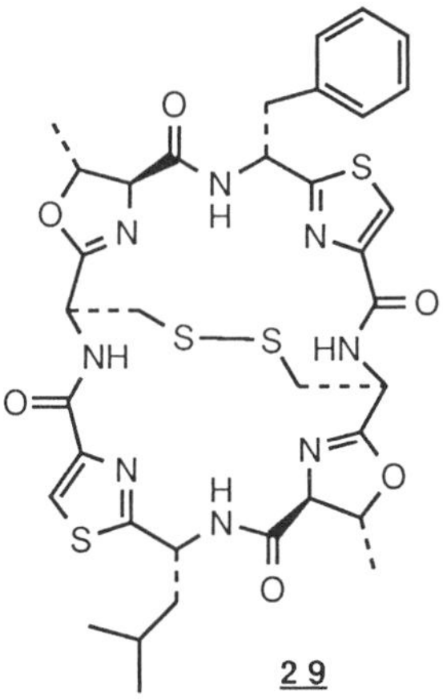

$\underline{29}$

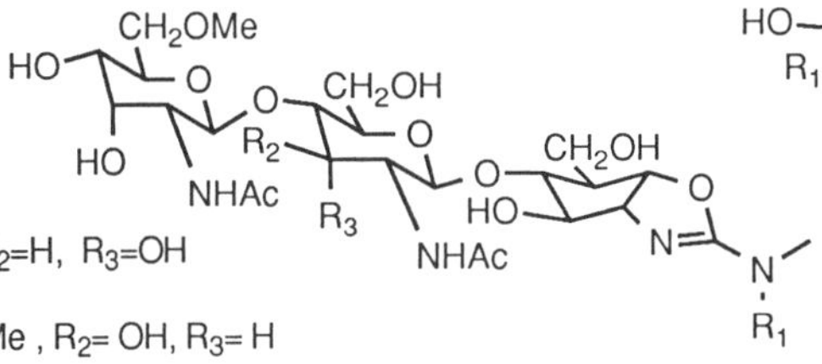

$\underline{30}$ $R_1=R_2=H$, $R_3=OH$

$\underline{31}$ $R_1=Me$, $R_2=OH$, $R_3=H$

$\underline{32}$ $R_1=R_3=H$, $R_2=OH$

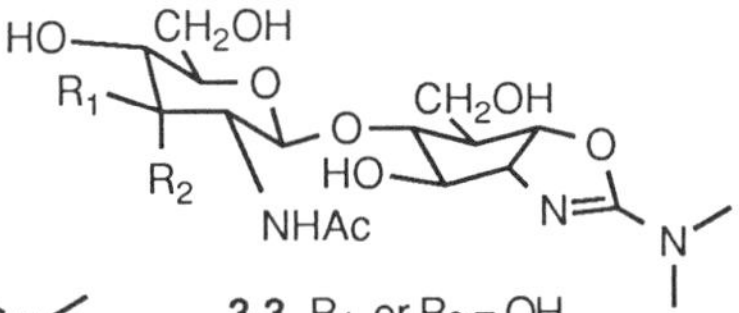

$\underline{33}$ R_1 or $R_2=OH$

Although synerazol ($\underline{34}$) had moderate *in vitro* activity against *Candida spp.* it was inactive in a mouse model possibly because of instability of the epoxide ring (129). Another unstable but very effective agent against a broad spectrum of yeasts and filamentous fungi with a profile similar to ketoconazole is restrictin ($\underline{35}$) (130-132). The dimethyl analog $\underline{36}$ was much less effective and the desglycine derivative $\underline{37}$ was inactive.

A45507 ($\underline{38}$-$\underline{40}$), a complex of weakly active anticandidal compounds produced by an unidentified mold, is also a weak fungal cell wall inhibitor which does not affect ^{14}C-glucose incorporation (133). Tjipanazoles ($\underline{41}$) are a mix of 15 new N-glycosides with moderate fungicidal activity against *A. flavus*, *T. mentagrophytes*, and *C. albicans* (134, 135). Two more antifungal agents from ectomycorrhizal fungi have been reported, pisolithin A (p-hydroxybenzoylformic acid) and pisolithin B ((R)-p-hydroxymandelic acid), but their inhibitory activities are limited to inhibiting mycelial growth of phytopathogens and other ectomycorrhizal fungi, and are without medicinal application (136).

Peptides/Proteins- Antifungal activity (believed to act by permeabilizing the fungal membrane) against *C. albicans* and *Trichoderma viride* has been identified in two small proteins (~20-23 kDa) from barley grain (137) similar to the potency of barley chitinase C and barley ribosome inactivating protein K (138), and the maize grain protein zeamatin (139).

Inhibitors of glucosamine-6-phosphate synthase delivered through peptide transport systems have been reported as having anti-candidal activity (140). The most potent of these antifungal dipeptides is L-norvalyl-FMDP (42) which has anti-candidal activity in a systemic mouse model (141). Such activity has been attributed to the inhibition of chitin, glucan (although not directly inhibiting glucan) and mannan -- the cumulative effect of inhibiting glucosamine-6-phosphate synthase. The surprising *in vitro* and especially *in vivo* activity (142) against *C. albicans* of cispentacin (43) led to an analog program (143) which showed that dipeptides such as 44 were just as efficacious as the parent provided the L- amino acid was used. Other cyclic amino acids or even the N,N-dimethyl derivative (R1=R2=Me) were inactive.

Antifungal peptides produced as a natural defense mechanism are being isolated from various sources. *A. giganteus* produces a 51 amino acid peptide rich in cysteine residues related to PLA$_2$ (144). Another cysteine-rich peptide, tracheal antimicrobial peptide (TAP), has been isolated from bovine trachea (145). This material shows a broad spectrum of antimicrobial as well as anticandidal activity. The mRNA encoding this peptide is more abundant in respiratory mucosa than in whole lung tissue. A family of histidine-rich proteins has been isolated from human saliva (146, 147). The exact mechanism of action is not known but histatins 1, 3 and 5 show different effects on different stages of growth of *C. albicans* (148, 149) and it has been speculated that the some of the activity is due to adoption of an α-helical conformation in the aqueous oral environment which allows interaction and permeabilization of the cell wall (150).

References

1.   M.D. Richardson, J. Antimicrob. Chemother., 28, 1 (1991).
2.   R.J. Hay, J. Antimicrob. Chemother., 28, 35 (1991).
3.   F.N. Vincent-Ballereau, O.N. Patey and C. Lafaix, Pharm. Weekbl., Sci. Ed., 13, 45 (1991).
4.   A. Polak and P.G. Hartman in "Progress in Drug Research, " Vol. 37, E. Jucker, Ed., Birkhauser Verlag, Basel, Switzerland, 1991, p. 181.
5.   D.W. Warnock, J. Antimicrob. Chemother., 28, 27 (1991).
6.   B. Leclef, P. Cerfontaine, J.M. Nicolas, H. Wantier and A. Trouet, EP 418 153 (1991).

7. J.M. Clark, R.R. Whitney, S.J. Olsen, R.J. George, M.R. Swerdel, L. Kunselman and D.P. Bonner, Antimicrob. Agents Chemother., 35, 615 (1991).
8. G. Lopez-Berestein, V. Fainstein, R. Hopfer, K. Mehta, M.P. Sullivan, M. Keating, M.G. Rosenblum, R. Mehta, M. Luna, E.M. Hersh, J. Reuben, R.L. Juliano and G.P. Bodey, J. Infect. Dis., 151, 704 (1985).
9. G. Lopez-Berestein, G.P. Bodey, V. Fainstein, M. Keating, L.S. Frankel, B. Zeluff, L. Gentry and K. Mehta, Arch. Int. Med., 149, 2533 (1989).
10. R.M. Fielding, A.W. Singer, L.H. Wang, S. Babbar and L.S. Guo, Antimicrob. Agents Chemother., 36, 299 (1992).
11. R.M. Fielding, P.C. Smith, L.H. Wang, J. Porter and L.S.S. Guo, Antimicrob. Agents Chemother., 35, 1208 (1991).
12. V. Joly, J. Bolard, L. Saint-Julien, C. Carbon and P. Yeni, Antimicrob. Agents Chemother., 36, 262 (1992).
13. K.V. Clemons and D.A. Stevens, Antimicrob. Agents Chemother., 35, 2144 (1991).
14. L.H. Hanson and D.A. Stevens, Antimicrob. Agents Chemother., 36, 486 (1992).
15. R.T. Mehta, T.J. Mcqueen, A. Keyhani and G. Lopez-Berestein, J. Infect. Dis., 164, 1003 (1991).
16. B. Dupont and E. Drouhet, Postgrad. Med. J., 55, 683 (1979).
17. I.J. Sud and D.S. Feingold, Antimicrob. Agents Chemother., 23, 185 (1983).
18. E. Ponce and J.C. Pechere, Eur. J. Microbiol. Infect. Dis., 9, 738 (1990).
19. H.-J. Schmitt, E.M. Bernard, F.F. Edwards and D. Armstrong, Mycoses, 34, 281 (1991).
20. W.H. Beggs, G.A. Sarosi and N.M. Steele, Antimicrob. Agents Chemother., 9, 863 (1976).
21. T.F. Patterson, D. George, D. Kordick, P. Miniter and V.T. Andriole, 31 st ICAAC, Sept. 29 (1991), Abs. No. 584
22. M.A. Petrou and T.R. Rogers, J. Antimicrob. Chemother., 27, 491 (1991).
23. A.M. Sugar, Antimicrob. Agents Chemother., 35, 1669 (1991).
24. J.W. Van 't Wout, I. Meynaar, I. Linde, R. Poell, H. Mattie and R. Van Furth, J. Antimicrob. Chemother., 25, 803 (1990).
25. R.T. Mehta, R.L. Hopfer, T. McQueen, R.L. Juliano and B.G. Lopez, Antimicrob. Agents Chemother., 31, 1901 (1987).
26. H. van Doorne and E.H. Bosch, Int. J. Pharm., 73, 43 (1991).
27. D.J. Woodbury and C. Miller, Biophys. J., 58, 833 (1990).
28. C. Mazumder, J. Basu, M. Kundu and P. Chakrabarti, Can. J. Microbiol., 36, 435 (1990).
29. M. Aperecida de Resende and F. Alterthum, Mycopathologia, 112, 165 (1990).
30. T. Komori and Y. Morimoto, J. Chromatog., 481, 416 (1989).
31. H. Decker, F. Walz, C. Bormann, H. Zaehner, H.P. Fiedler, H. Heitsch and W.A. Koenig, J. Antibiot., 43, 43 (1990).
32. T. Komori, J. Antibiot., 43, 778 (1990).
33. T. Komori, Y. Morimoto, M. Niwa and Y. Hirata, Tetrahedron Lett., 30, 3813 (1989).
34. T. Komori and Y. Morimoto, J. Antibiot., 43, 904 (1990).
35. J. Grzybowska and E. Borowski, J. Antibiot., 43, 907 (1990).
36. G. Etienne, E. Armau and G. Tiraby, J. Antibiot., 43, 199 (1990).
37. E. Roilides, T.J. Walsh, M. Rubin, D. Venzon and P.A. Pizzo, Antimicrob. Agents Chemother., 34, 196 (1990).
38. G.W. Sullivan, H.T. Carper and G.L. Mandell, Antimicrob. Agents Chemother., 36, 408 (1992).
39. M.A. Ghannoum, E.K.H. Abu, M.S. Motawy, H.M.A. Abu, A.S. Ibrahim and R.S. Criddle, Chemotherapy (Basel), 36, 308 (1990).
40. M.A. Ator, S.J. Schmidt, J.L. Adams, R.E. Dolle, L.I. Kruse, C.L. Frey and J.M. Barone, J. Med. Chem., 35, 100 (1992).
41. A.C. Oehlschlager, R.H. Angus, A.M. Pierce, H.D.J. Pierce and R. Srinivasan, Biochemistry, 23, 3582 (1984).
42. J. Ruiz-Herrera, Anton. Leeuwenhoek Int. J. Gen. Med., 60, 73 (1991).
43. M. Debono, B.J. Abbott, D.S. Fukuda, M. Barnhart, K.E. Willard, R.M. Molloy, K.H. Michel, J.R. Turner, T.F. Butler and A.H. Hunt, J. Antibiot., 42, 389 (1989).
44. M. Debono, B.J. Abbott, J.R. Turner, L.C. Howard, R.S. Gordee, A.S. Hunt, M. Barnhart, R.M. Molloy, K.E. Willard, D. Fukuda, T.F. Butler and D.J. Zeckner, Ann. N.Y. Acad. Sci., 544, 152 (1988)
45. K.A. McIntyre and J.N. Galgiani, Antimicrob. Agents Chemother., 33, 731 (1989).
46. K.A. McIntyre and J.N. Galgiani, Antimicrob. Agents Chemother., 33, 1095 (1989).
47. A. Huang, F. Edwards, E.M. Bernard, D. Armstrong and H.J. Schmitt, Eur. J. Clin. Microbiol. Infect. Dis., 9, 697 (1990).
48. C.S. Taft, T. Stark and C.P. Selitrennikoff, Antimicrob. Agents Chemother., 32, 1901 (1988).
49. J. Tang and T.R. Parr Jr., Antimicrob. Agents Chemother., 35, 99 (1991).
50. M. Pfaller, J. Riley and T. Koerner, Eur. J. Clin. Microbiol. Infect. Dis., 8, 1067 (1989).
51. C.S. Taft and C.P. Selitrennikoff, J. Antibiot., 43, 433 (1990).
52. J.R. Perfect, M.M. Hobbs, K.A. Wright and D.T. Durack, Antimicrob. Agents Chemother., 33, 1811 (1989).
53. A. Padula and H.F. Chambers, Antimicrob. Agents Chemother., 33, 1822 (1989).
54. D.W. Denning and D.A. Stevens, Antimicrob. Agents Chemother., 35, 1329 (1991).
55. J.W. Lee, P. Kelly, J. Lecciones, D. Coleman, R. Gordee, P.A. Pizzo and T.J. Walsh, Antimicrob. Agents Chemother., 34, 2240 (1990).

56. T.J. Walsh, J.W. Lee, P. Kelly, J. Bacher, J. Lecciones, V. Thomas, C. Lyman, D. Coleman, R. Gordee and P.A. Pizzo, Antimicrob. Agents Chemother., 35, 1321 (1991).
57. G.S. Hall, C. Myles, K.J. Pratt and J.A. Washington, Antimicrob. Agents Chemother., 32, 1331 (1988).
58. L.H. Hanson and D.A. Stevens, Antimicrob. Agents Chemother., 33, 1391 (1989).
59. M. Pfaller, R. Gordee, T. Gerarden, M. Yu and R. Wenzel, Eur. J. Clin. Microbiol. Infect. Dis., 8, 564 (1989).
60. L.H. Hanson, A.M. Perlman, K.V. Clemons and D.A. Stevens, Antimicrob. Agents Chemother., 35, 1334 (1991).
61. A.M. Sugar, L.Z. Goldani and M. Picard, Antimicrob. Agents Chemother., 35, 2128 (1991).
62. K.R. Smith, K.M. Lank, C.G. Cobbs, G.A. Cloud and W.E. Dismukes, Antimicrob. Agents Chemother., 34, 1619 (1990).
63. C.J. Morrison and D.A. Stevens, Antimicrob. Agents Chemother., 34, 746 (1990).
64. A.N. Bulo, S.F. Bradley and C.A. Kauffman, Mycoses, 32, 151 (1989).
65. J.N. Galgiani, S.H. Sun, K.V. Clemons and D.A. Stevens, J. Infect. Dis., 162, 944 (1990).
66. T. Meshulam, S.M. Levitz, R.D. Diamond and A.M. Sugar, J. Antimicrob. Chemother., 24, 741 (1989).
67. J.M. Balkovec, R.M. Black, M.L. Hammond, J.V. Heck, R.A. Zambias, G. Abruzzo, K. Bartizal, J. Puckett, C. Trainor, R. Schwartz, D.C. McFadden, K. Nollstadt, L.A. Pittarelli, M.A. Powles and D.M. Schmatz, 31st ICAAC, Sept. 29 (1991), Abs. No. 204
68. J.M. Balkovec, R.M. Black, M.L. Hammond, J.V. Heck, R.A. Zambias, G. Abruzzo, K. Bartizal, H. Kropp, C. Trainor, R.E. Schwartz, D.C. McFadden, K.H. Nollstadt, L.A. Pittarelli, M.A. Powles and D.M. Schmatz, J. Med. Chem., 35, 194 (1992).
69. K. Bartizal, G. Abruzzo, C. Trainor, J. Puckett, S. Ponticas, D. Krupa, D. Schmatz, K. Nollstadt, R. Schwartz, M. Hammond, J. Balkovec, R. Zambias and H. Kropp, 31st ICAAC, Sept. 29 (1991), Abs. No. 206
70. K. Bartizal, C. Gill, C. Renna, D. Shungu, G. Abruzzo, C. Trainor, J. Puckett, S. Ponticas, R. Schwartz, M. Hammond, J. Balkovec, R. Zambias and H. Kropp, 31st ICAAC, Sept. 29 (1991), Abs. No. 205
71. R. Hajdu, J.G. Sundelof, K. Bartizal, G. Abruzzo, C. Trainor, R. Thompson and H. Kropp, 31st ICAAC, Sept. 29 (1991), Abs. No. 209
72. A.A. Adefarati, R.A. Giacobbe, O.D. Hensens and J.S. Tkacz, J. Am. Chem. Soc., 113, 3542 (1991).
73. D.M. Schmatz, M.A. Romancheck, L.A. Pittarelli, R.E. Schwartz, R.A. Fromtling, K.H. Nollstadt, F.L. Vanmiddlesworth, K.E. Wilson and M.J. Turner, Proc. Natl. Acad. Sci. USA., 87, 5950 (1990).
74. T. Oki, M. Kakushima, M. Nishio, H. Kamei, M. Hirano, Y. Sawada and K. Masataka, J. Antibiot., 43, 1230 (1990).
75. T. Oki, O. Tenmyo, M. Hirano, K. Tomatsu and H. Kamei, J. Antibiot., 43, 763 (1990).
76. F. VanMiddlesworth, M.N. Omstead, D. Schmatz, K. Bartizal, R. Fromtling, G. Bills, K. Nollstadt, S. Honeycutt, M. Zweerink, G. Garrity and K. Wilson, J. Antibiot., 44, 45 (1991).
77. D. Loebenberg, A. Cacciapuoti, R. Parmegiani, E.L. Moss Jr., F. Menzel Jr., B. Antonacci, C. Norris, T. Yarosh-Tomaine, R.S. Hare and G.H. Miller, Antimicrob. Agents Chemother., 36, 498 (1992).
78. R. Allendoerfer, R.R. Yates, A.J. Marquis, D. Loebenberg, M.G. Rinaldi and J.R. Graybill, Antimicrob. Agents Chemother., 36, 217 (1992).
79. G. Pawelec, G. Ehninger, A. Rehbein, K. Schaudt and K. Jaschonek, Int. J. Pharmacol., 13, 299 (1991).
80. G. Pawelec, K. Jaschonek and G. Ehninger, Int. J. Pharmacol., 13, 875 (1991).
81. V. Vuddhakul, G.T. Mai, J.G. McCormack, W.K. Seow and Y.H. Thong, Int. J. Pharmacol., 12, 639 (1990).
82. D. Steinhilber, K. Jaschonek, J. Knospe, O. Morof and H.J. Roth, Arzneim.-Forsch./Drug Res., 40(II), 1260 (1990).
83. J. Van Cutsem, F. Van Gerven and P.A.J. Janssen, Antimicrob. Agents Chemother., 33, 2063 (1989).
84. J. Van Cutsem, F. Van Gerven and P.A.J. Janssen, Drugs of the Future, 14, 1187 (1989).
85. K.P. Fu, B. Foleno and A.J. Tobia, 31 st ICAAC, Sept. 29 (1991), Abs. No. 217
86. B. Foleno, K.P. Fu and A.J. Tobia, 31 st ICAAC, Sept. 29 (1991), Abs. No. 216
87. D.M. Isaacson, J. Lococo, J. Hilliard, K.P. Fu and A.J. Tobia, 31 st ICAAC, Sept. 29 (1991), Abs. No. 218
88. L. Angiolella, F. De Bernardis, C. Bromuro, F. Mondello, T. Ceddia and A. Cassone, J. Chemother., 2, 55 (1990).
89. K. Lazardi, J.A. Urbina and W. De Souza, Antimicrob. Agents Chemother., 35, 736 (1991).
90. J.A. Urbina, K. Lazardi, T. Aguirre, M.M. Piras and R. Piras, Antimicrob. Agents Chemother., 35, 730 (1991).
91. T. Konosu, T. Miyaoka, Y. Tajima and S. Oida, Chem. Pharm. Bull., 39, 2241 (1991).
92. T. Konosu, Y. Tajima, N. Takeda, T. Miyaoka, M. Kasahara, H. Yasuda and S. Oida, Chem. Pharm. Bull., 39, 2581 (1991).
93. M. Ogawa, H. Matsuda, H. Eto, T. Asaoka, T. Kuraishi, A. Iwasa, T. Nakashima and K. Yamaguchi, Chem. Pharm. Bull., 39, 2301 (1991).
94. A.G.M. Barrett and S.A. Lebold, J. Org. Chem., 56, 4875 (1991).
95. H. Decker, U. Pfefferle, C. Bormann, H. Zaehner, H.P. Fiedler, P.K.H. Van, M. Rieck and W.A. Konig, J. Antibiot., 44, 626 (1991).
96. E. Krainer, J.M. Becker and F. Naider, J. Med. Chem., 34, 174 (1991).
97. E. Cabib, Antimicrob. Agents Chemother., 35, 170 (1991).
98. E. Cabib, S.J. Silverman, J.A. Shaw, S. Das Gupta, H.-M. Park, J.T. Mullins, P.C. Mol and B. Bowers, Pure Appl. Chem., 63, 483 (1991).
99. K. Dickinson, V. Keer, C.A. Hitchcock and D.J. Adams, Biochim. Biophys. Acta, 1073, 177 (1991).
100. S. Milewski, F. Mignini and E. Borowski, J. Gen. Microbiol., 137, 2155 (1991).
101. J.C. Ribas, M. Diaz, A. Duran and P. Perez, J. Bact., 173, 3456 (1991).

102. C. Boone, S.S. Sommer, A. Hensel and H. Bussey, J. Cell. Biol., 110, 1833 (1990).
103. T. Roemer and H. Bussey, Proc. Natl. Acad. Sci. USA, 88, 11295 (1991).
104. R.F. Hector, B.L. Zimmer and D. Pappagianis, Antimicrob. Agents Chemother., 34, 587 (1990).
105. J.M. Hernández Molina, J. Llosá and A. Ventosa, Mycoses, 34, 323 (1991).
106. D. Guest and B. Grant, Biol. Rev., 66, 159 (1991).
107. I. Yalcin, E. Sener, T. Özden, S. Özden and A. Akin, Eur. J. Med. Chem., 25, 705 (1990).
108. E. Sener, I. Yalcin and E. Sungur, Quant. Struct.-Act. Relat., 10, 223 (1991).
109. M. Weidenboerner, H. Hindorf, H.C. Jha and P. Tsotsonos, Phytochemistry, 29, 1103 (1990).
110. M. Weidenboerner, H. Hindorf, H.C. Jha, P. Tsotsonos and H. Egge, Phytochemistry, 29, 801 (1990).
111. J.P. Monk and R.N. Brogden, Drugs, 42, 659 (1991).
112. E. Astorga, C.N. Cordero, Z.D. De Espinoza, R.F. Rojas, N. Zaias and J. Sefton, Curr. Ther. Res., 46, 1106 (1989).
113. J.R. Peterson, J.K. Zjawiony, C.D. Hufford, A.M. Clark and S. Liu, 31st ICAAC, Sept. 29 (1991), Abs. No. 214
114. S. Liu, B. Oguntimein, C.D. Hufford and A.M. Clark, Antimicrob. Agents Chemother., 34, 529 (1990).
115. J.R. Peterson, J.K. Zjawiony, A.M. Clark, C.D. Hufford, D.E. Graves and L.A. Walker, 31st ICAAC, Sept. 29 (1991), Abs. No. 215
116. P. McCarthy, T. Peterson and G. Gunawardana, 31st ICAAC, Sept. 29 (1991), Abs. No. 212
117. Y. Sawada, M. Nishio, H. Yamamoto, M. Hatori, T. Miyaki, M. Konishi and T. Oki, J. Antibiot., 43, 771 (1990).
118. M. Kakushima, S. Masuyoshi, M. Hirano, M. Shinoda, A. Ohta, H. Kamei and T. Oki, Antimicrob. Agents Chemother., 35, 2185 (1991).
119. K. Tomita, M. Nishio, K. Saitoh, H. Yamamoto, Y. Hoshino, H. Ohkuma, M. Konishi, T. Miyaka and T. Oki, J. Antibiot., 43, 755 (1990).
120. V.R. Hegde, M.G. Patel, H. Wittreich and V.P. Gullo, J. Org. Chem., 54, 2402 (1989).
121. K. Takesako, K. Ikai, F. Haruna, M. Endo, K. Shimanaka, E. Sono, T. Nakamura and I. Kato, J. Antibiot., 44, 919 (1991).
122. R. Hautzel, H. Anke and W.S. Sheldrick, J. Antibiot., 43, 1240 (1990).
123. N. Fusetani, K. Yasumuro, S. Matsunaga and K. Hashimoto, Tetrahedron Lett., 30, 2809 (1989).
124. T.M. Zabriskie, C.L. Mayne and C.M. Ireland, J. Am. Chem. Soc., 110, 7919 (1988).
125. D.G. Corley, R.E. Moore and V.J. Paul, J. Am. Chem. Soc., 110, 7920 (1988).
126. D.E. Williams, R.E. Moore and V.J. Paul, J. Nat. Prod., 52, 732 (1989).
127. D.W. Knight, G. Pattenden and D.E. Rippon, Synlett., 36 (1990).
128. Y. Nishimoto, S. Sakuda, S. Takayama and Y. Yamada, J. Antibiot., 44, 716 (1991).
129. O. Ando, H. Satake, M. Nakajima, A. Sato, T. Nakamura, T. Kinoshita, K. Furuya and T. Haneishi, J. Antibiot., 44, 382 (1991).
130. R.E. Schwartz, C. Dufresne, J.E. Flor, A.J. Kempf, K.E. Wilson, T. Lam, J. Onishi, J. Milligan, R.A. Fromtling, G.K. Abruzzo, R. Jenkins, K. Glazomitsky, G. Bills, L. Zitano, S. Mochales Del Val and M.N. Omstead, J. Antibiot., 44, 463 (1991).
131. O.D. Hensens, C.F. Wichmann, J.M. Liesch, F.L. VanMiddlesworth, K.E. Wilson and R.E. Schwartz, Tetrahedron, 47, 3915 (1991).
132. O.D. Hensens, J.M. Liesch, J. A. Milligan, S.M. Del Val, R.E. Schwartz and C. Wichmann, US 4,952,604 (1990).
133. D.S. Fukuda, W.M. Nakatsukasa, R.C. Yao, A.H. Hunt, R.S. Gordee, D.J. Zeckner and J.S. Mynderse, 31st ICAAC, Sept. 29 (1991), Abs. No. 211
134. G. Knübel, L.K. Larsen, R.E. Moore, I.A. Levine and G.M.L. Patterson, J. Antibiot., 43, 1236 (1990).
135. R. Bonjouklian, T.A. Smitka, L.E. Doolin, R.M. Molloy, M. Debono, S.A. Shaffer, R.E. Moore, J.B. Stewart and G.M.L. Patterson, Tetrahedron, 47, 7739 (1991).
136. H.H. Kope, Y.S. Tsantrizos, J.A. Fortin and K.K. Ogilvie, Can. J. Microbiol., 37, 258 (1991).
137. J. Hejgaard, S. Jacobsen and I. Svendsen, FEBS Lett., 291, 127 (1991).
138. R. Leah, H. Tommerup, I. Svendsen and J. Mundy, J. Biol. Chem., 266, 1564 (1991).
139. W.K. Roberts and C.P. Selitrennikoff, J. Gen. Microbiol., 136, 1771 (1990).
140. S. Milewski, R. Andruszkiewicz, L. Kasprzak, J. Mazerski, F. Mignini and E. Borowski, Antimicrob. Agents Chemother., 35, 36 (1991).
141. S. Milewski, H. Chmara, R. Andruszkiewicz, E. Borowski, M. Zaremba and J. Borowski, Drugs Exp. Clin. Res., 14, 461 (1988).
142. T. Oki, M. Hirano, K. Tomatsu, K.-I. Numata and H. Kamei, J. Antibiot., 42, 1756 (1989).
143. H. Ohki, Y. Inamoto, K. Kawabata, T. Kamimura and K. Sakane, J. Antibiot., 44, 546 (1991).
144. K. Nakaya, K. Omata, I. Okahashi, Y. Nakamura, H. Kolkenbrock and N. Ulbrich, Eur. J. Biochem., 193, 31 (1990).
145. G. Diamond, M. Zasloff, H. Eck, M. Brasseur, W.L. Maloy and C.L. Bevins, Proc. Natl. Acad. Sci. USA, 88, 3952 (1991).
146. F.G. Oppenheim, T. Xu, F.M. McMillan, S.M. Levitz, R.D. Diamond, G.D. Offner and R.F. Troxler, J. Biol. Chem., 263, 7472 (1988).
147. R.F. Troxler, G.D. Offner, T. Xu, J.C. Vanderspek and F.G. Oppenheim, J. Dent. Res., 69, 2 (1990).
148. T. Xu, E. Telser, R.F. Troxler and F.G. Oppenheim, J. Dent. Res., 69, 1717 (1990).
149. T. Xu, S.M. Levitz, R.D. Diamond and F.G. Oppenheim, Infect. Immun., 59, 2549 (1991).
150. P.A. Raj, M. Edgerton and M.J. Levine, J. Biol. Chem., 265, 3898 (1990).

Chapter 17. Progress in Antimicrobial Peptides

Sylvie E. Blondelle and Richard A. Houghten
Torrey Pines Institute for Molecular Studies, San Diego, CA 92121.

Introduction - In recent years, peptides have emerged as useful tools in all areas of biomedical research as well as effective therapeutics and immunodiagnostics. Over the past five years, nature has continued to serve as a common source in the search for novel peptides. Four of the main classes of "natural" antimicrobial peptides include the cecropins, magainins, defensins, and gramicidins. Research during the past two years has focused on the characterization of new natural antimicrobial peptides and the design of analogs with improved activity, as well as on the study of the relationships between their biological activity and secondary structure. These later studies have also provided insight into the design of totally synthetic amphipathic peptides which exhibit significant antimicrobial activity. Finally, a recent approach consisting of the systematic screening of peptide libraries represents a new source of antimicrobial peptides.

ANTIMICROBIAL PEPTIDES DERIVED FROM NATURAL SOURCES

Cecropins - Cecropins represent a class of antimicrobial peptides originally found in the humoral immune response of the North American silk moth *Hyalophora cecropia* (1). Three forms have been isolated (Fig. 1): cecropins A and B which show broad spectrum antimicrobial activities and cecropin D which is active only against a small number of gram-negative bacteria (1). The fragment cecropin D(9-37) shows no activity, while the substitution analog $[Q^3,L^4]$ cecropin D, which has amphipathicity similar to that of cecropin A, exhibits the same activity as cecropin D against *E. coli*. Furthermore, the hybrid cecropin A(1-11)D(12-37) shows an increased activity relative to both cecropin A and D (2). This enhanced activity was attributed to the presence of a more helical structure and a more hydrophobic C-terminal region relative to cecropin A, and to a more basic N-terminal region relative to cecropin D. Therefore, the low basicity at the N-terminal region of cecropin D appears to be the dominant factor in its lower activity (2).

Also of interest, cecropin D was found to modulate various biological functions of murine hematopoietic cells such as lymphocyte DNA synthesis and antibody production of B cells, while having no lymphocyte cytotoxic effects (3). In another study, a number of the intermediate precursor forms of the two cecropins A and B show activities similar to the native forms against *E. coli, P. aeruginosa*, and *B. megatorium*, while the full precursors are inactive and the Gly-extended cecropin A are less active against these bacteria (4).

In a search to improve the biological activity of the natural cecropins, hybrids of cecropin and melittin have been synthesized (5). Melittin was found to have a broader spectrum of activity than cecropin A, but much higher hemolytic activity. Three of the hybrids prepared, CA(1-24)M(1-13), CA(1-13)M(1-13), and M(1-13)CA(1-13), exhibit greater antimicrobial activity than cecropin A, especially against *S. aureus*, and no hemolytic activity. Furthermore, the most potent hybrid, CA(1-13)M(1-13), exhibits higher activity against *Plasmodium falciparum* than cecropin A or B. The lack of hemolytic activity observed for these hybrids, as well as the loss of hemolytic activity found for the inverted analog of melittin [M(16-26)M(1-13)], suggest that both the composition of the basic region and its location at the C-terminus are essential for lysis of the red blood cells. Two-dimensional

^{1}H-NMR of the hybrid CA(1-13)M(1-13) showed that, in common with the two parent peptides, the hybrid is made up of two α-helices joined by a hinge region (residues 13-15) (6-8). However, the melittin domain of this hybrid was found to form a more stable helix than the cecropin domain and the first three N-terminal residues are flexible and not included in the helix.

```
Hyalophora cecropia A:  KW   KLFKKIEKVGQNIRDGIIKAGPAVAVVGQATQIAK-NH₂
Hyalophora cecropia B:  KW   KVFKKIEKMGRNIRNGIVKAGPAIAVLGEAKAL-NH₂
Hyalophora cecropia D:  W    NPFKELEKVGQRVRDAVISAGPAVATVAQATALAK-NH₂
Bombyx mori         B1: RW   KIFKKIEKMGRNIRNGIVKAGPAIEVLGSAKAI-NH₂
Drosophila M.        A: GWLKKIGKKIERVGQHTRDATQLGIAQQAANVAATAR-NH₂
Porcine             P1: SWLSKTAKKLENSAK KR   ISEGIAIAIQGGPR
Andropin             : VFIDILDKVENAIHNAAQVGIGFAKPFEKLINPK-NH₂
CA(1-13)M(1-13)      : KW   KLFKKIEKVGQGIGAVLKVLTTGL-NH₂
Model peptide      MP1: KW   KLFKKIEKVAKKIKEAIEKALEAVAKLLKEAKEIAK-NH₂
Model peptide      MP2: KW   KLFKKIEKVAKKIKEAIEKALEAIAVLALALAL-NH₂
Model peptide      MP3: KW   KLFKKIEKVGRNGRNGIVKAGPAIAVLALALAL-NH₂
Model peptide      MP4: KW   KLFKKIEKV-NH₂
```

Figure 1. Cecropins sequences. The residues commonly found in the cecropins or related peptides and in the *Hyalophora* cecropins A and B are in bold. K:δ-hydroxylysine.

The mechanism of action of cecropins has been studied through the use of L- or D-amino acid-containing model peptides. For instance, four model peptides, termed MP1 to MP4 (Fig. 1), were designed to produce marked changes in the conformation of cecropins at particular sites of the molecules (9). The moderate antimicrobial activity found for the fully α-helical analog MP1 suggests that a complete amphipathic helix does not exclusively account for the activity of cecropins. The additional lack of amphipathicity of MP2 results in a loss in activity compared to MP1, although MP2 shows marked surface activity. In contrast, MP3 shows nearly equivalent activity to cecropins A and B, indicating the important role of a hinge region in the activity of the cecropins. Finally, the slight activity found for the shortened sequence MP4, as well as the lack of activity for the fragment cecropin A(9-37), confirm the requirement of the N-terminal segment 1-11 for lysis to occur (2). These results are consistent with a proposed mechanism of action which involves initial electrostatic interactions between the N-terminus and the membrane, followed by penetration of the hydrophobic C-terminus perpendicular to the membrane, and finally insertion of the amphipathic N-terminus into the membrane which results in the formation of positively charged channels through the membrane (10).

To determine if a specific chirality of the peptide is required for activity, cecropins and cecropin-melittin hybrids have been synthesized using D-amino acids (11). Against various bacteria such as *E. coli, P. aeruginosa, B. subtilis, S. aureus,* and *S. pyogenes,* the activity of the D-enantiomers is quantitatively equivalent to that of the L-enantiomers and none are hemolytic. In contrast, the D-enantiomers of the cecropin-melittin hybrids are less active than the corresponding L-enantiomers against the bloodstream form of the malaria parasite *Plasmodium falciparum.* These results suggest that chiral selectivity is not required for lysis of the bacterial cells, while both an achiral and chiral cell inactivation mechanism may occur in the case of *Plasmodium falciparum.*

Cecropins are now commonly found in the immune system of numerous other insects (12-14). For instance, six closely related cecropins have been isolated from immunized larval hemolymph of the silkworm *Bombyx mori.* One of these cecropins, cecropin B1 (Fig. 1), exhibits activity against both gram-negative and gram-positive bacteria, with the exception of *S. aureus* (15). A cecropin locus has been cloned from the immune response in *Drosophila melanogaster* (Fig. 1) (16). *Drosophila* cecropin A is at least as potent as the *Hyalophora* cecropin against gram-negative bacteria, but is less active against gram-positive organisms (17). Surprisingly, a cecropin-like peptide, named cecropin P1, has been isolated from pig small intestine (Fig. 1) (18). Both amide and carboxyl forms exhibit significant activity against *E. coli, Salmonella*, and *Acinetobacter*, with the amidated form being slightly more active. Little to no activity was found against gram-positive bacteria, other than *B. megatorium* (18). Cecropin P1 has been postulated to contribute to the regulation of *E. coli* in the upper intestine (19). These findings suggest that the cecropins are likely to be widely distributed in the animal kingdom.

While the *Drosophila* cecropin genes are expressed mainly in the fat body and hemocytes, a gene for a new cecropin-like peptide, named andropin, has been found in the *Drosophila* male reproductive tract (20). *Drosophila* cecropin A and andropin have similar precursor sequences and secondary structures but different mature sequences (Fig. 1). Andropin has moderate activity against gram-positive bacteria but little or no activity against gram-negative bacteria.

Magainins - Magainins consist of a family of basic antimicrobial peptides produced in the granular glands present in amphibian skin (Fig. 2). They exhibit a broad spectrum of antimicrobial activity including gram-positive and gram-negative bacteria, fungi, and protozoa (21,22). Enhancement of the amphipathicity and α-helicity (magainins B and G - Fig. 2) results in an increase in their antimicrobial and antiprotozoan activity (23-24). In contrast, the analog magainin H, which contains D-alanine at three positions in magainin 2 (Fig. 2), does not exhibit any antimicrobial or antiprotozoan activity (23,24). These results indicate that parallel relationships exist between the potential for helix formation and the antimicrobial or antiprotozoan activity. Magainin B and G, as well as magainin A (Fig. 2), which are more resistant to peptidase digestion, are at least 9-fold more potent than magainin 2 against hematopoietic tumor and solid tumor cells (25-26). Magainins A and G also show evidence of spermicidal activity which may occur through the disruption of the outer plasma membrane (27). While magainins 1 and 2 have no inhibitory effect on protein kinase C (PKC), the analog magainin B was found to be a potent inhibitor of PKC isozymes (28). In contrast to magainin 1, magainin 2 is readily phosphorylated by PKC. This would be expected to result in the reduction of its overall hydrophobicity and, in turn, of its membrane-disrupting effect on mammalian cells (28). Therefore magainin 2 has greater potential as a therapeutic agent than magainin 1.The D-magainins have similar antimicrobial activity combined with low hemolytic activity as do the L-enantiomers (11,29). As expected the D-enantiomers are much more resistant to proteolysis than the L-forms, providing a significantly enhanced therapeutic potential for the D-magainins (11,29). A systematic survey of other tissues of the adult frog has lead to the discovery of a group of antimicrobial peptides synthesized in the *Xenopus* stomach through a transcriptional process similar to that in the skin (30). Besides the known magainins, this group includes a novel peptide termed PGQ (Fig. 2) which shows antimicrobial and antifungal activity similar to that of the magainins; however, PGQ was also later found in skin extract (30). Magainins are therefore not found solely in the external mucosal surface of the frog. Indeed, magainin-like substances have been discovered in human submandibular and labial minor salivary glands, but the exact nature and role of these compounds are still under investigation (31).

As determined by Raman spectroscopy and NMR, the secondary structure of magainin 2 changes from unfolded in aqueous solution to helical when bound to lipid vesicles (32-33). To understand the mechanism of activity against gram-negative bacteria, the interactions between magainin 2 and lipopolysaccharide (LPS) present on the outer leaflet of the outer membrane of gram-negative cells were investigated by means of FT-IR spectroscopy (34). Thus, the addition of magainin to LPS from *S. typhimurium* causes a concentration-dependent and temperature-dependent disordering of the LPS, indicating effective binding of magainin. Furthermore, magainin 2 binds less effectively to bacterial mutants which contain lower concentrations of LPS (34), and the activity of magainin 2 is proportional to the LPS concentration (35,36). These results indicate that the activity of magainin is due to its ability to bind to the outer membrane.

$$\begin{array}{lll}
\text{Magainin 1} & : & \text{GIGKFLHSA}\mathbf{G}\text{KFGKAFVGEIMK}\mathbf{S}\text{-NH}_2 \\
\text{Magainin 2} & : & \text{GIGKFLHSA}\mathbf{K}\text{KFGKAFVGEIMN}\mathbf{S}\text{-NH}_2 \\
\text{Magainin A} & : & \boldsymbol{\beta}\mathbf{A}\text{IGKFLH}\mathbf{A}\mathbf{A}\text{KKF}\mathbf{A}\text{KAFV}\mathbf{A}\text{EIMNS-NH}_2 \\
\text{Magainin B} & : & \text{GIGKFLH}\mathbf{A}\mathbf{A}\text{KKF}\mathbf{A}\text{KAFV}\mathbf{A}\text{EIMNS-NH}_2 \\
\text{Magainin G} & : & \boldsymbol{\beta}\mathbf{A}\text{GIGKFLHSA}\mathbf{K}\text{KF}\mathbf{A}\text{KAFV}\mathbf{A}\text{EIMNS-NH}_2 \\
\text{Magainin H} & : & \text{GIGKFLHS}\mathbf{a}\text{KKF}\mathbf{a}\text{KAFV}\mathbf{a}\text{EIMNS-NH}_2 \\
\text{PGLa} & : & \text{GMASKAGAIAGKIAKVALKAL-NH}_2 \\
\text{PGQ} & : & \text{GVLSNVIGYLKKLGTGALNAVLKQ}
\end{array}$$

Figure 2. Sequences of magainins and analogs. The residues uncommon to the magainins are in bold, except for PGLa and PGQ which have little similarity to the magainins. βA: β-alanine; a: D-alanine.

The above results suggest that primary interaction between magainin 2 and the outer membrane is most likely electrostatic in nature. Magainin 2 was also found by Raman spectroscopy to bind to the surface of negatively charged liposomes without spontaneously penetrating the bilayer (32). Penetration into the bilayer, as well as the antimicrobial activity, are enhanced by mixing magainin 2 and PGLa, another antimicrobial peptide isolated from frog skin (Fig. 2) (32). Synergistic effects were also found between magainin 2 and the β-lactam antibiotic cefepime against $E.\ coli$, both $in\ vitro$ and $in\ vivo$ (37). Therefore, the cell surface of the β-lactam-altered bacteria may be more susceptible to cationic peptides such as magainins whose site of action is the inner bacterial membrane, as indicated below.

The peptide enhancement of the permeability in the lipid bilayer has been examined by measuring the leakage of a fluorescent dye from different types of small unilamellar vesicles (38). These analyses confirm the essential role of electrostatic interactions in the binding of magainin to lipids, and show that the membrane-perturbing activity is regulated by membrane fluidity. The penetration of magainins into the hydrophobic region of fluid membranes may be deeper than into rigid membrane, resulting in a disruption of the lipid organization. It was calculated that several hundred peptide molecules per cell are required for leakage to occur, which suggests a mechanism involving the accumulation of monomers electrostatically interacting with the membrane, resulting in leaky patches in the bilayer. The difference of one positive charge between magainins 1 and 2 may therefore explain the greater biological activity observed for magainin 2.

An endopeptidase that cleaves the magainins into half-peptides $in\ vivo$ has been isolated and characterized (39). Through the use of magainin analogs, this enzyme, termed magaininase, was found to cleave magainins only following their adoption of an amphipathic α-helical conformation. Since the resulting half-peptides are no longer biologically active, the $in\ vivo$ hydrolysis of magainins may be part of an inherent inactivating process (39). This hypothesis is under investigation.

Defensins - Defensins, a family of cationic peptides, were first identified in mammalian phagocytic cells. Although variable within and between species, defensin sequences all possess a specific array of cysteines connected by three disulfide bonds (Fig. 3) (40-44). Defensins exhibit broad activity against bacteria, fungi, tumor cells and viruses. RatNP-1, RatNP-2 and NP-1 were found to have similar activities against $S.\ typhimurium$, $S.\ aureus$, and *C. albicans*, and to be more potent than the less cationic defensins ratNP-3 and ratNP-4 (45). In contrast, human defensin HNP-1 is less active against *S. typhimurium* than any of the rat or rabbit defensins. *S. typhimurium* strains with mutations in either regulatory gene *pho*P or *pho*Q, as well as mutant strains missing the PhoP protein, are more sensitive to the rabbit defensins NP-1 and NP-2 (46).

Antimicrobial activity against clinical isolates of human ophthalmic pathogens has been reported for rabbit defensins NP-1 and NP-5 (47). While NP-1 exhibits bactericidal activity at 10μg/ml, NP-5 shows only bacteriostatic activity at 50μg/ml. These two defensins therefore exert antimicrobial activity through two distinct, yet undetermined, mechanisms. Defensins were also found to be involved in the control of the oral microorganisms present in the periodontium (48). Thus, HNP-1 and HNP-2 are active against *Capnocytophaga* species under both aerobic and anaerobic conditions and over a broad range of pH through a nonoxidative and oxygen-independent mechanism. In contrast, they fail to kill the two oral bacteria *A. actinomycetemcomitans* and *E. corrodens*, while rabbit defensin NP-1 exhibits significant potency and a wider range of activity against the oral microorganisms.

To understand the mechanism of defensin action, the crystal structure of HNP-3 was determined (49). In contrast to the cecropins and magainins, HNP-3 does not assume an α-helical structure but an antiparallel β-sheet conformation. The pattern of conserved residues of the defensin family (Fig. 3), as well as NMR studies of NP-5 (50), suggest similar conformations for all of the defensins. Similar to the cecropins and magainins, HNP-1 and NP-1 form voltage-dependent ion channels in lipid bilayer membranes at concentrations comparable to those required for *in vitro* antimicrobial activity (51). These results, as well as the lack of specificity of these channels, support channel-mediated membrane permeabilization as a broad-spectrum lytic mechanism.

```
Human       HNP-1      ACYCRIP ACIAGERRYGTCIYQGRLWAFCC
            HNP-2       CYCRIP ACIAGERRYGTCIYQGRLWAFCC
            HNP-3      DCYCRIP ACIAGERRYGTCIYQGRLWAFCC
            HNP-4      VCYCRLV FCRRTELRVGNCLIGGVSFTYCCTRV
Rabbit      NP-1      VVCACRRA LCLPRERRAGFCRIRGRIHPLCCRR
            NP-2      VVCACRRA LCLPLERRAGFCRIRGRIHPLCCRR
            NP-3A     GICACRRR FCPNSERFSGYCRVNGARYVRCCSRR
            NP-3B     GRCVCRKQLLCSYRERRIGDCKIRGVRFPFCCPR
            NP-4      VSCTCRRF SCGFGERASGSCTVNGVRHTLCCRR
            NP-5      VFCTCRGF LCGSGERASGSCTINGVRHTLCCRR
Rat         RatNP-1   VTCYCRRT RCGFRERLSGACGYRGRIYRLCCR
            RatNP-2   VTCYCRST RCGFRERLSGACGYRGRIYRLCCRR
            RatNP-3     CSCRTS SCRFGERLSGACRLNGRIYRLCC
            RatNP-4    ACYCRIG ACVSGERLTGACGLNGRIYRLCCR
Guinea pig  GPNP      RRCICTTR TCRFPYRRLGTCIFQNRVYTFCC
Mouse       cryptdin   VCYCRSR GCKGRERMNGTCRKGHLLYTLCCR
```

Figure 3. Sequences of the mammalian defensins. The characteristic residues are in bold.

The mechanism of HPN-1 defensin activity against tumor cells has been investigated by means of radiolabeled chromium or rubidium release assays (52,53). Small membrane channels are rapidly formed from HNP-1 binding to the plasma membrane. However, the continuing presence of defensin is required for lysis to occur. This second mechanism occurred after permeabilization was postulated to be mediated by DNA damage (53).

Rabbit defensins were found to be involved in the lysis of *T. pallidum* from syphilitic lesions (54). Activity is enhanced by the presence of complement in the serum, which suggests an activation of complement upon binding of defensins to the cells. Reduction and alkylation of NP-1 significantly alters the amphipathicity of NP-1 and results in a decrease in activity, illustrating the importance of the amphipathicity in the lytic effect of defensin against *T. pallidum*. In contrast to the antibacterial activity against gram-positive and gram-negative bacteria, the order of activity among the different rabbit defensins is independent of their net charges.

Defensins are not limited to mammals since defensin-like peptides have been isolated from the body fluid of the dipteran *Phormia terranovae* (55), in the larvae of the beetle *Zophobas atratus* (56), and in the embryonic cell line of *Sarcophaga peregrina*, named sapecin (57). A defensin-related 51-residue antimicrobial peptide has been characterized in the royal jelly of the honey bee, termed royalisin (58). Similar to the mammalian defensins, the insect defensins contain a motif of six cysteines (Fig. 4). These cysteines are spaced and the disulfides connected differently.

```
Phormia terranovae A  ATCDLLS    GTGINHSACAAHCLLRGNRGGYCNGKGVCVCRN
Phormia terranovae B  ATCDLLS    GTGINHSACAAHCLLRGNRGGYCNRKGVCVCRN
Zophobas atratus    B FTCDVLGFEIAGTKLNSAACGAHCLALGRRGGYCNSKSVCVCR
Zophobas atratus    C FTCDVLGFEIAGTKLNSAACGAHCLALGRTGGYCNSKSVCVCR
Sarcophaga  Sapecin  ATCDLLS    GTGINHSACAAHCLLRGNRGGYCNGKAVCVCRN
```

Figure 4. Sequences of the insect defensins. The common residues are in bold.

Gramicidins - Gramicidins are hydrophobic peptides produced by *Bacillus brevis* which exhibit antimicrobial activity primarily against gram-positive bacteria, but are known to be toxic to man. The hemolytic activity vanishes upon formylation of the tryptophan residues, although these analogs still specifically act on *Plasmodium falciparum* infected erythrocytes (59). The substitution of both ornithine and D-phenylalanine with alanine and α,β-dehydrophenylalanine respectively, in the cyclic gramicidin S (Fig. 5) results in an increase of the K^+ permeability of human erythrocytes and *S. aureus*, but not of *E. coli* (60). This activity is in contrast with the earlier belief that the amphipathicity of gramicidin S was important for its antimicrobial activity. The resistance to *E. coli* shows that the outer membrane forms a permeability barrier against hydrophobic antibiotics.

```
Gramicidin A:  HCO-VGAlAvVvWlWlWlW-NHCH2CH2OH
Gramicidin B:  HCO-VGAlAvVvWlFlWlW-NHCH2CH2OH
Gramicidin C:  HCO-VGAlAvVvWlYlWlW-NHCH2CH2OH
Gramicidin S:  cyclo (VOLfP)2
```

Figure 5. Gramicidins sequences. The D-amino acids are represented by the lower case letters. The uncommon residues at position 11 are in bold. O: ornithine.

Gramicidin channels are membrane-spanning structures that serve as models for the study of ion permeation through cell membranes (61). The channel structures of the three gramicidins A, B, and C are equivalent in bilayer systems (i.e., $\beta^{3,6}$ helix), while their conductance and duration properties vary (62). Kinetics of gramicidin A channel formation in lipid bilayers indicate that gramicidin permeates poorly across lipid bilayers and that the dissociation of dimers into monomers in the membrane is a very slow process (63-65). The dimer-monomer ratio, and in turn the conformation of gramicidin in the presence of phospholipids, is dependent on the solvents (64). In contrast, the behavior of the channels is independent of the nature of the solvent (66). The dipole moment along with the hydrogen bond forming capability of the tryptophan residues are two important features for the channel formation of gramicidin A. These four residues have the tendency to cluster toward the membrane-solution interface in an equivalent orientation to the channel axis (67). The role of the tryptophans was supported by the decrease of the channel-forming potency observed either for gramicidin B (which differs from gramicidin A by a phenylalanine at position 11) (Fig. 5) upon their formylation, or as the number of substitutions of tryptophan with phenylalanine increased (59,61,68). However, no correlation exists between the duration and conductance of the channel. In contrast, the alcohol function of the ethanolamine moiety has no influence on the conductance properties of the gramicidin channel since the presence of acyl groups on the C-terminus of gramicidin A resulted in similar single conductance channels (69). The channel lifetime increases with the length of the acyl chains, which may be due to stronger interactions between these chains and the hydrophobic part of the lipids.

Unrelated to the channel-forming potency, gramicidin A was found to induce both vesicle fusion and H_{II} phase formation (70). As with channel formation, N-formylation of the tryptophans reduces these properties. These findings suggest that small hydrophobic peptides might initiate a virus membrane fusion process by destabilizing the bilayers involved in this reaction.

Miscellaneous - A number of papers focused on the characterization of new classes of antimicrobial peptides (Fig. 6). For instance, abaecin, isolated from the hemolymph of immunized bees, shows antimicrobial activity; although moderate, its activity is substantially lower against gram-negative bacteria than the related apidaecin (71). Dermaseptin is the first peptide isolated from frog skin which inhibits the growth of a pathogenic mold and morphological alteration of its hyphae (72). Bombinin-like peptides (BLPs) from the skin of the asian toad *Bombina orientalis* show higher antimicrobial activity than magainin 2 and, in contrast to bombinin, have no hemolytic activity (73). A cysteine-rich peptide (TAP) with potent activity against *S. aureus, K. pneumonia, P. aeruginosa*, and *C. albicans* has been identified in the extracts of the bovine tracheal mucosa, while two proline-rich bactenecins (Bac5 and Bac7), which show high activity against gram-negative bacteria, have been isolated from the granule of bovine neutrophils (74,75). Nisin Z, a natural variant of the lantibiotic peptide nisin, that has higher antimicrobial activity compared to nisin, was found in *Lactococcus lactis* (76). Duramycins, another group of lantibiotics potently active against gram-positive bacteria, was isolated from various strains of *Streptoverticillium* and *Streptomyces* species (77,78). Two human serine protease, Cathepsin G and Granzyme B, possess homologous internal peptide sequences with broad-spectrum antimicrobial activities (HPQYNQR and HPAYNPK, respectively) (79).

```
Abaecin       YVPLPNVPQPGRRPFPTFPGQGPFNPKIKWPQGY
Dermaseptin   ALWKTMLKKLGTMALHAGKAALGAAADTISQGTQ
BLP-1         GIGASILSAGKSALKGLAKGLAEHFAN-NH2
TAP           NPVSCVRNKGICVPIRCPGSMKQIGTCVGRAVKCCRKK
Bac5          RFRPPIRRPPIRPPFYPPFRPPIRPPIFPPIRPPFRPPLRFP
Bac7          RRIRPRPPRLPRPRPRPLPFPRPGPRPIPRPLPFPRPGPRPIPRPLPFPRPGPRPIPRP
Nisin Z       I(Dhb)AI(Dha)LA(Abu)PGAK(Abu)GALMGANMK(Abu)A(Abu)ANASIHV(Dha)K
Duramycin     AKQAAAFGPF(Abu)FVA(HOAsp)GN(Abu)K
```
Figure 6. Sequences of new antimicrobial peptides. Dhb: dehydrobutyrine (β-methyldehydroalanine); Dha: dehydroalanine; Abu: α-aminobutyric acid; HOAsp: β-hydroxyaspartic acid

DESIGN AND DEVELOPMENT OF NEW ANTIMICROBIAL PEPTIDES

The fact that structural features such as amphipathicity and/or α-helicity are important for antimicrobial activity led to the design of potent synthetic antimicrobial peptides. Model amphipathic α-helical peptides composed of repeats of the LARL sequence exhibit potent broad spectrum antimicrobial activity (80-82). While the presence of a valeryl group at the N-terminal of Ac-(LARL)$_3$-NH$_2$ results in increased activity against gram-negative bacteria, the activity of this peptide against gram-positive bacteria significantly decreases upon the addition of a longer acyl group at the N-terminus (82). Their activity was found to depend on the degree of amphipathicity and/or helicity, since perturbation of the structure by introduction of a proline residue results in decreases in activity (80). These model peptides form ion-channels in lipid bilayers, with channel properties varying in the same order as the extent of their antimicrobial activity (81). Following RP-HPLC studies on the conformation of peptides when bound to the C-18 of the stationary phase, model amphipathic α-helical peptides composed of leucine and lysine peptides were found to have high activity against both gram-positive and gram-negative bacteria (83,84). Similarly, proline perturbation of this structure results in equivalent or lower activity depending on the position of proline within the sequence (84). In contrast, the introduction of a methionine sulfoxide residue improves the antimicrobial activity and decreases the hemolytic activity of the parent sequence (83). All of these results are well correlated with each analog retention time during RP-HPLC, confirming the importance of the amphipathicity and/or helicity for lysis to occur.

The development of new antimicrobial peptides through the isolation and characterization of active substances in insect, mammalian, or human tissues, or through the design of well defined structures, remains difficult and limited by the need to screen and synthesize large numbers of peptides. A unique and more general approach has been recently presented (85,86). This consists of systematically screening the antimicrobial activity of peptide libraries comprised of tens of millions of peptides. Thus, a number of 6-residue peptides such as Ac-RRWWRF-NH$_2$ and Ac-RRWWCR-NH$_2$, exhibits higher activity against *S. aureus* than the magainins or cecropins and shows no hemolytic activity.

Conclusion - Natural or synthetic model peptides represent a fertile source of novel chemotherapeutic agents. Although few of these compounds will be effectively used as drugs, they provide new classes of lead compounds as well as useful tools for studying the mechanism of action against bacterial cell membranes. With the development of peptide libraries, peptides are likely to come to the forefront of antibacterial research.

References

1. H.G. Boman and D. Hultmark, Annu. Rev. Microbiol., 41, 103 (1987).
2. J. Fink, R.B. Merrifield, I.A. Boman and H.G. Boman, J. Biol. Chem., 264, 6260 (1989).
3. Y. Okai and X. Qu, Immunol. Lett., 20, 127 (1989).
4. H.G. Boman, I.A. Boman, D. Andreu, Z. Li, R.B. Merrifield, G. Schlenstedt and R. Zimmermann, J. Biol. Chem., 264, 5852 (1989).
5. H.G. Boman, D. Wade, I.A. Boman, B. Wåhlin and R.B. Merrifield, FEBS Lett., 259, 103 (1989).
6. D. Sipos, K. Chandrasekhar, K. Arvidsson, Å. Engström and A. Ehrenberg, Eur. J. Biochem., 199, 285 (1991).
7. T.A. Holak, Å. Engström, P.J. Kraulis, G. Lindeberg, H. Bennich, T.A. Jones, A.M. Gronenborn and G.M. Clore, Biochemistry, 27, 7620 (1988).
8. B. Bazzo, M.J. Tappin, A. Pastore, T.S. Harvey, J.A. Carver and I.D. Campbell, Eur. J. Biochem., 173, 139 (1988).
9. J. Fink, I.A. Boman, H.G. Boman and R.B. Merrifield, Int. J. Pept. Protein Res., 33, 412 (1989).
10. B. Christensen, J. Fink, R.B. Merrifield and D. Mauzerall, Proc. Natl. Acad. Sci. USA, 80, 5072 (1988).
11. D. Wade, A. Boman, B. Wåhlin, C.M. Drain, D. Andreu, H.G. Boman and R.B. Merrifield, Proc. Natl. Acad. Sci. USA, 87, 4761 (1990).
12. H.G. Boman, I. Faye, G.H. Gudmunsson, J.Y. Lee and D.A. Lidholm, Eur. J. Biochem., 201, 23 (1991).
13. J.A. Hoffmann and D. Hoffmann, Res. Immunol., 141, 910 (1990).
14. P. Casteels, Res. Immunol., 141, 940 (1990).
15. I. Morishima, S. Suginaka, T. Ueno and H. Hirano, Comp. Biochem. Physiol., 95B, 551 (1990).
16. P. Kylsten, C. Samakovlis and D. Hultmark, Embo J., 9, 217 (1990).

17. C. Samakovlis, A. Krimbrell. P. Kylsten, Å. Engström and D. Hultmark, Embo J., 9, 2969 (1990).
18. J.Y. Lee, A. Boman, S. Chuanxin, M. Andersson, H. Jörnvall, V. Mutt and H.G. Boman, Proc. Natl. Acad. Sci. USA, 86, 9159 (1989).
19. J.M. Lundberg, A. Hemsen, M. Andersson, T. Hökfelt, V. Mutt and H.G. Boman, Acta Physiol. Scand., 141, 443 (1991).
20. C. Samakovlis, P. Kylsten, D.A. Kimbrell, Å. Engström and D. Hultmark, Embo J., 10, 163 (1991).
21. B. Berkowitz, C.L. Bevins and M.A. Zasloff, Biochem. Pharmacol., 39, 625 (1990).
22. H.G. Boman, Cell, 65, 205 (1991).
23. H.C. Chen, J.H. Brown, J.L. Morell and C.M. Huang, FEBS Lett, 236, 462 (1988).
24. C.M. Huang, H.C. Chen and C.H. Zierdt, Antimicrob. Agents Chemother., 34, 1824 (1990).
25. D. Juretic, H.C. Chen, J.H. Brown, J.L. Morell, R.W. Hendler and H.V. Westerhoff, FEBS Lett., 249, 219 (1989).
26. R.A. Crucian, J.L. Barker, M. Zasloff, H.C. Chen and O. Colamonici, Proc. Natl. Acad. Sci. USA, 88, 3792 (1991).
27. M.C. Edelstein, D.L. Fulgham, J.E. Gretz, N.J. Alexander, T.J. Bauer and D.F. Archer, Fertility Sterility, 55, 647 (1991).
28. H. Nakabayashi, J.H. Brown, J.L. Morell, H.C. Chen and K.P. Huang, FEBS Lett., 267, 135 (1990).
29. R. Bessalle, A. Kapitkosky, A. Gorea, I. Shalit and M. Fridkin, FEBS Lett., 274, 151 (1990).
30. K.S. Moore, L.C. Bevins, M.M. Brasseur, N. Tomassini, K. Turner, H. Eck and M. Zasloff, J. Biol. Chem., 266, 19851 (1991).
31. A. Wolff, J.E. Moreira, C.L. Bevins, A.R. Hand and P.C. Fox, J. Histochem. Cytochem., 38, 1531 (1990).
32. R.W. Williams, R. Starman, K.M.P. Taylor, K. Gable, T. Beeler, M. Zasloff and D. Covell, Biochemistry, 29, 4490 (1990).
33. D. Marion, M. Zasloff and A. Bax, FEBS Lett., 227, 21 (1988).
34. F.R. Rana, C.M. Sultany and J. Blazyk, FEBS Lett., 261, 464 (1990).
35. E.A. Macias, F.R. Rana, J. Blazyk and M.C. Modrzakowski, Can. J. Microbiol., 36, 582 (1990).
36. F.R. Rana, E.A. Macias, C.M. Sultany, M.C. Modrzakowski and J. Blazyk, Biochemistry, 30, 5858 (1991).
37. R.P. Darveau, M.D. Cunningham, C.L. Seachord, L. Cassiano-Clough, W.L. Cosand, J. Blake and C.S. Watkins, Antimicrobial. Agents Chemother., 35, 1153 (1991).
38. K. Matsuzaki, M. Harada, S. Funakoshi, N. Fujii and K. Miyajima, Biochim. Biophys. Acta, 1063, 162 (1991).
39. N.M. Resnick, W.L. Maloy, H.R. Guy and M. Zasloff, Cell, 66, 541 (1991).
40. R.I. Lehrer, T. Ganz and M.E. Selsted, ASM News, 56, 315 (1990).
41. R.I. Lehrer, T. Ganz, M.E. Selsted, B.M. Babior and J.T. Curnutte, Ann. Intern. Med., 109, 127 (1988).
42. T. Ganz, M.E. Selsted and R.I. Lehrer, Eur. J. Haematol., 44, 1 (1990).
43. R.I. Lehrer, T. Ganz and M.E. Selsted, Cell, 64, 229 (1991).
44. R.I. Lehrer and T. Ganz, Blood, 76, 2169 (1990).
45. P. Eisenhauer, S.S.S.L. Harwig, D. Szklarek, T. Ganz and R.I.Lehrer, Infect. Immun., 58, 3899 (1990).
46. S.I. Miller, W.S. Pulkkinen, M.E. Selsted and J.J. Mekalanos, Infect. Immun., 58, 3706 (1990).
47. J.S. Cullor, M.J. Mannis, C.J. Murphy, W.L. Smith, M.E. Selsted and T.W. Reid, Arch. Ophtalmol., 108, 861 (1990).
48. K.T. Miyasaki, A.L. Bodeau, T. Ganz, M.E. Selsted and R.I. Lehrer, Infect. Immun., 58, 3934 (1990).
49. C.P. Hill, J. Yee, M.E. Selsted and D. Eisenberg, Science, 251, 1481 (1991).
50. A. Pardi et al, J. Mol. Biol., 201, 625 (1988).
51. B.L. Kagan, M.E. Selsted, T. Ganz and R.I. Lehrer, Proc. Natl. Acad. Sci. USA, 87, 210 (1990).
52. A. Lichtenstein, J. Clin. Invest., 88, 93 (1991).
53. J.F. Gera and A. Lichtenstein, Cell. Immunol., 138, 108 (1991).
54. L.A. Borenstein, M.E. Selsted, R.I. Lehrer and J.N. Miller, Infect. Immun., 59, 1359 (1991).
55. J. Lambert, E. Keppi, J.L. Dimarcq, C. Wicker, J.M. Reichhart, B. Dunbar, P. Lepage, A.Van Dorsselaer, J. Hoffmann, J. Fothergill and D. Hoffmann, Proc. Natl. Acad. Sci. USA, 86, 262 (1989).
56. P. Bulet, S. Cociancich, J.L. Dimarcq, J. Lambert, J.M. Reichhart, D. Hoffmann, C. Hetru and A. Hoffmann, J. Biol. Chem., 266, 24520 (1991).
57. K. Matsuyama and S. Natori, J. Biol. Chem., 263, 17117 (1988).
58. S. Fujiwara, J. Imai, M. Fujiwara, T. Yaeshima, T. Kawashima and K. Kobayashi, J. Biol. Chem., 265, 11333 (1990).
59. G.N. Moll, V. Van den Eertwegh, H. Tournois, B. Roelofsen, J.A.F. Op den Kamp and L.L.M. Van Deenen, Biochim. Biophys. Acta, 1062, 206 (1991).
60. T. Katsu, K. Sanchika, M. Takahashi, Y. Kishimoto, Y. Fujita, H. Yoshitomi, M. Waki and Y. Shimohigashi, Chem. Pharm. Bull., 38, 2880 (1990).
61. B.A. Wallace, Ann. Rev. Biophys. Biophys. Chem., 19, 127 (1990).
62. D.B. Sawyer, L.P. Williams, W.L. Whaley, R.E. Koeppe II and O.S. Andersen, Biophys. J., 58, 1207 (1990).
63. A.M. O'Connell, R.E. Koeppe II and O.S. Andersen, Science, 250, 1256 (1990).
64. M.C. Bañó, L. Braco and C. Abad, Biochemistry, 30, 886 (1991).
65. P.L. Easton, J.F. Hinton and D.K. Newkirk, Biophys. J., 57, 63 (1990).

66. D.B. Sawyer, R.E. Koeppe II and O.S. Andersen, Biophys. J., 57, 515 (1990).
67. H. Michel and J. Deisenhofer, Curr. Top. Membr. Trans., 36, 53 (1990).
68. M.D. Becker, D.V. Greathouse, R.E. Koeppe II and O.S. Andersen, Biochemistry, 30, 8830 (1991).
69. A. Benayad, D. Benamar, N. Van Mau, G. Page and F. Heitz, Eur. Biophys. J., 20, 209 (1991).
70. H. Tournois, C.H.J.P. Fabrie, K.N.J. Burger, J. Mandersloot, P. Hilgers, H. Van Dalen, J. de Gier and B. de Kruijff, Biochemistry, 29, 8297 (1990).
71. P. Casteels, C. Ampe, L. Riviere, J. Van Damme, C. Elicone, M. Fleming, F. Jacobs and P. Tempst, Eur. J. Biochem., 187, 381 (1990).
72. A. Mor, V.H. Nguyen, A. Delfour, D. Migliore-Samour and P. Nicolas, Biochemistry, 30, 8824 (1991).
73. B.W. Gibson, D. Tang, R. Mandrell, M. Kelly and E.R. Spindel, J. Biol. Chem., 266, 23103 (1991).
74. G. Diamond, M. Zasloff, H. Eck, M. Brasseur, W.L. Maloy and C.L. Bevins, Proc. Natl. Acad. Sci. USA, 88, 3952 (1991).
75. R.W. Frank, R. Gennaro, K. Schneider, M. Przybylski and D. Romeo, J. Biol. Chem., 265, 18871 (1990).
76. J.W.M. Mulders, I.J. Boerrigter, H.S. Rollema, R.J. Siezen and W.M. de Vos, Eur. J. Biochem., 201, 581 (1991).
77. F. Hayashi, K. Nagashima, Y. Terui, Y. Kawamura, K. Matsumoto and H. Itazaki, J. Antibiotics, 43, 1421 (1990).
78. A. Fredenhagen, G. Fendrich, F. Märki, W. Märki, J. Gruner, F. Raschdorf and H.H. Peter, J. Antibiotics, 43, 1403 (1990).
79. W.M. Shafer, J. Pohl, V.C. Onunka, N. Bangalore and J. Travis, J. Biol. Chem., 266, 112 (1991).
80. S. Lee, N.G. Park, T. Kato, H. Aoyagi and T. Kato, Chem. Lett., 599 (1989).
81. K. Anzai, M. Hamasuna, H. Kadono, S. Lee, H. Aoyagi and Y. Kirino, Biochim. Biophys. Acta, 1064, 256 (1991).
82. H. Nakamura, H. Aoyagi, S. Lee, S. Ono, T. Kato, Y. Murata and G. Sugihara, Bull. Chem. Soc. Jpn., 63 1180 (1990).
83. S.E. Blondelle, K. Büttner and R.A. Houghten in "Peptides, Proc 21st European Peptide Symposium", E. Giralt and D. Andreu, Eds, Escom, 1991, p. 738.
84. K. Büttner, S.E. Blondelle, J.M. Ostresh and R.A. Houghten, Biopolymers, in press (1992).
85. R.A. Houghten and J.H. Cuervo in "Innovation and Perspectives in Solid Phase Synthesis", R. Epton, Ed, Solid Phase Conference Coordination, Ltd, Canterbury, 1992, in press.
86. R.A. Houghten, C. Pinilla, S.E. Blondelle, J.R. Appel, C.T. Dooley and J.H. Cuervo, Nature, 354, 84 (1991).

Chapter 18. Protein Tyrosine Kinases and Cancer

Ellen M. Dobrusin and David W. Fry
Parke-Davis Pharmaceutical Research Division
Warner-Lambert Company
Ann Arbor, Michigan 48105

<u>Introduction</u> - Recently, we have seen the beginning of a revolution in our knowledge and understanding of cellular signal transduction pathways that control and regulate a variety of cellular processes including the differentiation and proliferation of normal and malignant cells (1,2). Key among these pathways are those operating through growth factor receptors and protooncogene-encoded growth factor receptors. Protein kinase activity is among the most prominent enzymatic functions through which these receptors effect signalling. The complex interplay between tyrosine and serine/threonine kinases (and/or kinases and phosphatases) is not yet well understood. However, many growth factor receptors and oncogenes associated with tumor cells have been shown (or inferred through sequence homology) to have protein tyrosine kinase (PTK) activity (3). Isolation, purification, sequence analysis, and molecular cloning of these receptor PTKs has permitted the classification of receptor PTKs on the basis of their molecular structure (4). In this chapter, two PTKs are critically evaluated and evidence is provided for their association with transformation and malignancy. These findings inspire a new therapeutic approach to anticancer drug design based on selective interruption of the signalling pathways by PTK inhibitors and these are also reviewed.

TRANSFORMING POTENTIAL OF TYROSINE KINASES AND IMPLICATIONS IN CLINICAL MALIGNANCIES

This section will review existing evidence that implicates a role for specific PTKs in the initiation or progression of preclinical and clinical malignancy and dwell on abl and erbB-2 which currently are most relevant to this subject.

<u>abl Tyrosine Kinase</u> - c-abl encodes for a 145 Kd non-receptor tyrosine kinase. Its natural function remains unclear except for a likely role in signal transduction. While simple overexpression of this gene is insufficient for transformation of NIH 3T3 cells, the c-abl protooncogene can be mutated to become a transforming protein in numerous ways. These include deletion or mutation of critical regulatory domains such as deletion of N-terminal sequences including the SH3 region, linker insertion between the SH3 and SH2 domains (5,6), activating point mutations (7), or by N-terminal fusion with bulky heterologous sequences such as retroviral gag sequences in v-abl. Although not known at the time, c-abl has been implicated in several leukemic states through discovery of the Philadelphia chromosome (Ph^1). Ph^1 was the first recurring chromosome-specific abnormality to be associated with any type of malignancy. It is present in the leukemic cells of greater than 90% of chronic myeloid leukemia (CML) cases (>99% for those patients in blast crisis) and a significant number of acute lymphocytic leukemias (ALL) (8).

The molecular event underlying the formation of the Ph^1 is the juxtaposition of the abl proto-oncogene from chromosome 9 with the bcr gene on chromosome 22, forming a hybrid gene bcr/abl which encodes a fusion protein with activated abl tyrosine kinase activity (16). This molecule exists in two forms, reflecting the fact that some translocations occur in the first bcr intron while others occur within the originally designated breakpoint cluster region of bcr (9). This results in two forms of bcr/abl protein products, $p185^{bcr/abl}$ and $p210^{bcr/abl}$ (9,10). Both the p210 and p185 forms of the bcr/abl fusion protein have activated tyrosine-specific kinase activity (11); enzymatic activity is absolutely necessary for transformation (12). Direct comparisons between these two proteins show that the additional sequences contributed by bcr to the N-terminus of abl are necessary for its oncogenic properties and, depending on length, modulate both transforming activity and protein tyrosine kinase activity (11,13-15). In comparison to p210, the p185 bcr/abl protein is more efficient at transforming Rat-1 cells which correlates with its higher kinase activity and may account for its association with the more aggressive acute leukemias (11,16). The natural role of bcr and the mechanism by which it causes activation of abl tyrosine kinase when fused to this protooncogene is not understood. Some evidence, however, indicates that it is a novel serine/threonine kinase and that sequences essential for transformation by the bcr-abl oncogene bind to the abl SH2 regulatory domain (16,17). This raises the possibility that bcr may interfere with negative regulation of abl (18). Several reviews describing the genetics, functional domains and pathogenesis of v-abl, bcr, and bcr/abl have recently been published (19-21).

The most relevant preclinical studies of bcr/abl oncogenicity have been carried out in transgenic mice or in animals that have been transplanted with bone marrow carrying the gene of interest. A high percentage of transgenic mice carrying a bcr-v-abl construct, as well as their progeny, develop lymphoid tumors of both T and pre-B cell types (22). Transgenic mice expressing $p185^{bcr/abl}$ die of acute leukemia 10-58 days after birth (23). When the gene for $p210^{bcr/abl}$ is introduced into cultured murine bone marrow and then transplanted into a syngeneic strain of mice, about half of the mice that develop malignancies have a myeloproliferative disorder resembling human CML (24,25). Transformation in response to this gene also occurs in a wide variety of cell types including lymphoid, erythroid, macrophage and mast cell tumors (26). A comparative study in which either $p210^{bcr/abl}$ or $p185^{bcr/abl}$ was introduced into the bone marrow of mice shows that both genes induced a similar spectrum of diseases including granulocytic, myelomonocytic and lymphocytic leukemias. However, those mice receiving the smaller protein developed a more aggressive disease after a shorter latent period than animals treated with p210-infected cells (13). Similar studies have been performed in mice whose bone marrow has been infected with v-abl. These animals developed malignancies similar to, but distinct, from CML (27).

Philadelphia chromosome-positive blast cells have been isolated from patients with CML and cultured in the presence of synthetic antisense oligodeoxynucleotides complementary to two identified bcr-abl junctions (28). Leukemia colony formation was suppressed, whereas granulocyte-macrophage colony formation from normal marrow progenitors was unaffected. When equal proportions of normal marrow progenitors and blast cells were mixed and exposed to the oligodeoxynucleotides, the majority of residual colony forming cells were normal. These findings demonstrate the requirement for a functional bcr-abl gene in maintaining the leukemic phenotype in human CML and support the concept that inactivating the bcr-abl oncogene will selectively kill neoplastic tissue harboring this gene.

erbB-2 Tyrosine Kinase - The erbB-2 or HER-2 protooncogene is the normal human homolog for the neu oncogene, which was originally identified in nitrosourea induced neuroblastomas of the rat. It encodes for a 185 Kd protein which is structurally very similar to the EGF receptor and possesses an extracellular, transmembrane and cytoplasmic tyrosine kinase domain (29). The genetic sequences for the two proteins are 44% homologous (30). Despite the similarity of $p185^{erbB2}$ to the EGF receptor, it is quite distinct and does not bind EGF (31). Although a natural ligand has not been identified, several putative ligands that activate the tyrosine kinase activity of this protein have been found in the conditioned media from three different tumor or oncogene-transformed cells (29,32-35). Information to date with regard to the signaling properties and metabolism of $p185^{erbB2}$ has come from the use of these putative ligands, as well as from chimeric receptors and antibodies which trigger autophosphorylation and subsequent events (36-40).

Preclinical evidence implicating erbB-2 as a protooncogene was first obtained from the sequence of neu where it was found that its oncogenic properties were due to a single point mutation (val^{664} to glu) in the transmembrane region (41) and that this modification constitutively activated the tyrosine kinase activity of its gene product (38) perhaps by facilitating receptor dimerization (42). A similar mutation in the normal human gene, c-erbB-2, also exhibits highly transforming properties (43). Although this transforming mutation has not as yet been found clinically, simple overexpression of the normal human c-erbB-2 is sufficient to cause transformation and tumorigenicity (44-46).

Studies in transgenic animals that overexpress the $p185^{neu}$ provide evidence that breast tissue in particular is susceptible to the transforming properties of this oncogene. Transgenic mice uniformly expressing the MMTV/c-neu gene rapidly develop mammary adenocarcinomas that involve the entire epithelium in every gland, indicating that the activated c-neu oncogene appears to be sufficient to induce malignant transformation in mammary tissue in a single step (47). Studies in transgenic mice or rats have confirmed the susceptibility of mammary tissue to this gene (48,49).

Antibodies raised against the extracellular domain of $p185^{erb2}$ have been shown to inhibit growth of a number of breast and ovarian tumor cells lines (50) and a lung adenocarcinoma (39), as well as reverse the transformed phenotype in neu transfected but not ras-transfected cells (51,52).

Several *in vivo* studies have shown that antibodies that bind to $pp185^{neu}$ exert an antitumor effect. Neu transfected NIH-3T3 cells injected into nude mice cause rapidly growing progressive fibrosarcomas. Intravenous injection either immediately after tumor inoculation or as much as 7 days later with monoclonal antibodies that react with the extracellular domain inhibit tumor cell growth in a dose-dependent manner and extend the survival of treated mice (53,54). A combination of two monoclonals that are specific for different sites on the external portion of $p185^{neu}$ are synergistic causing cures in 9 of 17 mice (55). Studies in syngeneic BDIX rats carrying the original neu containing neuroblastoma cells give similar results. Mice that are immunized with vaccinia virus containing the sequence for $p185^{neu}$ develop substantial titers of antibody to p185 and fail to develop tumors when injected with neu transformed cells (56). Other *in vivo* studies have shown p185 antibody-mediated inhibition of the Murray breast carcinoma (a high expressor of erbB-2) (57) which had been implanted into the subrenal capsule of nude mice (50) and in another c-erbB-2 overproducing tumor grown as a xenograft (58).

Clinically, $p185^{erbB2}$ has generated extreme interest among oncologists both for its possible involvement in the development and progression of breast and possibly other cancers as well as its apparent predictive value in assessing prognosis and aggressive tumor behavior. No less than 40 studies have been conducted which assess the frequency of overexpression (relative to normal tissue of the same type) of $p185^{erbB2}$ in clinical tumor isolates from patients with a variety of cancers. Most of these clinical trials involve breast cancer. On the average, $p185^{erbB2}$ is overexpressed in approximately 30% (with a range of 8-44%) in breast cancer tissues (59-68). In those trials where $p185^{erbB2}$ was statistically evaluated as a prognostic indicator, most studies found that high expression of this protein in tumor tissue correlated with earlier relapse, shorter postrelapse survival and shorter overall survival (60-62,64-69). This correlation has been established most strongly for breast cancer patients where the disease has spread to the auxiliary nodes; however, as more patients are evaluated it has been established in node negative patients as well (60). Although not all studies are unanimous in these findings, some of the discrepancies may be due to low patient number or differences in the methodology for assessing expression of erbB-2 or its protein product (57,59,61,70). Overexpression of $p185^{erbB2}$ and, in some studies, correlation with poor prognosis has been established in other tumor types including ovarian, squamous cell carcinomas of the head, neck and lower female genital tract, gastric carcinoma, bladder and lung adenocarcinoma (71-75).

PTK INHIBITORS

<u>Introduction</u> - Implicit in the emerging understanding of the association of PTKs and transformation is the assumption that molecules with the ability to selectively inhibit the culpable PTK(s) will uncouple them from the signal transduction pathway. The specific disruption of this linkage is expected to result pharmacologically in the restoration of the normal regulation of cell growth and therapeutically in antitumor activity.

<u>Small Molecule Protein Tyrosine Kinase Inhibitors</u> - The complex relationships that exist between kinases in the signal transduction pathway make selectivity and mechanism of action critical issues in the search for nontoxic PTK inhibitors. For example, most of the flavone type natural products that inhibit PTKs compete for binding at the ATP site of the target kinase. Since the ATP binding site for ATP-dependent enzymes is highly conserved and ubiquitous, their potential for therapeutically useful, selective PTK inhibition without unacceptable toxicity has been questioned (76,77).

Surprisingly, the isoflavone genistein inhibits the PTK of the EGFR and $pp60^{v-src}$, EGF autophosphorylation, and EGF stimulated protein phosphorylation in A431 cells, but has relatively little effect on serine/threonine kinases such as cAMP-dependent protein kinase or PKC (78). Although it may discriminate between these kinases, the cellular effects of genistein may not be attributed solely to PTK inhibition, since it is also a potent inhibitor of DNA topoisomerase II (79,80). The flavone quercetin (1) inhibits $pp60^{v-src}$, but is far less selective for PTK than genistein since it also inhibits other ATP-binding enzymes including lipoxygenase, cyclooxygenase, ATPases, and PKC (81-83). Recent synthetic effort designed to achieve more potent and selective inhibitors based on these flavonoids has produced **2**, a competitive inhibitor of $p56^{lck}$ with high specificity for the inhibition of PTK over serine/threonine kinases (84).

<u>**1**</u> 3,3',4',5,7-OH
<u>**2**</u> 4' - NH_2, 6-OH

In contrast, erbstatin (3) is a tyrosine-like PTK inhibitor isolated from actinomycete that binds competitively at the substrate binding site of the EGF receptor (85). Among the PTK inhibitors discovered from natural product screening, erbstatin has thus far been the most useful prototype for synthetic drug design. Erbstatin inhibits autophosphorylation of the EGF receptor isolated from the membrane vesicle fraction of human epidermoid carcinoma A431 cells with an $IC_{50} = 3$ µM (86). Although erbstatin inhibits the *in vitro* EGF-dependent cell proliferation of A431 cells with an $IC_{50} = 20$ µM, its antitumor activity *in vivo* is tenuous since it can only be demonstrated with the simultaneous administration of foroxymithine, an iron chelator that protects erbstatin against oxidative decomposition catalyzed by ferric iron in serum (87). Using this protocol, erbstatin is weakly active with a T/C of 155%. Indeed, there have been few reports of *in vivo* activity for erbstatin and the paucity of such data may be attributable, in part, to the inherent instability of the phenol *in vivo* (87,88). A requirement for hydroxyls in tyrosine-like PTK inhibitors could, therefore, be a significant liability in preclinical *in vivo* evaluations and a potential impediment for further development.

The degree of inhibitory selectivity by erbstatin is the subject of some controversy in the literature. Although first reported to be highly selective for the EGF receptor PTK vs. serine/threonine kinases, a recent study indicates that erbstatin is only weakly selective relative to PKC (isoenzymes α, β, γ) and cAMP-dependent protein kinase where the IC_{50}'s for erbstatin were determined to be 17-28 µM and 0.78 µM, respectively (89). In this study, the mechanism of inhibition of PKC by erbstatin was shown to be competitive with ATP and not the protein substrate binding site as in the case of the EGF receptor TK. The ability of erbstatin to inhibit two distinct classes of kinases each by distinct mechanisms could be explained on the basis of an interplay between sequence homology and differences in the kinetic profile of substrate binding for the two kinases (90). The issues of inhibitor potency, kinase selectivity, and chemical instability have not yet been successfully addressed in the small series of closely related erbstatin analogs that have been reported (91,92). None achieve potency much below 1 µM and none retain activity without one or more aromatic hydroxyls.

The tyrphostins are the most promising and well studied of the recently designed PTK inhibitors based on erbstatin (77,93-96). They form a large series of p-hydroxy benzylidene malononitrile analogs, represented by **4-6**, designed to minimize nonspecific cytotoxicity and to selectively inhibit PTKs. Extensive SAR studies using conformationally restricted analogs (4) of the tyrphostins helped to define the discrete conformational requirement for optimal inhibition in which the nitrile and the catechol ring are fixed in a *cis*-coplanar orientation. It was also revealed through synthesis that most compounds lost significant inhibitory potency when the hydroxyl substituents were removed or replaced with other heteroatoms. A broad range of selectivity has been achieved among the various analogs. Many of the compounds are highly selective for inhibition of the EGF-TK as distinct from the highly homologous insulin receptor (Insr) kinase where their affinity for the receptor is much as 3 orders of magnitude weaker. For example, **5** is about 10^2-fold selective for inhibition of the PTKs of EGFR, PDGF, and $p210^{bcr-abl}$ relative to InsR (94). Amide **6** is $>10^2$-fold selective for inhibition of EGFR PTK compared to the highly homologous ErbB2/neu kinase (95). Although this selectivity falls in the less desirable direction, the ability of this class of inhibitors to discriminate between closely related PTKs is very promising from a therapeutic point of view.

Most of the tyrphostins bind competitively at the substrate binding site of the EGFR tyrosine kinase; competitive binding at the ATP binding domain by some analogs has been reported anecdotally, but these compounds were reported to maintain selectivity for PTKs (97). Some tyrphostins (i.e., **5**) selectively inhibit EGF-dependent autophosphorylation of the EGF receptor, EGF stimulated DNA synthesis and tyrosine phosphorylation of endogenous substrates in whole cells and these effects correlate well with their PTK inhibitory potency in a dose dependent fashion (98). Antiproliferative effects for EGF-dependent growth of A431 cells, HER14 cells, and human keratinocytes have been demonstrated for **5** (94,98,99). Similar effects of **3** in PDGF-dependent vascular smooth muscle cells have also been reported (100). *In vitro*, maximal inhibition requires 16-hour incubation implying that penetration into cells is slow. Extended incubation to 36 hours diminishes activity and suggests metabolism of the compound to an inactive form (98). *In vitro* inhibition of EGF or PDGF induced growth is completely reversible, with normal cell growth restored upon removal of drug (93,100). Minimal effect on EGF-independent growth has been observed. Some tyrphostins have also been shown to block phosphorylation of PLCγ (101) and PI turnover (102).

The most significant development in the tyrphostin series has been the evolution to second generation tyrphostins (**7,8**), which are devoid of hydroxyls, are metabolically stable and show *in vivo* activity (103). The inhibition of autophosphorylation by these compounds is sustained for more than 48 hours in HER14 cells. These compounds reversibly inhibited EGF-stimulated cell growth and DNA synthesis in MH-85 human squamous cell carcinoma *in vitro*. When tested *in vivo* (ip/ip), **8** inhibited tumor growth and significantly increased the life span of EGF-dependent MH-85 tumor bearing nude mice at a dose of 200 μg/mouse/injection treated twice daily for 5 weeks from the time of tumor inoculation. Similar activity is seen for the more potent dichloro analog **7**. This is the first report of *in vivo* activity for a nonhydroxylated tyrosine-like PTK inhibitor.

Pinneatannol is a naturally occurring tetrahydroxyl-*trans*-stilbene whose structure is also reminiscent of erbstatin and the tyrphostins (104). It inhibits $p56^{lck}$ PTK with an IC_{50} = 66 μM by competing at the peptide substrate binding site. The related phenylhydrazone analogs were also competitive inhibitors of comparable potency (105). A synthetic pyridylstilbene analog in which the hydroxyls are replaced with methoxyls was a competitive inhibitor of $p56^{lck}$ with respect to ATP, though the potency was very poor (IC_{50} = 178 μM). Surprisingly, the corresponding hydroxylated analog was even less potent (106).

Another inhibitor closely related to the tryphostins is ST-638 (**9**), one of a series of α-cyanocinnamamides with unspecified olefin geometry. *In vitro*, this compound selectively inhibits the PTKs of the EGF receptor and several protooncogene and oncogene tyrosine kinases including $pp60^{c\text{-}src}$ and $pp70^{gag\text{-}actin\text{-}v\text{-}fgr}$ (107-109). The inhibitor binds competitively at the binding site for exogenous substrate on the EGF and $p130^{gag\text{-}v\text{-}fps}$ PTKs, though the kinetic profile for the inhibition of EGF autophosphorylation by **9** is noncompetitive (109). Interestingly, these inhibitors are 5- to 100-fold selective for inhibition of $pp60^{c\text{-}src}$ versus $pp60^{v\text{-}src}$ in contrast to inhibitors binding at the

ATP site of these kinases (109). When tested *in vitro* in A431 cells at nontoxic concentrations of 25-100 µM, **7** selectively inhibited EGF-dependent phosphorylation of exogenous substrates in a dose dependent fashion (110). At 25 µM, EGF-induced autophosphorylation was inhibited by 26% whereas inhibition of the phosphorylation of lipocortin, an endogenous substrate, was almost completely inhibited. The inhibitor has no effect on EGF binding or EGF receptor internalization. No *in vivo* activity for this class of inhibitors has been reported.

The sulfonylbenzoyl nitrostyrene PTK inhibitors, exemplified by **10-13**, combine the structural features of a tyrosine substrate mimic like erbstatin with an ATP cofactor mimic to generate a potential bisubstrate type inhibitor (111).

	R_1	R_2	IC_{50}, µM:	EGF-R	v-abl	PKC	MK cells
10	H	OH		1.0	~100	>500	>50
11	2-OH	OH		0.054	27	500	>50
12	2-OH	OCH_3		1.8	>100	nt	5.7
13	H	$NHCH_3$		0.4	35	290	5.2

Previously, a similar strategy employing a tyrosine residue linked *via* a sulfonyl to an ATP mimic, 5'-[4-(fluorosulfonyl)benzoyl]adenosine, had produced moderately potent but nonselective kinase inhibitors that apparently bound indiscriminately at the ATP site common to both the p60^{v-abl} PTK and PKC (112,113). An extensive SAR for this series indicates that the p-nitro styrene is the PTK recognition element contributing PTK selectivity and addition of the sulfonyl benzoate moiety increases inhibition potency. Antiproliferation activity in MK mouse keratinocytes is improved by replacement of the carboxylate with an uncharged terminal ester or amide. Molecular modeling studies of a hypothetical transition state for the tyrosine phosphorylation reaction patterned from an MgAMP model predicted the increased potency for compounds bearing R_1 = OH attributable to enhanced cation chelation. *In vivo* studies with **13** in nude mice bearing A431 human epidermoid carcinoma showed inhibition of tumor growth (correlation with PTK activity was not established), but cumulative toxicity and metabolic cleavage of the sulfonate ester were noted.

It could be postulated that the "substrate-site" binding class of erbstatin-like PTK inhibitors with styrene structures might mimic a restricted conformation of tyrosine in a protein substrate as it presents itself to the kinase (114,115). This observation has also been made for a series of opiate structures which were tested for PTK activity (114). Prompted by this possibility, one study has examined a series of styrene inhibitors bearing several nonhydrolyzable, isosteric replacements for the phenolic phosphate ester to assess their effect on potency and selectivity (115). The requirements for substrate recognition must be much more complex since all were inactive in the EGFR and p56^{lck}TK assays. Another study of structure activity relationships for the styrene class of inhibitors reveals that selectivity and potency are controlled, in part, not only by the number of hydroxyls (82,116), but by contributions from both the pattern of aromatic hydroxylation and styrene functionalization (149,158).

Among the natural products with PTK inhibitory activity, lavendustin A is among the most potent *in vitro* with an IC_{50} (EGFR-PTK) = 0.012 µM (117). This microbial metabolite isolated from *Streptomyces griseolavendus* is selective for inhibition of PTK relative to serine/threonine kinase.

Despite its potent enzyme activity, the acid has very poor activity in cultured whole cell assays. The methyl ester of lavendustin A, designed to increase cellular penetration, has an improved IC_{50} [(EGFR-PTK) =   1.8 µM],   inhibits   EGFR   autophosphorylation,   EGFR   internalization, phosphatidylinositol (PI) kinase, and EGF-induced PI-turnover in A431 cells. The ester was cytotoxic to A431 cells, P388 murine leukemia cells, and NIH 3T3 cells *in vitro* with IC_{50}'s equal to 20.3 µM, 10.9 µM, and 8.6 µM, respectively. A detailed kinetic analysis of the inhibition of EGF-PTK by lavendustin A shows that it is a hyperbolic mixed type inhibitor with respect to both ATP and the peptide substrate. It is a slow, tight binding inhibitor that appears to bind in the kinase domain at a site distinct from the binding sites of ATP and peptide; this binding decreases the binding affinity of the kinase for both (118). In the same kinetic study, the inhibition by an analog of **12** was reduced to different extents in the presence of different peptide substrates derived from the EGFR and PLC.

Herbimycin A, a benzoquinoid ansamycin antibiotic isolated from *Streptomyces hygroscopicus*, is reported to inhibit the v-src and p210$^{bcr-abl}$ PTKs and reverse the transformed phenotype of cells infected by the v-src family of PTK oncogenes and this activity is correlated with the reduction in phosphotyrosine content in total cellular proteins (119). These effects are not observed in raf, ras, or myc transformed cells. The activity is inhibited by N-ethylmaleimide and iodoacetamide, thiol reagents that may act at sulfhydryls in the kinase (120,121). The precise mechanism of inhibition has been proposed and a total synthesis of herbimycin A has been reported (122,123). Another natural product, (+)-aeroplysinin-1, a metabolite from the marine sponge *Verongia aerophoba*, selectively and reversibly inhibits EGF-dependent proliferation of MCF-7 and ZR-75-1 breast cancer cells and EGF induced endocytosis of the EGF receptor (124). The inhibitory effects of this compound are selective for these tumor cell lines since no cytostatic effects are seen when normal human fibroblasts were treated at a 10-fold higher concentration. The EGF induced PTK activity of solubilized shed membrane vesicles from MCR-7 cells is inhibited by aeroplysinin at 0.5 µM.

<u>Small Peptide Substrates</u> - As a corollary to the notion that PTK inhibitors that bind to the ATP site of the kinase could be prone to indiscriminate binding, it is possible that the kinase specificity of inhibitors might best be achieved by incorporation of a component designed with consideration of the specific peptide substrate for the kinase. The substrate specificity and kinetics for a variety of PTKs and synthetic peptide substrates have been well summarized (125).

The feasibility of designing inhibitors from peptide substrates is supported by the observation that angiotensin is converted from a substrate for pp60^{src} PTK to an inhibitor by the amino acid substitution tyr→Δphe   (126). The use of small peptide substrates as a basis for PTK inhibitor design was recently reported for another angiotensin based decapeptide, conceived as a potential suicide substrate (127). This peptide and its diastereomer are very weak competitive inhibitors of p56^{lck}. Another peptide inhibitor approach based on consideration of the catalytic mechanism of PTKs incorporates tetrafluorotyrosine into gastrin analogs in place of tyrosine (128) in an effort to probe the effect of phenolic charge and nucleophilicity.   The gastrin analog incorporating L-tetrafluorotyrosine has an impressive K_i = 4 µM for inhibition of the InsR PTK. Interestingly, the D-analog has a different kinetic mechanism and a K_i = 20 µM.   Other peptide approaches incorporating modified tyrosines have explored nonhydrolyzable phosphotyrosine minics such as 4-phosphononphenylalanine (129) and phosphonomethylphenylalanine (130).

Very little information with respect to defining the secondary structural requirements for peptide substrates or inhibitors is known, though there is some evidence suggesting that ß-turns may be recognition elements (131,132). In contrast, there is a growing body of information utilizing synthetic peptides to decipher the recognition or specificity coded by sequence motifs in PTK substrate peptides (133-136).

<u>Future Directions</u> - Though the potential for tyrosine kinase inhibitors as therapeutic agents for the treatment of cancer has not yet been demonstrated, tyrosine kinase inhibitors are serving as valuable mechanistic tools in helping to unravel the role of aberrant cellular signal transduction in cancer. In the future, the search for potent, selective inhibitors will undoubtedly benefit from solution of the crystal structures of the PTKs. Additional information about the role of SH2 domains (137,138) in substrate recognition,   kinase regulation, and kinase/phosphatase cross talk will be of value in designing therapeutic agents. Very recent data demonstrate that it is possible to design PTK

inhibitors that selectively inhibit transforming $p210^{bcr-abl}$ relative to the normal abl protein, $p140^{c-abl}$ (139).

References

1. L.C. Cantley, K.R. Auger, C. Carpenter, R. Kapeller and S. Soltoff, Cell, 64, 281 (1991).
2. S. A. Aaronson, Science, 254, 1146 (1991).
3. A. Ullrich and J. Schlessinger, Cell, 61, 203 (1990).
4. A. F. Wilks, Progress in Growth Factor Res., 2, 97 (1990).
5. W.M. Franz, P. Berger and J.Y. J. Wang, EMBO, 8, 137 (1989).
6. P. Jackson and D. Baltimore, EMBO J., 8, 449 (1989).
7. S.K. Shore, S.L. Bogart and E.P. Reddy, Proc. Natl. Acad. Sci. USA, 87, 6502 (1990).
8. S.S. Clark, W.M. Crist and O.N. Witte, Annu. Rev. Med., 40, 113 (1989).
9. A. Hermans, N. Heisterkamp, M. von Lindern, S. van Baal, D. Meijer, D. van der Plas, L.M. Wiedemann, J. Groffen, D. Bootma and G. Grosveld, Cell, 51, 33 (1987).
10. E. Fainstein, C. Marcelle, A. Rosner, E. Canaani, R.P. Gale, O. Dreazen, S.D. Smith and C.M. Croce, Nature, 330, 386 (1987).
11. T.G. Lugo, A.-M. Pendergast, A.J. Muller and O.N. Witte, Science, 247, 1079 (1990).
12. R.W. Rees-Jones and S.P. Goff, J. Virol., 62, 978 (1988).
13. M. Kelliher, A. Knott, J. Mclaughlin, O.N. Witte and N. Rosenberg, Mol. and Cell. Biol., 11, 4710 (1991).
14. J.R. McWhirter and J.Y. J. and Wang, Mol. Cell. Biol., 11, 1553 (1991).
15. A.J. Muller, J.C. Young, A.-M. Pendergast, M. Pondel, N.R. Landau, D.R. Littman and O.N. Witte, Mol. Cell. Biol., 11, 1785 (1991).
16. A.M. Pendergast, A.J. Muller, M.H. Havlik, Y. Maru and O.N. Witte, Cell, 66, 161 (1991).
17. Y. Maru and O.N. Witte, Cell, 67, 459 (1991).
18. A.M. Pendergast, A.J. Muller, M.H. Havlik, R. Clark, F. McCormick and O.N. Witte, Proc. Nat. Acad. Sci. USA, 88, 5927 (1991).
19. A. Dobrovic, G.B. Peters and J.H. Ford, Chromosoma, 100, 479 (1991).
20. G.Q. Daley and Y. Benneriah, Advances in Cancer Res., 57, 151 (1991).
21. M.L. Campbell and R.B. Arlinghaus, Adv. Cancer Res., 57, 227 (1991).
22. I.K. Hariharan, A.W. Harris, H. Abud, E. Webb, S. Cory and J. Adams, Mol. Cell. Biol., 9, 2798 (1989).
23. N. Heisterkamp, G. Jenster, D. Zovich, P.K. Pattengale and J. Groffen, Nature, 344, 251 (1990).
24. G.Q. Daley, R.A. Van Etten and D. Baltimore, Science, 247, 824 (1990).
25. M.A. Kelliher, J. McLaughlin, O.N. Witte and N. Rosenberg, Proc. Natl. Acad. Sci. USA, 87, 6649 (1990).
26. A.G. Elefanty, I.K. Hariharan and S. Cory, EMBO, 9, 1069 (1990).
27. M.L. Scott, R.A. Van Etten, G.Q. Daley and D. Baltimore, Proc. Natl. Acad. Sci USA, 88, 6506 (1991).
28. C. Szczylik T. Skorski, N.C. Nicolaides, L. Manzella, L. Malagusmera, D. Venturelli, A.M. Gewirtz and B. Calabretta, Science, 253, 562 (1991).
29. Y. Yarden and E. Peles, Biochemistry, 30, 3543 (1991).
30. L.A. Maier, F.J. Xu, S. Hester, C.M. Boyer, S. Mckenzie, A.M. Bruskin, Y. Argon and R.C. Bast, Cancer Res., 51, 5369 (1991).
31. D.F. Stern, P.A. Hefferman and R.A. and Weinberg, Mol. Cell. Biol., 6, 1729 (1986).
32. Y. Yarden and R.A. Weinberg, Proc. Natl. Acad. Sci. USA, 86, 3179 (1989).
33. R. Lupu, R. Colomer, J. Sarup, M. Shepard, D. Slamon and M.E. Lippman, Science, 249, 1552 (1990).
34. J.G. Davis, J. Hamuro, C.Y. Shim, A. Samanta, M.I. Greene and K. Dobashi, Biochem. Biophys. Res. Comm., 179, 1536 (1991).
35. K. Dobashi, J.G. Davis, Y. Mikami, J.K. Freeman, J. Hamuro and M.I. Greene, Proc. Natl. Acad. Sci USA, 88, 8582 (1991).
36. J. Lee, T.J. Dull, I. Lax, J. Schlessinger and A. Ullrich, EMBO J., 8, 167(1989).
37. H. Lehvaslaiho, L. Lehtola, L. Sistonen and K. Alitalo, EMBO J., 8, 159 (1989).
38. Y. Yarden, Proc. Natl. Acad. Sci. USA, 87, 2569 (1990).
39. E. Tagliabue, F. Centis, M. Campiglio, A. Mastroianni, S. Martignone, R. Pellegrini, P. Casalini, C. Lanzi, S. Menard and M.I. Colnaghi, Int. J. Cancer, 47, 933 (1991).
40. G.K. Scott, J. Biol. Chem., 256, 14300 (1991).
41. C.I. Bargmann, M.-C. Hung and R.A. Weinberg, Cell, 45, 649 (1986).
42. M.J.E. Sternberg, Nature, 339, 587 (1989).
43. Y. Suda, S. Aizawa, Y. Furuta, T. Yagi, Y. Ikawa, K. Saitoh, Y. Yamada, K. Toyoshima and T. and Yamamoto, EMBO J., 9, 203 (1990).
44. H. Saya, S. Ara, P.S. Y. Lee, J. Ro and M.-C. Hung, Mol. Carinogen., 3, 198 (1990).
45. N.R. Lemoine, S. Staddon, C. Dickson, D.M. Barnes and W.J. Gullick, Oncogene, 5, 237 (1990).
46. R.M. Hudziak, J. Schlessinger and A. Ullrich, Proc. Natl. Acad. Sci. USA, 84, 7159 (1987).
47. W.J. Muller, E. Sinn, P.K. Pattengale, R. Wallace and P. Leder, Cell, 54, 105 (1988).
48. L. Bouchard, L. Lamarre, P.J. Tremblay and P. Jolicoew, Cell, 57, 931 (1989).

49. B.C. Wang, W.S. Kennan, M.J. Lindstrom and M.N. Gould, Cancer Research, 51, 5649 (1991).
50. H.M. Shepard, G.D. Lewis, J.C. Sarup, B.M. Fendly, D. Maneval, J. Mordenti, I. Figari, C.E. Kotts, M.A. Palladino and A. Ullrich, Journal of Clinical Immunology, 11, 117 (1991).
51. H.C. Maguire and M.I. Greene, Seminars in Oncology, 16, 148 (1989).
52. J.A. Drebin, V.C. Link, D.F. Stern, R.A. Weinberg and M.I. Greene, Cell, 41, 695 (1985).
53. J.A. Drebin, V.C. Link, R.A. Weinberg and M.I. Greene, Proc. Natl. Acad. Sci. USA, 83, 9129 (1986).
54. J.A. Drebin, V.C. Link and M.E. Greene, Oncogene, 2, 387 (1988).
55. J.A. Drebin, V.C. Link and M.I. Green, Oncogene, 2, 273 (1988).
56. R. Bernards, A. Destree, S. McKenzie, E. Gordon, R.A. Weinberg and D. Panicali, Proc. Natl. Acad. Sci USA, 84, 6854 (1987).
57. D.J. Slamon, W. Godolphin, L.A. Jones, J.A. Holt, S.G. Wong, D.E. Keith, W.J. Levin, S.G. Stuart, J. Udove and A. Ullrich, Science, 244, 707 (1989).
58. I. Stancovski, E. Hurwitz, O. Leitner, A. Ullrich, Y. Yarden and M. Sela, Proc. Natl. Acad. Sci. USA, 88, 8691 (1991).
59. G.M. Clark and W.L. McGuire, Cancer Res., 51, 944 (1991).
60. M.C. Paterson, K.D. Dietrich, J. Danyluk, A.H. G. Paterson, A.W. Lees, N. Jamil, J. Hanson, H. Jenkins, B.E.D. Krause and W.A. McBlain, Cancer Res., 51, 556 (1991).
61. W.J. Gullick, S.B. Love, C. Wright, D.M. Barnes, B. Gusterson, A.L. Harris and D.G. Altman, Br. J. Cancer, 63, 434 (1991).
62. T.J. Perren, Br. J. Cancer, 63, 328 (1991).
63. J. Lundy, A. Schuss, E.S. McCormack, S. Kramer and J.M. Sorvillo, Am. J. Pathol., 138, 1527 (1991).
64. H. Fukazawa, P.M. Li, C. Yamamoto, S. Mizuno and Y. Uehara, Biochem. Pharm., 42, 1661 (1991).
65. F. Rilke, M.I. Colnaghi, N. Cascinelli, S. Andreola, M.T. Baldini, R. Bufalino, G. Dellaporta, S. Menard, M.A. Pierotti and A. Testori, International Journal of Cancer, 49, 44 (1991).
66. S.M. Oreilly, D.M. Barnes, R.S. Camplejohn, J. Bartkova, W.M. Gregory and M.A. Richards, Brit. J. Cancer, 63, 444 (1991).
67. J. Winstanley, T. Cooke, G.D. Murray, A. Platt-Higgins, W.D. George, S. Holt, M. Myskov, A. Spedding, B.R. Barraclough and P.S. Rudland, Br. J. Cancer, 63, 447 (1991).
68. A.H. Mccann, P.A. Dervan, M. Oregan, M.B. Codd, W.J. Gullick, B.M. J. Tobin and D.N. Carney, Cancer Res., 51, 3296 (1991).
69. M. Tateishi, T. Ishida, T. Mitsudomi, S. Kaneko and K. Sugimachi, Eur. J. Cancer, 27, 1372 (1991).
70. S. Jain, M.I. Filipe, W.J. Gullick, J. Linehan and R.W. Morris, Int. J. Cancer, 48, 668 (1991).
71. F.L. Tyson, C.M. Boyer, R. Kaufman, K. Obriant, G. Cram, J.R. Crews, J.T. Soper, L. Daly, W.C. Fowler and J.S. Haskill, Am. J. of Ob. and Gyn., 165, 640 (1991).
72. N.R. Lemoine, S. Jain, F. Silvestre, C. Lopes, C.M. Hughes, E. Mclelland, W.J. Gullick and M.I. Filipe, Brit. J. Cancer, 64, 79 (1991).
73. A. Riviere, J. Becker and T. Loning, Cancer, 67, 2142 (1991).
74. R. Herbst, R. Lammers, J. Schlessinger and A. Ullrich, J. Biol. Chem., 266, 19908 (1991).
75. M. Tateishi, T. Ishida, T. Mitsudomi, S. Kaneko and K. Sugimachi, Eur. J. Cancer, 27, 1372 (1991).
76. G. Powis, Trends Phar. Sci., 12, 188 (1991).
77. A. Levizki, Biochem. Pharm., 40, (5), 913 (1990).
78. T. Akiyama, J. Ishida, S. Nakagawa, H. Ogawara, S. Watanabe, N. Itoh, M. Shibuya, Y. Fukami, J. Biol. Chem., 262, 5592 (1987).
79. N.M. Dean, M. Kanemista, A.L. Boynton, Biochem. Biophys. Res. Comm., 165, 795 (1989).
80. J. Markovitz, C. Linassier, P. Fosse, J. Couprie, A. Jaquemin-Sablon, J. B.Pecq, A. K. Larson, Cancer Res., 49, 5111 (1989).
81. Y. Graziani, E. Erickson, R.L. Erickson, Europ. J. Biochem., 135, 583 (1983).
82. M. Hagiwara, S. Inoue, T. Tanaka, K. Nunoki, M. Ito, H. Hidaka, Biochem. Pharm., 37, 2987, (1988).
83. Y. Graziani, R. Chayoth, N. Karny, R. Feldman, J. Levy, Biochim. Biophys. Acta, 714, 415 (1981).
84. M. Cushman, D. Magarathnam, D.L. Burg, R.L. Geahlen, J. Med. Chem., 34, 798 (1991).
85. M. Imoto, H. Umezawa, K. Isshiki, S. Kunimoto, T. Sawa, T. Takeuchi, H. Umezawa, J. Antibiot. (Tokyo), 40, 1471 (1987).
86. H. Umezawa, M. Imoto, T. Sawa, K. Isshiki, N. Matsuda, T. Uchida, H. Iinuma, M. Hamada, T. Takeuchi, J. Antibiot., 39, 170 (1986).
87. M. Imoto, K. Umezawa, K. Komuro, T. Sawa, T. Takeuchi, H. Umezawa, Jpn. J. Cancer Res. (Gann), 78, 329 (1987).
88. M. Toi, H. Mukaida, T. Wada, N. Hirabayashi, T. Hori, K. Umezawa, Eur. J. Cancer, 26, (6), 722 (1990).
89. W.R. Bishop, J. Petrin, L. Wang, U. Ramesh, R.J. Doll, Biochem. Pharm., 40, (9), 2129 (1990).
90. C. Erneux, S. Cohen, D.L. Garbers, J. Biol. Chem., 258, 4137 (1983).
91. T.R. Burke, Org. Prep. Procedure Int. 23, 127 (1991).
92. K. Isshiki, M. Imoto, T. Sawa, K. Umezawa, H. Umezawa, J. Antibiot. (Tokyo), 40, 1209 (1987)
93. P. Yaish, A. Gazit, C. Gilon, A. Levitzki, Science, 242, 933 (1988).
94. A. Gazit, P. Yaish, C. Gilon, A. Levitzki, J. Med. Chem., 32, 2344 (1989).

95.   A. Gazit, N. Osherov, I. Posner, P. Yaish, C. Gilon, A. Levitzki, J. Med. Chem., 34, 1896 (1991).
96.   A. Levitzki, A. Gazit, N. Osherov, I. Posner, C. Gilon, Methods in Enzymology (Protein Phosphorylation, Pt B), 201, 347 (1991).
97.   A. Levitzki, C. Gilon, Trends in Pharm. Sci., 12, 172 (1991).
98.   R. Lyall, A. Zilberstein, A. Gazit, C. Gilon, A. Levitzki, J. Schlessinger, J. Biol. Chem., 264, 14503 (1989).
99.   A. Dvir, Y. Milner, O. Chomsky, C. Gilon, A. Gazit, A. Levitzki, J. Cell Biol., 113, 857 (1991).
100. G. Bilder, J. Krawiec, K. McVety, A. Gazit, C. Gilon, R. Lyall, A. Zilberstein, A. Levitzki, M. Perrone, A. Schreiber, Am. J. Physiol. (Cell Physiol. 29), C721 (1991).
101. D. Margolis, S.G. Rhee, S. Felder, M. Mervic, R. Lyall, A. Levitzki, A. Ullrich, A. Zilberstein, J. Schlessinger, Cell, 57, 1101 (1989).
102. I. Posner, A. Gazit, C. Gilon, A. Levitzki, FEBS Lett., 257, 287 (1989).
103. T. Yoneda, R.M. Lyall, M.M. Alsina, P.E.Persons, A.P.Spada, A. Levitzki, A. Zilberstein, G.R. Mundy, Cancer Res. 51, 4430 (1991).
104. R.L. Geahlen, J.L. McLaughlin, Biochem. Biophys. Res. Comm., 165, 241 (1989).
105. M. Cushman, D. Nagarathnam, D. Gopal, R.L. Geahlen, Bioorg. Med. Chem. Lett., 1, 215 (1991).
106. M. Cushman, D. Nagarathnam, D. Gopal, R.L. Geahlen, Bioorg. Med. Chem. Lett., 1, 211 (1991).
107. T. Shiraishi, T. Domoto, N. Imai, Y. Shimada, Biochem. Biophys. Res. Comm., 147, 322 (1987).
108. T. Shiraishi, K. Kameyama, N. Imai, T. Domoto, I. Katsumi, Chem. Pharm. Bull., 36, 974 (1988).
109. T. Shiraishi, M.K. Owada, M. Tatsuka, T. Yamashita, T. Kaunaga, Cancer Res., 49, 2374 (1989).
110. T. Shiraishi, M.K. Owada, M. Tatsuka, Y. Fuse, T. Kakunaga, Jpn. J. Cancer Res., 81, 645 (1990).
111. P.M. Traxler, O. Wacker, H. L. Bach, J.F. Geissler, W. Kump, T. Meyer, U. Regenass, J.L. Roesel, N. Lydon, J. Med. Chem., 34, 2328 (1991).
112. C.H. Kruse, G.H. Kenneth, M.L. Pritchard, J.A. Feild, D.J. Reimann, R.G. Greig, R.G. Poste, J. Med. Chem., 31, 1762 (1988).
113. I. Baginski, B. Tocque, G. Colson, A. Zerial, Biochem. Biophys. Res. Commun., 165, 1324 (1989).
114. T.R. Burke, Z.H. Li, J.B. Bolen, M. Chapekar, Y. Gang, R.I. Gazer, K.C. Rice, V.E. Marquez, Biochem. Pharm., 41, R17 (1991).
115. T.R. Burke, Z.H. Li, J.B. Bolen, V.E. Marquez, J. Med Chem., 34, 1577 (1991).
116. T.R. Burke, Z.H. Li, J.B. Bolen, V.E. Marquez, Bioorg. Med. Chem. Lett., 1, 165 (1991).
117. T. Onoda, K. Isshiki, T. Takeuchi, K. Umezawa, Drugs Exptl. Clin. Res., XVI (6), 249 (1990).
118. C.Y. J. Hsu, P.E. Persons, A.P. Spada, R.A. Bednar, A. Levitzki, A. Zilberstein, J. Biol. Chem., 266, 21105, (1991).
119. Y. Uehara, Y. Murakami, S. Mizuno, S. Kawa, Virology, 164, 294 (1988).
120. H. Fukazawa, S. Mizuno, Y. Uehara, Biochem. Biophys. Res. Comm., 173, 276 (1990).
121. Y. Uehara, H. Fukazawa, Y. Murakami, S. Mizuno, Biochem. Res. Comm., 163, 803 (1989).
122. H. Fukazawa, P.M. Li, C. Yamamoto, S. Mizuno, Y. Uehara, Biochem. Pharm. 42, 1661 (1991).
123. M. Nakata, T. Osumi, A. Ueno, T. Kinura, T. Tamai, K. Tatsuta, Tet. Lett., 32, 6015 (1991).
124. M.H. Krueter, R.E. Leake, F. Rinaldi, W.M. Klieser, A. Maidhof, W.E.G. Muller, H.C. Schroder, Comp. Biochem. Physiol., 97B, 151 (1990).
125. R.L. Geahlen, M.L. Harrison, in "Peptides and Protein Phosphorylation," B.E. Kemp, Ed., CRC Press, 1990, p. 239.
126. T.W. Wong, A.R. Goldberg, J. Biol. Chem., 259, 3127 (1984).
127. M. Cushman, P. Chinnasamy, A.K. Chakraborti, J. Jurayj, R.L. Geahlen, R.D. Haugwitz, Int. J. Peptide Protein Res. 36, 538 (1990).
128. C.J. Yuan, S. Jakes, S. Elliott, D.J. Graves, J. Biol. Chem., 265, 16205 (1990).
129. K.S. Petrakis, T.L. Nagabhushan, J. Am. Chem. Soc., 109, 2831 (1987).
130. M. Cushman, E.-S. Lee, Tet. Lett., 33, 1193 (1992).
131. D.A. Tinker, E.A. Krebs, I.C. Feltham, V.S. Anathanarayanan, J. Biol. Chem., 263, 5024 (1988).
132. O. Marin, A. Donella-Deana, A.M. Brunati, S. Fischer, L.A. Pinna, J. Biol. Chem., 266, 17798 (1991).
133. T. Hunter, J. Biol. Chem., 257, 4843 (1982).
134. B.E. Kemp, R.B. Pearson, Trends Biochem. Sci., 15, 342 (1990).
135. C.Y.J. Hsu, D.R. Hurwitz, M. Mervic, A. Zilberstein, J. Biol. Chem., 266, 603 (1991).
136. H. Cheng, C.M.E. Litwin, D.M. Hwang, J.H.Wang, J. Biol. Chem., 266, 17919 (1991).
137. C.A. Koch, D. Anderson, M.F. Moran, C. Ellis, T. Pawson, Science, 252, 668 (1991).
138. C.H. Heldin, Trends Biochem. Sci., 16, 450 (1991).
139. M. Anafi, A. Gazit, C. Gilon, Y.B.-Neriah, A. Levitzki, J. Biol. Chem., 267, 4518 (1992).

SECTION IV. IMMUNOLOGY, ENDOCRINOLOGY AND METABOLIC DISEASES

Editor: William F. Michne, Sterling Winthrop Pharmaceutical Research Division
Rensselaer, NY 12144

Chapter 19. Monoclonal Antibodies in Therapy

George A. Heavner
Centocor, Inc., 200 Great Valley Parkway, Malvern, PA 19355

Introduction - The development in 1975 of the technology to produce monoclonal antibodies was the genesis of a flood of new and versatile molecules (1). Monoclonals have been extensively used in all phases of biological research, as *in vitro* and *in vivo* diagnostic agents and as part of rational drug design. Although they have the potential to be useful as therapeutic agents for a wide variety of indications, only three have been approved for human use: OKT3 for reversal of rejection in renal transplantation, HA-1A™ for treatment of gram negative septic shock and Myoscint® for imaging of non-viable myocardial tissue.

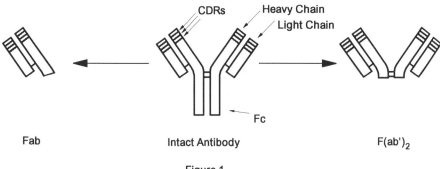

Figure 1

The structure and nomenclature of an antibody are shown in Figure 1. There are four different mechanisms of action for monoclonal antibodies. They can neutralize or block (antagonist) the action of endogenous molecules by binding to either receptors or ligands in such a way as to prevent the normal receptor-ligand interaction. They can function as agonists by binding to receptors and activating cells. They can clear materials from the system by coating and converting them into targets for phagocytic cells. They can kill cells directly with complement through ADCC (antibody dependent cellular cytotoxicity) or by acting as carriers of lethal agents such as high energy emitters or toxins.

The first monoclonal antibodies produced were murine. They are relatively easy to produce and constitute the majority of monoclonals generated. However, their immunogenicity makes them less attractive as *in vivo* therapeutic agents (2). Human monoclonals have been generated in order to overcome this immunogenicity. In addition to totally human monoclonals, chimeric antibodies containing the variable region of mouse antibodies with the constant region of human antibodies have been developed. These chimeric antibodies can still be immunogenic if the anti-antibody response is directed against the variable rather than the constant region. Immunogenicity of murine antibodies may be able to be reduced further by humanizing the antibody. Humanizing consists of transferring the CDRs (complementarity-determining regions that are responsible for the antibody binding) from a non-human monoclonal into a suitable human monoclonal (3,4). This approach is not straightforward since framework amino acids may influence conformation or be directly involved in binding. Since the exact mechanism of immunization is not known, there is the risk that the new "humanized" variable regions could be even more immunogenic than their murine counterparts. A new technique to decrease immunogenicity is veneering, which replaces the surface-exposed amino acids in a murine antibody with corresponding human residues, creating a pseodohuman surface on a murine antibody (5). Chimeric, humanized and human monoclonals generally are less immunogenic than murine antibodies and may have longer circulation times (6). In addition to the chimeric and humanized artificial constructs, bispecific antibodies have been designed where the variable binding region in each arm is specific for a different antigen. These artificial constructs are designed to bind to and bring into proximity different cells or receptors. A novel type of construct has been prepared by chemically linking the Fab portion of a mouse monoclonal antibody (OKT10, anti-CD38) to an Fc portion derived from human IgG to more effectively mediate ADCC with human effector cells (7).

Monoclonal antibodies are classified according to the antigen which they recognize. For example, antibodies which recognize an epitope on endotoxin are referred to anti-endotoxin antibodies while antibodies which recognize leukocyte cell surface antigens are referred to using the CD (cluster determinant) nomenclature (8). An important concept is that all monoclonal antibodies that recognize a specific molecule are not identical. They may vary not only in class and subtype but also in the specific site or epitope which they recognize. In general antibodies bind to epitopes 400-500 Å^2 in size, hence each molecule may have numerous epitopes. It is possible to have a number of monoclonal antibodies that bind to endotoxin but recognize totally distinct epitopes on the molecule. The clinical effectiveness of two such antibodies can be radically different although they both could be referred to as anti-endotoxin antibodies.

Since only a small portion of the antibody variable chains are involved in antigen recognition, antibody fragments can also be efficacious. Fragments being investigated include the bivalent $F(ab')_2$ and the monovalent Fab. In addition to these fragments, synthetic constructs containing the light and heavy variable regions (Fv fragments) and smaller peptide fragments (MRUs - minimum recognition units) containing CDRs and combinations have the potential for mimicking the activity of the parent monoclonal. The clinical effectiveness of an intact monoclonal versus a $F(ab')_2$ or other fragment may vary (9).

Monoclonal antibodies represent foreign proteins to the human immune system and the possibility exists of an immune response being generated against them. This can be pronounced in murine monoclonals where a strong HAMA (human anti-mouse antibody) response may be detected. Both antiidiotypic (recognizing the variable binding regions) and antiisotypic (recognizing the Fc chain) responses have been observed clinically. The precise nature of the response depends on the specific monoclonal, dose level and frequency of administration. The generation of an immune response to a monoclonal antibody can be either detrimental or beneficial (10). In cases where a strong immune response is generated, subsequent treatment with the antibody can result in boosting the immune response and rapid clearance. Anti-idiotypic antibodies (Ab_2) should possess similar molecular topology, but not necessarily sequence, to the antigen that is recognized by the monoclonal antibody (Ab_1). Anti-idiotypic antibodies could potentially be used to generate an immune response (anti-anti-idiotypic antibodies, Ab_3) and could serve as a means to vaccinate against the primary antigen.

Monoclonals are not without potential adverse reactions. Depending upon the antigen recognized, monoclonals can exert undesired effects such as T-cell activation or mediator release (11). Rapid clearance due to an immune response can be considered an adverse reaction if it affects efficacy.

Numerous articles and reviews have been published on monoclonal antibodies as potential therapeutic agents (12-14). The majority of these studies have been done using murine monoclonals in various animal models. Monoclonal antibodies can exhibit an extremely high species specificity. The murine monoclonal OKT3 recognizes the CD3 antigen only in human and selected primate species. Because of the high species specificity of monoclonal antibodies, animal models using monoclonals directed against cell surface antigens in animals can be used to demonstrate proof of concept, but unless those monoclonals can be shown to recognize comparable antigens within the human system they may have no clinical potential. As an example, a murine antiferritin monoclonal radioconjugate (QCI) showed tumor targeting in mouse but not in man (15) and the biodistribution of PK4S, an antibody against CEA, has a different pharmacokinetic profile in mice than it does in man (16). This review article will focus on monoclonal antibodies which have been investigated in clinical studies.

Transplantation - Rejection of grafts is a complex immunological process that can involve both cellular and humoral mechanisms (17). Numerous cell types are involved, including lymphocytes, plasma cells, monocytes and macrophages. An effective rejection response requires a competent immune system. For an effective cellular rejection, T-helper (CD4$^+$) cells appear to be required. Several reviews dealing with monoclonal antibodies used to treat or prevent graft rejection have recently been published (18-21). The majority of studies deal with OKT3, a murine monoclonal antibody, and the only one to be approved for human use for treating graft rejection in renal transplantation (22). OKT3 blocks T cell function by modulating the CD3-TCR complex. There are side reactions associated with the first dose that are related to cytokine release as the cells are activated as OKT3 binds. Both antiidiotypic and antiisotypic antibodies have been detected in OKT3 recipients. T lymphocyte activation has also been reported as might be expected with a powerful immunosuppressant (11). Increased infections have been noted with the conventional treatment protocol and increased incidence of lymphoproliferative disorders have been reported with prolonged use or reuse. Although directed against the CD3-TCR complex, OKT3 was successful in treating vascular (antibody mediated) as well as cellular rejection (23,24).

In addition to laboratory parameters such as decreased serum creatinine levels, indicating a functional renal graft, there are several end-point criteria for assessing the efficacy of graft rejection therapy. The simplest are the percent of surviving grafts and surviving patients. Less rigorous are the number and/or severity of early rejection episodes. OKT3 has been used in strategies both with and without concomitant cyclosporin treatment. In treating failed kidney grafts (rescue therapy), there was an increased graft survival when both drugs were used and a decreased HAMA response (25). Prophylactic treatment with OKT3 has also been evaluated (26). The protocol resulted in a 100% patient and graft survival at 18 months. Anti-OKT3 antibodies were detected in 50% of the patients. Despite the presence of both antiidiotypic and antiisotypic antibodies, a second course of therapy was beneficial in two heart transplant recipients.

As the only approved monoclonal antibody for treating graft rejection, OKT3 has become the standard against which to compare alternate treatments. The murine IgM T10B9.1A-31, directed against the $\alpha\beta$ T-cell receptor complex, was tested against OKT3 in a double blind trial in acute renal allograft rejection (27). Rejection was attenuated in all patients in both treatment groups. Initial side effects with T10B9.1A-31 were similar to OKT3 but milder and infectious complications and neurological side effects were less. HAMA responses were seen with both monoclonals. BMA031, another monoclonal directed against the T cell receptor, has been evaluated for treating renal rejection and shown to reverse steroid resistant cellular rejection but not vascular (28-29). Another murine anti-CD3 monoclonal (WT32) was used prophylactically in renal rejection (30). Although there was a delay in the onset time for first rejection, the overall incidence of rejection was no different than a non-antibody treated control group. Prophylactic treatment of kidney transplantation with anti-TAC (a monoclonal directed against the IL-2 receptor (IL-2R)) also gave delayed time to first rejection but no differences in patient or graft survival (31,32). IL-2R is an attractive target since it is only expressed on activated T cells. In an open study of five patients with renal allograft rejection, the murine anti-CD4 MAX.16H5 showed a rapid decrease in serum creatinine levels in all patients (33). No patient developed anti-murine antibodies. Campath-1G (recognizing CD52, an antigen present on all T- and B-lymphocytes and killing by ADCC) was used in 15 patients in an open study to reverse renal rejection (34). All reversal episodes were satisfactorily resolved. In contrast to OKT3, the immunosuppressive effect of Campath-1G remained for up to three weeks.

OKT3 has been used both for rescue therapy and prophylactically in heart transplantation. In prophylactic studies, OKT3 delayed the first rejection episodes but did not affect the frequency or severity of rejection or overall survival over conventional triple immunosuppressive therapy (prednisolone, azathioprine, cyclosporin) (35-37). Prophylactic trials have been conducted comparing OKT3 to various antithymocyte globulins. RATG (rabbit antithymocyte globulin) showed no differences in the overall number of rejection episodes compared to OKT3 (38). Interestingly, none of the OKT3 patients in this study developed anti-OKT3 antibodies. T lymphocyte recovery was slower in the RATG group, presumably because of the recognition of epitopes other than CD3. Antiidiotypic responses were seen with both OKT3 and ATS ,a polyclonal rabbit antithymocyte serum (39). Minnesota antilymphocyte globulin (MATG) was evaluated against OKT3 for rescue therapy in cardiac rejection with similar results to RATG (40). One individual has undergone five serial 14-day courses of OKT3 therapy in the course of two heart transplants (41). In each instance rejection was reversed with a decrease in $CD3^+$ cell levels. Low anti-OKT3 titers were only detected once, after the third course of therapy.

Liver graft rejections treated with OKT3 parallel the results obtained for kidney and heart transplants. Preexisting anti-OKT3 antibodies to not preclude successful retreatment (42). There is evidence in at least one clinical study that an anti-CD3-resistant subpopulation of T cells can emerge during prophylactic OKT3 therapy and may cause rejection (43). If so, then monitoring $CD3^+$ cells in circulation may not be an accurate predictor of events in the graft. As observed in treating other graft rejections, there appears to be an increased risk of viral infections in liver transplantation with OKT3 (44). In prophylactic studies in liver grafts, OKT3 showed no differences in graft or patient survival (45). The murine monoclonal BT 563 (CD25, IL-2R) was used prophylactically in an open study in liver transplantation (46). All patients developed anti-murine antibodies within ten days with corresponding drops in BT 563 levels. Another anti-IL-2R antibody, rat YTH-906, when used prophylactically, showed no differences compated to conventional triple immunotherapy (47). An immune response against the antibody was detected in 43% of the patients who received it.

Graft versus Host Disease - Graft versus host disease (GVHD) occurs when donor T lymphocytes become activated by recipient antigens and mount an immune response. During activation a variety of cell surface molecules not found on resting cells are expressed. GVHD may be treated at three stages: prophylactic treatment of the donor to deplete lymphocytes prior to graft removal, *ex vivo* treatment of the graft to remove or inactivate lymphocytes or, treatment (directed against lymphocytes, activation antigens or cytokines) of the recipient after engraftment. Prophylactic treatment of the donor has been demonstrated in animal systems (48).

Ex vivo lymphocyte depletion is more readily accomplished with bone marrow than in other organs and can be performed by several techniques. In one study, two different *ex vivo* T cell depletion techniques were compared (E-rosetting versus the monoclonal Campath IM with complement). Human T cells have a receptor for sheep erythrocytes which provides an easy method for their removal from bone marrow preparations. All bone marrow recipients were also treated with the murine anti-CD11a (LFA-1) antibody 25.3. The overall sustained engraftment rate was 72% versus a 26% historical control (49); however, engraftment with Campath IM depletion was lower that that obtained with E-rosetting. The same antibody used with non-T cell depleted bone marrow in 10 patients with acute GVHD showed partial, but transient and incomplete, responses in 8 patients (50). When bone marrow T cell depletion (measured by $CD5^+$ cells) was done using either murine anti-Leu1 plus complement or a ricin A-chain murine $F(ab')_2$ conjugate (ST1-IT), no difference in graft survival was observed (51). The anti-CD15 murine monoclonals Vln-D5, 8.27 and SMY15A were all equally effective when used with complement in *ex vivo* depletion of leukemic cells for autologous bone marrow transplantation (52). The same antibodies were also used without complement on magnetic beads, although cellular depletion was less effective. In addition to CD11a, several other antigens have been the target for treatment of GVHD in the recipient. In one patient a CD6 monoclonal (anti-T12) showed improvement in clincal parameters (53). Antibodies against the IL-2 receptor (IL-2R, CD25) have also been used to treat GVHD. Murine 2A3, specific for the 55 kD chain was used prophylactically in grafts involving unmodified bone marrow (54). Although circulating $CD25^+$ cells were saturated with antibody, the incidence of GVHD and survival was similar to controls. The murine CD25 antibody B-B10 showed increased survival when administered therapeutically to treat GVHD as did the murine anti-TNF-α antibody B-C7 (55).

Septic Shock - The principal cause of clinical manifestations of gram negative septic shock is the release of endotoxin (bacterial membrane lipopolysaccharides) during systemic infections with gram-negative bacteria. This initiates a cascade of events that can ultimately lead to adult respiratory distress syndrome (ARDS), acute renal failure (ARF), disseminated intravascular coagulation (DIC) and death. Principal cytokines in the septic shock cascade include tumor necrosis factor (TNF) and interleukin-1 (IL-1) (56,57). These molecules offer attractive targets for monoclonal antibodies to interrupt gram negative bacterial sepsis at the early stages in the septic shock cascade (58,59). The human IgM monoclonal antibody HA-1A, used for the treatment of gram-negative sepsis, binds selectively an epitope on the lipid A domain of endotoxin (60). In a double blind study, a single 100 mg dose of HA-1A reduced mortality by 39% in all patients with confirmed gram negative bacteremia (61). No patient had detectable anti-HA-1A antibodies. No anti-HA-1A antibodies were detected at doses up to 250 mg (62). A murine IgM monoclonal designated E5, also against an epitope of lipid A, was also evaluated in a double blind trial in patients with gram-negative sepsis (63). Two infusions of 2 mg/kg were given 24 hours apart. The difference in survival was not statistically significant between all E5 and placebo treated patients. There was an increase in survival over placebo in the subset of patients with gram-negative sepsis who were not in shock at study entry. An IgG anti-murine antibody response was detected in 47% of patients treated with E5. This confirms an earlier study with E5 where an immune response was seen at 2 mg/kg but not at 0.1 or 0.5 mg/kg (64).

Monoclonals against TNF-α have also been evaluated. In an open study CB0006, a murine IgG against recombinant human TNF-α was administered as a single dose infusion 0.4 or 2 mg/kg to 14 patients with gram negative septic shock (65). The HAMA response to CB0006 resembles that seen with E5.

Cancer - The majority of monoclonal antibodies in clinical study are directed against cancers. A number of reviews have appeared on the use of monoclonal therapy for cancer (66-73). Anticancer monoclonal therapy can be divided into four general strategies: 1) destruction of tumor cells via immune system mediation; 2) interdiction of tumor cell growth or differentiation; 3) delivery of antitumor agents to tumor cells: and 4) generation of an anti-antiidiotypic response. Clincal results with monoclonals that bind to tumor cells to enhance destruction by the immune system have yielded only limited results. A murine antibody (CLB-CD19) against CD19, an antigen present on the surface of both normal cells and maglinant B lymphocytes, was used in patients with progressive B cell non-Hodgkin's lymphoma at doses up to 1 g (74). Although the level of circulating tumor cells was temporarily reduced, there were no long term clinical effects. In a controlled trial involving 61 patients with resectable pancreatic cancer, the murine monoclonal 494/32, derived from mice immunized with the De-TA carcinoma cell line and recognizing a carbohydrate epitope, was used following surgery (75). No significant differences between treatment and control groups were observed. 17-1A, a murine monoclonal developed against the colorectal carcinoma cell line SW1038, was used with γ-INF to boost the immune response in patients with both pancreatic and metastatic colorectal adenocarcinoma (76,77). Although an improvement in natural cytotoxic activity was observed, there was no clinical efficacy at doses up to 1.6 g. γ-INF was also used with a mixture of four murine monoclonals against tumor cells (CO17-1A, CA19-9, BR55-2 and GA73-5) in patients with metastatic gastrointestinal adeoncarcinoma (78). Increased NK lytic activity and ADCC did not correlate with a clinical response.

Mycosis fungoides (MF) is a T-cell lymphoma that is composed primarily of $CD4^+$ cells. A chimeric anti-CD4 antibody was used to treat seven patients with MF at doses of from 10-80 mg twice a week for three weeks (79). No significant depletion of $CD4^+$or decreased T-cell proliferative responses were observed but there was suppression in mixed lymphocyte reactions. All patients had some transient clinical improvement lasting up to 12 weeks.

A chemically cross-linked, bispecific $F(ab')_2$ antibody fragment (ZCE/CHA) directed against carcino-embryonic antigen and an epitope on BLEDTA IV and EOTUBE (^{111}In chelates), gave a suppression of lymphocyte proliferation to mitogenic stimulation in patients with metastatic tumors; no clinical correlation was made (80). A monoclonal (Gd-2a) directed against Gd-2, disialoganglioside structures expressed in high concentration on neuroblastoma cells, has been used as an ^{131}I carrier for both imaging and therapy (81). A significant HAMA response was seen in all patients with no assessment of clinical efficacy.

A number of monoclonal antibodies have been used clinically in an attempt to initiate an idiotypic cascade. An antiidiotypic antibody (Ab_2) can mimic the topology of the initial tumor antigen. An immune response (Ab_3) mounted against Ab_2 can theoretically recognize the tumor antigen. Of 19 patients treated with MF11-30, a murine antiidiotypic antibody bearing the internal image of human high-molecular-weight-melanoma-associated antigen (HMW-MAA) and raised against anti-HMW-MAA 225.28, 16 developed antibodies that inhibited the binding of MF11-30 to 225.28 (82). Mixed clincal responses were observed. The $F(ab)_2$ fragment of OC125, recognizing the tumor associated antigen (TAA) CA125 of epithelial ovarian cancinoma, was capable of inducing an antiidiotypic network and a delayed clinical progression of ovarian cancer (83). A pool of 10 antiidiotypic antibodies raised against 17-1A in a patient treated with 17-1A was injected into other patients that have been previously treated with 17-1A, leading to the induction of anti-antiidiotypic antibodies (84). A chimeric 17-1A has been prepared which generated an immune response in only one of 16 patients (85). One patient with B cell chronic lymphocytic leukemia was treated with an antiidiotypic antibody and showed significant reduction in tumor size but not remission (86).

Radioimmunoconjugates using a variety of isotopes (including ^{131}I, ^{90}Y, ^{186}Re, ^{188}Re, ^{211}At and ^{212}Bi) have been used clinically to treat several different malignancies. The design and clinical use of these agents has been the subject of several reviews (87-90). Because of the variety of radionucleides and variation in doses given, it is difficult to compare the results of different clinical trials. The monoclonal HMFG1m developed against human milk fat globule and binding to a high molecular weight protein present on a variety of normal and neoplastic derivatives of glandular epithelium, was labeled with ^{131}I and used to treat seven patients with carcinomatous meningitis (91). Doses were from 30-60 mCi. Two patients showed positive clinical responses; however, a number of serious side effects were seen, including aseptic meningitis, seizures and myelosuppression. An ^{131}I labeled mixture of four monoclonals, anti-CEA FO23C5 and BW494/32 and anti-TAG B72.3 and AUA1, was used at isotope doses of 21-150 mCi to treat 18 patients with advanced and/or metastatic colorectal carcinoma. Out of 15 evaluable patients, two complete and two partial remissions were seen. Even in the presence of immunosuppressive therapy, all patients developed HAMA responses which altered the biodistribution of subsequent doses. The intact antibody or $F(ab')_2$ of LL2, a monoclonal having a high specificity for B cell lymphomas and leukemias was ^{131}I labeled and 5-58 mCi given to 16 patients with non-Hodgkin's lymphoma (92). The half life in blood was not significantly different between the IgG and $F(ab')_2$ Three out of eight patients having at least two doses developed a HAMA response after the second dose, but no change was seen in pharmacokinetics on repeated injections. In a series of five assessable therapy patients, there were two partial remissions, two mixed and one minor response.

Thrombosis - Arterial reocclusion can occur with current thrombolytic therapy using streptokinase, urokinase, t-PA (tissue plasminogen activator) or PTCA (percutaneous transluminal coronary angioplasty) (93). Clot formation is a result of the expression of competent GPIIb/IIIa fibrinogen receptors on platelets that become activated and subsequent receptor-mediated platelet aggregation. Demographic studies have shown a direct correlation between the number of competent GPIIb/IIIa receptors and both a decrease in platelet aggregation and the abolition of thrombus formation (94). The pharmacokinetics of the $F(ab')_2$ fragment of the blocking murine monoclonal 7E3 directed against GPIIb/IIIa was evaluated in 16 patients with unstable angina (95). A significant decrease in both free receptors and platelet aggregation was observed. Low titer HAMA responses were observed in two patients and one patient had an antibody response to 7E3- $F(ab')_2$ not inhibited by mouse IgG; however, no anamnestic responses were observed on repeated dosing. Patients treated with the murine 7E3-Fab fragment also showed decreased platelet aggregation with 25% developing a HAMA response (96). The chimeric 7E3-Fab fragment also showed comparable results in decreasing platelet aggregation when administered to patients prior to PTCA (97); all PTCA procedures were successful and no thrombocytopenia was observed. Chimeric 7E3-Fab was used at 0.1-0.25 mg/kg in 2 patients with myocardial infarction receiving t-PA (98).

Autoimmune Diseases - The major indication under clinical investigation with monoclonal antibodies in the area of autoimmune disorders is rheumatoid arthritis. Several reviews have addressed the use of monoclonal antibodies in the treatment of autoimmune disease and have identified specific target antigens (99,100). An opportunistic target is the CD4 antigen, an integral membrane glycoprotein on T helper cells that appears to act as a co-receptor with the antigen-T-cell-receptor complex and, as such, may be integrally involved with the pathogenesis of rheumatoid arthritis. Several anti-CD4 monoclonals have been used to treat rheumatoid arthritis in open studies. The murine monoclonal M-T151 was used at a dose of 10 mg/day for seven days to treat patients with active rheumatoid arthritis (101). There was a rapid depletion of $CD4^+$ cells that returned to normal levels within 24 hours after infusion. Clinical improvement was observed that lasted from four weeks to six months. Several patients developed a HAMA response. Retreatment of four patients after six months gave only transient clinical improvement lasting from three weeks to three months and a further increase in the HAMA titers in two of the patients. Patients treated with B-F5, a murine anti-CD4 antibody, showed clinical improvement, lasting from 1-12 months (102) A third murine anti-CD4 antibody, MAX.16H5, was given to patients with severe intractable rheumatoid arthritis (103). There was a marked depletion of $CD4^+$ cells during treatment that did not return to normal levels for two months. Low titer HAMA responses were detected in six of ten patients but this response did not reduce the efficacy of $CD4^+$ cell depletion. A 2-3 fold increase in HAMA levels was seen in four patients who underwent a second course of therapy. A murine anti-CD5 antibody conjugated to ricin A chain (CD5 Plus) was been used to treat 76 patients with rheumatoid arthritis (104). Clinical improvement, lasting up to 12 months was observed in some patients. An immune response to CD5 Plus was seen in 75/76 patients. Although an anamnestic response was seen in patients that were retreated, CD5 cell levels decreased and no change in serum half-life was observed.

One patient with relapsing polychondritis who did not respond to conventional immunosuppressive therapy was treated with the murine anti-CD4 monoclonal MT-412 (105). $CD4^+$ cells were rapidly depleted but returned to baseline levels within 24 hours. Treatment led to disease regression for six months. Although the patient had a positive HAMA response, retreatment gave repeated clinical improvement. The anti-CD4 chimeric M-T412 (cM-T412) was used to treat severe generalized pustular psoriasis that had developed following oral administration of steroids (106). Clinical improvement correlated with some laboratory parameters.

Other Indications - Although generally administered parenterally for systemic effect, four monoclonals (Guy's 1, 9, 11 and 13) specific for the mutans group of streptococci were used topically in the mouth to prevent recolonization by Strep. mutans (107). Treated subjects remained free of Strep. mutans for up to two years while control subjects started recolonization within days.

Imaging - Monoclonal antibodies can be used clinically to carry toxins or high energy emitters for therapeutic cellular deletion. This same target specificity makes them ideal imaging agents as a diagnostic adjunct to therapy. Myoscint, an ^{111}In labeled antimyosin antibody that detects myocyte necrosis associated with myocarditis, was used in cardiac transplant patients to assess myocardial damage (108). Decreased antibody uptake was associated with an uneventful clinical course, whereas persistent uptake correlated with rejection-related complications. ^{99m}Tc labeled Fab′ of T2G1s and ^{111}In Fab 59D8, both of which recognize the N-terminal domain of fibrin that is only available on newly formed acute thrombi, have been used successfully to rapidly image deep venous thrombosis (109). The ^{99m}Tc labeled Fab fragment of C22A, against the fibrin monomer beta chain, was evaluated against contrast venography to image deep venous thrombi with an imaging sensitivity of 85% (110). No negative contrast venography had a positive antifibrin antibody scan. The humanized monoclonal Hu2PLAP recognizing human placental alkaline phosphatase and reacting with a wide variety of neoplasms, was radiolabeled with ^{111}In through the coupled chelator DOTA (111). Successful tumor imaging was seen in four of seven patients. No patient developed antibodies against Hu2PLAP but three of six patients tested showed an immune response against DOTA.

Conclusions - Monoclonal antibodies will occupy an increasingly important position in the physician's armamentarium. Small scale clinical trials with monoclonal antibodies may be highly variable since recognition of the desired antigen is not sufficient in itself for successful therapy. Where successive clinical studies have been done with the same antibody, the results are not always consistent. This may be due to the combination of patient response variability and small patient numbers or other less readily characterizable factors. For those antibodies where large multicenter trials have been done, OKT3 and HA-1A, clinical efficacy is clear. Murine antibodies can generate HAMA responses, even in profoundly immunosuppressed patients; although in clinical studies with small patient populations, the percent HAMA responses can be highly variable, even with the same antibody. Where no HAMA response has been detected, lack of immunogenicity should be considered tentative, pending confirmation in large scale trials. Even where a HAMA response occurs, its effect on subsequent doses is not always predictable. In those cases such as 7E3 where a murine and chimeric have been clinically tested, the chimeric is less immunogenic. In general, monoclonal antibodies, unlike conventional drugs, are safe, can be targeted to specific epitopes and are relatively easy to produce. Monoclonal antibody therapy will play a large role in clinical diagnosis and therapy in the future.

References

1. G. Kohler and C. Milstein, Nature, 256, 495 (1975).
2. M.S. Co and C. Queen, Nature, 351, 501 (1991).
3. P.R. Tempest, P. Bremmer, M. Lambert, G. Taylor, J.M. Furze, F.J. Carr and W.J. Harris, Biotechnology, 9, 266, (1991).
4. L. Riechmann, M. Clark, H. Waldmann and G. Winter, Nature, 332 24 (1988).
5.. J.A. DeMartino, B.L. Daugherty, M-F. Law, G.C. Cuca, K. Alves, M. Silberklang and G.E. Mark, Abstracts of the Second Annual IBC International Conference on Antibody Engineering, 1991.
6. A.F. LoBuglio, R.H. Wheeler, J. Trang, A. Haynes, K. Rogers, E.B. Harvey, L. Sun, J. Ghrayeb and M.B. Khazeli, Proc.Natl.Acad.Sci.USA, 86, 4220 (1989).
7. F.K. Stevenson, A.J. Bell, R. Cusack, T.J. Hamblin, C.J. Slade, M.D. Spellerberg and G.T. Stevenson, Blood, 77, 1071 (1991).
8. W.N. Erber, Pathology, 22, 61, (1990).
9. F. Buchegger, A. Pèlegrin, B. Delaloye, A. Bischof-Delaloye and J-.P. Mach, J.Nucl.Med., 31, 1035 (1990).
10. A. Nisonoff, J.Immunol., 147, 2429 (1991).
11 J.D.I. Ellenhorn, E.S. Woodle, J.R. Thistlethwaite and J.A. Buestone, Cur.Surg., 47, 458 (1990).
12. "Therapeutic Monoclonal Antibodies," C.A.K. Borrebaeck and J.W. Larrick, Ed., Stockton Press, New York, N.Y., 1990.
13. T.A. Waldman, Science, 252, 1657 (1991).
14. H. Chmel, Comprehensive Therapy, 16, 12 (1990).
15. H.W. Vriesendorp, J.M. Herpst, M.A. Gernack, J.L. Klein, P.K. Leichner, D.M. Loudenslager and S.E. Order, J.Clin.Oncol. 9, 918 (1991).
16. R.H.J. Begent and R.B. Pedley, Cancer Treat.Rev, 17, 373 (1990).
17. I.V. Hutchinson, Current Opinions in Immunology, 3, 722 (1991).
18. L. Chatenoud, C. Ferran and J.-F. Bach, Adv. Nephrol Necker Hosp., 20, 281 (1991).
19. P.J. Morris, Transplant.Proc.,23, 2133 (1991).
20. J. Dantal and J.-P. Soulillou, Current Opinions in Immunology, 3, 740 (1991).
21. P.J. Friend, Current Opinions in Immunology, 3, 859 (1991).
22. D.J. Norman and M.R. Leone, Pediatric Nephrol., 5, 130 (1991).
23. T.J. Schroeder, M.A. Weiss, R.D. Smith, G.W. Stephens, M. Carey and M.R. First, Transplant.Proc., 23, 1043 (1991).
24. T.J. Schroeder, M.A Weiss, R.D. Smith, G.W. Stephens and M.R. First, Transplantation, 51, 312 (1991).
25. J.A. Schulak, J.T. Mayes and D.E. Hricik, Transplant.Proc., 23, 2119 (1991).
26. W. Weimar, C.J. Hesse, L.M.B. Vaessen, G.F.J. Hendriks, M.H.P.M. Jutte and J. Jeekel, Biotherapy, 2, 267 (1990).
27. T.H. Waid, B.A. Lucas, J.S. THompson, L.C. Munch, S. Brown, R. Kryscio, R. Prebeck, M.A. VanHoy and D. Jezek, Transplant.Proc., 23, 1062 (1991).
28. P.F. Pfeffer, A. Jakobsen, D. Albrechtsen, G. Sødal, I. Brekke, Ø. Bentdal, T. Leivestad, P. Fauchald and A. Flatmark, Transplant.Proc., 23, 1099 (1991).
29. G. Hillebrand, U. Dendorfer, H.E. Feucht, C. Brockmeyer, J.M. Gokel, W.D. Illner, D. Abendroth, C. Kasper and W. Land, Transplant.Proc., 23, 1092 (1991).
30. L.A..M. Frenken, A.J. Hoitsma, W.J.M Tax and R.A.P. Koene, Transplant.Proc., 23, 1072 (1991).
31. R.L. Kirkman, M.E. Shapiro, C.B. Carpenter, D.B. McKay, E.L. Milford, E.L. Ramos, N.L. Tilney, T.A. Waldmann, C.E. Zimmerman and T.B. Strom, Transplantation, 51, 107 (1991).
32. R.L. Kirkman, M.E. Shapiro, C.B. Carpenter, D.B. McKay, E.L. Milford, E.L. Ramos, N.L. Tilney, T.A. Waldmann, C.E. Zimmerman and T.B. Strom, Transplant.Proc., 23, 1066 (1991).
33. P. Reinke, H. Miller, E. Fietze, D. Herberger, H.-D. Volk, K. Neuhaus, J. Herberger, R. v. Baehr and F. Emmrich, Lancet, 338, 702 (1991).
34. P.J. Friend, H. Waldkann, G. Hale, S. Cobbold, P. Rebello, S. Thiru, N.V. Jamieson, P.S. Johnston and R.Y. Caine, Transplant.Proc., 23, 2253 (1991).
35. M.L. Barr, J.A. Sanchez, L.A. Seche, L.L. Schulman, C.R. Smith and E.A. Rose, Circulation, 82 (suppl. IV), IV-291 (1990).

36. A.H.M.M. Balk, M.L. Simoons, N.H.P.M. Jutte, M.L. Brouwer, K. Meeter, B. Mochtar and W. Weimar, Clinical Transplantation, 5, 301 (1991).
37. L.A. Pulpón, P. Dominguez, M. Cháfer, J. Segovia, M.G. Crespo, G. Pradas, R. Burgos and D. Figuera, Transplant.Proc., 22, 2319 (1990).
38. G. Ippoliti, M. Negri, P. Abelli, B. Rovati, L. Di Franco, P. Grossi, L. Minoli, E. Ascari and M. Viganó, Transplant.Proc., 23, 2272 (1991).
39. W.H. Frist, W.H. Merrill, T.E. Eastburn, J.B Atkinson, J.R. Stewart, J.W. Hamon, Jr. and H.W. Bender, Jr., J.Heart Transplant., 9, 489 (1990).
40. G.M. Deeb, S.F. Bolling, C.N. Steimle, J.E. Dawe, A.L. McKay and A.M. Richardson, Transplantation, 1, 180 (1991).
41. W.H. Frist, E.B. Gerhardt, W.H. Merrill, J.B. Atkinson, T.E. Eastburn, J.R. Stewart, J.W. Hammon, Jr. and Harvey W. Bender, Jr., J.Heart Transplant., 9, 724 (1990).
42. E.S. Woodle, J.R. Thistlethwaite, Jr., J.C. Emond, P.F. Whitington, D.D. Black, P.P. Aran, A.L. Baker, F.P. Stuart and C.E. Broelsch, Transplantation, 51, 1207 (1991).
43. F.R. Sutherland, M. Aboujaoude, M.J. White, J. Yamada, C. Ghent, D. Grant, W. Wall, B. Garcia, R. Mazaheri and A.I. Lazarovits, Clinical Immunol.Immunopathol., 60, 40 (1991).
44. J.S. Bowman, M. Green, V.P. Scantlebury, S. Todo, A. Tzakis, S. Iwatsuki, L. Douglas and T.E. Starzl, Clinical Transplant., 5, 294 (1991).
45. S.V. McDiarmid, M.J. Millis, P.I Terasaki, M.E. Ament and R.W. Busuttil, Dig.Dis. Sci., 36, 1418 (1991).
46. G. Otto, J. Thies, D. Kabelitz, W.J. Hofmann, Ch. Herfarth and S. Meuer, Transplant.Proc., 23, 1387 (1991).
47. P.J. Friend, H. Waldmann, S. Cobbold, H. Tighe, P. Rebello, D. Wight, W. Gore, S. Pollard, S. Lim, P. Johnston, R. Williams, J. O'Grady and R.Y. Caine, Transplant.Proc., 23, 1390 (1991).
48. D. Shaffer, M.A. Simpson, E.L. Milford, R. Gottschalk, J.P. Kut, T. Maki and A.P. Monaco, Transplant.Proc., 23, 679 (1991).
49. A. Fischer, W. Friedrich, A. Fasth, S. Blanche, F. Le Deist, D. Girault, F. Veber, J. Vossen, M. Lopez, C. Griscelli and M. Hirn, Blood, 77, 249 (1991).
50. A.M. Stoppa, D. Maraninchi, D. Blaise, P. Viens, M. Hirn, D. Olive, J. Reiffers, N. Milpied, M.H. Gaspard and C. Mawas, Transplant.Int., 4, 3 (1991).
51. J.H. Antin, B.E. Bierer, B.R. Smith, J.Ferrara. E.C. Guinan, C. Sieff, D.E. Golan, R.M. Macklis, N.J. Tarbell, E.Lynch, T.A. Reichert, H. Blythman, C. Bouloux, J.M. Rappeport, S.J. Burakoff and H.J. Weinstein, Blood, 78, 2139 (1991).
52. D. Guyotat, Z.H. Shi, L. Campos, M. Rabat, S. Bonnier, P. Poncelet, J.C. Laurent and D. Fiere, Bone Marrow Transplantation, 6, 385 (1990).
53. M.D. Blomquist, M. Boggards, I.C. Guerra Hanson, H.M. Rosenblatt, M.S. Pollack, E. Hawkins, J. Ritz and W.T. Shearer, J.Allergy Clin.Immunol., 87, 1029 (1991).
54. C. Anasetti, P.J. Martin, R. Storb, F.R. Appelbaum, P.G. Beatty, E. Calori, J. Davis, K. Doney. T. Reichert, P. Stewart, K.M. Sullivan, E.D. Thomas, R.P. Witherspoon and J.A. Hansen, Bone Marrow Transplant., 7, 375 (1991).
55. P. Herve, P. Bordigoni, J.Y Cahn, M. Flesch, P. Tiberghien, E. Racadot, N. Milpied, J.P. Bergerat, E. Plouvier, C. Roche and J. Wijdenes, Transplant.Proc., 23, 1692 (1991).
56. K.J. Tracey, Circ.Shock, 35, 123 (1991).
57. R. C. Bone, Crit.Care, 100, 802 (1991).
58. J.D. Baumgartner, Eur.J.Clin.Microbiol.Infect.Dis., 9, 711 (1990).
59. D.L. Dunn, J.Trauma, 30, S100 (1990).
60. E.J. Ziegler, C.J. Fisher, Jr., C.L. Sprung, C.R. Smith, R.C. Straub, J.C. Sadoff, R.P. Dellinger, N.Engl.J.Med., 325, 279 (1991).
61. E.J. Ziegler, C.J. Fisher, Jr., C.L. Sprung, R.C. Straube, J.C. Sadoff, G.E. Foulke, C.H. Wortel, M.P. Fink., R.P. Dellinger, N.N.H. Teng, I.E. Allen, H.J. Berger, G.L. Knatterud, A.F. LoBuglio, C.R. Smith, N.Engl.J.Med., 324, 429 (1991).
62. C.J. Fisher, Jr., J. Zimmerman, M.B. Khazaeli, T.E. Albertson, R.P. Dellinger, E.A. Panacek, G.E. Foulke, C. Dating, C.R. Smith, A.F. LoBuglio, Crit.Care Med., 18, 1311 (1990).
63. R.L. Greenman, R.M.H. Schein, M.A. Martin, R.P. Wenzel, N.R. MacIntyre, G. Emmanuel, R.B. Kohler, M. McCarthy, J. Plouffe, J.A. Russell, J.Am.Med.Assoc., 266, 1097 (1991).
64. L.S. Young, R. Gascon, S. Alam, L.E.M. Bermudez, Rev.Infect.Dis., II 7, S1564 (1989).
65. A.R. Exley, J. Cohen, W. Buurman, R. Owen, G. Hanson, J. Lumley, J.M. Aulakh, M. Bodmer, A. Riddell, S. Stephens, M. Perry, Lancet, 335, 1274 (1990).
66. A.S. Freeman, A. Pedrazzini and L.M. Madler, Cancer Investigation 9, 69 (1991).
67. R. Levy and R.A. Miller, J.Natl.Cancer Inst.Monogr., 10, 61 (1990).
68. H. Mellstedt, J.E. Frodin and G. Masucci, Oncology, 3, 25 (1989).
69. R. Stein and D.M. Goldenberg, Chest, 99, 1466 (1991).
70. J.J. DiMaggio, D.A. Scheinberg and A.N. Houghton in "Cancer Chemotherapy and Biological Response Modifiers Annual 11", H.M. Pinedo, B.A Chabner and D.L.Longo Eds., Elsevier Science Publishers B.V., 1990, p. 177.
71. R.W. Baldwin, V.S. Byers and M.V Pimm in "Cancer Chemotherapy and Biological Response Modifiers Annual 10", H.M. Pinedo, D.L.Longo and B.A Chabner Eds., Elsevier Science Publishers B.V., 1988, p. 397.
72. S. Ferrone, Current Opinions in Oncology, 2, 1146 (1990).
73. H. Mellstedt, J.-E. Frödin, G.Masucci, P. Ragnhammer, J. Fagerberg, A.-L. Hjelm, J. Shetye, P. Wersäll and A. Österborg, Seminars in Oncology, 18, 462 (1991).
74. A. Hekman, A. Honselaar, W.M. Vuist, J.J. Sein, S. Rodenhuis, W.W. ten Bokkel Huinink, R. Somers, P. Rumke and C.J Melief, Cancer Immunol.Immunother., 32, 364 (1991).

75. M. Böchler, H. Friess, K.-H. Schultheiss, C. Gebhardt, R. Köbel, K.-H. Muhrer, M. Winkelmann, T. Wagener, R. Klapdor, M. Kaul, Günther Möler, G. Schulz and H.G. Beger, Cancer, $\underline{68}$, 1505 (1991).

76. M.A. Tempero, C. Sivinski, Z. Steplewski, E. Harvey, L. Lkassen and H.D. Kay, J. Clinical Oncology, $\underline{8}$, 2019 (1990).

77. M.N. Saleh, A.F. LoBuglio, R.H. Wheeler, K.J. Rogers, A. Haynes, J.Y Lee and M.B. Khazaeli, Cancer Immunol.Immunother., $\underline{32}$, 185 (1990).

78. H.M Blottiere, J.Y. Douillard, H. Koprowski and Z. Steplewski, Cancer Immunol.Immunother., $\underline{32}$, 29 (1990).

79. S.J. Knox, R. Levy, S. Hodgkinson, R. Bell, S. Brown, G.S. Wood, R. Hoppe, E.A. Abel, L. Steinman, R.G. Berger, C. Gaiser, G Young,, J. Blndl, A. Hanman and T. Reichert, Blood, $\underline{77}$, 20 (1991).

80. D.S. Gridley and D.R. Stickney, Clin.Exp.Immunol., $\underline{84}$, 289 (1991).

81. P. Reuland, R. Handgretinger, H. Smykowsky, R. Dopfer, T. Lkingebiel, B.M. Miller, R.A. Reisfeld, S. Gallagher, E. Koscelniak, J. Treuner, D. Niethammer and U. Feine, Int.J.Rad.Appl.Instrum.[B], $\underline{18}$, 121 (1991).

82. A. Mittelman, Z.J. Chen, T. Kageshita, H. Yang. M. Yamada, P. Baskind, N. Goldberg, C. Puccio, T. Ahmed, Z. Arlin and S. Ferrone, J.Clin.Invest.,$\underline{86}$, 2136 (1990).

83. U. Wagner, J. Reinsberg and D. Krebs, Geburtshilfe Frauenheilkd, $\underline{50}$, 785 (1990).

84. H. Mellstedt, J.E. Frodin. P. Biberfeld, J. Fagerberg, R. Giscombe, A. Hernandex, G. Masucci, S.L. Li and M. Steinitz, Int.J.Cancer, $\underline{48}$, 344 (1991).

85. A. LoBuglio, M. Khazaeli, R. Meredith, M. Salter, M. Hardin, E. Plott and C. Russell, Abstracts of Advances in the Applications of Monoclonal Antibodies in Clinical Oncology, 7th International Meeting, 1990.

86. W. Allebes, R. Knops, M. Herold, C. Huber, C. Haanen and P. Capel, Leuk.Res., $\underline{15}$, 215 (1991).

87. J.F. Eary, Ann. Oncol., $\underline{2 \text{ Suppl. 2}}$, 187 (1991).

88. K.E. Beritton, S.J. Mather and M. Granowska, Nucl.Med.Commun., $\underline{12}$, 333 (1991).

89. G.A. Koppel, Bioconjugate Chemistry, $\underline{1}$, 13 (1990).

90. J.F. Eary, Int.J.Rad.Appl.Instrum.[B], $\underline{15}$, 105 (1991).

91. R.P. Moseley, J.C. Benjamin, R.D. Ashpole, N.M. Sullivan, J.A. Bullimore, H.B. Coakham and J.T. Kemshead, J.Neurol.Neurosurg.Psychiatry, $\underline{54}$, 260 (1991).

92. D.M. Goldenberg, J.A. Horowitz, R.M Sharkey, T.C Hall, S. Murthy, H. Goldenberg, R.E. Lee, R. Stein, J.A. Slegel, D. O. Izon, K. Burger, L.C. Swayne, E. Belisle, H.J Hansen and C.M Pinsky, J.Clin.Oncol., $\underline{9}$, 548 (1991).

93. D. Collen, H.R. Lijnen and H.K Gold, Prog.Cardiovas.Dis., $\underline{XXXIV}$, 101 (1991).

94. B.S.Coller, L.E Scudder, J. Beer, Herman K. Gold, J.D. Foots, J. Cavagnaro, R. Jordan, C. Wagner, J. Iuliucci, D. Knight, J. Ghrayeb, C. Smith, H.F. Weisman and H. Berger, Ann.N.Y.Acad.Sci., $\underline{614}$, 193 (1991).

95. H.L. Gold, L.W. Gimple, T. Yasuda, R.C. Leinbach, W. Werner, R. Hold, R. Jordan, D. Collen and B.S. Coller, J.Clin.Invest.,$\underline{86}$, 651 (1990).

96. S. Bhattacharya, I. Mackie, S. Machin, P. Leese, T. Schaible, R. Jordan, C. Smith, H. Berger and A. Lahiri, Circulation, $\underline{82}$,(Suppl. III) III-2392, (1990).

97. J.E. Tcheng, N.S. Kleinman, M.J. Miller, D.C. Sane, A.L. Wang and H.F. Weisman, Circulation, $\underline{84}$ (Suppl. II) II-2344 (1991).

98. N.S. Kleinman, E. M. Ohman, D.J. Keriakes, S.G. Ellis, H.F. Weisman and E.J. Topol, Circulation, $\underline{84}$ (Suppl. II), II-2076 (1991).

99. L. Steinman, J.Clin.Immunol., $\underline{10}$, 30S (1990).

100. M.K. Waldor in "Cell Surface Antigen Thy-1, Immunology, Neurology, and Therapeutic Applications", A.E. Reif and M. Schlesinger Eds., Marcel Dekker, Inc, New York, NY,1989, p. 573.

101. C. Reiter, B. Kakavand, E.P. Rieber, M. Schattenkirchner, G. Riethmüller and K. Krüger, Arthritis Rheum., $\underline{34}$, 525 (1991).

102. D. Wendling, J. Wijdenes, E. Racadot and B. Morel-Fourrier, J. Rheumatol., $\underline{18}$, 325 (1991).

103. G. Horneff, G.R. Burmester, F. Emmrich and J.R. Kalden, Arthritis Rheumatol., $\underline{34}$, 129(1991).

104. V. Strand, P.E. Lipsky, G. Cannon and L. Calabrese. Abstracts of the Americal College of Rheumatology 55th Annual Scientific Meeting, 1991.

105. P.A. van der Lubbe, A.M. Miltenburg and F.C. Breedveld, Lancet, $\underline{337}$, 1349 (1991).

106. J. Prinz, O. Braun-Falco, M. Meurer, P. Daddona, C. Reiter, P. Rieber and G. Riethmüller, Lancet, $\underline{338}$, 320 (1991).

107. J.K-.C. Ma and T. Lehner, Archs. Oral Biol., $\underline{35}$, 115S (1990).

108. I. Carrió, L. Berná, M. Estorch, G. Torres, C. Duncker, M. Gallester and D. Obrador, Abstracts of the 38th Annual Meeting of the Society of Nuclear Medicine, 1991.

109. T.S. Schaible and A. Alavi, Semin.Nucl.Med., $\underline{XXI}$, 313 (1991).

110. P.De Faucal, P. Peltier, B. Planchon, B. Dupas, M-.D. Touze, D. Baron, T. Scaible, H.J. Berger and J-.F. Chatal, J.Nucl.Med., $\underline{32}$, 785 (1991).

111. V. Hird, M. Verhoeyen, R.A. Badley, D. Price, D. Snook, C. Kosmas, C. Gooden, A. Bamias, C. Meares, J.P. Lavender and A.A. Epenetos, Br. J.Cancer, $\underline{64}$, 911 (1991).

Chapter 20. Techniques for Determining Epitopes for Antibodies and T-cell Receptors

David C. Benjamin and Samuel S. Perdue
Department of Microbiology
Health Sciences Center, University of Virginia
Charlottesville, VA 22908

The immune response is characterized by its high degree of specificity for antigen. Over the last 10 years it has become increasingly apparent that this specificity occurs at three levels and involves several evolutionarily-related cell-surface receptors and secreted molecules that differ fundamentally in the antigenic structures each recognizes. These specific recognition molecules are antibody (both as a B cell surface receptor and as secreted immunoglobulin), the class I and class II cell surface receptors which are encoded by genes within the Major Histocompatibility Complex (MHC), and the T cell receptor. Over this same period, major progress has been made in defining the structures of each of these macromolecular recognition molecules and the structures of the antigenic sites they recognize. Antibodies recognize a wide variety of chemical structures including proteins, peptides, lipids, DNA, RNA, carbohydrates and small organic molecules. In contrast, the class I and class II MHC molecules and the T cell receptor are restricted to recognizing peptides derived from protein antigens and, in some cases, small organic molecules covalently associated with peptides. The MHC molecules bind antigen-derived peptides whereas the T cell receptor interacts with these MHC-bound peptides on antigen-presenting cells. The antigenic site recognized by antibody is referred to as the B cell epitope, the site on the peptide recognized by the MHC molecule is called the agretope (for antigen recognition), and the site on the MHC-bound peptide recognized by the T cell receptor is called the T cell epitope. An understanding of the structural characteristics of antigenic sites could result in simplification of vaccine development, new methods for detecting antibodies and/or T cells specific for a given antigen, rational design of catalytic antibodies (abzymes) and peptide inhibitors, and an understanding of the structural basis of antibody-mediated inhibition of biological processes. This chapter presents a summary of many of the methods currently being used to define the locations and structural properties of antigenic sites on proteins.

B CELL EPITOPES

Anti-protein antibodies are all conformationally specific (1). The conformation recognized may be comprised of amino acids from a relatively short, continuous segment of the polypeptide chain, or it may be comprised of amino acids from two or more regions which are separate from each other in the primary structure of the polypeptide chain but which are brought together on the surface of the protein during folding. The former, and conformationally more simple, structure is referred to as a *continuous epitope* (or as a linear, segmental , or sequential epitope) and may be found on the folded protein molecule, the unfolded polypeptide chain, or a protein-derived short peptide . The latter, more complex antigenic structure, is referred to as a *discontinuous epitope* (or as a conformational or topographically assembled epitope). The terms continuous and discontinuous will be used throughout this chapter.

Two other descriptive terms are in common use (2). A *structural epitope* is a three-dimensional structure defined by X-ray crystallographic analysis of an antigen-antibody complex. It includes all atoms of the antigen which are in contact with antibody atoms using standard interatomic distances. A *functional epitope* is defined strictly in functional terms (e.g., by crossreactivity of an antibody with variant antigen molecules having one or more amino acid differences within the antigenic site). The functional epitope usually consists of fewer amino acids than the structural epitope most probably because there are far too few variants available for study. For example, a structural epitope was defined by the crystal structure of a complex between the monoclonal antibody (mAb) HyHEL-5 and the antigen Hen Egg-white Lysozyme (HEL), and is comprised of 14 amino acids from three distant segments of the polypeptide chain (3). The functional epitope defined by interaction of the HyHEL-5 mAb with a series of evolutionary lysozyme variants consists of but three amino acids (from two different segments) each of which was missing in one or more of the variant lysozymes used (4).

When discussing antigen structure one must be careful to state clearly whether the epitope is continuous or discontinuous, whether it is defined by structural or functional methods, and what functional assay (ELISA, RIA, solid-phase, in solution inhibition assays, etc) was used to define it.

Fragments and Peptides - Cleavage of proteins to obtain antigenically active fragments has been done since early this century (5). The use of fragments to map epitopes became widespread in the 1960s and 1970s following reports of experiments that utilized chemical and/or proteolytic digestion of proteins to demonstrate that protein antigens possess multiple, distinct epitopes (6-8). Overlapping fragments isolated from proteolytic digests may be used as a starting point in mapping the position of B cell epitopes on the parent macromolecule (9, 10). Once the general location of the epitope is determined synthetic peptides can be produced and used to define the minimal continuous epitope recognized by a particular polyclonal antibody. For example, for one or more continuous epitopes on sperm whale myoglobin, it was shown that the length of the peptide required for optimal reactivity with different antisera varied by no more than a single amino acid at the amino- and/or carboxyl-terminal ends (9).

Fragments generated by chemical, enzymatic, or recombinant DNA methods still represent a viable initial approach to mapping epitopes, and are useful for isolating restricted populations of antibody from polyclonal antisera. For discontinuous epitopes, the use of fragments may be the only method by which the epitope can be located to one or more regions of a protein antigen. This is not a small point since the majority of antibodies produced in response to native folded protein antigens recognize discontinuous epitopes (1,11-13). Indeed, all structural epitopes reported to date are of the discontinuous type (3, 14-17). Reaction of some antibodies, that were made to native antigens, with small peptides (vide infra) may occur because the peptide represents a portion of the native epitope and can react with the antibody with a sufficiently high affinity to be detected in the assay (18, 19). Other peptides, also representing portions of the discontinuous epitope, may not react with sufficient affinity to be detected. Thus, when mapping an epitope using small peptides, one should remember that it probably represents only one of several regions on the protein that contributes to the total epitope.

However, the use of synthetic peptides can give useful information. For example, if one wants to use the antibody to detect the presence of a given antigen in a biological specimen, it does not matter whether the epitope is discontinuous or continuous as long as the reaction with antibody is specific. On the other hand, if one is using information obtained from antibody reactions with peptides to determine the sequence recognized in anticipation of producing a synthetic vaccine, then one must consider the very real possibility that the anti-peptide antibody produced will not react with the native protein and thus be ineffective as a vaccine (19).

In recent years, the use of synthetic peptides to map epitopes recognized by both polyclonal and monoclonal antibodies and to determine their structural properties has gained considerable momentum. Part of the renewed interest stems from a desire to produce synthetic vaccines and part is due to the development of new methods for rapidly and simultaneously synthesizing a large number of peptides. Synthesis of peptides on solid supports such as polyethylene rods or pins has allowed the screening of a large number of peptides for antigenic activity (20-24). This method was used with polyclonal antisera to define nine continuous sites on myohemerythrin (20). Replacement nets were constructed to measure the relative importance of each amino acid within the epitope. Each net consists of a collection of peptides wherein each amino acid, of a hexapeptide representing a continuous epitope, was replaced in turn by each of the other 19 possible amino acids (21). Residues were considered critical to antibody binding if no substitutions or only a few substitutions were tolerated. For example, one continuous site was found to be comprised of residues 4-9 of myohemerythrin with the sequence, I^4PEPYV^9 and the residues, Glu^6, Tyr^8, and Val^9 were critical for antibody binding. Therefore, the sequence of this continuous epitope could be represented as

XXEXYV where X represents a replaceable residue. Different antisera see different, yet overlapping, sets of continuous epitopes and a given residue within a site might be critical or replaceable depending on the antiserum being used (21,22).

A similar study was performed using both murine monoclonal antibodies specific for the major Mycobacterium bovis secreted protein MPB70 and polyclonal antisera obtained from M. bovis infected cattle (13). Competitive inhibition assays showed six mAb defining three antigenic sites with two mAb recognizing each site. Four of the six mAb did not react with any of the complete set of overlapping heptapeptides, indicating they recognized only discontinuous epitopes. The other two mAb recognized different continuous epitopes defined by the peptides $P^{16}TGPASV^{22}$ and $S^{26}QDPVAV^{32}$. Replacement net analyses found these two peptides to have three and four critical residues, respectively, and could be represented by the generic sequences XXGXXSV and XQDPVXX. In contrast to the polyclonal response of mice to myohemerythrin, infected cattle see only a single continuous epitope on MPB70. This epitope, with the sequence $S^{35}NNPEL^{40}$, was nearby the continuous epitopes seen by the murine mAb to MPB70 and could be represented by the XNXPXX generic sequence. Although there was only a single epitope recognized by sera from infected cattle, one cannot consider this epitope to be immunodominant since neither the amount of antibody directed to it nor the number of discontinuous epitopes recognized by the same antisera is known.

More recently, a variation on the pin technology has been introduced (25). This method involves creating a large peptide library consisting of millions of plastic beads with each bead containing a single peptide and with the complete collection representing the total number of possible random peptides in roughly equal proportions. A rapid ELISA-based screening assay can be used and a positive reaction involving a few highly colored beads against a colorless background is observed. The colored bead is removed and the sequence of the peptide on the bead determined in a microsequencer. The validity of this method was tested using an anti-β-endorphin mAb with specificity for the sequence YGGFL. Six reactive beads were isolated from a 2 million pentapeptide bead library. Three of these peptides, with the sequences YGGFQ, YGGFA, and YGGFT, had essentially the same affinity for the antibody as did the native peptide. The other three peptides (with sequences of YGGLS, YGALQ, and YGGMQ) had considerably lower affinities for the mAb. This method can be used to synthesize overlapping sets of peptides representing only the polypeptide sequence, replacement nets for determining the contribution of single amino acids to binding, or large random libraries representing the universe of peptides of a given length. However, once a peptide sequence is determined from a reactive bead, an independent analysis must be done to confirm the specificity of the original reaction with antibody. For example, one must use unrelated antibodies as negative controls and assess the ability of the peptide to inhibit reaction of the antibody with intact protein antigen.

Prediction Algorithms -- A number of attempts have been made to correlate chemical, physical or structural properties of proteins or peptides with antigenicity. Algorithms based on hydrophilicity (26), solvent accessibility (27), protrusion (28), and main chain segmental mobility (29, 30) have all been used with some success. However, each is merely an indicator of surface position and none is an infallible predictor of antigenicity. Still, the use of a combination of these methods may lead to a good first approximation, e.g., for a protein of known sequence but unknown three dimensional structure a combination of hydrophilicity and β-turn prediction may indicate that a given sequence is on the surface of the protein and is in a relatively continuous exposed and protruding segment. Once such a segment is identified, a peptide can be synthesized and used to produce antibodies which, in turn, are assessed for their ability to react with the peptide and with the intact protein.

Species Variants - Reactivity of an antibody, produced to a given protein, with evolutionarily-related species variants has been an extremely productive and valuable tool for mapping epitopes (1). The reactivity of the antibody with each member of a family of related proteins is determined. The extent of reactivity (or lack thereof) is then correlated with differences in amino acid sequence between the protein to which the antibody was made and each of the other family members. Given the absence of large conformational differences between closely-related proteins, one assumes differences in amino acids within the epitope are responsible for changes in reactivity. This method has been used with a wide variety of proteins including, cytochromes, lysozymes, myoglobins, serum albumins, etc. (1, and references therein). The success of this method was confirmed when studies using species variants to map functional epitopes (4,14) accurately predicted the surface position of each of the three structural epitopes on hen egg-white lysozyme (HEL) determined by X-ray diffraction of HEL-anti-HEL complexes (3,14,16). The major, and perhaps only, drawback to this method is the lack of a sufficient number of species variants to accurately map the entire epitope.

Antibody Protection - Reaction of antibody with antigen can protect side-chains within the epitope from chemical alteration. This approach has been used to map a discontinuous epitope on the surface of horse cytochrome-c (31). Antibody-bound and free cytochrome-c were trace labeled with ^{3}H-acetic anhydride and then fully acetylated with non-radioactive anhydride. This labeled material was mixed with ^{14}C-cytochrome-c, the mixture proteolyzed and the peptide mixture fractionated on HPLC. The peptides were then sequenced and the degree of acetylation during trace-labeling was determined by measuring the ^{3}H/^{14}C ratio. The two residues found to be protected, Lys^{60} and Lys^{99} are within 7 Å of each other on the surface of the folded cytochrome-c molecule.

Antibody protection of antigen from digestion with proteolytic enzymes can also be used to map epitopes (32). Antibody-bound and free horse heart cytochrome-c (the latter mixed with an unrelated antibody) were digested with trypsin for varying periods of time and the resulting peptide mixtures fractionated on HPLC. Differences in the peptide elution profiles of the two digests represent peptides protected by antibody during digestion. Isolation and characterization of these peptides define amino acids that may contribute to the epitope. For one mAb, peptides from two distinct regions of the polypeptide chain were protected. One of these peptides contained a residue previously identified to be within the epitope. The carbonyl oxygen atom of this residue is hydrogen-bonded to a residue contained in the second protected segment in the intact, folded protein.

More recently, a method using 2-D nuclear magnetic resonance (NMR) to map and define antigenic sites on small proteins in solution has been described (33). By determining the exchange rate (by NMR) of amide-hydrogens in the presence and absence of a monoclonal antibody, a region, on the surface of cytochrome-c, was found to contain the protected residues, their hydrogen bond acceptors, and several residues that had previously been shown to lie within the epitope defined by the mAb used (vide supra). In addition, the solvent accessible surface area of this region was of approximately the same size and general shape of other protein epitopes defined by X-ray diffraction methods.. However, this study suffered from the fact that the crystal structure of this mAb-cytochrome-c complex had not been determined and thus a direct comparison between the NMR and x-ray diffraction analyses could not be made. A subsequent study (34) using the same methods and an antigen (HEL)-mAb (HyHEL-5) complex for which the crystal structure had been reported (3), showed that residues outside the epitope can be significantly protected and that protection effects can be seen in residues quite distant from the epitope. Thus, in the absence of other information, this method may not be generally applicable to the study of antigen structure.

Recombinant DNA Methods - Potentially the most powerful tool for studying antigen structure is recombinant DNA. The ability to systematically manipulate the DNA sequences encoding protein and measure the effect on antigen structure has already provided interesting information. In eukaryotes the DNA encoding protein does not exist as contiguous sequences on the genome: rather, it is divided into discreet coding sections called *exons* (for *ex*pressed sequences) which are separated on the genome by non-coding DNA segments called *introns* (for *int*ervening sequences). Quite often exons correspond to the structural domains of proteins. Exon shuffling is a recombinant DNA method whereby individual exons are exchanged between the genes encoding related proteins. For example, the exons of the genes encoding class I MHC molecules correspond to the structural domains of the protein. In order to map epitopes detected by alloantisera (or mAb) a number of studies exchanged (shuffled) exons between genes encoding different Class I molecules within a species or between genes encoding the same class I molecule across species. The homo- and/or hetero-hybrid molecule was then reacted with the alloantibody specific for one of the parent proteins. Loss of reactivity with the antibody upon loss of a given domain by the donor antigen molecule together with gain of reactivity by the recipient molecule localized the epitope to the exchanged domain. This method has been used to map epitopes to domains (35), interdomain epitopes that require amino acids from more than one domain (36), and epitopes to regions within domains (37).

A related method involves the use of cDNA rather than genomic DNA as the starting point. As opposed to genomic DNA, the coding sequences on mRNA are contiguous. Therefore a cDNA copy of the mRNA can be manipulated without regard to exon or domain boundaries. A method, called scanning mutagenesis, was used to map the epitopes on human growth hormone by systematically exchanging sequences between cDNAs encoding growth hormone from several species (38).

Deletion of segments of DNA encoding the viral oncogene product $pp60^{src}$ and examining the resultant mutant oncogene products for antibody reactivity resulted in the localization of epitopes recognized by three groups of mAb (39). Randomly cloning small fragments of DNA, produced by sonication or partial digestion with DNase I, into a suitable expression vector can be used to produce a large library of recombinant organisms producing fragments of the antigenic protein. Screening this library with the appropriate antibody together with DNA sequence analysis, an accurate location of the epitope can be determined (40). Chemical and/or oligonucleotide directed mutagenesis has been used to generate single amino acid variants leading to mapping of antigenic sites on a variety of proteins (41-43). Site-directed mutagenesis using oligonucleotides fully redundant at individual codons has permitted the construction of a library of site-directed mutants at specific residues within an epitope on the surface of staphylococcal nuclease (43). Analysis of these mutants has shown that one or more (but not all) changes at any given residue may affect antibody binding suggesting that all residues within an epitope are important. These studies suggest that, given a sufficient number of variants, all residues within a structural epitope might be functionally defined and that the distinction between structural and functional epitopes may be artificial. In addition, these studies suggest that the importance of any given residue to antibody recognition and strength of binding is not an inherent property of that particular residue but rather is unique to a particular antigen-antibody pair. Thus, each antibody of many that recognize the same epitope offers a unique solution to a chemical problem.

An interesting alternative to the random synthetic peptide library method discussed above is the production of very large libraries using bacteriophage vectors into which random oligonucleotide sequences have been introduced (44-46). In one version of this method, each phage displays a single peptide sequence (encoded by the inserted oligonucleotide) on the virion surface as part of the viral surface protein. Antibody (or other binding ligand) is then used to affinity purify the bacteriophage that possess the peptides. Subsequent amplification and purification of the individual, affinity-purified, phage permits characterization of the peptide by sequencing the appropriate coding region in the viral DNA.

X-ray Diffraction - The ultimate method for defining the epitope which does not rely on peculiarities of the assay method and interpretation of negative binding data (low binding constant etc.) is x-ray diffraction analysis of antigen-antibody complexes. This method has been used with protein antigens (3, 14-17), peptides (47), carbohydrates (48) and haptens (49-51). The results with proteins clearly demonstrate that discontinuous epitopes do exist -- in fact all the structural epitopes defined by x-ray diffraction are discontinuous. They also show that the conformation of a peptide when bound to antibody can be very different from the same region as part of the intact native protein (47) emphasizing the need to carefully interpret experimental data resulting from use of peptides to map and define epitopes.

AGRETOPES AND T CELL EPITOPES

Antigen recognition by T cells differs from that described for antibodies in that the T cell receptor (TCR) does not recognize native protein structure. Rather, the TCR binds specifically to an antigen-derived peptide which in turn is bound to a self MHC-encoded surface glycoprotein, and the TCR must recognize residues on both the peptide and the MHC protein in order for T cell activation to occur. Therefore, TCR-antigen recognition differs from antibody-antigen interactions in that TCRs must recognize a complex comprised of two distinct protein entities. The requirement for peptide binding to an MHC protein prior to immune recognition presents an additional consideration for those who wish to identify antigenic regions of a protein: any peptide that induces a T cell response must possess not only the appropriate epitope (the residues that bind the TCR), but also the proper agretope (the residues which interact with the MHC). In addition, the amino acids contained within the epitope and agretope must be ordered in a way that is not detrimental to the proper presentation of the peptide. Furthermore, since the antigen must undergo processing prior to presentation (see 52 for review), residues flanking the final peptide must be taken into consideration as it is unknown how important these regions are in effective processing and proteolysis (53-54). Consequently, the problems associated with determining T cell epitopes are considerable.

TCR and MHC Structure - The TCR is a heterodimer composed of two transmembrane glycoproteins. The three-dimensional conformation of the TCR is unknown, but sequence analysis and the generation of immunoglobulin-TCR chimeras (55) has provided evidence that the TCR is similar in structure to antibody molecules. Comparison of the amino acid sequence with that of numerous immunoglobulins reveals a high degree of homology between TCRs and antibodies, and comparisons between different TCRs indicate the presence of defined constant and variable domains, the latter containing distinct regions of hypervariability (56-57). In immunoglobulins, these domains are called CDRs (complementarity determining regions), and in this review the same designation will also be used for the equivalent regions of the TCR. Given the similarity between TCRs and antibodies, it is likely that TCRs recognize their antigens in a manner at least similar to that of an antibody, with the difference being that the residues that are brought together at the surface of the antigen complex come from different proteins.

More is known about MHC protein structure, as crystal structures of three class I MHC molecules have been determined (58-61). The class I molecule is a dimer composed of a membrane- bound heavy chain in a non-covalent association with a molecule of β_2-microglobulin. The heavy chain contains three domains: the membrane-proximal $\alpha3$ domain adopts an immunoglobulin fold, while the outer $\alpha1$ and $\alpha2$ domains together form two antiparallel α helices atop a ß-sheet. The β_2-microglobulin and $\alpha3$ domains form a "base" upon which rests the ß-sheet platform and α-helices. The helices and platform together form a groove, with the walls of the groove being the α-helices and the bottom being the ß-sheet platform. Class II MHC molecules are composed of an α/β heterodimer of transmembrane glycoproteins, with each chain possessing two extracellular domains. Based on the information obtained from the class I diffraction studies and analysis of class II sequence, a model has been proposed for class II (62). According to the model, the $\alpha2$ and ß2 domains also adopt immunoglobulin folds, while the $\alpha1$ and ß1 domains together form a structure similar to that seen for class I molecules (with one helix provided by each chain), thereby also forming a groove between two α-helices. It has long been thought that the processed peptide rests within the cleft between the α-helices, and the recent crystallography of the human class I product HLA-B27 reveals the presence of an identifiable peptide within the binding groove (61). Therefore, antigenic peptides must possess the characteristics necessary for binding in the MHC cleft, yet still expose the proper surface for T cell recognition.

A model has been proposed where the first and second hypervariable domains (or CDRs) of a TCR interact with residues on the MHC, while CDR3 interacts with the bound peptide (63). This model is based on the high degree of homology between TCRs and antibodies, and on the degree of variability within the TCR and MHC molecules. The distance between the α-helices on the MHC molecule is equivalent to that seen between the heavy and light chain CDRs 1 and 2 of an antibody. Furthermore, the highest degree of polymorphism within the TCR is located in the CDR3-equivalent region, while the first and second CDRs are not as variable. Thus, according to the model, CDRs 1 and 2 bind residues on the MHC helices, while the more variable CDR3 interacts with the large variety of antigenic peptides that could potentially bind in the groove. However, until the crystal structure of a TCR is determined, the precise conformation of the molecule is uncertain.

Species variants - Early investigations into the nature of T cell-inducing peptides utilized evolutionary variants of different proteins. Variants of myoglobin (64-65), lysozyme (66-67), and ovalbumin (68) were three of the initial proteins used in these studies. Species variants of these proteins demonstrated varying degrees of reactivity, and sequence comparisons between the variant proteins allowed the determination of regions believed to be responsible for T cell induction. The proteins were subjected to chemical or proteolytic cleavage and the resulting fragments were tested for their ability to stimulate T cells when presented in context with the appropriate MHC molecule. A subset of the protein fragments were shown to be capable of activating T cells, thereby allowing localization of the epitope to a particular region of the protein antigen. This methodology has increased our understanding of T cell epitopes, but is limited in that it indicates only the general location of the epitope within the protein. In order to determine the minimal antigenic peptide, further studies using mutational analysis and synthetic peptides are necessary.

Synthetic peptides - Many of the peptide epitopes reported to date have been identified using synthetic peptides corresponding to different regions of a protein antigen. Once the epitope is localized to a particular region of a protein, overlapping peptides are synthesized that span the site and are tested for their ability to prime target cells for T cell recognition. Using this methodology, the location of the T cell epitope can be further elucidated. This procedure has been used extensively to determine T cell-inducing peptides from a variety of proteins, including recent studies of sperm whale myoglobin (69), HIV gp160 (70), influenza hemagglutinin (71) and nucleoprotein (72), and the phospholipid protein linked with autoimmune encephalitis (73).

One problem with using synthetic peptides in determining T cell epitopes is the practical considerations of the simultaneous generation of a large panel of peptides for use. Considerable progress has been made in this area in recent years, notably the pin synthesis described above. A modification of this methodology has been described in which peptides are attached to polyethylene pins by a linker that is cleavable at physiological pH (74). Thus, once the peptides are synthesized they can undergo deprotection and removal of contaminants under acid or near neutral (pH 5.0) conditions, followed by removal from the pin under physiological pH. This pin technology has been used to generate overlapping peptides that together span a putative eight-residue determinant for tetanus toxoid. These peptides were then tested for their ability to stimulate the appropriate T cell clone. The peptides demonstrated functional T cell-inducing activity in most cases, though some of the longer peptides that included the octamer region failed to cause proliferation. Analysis of peptides produced under these conditions may prove to be useful in the examination of T cell-inducing peptides. A potential drawback to this technique is that cleavage of the linker through diketopiperazine formation yields a peptide that remains attached to the *cyclo*(Lys-Pro) moiety of the linker, thereby producing a peptide with a bulky, uncharged C-terminus. Elimination of a potential charge or addition of a bulky group on the end of the peptide may have significant consequences in the binding of the peptide to MHC or in TCR recognition, as binding of a charged lysine at the C-terminus to the MHC molecule has been documented, and some studies have demonstrated a preference for a positively charged C terminus (61,75).

Prediction algorithms - Empirical analysis of the many reported immunogenic peptides has resulted in the development of two algorithms, one based on primary sequence and the other based on secondary structure, for predicting regions of a protein that can stimulate a T cell response. These predictions are based on the sequence or structure of a native protein, not the MHC-bound peptide, and do necessarily attempt to explain the structure of the peptide as presented to a TCR. The requirements for processing and the attendant proteolytic cleavage of the protein may require the presence of certain motifs that play no actual role in peptide binding or recognition.

A common sequence motif consisting of a glycine or charged residue followed by two hydrophobic amino acids has been identified (76). In the analysis of 57 immunogenic peptide sequences, 84% possessed this motif. Furthermore, when the peptides were sorted according to allelic specificity, subpatterns could be determined centered around the main motif. By testing the model with synthetic peptides corresponding to predicted regions of five different proteins, they were able to accurately predict 11 epitopes. Additional applications of the model have also demonstrated some success (77).

A second prediction algorithm is based not on primary sequence, but on secondary structure. It has been proposed that potential peptides must be able to adopt an amphipathic α-helical conformation in which one face of the helix is hydrophilic and one is hydrophobic (78-79). T cell-inducing peptides were predicted by locating regions that could potentially form an α-helix, and then using a hydrophobicity scale to determine potential amphipathic regions. The algorithm was tested by comparing predicted regions with a panel of 23 known immunogenic peptides. Using this model, 78% of these peptides contained the motif. Later use of the algorithm has been successful in predicting sites in several proteins, providing promising potential for utilization in the generation of synthetic vaccines (70,80-81) or immunotherapeutics (82). It is likely that both models will remain useful in predicting potential antigenic sites, as both have been utilized to a degree of success. Furthermore, the two models are not mutually exclusive, since the former motif is capable of fitting into an amphipathic helix. However, this motif can also exist in ß-strands and is therefore not constrained to form a helix turn.

Acid extraction - The use of synthetic peptides to mimic natural peptides in T cell stimulation assays has produced important information regarding the characteristics required for MHC binding and TCR recognition. However, the information provided by these studies alone fails to reveal the nature of naturally-processed peptides, and therefore sequence and structural requirements for peptide binding can only be inferred. Recent investigations involving the extraction of naturally processed peptides from MHC molecules promise to provide greater insight into the nature of T cell peptides (reviewed in 83). Whole cell homogenates are acid extracted and the large molecular weight material is precipitated. Smaller material ($M_r < 5000$) is then purified by reverse phase HPLC and the fractions are tested for their ability to stimulate T cells. Any fractions that are able to stimulate are then compared to synthetic peptides, in both biochemical and immunological assays in order to determine the nature of the peptide. Using this technique, several naturally-processed T cell-inducing peptides from both self and viral proteins have been isolated (84-87).

A modification to this acid extraction methodology allows one to extract only those peptides bound to MHC molecules. Following digestion of the cells, the MHC molecules are isolated either by immunoprecipitation or antibody affinity chromatography (75,88). Bound peptides are then eluted from the MHC molecules at low pH as before, purified through reversed-phase HPLC, and checked for their ability to stimulate T cells. Direct extraction of peptide from both class I (75) and class II (88) MHC molecules has been reported. Not only does this method yield naturally processed peptides for study, but direct sequencing of some peptides is also possible (88-89).

Perhaps the most interesting feature discovered in these peptide-extraction studies is the presence of allele-specific motifs. A sequence motif that is highly conserved in peptides bound to the murine class I allele H-2K^d has been identified (87). After isolating a nonapeptide using the above technique, its sequence was compared to that of other known sequences containing a K^d-reactive peptide, and found that all contained a tyrosine residue. As the tyrosine residue in the naturally-processed peptide was located in the second position, each known sequence was aligned so that their tyrosines were in the same position. The analysis revealed a motif of a conserved tyrosine at position 2, an amino acid bearing a side chain methyl group at position 9, and a substantial degree of conservation at positions 3, 5, and 6. Residues 1, 4, 7, and 8 were highly variable. This allele-specific motif has been used to predict a T cell-inducing peptide of the bacterial protein lyseriolysin, LLO (90). The sequence of LLO indicates five regions that fit the K^d motif. Nonamers corresponding to these motif regions were synthesized (together representing only 18% of the molecule) and examined for CTL targeting. One peptide, LLO91-99, was found to comigrate with the naturally-processed peptide. Another motif was discovered for HLA-B27, which contains a conserved arginine at position 2, a hydrophobic residue at position 3, a nonpolar or small polar residue at position 6, and a preference for a positive charge at positions 1 and 9. Once again, positions 4, 7 and 8 are highly variable (75). The presence of these allele-specific motifs correlates well with the crystal structure of HLA-B27, revealing orientations of the residues in the bound peptide that are consistent with the data above (61,86). In the bound peptide, the side chains located at positions 4 and 8 point up, while those at positions 3, 7, and 9 point down, and the position 2 side chain extending into the "45" pocket (59-60). Amino acids 5 and 6 contain side chains that point across the width of the cleft, and although there is no electron density corresponding to position 1, modeling indicates that a long side chain could extend up out of the groove. This information supports the motif mentioned above, for the conserved residues are all located at positions extending into the cleft, where they would interact with allele-specific residues of the MHC, while the variable residues, with the exception of position 7, extend outward. In this instance, residues 5 and 6 could theoretically interact with either the MHC or TCR.

References

1. D.C. Benjamin, J.A. Berzofsky, I.J. East, F.R.N. Gurd, C. Hannum, S.J. Leach, E. Margoliash, J.G. Michael, A. Miller, E.M. Prager, M. Reichlin, E.E. Sercarz,S.J. Smith-Gill, P.E. Todd, and A.C. Wilson, Ann.Rev.Immunol., $\underline{2}$, 67 (1984).

2. S.J. Smith-Gill, M.A. Newman, C.P. Mallett, L.N.W. Kam-Morgan, J.F. Kirsch, R.J. Poljak, R.E. Burcolleri and J. Novotny, FASEB J. $\underline{4}$, A2086 (1990).

3. S. Sheriff, E.W. Silverton, E.A. Padlan, G.H. Cohen, S.J. Smith-Gill, B.C. Finzel, and D.R. Davies, Proc.Nat.Acad.Sci.USA, $\underline{84}$, 8075 (1987).

4. S.J. Smith-Gill, T.B. Lavoie, and C.R. Mainhart, J.Immunol. $\underline{133}$, 384 (1984).

5. S.B. Hooker and W.C. Boyd, J.Immunol., $\underline{26}$, 469 (1934).

6. M. Kaminski, Progr. Allergy, $\underline{9}$, 79 (1965).

7. M. J. Crumpton and J. M. Wilkinson, Biochem.J., $\underline{94}$. 545 (1965).

8. D.C. Benjamin and W.O. Weigle, Immunochem., $\underline{8,}$ 1208 (1971).

9. M.Z. Atassi, Immunochem., $\underline{12}$, 423 (1975).

10. M. Z. Atassi, Immunochem., $\underline{15}$, 909 (1978).

11. D.J. Barlow, M.S. Edwards, and J.M. Thornton, Nature, $\underline{322}$, 747 (1986).

12. G. Lando and M. Reichlin, J.Immunol., $\underline{129}$, 212 (1982).

13. A.J. Radford, P.R. Wood, H. Billman-Jacobe, H.M. Geysen, T.J. Mason, and G. Tribbick, J.Gen.Microbiol., $\underline{136}$, 265 (1990).

14. A.G. Amit, R.A. Mariuzza, S.E.V. Phillips, and R.L. Poljak, Science, $\underline{233}$, 747 (1986).

15. P.M. Colman, W.G. Laver, J.N. Varghese, A.T. Baker, P.A.Tulloch, G.M. Air, and R.G. Webster, Nature, $\underline{326}$, 358 (1987).

16. E.A. Padlan, E.W. Silverton, S. Sheriff, G.H. Cohen, S.J. Smith-Gill, and D.R. Davies, Proc.Natl. Acad.Sci.USA, $\underline{86}$, 5938 (1989).

17. F.A. Bentley, G. Boulot, M.M. Riottot, and R.J. Poljak, Nature, $\underline{348}$, 254 (1990).

18. R. Jemmerson, Proc.Nat.Acad.Sci.USA, 84, 9180 (1987).

19. B.D. Spangler, J.Immunol., $\underline{146}$, 1591 (1991).

20. H.M. Geysen, J.A. Tainer, S.J. Rodes, T.J. Mason, H. Alexander, E.D. Getzoff, and R.A. Lerner, Science, $\underline{235}$, 1184 (1987).

21. E.D. Getzoff, H.M. Geysen, S.J. Rodda, H. Alexander, J.A. Tainer, and R.A. Lerner, Science, $\underline{235,}$ 1191 (1987).

22. E.D. Getzoff, J.A. Tainer, R.A. Lerner, and H.M. Geysen, Adv.Immunol., $\underline{43}$, 1 (1988).

23. H.M. Geysen, S.J. Rodda, T.J. Mason, G. Tribbick, and P.G. Schoofs, J.Immunol.Meth., $\underline{102}$, 259 (1987).

24. H.M. Geysen, Southeast Asian J.Trop.Med. and Pub. Health, $\underline{21}$, 523 (1990).

25. K.S. Lam, S.E. Salmon, E.M. Hersh, V.J. Hruby, W.M. Kazmierski, and R.J. Knapp, Nature, $\underline{354}$, 82 (1991).

26. T.P. Hopp, and K.R. Woods, Proc.Nat.Acad.Sci.USA, $\underline{78}$, 824 (1981).

27. J. Novotny, M. Handschumacher, E. Haber, R.E. Brucolleri, W.B. Carlson, D.W. Fanning, J.A. Smith, and G.D. Rose, Proc.Nat.Acad.Sci.(USA), $\underline{83}$, 226 (1986).

28. E. Westoff, Nature, $\underline{311}$, 123 (1984).

29. M.H.V. Van Regenmortel, J.P. Briand, Z. Al Moudallal, D. Altschuh, and E. Westhof in "Current Communications in Molecular Biology: Immune Recognition of Protein Antigens," W.G. Laver and G.M. Air, Ed., Cold Spring Harbor Laboratory Press, Cold Spring Harbor, N.Y., 1985, p. 181.

30. J.M. Thornton, EMBO J., $\underline{5}$, 409 (1986).

31. A. Burnens, S. Demotz, G. Corradin, H. Binz, and H.R. Bosshard, Science, $\underline{235}$, 780 (1987).

32. R. Jemmerson and Y. Paterson, Science, $\underline{232}$, 1001 (1986).

33. Y. Paterson, S.W. Englander, and H. Roder, Science, $\underline{249}$, 755 (1990).

34. D.C. Benjamin, D.C. Williams, Jr., S.J. Smith-Gill, and G. S. Rule, Proc.Nat.Acad.Sci.USA, in press.

35. K. Ozato, G.A. Evans, D.H. Marguiles, J.G. Seidman, and R.B. Levy, Transplant.Proc., $\underline{15}$, 2074 (1983).

36. I. Stroynouski, S.S. Clark, L.A. Henderson, L.E. Hood, M. McMillan, and J. Forman, J.Immunol., $\underline{135}$, 2160 (1985).

37. A. Toubert, C. Raffoux, J. Boretto, J. Sire, R. Sodoyen, S.R. Thurau, P. Amor, J. Colombani, F.A. Lenmonier, and B.R. Jordan, J.Immunol., $\underline{141}$, 2503 (1988).

38. B.C. Cuningham, P. Jhurani, P. Ng, and J.A. Wells, Science, $\underline{243}$, 1330 (1989).

39. S.J. Parsons, D.J. McCarley, V.W. Raymond, and J.T. Parsons, J.Virol., $\underline{59}$, 755 (1986).

40. R.A. Young and R.W. Davis, Science, $\underline{222}$, 778 (1983).

41. A.M. Smith, M. P. Woodward, C.W. Hershey, E.D. Hershey and D.C. Benjamin, J.Immunol., $\underline{146,}$ 1254 (1991).

42. D.C. Benjamin, Intl.Rev.Immunol., $\underline{7}$, 149 (1991).

43. A.M. Smith and D.C. Benjamin, J.Immunol., $\underline{146}$, 1259 (1991).

44.   J.K. Scott and G.P. Smith, Science, 249, 386 (1990).
45.   S.E. Cwirla, E.A. Peters, R.W. Barrett and W. J. Dower, Proc.Nat.Acad.Sci.USA, 87, 6378 (1990).
46.   J.J. Devlin, L.C. Panganiban and P.E. Devlin, Science, 249, 386 (1990).
47.   R.L. Stanfield, T.M. Fieser, R.A. Lerner, and I.A. Wilson, Science, 248, 712 (1990).
48.   M. Cygler, D.R. Rose, and D.R. Bundle, Science, 253, 442 (1991).
49.   Y. Satow, G.H. Cohen, E.A. Padlan, and D.R. Davies, J.Mol.Biol., 190, 593 (1986).
50.   J.N. Herron, X.M. He, M.L. Mason, E.W. Voss and A.B. Edmundson, Proteins, 5, 271 (1989).
51.   A.T. Brunger, D.J. Leahy, T.R. hynes, and R.O. Fox, J.Mol.Biol., 221, 239 (1991).
52.   T.P. Levine and B.M. Chain, Crit.Rev.Biochem.Mol.Biol., 26, 439 (1991).
53.   M. Del Val, H.-J. Schlicht, T. Ruppert, M.J. Reddehase, and U.H. Koszinowski, Cell, 66, 1145 (1991).
54.   H. Bhayani, F.R. Carbone, and Y. Paterson, J. Immunol., 141, 377 (1988).
55.   M.L.B. Becker, R. Near, M. Mudgett-Hunter, M.N. Margolies, R.T. Kubo, J. Kaye, and S.M. Hedrick, Cell, 58, 911 (1989).
56.   C. Chothia, D.R. Boswell, and A.M. Lesk, EMBO J., 7, 3745 (1988).
57.   S.M. Hedrick, E.A. Nielsen, J. Kavaler, D.I. Cohen, and M.M. Davis, Nature, 308, 153 (1984).
58.   P.J. Bjorkman, M.A. Saper, B. Samraoui, W.S. Bennett, J.L. Strominger, and D.C. Wiley, Nature, 329, 506 (1987).
59.   M.A. Saper, P.J. Bjorkman, and D.C. Wiley, J.Mol.Biol., 219, 277 (1991).
60.   T.P.G. Garrett, M.A. Saper, P.J. Bjorkman, J.L. Strominger, and D.C. Wiley, Nature, 342, 692 (1989).
61.   D.R. Madden, J.C. Gorga, J.L. Strominger, and D.C. Wiley, Nature, 353, 321 (1991).
62.   J.H. Brown, T. Jardetsky, M.A. Saper, B. Samraoui, P.J. Bjorkman, and D.C. Wiley, Nature, 332, 845 (1988).
63.   P.J. Bjorkman and M.M. Davis in "Symposia on Quantitative Biology," Vol. LIV, Cold Spring Harbor Laboratory Press, Cold Spring Harbor, N.Y., 1989, p. 365.
64.   I. Berkower, L.A. Matis, G.K. Buckenmeyer, F.R.N. Gurd, D.L. Longo, and J.A. Berzofsky, J.Immunol., 132, 1370 (1984).
65.   I. Berkower, H. Kawamura, L.A. Matis, and J.A. Berzofsky, J.Immunol., 135, 2628 (1985).
66.   F. Manca, J.A. Clarke, A. Miller, E.E. Sercarz, and N. Shastri, J.Immunol., 133, 2075 (1984).
67.   P.M. Allen, B.P. Babbitt, and E.R. Unanue, Immunol.Rev., 98, 171 (1987)..
68.   R. Shimonkevitz, J. Kappler, P. Marrack, and H. Grey, J.Exp.Med., 158, 303 (1983).
69.   J.E. Kim, M. Kojima, R. Houghten, C.D. Pendleton, J.L. Cornette, C. DeLisi, and J.A. Berzofsky, Mol. Immunol., 27, 941 (1990).
70.   K.B. Cease, H. Margalit, J.L. Cornette, S.D. Putney, W.G. Robey, C. Ouyand, H.Z. Streicher, P.J. Fischinger, R.C. Gallo, C. DeLisi, and J.A. Berzofsky, Proc.Natl.Acad.Sci.USA, 84, 4249 (1987).
71.   L.R. Brown, N.R. Nygard, M.B. Graham, C. Bono, V.L. Braciale, J. Gorka, B.D. Schwartz, and T.J. Braciale, J.Immunol., 147, 2677 (1991).
72.   S.J. Brett, J. Blau, C.M. Hughes-Jenkins, J. Rhodes, F.Y. Liew, and J.P. Tite, J.Immunol., 147, 984 (1991).
73.   R.H. Whitham, R.E. Jones, G.A. Hashim, C.M. Hoy, R.-Y. Wang, A.A. Vandenbark, and H. Offner, J.Immunol., 147, 3803 (1991).
74.   N.J. Maeji, A.M. Bray, H.M. Geysen, J.Immunol.Meth., 134, 23 (1990).
75.   T.S. Jardetsky, W.S. Lane, R.A. Robinson, D.R. Madden, and D.C. Wiley, Nature, 353, 326 (1991).
76.   J.B. Rothbard and W.R. Taylor, EMBO J., 7, 93 (1988)
77.   J.R. Lamb, J. Ivanyi, A.D.M. Rees, J.B. Rothbard, K. Howland, R.A. Young, and D.B. Young, EMBO J., 6, 1245 (1987).
78.   H. Margalit, J.L. Spouge, J.L. Cornette, K.B. Cease, C.L. DeLisi, and J.A. Berzofsky, J.Immunol., 138, 2213 (1987).
79.   C. DeLisi and J.A. Berzofsky, Proc.Natl.Acad.Sci.USA, 82, 7048 (1985).
80.   J.A. Berzofsky, Sem.Immunol., 3, 203 (1991).
81.   R. Arnon, Mol.Immunol., 28, 209 (1991).
82.   V.C. Schad, R.D. Garman, and J.L. Greenstein, Sem.Immunol., 3, 217 (1991).
83.   O. Rötzsche and K. Falk, Immunol.Today, 12, 447 (1991).
84.   O. Rötzsche, K. Falk, H.-J. Wallny, S. Faath, and H.-G. Rammensee, Science, 249, 283 (1990).
85.   O. Rötzsche, K. Falk, K. Deres, H. Schild, M. Norda, J. Metzger, G. Jung, and H.-G. Rammensee, Nature, 348, 252 (1990).
86.   G.M. Van Bleek, and S.G. Nathenson, Nature, 348, 213 (1990).
87.   K. Falk, O. Rötzschke, K. Deres, J. Metzger, G. Jung, and H.G. Rammensee, J.Exp.Med., 174, 425 (1991).
88.   A.Y. Rudensky, P. Preston-Hurlburt, S.-C. Hong, A. Barlow, and C.A. Janeway, Jr., Nature, 353, 622 (1991).
89.   D.F. Hunt, R.A. Henderson, J. Shabanowitz, K. Sakaguchi, H. Michel, N. Sevilir, A. Cox, E. Appella, and V.H. Engelhard, Science, Science, 255, 1261 (1992).
90.   E.G. Pamer, J.T. Harty, and M.J. Bevan, Nature, 353, 852 (1991).

Chapter 21. Potential Therapeutic Modifiers of the Complement Cascade

William K. Hagmann
Merck Research Laboratories, Rahway, NJ 07065

Robert D. Sindelar
Department of Medicinal Chemistry, The University of Mississippi, University, MS 38677

Introduction - The human complement cascade, a system of approximately 30 plasma and membrane protein components, factors, regulators, and receptors, plays three integral roles in host defense mechanisms: 1) recruiting phagocytic cells to the site of complement activation; 2) targeting invading microorganisms and other antigens to distinct surface complement receptor-bearing cells of the immune system; and 3) causing invading cell membrane damage and lysis. In addition to the immune response, components of the complement cascade interact directly or indirectly with other important multi-component systems including cell adhesion and coagulation. While host protective in purpose, the complement cascade is also an important mediator of the acute inflammatory response in many autoimmune and inflammatory diseases. Specific modifiers of the complement cascade would therefore be expected to be potentially useful therapeutic agents altering both the inflammatory and immune functions of complement. *Annual Reports in Medicinal Chemistry* has previously reviewed the complement cascade in 1972 (1) and complement inhibitors in 1980 (2). This chapter will offer a brief summary of the literature published in the last four years in this area .

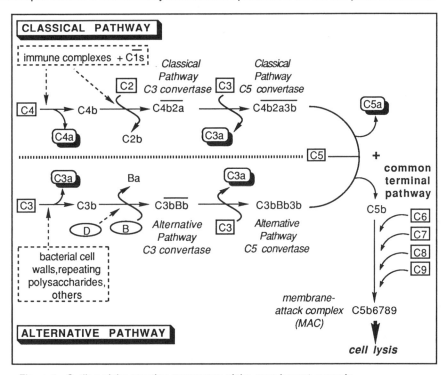

Figure 1. Outline of the reaction sequences of the complement cascade.

Complement cascade - Activation of the complement cascade (Fig. 1) via the immune complex-dependent classical pathway (utilizing components C1q, C1r, C1s, C2 and C4) or via the immune complex-independent alternative pathway (utilizing component C3 and Factors B and D) results in the sequential conversion of proenzymes to active enzymes, the formation of the pathway specific complex C3 and C5 convertases and the release of the biologically-active anaphylatoxins C3a, C4a and C5a. The common terminal pathway, initiated upon generation of protein C5b originating from either classical or alternative pathway, leads to the formation of the large protein membrane attack complex (MAC). Among the recent general reviews of the complement cascade is a book (3), a special September 1991 issue to *Immunology today* (4-14) and numerous journal reviews (15-17). Additional reviews have recently appeared on complement activation (18-20), complement regulation (21, 22), the anaphylatoxins (23-25), the membrane attack complex and its formation (26-32) and the link between blood coagulation and complement (33).

Complement receptors - Important discoveries now being made about the cellular receptors regulating the complement cascade can be expected to provide a mechanistic basis for future modulator drug design. Several excellent reviews have recently appeared (34-37). In general, these receptors are structures with long amino acid repeats, proteins anchored to membranes by 1,2-diacylglycerol, or heterodimers with a ligand-binding preference of Arg-Gly-Asn. Table 1 summarizes the known cell surface molecules reacting with complement components.

Table 1. Complement Receptors (34, 36)

Receptor Type *	Specificity	Receptor Type *	Specificity
B	Ba, Bb	CR2 (CD21)	iC3b, C3dg/C3d, EBV gp350
C1qR	C1q collagen tail	CR3 (CD11b/CD18)	iC3b, C3d, β–glucan, lipopolysaccharide
C3aR	C3a, C4a	CR4 (CD11c/CD18 p150,95)	iC3b, endothelial cells
C3eR	C3dk, C3e	DAF (CD55)	C3bBb, C4b2a
C5aR	C5a, C5a-des-Arg	HRF	Factor H
CR1 (CD35)	C4b, C3b, iC3b, C3c	MCP (CD46)	C3b, iC3b, C4b

(* CnR or CRn = complement receptor for n, DAF = decay accelerating factor, HRF = homologous restriction factor, MCP = membrane cofactor protein)

COMPLEMENT AND DISEASE

Reviews of the involvement of complement in disease states offer a recent update of the clinical manifestations of complement component deficiencies and excessive activities (38-40). Although rare, complement deficiencies provide the opportunity to delineate the role of individual components of the complement cascade (6). Table 2 represents those diseases characterized by deficiencies of one or more components of the complement system.

Table 2. Diseases Associated with Complement Deficiencies (6)

Component	Disease association
Classical pathway proteins	systemic lupus erythematosus (SLE)
C1q	pyogenic infections including meningitis
C1r / C1s	
C4	
C2	
Alternative pathway proteins	neisserial and pyogenic infections
Factor D	glomerulonephritis
Properdin	SLE
C3	hemolytic uremic syndrome
Factor H	
Factor I	
Terminal components	neisserial infections
C5, C6, C7, C8, C9	

As the principle biological role for the complement cascade is the neutralization and removal of invading pathogens, it is clear that it is an important part in the immune system. As seen above, deficiencies of many components allows recurring infections to occur. Diseases characterized by excessive complement activity lead to many inflammatory conditions. Diseases, many of inflammatory origin, where one or more components of the complement cascade are implicated are listed in Table 3. The specific disease will often be a result of the activation of complement in a particular tissue or organ or may be of systemic origins. The role of complement in these diseases is often surmised from the deposition of complement components in or near the sites of disease activity. Therapeutic invention of the various components of the complement cascade in these inflammatory diseases would probably not address the underlying etiology of the disease but rather would provide symptomatic relief unless there were feedback mechanisms involved. In those diseases where the pathology is acute (eg., ARDS or myocardial infarct), immediate relief of symptoms could be life saving and prevent permanent damage.

Table 3. Diseases Associated with Excessive or Uncontrolled Complement Activity

Kidney	Various nephritises	Dermatological	Bullous diseases
			SLE, DLE
Systemic	SLE		Acne
	Rheumatoid arthritis		Psoriasis
	Vasculitis		Lichen planus
Neurological / Muscular	Myasthenia gravis	Post traumatic	ARDS
	Multiple sclerosis		
	Guillain-Barre syndrome	Pulmonary	Chronic bronchitis
	Alzheimer's syndrome		Bronchial asthma
	Polymyositis		Sarcoidosis
	Muscular dystrophy		
		Cardiovascular	Myocardial infarct
Endocrine	Graves' disease		
	Hashimoto's thyroiditis	Other	Inflam. bowel disease
			Hemolytic anemia

INHIBITORS OF THE COMPLEMENT CASCADE

Reviews - Since the review of complement inhibitors appeared in *Annual Reports in Medicinal Chemistry* (2), several now dated summaries of the literature regarding modulation of the complement cascade have been published. The reviews summarized the literature on the basis of the stage of regulation during the cascade (41), the category of complement inhibitors by chemical class (42), or by classical or alternative pathway intervention (43). The following overview of classes of complement inhibitors will focus on the literature appearing in the last four years. References assuring a historical perspective of these inhibitors will be provided as appropriate.

Microbial- and plant-derived inhibitors and analogs - Complestatin (1), a peptide-like compound containing the two unusual amino acids D-(-)-4-hydroxyphenylglycine and D-(-)-3,5-dichloro-4-hydroxyphenylglycine has been shown to be a potent inhibitor of complement (44). Isolated from the mycelium of a strain of *Streptomyces lavendulae* (SANK 60477), the structure of 1 was determined on the basis of heteronuclear multiple bond correlation NMR spectroscopy (45). Found not to be a general protease inhibitor, compound 1 did inhibit 50% the classical pathway guinea pig and human complement-mediated hemolysis of sensitized sheep erythrocytes at 0.4 and 0.7 μg/ml, respectively. *In vivo*, 1 prevents the convulsion and death associated with anaphylatoxin C3a and C5a release in typical systemic anaphylactic shock. Without details and references, a later paper from an affiliated group states that 1 inhibits both the classical pathway of complement activation by inhibiting the formation of the C4b2b complex through binding to C2 and the alternative pathway by preventing the formation of the C3bBb complex by binding to factor B (46). Of interest, this paper details the inhibition of HIV-1 infection by 1 *in vitro*, possibly by the modulation of a cell-cell or cell-virion interaction.

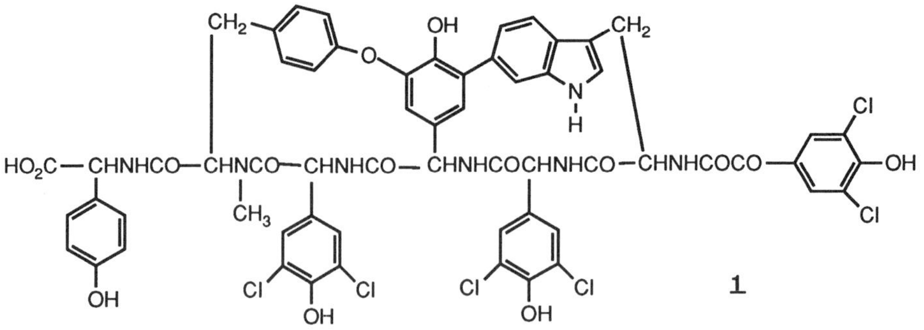

1

A series of 20 partial analogs of the complement inhibitory fungal metabolite isolated from *Stachybotrys complementi* (nov. sp. 76) (47, 48), K-76 (**2**) and its carboxylic acid derivative (K-76-COOH, **3**), has recently been synthesized and evaluated (49, 50). Compound **3** served as a model for the design of the partial analogs based on its reported inhibition of both the classical pathway and the alternative pathway *in vitro* (51-55) and *in vivo* (48, 56, 57). Compared to parent **3** (580 μM), the simple BCD-ring analog **4** (400 μM) was equipotent at inhibiting 50% the classical pathway human complement-mediated hemolysis of sensitized sheep erythrocytes, while regioisomer **5** (1300 μM) was less potent. ACD-ring analog **6** (300 μM) was twice as potent as that prepared from the enantiomer (700 μM).

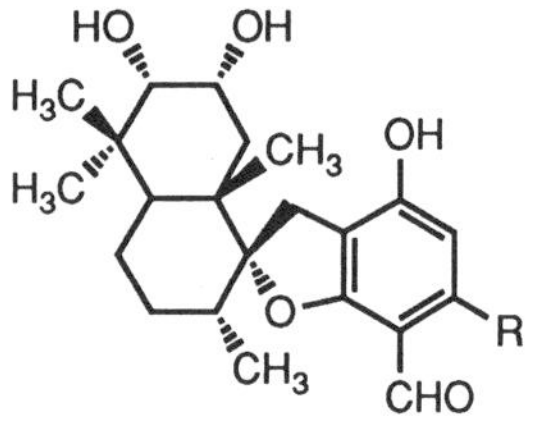

2, R = CHO

3, R = COOH

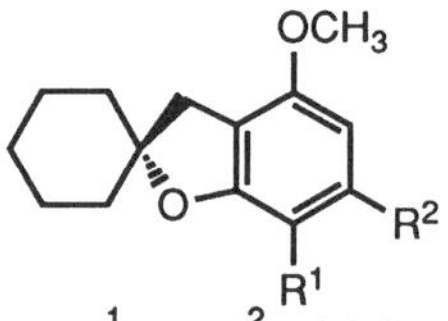

4, R^1 = H, R^2 = COOH

5, R^1 = COOH, R^2 = H

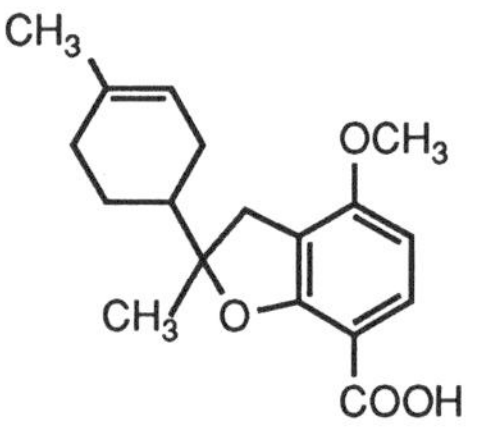

6

Rosmarinic acid (**7**), an ester extracted and purified from the labiatae *Rosmarinus officinalis* and *Melissa officinalis*, is reported to have *in vitro* inhibitory activity against the lytic sequence of complement by blocking classical pathway C3 convertase (10 μM) (58) and the cobra venom factor-induced (CVF-induced) alternative pathway enzyme (1mM) (59). A recent study examining the effect of **7** on individual complement components and pathways *in vitro* suggests that the complement modulation occurs principally through inhibition of C5 convertase by an unknown mechanism (60). *In vivo*, **7** blocks several complement-mediated inflammatory processes including CVF-induced paw edema (0.316-3.16 mg/kg i.m.) and ovalbumin/anti-ovalbumin-mediated passive cutaneous anaphylaxis in rats (1-100 mg/kg i.m.) (58). Additionally, **7** inhibits complement activation-induced prostacyclin synthesis in rabbit peritoneal tissue (59, 61). It has been reported that **7** has the complement-independent effects of inhibition of elastase (58) and scavenging oxygen free radicals (62,63)

Two structurally-related polypeptide antibiotics isolated over four decades ago, polymyxin B sulfate and colistin sulfate have recently been reported to inhibit the early stages of the classical pathway of human complement (64, 65). The inhibition is reversible since hemolytic activity was completely restored after dialysis.

Polyionic substances - Numerous studies have been published detailing the modulation of complement activation by polyanionic compounds such as heparin and heparin fragments and polycationic substances such as protamine sulfate (66-73). These reports support a mechanism of inhibition resulting from preventing interactions of C1, C4 and C2 during the formation of the classical pathway C3 convertase and blocking the binding site for B on C3b, thus hindering the formation of the alternative pathway C3 convertase. A minimal structure of the heparin molecule necessary for activity is a pentasaccharide with potency increasing with increasing sulfate content (74) as is also

$\underline{7}$ $\underline{8}$

observed with chondroitin sulfates (75). Additionally, this relationship is reported
for the dose-dependent inhibition of immune complex binding to human CR_1 studied with heparin, dextran sulfate, chondroitin sulfate, and suramin (76). Complement inhibition is a common activity of highly charged substances with polyanions preferentially inhibiting the alternative pathway while polycations preferentially inhibit the classical pathway (73). For example, heparin at 1.0 µg/10^7 cellular intermediates inhibited alternative pathway and classical pathway by 77% and 14%, respectively, whereas protamine sulfate at 0.25-1.0 µg/10^7 cellular intermediates inhibited these pathways at 34% and 98%, respectively. Inhibition of the lytic activity of cytolytic T-lymphocytes and the late complement components by proteoglycans and other polyions was studied in relation to homology with LDL-LDL receptor interactions (77).

A novel small molecule polyanionic substance, $\underline{8}$ resulting from modification of the complement inhibitory dye Chlorazol Fast Pink B blocks early steps in both the classical and alternative pathways (78, 79). Compound $\underline{8}$ is reported to suppress the vascular injury to endothelium and basal lamina at the site of the Arthus reaction when injected in rabbits at 2 x 24 mg/kg (80).

Small molecular weight inhibitors - Literature reports of small molecule complement inhibitors over the past four years have been very limited. Studies continue on the anticomplement benzamidine drug nafamostat mesylate (FUT-175, $\underline{9}$), a general protease inhibitor marketed in Japan for use in the treatment of acute pancreatitis. In vitro, $\underline{9}$ inhibits by 50% the guinea pig classical pathway-mediated hemolysis of sensitized sheep erythrocytes at 6.9 X 10^{-8} M and the human alternative pathway-mediated hemolysis at 5.1X 10^{-7} M (81, 82). In vivo, $\underline{9}$ prevents Forssman shock in guinea pigs, the Arthus reaction in rats, endotoxin shock in mice (83) and lupus nephritis in a mouse model (84). The rationale for studying the use of $\underline{9}$ to block complement activation during cardiopulmonary bypass has been reviewed (85). Compound $\underline{9}$ inhibited the increase in hemolytic complement in a dose-dependent manner in adjuvant arthritic rats (86).

$\underline{9}$

The synthetic dipeptide N-acetyl-L-aspartyl glutamic acid, detected as traces in
human brains, was found to inhibit hemolytic complement at quite high doses for both the classical pathway (2-6.10 X 10^{-3} M) and alternative pathway (2-10 X 10^{-3} M). Examples do exist of certain peptides blocking complement. For example, leupeptin (acetyl-Leu-Leu-Arg) suppresses C1s esterolysis to 60% at 0.03 mM (87). SAR studies of synthetic peptide blockers of complement serine proteases functionally equivalent to the cleavage site of C3 and C4, protease inhibitor sequences in serpins or the C-terminus of antithrombin III (AT III) have been reported (88, 89). A peptide corresponding to the AA sequence 382-420 of AT III caused significant inhibition of hemolysis in both pathways at a concentration of 5 µM.

Macromolecular inhibitors - One of the most active research areas in complement cascade modulation is the discovery of promising macromolecular agents, concurrently contributing to the definition of the therapeutic potential of such modulators in complement-dependent autoimmune and inflammatory diseases. Five endogenous complement system regulatory proteins, members of the regulators of complement activation (RCA) family (90), inhibit the proteolytic enzymes within the cascade that activate C3 and C5. The RCA family includes the plasma proteins, factor H (91, 92) and C4-binding protein (93, 94), and the three membrane receptor proteins, CR1 (95, 96), decay-accelerating factor (DAF) (97) and membrane cofactor protein (MCP) (98). Each protein, coupled with rDNA technology, offers the promise of new leads for the design of complement cascade modulators. Recombinant soluble CR1 (sCR1), the 200,000 molecular weight truncated form of CR1 lacking the transmembrane and cytoplasmic domains, has been shown to block both pathways of complement activation in human serum in nanomolar concentrations (99). Without interfering with healing, infarction size was reduced 44% and leukocyte infiltration to the site was diminished by sCR1 (bolus 1 mg I.V.) in a rat model of reperfusion injury of ischemic myocardium. In a dose-dependent manner, sCR1 reduced the vasculititis and the immunohistologic localization of neoantigen deposits in a rat reversed passive Arthus reaction model with a minimum effective reduction at 1 μg/dermal site (100). In addition, a short consensus repeat sequence of CR1 was fused to the heavy chains of a murine antibody to produce a chimeric protein able to inhibit the activation of the alternative patway (101). DAF is reported to inhibit in a dose-dependent manner, the cytotoxicity of porcine endothelial cells in the prevention of complement-mediated tissue injury *in vitro* in a porcine to human hyperacute rejection of discordant xenografts model (102). A mechanistic study detailed the preferential inactivation of alternative pathway C5 convertase by Factor I and MCP (103).

Structurally-defined gangliosides containing modified sialic acid residues were found to inhibit activation of the alternative pathway of human complement (104). Numerous polysaccharides reported to inhibit hemolytic complement have been isolated from plants, structurally characterized to varying degrees and chemically modified (105-111).

Complement inhibiting substances or extracts - Complement inhibitory activity observed in extracts may serve as a lead to novel chemical structures upon isolation and structural elucidation. The following studies have recently been reported in the literature: an inhibitor from *Aspergillus fumigatus* (112); inhibitory fractions from the Japanese herbal medicine "Juzen-Taiho-To" (113); inhibitory activity of *Jatropha multifida* (114); and inhibitors from spices and pokeweed fruits (115-117).

MODULATORS OF COMPLEMENT CASCADE PRODUCTS

C3a (23) - A comparison of the two-dimensional ^{1}H NMR solution structure of des-Arg^{77}-C3a (118) with the X-ray structure (119) agreed very well in the core region composed of three tightly packed α-helices, but the differences noted included a more helical structure in the amino-terminus and a loss of structure beyond residue 66. The loss of the carboxy-terminal Arg residue from C3a eliminates its biological activities, although its conformation is unaffected (120). A 21-residue peptide corresponding to the carboxy terminus elicits the same activities in a guinea pig ileum contraction assay as does intact C3a (121). The minimal active peptide in the C-terminal region of guinea pig C3a was found to be LGLAR (122) having 10^3-10^4 times less activity in the ileum contraction or platelet ATP-release assays than the parent C3a. This C-terminal sequence has formed the basis for much of the subsequent analog work to prepare more potent agonists of C3a activity and explore the binding requirements for the C3a receptor.

A thirteen amino acid peptide having 6-(4-azido-2-nitrophenylamino)hexanoate (Nap-Ahx) on the N-terminus (Nap-Ahx-YRRGRAAALGLAR) was six times more active than C3a in the guinea pig platelet assay (123, 124). An even shorter active sequence containing only three amino acids (Fmoc-Ahx-LAR) was prepared that had ~300 times less activity than C3a in the platelet assay and suggested a specific hydrophobic region on the C3a receptor (125). Exceptionally potent peptide agonists were prepared by optimizing the relative positions of the carboxy-terminal binding component (LGLAR) and an N-terminal hydrophobic unit (WW) (126). The most potent agonist (WWGKKYRASKLGLAR) was 12-15 times more potent than C3a in a guinea pig platelet aggregation assay.

C5a - A three dimensional model for the C5a structure was developed from the X-ray crystal structure of the homologous anaphylatoxin C3a (36% identity) (127). The tertiary structure of human C5a was determined by ^{1}H NMR (128) and by heteronuclear ^{1}H-^{15}N NMR (129). The structure was found in good agreement with the proposed model in the region of the N-terminal region and the core but the important carboxy-terminal was disordered (130). Similar structures were obtained by ^{1}H NMR for bovine C5a (131) and porcine des-Arg^{74}-C5a (132). Site-directed mutagenesis studies identified the following residues as being important for C5a binding to its receptor: Lys^{68}, Leu^{72} and

Arg^{74} in the carboxy-terminus, Arg^{40} (and/or Ile^{41}) in the core region, and His^{15} (implied by a conformational change induced by a mutation at Ala^{26}) in the N-terminus (133).

A series of peptides were prepared to systematically survey various regions of the C5a molecule in order to identify those areas important for binding to the C5a receptor (134). Like C3a, the carboxy-terminal peptide 59-74(SQLRANISHKDMQLGR) was the only one prepared that effectively blocked binding of radiolabeled C5a to its receptor (>20% inhibition at 0.1-1 mM). Further investigation of this region identified the terminal octapeptide (HKDMQLGR) as the minimal active sequence (K_i = 300 µM vs K_i = 0.00016 µM for C5a). Optimization of this sequence yielded an octapeptide (FKA-Cha-AL-(N-Me)dA-R) with excellent affinity for the C5a receptor (Ki = 0.012 µM) (135, 136). The Phe for His replacement at the N-terminus alone improved binding affinity by nearly three orders of magnitude (137). Chemokinesis data suggested that these peptides are all agonists of the C5a receptor.

There are few examples of nonpeptidyl antagonists of the C5a receptor. The C5a molecule is highly cationic which has suggested searching for mimics that might interact with the presumedly anionic-rich C5a receptor. Poly-L-Arg and protamine were found to inhibit C5a-induced leukocyte responses including histamine release from basophils and chemiluminescence and β-glucuronidase from neutrophils in the range of 5-20 µg/mL (138). A similar strategy of searching among cationic structures for C5a antagonists yielded substituted 4,6-diaminoquinolines $\underline{10}$ and $\underline{11}$ with IC_{50}'s = 3.3 and 12 µg/mL, respectively for inhibiting radiolabeled C5a binding to its receptor (139). Attempts to enhance these potencies by chemical modification revealed a very narrow profile for effective C5a receptor antagonism. A heterocyclic antiinflammatory agent AA-2379 ($\underline{12}$) that is active in Arthus reactions is reported to be a weak inhibitor (10-100 µM) of C5a- and fMLF-induced degranulation of rat neutrophils (140).

$\underline{10}$ $\underline{11}$

$\underline{12}$

A C5a antagonist L-156,602, a cyclic hexadepsipeptide containing five "unnatural" amino acids and a fourteen carbon side chain, was isolated from a *Streptomyces* fermentation and its structure, determined by ^1H NMR and X-ray crystallography, was found to be identical to a previously isolated antibiotic (141). The successful total synthesis of this novel cyclic hexadepsipeptide (142) as well as direct chemical modification of the natural product (143) provided the methodology for an extensive examination its binding requirements to the C5a receptor. The N-hydroxyl groups were unimportant but each of the amino acids in the cyclic peptide as well as the fourteen carbon side chain were crucial for effectively inhibiting C5a binding (144).

Understanding of the C5a receptor is rapidly increasing. Its induction and expression on neutrophils is regulated independently of other receptors such as fMLF, CR1, CR3 and Fc (145, 146). The receptor has been isolated as a tight complex with a G-protein confirming it as a member of the G-linked receptor family (147). The C5a receptor has also been cloned and expressed and its deduced amino acid sequence places it among the rhodopsin superfamily of membrane-bound receptors with seven membrane spanning sequences (148). A novel recombinant C5a molecule with a 19-residue addition on the N-terminus that can be recognized by antibodies when bound to its receptor provides a useful probe of the receptor molecule on cell surfaces (149).

References

1. H.R. Colten in *Ann.Repts. Med.Chem.*, R.V. Heinzelman, Ed., Academic Press, New York, 7, 228 (1972).
2. R.A. Patrick and R.E. Johnson in *Ann.Repts.Med.Chem.*, H-J Hess, Ed., Academic Press, New York, 15, 193 (1980).
3. "The Complement System", K. Rother and G. O. Till, Eds., Springer-Verlag, Berlin, 1988.
4. T. Kinoshita, Immunol.Today, 12, 291 (1991).
5. T.C. Farries and J.P. Atkinson, Immunol.Today, 12, 295 (1991).
6. B.P. Morgan and M.J. Walport, Immunol.Today, 12, 301 (1991).
7. R.B. Sim and K.B.M. Reid, Immunol.Today, 12, 307 (1991).
8. P.J. Lachmann, Immunol.Today, 12, 312 (1991).
9. A.F. Esser, Immunol.Today, 12, 316 (1991).
10. S. Bhakdi and J. Tranum-Jensen, Immunol.Today, 12, 318 (1991).
11. M.M. Frank and L.F. Fries, Immunol.Today, 12, 322 (1991).
12. N.R. Cooper, Immunol.Today, 12, 327 (1991).
13. A. Erdei, G. Fust and J. Gergely, Immunol.Today, 12, 332 (1991).
14. P.J. Sims and T. Wiedmer, Immunol.Today, 12, 338 (1991).
15. M.M. Frank, J.AllergyClin.Immunol., 84, 411 (1989).
16. H. Muller-Eberhard, Ann.Rev.Biochem., 57, 321 (1988)
17. P.J. Lachmann, Biochem.Soc.Trans., 18, 1143 (1990).
18. T.C. Farries, K.L.K. Steuer and J.P. Atkinson, Immunol.Today, 11, 78 (1990).
19. J.E. Volanakis, Curr.Top.Microbiol.Immunol., 153, 1 (1990).
20. W.P. Kolb, P.R. Morrow and J.D. Tamerius, Complement Inflamm., 6, 175 (1989)
21. S.R. Barnum, Immunol.Res. (Basel), 10, 28 (1991).
22. P.J. Lachmann, Biochem.Soc.Trans., 18, 1159 (1990).
23. T.E. Hugli, Curr.Top.Microbiol.Immunol., 153, 181 (1990).
24. J.A. Schifferli and J.P. Paccaud, Complement Inflamm., 6, 19 (1989)
25. J. Janatova, Methods Enzymol., 162 (Immunochem. Tech., Pt. L), 579 (1988).
26. K.K. Stanley, Curr.Top.Microbiol.Immunol., 140, 49 (1989).
27. J.M. Stodetz, Curr.Top.Microbiol.Immunol., 140, 19 (1989).
28. A.F. Esser, Year Immunol. 90, 6, 229 (1989).
29. B.P. Morgan, Biochem.J., 264, 1 (1989).
30. B.P. Morgan, Complement Inflamm., 6, 104 (1989).
31. A.P. Dalmasso, R.J. Falk and L. Raij, Complement Inflamm., 6, 36 (1989).
32. T.E. Mollnes, T. Lea and J. Tschopp, Complement Inflamm., 6, 223 (1989).
33. M. Hessing, Biochem.J., 277, 581 (1991).
34. J.D. Becherer, J. Alsenz, C. Servis, B.L. Myones and J.D. Lambris, Complement Inflamm., 6, 142 (1989).
35. J. M. Ahearn and D.T. Fearon, Adv.Immunol., 46, 183 (1989).
36. G.D. Ross, Curr.Opinion Immunol., 2, 50 (1989).
37. M.P. Dierich, T.F. Schultz, A. Eigentler, H. Huemer and W. Schaeble, Mol.Immunol., 25, 1043 (1988).
38. B.P. Morgan, "Complement: Clinical Aspects and Relevance to Disease", Academic Press, New York, N.Y., 1990.
39. J.E. Figueroa and P. Densen, Clin.Microbiol.Rev., 4, 359 (1991).
40. W.R. Bartholomew and T.C. Shanahan, Immunol.Ser., 52, 33 (1990).
41. A.M. Reynard, J.Immunopharmacol., 2, 1 (1980).
42. S.S. Asghar, Pharmacol.Rev., 36, 223 (1984).
43. S. Fujii and T. Aoyama, Drugs Future, 9, 849 (1984).
44. I. Kaneko, K. Kamoshida and S. Takahashi, J.Antibiot., 42, 236 (1989).
45. H. Seto, T. Fujioka, K. Furihata, I. Kaneko and S. Takahashi, Tet.Lett., 30, 4987 (1989).
46. K. Momota, I. Kaneko, S. Kimura, K. Mitamura and K. Shimada, Biochem.Biophys.Res.Commun., 179 243 (1991).
47. H. Kaise, M. Shinohara, W. Miyazaki, T. Izawa, Y. Nakano, M. Sugawara and K. Sugiura, J.Chem.Soc., Chem.Commun., 726 (1979).
48. W. Miyazaki, H. Tamoaka, M. Shinohara, H. Kaise, T. Izawa, Y. Nakano, T. Kinoshita, K. Hong and K. Inoue, Microbiol.Immunol., 24, 1091 (1980).
49. B.J. Bradbury and R.D. Sindelar, J.Heterocycl.Chem., 26, 1827 (1989).
50. R.D. Sindelar, B.J. Bradbury, T.S. Kaufman, C. Lee, S. Scesney and H.C. Marsh, Jr., Abs.Pap.200thNatl.ACS Mtg., MED178 (1990)
51. K. Hong, T. Kinoshita, W. Miyazaki, T. Izawa and K. Inoue, J.Immunol., 122, 2418 (1979).
52. K. Hong, T. Kinoshita, H. Kitajima and K. Inoue, J.Immunol., 127, 104 (1981) .
53. K. Hong, T. Kinoshita and K. Inoue, J.Immunol., 127, 109 (1981).
54. D. Hudig, D. Redelman, L. Minning and K. Carine, J.Immunol., 133, 408 (1984).
55. D. Redelman and D. Hudig, Cell.Immunol., 88, 16 (1984).

56. W. Miyazaki, T. Izawa, Y. Nakano, M. Shinohara, K. Hing, T. Kinoshita and K. Inoue, Complement, 1, 134 (1984).
57. S. Konno and S. Tsurufuji, Japan.J.Pharmacol., 38, 116 (1985).
58. W. Englberger, U. Hadding, E. Etschenberg, E. Graf, S. Leyck, J. Winkelmann and M.J. Parnham, Int.J.Immunopharmacol., 10, 729 (1988).
59. M. Rampart, J. R. Beetens, H. Bult, A. G. Herman, M.J. Parnham, and J. Winkelmann, Biochem.Pharmacol., 35, 1397 (1986).
60. P.W. Peake, B.A. Pussell, P. Martyn, V. Timmermans, and J.A. Charlesworth, Int.J.Immunopharmacol., 13, 853 (1991).
61. H. Bult, A.G. Herman and M. Rampart, Br.J.Pharmacol., 84, 317 (1985).
62. J.K.S. Nuytinck, R.J.A. Goris, E.S. Kalter and P.H.M. Schillings, Agts.Actns., 17, 373 (1985).
63. K.P.M. Van Kessel, E.S. Kalter and J. Verhoef, Agts.Actns., 17, 375 (1985).
64. S.S. Asghar, T. Boot and H. J. van der Helm, Biochem.Pharmacol., 36, 2927 (1987).
65. S.S. Asghar, A. de Koster and H. J. van der Helm, Biochem.Pharmacol., 35, 2917 (1986).
66. E.E. Ecker and P. Gross, J.Infect.Dis., 44, 250 (1929).
67. E. Raepple, H.U. Hill and M. Loos, Immunochem., 13, 251 (1976).
68. M. Loos, J.E. Volanakis and R.M. Stroud, Immunochem., 13, 257 (1976).
69. J.M. Weiler, R.W. Burt, D.T. Fearon and K.F. Austen, J.Exp.Med., 147, 409 (1978).
70. M.D. Kazatchkine, D.T. Fearon, D.D. Metcalfe, R.D. Rosenberg and K.F. Austen, J.Clin.Invest., 67, 223 (1981).
71. W.E. Hennink, J.P.A.M. Klerx, H. van Dijk and J. Feijen, Thromb.Res., 36, 281 (1984).
72. M.D. Sharath, Z.M. Merchant, Y.S. Kim, K.G. Rice, R.J. Linhardt and J.M. Weiler, Immunopharmacol., 9, 73 (1985).
73. J.M. Weiler and R.J. Linhardt, Immunopharmacol., 17, 65 (1989).
74. F. Maillet, M. Petitou, J. Choay and M.D. Kazatchkine, Mol.Immunol., 25, 917 (1988)
75. M. Biffoni and E. Paroli, Drugs Exptl.Clin.Res., 17, 35 (1991).
76. H.H. Jepsen, H.-P.T. Ekre and S.-E. Svehag, Int.J.Immunopharmacol., 9, 587 (1987).
77. J. Tschopp and D. Masson, Mol.Immunol., 24, 907 (1987).
78. R.B. Conrow, N. Bauman, J.A. Brockman and S. Bernstein, J.Med.Chem., 23, 242 (1980).
79. N. Bauman and J.A. Brockman, J.Immunol., 120, 1764 (1978).
80. M.Y.M. Abdel Mawla, K.P. Dingemans, M. Amer., G.T. Venneker, M. van Meegen and S.S. Asghar, Complement Inflamm., 8, 50 (1991).
81. S. Fujii and Y. Hitomi, Biochem.Biophys.Acta, 661, 345 (1981).
82. T. Aoyama, Y. Ino, M. Ozeki, M. Oda, T. Sato, Y. Koshiyama, S. Suzuki and M. Fujita, Japan.J.Pharmacol., 35, 203 (1984).
83. Y. Hitomi and S. Fujii, Int.Arch.Allergy Appl.Immunol., 69, 262 (1982).
84. S. Ikehara, K. Shimamura, T. Aoyama, S. Fujii and Y. Hamashima, Immunol., 55, 595 (1985).
85. Y. Miyamoto, H. Matsuda and Y. Kawashima, J.Biomater.Appl., 4, 56 (1989).
86. Y. Ino, T. Sato, Y. Koshiyama, K. Suzuki, M. Oda and M. Iwaki, Gen.Pharmacol., 18, 513 (1987).
87. Y. Takeda, Y. Arimoto, H. Mineda and A. Takada, Immunol., 34, 509 (1978).
88. G.I. Glover, C.S. Schasteen, W.-S. Liu and R.P. Levine, Mol.Immunol., 25, 1261 (1988).
89. C.S. Schasteen, S.A. McLafferty, G.I. Glover, C.Y. Han, J.C. Mayden, W.-S. Liu and R.P. Levine, Mol. Immunol., 25, 1269 (1988).
90. D. Hourcade, V.M. Holers and J.P. Atkinson, Adv.Immunol., 45, 381 (1989).
91. K. Whaley and S. Ruddy, J.Exp.Med., 144, 1147 (1976).
92. J. M. Weiler, M.R. Daha, K.F. Austen, D.T. Fearon, Proc.Natl.Acad.Sci.U.S.A., 73, 3268 (1976).
93. S. Shiraishi and R.M. Stroud, Immunochem., 12, 935 (1975).
94. I. Gigli, T. Fujita and V. Nussenzweig, Proc.Natl.Acad.Sci.U.S.A., 76, 6596 (1979).
95. D.T. Fearon, Proc.Natl.Acad.Sci.U.S.A., 76, 5867 (1979).
96. K. Iida and V. Nussenzweig, J.Exp.Med., 153, 1138 (1981).
97. A. Nicholson-Weller, J. Burge, D.T. Fearon and K.F. Austen, J.Immunol., 129, 184 (1982).
98. T. Seya, J. Turner and J.P. Atkinson, J.Exp.Med., 163, 837 (1985).
99. H.F. Weisman, T. Bartow, M. K. Leppo, H.C. Marsh, Jr., G.R. Carson, M.F. Concino, M.P. Boyle, K.H. Roux, M.L. Weisfeldt and D.T. Fearon, Science, 249, 146 (1990).
100. C.G. Yeh, H.C. Marsh, Jr., G.R. Carson, L. Berman, M.F. Concino, S.M. Scesney, R.E. Kuestner, R. Skibbens, K.A. Donahue and S.H. Ip, J.Immunol., 146, 250(1991).
101. K.R. Kalli, P. Hsu, T.J. Bartow, J.M. Ahearn, A.K. Matsumoto, L.B. Klickstein and D.T. Fearon, J.Exp.Med., 174 1451 (1991).
102. A.P. Dalmasso, G.M. Vercellotti, J.L. Platt and F.H. Bach, Transplantation, 52, 530 (1991).
103. T. Seya, M. Okada, M. Matsumoto, K. Hong, T. Kinoshita and J.P. Atkinson, Mol.Immunol., 28, 1137 (1991).
104. M.T. Michalek, C. Mold and E.G. Bremer, J.Immunol., 140, 1588 (1988).
105. H. Yamada, H. Kiyohara, J.-C. Cyong, and Y. Otsuka, Carbohyd.Res., 159, 275 (1987).
106. H. Yamada, K.-S. Ra, H. Kiyohara, J.-C. Cyong, H.C. Yang and Y. Otsuka, Phytochem., 27, 3163 (1988).

107. H. Kiyohara, J.-C. Cyong and H. Yamada, Carbohyd.Res., 193, 193 (1989).
108. H. Kiyohara, J.-C. Cyong and H. Yamada, Carbohyd.Res., 193, 201 (1989).
109. Q. Gao, H. Kiyohara, J.-C. Cyong and H. Yamada, Planta Med., 55, 9 (1989).
110. R. Gonda, M. Tomoda and N. Shimizu, Carbohyd.Res., 198, 323 (1990).
111. J.-F. Zhao, H. Kiyohara and H. Yamada, Carbohyd.Res., 219, 149 (1991).
112. R.G. Washburn, D.J. DeHart, D.E. Agwu, B.J. Bryant-Varela and N.C. Julian, Infect.Immunity, 58, 3508 (1990).
113. H. Yamada, H. Kiyohara, J.-C. Cyong, N. Takemoto, Y. Komatsu, H. Kawamura, M. Aburada and E. Hosoya, Planta Med., 56, 386 (1990).
114. S. Kosasi, L.A. Hart, H. van Dijk and R.P. Labadie, J.Ethnopharmacol., 27, 81 (1989).
115. G.G. Gancevici and C. Popescu, Arch.Roum.Pathol.Exp.Microbiol., 46, 47 (1987).
116. G.G. Gancevici and C. Popescu, Arch.Roum.Pathol.Exp.Microbiol., 46, 321 (1987).
117. C. Popescu, M. Croitoru and G.G. Gancevici, Arch.Roum.Pathol.Exp.Microbiol., 47, 37 (1988).
118. D.G. Nettesheim, R.P. Edalji, K.W. Mollison, J. Greer and E.R.P. Zuiderweg, Proc.Natl.Acad.Sci.USA, 85, 5036 (1988).
119. R. Huber, H. Scholze, E.P. Paques and J. Deisenhofer, Hoppe-Seyler's Z.Physiol.Chem., 361, 1389 (1980).
120. T. E. Hugli, W.T. Morgan and H.J. Muller-Eberhard, J.Biol.Chem., 250, 1479 (1975).
121. R. Huey, C.M. Bloor, D.S. Kawahara and T.E. Hugli, Am.J. Pathol, 112, 48 (1983).
122. L.H. Caporale, P.S. Tippett, B.W. Erickson and T.E. Hugli, J.Biol.Chem., 255, 10758 (1980).
123. R. Gerardy-Schahn, D. Ambrosius, M. Casaretto, J. Grotzinger, D. Saunders, A. Wollmer, D. Brandenburg and D. Bitter-Suermann, Biochem.J., 255, 209 (1988).
124. D. Ambrosius, M. Casaretto, R. Gerardy-Schahn, D. Saunders, D. Brandenburg and H. Zahn, Biol.Chem.Hoppe-Seyler, 370, 217 (1989)
125. J. Kohl, M. Casaretto, M. Gier, G. Karwath, C. Gietz, W. Bautsch, D. Saunders and D. Bitter-Suermann, Eur.J.Immunol., 20, 1463 (1990)
126. J.A. Ember, N.L. Johansen and T.E. Hugli, Biochem., 30, 3603 (1991).
127. J. Greer, Science, 228, 1055 (1985).
128. E.R.P. Zuiderweg, D.G. Nettesheim, K.M. Mollison and G.W. Carter, Biochem., 28, 172 (1989).
129. E.R.P. Zuiderweg and S.W.Fesik, Biochem., 28, 2387 (1989).
130. E.R.P. Zuiderweg, J. Henkin, K.W. Mollison, G.W.Carter and J. Greer, Proteins: Structure, Function, and Genetics, 3, 139 (1988).
131. J. Zarbock, R. Gennaro, D. Romeo, G.M. Clore and A.M. Gronenborn, FEBS Lett., 238, 239 (1988).
132. M.P. Williamson and V.S. Madison, Biochem., 29, 2895 (1990).
133. K.W. Mollison, W. Mandecki, E.R.P. Zuiderweg, L. Fayer, T.A. Fey, R.A. Krause, R.G. Conway, L. Miller, R.P. Edalji, M.A. Shallcross, B. Lane, J.L. Fox, J. Greer and G.W.Carter, Proc.Natl.Acad.Sci.USA, 86, 292 (1989).
134. M. Kawai, D.A. Quincy, B. Lane, K.W. Mollison, J.R. Luly and G.W. Carter, J.Med.Chem., 34, 2068 (1991).
135. M. Kawai, Y.S. Or, P.E. Wiedeman, J.R. Luly and M.P. Moyer, Pat. WO 90/09162 (1990).
136. M. Kawai, D.A. Quincy, B. Lane, K.W. Mollison, Y.-S. Or, J.R. Luly and G.W. Carter, J.Med.Chem., 35, accepted for publication (1992).
137. Y.-S. Or, R.F. Clark, B. Lane, K.W. Mollison, G.W. Carter and J.R. Luly, J.Med.Chem., 35, 402 (1992).
138. U.B. Olsen, J. Selmer and J.-U. Kahl, Complement, 5, 153 (1988).
139. T.J. Lanza, P.L. Durette, T. Rollins, S. Siciliano, D.N. Cianciarulo, S.V. Kobayashi, C.G. Caldwell, M.S. Springer and W.K. Hagmann, J.Med.Chem., 35, 252 (1992).
140. H. Makino, T. Naka, T. Saijo and Y. Maki, Agts.Actns., 25, 326 (1988).
141. O.D. Hensens, R.P. Borris, L.R. Koupal, C.G. Caldwell, S.A. Currie, A.A. Haidri, C.F. Homnick, S.S. Honeycutt, S.M. Lindenmayer, C.D. Schwartz, B.A. Weissberger, H.B. Woodruff, D.L. Zink, L. Zitano, J.M. Fieldhouse, T. Rollins, M.S. Springer and J.P. Springer, J.Antibiot., 44, 249 (1991).
142. P.L. Durette, F. Baker, P.L Barker, J. Boger, S.S. Bondy, M.L. Hammond, T.J. Lanza, A.A. Pessolano and C.G. Caldwell, Tet.Lett., 31, 1237 (1990).
143. I.E. Kopka, Tet.Lett., 33, 4711 (1990).
144. P.L. Durette, Abs.Pap.200th Natl.ACS Mtg., MEDI88 (1990).
145. W. Zimmerli, A.-M. Reber and C.A. Dahinden, J.Infect.Dis., 161, 242 (1990).
146. D.E. Van Epps, J.G. Bender, S.J. Simpson and D.E. Chenoweth, J.Leuk.Biol., 47, 519 (1990).
147. T.E. Rollins, S. Siciliano, S. Kobayashi, D.N. Cianciarulo, V. Bonilla-Argudo, K. Collier and M.S. Springer, Proc.Natl.Acad.Sci.USA, 88, 971 (1991).
148. N.P. Gerard and C. Gerard, Nature, 349, 614 (1991).
149. N.P. Gerard and C. Gerard, Biochem., 29, 9274 (1990).

Chapter 22. Cytokine Modulation as a Medicinal Chemistry Target

Kelvin Cooper and Hiroko Masamune
Central Research Division, Pfizer Inc, Groton, CT 06340

Introduction — The cytokines are the protein hormones that orchestrate leukocyte function during the active phases of natural and specific immunity in addition to regulating immune and inflammatory responses. Two fundamental features are clearly emerging as characteristic of all the cytokines: 1) pleiotropy — each cytokine has more than one biological function, and 2) redundancy — each function is mediated by more than one cytokine. Furthermore, each cytokine is produced by and acts on multiple cell types, often influencing the function of other cytokines. Thus, these far ranging activities have made cytokines challenging targets for drug research. This chapter will review recent progress made in the discovery of modulators of cytokine function with emphasis on the activity of small molecules. Clinical use of specific biologicals such as antibodies, soluble receptors, or recombinant cytokines will also be covered. The cytokine receptors were reviewed in Volume 26 of this series (1).

Interleukin 1 — IL-1 exists in two different forms, IL-1α and IL-1β, and is primarily produced by macrophages. It plays a central role in both inflammatory and immunological responses and has been implicated in a wide range of human diseases such as hemodynamic shock, arthritis, inflammatory bowel disease, and lethal sepsis. The structure and biological functions of IL-1 have been reviewed (2). A wide range of small molecules has been found to modulate IL-1 synthesis or release and was reviewed in volume 25 of this series (3). This section will update and extend that review.

$$\underline{1} \qquad\qquad \underline{2} \qquad\qquad \underline{3}\ R = CH_3S \qquad \underline{4}\ R = CH_3S\,O$$

Pentamidine $\underline{1}$ blocks (IC$_{50}$ = 1 μM) the post translational modification of proIL-1α to the mature 17 kDa form in lipopolysaccharide (LPS)-stimulated mouse macrophages without affecting phagocytosis or Ia antigen expression (4). IX 207-887 $\underline{2}$ inhibits release, but not synthesis of IL-1 from human monocytes and mouse macrophages with IC$_{50}$'s of 30 and 60 μM, respectively, without affecting secretion of IL-6 and tumor necrosis factor (TNF) (5). SKF 105,809 $\underline{3}$ is converted *in vivo* to the combined cyclooxygenase/lipoxygenase (CO/LO) inhibitor SKF 105,561 $\underline{4}$ and shows unique anti-inflammatory activity in murine models that are normally resistant to CO inhibitors (6). This activity is due to its *in vitro* inhibition of IL-1 production (7), although the mechanism of action is unknown. A series of naphthalene propenoic acids $\underline{5}$ blocks IL-1 generation *in vitro* in LPS-stimulated human monocytes and rat macrophages, and *in vivo* in a rat CMC-LPS air-pouch model (8). Acetoxy E 5090 $\underline{6}$ serves as a prodrug for the active hydroxy derivative $\underline{7}$, the most potent compound in the series

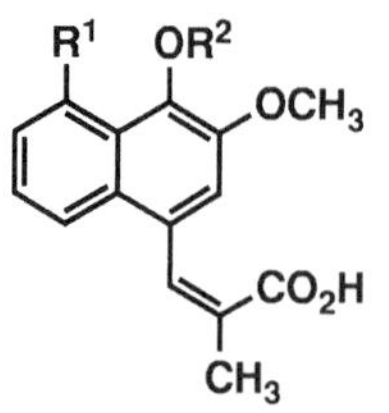

5
6 R^1 = CH$_3$, R^2 = OAc
7 R^1 = CH$_3$, R^2 = OH

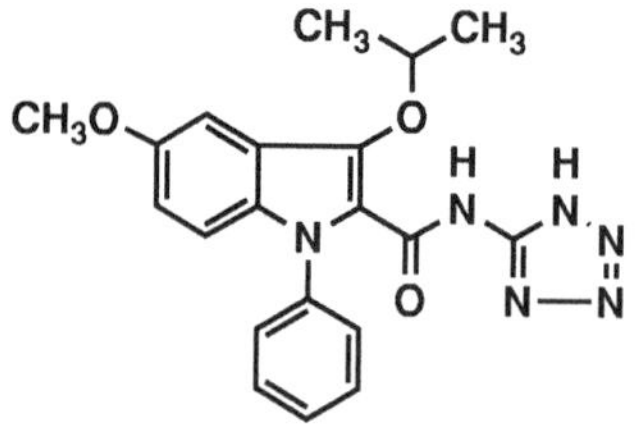

8

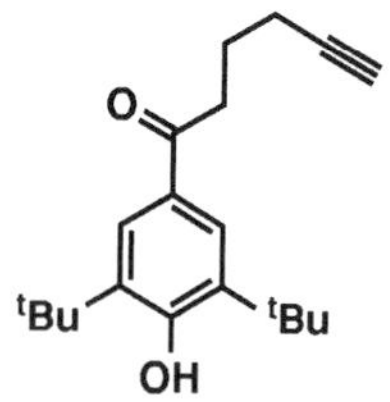

9

(9). The anti-allergy compound CI-949 **8**, an inhibitor of histamine, leukotriene and thromboxane production from human mast cells, has also been shown to be a weak inhibitor of IL-1 release from human peripheral blood leukocytes (10). In contrast, the CO/LO inhibitor, tebufelone **9**, enhances LPS-stimulated IL-1 and TNF production by human peripheral blood mononuclear cells at concentrations similar to those which block the production of leukotriene (LT) synthesis. The mechanism of increased production is a post translational event since IL-1 or TNF mRNA levels were not affected (11). The novel steroid, mometasone furoate **10** is an inhibitor of IL-1, IL-6, and TNF production in LPS-stimulated mouse leukemia cells, being 100-200X more potent than dexamethasone **11** (12).

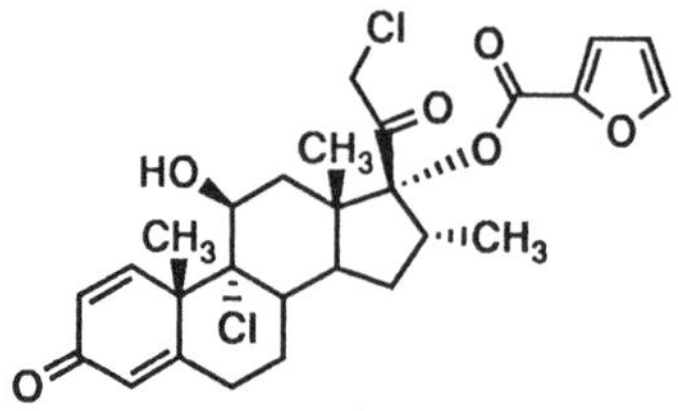

10

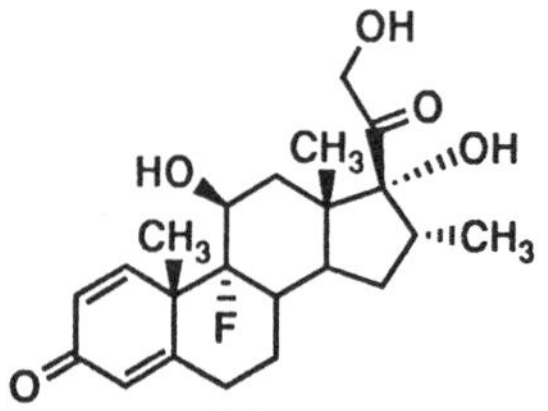

11

In addition to their direct effect on IL-1 production, gold sodium thiomalate **12** and auranofin **13** inhibit γ-interferon (γ-IFN) enhancement of IL-1 production (13). The immunosuppressant FK506 **16** (vide infra), in addition to its effect on T-cell responses, blocks the synthesis and production of IL-1 from LPS-stimulated human monocytes and alveolar macrophages (14). The protease which cleaves proIL-1β to the mature 17 kDa form represents an attractive target for drug intervention. The substrate specificity of this enzyme, IL-1 converting enzyme (ICE), has been determined (15) and patent applications on the sequence, as well as on a series of peptidic inhibitors **14**, have been filed (16). Human derived IL-1 receptor antagonist (IL-1ra), a 152 amino acid peptide (17), binds to the receptor with high affinity (Kd = 150 pM)(18).

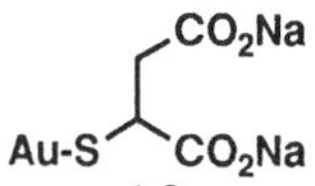

12

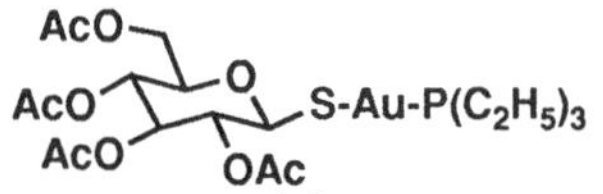

13

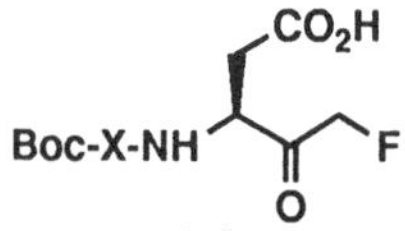

14

Interleukin 2 — IL-2 plays an integral role in the clonal expansion of T-lymphocytes after antigen stimulation (19, 20). In addition, it stimulates the growth and differentiation of a variety of other lymphocytes, including B cells, natural killer (NK) cells, and lymphokine-activated killer (LAK) cells. Il-2 exists as a 15 kDa monomeric glycoprotein and recently, an FDA advisory committee recommended approval for Chiron/Cetus' recombinant IL-2 (Proleukin) for the treatment of renal cell carcinoma, an often fatal disease (21). The efficacy of IL-2 has been investigated in a number of other cancerous states (22), but unfortunately, it appears that patients receiving high doses of IL-2 suffer from a number of serious side effects (23), as well as experiencing disease relapse (24). Several approaches are being explored to decrease the limiting toxicity of IL-2, including administering low-dose IL-2 (25), combining IL-2 with ibuprofen (26), and examining the possible mechanisms of toxicity (27). The IL-2 receptor (IL-2R) is also the focus of much research (28).

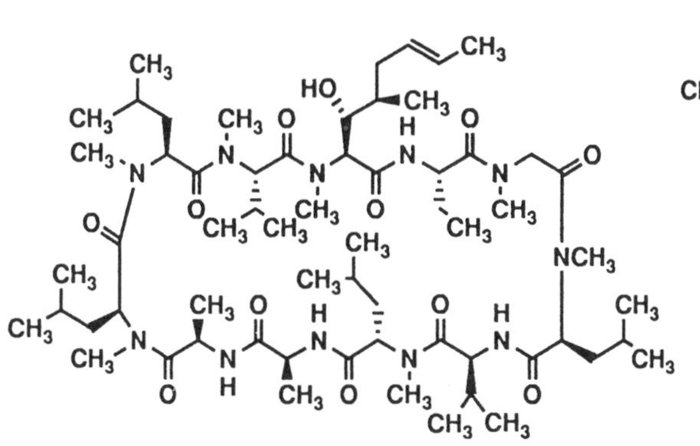

15

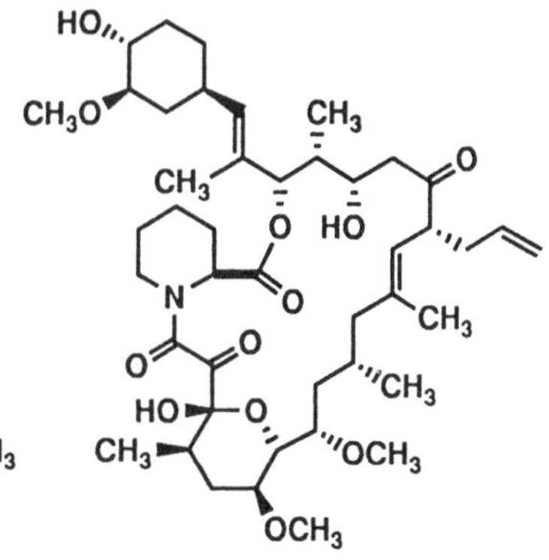

16

Although the direct involvement of IL-2 has been difficult to assess (19), it is implicated in a number of autoimmune diseases such as rheumatoid arthritis (RA) and in patients undergoing organ transplants. Inhibition of IL-2 production has thus been a major objective of drug discovery and is thought to be one of the modes of action of cyclosporin A **15**, the current drug of choice in anti-rejection therapy (29). Unfortunately, the use of **15** is dose-limiting due to systemic toxicity, particularly nephrotoxicity. Another macrolide of interest is FK506 **16**, which appears to be 10-100X more potent that **15** (30) and appears to inhibit IL-2 production, even though it binds to a different cytosolic receptor than **15** (31). Early clinical data on **16** suggest that it is also an effective immunosuppressant with a side effect profile which may limit its use (32).

A drug which is being developed as therapy for RA, bucillamine **17**, has been found to inhibit the production of IL-2 by 50% at 100 μM (33). Similar effects were seen with sulphasalazine **18**, an anti-rheumatic drug used to treat ulcerative colitis (34). Dexamethasone **11**, inhibits IL-2 production at much lower concentrations (35). Compound **11** also has a significant effect on IL-2R production and IL-2 mRNA levels.

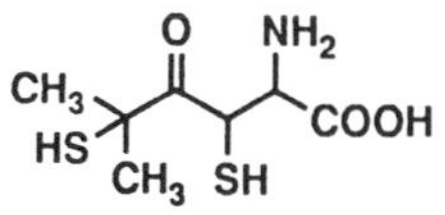

17

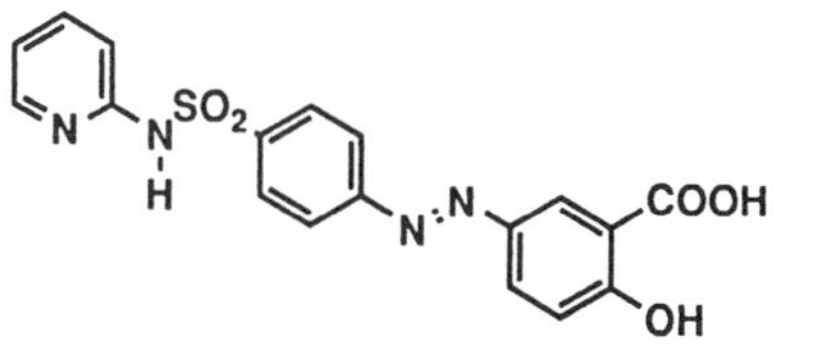

18

There have been several other compounds that have been examined pre-clinically for immuno-suppressive effects. Recent reports suggest that protein kinase C (PKC) is directly involved in the production of IL-2 (36); accordingly, H-7 **19**, a PKC inhibitor, was found to suppress IL-2 production by 70% at 10 μM (37). CI-959 **20**, an anti-allergy compound, blocks the release of IL-2 from con-canavalin A-stimulated rat splenocytes (IC_{50} = 19.1 μM) and human lymphocytes (IC_{50} = 23.1 μM) (38). This effect is considerably less potent than that produced with **11** or **15**, albeit comparable to NSAID's such as indomethacin. 1,25-Dihydroxyvitamin D_3 **21**, an immunomodulatory hormone, also shows potent effects on IL-2 production (39).

In contrast, up-regulation of IL-2 can also be induced with small molecules. Ciprofloxacin **22**, a quinolone antibacterial, significantly increases IL-2 production in phytohemagglutinin-stimulated human peripheral blood lymphoctes (3-5X increase at 50-100 μg/ml) (40). Increased IL-2 production in the spleen of tumor-bearing mice is likewise seen upon treatment by RS-0481 **23** (41).

19   **20**   **21**

Interleukin 3 — IL-3 is a 23 kDa glycoprotein whose general function is to induce proliferation of early hematopoietic progenitor cells (19, 42). It is closely related to granulocyte-macrophage colony stimulating factor (GM-CSF) and often has similar activity. One of the modes of action attributed to IL-3 and GM-CSF is to enhance histamine synthesis in the progenitor cells (42). Histamine plays a role not only in allergic reactions, but also in signalling between leukocytes involved in immunologically mediated inflammation. It has been suggested that histamine is an important mediator in inducible hematopoiesis and as such, is the rationale for the immunomodulatory effects of such H_2 receptor antagonists as cimetidine **24** (43). Other features of IL-3 activity include downregulation of the expression of IL-2R (44) and the involvement of tyrosine kinases (45).

22   **23**   **24**

Interleukin 4 — IL-4 is a 20 kDa peptide, produced principally by T cells, which plays a key role in B-cell growth and is the sole cytokine responsible for initiation of IgE synthesis, implicating it in allergic diseases and parasitic infections. Its immunoregulatory functions and potential application in cancer therapy have been reviewed (46, 47). Several structure function studies have been carried out on human IL-4. Using secondary structure predictions, site directed mutagenesis and CD spectroscopy, a model of the protein has been generated and predicts an all parallel four helix globular protein with two overhand loop connections (48). The protein contains three disulphide links but mutagenesis studies show that only one bridge, Cys46 to Cys99, is essential for biological activity (49). Polyclonal antibodies to a series of segments covering the entire sequence have been used to attempt to identify the receptor binding region of the peptide (50). Antibodies to Leu52-Cys65 and Lys61-Phe82 bind the most effectively to IL-4. In addition, an anti-idiotypic antibody to the Lys61-Phe82 sequence antagonized IL-4 binding to its receptor without binding to IL-4 which suggests that receptor binding is localized to that region. Two small segments of IL-4, Ala70-Arg88 and Asn89-Glu122, show agonist activity in a fibroblast chemotaxis assay, with the former showing the most potent effects (51).

Few small molecules have been reported to show modulatory effects on IL-4 activity. IPD-1151T **25** has been reported to inhibit mitogen-induced IL-4 production (52). Dexamethasone **11** treatment of a homogeneous T cell line simultaneously inhibits IL-2 production and stimulates IL-4 production from the antigen-stimulated cells and this modulatory effect can be inhibited by the steroid antagonist RU486 (53). Another steroid, dihydrotestosterone **26**, inhibits IL-4 production, as well as IL-5 and γ-INF production, by anti-CD$_3$-stimulated mouse T cells. Although testosterone has no effect on isolated T-cells, in the presence of macrophages which possess 5α-reductase, it also inhibits IL-4 production (54). Cyclosporin A **15** inhibits both IL-4 and IL-4 receptor expression in anti-CD$_3$-stimulated human peripheral blood mononuclear cells (55).

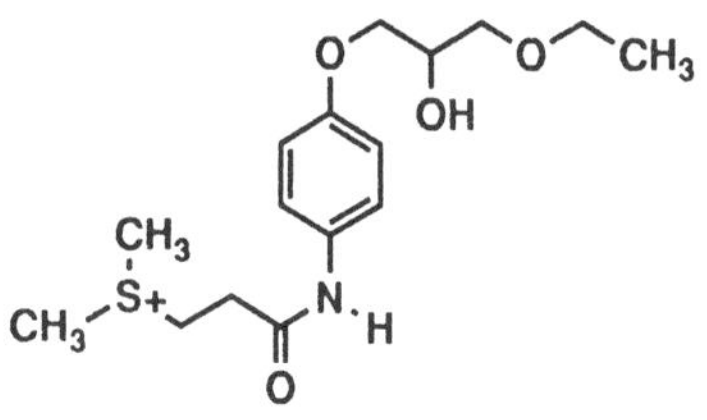

25 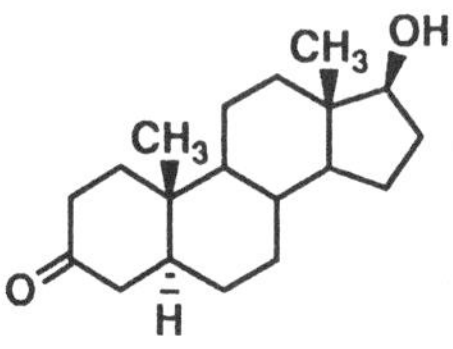   26 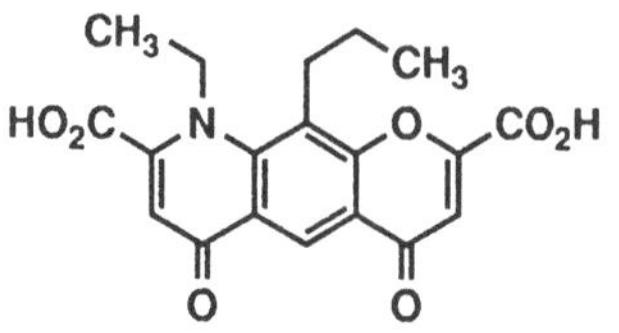   27

Interleukin 5 — IL-5 is produced principally by T cells and plays a key role in eosinophil and B-cell proliferation and differentiation. Hence, this cytokine is also thought to have a role in allergic diseases and parasitic infections (56). Structure function studies of human/mouse chimeric IL-5 molecules have defined the C-terminus as the receptor binding region and together with the dimer studies suggest that IL-5 may function as a bifunctional ligand inducing cross linking in two IL-5 receptors (57). In agreement with the chimeric protein work, C-terminus truncated human IL-5 failed to induce B cell differentiation (58). No selective, small molecule modulators of IL-5 have been disclosed, although the inhibition of IL-5-induced eosinophil survival is blocked by high concentrations of nedocromil **27** (59).

Interleukin 6 — IL-6 is a 26 kDa peptide produced by macrophages, T cells, and fibroblasts. Along with IL-1 and TNF, IL-6 is a major mediator of the acute phase response to infection or injury. Its abnormal production is thought to play a key role in inflammatory and autoimmune diseases (60). In addition, the potential uses of IL-6 and antagonists of IL-6 as therapeutants have been reviewed (61). A variety of structure function studies on human IL-6 have been performed allowing some predictions about the receptor binding region to be made. More recently, a combination of nmr and site-directed mutagenesis studies support a role for Met162 in receptor binding and additionally suggest that Leu159 and Leu166 are also important for binding (62).

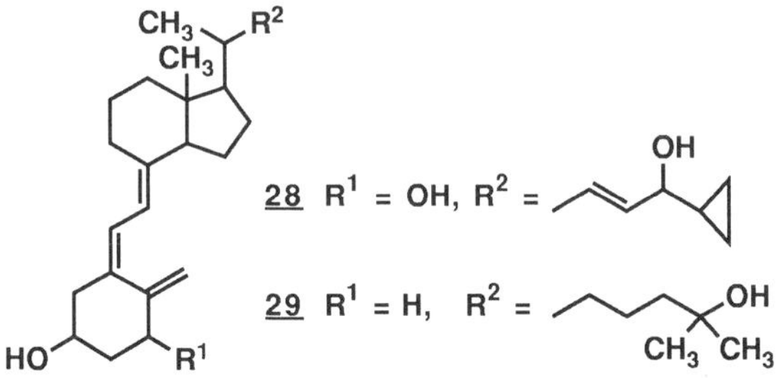

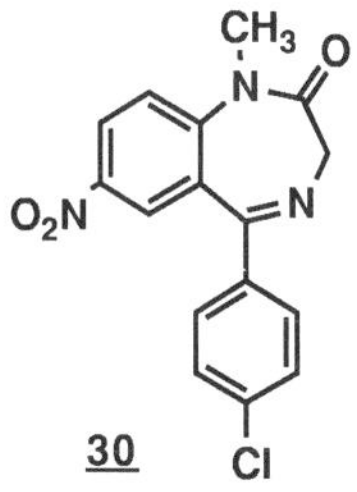

28 R^1 = OH, R^2 =

29 R^1 = H, R^2 =

30

Dihydroxyvitamin D_3 **21** and the analog MC 903 **28** inhibit the production of IL-6 by LPS-stimulated human mononuclear cells (63), whereas hydroxyvitamin D_3 **29** has no effect, suggesting a specific action through the vitamin D_3 receptor. Additionally, **28** has been shown to reduce the levels of IL-6 in psoriatic lesions (64). Dexamethasone and cortisol inhibit the production of IL-6 from LPS-stimulated mouse macrophages and human monocytes, endothelial cells and fibroblasts *via* stimulation of the glucocorticoid receptor (65). Prostaglandin E_2 (PGE_2) is a potent inhibitor of IL-6 production in endotoxin-stimulated Kupffer cells and thus the CO inhibitor indomethacin can enhance IL-6 production (66). Similarly, treatment with ibuprofen prior to endotoxin adminstration leads to elevated levels of IL-6 as well as elevated levels of TNF-α, presumably by inhibition of PGE_2 synthesis (67). *In vivo* treatment with the peripheral benzodiazepine Ro5-4864 **30**, blocks the *ex vivo* production of IL-1, IL-6, and TNF-α by LPS-stimulated mouse spleen macrophages. In contrast, peritoneal macrophages from the same mice are unaffected (68). Retinoic acid and menthol inhibit IL-6 receptor (IL-6R) expression in human leucocytes *via* down regulation of IL-6R mRNA; this effect may explain their anti-proliferative effects (69).

Interleukin 7 — IL-7 is a 25 kDa protein derived from stromal cells. It stimulates the proliferation of pre-B cells, thymocytes, and mature T cells (70, 71), as well as stimulating the tumoricidal activity of monocytes (72) and the killing ability of cytotoxic T lymphocytes (73). The IL-7 receptor has recently been cloned and characterized, showing it to be a member of a receptor superfamily (74). In addition to the membrane bound receptor forms, a cDNA was isolated that encoded for a soluble form of the IL-7 receptor. IL-7 has also been shown to be effective as a therapeutic agent *in vivo*, in cyclophosphamide-induced lymphopenia in mice (75). Treatment with 200 mg/kg i.p. of cyclophospha-mide causes a substantial decrease in the cellularity of the bone marrow and lymphoid organs as reflected by a drop in T cell and B cell numbers. Administration of IL-7 (500 ng, s.c.) results in a more rapid return to the normal range of cellularity in the spleen and mesenteric lymph node but not in the thymus or bone marrow.

Interleukin 8 — IL-8 is a monocyte-derived chemotactic factor for neutrophils and hence, has also been described as Neutrophil Activating Factor (NAF), Neutrophil Activating Peptide (NAP), and Neutrophil Chemotactic Factor (NCF). It is a 72 amino acid peptide and in addition to its chemotactic ability for neutrophils, it also induces neutrophil activation (76, 77). IL-8 acts as a "later stage" chemotactic factor, on the order of hours, after its expression is induced by other cytokines (78). Elevated levels of IL-8 have been determined in a number of disease states, such as rheumatoid arthritis (79), chronic airway diseases (80), and idiopathic pulmonary fibrosis (80).

The role of calcium in IL-8 induced migration of lymphocytes was investigated using a series of calcium channel antagonists (81). Verapamil, nifedipine, and diltiazem are all potent inhibitors of stimulated migration (IC_{50}'s of 10 nM, 60 nM, and 10 nM, respectively), implying a role for calcium channel activation. IL-8 induced migration is also inhibited by PKC inhibitors H-7 **19**, Ro series **31**, and sphingosine **32** (82). An endogenous immunomodulator, dihydroxyvitamin D_3 **21**, inhibits IL-1α induced IL-8 production and mRNA expression in keratinocytes, fibroblasts, and peripheral blood monocytes, but not in endothelial cells (83). Antiinflammatory agents have also been found to be efficacious; the 5-LO/CO inhibitor BI-L-93BS **33** inhibits the release of IL-8 from LPS-stimulated THP-1 cells (84). This compound has an ED_{50} of 16.5 mg/kg in the rat carrageenan-induced paw edema model (85).

3 1   **3 2**   **3 3**

Interleukin 10 — First described as Cytokine Synthesis Inhibitory Factor, IL-10 is a 35 kDa homodimeric protein which is secreted by T_H2 (helper T) cells and mast cells. As its earlier name implies, IL-10 is involved in the cross-regulation between T_H1 and T_H2 cells by inhibiting cytokine synthesis by T_H1 clones (86). This cross-regulation is important since T_H1 and T_H2 activities are often mutually exclusive (delayed type hypersensitivity and antibody responses, respectively). In addition, IL-10 appears to be a growth cofactor for mature and immature T cells (87), a cytotoxic T cell differentiation factor (88), and in conjunction with IL-3 and/or IL-4, a stimulatory factor for mast cells (89). Interestingly, it was discovered that the BCRF1 gene of the Epstein Barr virus (EBV) has high homology to human IL-10 (90). Recombinant BCRF1 protein inhibits interferon production by human T cells stimulated either with phytohemagglutinin, CD_3 antibodies, or IL-2, a property shared by IL-10. One of the actions of γ-IFN is that it inhibits viral replication and in synergy with tumor necrosis factor, selectively kills virus-infected cells. This suggests that EBV has captured and conserved the mammalian genome for the purpose of reducing the anti-viral effects of the host immune response.

Tumor Necrosis Factor-α — TNF-α is a 157 amino acid peptide produced by macrophages, keratinocytes and T cells. This cytokine plays a central role in host defense mechanisms and has been implicated in shock, chronic inflammation, and other autoimmune diseases. Its structure, function, and potential clinical uses have been reviewed (91, 92, 93).

Histamine inhibits LPS-stimulated TNF-α synthesis in human peripheral blood mononuclear cells *via* H_2-receptor mediated TNF-α mRNA reduction, and suggests that local release of histamine may limit TNF-α release (94). Thalidomide **34**, used as a treatment for erythema nodosum leprosum, has been found to reduce the production of TNF-α in LPS-stimulated human monocytes (95). Compound **34** has no effect on general protein synthesis and also does not inhibit IL-1 or GM-CSF production. Dexamethasone **11** inhibits TNF-α production from LPS-stimulated Kupffer cells, an effect which can be blocked with the steroid antagonist RU 486 (96). Cyclosporin A **15** inhibits TNF-α release from mouse macrophages, without suppresion of TNF-α mRNA levels or reduction of intracellular TNF-α (97). The methyl xanthine pentoxifylline **35**, a known *in vitro* inhibitor of TNF-α production, has been shown to protect animals from LPS-induced TNF-α production and subsequent death (98). In similar studies in man, **35** inhibits the increase in serum concentrations of TNF-α in endotoxin treated volunteers (99). However, the acute rise in IL-6 levels is unaffected, indicating a selective action of **35**. The mechanism by which **35** works has been linked to its phosphodiesterase (PDE) inhibitory activity, and this is supported by the ability of the PDE inhibitors theophylline **36** and isobutylmethyl-xanthine **37** to also suppress TNF-α production from LPS-stimulated mouse macrophages (100). Compound **35** is also capable of reversing TNF-α-induced inhibiton of f-Met-Leu-Phe-mediated lympho-cyte chemotaxis (101). TNF-α upregulation of adhesion molecule (ICAM-1) expression on human umbilical vein endothelial cells is mediated at least partially by PKC's, since the inhibitors staurosporinone **38** and H-7 **19** are inhibitory (102).

35 $R^1 = R^2 = CH_3$, $R^3 = $ (structure)
36 $R^1 = H$, $R^2 = R^3 = CH_3$
37 $R^1 = H$, $R^3 = CH_3$, $R^2 = $ (structure)

34

38

Granulocyte-Macrophage and Granulocyte Colony Stimulating Factors — The colony stimulating factors (CSF's) are a family of regulatory glycoproteins that stimulate the production of lineage specific cells from committed hematopoietic progenitor cells. The effects of GM-CSF and G-CSF are far ranging and affect virtually every function of the granulocyte and macrophage that has been studied (103, 104). In addition, recent evidence suggests a role for GM-CSF in eosinophil chemotaxis (105). GM-CSF has been detected in a number of disease states, such as RA (106) and psoriasis (107). For neutropenic disease states, however, recombinant CSF's are being investigated as therapeutic agents because of their stimulatory effects (108, 109, 110). Because of the ability of multiple cytokines to accelerate hematopoiesis and known synergism *in vitro*, fusion proteins such as GM-CSF/IL-3 are also being developed (111).

A synthetic peptide, SKF-107,647 **39**, has been found to be efficacious in stimulating general colony stimulating activity (CSA) (112). Single injections of 1-10 ng/kg of **39** increase CSA levels in mice by 400-600%, and multiple doses of 10 ng/kg qd x 7 increase peritoneal macrophage super-oxide production and candidacidal activity. In another approach, significant enhancement of GM-CSF-stimulated neutrophil responses are observed by the preincubation of whole blood with H-7 **19**, suggesting that a protein kinase-mediated phosphorylation step is important in the down-regulation of neutrophil responses to GM-CSF (113).

pGluGluAsp—(structure)—LysOH

39

40

When alloreactive T helper cell clones are treated with cyclosporin A **15**, even in the presence of exogenous IL-2, GM-CSF secretion is inhibited (114). Coinciding with the observation of leukotrienes *in vivo* in patients being treated with GM-CSF (115), the cytotoxicity in monocytes which is elicited by GM-CSF is found to be augmented by indomethacin, a CO inhibitor, and found to be suppressed by nordihydroguaiaretic acid $\underline{40}$, an LO inhibitor (116). Dexamethasone $\underline{11}$ inhibits the prolongation of eosinophil survival caused by GM-CSF (117).

Interferon — IFN is a generic term for a family of proteins with a multitude of effects including virus inhibition, oncogene suppression, and slowing of cell proliferation. The IFN's are comprised of α-IFN, β-IFN, γ-IFN and ω-IFN, each with several subtypes which bind to at least two diferent types of receptors (118). In chronic viral infections, such as hepatitis B, α-IFN has shown positive benefit, but is usually ineffective in acute viral infection (119). Both β- and γ-IFN have been extensively tested in the treatment of various cancers. However, the outcome is variable with the best results obtained in therapy of residual malignancies (120, 121). Much work has been done on the mapping of the effector regions of the IFN's using synthetic fragments and epitope mapping with monoclonal antibodies. Structure function studies with α-IFN have recently been summarized and extended (122). When IFN was last reviewed in this series, a considerable number of interferon inducers were described (123). However, the early promise of this area has not come to fruition; very few reports of these agents having appeared in recent years. Nevertheless, one such agent continues to be of interest; CL 246,738 $\underline{41}$ is a potent inducer of IFN *in vivo*, elevating circulating levels of IFN and activating natural killer cells. Compound $\underline{41}$ is currently in Phase I clinical trials (124).

$$\underline{4\,1}$$

Conclusion — The controlled modulation of cytokine function presents an enormous challenge to the medicinal chemist. Some progress has been made in defining agents which interfere with cytokine activity but with few exceptions, these compounds have not been designed for that purpose. This has meant that selectivity for modulation of an individual cytokine is rare, which in turn has hampered the execution of pharmacological experiments which define the precise roles of the cytokines. Without doubt, the future lies with purpose-designed antagonists, agonists, inhibitors, or enhancers. These studies will provide not only the pharmacological tools but also the future drugs for the treatment of immunological, inflammatory, and oncological diseases.

References

1. D.L. Urdal, Annu.Rep.Med.Chem., $\underline{26}$, 221 (1991).
2. C.A. Dinarello, Blood, $\underline{77}$, 1627 (1991).
3. P.E. Bender and J.C. Lee, Ann.Rep.Med.Chem., $\underline{25}$, 185 (1990).
4. G.J. Rosenthal, E.Corsini, W.A. Craig, C.E. Comment, and M.I. Luster, Tox.Appl.Pharmacol., $\underline{107}$, 555 (1991).
5. J. Schnyder, P. Bollinger, and T. Payne, Agents Actions, $\underline{30}$, 350 (1990).
6. D.E. Griswold, P.J. Marshall, J.C. Lee, E.F. Webb, M.L. Hillegass, J. Wartell, J. Newton Jr., and N. Hanna, Biochem.Pharmacol., $\underline{42}$, 825 (1991).
7. P.J. Marshall, D.E. Griswold, J. Breton, E.F. Webb, L.M. Hillegass, H.M. Sarau, J. Newton Jr., J.C. Lee, P.E. Bender, and N. Hanna, Biochem.Pharmacol., $\underline{42}$, 813 (1991).
8. M. Tanaka, K. Chiba, M. Okita, T. Kaneko, K. Tagami, S. Hibi, Y. Okamoto, H. Shirota, M. Goto, H. Sakurai, Y. Machida, and I. Yamatsu, J.Med.Chem., $\underline{34}$, 2647 (1991).
9. M. Goto, K. Chiba, R. Hashida, and H. Shirota, Agents Actions, $\underline{32}$, (suppl) 225 (1991).
10. R.B. Gilbertsen, K.M. Cullinen, D.J. Wilburn, M.K. Dong, and M.C. Conroy, Agents Actions, $\underline{27}$, 303 (1989).
11. S.P. Sirko, R. Schindler, M.J. Doyle, S.M. Weisman, and C.A. Dinarello, Eur.J.Immunol., $\underline{21}$, 243 (1991).
12. B.E. Barton, J.P.Jakway, S.R. Smith, and M.I. Siegel, Immunopharmacol.Immunotox., $\underline{13}$, 251 (1991).
13. M. Harth, G.A. McCain, and K. Cousin, Immunopharm., $\underline{20}$, 125 (1990).
14. N. Keicho, S. Sawada, K. Kitamura, H. Yotsumoto, and F. Takaku, Cell. Immunol., $\underline{132}$, 285 (1991).
15. A.D. Howard, M.J. Kostura, N. Thornberry, G.J.F. Ding, G. Limjuco, J. Weidner, J.P. Salley, K.A. Hogquist, D.D. Chaplin, R.A. Mumford, J.A. Schmidt, and M.J. Tocci, J.Immunol., $\underline{147}$, 2964 (1991).
16. R.A. Black, P.R. Sleath, and R.S. Kronheim, World Patent 9,115,577 (1991).
17. W.P. Arend, Prog.Growth Factor Res., $\underline{2}$, 193 (1990).
18. D.J. Dripps, B.J. Brandhuber, R.C. Thompson, and S.P. Eisenberg, J.Biol. Chem., $\underline{266}$, 10331 (1991).
19. J. Boger and J.A. Schmidt, Annu. Rep. Med. Chem., $\underline{23}$, 171 (1988).
20. A. Rebollo, D. De Groote, M. Baudrihay, J. Theze, and D.L. Jankovic, Mol. Immunol., $\underline{29}$, 119 (1992).

21. SCRIP, 1686, 20 (1992).
22. M.T. Lotze, Clin. Immunol. Immunopath., 62, S47 (1992).
23. R. Rahman, Z. Bernstein, L. Vaickus, R. Penetrante, S. Arbuck, I. Kopec, D. Vesper, H.O. Douglass, and K.A. Foon, J.Immunother., 10, 221 (1991).
24. R.M. Sherry, S.A. Rosenberg, and J.C. Yang, J. Immunother., 10, 371 (1991).
25. R. J. Soiffer, C. Murray, K. Cochran, C. Cameron, E. Wang, P.W. Schow, J.F. Daley, and J. Ritz, Blood, 79, 517 (1992).
26. T.J. Eberlein, D.D. Schoof, H.R. Michie, A.F. Massaro, U. Burger, D.W. Wilmore, R.E. Wilson, Arch. Surg., 124, 542 (1989).
27. A.L. Sagone, R.M. Husney, P.L. Triozzi, and J. Rinehart, Blood, 78, 2931 (1991).
28. T.A. Waldmann, J. Biol. Chem., 266, 2681 (1991).
29. Drugs of the Future, 16, 208 (1991).
30. D.J. Henderson, I. Naya, R.V. Bundick, G.M. Smith, and J.A. Schmidt, Immunol., 73, 316 (1991).
31. M.J. Tocci, D.A. Matkovich, K.A. Collier, P. Kwok, F. Dumont, S. Lin, S. Degudicibus, J.J. Siekierka, J. Chin, and N.I. Hutchinson, J. Immunol., 143, 718 (1989).
32. Transplantation Proceedings, 23, 2709 (1991).
33. M. Sasano, K. Nakata, S. Mita, Int. J. Immunother., 7, 87 (1991).
34. M. Fujiwara, K. Mitsui, and I. Yamamoto, Japan. J. Pharmacol., 54, 121 (1990).
35. D.T. Boumpas, E.D. Anastassiou, S.A. Older, G.D. Tsokos, D.L. Nelson, and J.E. Balow, J. Clin. Invest., 87, 1739 (1991).
36. J. F. Modiano, R. Kolp, R.J. Lamb, and P.C. Nowell, J. Biol. Chem., 266, 10552 (1991).
37. D. Atluru, S. Polam, S. Atluru, and G.E. Woloschak, Cell. Immunol., 129, 310 (1990).
38. M.K. Dong, D.J. Wilburn, M.C. Conroy, and R.B. Gilbertsen, Agents Actions, 34, 53 (1991).
39   W.F.C. Rigby, B.J. Hamilton, and M.G. Waugh, Cell. Immunol., 125, 396 (1990).
40. T. Zehavi-Willner and I. Shalit, Lymphokine Res., 8, 35 (1989).
41. S. Kurakata, M. Tomatsu, M. Arai, H. Arai, A. Hishinuma, H. Kohno, K. Kitamura, T. Kobayashi, and K. Nomoto, Cancer Immunol., Immunother., 33, 71 (1991).
42. M. Dy and E. Schneider, Eur. Cytokine Netw., 2, 153 (1991).
43. D.M. Sahasrabudhe, C.S. McCune, R.W. O'Donnell, and E.C. Henshaw, J. Immunol., 138, 2760 (1987).
44. R. Onishi, T. Ishikawa, T. Kodaka, M. Okuma, and T. Uchiyama, Blood, 78, 2908 (1991).
45. T. Satoh, Y. Uehara, and Y. Kaziro, J. Biol. Chem., 267, 2537 (1992).
46. W.E. Paul, Blood, 77, 1859 (1991).
47. D.W. Maher, I. Davis, A.W. Boyd, and G. Morstyn, Prog. Growth Factor Res., 3, 43 (1991).
48. B.M. Curtis, S.R. Presnell, S. Srinivasan, H. Sassenfeld, R. Klinke, E. Jeffery, D. Cosman, C.J. March, and F.E. Cohen, Proteins, 11, 111 (1991)..
49. N. Kruse, T. Lehrnbecher, and W. Sebald, FEBS Lett., 286, 58 (1991).
50. L. Ramanathan, G.F. Seelig and P.P. Trotta, World Patent 9,109,059 (1991).
51. A.E. Postlethwaite and J.M. Seyer, J.Clin.Invest., 87, 2147 (1991).
52. Drugs of the Future, 15, 1136 (1990).
53. R.A. Daynes and B.A. Araneo, Eur.J.Immunol., 19, 2319 (1989).
54. B.A. Araneo, T. Dowell, M. Diegel, and R.A. Daynes, Blood, 78, 688 (1991).
55. B.M.J. Foxwell, G. Woerly, and B. Ryffel, Eur.J.Immunol., 20, 1185 (1990).
56. K. Takatsu and A. Tominaga, Prog.Growth Factor Res., 3, 87 (1991).
57. A.N.J. McKenzie, S.C. Barry, M. Strath, and C.J. Sanderson, EMBO J., 10, 1193 (1991).
58. S. Kodama, N. Tsuruoka, and M. Tsujimoto, Biochem.Biophys.Res.Comm., 178, 514 (1991).
59. B. Resler, J.B. Sedgwick, and W.W. Busse, J.Allergy Clin.Immunol., 89, 235 (1992).
60. T. Hirano, Clin.Immunol.Immunopath., 62, S60 (1992).
61. J. Bauer and F. Herrmann, Ann.Hematol., 62, 203 (1991).
62. C. Nishimura, K. Futatsugi, K. Yasukawa, T. Kishimoto, and Y. Arata, FEBS Lett., 281, 167 (1991).
63. K. Müller, M. Diamant, and K. Bendtzen, Immunol.Lett., 28, 115 (1991).
64. A. Oxholm, P. Oxholm, B. Staberg, and K. Bendtzen, Acta Derm.Venerol., 69, 385 (1989).
65. A. Waage, G. Slupphaug, and R. Shalaby, Eur.J.Immunol., 20, 2439 (1990).
66. M.P. Callery, M.J. Mangino, T. Kamei, and M.W. Flye, J.Surg.Res., 48, 523 (1990).
67. G.A. Spinas, D. Bloesch, U. Keller, W. Zimmerli, and S. Cammisuli, J.Infect.Dis., 163, 89 (1991).
68. F. Zavala, V. Taupin, and B. Descamps-Latscha, J.Pharmacol.Exp.Ther., 255, 442 (1990).
69. N. Sidell, T. Taga, T. Hirano, T. Kishimoto, and A. Saxon, J.Immunol., 146, 3809 (1991).
70. M. Masuda, T. Motoji, K. Oshimi, and H. Mizoguchi, Leukemia Lymphoma, 5, 231 (1991).
71. A. Herbelin, F. Machavoine, E. Schneider, M. Papiernik, and M. Dy, J. Immunol., 148, 99 (1992).
72. M.R. Alderson, T.W. Tough, S.F. Ziegler, and K.H. Grabstein, J. Exp. Med., 173, 923 (1991).
73. C.J. Hickman, J.A. Crim, H.S. Mostowski, and J.P. Siegel, J. Immunol., 145, 2415 (1990).
74. R.G. Goodwin, D. Friend, S.F. Ziegler, R. Jerzy, B.A. Falk, S. Gimpel, D. Cosman, S.K. Dower, C.J. March, A.E. Namen, and L.S. Park, Cell, 60, 941 (1990).
75. P.J. Morrissey, P. Conlon, S. Braddy, D.E. Williams, A.E. Namen, and D.Y. Mochizuki, J. Immunol., 146, 1547 (1991).

76. I. Colditz, R. Zwahlen, B. Dewald, and M. Baggiolini, Am. J. Pathol., 134, 755 (1989).
77. M. Sticherling, J.M. Schroder, and E. Christophers, J. Immunol., 143, 1628 (1989).
78. C.E.M. Griffiths, J.N.W.N. Barker, S. Kunkel, and B.J. Nickoloff, Br. J. Dermat., 124, 519 (1991).
79. M. Seitz, B. Dewald, N. Gerber, and M. Baggiolini, J. Clin. Invest., 87, 463 (1991).
80. T. Ozaki, H. Hayashi, K. Tani, F. Ogushi, S. Yasuoka, and T. Ogura, Am. Rev. Respir. Dis., 145, 85 (1992).
81. K.B. Bacon, J. Westwick, and R.D.R. Camp, Biochem. Biophys. Res. Commun., 165, 349 (1989).
82. K.B. Bacon and R.D.R. Camp, Biochem. Biophys. Res. Commun., 169, 1099 (1990).
83. C.G. Larsen, M. Kristensen, K. Paludan, B. Deleuran, M.K. Thomsen, C. Zachariae, K. Kragballe, K. Matsushima, and K. Thestrup-Pedersen, Biochem. Biophys. Res. Commun., 176, 1020 (1991).
84. C.A. Homon, R. DeLeon, R. Mucci, and P.R. Farina, 1st Intern. Cong. Inflamm., Barcelona, June 17-22, Abs 003 (1990).
85. E.S. Lazer, H.C. Wong, G.J. Possanza, A.G. Graham, and P.R. Farina, J. Med. Chem., 32, 100 (1989).
86. T.R. Mosmann, Ann. NYAS, 628, 337 (1991).
87. I.A. MacNeil, T. Suda, K.W. Moore, T.R. Mosmann, and A. Zlotnik, J. Immunol., 145, 4167 (1990).
88. W. Chen and A. Zlotnick, J. Immunol., 147, 528 (1991).
89. L. Thompson-Snipes, V. Dhar, M.W. Bond, T.R. Mosmann, K.W. Moore, and D. M. Rennick, J. Exp. Med., 173, 507 (1991).
90. D.H. Hsu, R. deWaal Malefyt, D.F. Fiorentino, M.N. Dang, P. Vieira, J. DeVries, H. Spits, T. Mosmann, and K.W. Moore, Science, 250, 830 (1990).
91. M.A. Clark, P.L. Simon, M.-J. Chen, and J.S. Bomalaski, Annu.Rep.Med.Chem., 22, 235 (1987).
92. W. Fiers, FEBS Lett., 285, 199 (1991).
93. C.E. Spooner, N.P. Markowitz, and L.D. Saravolatz, Clin.Immunol.Immunopath., 62, S11 (1992).
94. E. Vannier, L.C. Miller, and C.A. Dinarello, J.Exp.Med., 174, 281 (1991).
95. E. Sampaio, E.N. Sarno, R. Galilly, Z.A. Cohn, and G. Kaplan, J.Exp.Med., 173, 699 (1991).
96. W.H. Kutteh, W.E. Rainey, and B.R. Carr, J.Clin.Endocrin.Metab., 73, 296 (1991).
97. D.T. Nguyen, M.K. Eskandari, L.E. DeForge, C.L. Raiford, R.M. Strieter, S.L. Kunkel, and D.G. Remick, J.Immunol., 144, 3822 (1990).
98. P. Noel, S. Nelson, R. Bokulic, G. Bagby, H. Lippton, G. Lipscomb, and W. Summer, Life Sci., 47, 1023 (1990).
99. P. Zabel, D.T. Wolter, M.M. Schönharting, and U.F. Schade, Lancet, 1474 (1989).
100. S. Endres, H.-J. Fülle, B. Sinha, D. Stoll, C.A. Dinarello, R. Gerzer, and P.C. Weber, Immunol., 72, 56 (1991).
101. G.W. Sullivan, J. Linden, E.L Hewlett, H.T. Carper, J.B. Hylton, and G.L. Mandell, J.Immunol., 145, 1537 (1990).
102. T.A. Lane, G.E. Lamkin, and E.V. Wancewicz, Biochem.Biophys.Res.Comm., 172, 1273 (1990).
103. W.W. Grosh and P. Quesenberry, Clin. Immunol. Immunopath., 62, S25 (1992).
104. G.D. Demetri and J.D. Griffin, Blood, 78, 2791 (1991).
105. P.F. Weller, Clin. Immunol. Immunopath., 62, S55 (1992).
106. J.M. Alvaro-Gracia, N.J. Zvaifler, C.B. Brown, K. Kaushansky, and G.S. Firestein, J. Immunol., 146, 3365 (1991).
107. H. Takematsu and H. Tagami, Dermatologica, 181, 16 (1990).
108. C.H. Spiridonidis, Drugs of Today, 27, 503 (1991).
109. J.D. Levine, J.D. Allan, J.H. Tessitore, N. Falcone, F. Galasso, R.J. Israel, and J.E. Groopmen, Blood, 78, 3148 (1991).
110. L.M. Hollingshead and K.L. Goa, Drugs, 42, 300 (1991).
111. B.M. Curtis, D.E. Williams, H.E. Broxmeyer, J. Dunn, T. Farrah, E. Jeffery, W. Clevenger, P. de Roos, U. Martin, D. Friend, V. Craig, R. Gayle, V. Price, D. Cosman, C.J. March, and L.S. Park, Proc. Natl. Acad. Sci. USA, 88, 5809 (1991).
112. L.M. Pelus, P. Bhatnagar, C. Frey, P. DeMarsh, and A.G. King, Exp. Hematol., 19, 488 (1991).
113. A. Khwaja, P.J. Roberts, H.M. Jones, K. Yong, M.S. Jaswon, and D.C. Linch, Blood, 76, 996 (1990).
114. G. Pawelec, A. Rehbein, K. Katrilaka, I. Balko, and F.W. Busch, Immunopharmacol., 16, 207 (1988).
115. C. Denzlinger, A. Kapp, M. Grimberg, H.H. Gerhartz, and W. Wilmanns, Blood, 76, 1765 (1990).
116. D.P. Braun, K.P. Siziopikou, L.C. Casey, and J.E. Harris, Cancer Immunol. Immunother., 32, 55 (1990).
117. A.M. Lamas, O.G. Leon, and R.P. Schleimer, J. Immunol., 147, 254 (1991).
118. E.C. Borden, Clin.Immunol.Immunopath., 62, S18 (1992).
119. N.B. Finter, S. Chapman, P. Dowd, J.M. Johnston, V. Manna, N. Sarantis, N. Sheron, G. Scott, S. Phua, and P.B. Tatum, Drugs, 42, 749 (1991).
120. A.M. Leberati, M. Horisberger, M. Schippa, F. Di Clemente, M. Fizzotti, S. Filippo, M.G. Proietti, S. Arzano, P. Berruto, L. Palmisano, and S. Cinieri, Cancer Immunol.Immunother., 34, 115 (1991).
121. C.H.Spiridonidis, Drugs of Today, 27, 349 (1991).
122. P. Kontsek, L. Borecky, E. Kontsekova, I. Macikova, A. Kolcunova, M. Novak, and V. Krchnak, Mol.Immunol., 28, 1289 (1991).
123. W. Wierenga, Annu.Rep.Med.Chem., 17, 151 (1982).
124. V. Ruszala-Mallon, J. Silva, A.L. Lumanglas, F.E. Durr, and B.S. Wang, Int.J.Immunopharmacol., 13, 913 (1991).

Chapter 23. Recent Advances in the Discovery and Development of Potential Antidiabetic Agents.

Jerry R. Colca and Steven P. Tanis
The Upjohn Company, Kalamazoo, MI 49001

Introduction - Non-insulin-dependent diabetes mellitus (NIDDM) is a common metabolic disorder affecting approximately 5% of the general population of industrialized nations. Although perhaps of heterologous origins, the disease is genetically transmitted. Incidence varies in selected populations ranging up to 50% in some minority populations. NIDDM accounts for greater than 90% of the cases of diabetes mellitus. The estimated economic burden of diabetes is in excess of 20 billion dollars per year in the United States alone (1). Over 95% of the costs are incurred as a result of the numerous sequelae of the disease. These complications include macrovascular disease, resulting in cardiovascular and cerebrovascular disturbances; microvascular disease, resulting in renal failure and retinopathy; and numerous dysfunctions of the autonomic nervous system. It is generally believed that a therapy that adequately corrected the metabolic disturbances that occur in NIDDM should alleviate these complications (2). To understand the complexities required for the such therapy, it is necessary to discuss the nature and possible etiology of the metabolic disturbances that occur in NIDDM.

Unlike insulin-dependent diabetes mellitus (IDDM) in which the source of endogenous insulin (pancreatic β-cells) is destroyed via an autoimmune attack on the pancreatic β-cells, NIDDM is not accompanied by an absolute deficiency in circulating insulin (3-6). The hallmark of this condition is a reduced responsiveness of the various tissues to the actions of insulin. In a majority of cases, "insulin resistance" appears to precede the onset of overt symptoms (e.g. elevated fasting glucose) by many years (7). Once overt symptoms of the disease occur, dysfunction of pancreatic insulin secretion are measurable (8,9). The general consensus is that the combination of tissue insulin resistance and the pancreatic secretory defects conspire to produce the decompensated regulation of metabolism that is diagnosed as NIDDM. Potential points of therapeutic intervention include augmentation of pancreatic β-cell secretion or alteration of the target tissue responsiveness to insulin.

Factors Affecting the Secretion of Insulin - Several possible mechanisms exist whereby insulin secretion may be augmented resulting in reduced circulating glucose concentrations. It is not clear, however, that any of the currently proposed mechanisms actually corrects the specific secretory defect(s) that occur in NIDDM. Early secretagogues were patterned after sulfonylureas that were found empirically by their ability to lower circulating glucose by increasing the secretion of insulin. Recent evidence suggests that the mechanism of action of these compounds is the inhibition of the ATP-sensitive potassium channel located in the pancreatic β-cells. Augmented secretion thus occurs secondary to depolarization of the β-cell by this mechanism followed by the entry of Ca²⁺ through voltage dependent channels (10). This mechanism suggests the possibility for novel agents that might more specifically augment the secretion of insulin in this fashion. Mechanism-based development of stimulants of insulin secretion might also be based upon the observations that insulin secretion can be inhibited by the sympathetic nervous system, through α2 adrenoreceptors (11). The stimulation of the α2 adrenoreceptor with a number of agonists has been reported to produce a hyperglycemic response (11) through an inhibition of insulin secretion. The selection and application of α2 selective adrenorecptor antagonists might potentiate glucose mediated insulin release and effect hyperglycemia. Midaglizole (1,12-16), idazoxan (2,17-19), and SL 84.0418, (3, 20,21) have been reported to be peripherally active α2 adrenoreceptor antagonists which are antihyperglycemic agents in a variety of models of NIDDM. However, 2 was found not to alter insulin release, in response to a glucose infusion, in either normal volunteers or patients with NIDDM raising possible questions regarding this method of intervention (22).

ANNUAL REPORTS IN MEDICINAL CHEMISTRY—27   219   Copyright © 1992 by Academic Press, Inc.
All rights of reproduction in any form reserved.

Another class of novel insulin secretagogues, represented by phenylalanine derivative A-4166 (<u>4</u>) has been reported (23-26). The *trans* substituted cyclohexane moiety and the unnatural-R-absolute stereochemistry at the phenylalanine center (24-26) were crucial to the observed blood glucose lowering activity in KK mice. Compounds such as <u>4</u> seem to augment the secretion of insulin in a manner different than sulfonylureas which might reduce the risk of uncontrolled hypoglycemia (23-26). A-4166 results in a more rapid on and off augmentation of insulin output and, unlike sulfonylureas, does not appear to inhibit the secretion of the counterregulatory hormone, glucagon (23-26). The mechanism by which these compounds stimulate insulin secretion has not yet been established.

Glimeperide (<u>5</u>, 27) represents the newest class of sulfonylurea-type compounds that are undergoing development for possible use in the treatment of NIDDM. A series of thiopyranopyrimidines (28,29), characterized by MTP-3115 (<u>6</u>, 28) has been reported to lower blood glucose in normal and diabetic (VY/WfL-A^{vy}/a) mice by a mechanism which might involve stimulation of insulin secretion.

Another possible mechanism by which insulin secretion might be augmented is by the actions of the incretin factors released from the gut. These are thought to regulate the release of insulin in response to oral stimuli. The physiological regulator of this process appears to be a fragment of glucagon-like peptide 1 (30). This insulinotropin, or mimics of this peptide, may have some utility in the treatment of NIDDM (31).

<u>Factors Affecting Insulin Sensitivity-</u> The mechanism of the reduced tissue sensitivity to insulin, or insulin resistance that occurs in NIDDM is unknown (5). However, it is possible that more than one mechanism may be involved. For this reason the discovery of compounds that influence the sensitivity of the target tissues to insulin has been by a purely empirical approach. Murine models of insulin resistance have led to the discovery of compounds such as the thiazolidinedione, ciglitazone (<u>7</u>, 32-34). Analysis of the structure activity relationships of related compounds have given rise to at least three entities that are in clinical development, pioglitazone (<u>8</u>, 35), CS-045 (<u>9</u>, 36-38), and englitazone (<u>10</u>, 39,40). Further analysis of related compounds has led to some variations on the basic structural theme which seem to act by a similar mechanism (*vide infra*).

In general, compounds related to <u>7</u> appear to augment the pleiotropic effects of insulin in all of the target tissues known to respond to the hormone. Augmented actions of insulin have been shown to occur in muscle, adipose, and liver by studies *in vivo* as well as *ex vivo* (34). As a result of the improved responsiveness of the target tissues to insulin, circulating levels of the hormone are reduced allowing pancreatic stores of insulin to increase (34). It has been reported that treatment of diabetic KKA^y mice with <u>8</u>, causes the levels of the insulin regulated glucose transporter to increase in the tissues where insulin is known to produce this effect (adipose and muscle) while circulating insulin levels are reduced (41,42). That these effects occur secondary to augmented insulin action is demonstrated by the fact that pioglitazone (**8**)-augmented glucose transport did not occur in animals that were deficient in insulin, but was observed when insulin was supplied (41). Englitazone (**10**) has also been reported to have a similar effect on glucose transporter expression in 3T3-L1 adipocytes (39,43). Other effects of compounds such as **8** (44-47), **9** (37), and **10** (40, 48), such as reductions in circulating glucose, triglycerides and cholesterol, are dependent on the presence of circulating insulin (44). Recent evidence indicates that this includes improvements in the ratio of HDL-cholesterol to total cholesterol (44) as well as reductions in blood pressure (49). Thus it appears that compounds which augment the physiological actions of insulin and reduce circulating insulin levels may impact many metabolic consequences that have been linked to insulin insensitivity and hyperinsulinemia (3).

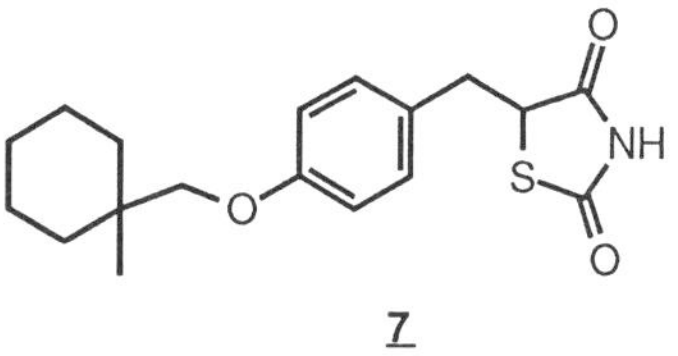

7

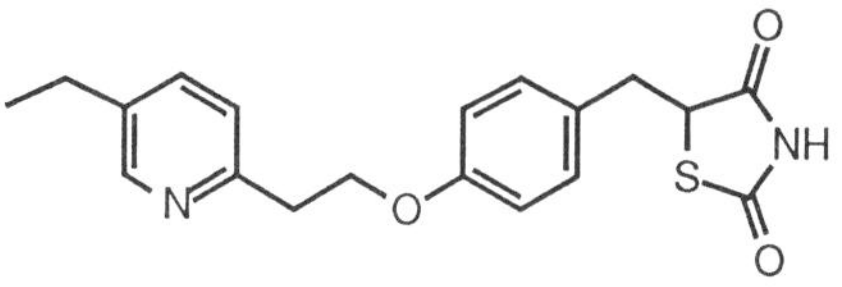

8

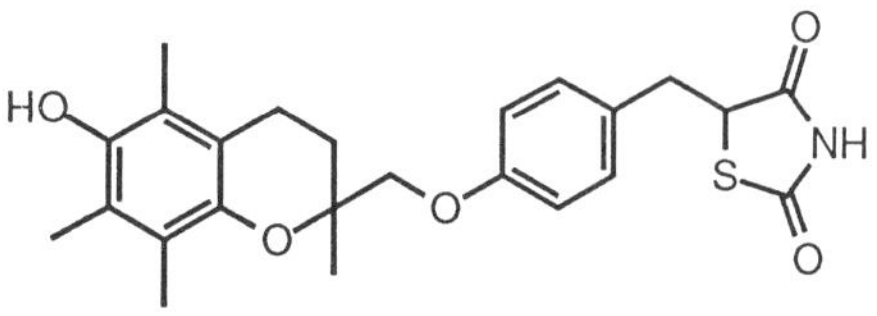

9

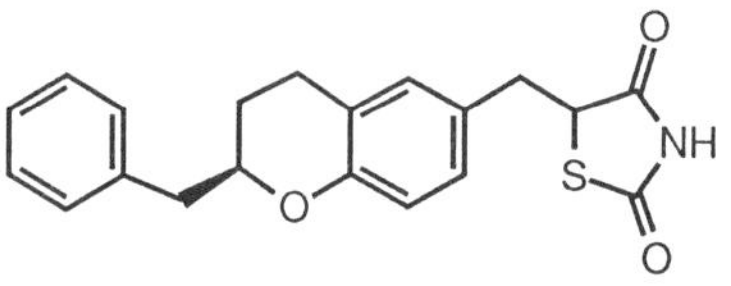

1 0

As a result of the interesting activities exhibited by compounds such as **7-10** a great deal of effort has been directed toward the thiazolidinedione-like insulin sensitizing agents (40,47,50-63). These efforts have resulted in a clearer definition of the structural requirements for action. Of interest are the reports that alterations in the benzyl-thiazolidinedione moiety, which typifies many of these molecules, can lead to active antihyperglycemic agents such as the naphthalenyl-sulfonyl-thiazolidinedione AY-31637 (**11**) which appears equipotent to **7** (57). Replacement of the thiazolidinedione with an oxazolidinedione, and concomitant introduction of a remote oxazole moiety, as in **12** (59), affords molecules which are more potent than **7**. Substitution of the thiazolidinedione with an oxathiadiazole-oxide, as in **13**, provides compounds which are more potent than **7**, as measured by postprandial glucose reduction in the db/db mouse (62).

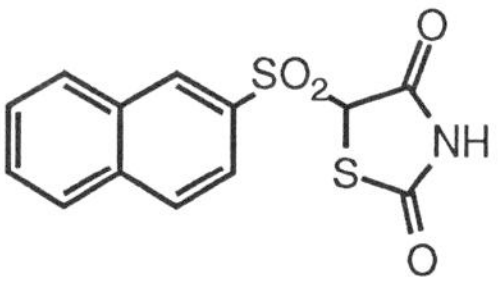

11

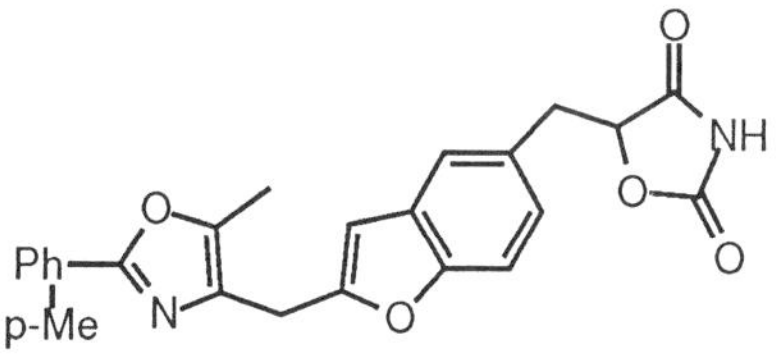

12

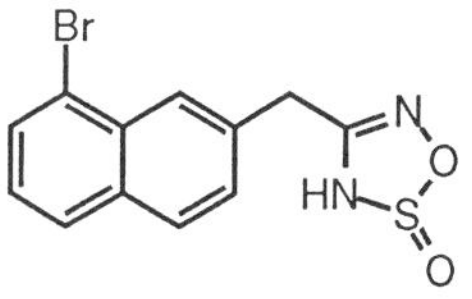

13

Information as to the specific biochemical action of compounds such as **7-13**, may give rise to mechanism-based screens that should allow broader evaluation of chemical classes that might act by this mechanism.

An additional class of compounds designed as ciglitazone (**7**) analogs, represented by the perfluoroalkyl-tetrazole **14**, had been previously described as thiazolidinedione-like insulin sensitizing agents (64, 65). A more recent report (66) describes an extensive SAR examination of this class which suggests that the antihyperglycemic activity associated with these compounds is mainly dependent upon the perfluoroalkyl chain, as compounds **15** and **16** also exhibit blood glucose lowering activity in the ob/ob mouse. The lack of dependence of the antidiabetic action upon the presence of the tetrazole thiazolidinedione-equivalent as well as the separation in the lowering of plasma glucose and insulin suggests that independent mechanisms are involved (66).

14   **15**   **16**

The sulfonamidobenzamide M&B 39890A (**17**) lowered plasma glucose and insulin in fed, diabetic VY/WfL-A^{vy}/a mice by improving insulin sensitivity (67). Compound **17** was inactive in normal mice and streptozotocin-rats.   The 17-amino steroid **18** lowers fed blood glucose in the insulin resistant KKa^y mouse (68,69). Given the profound insulin resistance of this test animal, it appears likely that **18** lowers blood glucose by reducing insulin resistance. A series of phenacylimidizolium halides, such as **19** (70), was observed to lower blood glucose and improve glucose tolerance in the insulin resistant, diabetic VY/WfL-A^{vy}/a mouse. This action has been attributed (71) to a stabilization of hepatic glycogen *via* the activation of glycogen synthase and the inhibition of glycogen phosphorylase.

17   CF$_3$ **18**   **19**

A large series of phenyl-guanidines and phenyl-amidines, such as **20**, have improved glucose tolerance in glucose-primed rats (72). The somewhat related biguanide-like compound **21**, which inhibits monosaccharide uptake, reduces gluconeogenesis, and increases peripheral glucose utilization, is reported to improve glucose tolerance in the rat and appears equipotent to phenformin (73).

20   **21**

Thermogenic β$_3$-Agonists - The effect of the sympathetic nervous system on heat generation from brown adipose cells is mediated by a specific adrenoreceptor which has been termed β$_3$ (74-76). Thus, it is theoretically possible to selectively access the beneficial effects of interaction with this receptor without affecting the various other physiological consequences of sympathetic activation of β$_1$ (heart) and β$_2$ (muscle, trachea, uterus) receptors (77). Activation of the thermogenic system in this manner has been shown to result in weight loss specifically from fat stores. The majority of patients with NIDDM are overweight and it has been hypothesized that this obesity contributes to the development of insulin resistance (78). Loss of body weight can lead to an improvement in the control of blood glucose (79). Thus, it is reasonable to assume, that reduced obesity with thermogenic agonists would improve metabolic control. Furthermore, for reasons not well understood, thermogenic β_3 agonists may have additional effects on insulin sensitivity which are independent of weight loss (80).

22   **23**

The β_3-selective agonist BRL 37344 (**22**) (400X *vs* atrial β_1, 20X *vs* trachial β_2) was found to be a potent full agonist on rat adipocytes, but was a partial agonist when examined in a human adipocyte system (81). ICI D7114 (**23**), a selective β_3-agonist, activated rat brown adipose tissue at doses which did not induce tremor (82). Compound **23** produced weight loss, a lowering of triglycerides, decreased fasting plasma glucose and improved glucose tolerance in the Zucker rat (83). The main impediment to the successful development of these compounds will likely be adequate selectivity with regard to the β_3 receptor. It also remains to be demonstrated that the selectivity of the various compounds, demonstrated in rodents, will be transferred to the human receptors (84).

Inhibition of Gluconeogenesis- As discussed previously, the imbalance of metabolic control that occurs in NIDDM may be considered to occur from failure of both the pancreas in terms of the adequate secretion of insulin, and the target tissues in terms of the ability to respond to insulin. One of the consequences of these deficits is an increased production of glucose by the liver. Although this hepatic defect could be repaired by a correction of the physiological control mechanisms, it is also possible to directly inhibit hepatic glucose production. For example, the inhibition of fatty acid oxidation may lower cellular acetylCoA such that pyruvate carboxylase, a key gluconeogenetic enzyme, is not fully activated (85). Alternate strategies for inhibiting gluconeogenesis, including the inhibition of fructose-1,6-diphosphatase (86) are possible (87). This approach is attractive from the point of view of drug design; however, substantial inhibition of hepatic gluconeogenesis could be extremely dangerous during periods of fasting (85).

Amylin- Recently it was suggested that a putative peptide hormone amylin, released in concert with insulin from the pancreatic β–cells, may be the cause of a portion of the physiological imbalance in insulin secretion and action that occurs in NIDDM (88). Although controversial, evidence has been presented that this peptide can inhibit the secretion of insulin and inhibit some actions of insulin in skeletal muscle (89). Therefore, antagonists of amylin secretion/action might be expected to be useful in the treatment of NIDDM. Interestingly **7** has been shown to reduce circulating levels of amylin in concert with reductions in circulating insulin (90). Sulfonylureas, however, lower blood glucose and elevate circulating amylin (91). There is no evidence that amylin levels are elevated over insulin concentrations in NIDDM. The generation of successful receptor-level antagonists should help clarify this situation.

Insulin Signal Transduction - The insensitivity of the target cells to insulin that occurs in NIDDM is thought to result from a reduced transduction of the insulin signal from the insulin receptor. This transduction can apparently be increased by treatment with such compounds as thiazolidinediones, as well as other insulin sensitizers, and β_3 agonists. The biochemical systems that are affected by these treatments remain to be defined. Another possible approach for drug discovery might be a specific interaction with elements of the insulin transduction cascade. Unfortunately, in spite of extensive examination, the biochemical events involved in the transduction of insulin action remain to be elucidated (92, 93). Evidence has emerged that some of the actions of insulin may occur secondary to the generation of second messengers or mediators from the cell surface. Hence, some but not all of the actions of insulin may be mimicked by the addition of water soluble compounds liberated following the treatment either of whole cells or membranes with insulin (94, 95). Although the chemical structure(s) of the mediator(s) remain to be completely described there is evidence that a class of inositol-containing mediators may be generated by insulin activation of a phosphoinositol-specific phospholipase C. There appear to be differences in the physical nature of these water soluble mediators depending on the cell type utilized by the various investigators. These mediators may also represent different bifurcations in insulin signaling pathways (95). The proof of this concept awaits the identification of these compounds and proof of action by synthesis. Thus it may be possible to selectively affect (stimulate or inhibit) particular actions of insulin. This could have important implications to the treatment of obesity as well as NIDDM.

CONCLUSION

The consequences of inadequate control of NIDDM have an enormous medical, emotional, and economic impact. Further definition of the problems associated with the disease and recent observations in the area of drug discovery, particularly with the advent of insulin sensitizing compounds, are facilitating the discovery and development of novel agents. These will likely lead to more diverse therapeutic alternatives. Of particular interest, with the insulin sensitizing compounds, is the apparent wide-spread effectiveness on a number of parameters that are associated not only with NIDDM but with poor insulin sensitivity and consequent high circulating insulin concentrations in individuals without clinical diabetes. This phenomenon is known to be associated with reduced HDL-cholesterol, and elevated triglycerides and blood pressure and is believed to contribute to the premature atherosclerosis and coronary heart disease that occurs in these patients whether or not they express the manifestations of altered carbohydrate metabolism that leads to the diagnosis of impaired glucose tolerance or NIDDM. As discussed above, several studies have shown that thiazolidinediones can elevate HDL cholesterol (reducing the ratio of total cholesterol/HDL-cholesterol) (37, 44). Further, it has been reported that these compounds can lower blood pressure in animals (49). It remains to be seen whether these effects will significantly affect the vascular pathology associated with these metabolic disturbances. Nevertheless it is clear that these findings, together with the definition of the mechanisms involved, will make this area a fruitful field for the efforts of medicinal chemists.

References

1. E.D. Bransome, Diabetes Care, $\underline{15}$ (Suppl. 1), 1 (1991).
2. R.A. DeFronzo, E. Ferrannini, and V. Koivisto, Am. J. Med., $\underline{74}$, 52 (1983).
3. G.M. Reaven, Diabetes, 37, 1595 (1988).
4. R.A. DeFronzo, Diabetes, $\underline{37}$, 667 (1988).
5. J.M. Olefsky, W.T. Garvey, R.R., Henry, S. Matthei, and G.R. Freidenberg, Am. J. Med., $\underline{85}$, 86 (1988).
6. R.A. DeFronzo, R.C. Bonadonna, and E. Ferrannini, Diabetes Care, $\underline{15}$, 318 (1992).
7. J.H. Warram, B.C. Martin, A.S. Krolewski, J.S. Soeldner, and C.R. Kahn, Ann. Intern. Med., $\underline{113}$, 909 (1990).
8. S.E. Kahn, and D. Porte, Am. J. Med., $\underline{85}$, 4 (1988).
9. R. Taylor, Br. Med. Bull., $\underline{45}$, 73 (1989).
10. A.E. Boyd 3rd, B.L. Aguilar, J. Bryan, D.L. Kunze, L. Moss, D.A. Nelson, A.S. Rajan, H. Raef, H.D. Xiang, and G.C. Yaney, Recent Prog. Horm. Res., $\underline{47}$, 299 (1991).
11. I. Angel, R. Niddam, and S.Z. Langer, J. Pharmacol. Exp. Ther., $\underline{254}$, 877 (1990).
12. K. Yamanaka, S. Kigoshi, and I. Muramatsu, Eur. J. Pharmacol., $\underline{106}$, 625 (1984).
13. S. Kawazu, M. Suzuki, K. Negishi, T. Watanabe, and J. Ishii, Diabetes, $\underline{36}$, 216 (1987).
14. S. Kawazu, M. Suzuki, K. Negishi, J. Ishii, H. Sando, H. Katagiri, Y. Kanazawa, S. Yamanouchi, Y. Akanuma, H. Kajinuma, K. Suzuki, K. Watanabe, T. Itoh, T. Kobayashi, and K. Kosaka, Diabetes, $\underline{36}$, 221 (1987).
15. K. Kameda, I. Koyama, and Y. Abiko, Biochim. Biophys. Acta, $\underline{677}$, 263 (1981).
16. G. Koh, Y. Seino, K. Tsuda, S. Nishi, H. Ishida, J. Takeda, H. Fukumoto, T. Taminato, and H. Imura, Life Sci., $\underline{40}$, 1113 (1987).
17. C.B. Chapleo, P.L. Myers, R.C.M. Butler, J.C. Doxey, A.G. Roach, and C.F.C. Roach, J. Med. Chem., $\underline{26}$, 823 (1983).
18. C.B. Chapleo, P.L. Myers, R.C.M. Butler, J.A. Davis, J.C. Doxey, S.D. Higgins, M. Myers, A.G. Roach, C.F.C. Smith, M.R. Stillings, and A.P. Welbourn, J. Med. Chem., $\underline{27}$, 570 (1984).

19. G.P. Fagan, C.P. Chapleo, A.C. Lane, M. Myers, A.G. Roach, C.F.C. Roach, M.R. Stillings, and A.P. Welbourn, J. Med. Chem., 31, 944 (1988).
20. D. Biggs, S. Langer, C. Morel, and M. Servin, U.S. Patent 4,904,684 (1990).
21. I. Angel, H. Schoemaker, R. Azerhad, S. Bidet, N. Duval, and S.Z. Langer, Int. J. Obesity, 14 (Suppl. 2), 50 (Abstract IF12) (1990).
22. C.G. Ostenson, J. Pigon, J.C. Doxey, S. Efendic, J. Clin. Endocrin. Metab., 67, 1054 (1988).
23. H. Shinkai, K. Toi, I. Kumashiro, Y. Seto, M. Fukuma, K. Dan, and S. Toyoshima, J. Med. Chem., 31, 2092 (1988).
24. H. Shinkai, M. Nishikawa, Y. Sato, K. Toi, I. Kumashiro, Y. Seto, M. Fukuma, K. Dan, and S. Toyoshima, J. Med. Chem., 32, 1436 (1989).
25. H. Shinkai and Y. Sato in "New Antidiabetic Drugs", C.J.Bailey, P.R. Flatt, Eds., Smith-Gordon, London, 1990, p. 249.
26. Y. Sato, M. Nishikawa, H. Shinkai, and E. Sukegawa, Diabetes Res. Clin. Practice, in press (1992).
27. K. Geisen, Arzneim-Forsch/Drug Res., 38, 1120 (1988).
28. T.T. Yen, N.B. Dininger, C.L. Broderick, and A.M. Gill, Arch. Int. Pharmacodyn., 308, 137 (1990).
29. R. Saperstein, R.M. Tolman, and E.E. Slater, U.S. Patent 4,978,667 (1990).
30. B. Kreymann, M.A. Gitatei, G. Williams, and S.R. Bloom, Lancet II, 1300 (1987).
31. H.C. Fehmann, and J.F. Habener, Endocrinology, 130, 159 (1992).
32. K.E. Steiner, and E.L. Lien, Prog. Med. Chem., 24, 209 (1987).
33. E.R. Larsen, D.A. Clark, and R.W. Stevenson, Ann. Rep. Med. Chem., 25, 205 (1989).
34. J.R. Colca and D.R. Morton in "New Antidiabetic Drugs", C.J.Bailey, P.R. Flatt, Eds., Smith-Gordon, London, 1990, p. 255.
35. Drugs of the Future, 15, 1080 (1990).
36. Drugs of the Future, 16, 853 (1991).
37. S.L. Suter, J.J. Nolan, P. Wallace, B. Gumbiner, and J.M. Olefsky, Diabetes Care, 15, 193 (1992).
38. T. Kuzuya, Y. Iwamoto, K. Kosaka, K. Takebe, T. Yamanouchi, M. Kasuga, H. Kajinuma, Y. Akanuma, S. Yoshida, and Y. Shigeta, Diabetes Res. Clin. Pract., 11, 147 (1991).
39. R.W. Stevenson, R.K. McPherson, P.E. Genereux, B.H. Danbury, and D.K. Kreutter, Metabolism, 40, 1268 (1991).
40. D. A. Clark, S. W. Goldestein, R. A. Volkmann, J. F. Eggler, G. F. Holland, B. Hulin, R.W. Stevenson, D.K. Kreutter, E.M. Gibbs, M.N. Krupp, P. Merrigan, P.L. Kelbaugh, E.G. Andrews, D.L. Tickner, R.T. Suleske, C.H. Lamphere, F.J. Rajeckas, W.H. Kappeler, R.E. McDermott, N.J. Hutson, and M.R. Johnson, J. Med. Chem., 34, 319 (1991).
41. C. Hofmann, K. Lorenz, and J.R. Colca, Endocrinol., 129, 1915 (1991).
42. C.A. Hofmann, C.W. Edwards, III, R.M. Hillman, and J.R. Colca, Endocrinol., 130, 735(1992).
43. E.M. Gibbs, P.E. Genereux, D.K. Kreutter, Diabetologia, 32, 491A (1989).
44. J.R. Colca, C.F. Dailey, B.J. Palazuk, R.M. Hillman, D.M. Dinh, G.W. Melchior, and C.H. Spilman, Diabetes, 40, 1669 (1991).
45. H. Ikeda, S. Taketomi, Y. Sugiyama, T. Sodha, K. Meguro, and T. Fujita, Arzneim-Forsch/Drug Res., 40, 156 (1990).
46. Y. Sugiyama, S. Taketomi, Y. Shimura, H. Ikeda, and T. Fujita, Arzneim-Forsch/Drug Res., 40, 263 (1990).
47. Y. Sugiyama, Y. Shimura, and H. Ikeda, Arzneim-Forsch/Drug Res., 40, 436 (1990).
48. R.W. Stevenson, N.J. Hutson, M.N. Krupp, R.M. Volkmann, G.F. Holland, J.F. Eggler, D.A. Clark, R.K.McPherson, K.L. Hall, B.H. Danbury, E.M. Gibbs, and D.K. Kreutter., Diabetes, 39, 1218 (1990).
49. R.K. Dubey, T.A. Kotchen, M.A. Boegehold, and H.Y. Zhang, FASEB, abstract in press (1992).
50. Y. Momose, K. Meguro, H. Ikeda, C. Hatanaga, S. Ot, and T. Sohda, Chem. Pharm. Bull., 39, 1440 (1991).
51. T. Sohda, Y. Momose, K. Meguro, Y. Kawamatsu, Y. Sugiyama, and H. Ikeda, Arzneim-Forsch/Drug Res., 40, 37 (1990).
52. D.A. Clark, S.W. Goldstein, and B. Hulin, U.S. Patent 5,036,079 (1991).
53. D.A. Clark, S.W. Goldstein, and B. Hulin, WO 9103474 (1991).
54. D.A. Clark, and M.R. Johnson, U.S. Patent 4,968,707 (1990).
55. B.C.C. Cantello, and P.T. Duff, European Patent 415,605A (1991).
56. R.M. Hindley, R. Southgate, and P.T. Duff, European Patent 419,035A (1991).
57. A. Zask, I. Jirkovsky, J.W. Nowicki, and M.L. McCaleb, J. Med. Chem., 33, 1418 (1990).
58. A. Zask, and I. Jirkovsky, U.S. Patent 4,997,948 (1991).

59. R.L. Dow, B.M. Bechle, T.T. Chou, D.A. Clark, B. Hulin, and R.W. Stevenson, J. Med. Chem., 34, 1538 (1991).

60. S.W. Goldstein, U.S. Patent 5,037,842 (1991).

61. J.W. Ellingboe, J.F. Bagli, and T.R. Alessi, U.S. Patent 4,966,975 (1990).

62. T.R. Alessi, T.M. Dolak, J.W. Ellingboe, and L.J. Lombardo, U.S. Patent 4,897,405 (1990).

63. T.R. Alessi, and T.M. Dolak, U.S. Patent 4,895,862 (1990).

64. K.L. Kees, R.S. Cheeseman, D.H. Prozialeck, and K.E. Steiner, J. Med. Chem., 32, 11 (1989).

65. K.E. Steiner, M.L. McCaleb and K.L. Kees in "New Antidiabetic Drugs", C.J.Bailey, P.R. Flatt, Eds., Smith-Gordon, London, 1990, p. 237.

66. K.L. Kees, T.M. Smith, M.L. McCaleb, D.H. Prozialeck, R.S. Cheeseman, T.E. Christos, W.C. Patt, and K.E. Steiner, J. Med. Chem., 35, 944 (1992).

67. T.T. Yen, K.K. Schmiegel, G. Gold, G.D. Williams, N.B. Dininger, C.L. Broderick, and A.M. Gill, Arch. Int. Pharmacodyn., 310, 162 (1991).

68. G.A. Youngdale, WO 8909224 (1989).

69. R.A. Johnson, G.L. Bundy, G.A. Youngdale, D.R. Morton, and D.P. Wallach, U.S. Patent 4,917,826 (1990).

70. S.J. Dominianni, and T.T. Yen, J. Med. Chem., 32, 2301 (1989).

71. T.T. Yen, S.J. Dominianni, R.A. Harris, and T.W. Stephens in "New Antidiabetic Drugs", C.J.Bailey, P.R. Flatt, Eds., Smith-Gordon, London, 1990, p. 245.

72. G. Balasubramanian, GB 2,226,562A (1990).

73. A.B. Reitz, R.W. Tuman, C.S. Marchione, A.D. Jordan Jr., C.R. Bowden, and B.E. Maryanoff, J. Med. Chem., 32, 2110 (1989).

74. J.R.S. Arch, A.T. Ainsworth, M.A. Cawthorne, V. Piercy, M.V. Sennitt, V.E. Thody, C. Wilson, and S. Wilson, Nature, 309, 163 (1984).

75. C.S. Wilson, V. Wilson, M. Piercy, and M.V. Sennitt, J.R.S. Arch, Eur. J. Pharmacol., 100, 309 (1984).

76. P.L. Thurlby, and R.D.M. Ellis, Can. J. Physiol. Pharmacol., 64, 1111 (1986).

77. C. Henry, A. Buckert, Y. Schutz, E. Jequeir, and J.P. Felber, Int. J. Obesity, 12, 227 (1988).

78. A. Golay, J-P. Felber, E. Jequier, R.A. DeFronzo, and E. Ferrannini, Diabet./Metab. Rev., 4, 727 (1988).

79. N.B. Watts, R.G. Spanheimer, M. DiGirolamo, S.S.P. Gebhart, V.C. Musey, Y.K. Siddiq, and L.S. Phillips, Arch. Int. Med., 150, 803 (1990).

80. J. Challiss, L. Budohoski, E.A. Newsholme, M.V. Sennitt, and M.A. Cawthorne, Biochem. Biophys. Res. Comm., 128, 928 (1985).

81. C. Hollenga, M. Haas, J.T. Deinum, and J. Zaagsma, Horm. Metabol. Res. 22, 171 (1990).

82. B.R. Holloway, R. Howe, B.S. Rao, and D. Stribling, Int. J. Obesity, 14 (Suppl. 2), 51 (Abstract IF-15) (1990).

83. B.R. Holloway, R. Mayers, J. Jackson, and M. Briscoe, Int. J. Obesity, 14, (Suppl. 2), 142 (Abstract 2P-85) (1990).

84. J. Zaagsma, and S.R. Nahorski, TIPS, 11, 3 (1990).

85. P.L. Selby, and H.S.A. Sherratt, TIPS, 10, 495 (1989).

86. M. von-Herrath, and H. Holzer, Z. Lebensm. Unters. Forsil., 186, 427 (1988).

87. D.M. Regen and S.J. Pilkis, in "Hormonal Control of Gluconeogenesis", Vol. III, N. Kraus-Friedmann, Ed., CRC Press, Boca Raton, Florida, 1986, p. 101.

88. B. Leighton, and G.J.S. Cooper, Nature, 335, 632 (1988).

89. D.A. Young, R.O. Deems, R.W. Deacon, R.H. McIntosh, and J.E. Foley, Amer. J. Physiol., 259, E457 (1990).

90. A. M. Gill and T. T. Yen, Life Sciences, 48, 703 (1990).

91. C.X. Moore, and G.J.S. Cooper, Biochem. Biophys. Res. Comm., 179, 1 (1991).

92. R. Taylor, Clin. Endocrinol., 34, 159 (1991).

93. H.U. Haring, Diabetologia, 34, 848 (1991).

94. M.D. Houslay, M.J.O. Wakelman, and N.J. Pyne, TIBS II, 393 (1986).

95. A.R. Saltiel, J. Bioenerg. Biomembranes, 23, 29 (1991).

SECTION V. Topics In Biology

Editor: Kenneth B. Seamon
Food and Drug Administration
Bethesda MD 20892

Chapter 24. The Role of Homeobox Genes In Vertebrate Embryonic Development

Kathleen A. Mahon[1] and Milan Jamrich[2]
[1]Laboratory of Mammalian Genes and Development, NICHD, NIH
and [2]Laboratory of Molecular Pharmacology, Center for Biologics Evaluation and Research, FDA,
8800-Rockville Pike Bethesda, MD 20892

Introduction - Despite their obvious differences in morphogenesis, recent molecular evidence has indicated that the development of invertebrate and vertebrate embryos share many common genetic features. Embryogenesis proceeds through a system of genetic controls that act upon the symmetrical fertilized egg to progressively transform it into a complex multicellular embryo with distinct anteroposterior, dorsoventral, and proximodistal axes. This genetic system also provides cells with information about their structural and positional identity resulting in the reproducible morphological and physiological makeup of the given species. Embryonic cells form ordered spatial arrangements resulting in familiar structures such as arms, legs, tails and heads. This process is called pattern formation.

A large number of genes involved in the process of patterning have been identified over the past 10 years. Many of these genes encode transcriptional regulators that appear to provide positional cues during embryonic development (1). Most interestingly, many of these genes share a highly conserved 180 bp DNA sequence termed the "homeobox". The homeobox encodes a 60 amino acid DNA-binding region of the helix-turn-helix type, the so-called homeodomain, similar to the domain found in many bacterial DNA-binding proteins (2). The homeobox was originally found in Drosophila developmental control genes (3,4). Shortly after it was realized that the homeobox was conserved in other animal species (5), an intense effort was initiated to identify additional family members. To date, homeobox genes have been isolated from widely isolated and diverse representatives of the animal kingdom, e.g. coelenterates, Drosophila, worms, frogs, mice, and man (5-11). Recently, these genes have also been identified in plants (12,13). Presumably they perform the same function in all of these species - the regulation of transcription during the process of pattern formation. Their developmental roles have been best illustrated in Drosophila, but there is an increasing body of evidence that they have the same function in vertebrates.

An understanding of pattern formation and the genes involved in it would be highly desirable, not only from the theoretical standpoint, but also from a medical perspective since hundreds of thousands of children are born yearly in the USA alone with some kind of malformation due to incorrect embryonic patterning. It is conceivable that at least some of these malformations are due to the aberrant expression of homeobox genes. In this chapter, we will concentrate on the progress achieved in this field with special emphasis on three organisms - Drosophila, Xenopus, and mouse. Each of these developmental systems offers unique advantages for the experimental study of pattern formation.

Drosophilia Homeobox Genes - Embryonic development in Drosophila is known to be controlled by a regulatory cascade of gene expression that acts to subdivide the embryo into progressively smaller developmental units. At the apex of the regulatory hierarchy are genes whose products are expressed maternally and deposited in the unfertilized egg. These genes, such as bicoid and dorsal, determine the absolute polarity of the anteroposterior and dorsoventral axes of the embryo respectively (14-16). Their gene products then act upon the genes at intermediate positions in the hierarchy, which specify the patterning and polarity the basic regions and segments of the fly along these axes. However, the fine tuning of embryonic patterning, particularly with respect to the anteroposterior axis, is under the control of the homeotic selector genes, which determine segment identity, i.e., whether a head, thoracic or abdominal segment develops. These genes, originally identified by dramatic mutations that "transform" one body structure into that of an analogous structure from a neighboring segment (the so-called homeotic transformations), are present in two linked complexes, the Bithorax and Antennapedia gene complexes. For example, a series of mutations in the Bithorax complex will transform the third thoracic segment into another second thoracic segment creating a fly with four, instead of, two wings. Similarly, certain genetic mutations in the Antennapedia complex can cause formation of legs instead of antennae on the head of the fruitfly.

It is now clear that the majority of genes in the regulatory hierarchy encode putative transcriptional regulators, and as such, are believed to activate or repress batteries of structural genes involved in morphogenesis. There is also a great deal of genetic and molecular evidence that these transcription factors can regulate each other (1). Most of these genes contain the homeobox DNA binding motif, although other classes of transcription factors are also represented. However, among the homeobox genes, several sub-classes of homeobox sequences can be ascertained based on sequence similarity (17). Most of the genes of the Antennapedia and Bithorax gene complexes encode homeodomain proteins of the so-called Antennapedia (Antp) class. Genes from other tiers of the regulatory hierarchy, such as engrailed, paired and bicoid encode clearly divergent homeodomains.

One of the hallmarks of homeobox gene activity in flies (as well as in vertebrates) is region specific expression. The Antennapedia and Bithorax complex genes are expressed in distinct domains along the anteroposterior axis (18). Generally, these expression domains coincide with the regions or segments that are defective or missing in corresponding mutants of these genes. The very detailed genetic and molecular studies in this invertebrate system revealed for the first time how embryonic development proceeds, and have laid the conceptual framework for current hypotheses about the molecular basis of development in higher organisms.

<u>VERTEBRATE HOMEOBOX GENES</u>

Organization of Homeobox Genes - The first vertebrate homeobox containing gene to be identified was isolated from Xenopus (6). Subsequently, a large number of genes were isolated from three vertebrate species - Xenopus, mouse, and man (11,17,19). Because of the relative ease in mapping genes in mouse and man, much of what is known about the genomic and chromosomal organization of these genes has been obtained from these species. Approximately 40 mammalian genes have been isolated that contain homeobox sequences most similar to that of the Drosophila Antp gene. These genes are collectively known as Hox genes and represent the largest group of vertebrate homeobox genes known to date.

Hox genes are organized in four chromosomal clusters in both mouse and man. The clusters appear to have evolved by duplication and divergence from an ancestral gene. Sequence comparison has shown that the genes from one cluster can be aligned by virtue of their homeobox sequences with genes at identical positions in other clusters (Figure 1). The conservation of this clustered arrangement is even more intriguing when one considers that the same linear order is maintained in the homeotic Bithorax and Antennapedia gene complexes of Drosophila (Figure 1) (14,20,21). Recently, it has been shown that Antp-class homeobox genes of the nematode C. elegans are also clustered in this same linear arrangement (9). Due to the lack of genetics and chromosomal mapping techniques in Xenopus, it is currently not known if the homeobox genes are clustered in this organism. However, it seems unlikely that they would not be, given the extreme evolutionary conservation of this colinear arrangement that includes animals separated by more than 500 million years.

The remarkable conservation of homeobox genomic organization raises some interesting evolutionary questions and provides clues concerning the regulation of these genes in development. There is growing evidence that the clustered arrangement is necessary for the proper temporal and spatial expression of these genes. It has been shown that there are multiple control elements necessary for proper expression of these genes which can be interspersed or overlap with regulatory elements from other adjacent genes in the cluster (22,23), thus explaining why the colinearity of the cluster would be evolutionarily maintained. Progressive "opening" of chromosomal domains has also been suggested as a mechanism for the sequential expression of the genes along the cluster (24).

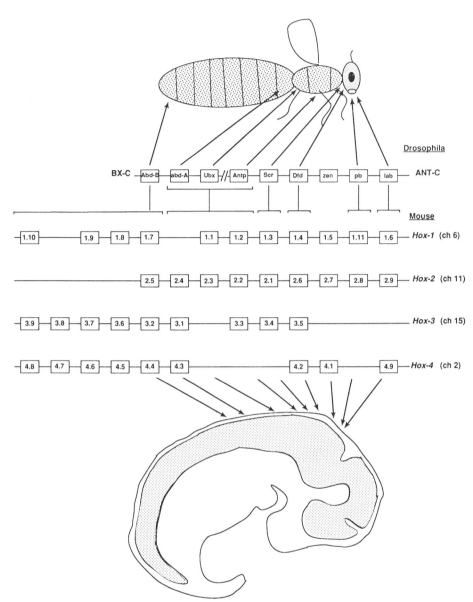

Figure 1. Chromosomal organization of homeobox gene clusters in <u>Drosophila</u> and mouse. The homeobox genes of the fly and the mouse are clustered, and are aligned vertically based on sequence comparisons of their homeodomain regions. Each linear array represents the cluster on one chromosome, whereas each box represents a different homeobox gene within the cluster. The top cluster is from the <u>Drosophila</u> <u>Antennapedia</u> (ANT-C) and <u>Bithorax</u> (BX-C) complexes. They are aligned with the four murine gene clusters, <u>Hox</u>-1, <u>Hox</u>-2, <u>Hox</u>-3 and <u>Hox</u>-4. Their chromosomal locations are given in parentheses. The names of individual genes within each cluster are given within the boxes. There is a correlation between the linear organization of the genes and the expression domains along the anteroposterior axis in both organisms. This is indicated by the diagrams of the fly and mouse embryo at the top and bottom of the figure respectively. The arrows indicate the approximate anterior or rostral boundary of expression of each gene. For the mouse, the arrows indicate boundaries in the CNS for all genes at analogous positions in each cluster. (For example, <u>Hox</u>-1.4, 2.6, 3.5, and 4.2 have similar expression boundaries). This figure is based on Graham et al. (21), and McGinnis and Krumlauf (11).

Expression During Development - Like their Drosophila counterparts in the Antennapedia and Bithorax complexes, the murine Hox genes are expressed in spatially restricted domains along the anteroposterior axis during embryogenesis, most notably in the developing central nervous system (CNS) and the paraxial mesoderm (8,19,11). Extensive in situ hybridization studies have shown that each Hox gene has a distinct expression domain in the CNS with a precise but different rostral expression boundary. Thus the Hox genes are expressed in overlapping, yet distinct, expression domains along the axis (see Figure 1). The most rostral extent of Hox gene expression is seen in the hindbrain, where Hox expression boundaries correspond to rhombomeric constrictions in the neuroepithelium (25-27). These expression patterns are reminiscent of the segmental expression domains of many Drosophila homeotic genes, and are consistent with a role for the Hox genes in the determination of regional identity along the anteroposterior axis during development in the same way the homeotic selector genes act in Drosophila development.

Many Hox gene family members are differentially expressed in the vertebrate limb bud, often with a graded pattern of expression (28-32). As in the CNS, Hox expression domains overlap, yet are distinct for each gene along each of the three axes apparent in the limb - anteroposterior, dorsoventral, and proximodistal (30). These intriguing expression patterns clearly implicate the Hox genes in the patterning of the limb through both differential and combinatorial interactions of Hox gene products.

Surprisingly, the activation of the Hox genes along the anteroposterior axis (and proximodistal axis in the case of the limb bud) corresponds to the linear arrangement of the genes within the cluster as shown in Figure 1 (20,21,30,33,34). Genes at the 3' end of the cluster are expressed more anteriorly, whereas genes at the 5' end are expressed more posteriorly. The correspondence of the order of the genes in the cluster with their order of expression along the axis has long been recognized for the Antennapedia and Bithorax gene complexes in Drosophila (35,14), again underscoring the extreme structural and functional conservation of this gene family.

Other Classes of Vertebrate Homeobox Genes - While the Antp-like Hox homeobox genes appear to be counterparts of the homeotic selector genes of Drosophila, an important question remains: Is there a regulatory hierarchy at work in vertebrates as there is in Drosophila that sets up embryonic patterning in the very early embryo? In addition to the Antp class of homeobox genes, there appear to be many divergent classes of homeobox genes in vertebrates as well as in Drosophila. These include homologs of the Drosophila engrailed (36), caudal (37), paired (38,39), even-skipped (40), and Distal-less (41-43) genes, as well new, undescribed homeobox sequences (44,45). In general, these genes are expressed in region specific domains quite distinct from those of the Hox genes, suggesting that they provide positional information to other regions of the developing embryo.

For example, in the CNS, the Hox genes are expressed in the hindbrain and neural tube, the engrailed genes, en-1 and en-2, are expressed in the midbrain (46), and the Distal-less genes, Dlx-1 and Dlx-2, are expressed in the forebrain (41,42,43,47). Thus the CNS can be spatially subdivided according homeobox expression domains. In addition, a number of homeobox genes have been isolated that are temporally expressed prior to the onset of Hox gene expression. Thus it is likely that a regulatory hierarchy akin to that in Drosophila is responsible for the generation of embryonic form in vertebrates as well.

DEVELOPMENTAL FUNCTION OF VERTEBRATE HOMEOBOX GENES

Homeobox Genes, Growth Factors, and Embryonic Induction - The discovery of homeobox genes came at a time when frog developmental biologists were just beginning to understand, at the molecular level, the nature of mesoderm inducing factors. It had been known for some time in frogs that mesodermally derived structures, such as muscle and notochord, can be formed from uncommitted ectodermal cells under the influence of mesoderm inducing factors. These factors have been identified as belonging to the group of FGF and TGF-β like growth factors (48-51). Whereas FGF and TGF-β were able to induce formation of mesoderm from ectoderm, none of these was able to initiate correct patterning these explants. Finally, activin, a TGF-β like growth factor, was found not only to be able to induce mesoderm but also to initiate correct patterning leading to the formation of a mini-embryo (52-55). For the first time, a connection was made between an isolated growth factor and the initial steps of pattern formation. Subsequently, a search for homeobox genes that have a primary response to activin induction was intensively pursued. To date, a number of homeobox genes have been found to respond to activin induction. The first gene of this kind to be isolated, Mix-1, can be induced in ectodermal explants by activin even in the absence of protein synthesis (56). Consequently, the homeobox genes goosecoid (57,58) and Xlim-1 (59) have been shown to have similar properties.

"Mutational" Analysis of Vertebrate Homeobox Gene Function - Mutant analysis has dramatically demonstrated the absolute requirement of functional homeobox gene activity for proper embryonic development in Drosophila (14,15). Although their extreme conservation in organization and expression provides a great deal of circumstantial evidence that the vertebrate homeobox genes are important for embryonic development, direct evidence for their developmental roles has only recently been forthcoming. The lack of known mutants of vertebrate Hox genes has not hampered researchers, who have utilized a combination of molecular biology and experimental embryology to generate both dominant "gain of function" and recessive "loss-of-function" mutant phenotypes in both frog and mouse.

Xenopus - In Xenopus, functional aspects of homeobox gene activity have been studied by ectopic expression or overexpression of these genes or by ablation of their gene products in embryos. Ectopic expression of homeobox genes in Xenopus can be achieved by microinjection of synthetic mRNA into 2-16 cell stage embryos. Overexpression of Xhox 1A, a gene normally expressed in the developing somites, induces abnormalities in somite segmentation (60), suggesting that the correct regulation of the Xhox 1A gene is important for the normal segmentation in early embryos. Overexpression of a posterior restricted gene, Xhox-3 causes posteriorization of the embryo; i.e. a loss of anterior structures (61). It has also been shown that overexpression of a posterior specific homeobox gene, in this case XlHbox-6 (equivalent to murine Hox-2.5), may confer posterior character to microinjected embryos (57).

Xenopus homeobox genes have been shown to be involved in pattern formation (58). It was demonstrated that ectopic expression of the goosecoid homeobox gene can induce a secondary axis in Xenopus embryos. Functional ablation of homeobox gene products by antibodies was also attempted (62). Embryos injected with antibodies against XlHbox-1 (equivalent to murine Hox-3.3) did show malformations in the region of the neural tube normally expressing this gene.

Table 1. Gain of Function Phenotypes of Homeobox Genes in Flies, Frogs, and Mouse

GENE	PHENOTYPE	Ref
Dfd (Drosophila)	Homeotic transformation of the labial and thoracic epidermis to maxillary, head involution defects	81-83
Antp (Drosophila)	Partial homeotic transformation of head segment epidermis to thoracic epidermis	84,85
X1Hbox1 (Xenopus)	Alteration of the anterior spinal cord, expanded hindbrain	62
Goosecoid (Xenopus)	Formation of a secondary axis	58
Hox-1.1 (Mouse)	Dominant lethal, craniofacial abnormalities, changes in axial skeleton	64,65
Hox-1.4 (Mouse)	Dominant lethal, hyperproliferation of the colon, resembles Hirschsprung's disease	63

Mouse - "Gain-of-function" phenotypes (see Table 1) have been similarly generated in mice, with ectopic expression and overexpression of homeobox gene constructs being achieved through the use of transgenic technology. Overexpression of a Hox-1.4 gene under the control of its own promoter led to abnormalities of the gastrointestinal tract, the so-called "megacolon" phenotype (63). Ectopic expression of the Hox-1.1 gene, a homeobox gene normally expressed in the lower cervical and thoracic region of the neural tube and prevertebral elements respectively, resulted in profound cranio-facial defects and other morphological abnormalities in the head (64). Interestingly, the regions most affected were rostral to the region normally expressing Hox-1.1. In many cases there appeared to be a anterior to posterior transformation of vertebrae, that is, cervical vertebrae C1-C2 often resembled more caudal vertebral elements (65). These phenotypic abnormalities are reminiscent of the transformations characteristic of Drosophila homeotic mutants. Very similar malformations of the vertebrae can also be obtained through treatment of normal embryos with retinoic acid, a morphogen known to alter the regional expression of the Hox genes. These abnormalities are accompanied by characteristic shifts in Hox gene expression domains along the axis (66).

"Loss-of-function" phenotypes (see Table 2) can be obtained through the use of embryonic stem (ES) cells, in which genes can be mutated via homologous recombination. These pluripotent ES cells can then be used to generate chimeric mice. By breeding these chimeric mice the mutated cells can be passed through the germline, ultimately leading to the generation of mice homozygous for the respective mutation. Two homeobox genes, Hox-1.5 and Hox-1.6, have been targeted in this way (67,68). Both mutants die shortly after birth and are characterized by profound malformations of the head and thorax, consistent with the hypothesis that Hox genes are providing positional cues for proper patterning. However, although these genes have largely overlapping expression domains in normal embryos, there are distinct differences in the malformations observed, suggesting that Hox-1.5 and Hox-1.6 play distinct roles in morphogenesis.

Table 2. Loss of Function Phenotypes of Homeobox Gene in Flies, Frogs, and Mouse

GENE	PHENOTYPE	Ref
Dfd (Drosophila)	Deletion of epidermal structures derived from ventral maxillary and mandibular segments	81-83
Scr (Drosophila)	Homeotic transformation of anterior epidermis of the thoracic segment 1 to thoracic segment 2	85,86
X1Hbox1 (Xenopus)	Alteration of the anterior spinal cord, enlarged hindbrain, defects in neural crest cells	62
Hox-1.5 (Mouse)	Neonatal letha., defects in head and thorax, missing thymus, altered heart and circulatory system	67
Hox-1.6 (Mouse)	Neonatal lethal, defects in head mesoderm, neural crest, and cranial sensory ganglia	68

These "gain of function" and "loss-of-function" studies clearly illustrate the developmental importance of homeobox gene activity for proper vertebrate morphogenesis. However, the elaboration of their exact roles remains to be elucidated. It seems clear that, in general, precise homeotic transformations are more the exception that the rule among vertebrate homeobox "mutants". This is most likely due to the extreme dependence of vertebrate development on cell and tissue interactions as opposed to the lineage dependent development characteristic of invertebrates.

Homeobox Genes in Human Development and Disease - Could mutations in homeobox genes be responsible for human malformations? Although naturally occurring mutations in Antp-like Hox genes have not been identified in the mouse, there has been at least one candidate among human patients with developmental anomalies that might be due to aberrant Hox gene expression. A patient has been described with campomelic dysplasia, a syndrome characterized by malformations of the face, thorax, and axial and appendicular skeleton (69). This condition is phenotypically similar to the experimentally induced mutations of Hox genes in mice. This patient bore a paracentric inversion in the vicinity of the Hox 2 cluster on chromosome 17, raising the intriguing possibility that Hox gene misexpression is responsible for the malformations seen.

There are, however, several naturally occurring mutations that have been identified in non-Antp class homeobox genes. A mutation in the homeobox of the pituitary-specific Pit-1/GHF-1 gene has been found in the Snell's dwarf (dw) mouse mutant (70,71). These mice are characterized by abnormal anterior pituitary function, due to lack of growth hormone production. Pit-1 is known to transcriptionally activate growth hormone (72,73). At least two known mouse mutations affecting morphogenesis have been shown to bear mutations in homeobox genes. The Splotch (Sp^{2H}) and

Small eye (Sey) mutations carry defective Pax-3 and Pax-6 homeobox genes respectively (74,75). (These Pax genes contain a paired-type homeobox as well as another putative DNA binding domain, the paired domain). Interestingly, Sp^{2H} and Sey appear to be homologues of the Waardenburg syndrome (WS) and aniridia (AN) loci in man respectively. Waardenburg's syndrome is characterized by deafness and pigmentary disturbances, probably the result of defective neural crest, while mouse Sp^{2H} mutants have defective neural tube closure. Both AN and Sey mutations involve defects in eye development. Mutations in the Pax-3 gene have been found in WS patients (76,77), and the AN locus has been positionally cloned and does in fact correspond to the Pax-6 gene (78).

There is also convincing evidence that aberrant homeobox gene expression can contribute to the genesis or progression of cancer. Transcriptional activation of Hox-2.4 by proviral insertion has led to the oncogenic transformation in mouse fibroblasts (79). Similarly, 25% of pediatric B-cell acute lymphoblastic leukemias contain a translocation that activates a novel homeobox gene, Pbx-1 (80). These data provide intriguing clues concerning the developmental role of the homeobox genes in vertebrates. It is likely that the further analysis of homeobox genes will implicate them in human disease and developmental malformation.

References

1.   M. Levine and T. Hoey, Cell, 55, 537 (1988).
2.   C.O. Pabo and R.T. Sauer, Annu. Rev. Biochem., 53, 292 (1984).
3.   W. McGinnis, M.S. Levine, E. Hafen, A. Kuroiwa and W.J. Gehring, Nature, 308, 428 (1984).
4.   M.P. Scott and A.J. Weiner, Proc. Natl. Acad. Sci. U.S.A., 81, 4115 (1984).
5.   W. McGinnis, R.L. Garber, J. Wirz, A. Kuroiwa and W.J. Gehring, Cell, 37, 403 (1984).
6.   A.E. Carrasco, W. McGinnis, W.J. Gehring and E.M. De Robertis, Cell, 37, 409 (1984).
7.   M. Rabin, C.P. Hart, A. Ferguson-Smith, W. McGinnis, M. Levine and F.H. Ruddle, Nature, 314, 175 (1985).
8.   P.W. Holland and B.L. Hogan, Genes Dev., 2, 773 (1988).
9.   C. Kenyon and B. Wang, Science, 253, 517 (1991).
10.   J. Marx, Science, 255, 399, (1992).
11.   W. McGinnis and R. Krumlauf, Cell, 68, 283 (1992).
12.   E. Vollbrecht, B. Veit, N. Sinha and S. Hake, Nature, 350, 241 (1991).
13.   I. Ruberti, G. Sessa, S. Lucchetti and G. Morelli, EMBO J., 10, 1787 (1991).
14.   M.E. Akam, Development, 101, 1 (1987).
15.   P.W. Ingham, Nature, 335, 25 (1988).
16.   D. St. Johnston and Ch. Nusslein-Volhard, Cell, 68, 201 (1992).
17.   M.P. Scott, J.W. Tamkun and G.W. Hartzell III, Biochim. Biophys. Acta, 989, 25 (1989).
18.   K. Harding, C. Wedeen, W. McGinnis, and M. Levine, Science, 229, 1236 (1985).
19.   M. Kessel and P. Gruss, Science, 249, 374 (1990).
20.   D. Duboule and P. Dolle, EMBO J., 8, 1497 (1989).
21.   A. Graham, N. Papalopulu and R. Krumlauf, Cell, 57, 367 (1989).
22.   A. Puschel, R. Balling and P. Gruss, Development, 108, 435 (1990).
23.   J. Whiting, H. Marshall, M. Cook, R. Krumlauf, P.W.J. Rigby, D. Scott and R.K. Allemann, Genes and Devel., 5, 2048 (1991).
24.   S.J. Gaunt and P.B. Singh, Trends in Genetics, 6, 208 (1990).
25.   P. Murphy and R.E. Hill, Development, 111, 61 (1991).
26.   D. Wilkinson, S. Bhatt, M. Cook, E. Boncinelli and R. Krumlauf, Nature, 341, 405 (1989).
27.   P. Hunt and R. Krumlauf, Cell, 66, 1075 (1991).
28.   C.V.E. Wright, K.W.Y. Cho, J. Hardwicke, R.H. Collins and E.M. DeRobertis, Cell, 59, 81 (1989).
29.   K.A. Mahon, H. Westphal, and P. Gruss, Development, 104, 187 (1988).
30.   P. Dolle, J.-C. Izpisua-Belmonte, H. Falkenstein, A. Renucci and D. Duboule, Nature, 342, 767 (1989).
31.   S. Mackem and K.A. Mahon, Development, 112, 791 (1991).
32.   T. Nohno, S. Noji, E. Koyama, K. Ohyama, F. Myokai, A. Kuroiwa, T. Saito and S. Taniguchi, Cell 64, 1197 (1991).
33.   S. Gaunt, P. Sharpe and D. Duboule, Development, 104, 169 (1988).
34.   G.R. Dressler and P. Gruss, Differentiation, 41, 193 (1989).
35.   E.B. Lewis, Nature, 276, 565 (1978).
36.   A.L. Joyner and G.R. Martin, Genes Dev. 1, 29 (1987).
37.   P. Duprey, K. Chowdhury, G.R. Dressler, R. Balling, D. Simon, J.-L. Guenet and P. Gruss, Genes Dev.

2, 1647 (1988).

38. K. Kongsuwan, E. Webb, P. Housiaux and J.M. Adams, EMBO J, 7, 2131 (1988).

39. C. Walther, J.L. Guenet, D. Simon, U. Deutsch, B. Jostes, M.D. Goulding, D. Plachov, R. Balling, P. Gruss, Genomics, 11, 424 (1991).

40. H. Bastian and P. Gruss, EMBO J., 9, 1839 (1990).

41. M. Price, M. Lemaistre, M. Pischetola, R. Di Lauro and D. Duboule, Nature, 351, 748 (1991).

42. M.H. Porteus, A. Bulfone, R.D. Ciaranello and J.L.R. Rubenstein, Neuron, 7, 221 (1991).

43. G. Robinson, S. Wray and K. Mahon, New Biologist, 3, 1183 (1991).

44. M.T. Murtha, J.F. Leckman and F.H. Ruddle, PNAS 88, 10711 (1991).

45. G. Singh, S. Kaur, J.L. Stock, N.A. Jenkins, D.J. Gilbert, N.G. Copeland and S.S. Potter, Proc. Natl. Acad. Sci. USA, 88, 10706 (1991).

46. C.A. Davis and A.L. Joyner, Genes Dev. 2, 1736 (1988).

47. M. Dirksen, S. Mackem and M. Jamrich (Neuron) in press.

48. J.M.W. Slack, B.G. Darlington, J.K. Heath and S.F. Godsave, Nature, 326, 197 (1987).

49. D. Kimelman and M. Kirschner, Cell, 51, 869 (1987).

50. J.C. Smith, Development, 105, 665 (1989).

51. A.P. Roberts, P. Kondaiah, F. Rosa, S. Watanabe, P. Good, D. Danielpour, N.S. Roche, M.L. Rebbert, I.B. Dawid and M.B. Sporn, Growth Factors, 3, 277 (1990).

52. M. Asashima, H. Nakano, K. Shimada, K. Kinoshita, K. Ishii, H. Shibai and N. Ueno, Roux's Arch. Dev. Biol., 198, 330 (1990).

53. J.C. Smith, B.M.J. Price, K. Van Nimmen and D. Huylebroeck, Nature, 345, 729 (1990).

54. S. Sokol, G.G. Wong and D.A. Melton, Science, 249, 561 (1990). 55.G. Thomsen, T. Woolf, M. Whitman, S. Sokol, J. Vaughan, W. Vale and D.A. Melton, Cell, 63, 485 (1990).

56. F.M. Rosa, Cell, 57, 965 (1989).

57. K.W.Y. Cho, E.A. Morita, C.V.E. Wright and E.M. DeRobertis, Cell, 65, 55 (1991).

58. K.W. Cho, B. Blumberg, H. Steinbeisser and E.M. DeRobertis, Cell, 67, 1111 (1991).

59. M. Taira, M. Jamrich, P.J. Good and I.B. Dawid, Genes Dev., 6, 356 (1992).

60. R.P. Harvey and D.A. Melton, Cell, 53, 687 (1988).

61. A. Ruiz i Altaba and D.A. Melton, Cell, 57, 317 (1989).

62. C.V.E. Wright, K.W.Y. Cho, J. Hardwicke, R.H. Collins and E.M. DeRobertis, Cell, 59, 81 (1989).

63. D.J. Wolgemuth, R.R. Behringer, M.P. Mostoller, R.L. Brinster and R.D. Palmiter, Nature, 337, 464 (1989).

64. R. Balling, G. Mutter, P. Gruss and M. Kessel, Cell, 58, 337 (1989).

65. M. Kessel, R. Balling and P. Gruss, Cell, 61, 301 (1990).

66. M. Kessel and P. Gruss, Cell, 67, 89 (1991).

67. D. Chisaka and M.R. Capecchi, Nature, 350, 473 (1991).

68. T. Lufkin, A. Dierich, M. LeMeur, M. Mark and P. Chambon, Cell, 66, 1105 (1991).

69. R. Maraia, H.M. Saal and D. Wangsa, Clinical Genetics, 39, 401 (1991).

70. S. Li, E.B. Crenshaw, E.J. Rawson, D.M. Simmons, L.W. Swanson, and M.G. Rosenfeld, Nature, 347, 528 (1990).

71. S.A. Camper, T.L. Saunders, R.W. Katz and R.H. Reeves, Genomics, 8, 586 (1990).

72. M. Bodner, J.L. Castrillo, L.E. Theill, T. Deerinck, M. Ellisman, M. Karin, Cell, 55, 505 (1988).

73. H.A. Ingraham, R. Chen, H.J. Mangalam, H.P. Elsholtz, S.E. Flynn, C.R. Lin, D.M. Simmons, L. Swanson and M.G. Rosenfeld, Cell, 55, 519 (1988).

74. D.J. Epstein, M. Vekemans and P. Gros, Cell, 67, 767 (1991).

75. R.E. Hill, J. Favor, B.L.M. Hogan, C.C.T. Ton, G.F. Saunders, I.M. Hanson, J. Prosser, T. Jordan, N.D. Hastie and V. van Heyningen, Nature, 354, 522 (1991).

76. M. Tassabehji, A.P. Read, V.E. Newton, R. Harris, R. Balling, P. Gruss and T. Strachan, Nature, 355, 635 (1991).

77. C.T. Baldwin, C.F. Hoth, J.A. Amos, E.O. da-Silva and A. Milunsky, Nature, 355, 637 (1992).

78. C.C.T. Ton, H. Hirvonen, H. Miwa, M.M. Well, P. Monaghan, T. Jordan, V. van Heyningen, N.D. Hastle, H. Meijers-Heijboer, M. Drechsler, B. Royer-Pokora, F. Collins, A. Swaroop, L.C. Strong, and G.F. Saunders, Cell, 67, 1059 (1991).

79. D. Aberdam, V. Negreanu, L. Sachs and C. Blatt, MCB, 11, 554 (1991).

80. M.P. Kamps, A.T. Look and D. Baltimore, Genes and Dev., 5, 358 (1991).

81. V.K.L. Merrill, F.R. Turner, and T.C. Kauffman, Dev. Biol., 122, 379 (1987).

82. M. Regulski, N. McGinnis, R. Chadwick, and W. McGinnis, EMBO J., 6, 767 (1987).

83. M.A. Kuziora and W. McGinnis, Cell, 55, 477 (1988).

84. G. Gibson and W.J. Gehring, Development, 102, 657 (1988).

85. B.T. Wakimoto and T.C. Kaufman, Dev. Biol., 81, 51 (1981).

86. G. Gibson, A. Schier, P. Le Motte, and W.J. Gehring, Cell, 62, 1087 (1990).

Chapter 25. Emerging Therapies for Cystic Fibrosis

Alan E. Smith
GENZYME Corporation
One Mountain Road
Framingham, Massachusetts   01701

<u>Introduction</u> - Therapies to treat any given disease can be developed based on observation of the symptoms and course of the disease or based on knowledge of the underlying molecular defect. Often therapies in the first category are only palliative, whereas those in the second attempt to target the basic problem. Although this latter course is ultimately preferable, the molecular basis of only a few diseases is known. With the isolation of the gene responsible for cystic fibrosis (CF) in 1989 (1-3), the development of therapies for this disease have now reached the fascinating period where new treatments are beginning to emerge at the same time as several more traditional approaches are also showing signs of promise.

<u>Status of the Disease</u> - CF is the most common lethal, recessive genetic disease of the Caucasian population (4). The disease is manifest in many secretory epithelial surfaces of the body, including the airways, sweat glands, pancreas and gut (5-6). Biochemical and electrophysiological studies on normal and CF epithelia have identified a number of differences between them, particularly abnormalities in Cl^- secretion, especially in sweat gland ducts and airway epithelia and in Na^+ absorption (5-8).

The most life threatening aspects of CF involve pulmonary events, including the production of thick, purulent mucus, infection and colonization of the mucus and resulting inflammatory responses which lead to cellular and tissue damage. Because of pancreatic and gut involvement, CF patients often also suffer nutritional deprivation (9). Survival times for CF patients have improved significantly in recent decades based in part on use of dietary supplements such as pancreatic enzymes (9,10) and in part on aggressive attempts to keep the airways functional by a combination of pulmonary percussion (11,12) and treatment with antibiotics (13). Typically, however, by age thirty, bacterial infections of the airways become difficult to erradicate, and airway blockage and damage progress, such that pulmonary dysfunction is the eventual likely cause of mortality.

Most early efforts to develop new therapies to treat CF have focused on the pulmonary complications. In some extreme cases, lung transplantation has been attempted (14). Since CF mucus consists of a surprisingly high percentage of DNA derived from lysed neutrophils, another approach has been to develop recombinant human DNase (15). Treatment of patients with aerosolized enzyme appears, in early trials, to be effective in reducing the viscosity of mucus. This should be helpful in clearing the airways of obstruction and perhaps in reducing infections. In an attempt to limit damage caused by excess of neutrophil derived elastase, protease inhibitors have been used. Aerosolized alpha-1-antitrypsin has been used to deliver enzyme activity to the lung of CF patients (16). Recombinant soluble leukocyte protease inhibitor (SLPI) is also being tested in CF patients. Another approach would be the use of agents to inhibit the action of oxidants derived from neutrophils. Although biochemical parameters have been measured successfully, the longer term beneficial effects of these treatments is yet to be clearly established.

Using a very different rational, other investigators have attempted to use pharmacological agents to reverse the abnormal Na^+ absorption in CF airways. Assuming that some of the symptoms of the disease such as altered mucus composition reflect reduced hydration resulting in turn from defective ion transport, treatment with drugs that might correct ion balance could be beneficial. Trials are in progress with aerosolized versions of the drug amiloride and initial results indicate that disease progression, as measured by lung function tests, can be reduced (17,18). Preliminary trials to test the ability of nucleotides such as ATP or UTP to stimulate the purinergic receptor and thereby correct the ion balance by an alternative pathway also appear promising (19).

<u>The CF Gene and Gene Product</u> - The identification of the gene associated with cystic fibrosis allows new therapies to be developed based on knowledge of the gene and its gene product and on studies of the molecular basis of the disease made possible by the availability of these reagents. The CF gene is located in region q31 of chromosome 7. It comprises about 240 kb and contains at least 27 exons. The mRNA is about 6.5 Kd. The gene encodes a protein named cystic fibrosis transmembrane conductance regulator or CFTR. The protein comprises 1,480 amino acids, contains 2 potential glycosylation sites and 12 hydrophobic sequences considered to be transmembrane segments (Figure 1). By sequence homology with other proteins, CFTR contains two nucleotide binding domains (NBD) and it appears to be related to several other proteins containing two copies of a unit of six transmembrane domains and an NBD, such as multidrug resistance protein (mdr) and the yeast mating factor transporter Ste6. Between the two repeats, CFTR contains a large R-domain with multiple potential phosphorylation sites (1,2).

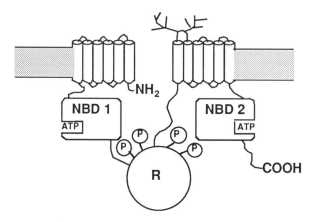

Figure 1. Proposed membrane topology and structure of CFTR. The nucleotide binding sites (**NBD1,2**), the regulatory domain (**R**), and potential sites for phosphorylation (**P**) are indicated.

Several antibodies to CFTR have been isolated and these have allowed identification of the protein (20-27). The protein has been expressed in a variety of recombinant cells (20,21,25). Importantly, expression of CFTR in cells derived from CF patients corrects the defect in Cl⁻ transport characteristic of such cells (28,29). This single and simple experiment establishes the feasibility of reversing the effects of the abnormal gene in CF patients. Expression of CFTR in a variety of recombinant cells including insect cells and frog oocytes results in the appearance of cAMP regulated Cl⁻ channel activity (21, 30-32). The simplest interpretation of this result is that CFTR itself is a Cl⁻ channel and this hypothesis has been tested by mutagenesis of residues in the transmembrane segments of CFTR (33). The finding that two such mutants in which charged residues within the transmembrane domains were changed had altered ion selectivity for Br⁻ Cl⁻, and I⁻ argues that CFTR is a Cl⁻ channel. It does not exclude the possibility that CFTR also has other functions. Electrophysiological experiments on the properties of CFTR expressed in recombinant cells indicate that the Cl⁻ channel is regulated by phosphorylation (34-35).

Extensive mutagenesis of CFTR, followed by electrophysiological studies of the variant protein indicate that the R-domain acts to regulate passage of Cl⁻ through the channel since a mutant lacking part of this region is constitutively active (36). ATP is required for the activity of CFTR both for binding at the NBDs (37) and for phosphorylation of 4 serine residues in the R-domain by cAMP dependent protein kinase A (38). Perhaps the R-domain functions as a molecular plug and movement of the plug requires both phosphorylation and hydrolysis of ATP to provide energy.

Studies on the biosynthesis of CFTR indicate that the protein is only inefficiently processed to its mature, fully glycosylated form (39). In recombinant cells, several mutants including the common ΔF508 mutation that accounts for over 70% of all CF alleles fail completely to mature (39,40). One hypothesis to account for this data is that some mutant CFTR species fail to fold correctly and in consequence are recognized by a cellular quality control mechanism located in the endoplasmic reticulum and are degraded (25,39-42). If this hypothesis is correct, and so far data particularly using non recombinant cells are conflicting (26, 27, 43-45), then CF in most cases is caused, at least in part, by the failure of CFTR to traffick to the correct cellular location. Although CFTR can be measured in the apical surface of the epithelial cells, it is not established whether it is also present and functions at other cellular locations.

EMERGING THERAPIES BASED ON KNOWLEDGE OF THE CF GENE AND GENE PRODUCT

For the purposes of the following discussion, we will assume the following hypotheses based on the data reviewed above: CF is caused by mutations in the gene encoding CFTR; such mutations either inactivate the activity of CFTR or cause it to be mislocated within the cell and to be degraded more rapidly than wild type; and one of the functions of CFTR is as a Cl^- channel. Three general approaches to develop therapies for CF can be envisaged based on this knowledge; pharmacological approaches designed to bypass the effects of malfunctioning CFTR or to relocate it to its correct cellular position; gene therapy, designed to augment the defective CF gene and protein replacement therapy designed to deliver the wild type CFTR protein to cells lacking it.

PHARMACOLOGICAL APPROACHES

Pharmacological approaches to CF are least impacted by knowledge of CFTR and its associated biochemical functions. Attempts to exploit pharmacological means to alter salt balance using amiloride and ATP/UTP predate the knowledge that CFTR is a Cl^- channel (17-19). Alternatives to these agents would be the use of pharmaceuticals to activate other Cl^- channels present in airway epithelia (46-48), or of membrane active agents that form pores (49,50). As with all pharmacological agents, issues such as drug toxicity and dosing will be important. A more fundamental consideration is whether the Cl^- channel activity associated with CFTR is the crucial property that leads to the disease state. Perhaps there is another, as yet, unidentified component of the CFTR system and this is the key regulator. If this were the case, it is possible that a pharmacological approach based on Cl^- transport might successfully adjust ion balance but still not relieve the fundamental physiological problem.

The finding that $\Delta F508$ CFTR may have altered intracellular trafficking properties (39), and the subsequent demonstration that under certain circumstances $\Delta F508$ CFTR has some low level of Cl^- channel activity (25,44) suggests another pharmacological approach; the development of agents able to relocate CFTR to the appropriate cellular destination (39). Agents able to promote folding, inhibit proteases or redistribute proteins within cells might be effective. Aside from protein foldases and nonspecific protease inhibitors (51-52), few agents with the desired properties are known. Nevertheless, it is not inconceivable that they exist, since inhibitors, rather than promoters, of intracellular trafficking and glycosylation pathways are known (53). However, nonspecific interference with intracellular degradation, trafficking or quality control mechanisms might be expected to have deleterious general effects on cellular physiology.

PROTEIN REPLACEMENT THERAPY

The delivery of wild type protein to a patient lacking the associated biochemical activity is the present treatment of choice for many protein deficiency diseases, especially genetic diseases. Such treatment directly augments the missing activity. Proteins purified from human sources (such as blood or placenta) and more recently produced by recombinant means, have been used successfully. For example, various purified blood clotting factors have been used to treat hemophilias; alpha-1-antitrypsin has been used to treat hereditary emphysema; adenosine deaminase to treat severe combined immunodeficiency disease; recombinant growth hormone to treat hereditary dwarfism. Insulin, of course, has been used for over half a century to treat different forms of diabetes, most recently with recombinant protein. Glucocerebrosidase, to treat Gaucher's disease, is a good example of a recently developed protein replacement therapy (54,55). The protein was initially purified from human material but a recombinant version is now being tested. Interestingly, the protein requires substantial modification to alter its native carbohydrate structure and thereby target it to the mannose receptor of macrophage cells (55). One advantage of protein therapy in terms of drug development is that since such treatments are well characterized and well tolerated, the regulatory pathway for such agents is now relatively conventional.

The concept of protein replacement therapy for CF is simple; a preparation of highly purified recombinant CFTR formulated in some fusogenic liposome or reassembled virus carrier delivered to the airways by installation or aerosol. Assuming that such a concept were possible, the hope would be that sufficient active CFTR would fuse into airway cells in the correct orientation so as to correct the physiological defect, irrespective of whether it is Cl^- channel activity or some other function that is the major contributor to the disease state.

Production of a Recombinant Membrane Protein - CFTR is not a soluble protein of the type that has been used for previous protein replacement therapies or for other therapeutic uses. Examples of soluble proteins that can be made in large quantities by recombinant means include insulin, growth hormone, erythropoietin and tissue plasminogen activator. Because such proteins are secreted, the issue of protein activity or toxicity within the recombinant cell seldom arises and very high levels of expression can be achieved with 50 mgs protein/L or 2000 molecules/sec/cell being possible (56). By contrast, there may be a limit to the amount of a membrane protein with biochemical activity that can be expressed in a recombinant cell. There are reports in the literature of 10^5 - 10^6 molecules/cell but this may represent an upper limit (57).

CFTR has been expressed in a variety of recombinant cells, some of which would be suitable for commercial production. The maximum reported expression levels in mammalian cells are approximately 10^5 molecules/cell (41). Insect cells make large amounts of CFTR but much of the protein appears to be insoluble (21). Recombinant cells with these modest expression levels could be scaled up to be the basis of a commercial process, but the process might be inefficient. As a radically different approach, CFTR has been expressed in the mammary glands of transgenic mice under the direction of a caesin promoter (58). The advantage of this system is that nascent CFTR arriving at the plasma membrane is budded off in the process of apocrine secretion utilized in the production of milk fat globules. As the milk fat globules exit the cell, they are surrounded by a layer of plasma membrane including, in this case, CFTR. By using transgenic rabbits or goats, it should be possible to make virtually unlimited amounts of CFTR.

Assays - Another difficulty at present in the production, solubilization and purification of CFTR is that there are no simple functional assays for the protein in vitro. This means that purification and solubilization procedures that could damage or inactivate the protein are difficult to assess.

Purification and Formulation - The purification of membrane proteins is generally more difficult than soluble proteins. The protein will require solubilization in detergents and purification under such conditions. It is likely that the process will be less efficient than for a soluble protein. Being a membrane protein, CFTR will require formulation in a suitably lipophilic vehicle. This may be a proteoliposome or perhaps a reconstituted virosome.

Fusion - The fusion of CFTR from its delivery vehicle into the membrane of recipient cells is likely to be an inefficient process. Delivery of a variety of proteins into cell membranes has been reported but the efficiency is usually low and difficult to quantitate accurately (59-62). It is unlikely that this process will exceed 1% of the added protein, particularly when applied to delivery to the airway cells of an animal rather than tissue culture cells. Delivery itself may also be problematic in the sense that aerosolization, for example, may cause damage to the protein. Particular attention will need to be paid to the orientation of the CFTR protein in the delivery vehicle and recipient cell. Only correctly oriented protein will be subject to appropriate regulation by cAMP and consequently able to function. The insertion of CFTR into a liposome may or may not favor one or other orientation and techniques to purify the appropriate preparation may be required. Another aspect of the efficiency of fusion relates to the percentage of cells in any given epithelial surface that need to be repaired to restore function. Experiments in which cells containing retroviruses expressing CFTR or a marker protein were mixed in different proportions followed by measurements of transepithelial conductance indicate that surprisingly few cells perhaps only 5%-10%, need to be complemented to restore normal current (63). Cells adjacent to a complemented cell are coupled to it by gap junctions and perhaps this allows movement between cells of low molecular weight components such as Cl^-.

Targeting - The human airway contains a variety of different cell types. Efforts to date to detect CFTR protein in such cells have been unsuccessful probably because the amounts of mRNA and protein are very low (64,65). It is not clear, therefore, whether CFTR expression is limited to a subset of airway cells or expressed in all. It once seemed possible that the expression of CFTR in some cells of the airways might be deleterious and were this the case, a very specific targeting process would be required and accurate dosing regimes might be needed. Recent experiments in which transgenic mice have been produced that express CFTR in most if not all cells of the tracheo-bronchial tree indicate that high level of expression in a wide variety of mouse cells is not harmful (66). Thus, even if CFTR is not normally present in all cells of the airway nor present in high amounts, were this the case following drug treatment, it might not be problematic. The results with the transgenic animals would imply that CFTR is not toxic and would also imply that targeting of protein replacement therapy for CF need not be especially sophisticated.

Immunogenicity - Since proteins turn over with a given half-life, protein therapy would need to be repeated at regular intervals. The potential immunogenicity of the reagents used will, therefore, require special attention. CFTR itself and the lipid components of liposomes should not present problems, unless some patients express absolutely no CFTR and recognize introduced protein as foreign. However, any agents added to the formulation to aid fusion, cell binding or targeting, however, could prove problematic.

Clinical Challenges - The clinical challenges for human protein therapy for CF revolve around introducing a protein-containing formulation into the airway followed by incorporation in the correct orientation of the introduced protein into the membrane of epithelial cells. The cells may be inaccessible due to mucus build up. The environment may be hostile because of the presence of oxidants and proteases derived from neutrophils attracted to the location by infection. The incorporation of the protein involving the fusion process between introduced formulation and cellular membrane will probably be very inefficient. Furthermore, the reagents used to promote fusion and possibly targeting may be immunogenic upon multiple application. In more severe cases of the disease, airway cell damage may be so advanced as to be irrepairable.

There is as yet no CF animal model to assess the likely efficacy of potential treatments (67) nor are there simple clinical end points to assess efficacy. It is clear, therefore, that even when the technical aspects of developing a CF protein therapy are addressed considerable effort will be needed to decide the most appropriate patient population in whom to attempt early clinical testing and to determine the clinical parameters most likely to be informative in terms of clinical, rather than biochemical, efficacy.

Protein Replacement: Present Status - To date CFTR has been expressed in a variety of recombinant cells, some of which can be readily scaled up (41). Purification of CFTR from such sources has been achieved to a limited extent. CFTR from washed membrane vesicles isolated from recombinant cells has been transferred in active form to lipid bilayer preparations (68). The next significant milestone is to establish the feasibility of transfer of partially purified CFTR into recipient cells from a CF patient followed by measurements of membrane function. Such an experiment will establish the feasibility of protein replacement therapy. Once established, scale up and optimization of the purification and fusion processes will occur, formulations to deliver the protein to airway cells of animals will be developed and safety aspects of the proposed therapy will be addressed prior to initial clinical testing.

GENE THERAPY

Although the potential of gene therapy to treat genetic diseases has been appreciated for many years, it is only in the 1990's that such approaches have become practical with the treatment of two patients with adenosine deaminase deficiency. The protocol consists of removing lymphocytes from the patients, stimulating them to grow in tissue culture, infecting them with an appropriately engineered retrovirus followed by reintroduction of the cells into the patient (69). Initial results of treatment are very encouraging. With the approval of a number of other human gene therapy protocols for limited clinical use, and with the rapid demonstration of the feasibility of complementing the CF defect by gene transfer gene therapy for CF appears a very viable option (28,29).

The concept of gene replacement therapy is also very simple; a preparation of CFTR coding sequences in some suitable vector in a viral or other carrier delivered directly to the airway cells of CF patients. Since, at present, we do not know the identity of the stem cell for air epithelial cells, nor do most potential treatments incorporate DNA into the host cell genome, the first generation of CF gene therapy is likely to be transient and require repeated delivery. Eventually, however, such an approach may offer a cure in that were DNA incorporated into airway stem cells, all subsequent generations of such cells would make authentic CFTR from the integrated sequences and should correct the physiological defect almost irrespective of the biochemical basis of the action of CFTR. Although simple in concept, scientific and clinical problems also face approaches to gene therapy, not least of these being that CF requires an in vivo approach while all gene therapy treatments to date have involved ex vivo treatment of patient cells followed by reintroduction.

Efficiency of Delivery - Methods to introduce DNA into cells are generally inefficient. Since viruses have evolved very efficient means to introduce their nucleic acid into cells, many approaches to gene therapy make use of engineered defective viruses. In the case of CF, viruses show much promise in terms of efficiency but at the same time the need to use them in vivo raises new problems. The use of simple DNA plasmid constructs containing minimal additional DNA, on the other hand, also has attractive features. Use of such plasmids, however, is often very inefficient and normally results in transient protein expression.

DNA Integration - The integration of introduced DNA into the host chromosome has advantages in that such DNA will be passed to daughter cells. In some circumstances, integrated DNA may also lead to high or more sustained expression. However, integration often, perhaps always, requires cellular DNA replication in order to occur. This is certainly the case with the present generation of retroviruses. This limits the use of such viruses to circumstances where cell division occurs in a high proportion of cells. For cells cultured in vitro, this is seldom a problem, however, the cells of the airway are reported to divide only infrequently (70). The use of retroviruses in CF will probably require damaging the airways (by agents such as SO_2 or O_3) to induce cell division. This may prove impracticable in CF patients.

Even if DNA integration could be achieved using viruses, the human genome contains many elements involved in the regulation of cellular growth, of which it is likely only a small fraction are presently identified. By integrating adjacent to an element such as a proto-oncogene or an anti-oncogene, activation or inactivation of that element could occur, leading to uncontrolled growth of the altered cell. It is considered likely that several such activation/inactivation steps are usually required in any one cell to induce uncontrolled proliferation (71) and this may reduce somewhat this potential risk. On the other hand, insertional mutagenesis leading to tumor formation is certainly known in animals with some nondefective retroviruses (72,73) and the large numbers of potential integrations occuring during the lifetime of a patient treated repeatedly in vivo with retroviruses must raise concerns on the safety of such a procedure.

Safety - In addition to the potential problems associated with viral DNA integration, a number of additional safety issues arise. Many patients may have preexisting antibodies to some of the viruses that are candidates for vectors, for example, adenoviruses. In addition, repeated use of such vectors might induce an immune response. Fortunately, the fact that defective viral vectors will be used may mean that this is less of a problem in the sense that the vectors will not lead to productive viral life cycles generating infected cells, cell lysis or large numbers of progeny viruses. Other issues associated with the use of viruses are the possibility of recombination with related viruses naturally infecting the treated patient, complementation of the viral defects by simultaneous expression of wild type virus proteins and containment of aerosols of the engineered viruses.

Clinical Challenges - Gene therapy approaches to CF will face many of the same clinical challenges as protein therapy. These include the inaccessibility of airway epithelium caused by mucus build-up and the hostile nature of the environment in CF airways which may inactivate viruses/vectors. Elements of the vector carriers may be immunogenic and introduction of the DNA may be inefficient. These problems, as with protein therapy, are exacerbated by the absence of good animal model for the disease nor a simple clinical end point to measure the efficacy of treatment. Gene therapy faces additional safety considerations, mentioned above, if engineered viruses are used. CF may be the first clinical use of gene therapy in vivo rather than ex vivo and this too will mean that such procedures will attract additional regulatory scrutiny.

CF GENE THERAPY - POSSIBLE OPTIONS AND PRESENT STATUS

Retroviruses - Although defective retroviruses are the best characterized system and the only one approved for human gene therapy (74), the major issue in relation to CF is the requirement for dividing cells to achieve DNA integration and gene expression. Were conditions found to induce airway cell division, the in vivo use of retroviruses, especially if repeated over many years, would necessitate assessment of the safety aspects of insertional mutagenesis in this context.

Adenoviruses - Defective adenoviruses at present appear a potentially very promising approach to CF gene therapy (75). CFTR expression following infection of cotton rats with an E1/E3 deleted adenovirus has been detected for periods up to six weeks (76). Since adenoviruses have been administered orally to large numbers of military recruits for vaccination purposes, some human safety data, albeit distant, already exists (77). Furthermore, extensive studies to attempt to establish adenovirus as a causative agent in human cancer were uniformly negative (78). However, the defective adenovirus system is as yet poorly studied. We know little of the potential immunogenicity of repeated exposure to defective adenovirus. Additional potential problems include complementation, recombination and containment. All of these issues are likely to be intensively studied in the very near future.

Adeno-Associated Virus - AAV is a naturally occuring defective virus that requires other viruses such as adenoviruses as helper (79). It is also one of the few viruses that may integrate its DNA into non-dividing cells, although this is not yet certain. Vectors containing as little as 300 base pairs of AAV can be packaged and can integrate, but space for exogenous DNA is limited to about 4.5 kb. CFTR DNA may be towards the upper limit of packaging. Furthermore, the packing process itself is presently inefficient and safety issues such as immunogenicity, complementation and containment will also apply to AAV. Nevertheless, this system is sufficiently promising to warrant further study.

Plasmid DNA - Naked plasmid can be introduced into muscle cells by injection into the tissue. Expression can extend over many months but the number of positive cells is low (80). Cationic lipids aid introduction of DNA into some cells in culture (81). Injection of cation lipid plasmid DNA complex into the circulation of mice leads to expression in lung (82). Installation of cationic lipid plasmid DNA into lung also leads to expression in epithelial cells but the efficiency of expression is relatively low and transient (83). One advantage of the use of plasmid DNA is that it can be introduced into non replicating cells. However, the use of plasmid DNA in the CF airway environment containing high concentrations of endogenous DNA seems problematic.

Receptor Mediated Entry - In an effort to improve the efficiency of plasmid DNA uptake, attempts have been made to utilize receptor-mediated endocytosis as an entry mechanism and to protect DNA in complexes with polylysine (84). One problem with this approach is that the incoming plasmid DNA enters the pathway leading from endosome to lysosome, where much incoming material is degraded. This, to some extent, negates the advantage of this approach. One solution to this problem is the use of transferin DNA-polylysine complexes linked to adenovirus capsids (85). The latter enter efficiently but have the added advantage of naturally disrupting the endosome thereby shuttling it to the lysosome. This approach has promise but at present is relatively transient and suffers from the same potential problems of immunogenicity as other adenovirus based methods.

CONCLUSIONS

Progress on understanding the molecular basis of CF has been extremely encouraging over the time period since the gene was identified. Approaches to therapy based on knowledge of the gene have the advantage of being completely rational and there is a high expectation that they will eventually be efficacious. Present efforts are aimed at establishing that they are technically feasible, but once that is done, progress in optimizing and scaling up the procedures to enable safety studies and early clinical testing can be expected to be very rapid. At this stage, there is sufficient uncertainty, both technical and regulatory and in terms of ultimate clinical use, to warrant work on both protein based and gene based approaches and on many different vectors. However, there is every reason to hope that successful therapies based on this work will be available by the end of the present decade.

REFERENCES

1. J. Rommens, M. Iannuzi, B. Kerem, M. Drumm, G. Melmer, M. Dean, R. Rozmahel, J. Cole, D. Kennedy, N. Hidaka, M. Zsiga, M. Buchwald, J.R. Riordan, L. Tsui, F. Collins, Science, 245, 1059 (1989).
2. J. Riordan, J. Rommens, B. Kerem, N. Alon, R. Rozmahel, Z. Grzelczack, J. Zielenski, S. Lok, N. Plavsic, J. Chou, M. Drumm, M. Iannuz, F. Collins, L. Tsui, Science, 245, 1066 (1989).
3. B. Karem, R. Rommens, J. Buchanan, D. Markiewicz, T. Cox, A. Chakravarti, M. Buchwald, L. Tsui, Science, 245, 1073 (1989).
4. T. Boat, M. Welsh, A. Beaudet, Cystic fibrosis, C. Scriver, A. Beaudet, W. Sly, D. Valle, eds, New York, McGraw Hill, p. 2649 (1989).
5. M. Welsh, FASEB J., 4, 2718 (1990).
6. P. Quinton, FASEB J., 4, 2709 (1990).
7. W.T. Gerson, P. Swan, A. Walker, Nutr. Rev., 45, 353 (1987).
8. M. Knowles, J. Gatzy, R. Boucher, J. Clin. Invest., 71, 1410 (1983).

9. P. Quinton, Nature, 301, 421 (1983).
10. S.H. Michel, D.H. Mueller, Topics In Clin. Nutr., 4, 46 (1989).
11. M.S. Zach, B. Oberwaldner, Infection, 15, 381 (1987).
12. J. Reisman, B. Rivington-Law, M. Corey, J. Marotte, E. Wannamaker, D. Harcourt, H. Levison, J. Pediatrics, 113, 632 (1988).
13. A. Smith, B. Ramsey, D. Hedges. B. Hack, J. Williams-Warren, A. Weber, E. Gore, G. Redding, Pediatr. Pulmon., 7, 265 (1989).
14. S. Marshall, N. Lewiston, V. Starnes J. Theodore, Chest 98, 1488 (1990).
15. S. Shak, D. Capon, R. Hellmis, S. Marsters, C. Baker, Proc. Natl. Acad. Sci. USA, 87, 9188 (1990).
16. N. McElvaney, R. Hubbard, P. Birrer, M. Chernick, D. Caplan, M. Frank, R. Crystal, The Lancet, 337, 392 (1991).
17. E. App, M. King, R. Helfesrieder, D. Kohler, H. Matthys. Am. Rev. Respir. Dis., 141, 605 (1990).
18. M. Knowles, N. Church, W. Waltner, J. Yankaskas, P. Gilligan, M. King, L. Edwards, R. Helms R. Boucher, N. Engl. J. Med., 322, 1189 (1990).
19. M. Knowles, L. Clarke, R. Boucher, N. Engl. J. Med., 325, 533 (1991).
20. R. Gregory, S. Cheng, D. Rich, J. Marshall, S. Paul, K. Hehir, L. Ostedgaard, K. Klinger, M. Welsh, A. Smith, Nature, 347, 382 (1990).
21. N. Kartner, J. Hanrahan, T. Jensen, A. Naismith, S. Sun, C. Ackerley, E. Reyes, L. Tsui, J. Rommens, C. Bear, J. Riordan, Cell, 54, 681 (1991).
22. I. Crawford, P. Maloney, P. Zeitlin, W. Guggino, S. Hyde, H. Turley, K. Gatter, A. Harris, C. Higgins, Proc. Natl. Acad. Sci. USA, 88, 9262 (1991).
23. C. Marino, L. Matovcik, F. Gorelick, J. Cohn, J. Clin. Invest., 88, 712 (1991).
24. A. Hoogeveen, J. Keulemans, R. Willemsen, B. Scholtf, J. Human, M.J. Edixhoven, H. DeJonge, H. Galiaard, Exp. Cell Res. 193, 435 (1991).
25. W. Dalemans, P. Barbry, G. Champigny, S. Jallat, K. Dott, D. Dreyer, R. Crystal, A. Pavirani, J. Lecocq, M. Lazdunski, Nature, 354, 526 (1991).
26. B. Sarkadi, D. Bauzon, W. Huckle, H. Earp, A. Berry, H. Suchindran, E. Price, J. Olsen, R. Boucher, G. Scarborough, J. Biol. Chem., 267, 2087 (1992).
27. P. Zeitlin, I. Crawford, C. Lu, S. Woel, M. Cohen, M. Donowitz, M. Montrose, A. Hamosh, G. Cutting, D. Gruenert, R. Huganir, P. Maloney W. Guggino, Proc. Natl. Acad. Sci. USA, 89, 344 (1992).
28. D. Rich, M. Anderson, R. Gregory, S. Cheng, S. Paul, D. Jefferson, J. McCann, K. Klinger, A. Smith, M. Welsh, Nature, 347, 358 (1990).
29. M. Drumm, H. Pope, W. Cliff, J. Rommens, S. Marvin, L. Tsui, F. Collins, R. Frizazel, J. Wilson, Cell 62, 1227 (1990).
30. M. Anderson, D. Rich, R. Gregory, A. Smith, M. Welsh, Science, 251, 679 (1991).
31. J. Rommens, S. Dho, C. Bear, N. Kartner, D. Kennedy, J. Riordan, L. Tsui, J. Foskett, Proc. Natl. Acad. Sci. USA, 88, 7500 (1991).
32. C. Bear, F. Duguay, A. Naismith, N. Kartner, J. Hanrahan, J. Riordan, J. Biol. Chem., 266, 19142 (1991).
33. M. Anderson, D. Rich, R. Gregory, A. Smith, M. Welsh, Science, 251, 679 (1991).
34. H. Berger, M. Anderson, R. Gregory, S. Thompson, P. Howard, R. Maurer, R. Mulligan, A. Smith, M. Welsh, J. Clin. Invest., 88, 1422 (1991).
35. J. Tabcharani, X. Chang, J. Riordan, J. Hanrahan, Nature 352, 628 (1991).
36. D. Rich, R. Gregory, M. Anderson, P. Manavalan, A. Smith, M. Welsh, Science, 253, 205 (1991).
37. M. Anderson, H. Berger, D. Rich, R. Gregory, A. Smith, M. Welsh, Cell, 67, 775-784 (1991).
38. S. Cheng, D. Rich, J. Marshall, R. Gregory, M. Welsh, A. Smith, Cell, 66, 1027 (1991).
39. S. Cheng, R. Gregory, J. Marshall, S. Paul, D. Souza, G. White, C. O'Riordan, A. Smith, Cell, 63, 827 (1990).
40. R. Gregory, D. Rich, S. Cheng, D. Souza, S. Paul, P. Manavalan, M. Anderson, M. Welsh, A. Smith, Mol. Cell Biol., 11, 3886 (1991).
41. S.Cheng, R. Gregory, J. Amara, D. Rich, M. Anderson, M. Welsh, A. Smith, In: Current Topics in Cystic Fibrosis Vol. 1., Ed. J. Dodge, D.J. Brock & J.M. Widdicome, John Wiley, Chichester (1992) (in press).
42. J. Amara, S. Cheng, A. Smith, In: Trends in Cell Biology (1992) (in press).
43. G. Denning, L. Ostedgaard, S. Cheng, A. Smith, M. Welsh, J. Clin. Invest., 89, 339 (1992).
44. M. Drumm, D. Wilkinson, L. Smit, R. Worrell, T. Strong, R. Frizzell, D. Dawson, F. Collins, Science, 254, 1797 (1991).
45. M. Welsh, G. Denning, L. Ostedgaard, Abnormal Localization of the Cystic Fibrosis Transmembrane Conductance Regulator in Primary Cultures of CF Airway Epithelial Cells, (Submitted for publication).
46. R. Worrell, R. Frizzell, Am. J. Physiol., 260, C877 (1991).
47. R. Frizzell, D. Halm, In: Current Topics in Membranes and Transport (Academic Press), 247 (1990).
48. J. Wagner, A. Cozens, H. Schulman, D. Gruenert, L. Stryer P. Gardner, Nature, 349, 793 (1991).
49. S. Harshman, P. Boquet, E. Dufont, J. Alouf, C. Montecucco, E. Papini, J. Biol. Chem. 264, 14978 (1989).
50. N. Willumsen, R. Boucher, Am. J. Physiol., 256, C226 (1989).
51. N. Takahashi, T. Hayano, M. Suzuki, Nature 337, 473 (1989).
52. A. Stamnes, B. Shieh, L. Chuman, G. Harris, C. Zuker, Cell, 65, 219 (1991).
53. J. Lippincott-Schwartz, L. Yuan, J. Bonifacino, R. Klausner, Cell, 56, 801 (1989).

54. N. Barton, F. Furbish, G. Murray, M. Garfield, R. Brady, Proc. Natl. Acad. Sci., 87; 1913 (1990).
55. N. Barton, R. Brady, J. Dambrosia, A. DiBiscelglie, S. Doppelt, S. Hill, H. Mankin, G. Murray, R. Parker, C. Argoff, R. Grewai, K-T Yu, N. Engl. J. Med., 324, 1464 (1991).
56. R. Kaufman, Methods in Enzymology, 185, 487 (1990).
57. H-Y Wang, L. Lipfert, C. Malbon, S. Bahouth, J. Biol. Chem., 264, 14424 (1989).
58. P. DiTullio, S. Cheng, J. Marshall, R. Gregory, K. Ebert, H. Meade, A. Smith, Bio/Technology 10, 74 (1992).
59. M. Schramm, Proc. Natl. Acad. Sci. USA, 76, 1174 (1979).
60. D. Doyle, E. Hou, R. Warren, J. Biol. Chem., 254, 6853 (1979).
61. A. Prujansky-Jakobovits, D. Volsky, A. Loyter, N. Sharon, Proc. Natl. Acad. Sci. USA, 77, 7247 (1980).
62. A. Newton, W. Heustis, Biochemistry, 427, 4655 (1988).
63. L. Johnson, J. Olsen, B. Sarkardi, K. Moore, R. Swanstrom, R. Boucher, (personal communication) (1991).
64. B. Trapnell, C-S Chu, P. Paakko, T. Banks, K. Yoshimura, V. Ferrans, M. Chernick, R. Crystal, Proc. Natl. Acad. Sci. USA, 88, 6565 (1991).
65. A.E.O. Trezise, M. Buchwald, Nature, 353, 434 (1991).
66. J. Whitsett, C. Dey, B. Stripp, K. Wikenheiser, S. Wert, R. Gregory, A. Smith, J. Cohen, J. Wilson, J. Engelhardt, (submitted for publication).
67. B.H. Koller, H-S Kim, A.M. Latour, K. Brigman, R.C. Boucher, Jr., P. Scambler, B. Wainwright, O. Smithies, Proc. Natl. Acad. Sci. USA, 88, 10730 (1991).
68. B. Tilly, M. Winter, L. Ostedgaard, C. O'Riordan, A. Smith, M. Welsh, J. Bio. Chem., (in press).
69. P.Kantoff, A. Gillio. J. McLachlin, C. Bordignon, M. Eglitis, N.A. Kernan, R.C. Moen, D.B. Kohn, S-F Yu, E. Karson, S. Karlsson, J.A. Zwiebel, E. Gilboa, R.M. Blaese, A. Nienhuis, R.J.O'Reilly, W.F. Anderson, J. Exp. Med., 166, 219 (1987).
70. O. Kawanami, V. Ferrans, R. Crystal, An. Rev. Respir. Dis., 120, 595 (1979).
71. R. Weinberg, Cancer Research, 49, 3713 (1989).
72. W. Haywood, G. Neel, S.M. Astrin, Nature, 290, 475 (1981).
73. G. Payne, J.M. Bishop, H.E. Varmus, Nature, 295, 209 (1982).
74. A. Miller, Blood, 76, 271 (1990).
75. K. Berkner, BioTechniques, 6, 616 (1988).
76. M. Rosenfeld, K. Yoshimura, B. Trapnell, K. Yoneyama, E. Rosenthal, W. Dalemans, M. Fukayama, J. Bargon, L. Stier, L. Stratford-Perricaudet, M. Perricaudet, W. Guggino, A. Pavirani, J.P.Lecocq, R. Crystal, Cell, 68, 143 (1992).
77. T. Smith, E. Buescher, F. Top, Jr., W. Altemeier, J. McCown, J. Inf. Dis., 122, 239 (1970).
78. M. Green, W.S.M. Wold, J.K. Mackey, P. Rigden, Proc. Natl. Acad. Sci. USA, 76, 6606 (1979).
79. N. Muzyczka, In: Current topics in microbiology and immunology, 158, 97 (1992).
80. J. Wolff, R. Malone, P. Williams, W. Chong, G. Acsadi, A. Jani, P.L. Felgner, Science, 247, 1465 (1990).
81. P.L. Felgner, G.M. Ringold, Nature, 337, 387 (1989).
82. K. Brigham, B. Meyrick, B. Christman, M. Magnuson, G. King, L. Berry, Jr., Am. J. Med. Sci., 298, 278 (1989).
83. T. Hazinski, P. Ladd, C. DeMatteo, Am. J. Respir., Cell Mol. Biol., 4, 206 (1991).
84. G. Wu, C.H. Wu, J. Biol. Chem., 263, 14621 (1988).
85. D.T. Curiel, S. Agarwal, E. Wagner, M. Cotten, Proc. Natl. Acad. Sci. USA, 88, 8850 (1991).

Chapter 26. The Guanylyl Cyclase Family of Receptor/Enzymes

Hans-Jürgen Fülle and David L. Garbers
Howard Hughes Medical Institute and Department of Pharmacology
University of Texas Southwestern Medical Center at Dallas
Dallas, TX 75235-9050

Introduction - Guanylyl cyclases [GTP pyrophosphate-lyase (cyclizing), EC 4.6.1.2] constitute a growing family of proteins that catalyze the enzymatic formation of cyclic guanosine 3',5'-monophosphate (cGMP) and pyrophosphate from guanosine trisphosphate (GTP). Since the discovery of cGMP in 1963 (1), this intracellular signaling molecule has been found in nearly all organs, tissues and cells that have been investigated (2). cGMP participates as a second messenger in vascular smooth muscle relaxation, platelet aggregation and retinal phototransduction. It is probably involved in many other cellular functions acting via specific ion channels, phosphodiesterases and protein kinases (2,3). Diverse forms of guanylyl cyclases have been identified as receptors for such different agents as sea urchin egg peptides, mammalian peptides with natriuretic properties, bacterial enterotoxins, endogenously formed nitric oxide and nitrovasodilatory drugs (4-7).

Guanylyl cyclase was first described in mammalian tissues as an enzyme different from adenylyl cyclase (8-10). Soon it became clear that different proteins (based on solubility properties, immunological cross-reactivities, kinetic parameters, responses to effector molecules and subunit composition) accounted for the guanylyl cyclase activities found in particulate and soluble fractions of tissue homogenates and cell lysates (11-13). Most tissues contain detergent-soluble particulate guanylyl cyclases aside from cytoplasmic forms. Detergent-insoluble particulate guanylyl cyclases were found in retinal rod outer segments, intestinal cells and in protozoans. Several guanylyl cyclases were purified to apparent homogeneity (14-27). cDNAs were obtained encoding some of the purified, and other, unknown forms of guanylyl cyclases (28-43). The identification of different guanylyl cyclase isoenzymes based on their primary structure established the expanding guanylyl cyclase family of receptor/enzymes.

SEA URCHIN MEMBRANE GUANYLYL CYCLASE

Guanylyl cyclase was first purified from sea urchin spermatozoa (14,15). These cells are a very rich source of guanylyl cyclase and enzyme activity is located almost entirely in the particulate fraction. Characteristics of this membrane-bound enzyme are very similar to those of mammalian tissues with respect to size, kinetic properties, immunogenic determinants and regulation by extracellular peptides (4,44). Based on partial amino acid sequence from the purified protein, the first mRNAs encoding plasma membrane forms of guanylyl cyclase were cloned and sequenced from a sea urchin testis cDNA library (28,29). Sea urchin sperm guanylyl cyclase displays a series of domains that are also present in mammalian membrane-associated peptide receptor/guanylyl cyclases (Figure 1).

Primary Structure - cGMP is involved in the regulation of metabolism and motility of sea urchin spermatozoa although its molecular site(s) of action are not known. Several small peptides from egg-conditioned media, e.g. speract or resact, cause a species-specific, marked increase of cGMP levels in spermatozoa by activation of a particulate guanylyl cyclase (44,45). Results from cross-linking studies suggested that guanylyl cyclase may directly function as a receptor for some or all of these egg peptides (46). The cDNA for guanylyl cyclases from two different sea urchin species, Arbacia punctulata and Strongylocentrotus purpuratus, encodes intrinsic membrane proteins with an extracellular, amino-terminal domain, separated by a single putative membrane-spanning domain, from an intracellular, carboxy-terminal domain (28,29). The predicted amino acid sequences of these sea urchin guanylyl cyclases show a 77% overall identity. The extracellular putative egg peptide binding domains are more diverse than the intracellular regions (29). Just within the transmembrane domain, an apparent signature domain of all membrane-bound guanylyl cyclases exists, showing homology to the catalytic domains found in protein kinases (47).

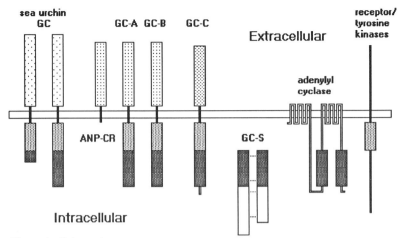

Figure 1. Schematic representation of the general topology of the guanylyl cyclase (GC) family of receptor/enzymes and related proteins. Areas with similar shading indicate homologous regions of sea urchin guanylyl cyclases (from *Arbacia punctulata* and *Strongylocentrotus purpuratus*), the ANPC receptor (ANPCR), GC-A, GC-B, GC-C, the heterodimeric GC-S, bovine brain adenylyl cyclase and the family of receptor/tyrosine kinases. For further explanations and references see text.

THE GUANYLYL CYCLASE-COUPLED PEPTIDE RECEPTORS (GC-A AND GC-B)

Membrane-bound guanylyl cyclase A (GC-A) and guanylyl cyclase B (GC-B) represent the two identified mammalian guanylyl cyclases that are directly linked to cell surface receptors for natriuretic peptide hormones. Transmembrane signaling through a single protein is achieved by the direct coupling of the extracellular binding of ligands and the intracellular production of cGMP. Associated regulatory proteins may exist, although none have yet been discovered. The topology of the membrane forms of guanylyl cyclase is similar to other signal-transducing receptor/enzymes, *e.g.* the growth factor receptor-coupled protein tyrosine kinases (47,48) or the receptor-linked protein tyrosine phosphatases (49). Based on the predicted amino acid sequences, four distinct domains can be assigned to the membrane-bound guanylyl cyclases: an amino-terminal extracellular ligand binding domain is coupled via a single transmembrane domain to an intracellular regulatory protein kinase-like domain and a carboxy-terminal guanylyl cyclase catalytic domain (Figure 1).

Natriuretic Peptides and ANP-C Receptor - In 1981, atrial natriuretic peptide (ANP) was isolated from secretory granula of specialized myoendocrine cells in rat cardiac atria as the first member of a new family of natriuretic, diuretic and vasodilatory peptide hormones (50). Meanwhile, several other related peptides were discovered, *e.g.* brain natriuretic peptide (BNP) and C-type natriuretic peptide (CNP). Synthesis of these peptides is not restricted to the heart but occurs also in brain and various other tissues. Natriuretic peptides directly inhibit renin and aldosterone synthesis, inhibit the release of vasopressin, and increase Leydig cell testosterone synthesis (51). Initially, it was shown that ANP induces a rapid increase in intracellular cGMP levels in various mammalian intact cells and tissues (52-54) and that it activates a guanylyl cyclase in the particulate fraction of broken cell preparations (54,55). Subsequently, two types of cell surface receptors specific for ANP were identified by binding and chemical cross-linking experiments, and by receptor purification. Whereas the higher apparent molecular weight (M_r) form, a monomeric 130 kDa protein, was shown to directly couple to guanylyl cyclase activation and is now called guanylyl cyclase A (GC-A), another, more abundant ANP receptor, a homodimer of 66 kDa subunits, is apparently not linked to guanylyl cyclase (18,56-58). The latter has been designated as a silent ANP receptor or ANP-clearance receptor (ANP-CR), since it was suggested that its physiological function was to bind and remove ANP or various truncated ANP analogues by receptor internalization from the circulation (58,59). The ANP-CR protein has been purified (59,60) and its mRNA cloned and sequenced (61). Its extracellular ligand binding domain is about 33% identical to GC-A. It shows a single transmembrane domain as GC-A, but only 37 amino acids are located within the cytoplasm (61). Therefore, ANP-CR is regarded as a truncated member of the guanylyl cyclase family of receptor/enzymes (6). The small cytoplasmatic region does not exclude other signaling functions of this low M_r ANP receptor. It may mediate inhibition of adenylyl cyclase by ANP (62,63), possibly by interaction with an inhibitory guanine nucleotide-binding regulatory protein (64). Stimulation of phospholipid hydrolysis via ANP-CR has also been suggested (65).

Guanylyl Cyclase A - During the mid and late 1980s, several reports described the co-purification of ANP binding and membrane-bound guanylyl cyclase activities in a M_r 120-180 kDa protein suggesting that both functions might reside within this single protein (16-19). However, the low amounts of purified protein and the fact that the sensitivity of the enzyme to ANP stimulation was lost in all protocols during the purification, did not allow confirmation that the high M_r ANP receptor and membrane-bound guanylyl cyclase were same. This was subsequently achieved by the isolation, sequencing and expression of a cDNA clone termed guanylyl cyclase A (GC-A) encoding the enzyme from rat brain (30). The transfection of this cDNA into cultured mammalian cells resulted in expression of increased specific binding of ANP and increased ANP-stimulatable guanylyl cyclase activity. The ligand binding properties were characteristic for the high M_r ANP receptor showing a higher affinity for ANP than for truncated ANP analogues (30). By contrast, the low M_r ANP-CR has less stringent constraints on ligand structure and binds truncated ANP analogues with high affinity, too (57,66). Cross-linking experiments with expressed GC-A and radiolabeled ANP identified a single M_r 130 kDa protein band in SDS polyacrylamide gel electrophoresis coinciding with the reported mobility of the high M_r ANP receptor (16,18,30,57). The cloning and expression of homologous mRNAs of human origin (31) and, later, cloning of GC-A from mouse Leydig cells (32) provided definitive evidence that GC-A is a cell surface receptor for ANP (Table 1).

Guanylyl Cyclase B - The conserved 3' region of the GC-A clone was used as a probe to screen other cDNA libraries to identify related members of the plasma membrane subfamily of guanylyl cyclases. Corresponding clones homologous to GC-A were isolated from rat brain and human and porcine tissues (33,34). The predicted amino acid sequences of the catalytic domain showed the highest identity (91%), however the protein kinase-like domain (72%) and the extracellular domain (43%) were also homologous (33). These clones encoded a second type of natriuretic receptor/guanylyl cyclase, guanylyl cyclase B (GC-B), with different binding specificities for natriuretic peptides (Figure 1). Although GC-A- and GC-B-transfected cells specifically bound ANP and BNP, the relative potencies of these natriuretic peptides were markedly different. Half-maximal cGMP-elevating concentrations were 3 nM rat ANP and 25 nM rat BNP in rat GC-A-transfected cells but 25 μM rat ANP and 10 μM rat BNP in rat GC-B-transfected cells (33). The non-physiologically high concentrations of ANP and BNP that were required to activate GC-B suggested that a more potent natural ligand for this receptor/guanylyl cyclase might exist (33). Eventually, a new member of the natriuretic peptide hormone family, C-type natriuretic peptide (CNP), was discovered in porcine brain (67,68). CNP selectively binds and potently activates cloned GC-B expressed in mammalian cells at physiologically relevant concentrations but is very ineffective in both binding to and stimulating GC-A (67,68). Therefore, CNP may be a natural ligand for GC-B (68).

The existence of multiple natriuretic peptides and multiple natriuretic peptide receptor/guanylyl cyclases suggests a rather complex physiology for these hormone systems. ANP and CNP as well as GC-A and GC-B are highly conserved among different species. The primary structure of BNP, in contrast, varies considerably across species as does its potency to activate GC-A and GC-B. The reason for this is unknown, but it is conceivable that BNP may be the natural ligand of a yet unidentified member of the receptor/guanylyl cyclase family. The tissue distribution of GC-A, GC-B and ANP-CR mRNA in monkey was assayed by in situ hybridization (69). GC-A mRNA was predominantly found in kidney, adrenal cortex, pituitary, cerebellum and in the endothelium of the heart. GC-B expression was detected only in neural and neural crest-derived tissues, whereas ANP-C receptor mRNA was most prominent in myocardium of the right atrium and also evident in kidney, adrenal cortex and medulla and in brain cortex and cerebellum (69). Since both GC-B and CNP were predominantly found in the central nervous system, this ligand-receptor pair may be responsible for central hormonal regulation of body fluid homeostasis and complement the ANP/GC-A system that is localized in peripheral and central tissues.

Regulation of Membrane-Bound Guanylyl Cyclases - The two intracellular domains, the protein kinase-like domain and the putative catalytic domain at the carboxy terminus are distinctive elements in the topology of the sea urchin guanylyl cyclases and of mammalian GC-A and GC-B (Figure 1). Originally, it was not clear which of these regions would actually bind GTP and catalyze cGMP formation, and how they would participate in signal transduction (28,30). Sequence comparisons between soluble guanylyl cyclase and bovine brain adenylyl cyclase indicated that the distal carboxy-terminal regions of the membrane forms may represent the catalytically active domain

(37,38,70). To test this hypothesis the intracellular domains of GC-A were analyzed by deletion mutagenesis. Complete loss of guanylyl cyclase activity was seen after deletion of the distal carboxy-terminal domain (71). Expression in bacteria of this region alone was sufficient for guanylyl cyclase activity and provided direct evidence that the distal carboxy-terminal domain of GC-A is solely responsible for cGMP formation (72). With more guanylyl and adenylyl cyclases cloned and sequenced it has been possible to propose a cyclase consensus sequence (4).

The protein kinase-homologous domain of membrane-bound guanylyl cyclases stretches over about 250 amino acids (28). It does not encode guanylyl cyclase activity and no data yet show that it contains phosphotransferase activity. The deletion of this domain instead has resulted in expression of constitutively high guanylyl cyclase activity (71). The mutant form of GC-A was no longer sensitive to the stimulating effects of ANP, although ANP binding was not inhibited (71). The protein kinase-like domain, thus, is required for transmembrane signal transduction. It was suggested that this domain could act as a negative regulatory element, suppressing the maximal catalytic activity in the basal state. According to this hypothesis the binding of ANP and a conformational change of the receptor would trigger a de-inhibition of the catalytic domain which would be mediated by the protein kinase-like domain (71).

ATP plays an important regulatory role in the signal transduction through natriuretic receptor/guanylyl cyclases. ATP and some nonhydrolyzable analogues potentiate the effect of ANP on membrane-bound guanylyl cyclase activity but the mechanism for this is still unclear (73-76). Binding of ANP was reduced by ATP according to one report (77) but not altered according to another (73). In contrast, ANP binding to ANP-CR from a fibroblast cell line seemed to be slightly enhanced by ATP (78). When expressed in a baculovirus system lacking endogenous ANP receptors, the activation of rat GC-A by ANP strictly required the presence of adenine nucleotides; ATP did not affect basal enzyme activity (79). Nonhydrolyzable nucleotide analogues were also effective, and the effect was reversible, suggesting that ATP acted via an allosteric site and not as a substrate in a phosphorylation reaction (79). Such an allosteric site had been previously proposed since one uniform characteristic of all membrane-bound guanylyl cyclases has been their apparent positive cooperativity (4-7). The protein kinase-like domain is a potential candidate for such a regulatory ATP binding site, although cooperative interactions due to dimerization or other higher-ordered structures is a more likely explanation. The potentiating effect of ATP on guanylyl cyclase activity was lost after deletion of this domain from GC-A (71) whereas the modulating effect on ANP binding was still observed in an affinity-purified receptor/guanylyl cyclase preparation (80). The receptor/guanylyl cyclases appear to contain a cell surface receptor, a nucleotide-binding regulatory protein and an enzyme in one molecule (4,6).

Several functional aspects of membrane-bound receptor/cyclases are currently under investigation. For example, sea urchin guanylyl cyclase is highly phosphorylated in its basal state (4,45). After ligand-induced enzyme stimulation, rapid dephosphorylation occurs and leads to desensitization (44). Preliminary findings indicate that mammalian GC-A exists as a phosphoprotein and that dephosphorylation of GC-A also results in a decrease of ANP-stimulatable enzyme activity (81). Others addressed the existence of higher-order structures of receptor/cyclase. Oligomerization is one model to explain the positive enzyme cooperativity observed in the basal state (6). Evidence has been obtained for homodimer formation and interaction of the catalytic sites by overproducing the carboxy-terminal catalytic domain of GC-A in a bacterial expression system (82). The characterization of ANP receptor/guanylyl cyclase by the use of ^{125}I-ANP cross-linking of bovine adrenal cortex membranes indicated a disulfide-linked tetrameric structure (SDS/PAGE profile under non-reducing conditions). The use of ^{125}I-ANP to trace the position of the receptor precluded to determine whether or not aggregation occurred in response to ANP addition (83).

THE GUANYLYL CYCLASE-COUPLED HEAT-STABLE ENTEROTOXIN RECEPTOR (GC-C)

GC-C represents a subtype of particulate guanylyl cyclases that is predominantly located in the intestinal mucosa. GC-C acts as a cell surface receptor for heat-stable enterotoxins (ST) from *Escherichia coli* and other pathogenic bacteria (Table 1). These small peptides bind to the extracellular domain of GC-C and directly stimulate enzyme activity (35). The prominent increase in cGMP formation affects chloride and other ion transport mechanisms in small and large intestine resulting in acute diarrhea (84,85). The intestinal mucosa ST-receptor/GC-C has not been purified and characterized in its native form. This may be due to the fact that this form is difficult to solubilize by nonionic detergent treatment and it is assumed that it is associated with cytoskeletal structures (86). There is a high content of proteases in intestinal mucosa which could be one explanation for the identification of several apparent ST binding proteins not coupled to guanylyl cyclase as determined by chemical cross-linking and measurement of enzyme activity (87,88).

Primary Structure - A cDNA clone representing GC-C has been isolated from a rat intestinal mucosa library (35). GC-C is closely related to the natriuretic peptide receptor-type GC-A and GC-B. From its deduced amino acid sequence a polypeptide of 114 kDa is predicted that fits the general topological model of membrane guanylyl cyclases (Figure 1). Like GC-A and GC-B, GC-C has an intracellular kinase-like domain and a carboxyl-terminal putative catalytic domain. Additionally, there is an extended carboxyl tail which is rich in uncharged polar amino acids and could function as a means to anchor GC-C to the cytoskeleton (35). The sequence homology of GC-C with GC-A or GC-B is considerably lower than between the natriuretic peptide receptor/cyclases. GC-A and GC-B are 43% identical in their extracellular binding domains compared to only 10% identity between GC-C and either of these forms. GC-C does not bind natriuretic peptides. The intracellular kinase-like and catalytic domains of GC-A and GC-B are 72% and 91% identical, respectively. Compared to either of these forms GC-C is 35% identical in the kinase-like region and 55% in the catalytic domain (35). A stretch of amino acids (Asp-Asp-Asp-Arg) located intracellularly close to the transmembrane domain is very similar to the proteolytic cleavage site of enterokinase (Asp-Asp-Asp-Asp-Lys-X COOH). This may have some significance with respect to the sensitivity of GC-C to proteolysis, taking into account the rare occurrence of this sequence in proteins, and the high amount of enterokinase in the intestine.

A cDNA clone that is 81% identical to GC-C was derived from the human colon carcinoma cell line T84 (36) and guanylyl cyclase-coupled ST receptors were described in epithelium cells of different tissues of opossum (89). ST receptors lacking intrinsic guanylyl cyclase activity and a related guanylyl cyclase cDNA clone from small intestinal mucosa have also been reported (35,87). These observations suggest that a diverse subfamily of GC-C-type proteins might exist. Recently, a 15-amino acid peptide with a high degree of homology to ST was purified from rat intestinum (90). In T84 cells, it stimulates increases in cGMP levels and displaces ST binding. This peptide, termed guanylin, is proposed to be an endogenous activator of intestinal guanylyl cyclase (90). The physiologic and pathophysiologic role of guanylin and GC-C in epithelial ion transport has yet to be defined.

Table 1.   The Guanylyl Cyclase Family of Receptor/Enzymes

FORM	RECEPTOR	ACTIVATORS*
Sea urchin GC	egg peptide receptor	Resact; Speract
Mammalian ANP-CR	natriuretic peptide receptor	ANP=CNP>BNP
Mammalian GC-A	natriuretic peptide receptor	ANP>BNP>>CNP
Mammalian GC-B	natriuretic peptide receptor	CNP>>ANP=BNP
Mammalian GC-C	bacterial enterotoxin receptor	ST, guanylin?
Vertebrate retinal GC	Ca^{2+}-regulated <u>via</u> recoverin	recoverin
Protozoan GC	Ca^{2+}/calmodulin-activated	Ca^{2+}/calmodulin
Mammalian GC-S	nitric oxide (NO) receptor	NO, nitrovasodilatory drugs

* The peptides with natriuretic properties possess a conserved 17-residue disulfide ring but do not necessarily function as natriuretic hormones under physiological conditions.

CALCIUM-REGULATED GUANYLYL CYCLASES

Guanylyl cyclases from mammalian retina and from ciliary membranes of protozoans are characterized by an apparent calcium regulation of their enzyme activity, and by their localization in the particulate or cytoskeleton fraction of cell homogenates (91,92). At present however, no primary sequence data for these forms are available (Table 1).

Retinal Guanylyl Cyclase - Guanylyl cyclase plays a crucial role in visual transduction: The light-induced hydrolysis of cGMP leads to the closure of cGMP-gated cationic channels which in the dark state are kept open by cGMP. This results in a decrease in intracellular Ca^{2+} concentration and ultimately in activation of rod guanylyl cyclase and resynthesis of cGMP (93,94). The resumed production of cGMP then results in reopening of the cationic channels and in a consequent recovery of the dark state with intracellular Ca^{2+} concentrations of 300-400 nM. Rod guanylyl cyclase is resistant to solubilization by nonionic detergents under mild conditions (20-22,95,96) and seems to be associated with axonemal structures (97,98). It may also be present in disk membranes (95,99). Guanylyl cyclase from bovine rod photoreceptors was purified to or near to apparent homogeneity only in 1991 (20-22). There exists, however, some discrepancy concerning the characteristics of the purified proteins: two of the reports estimate a M_r of 110-112 kDa (20,21), the third a M_r of 60 kDa (22).

A retinal-specific accessory Ca^{2+}-binding protein of 26 kDa, which mediates the effects of Ca^{2+} on guanylyl cyclase activity in bovine rods (91), has been purified (100), cloned and designated as recoverin (101). Recoverin is a member of a novel family of soluble Ca^{2+}-binding proteins distinct from calmodulin, which also includes visinin found in chicken cones (100,101). In its Ca^{2+}-free form (below 200 nM Ca^{2+}) recoverin activates rod guanylyl cyclase cooperatively while it is not active at the higher Ca^{2+} concentrations of the dark state (100). In the absence of recoverin, the photoreceptor guanylyl cyclase exhibits only a low basal activity and is insensitive to Ca^{2+} (20,100).

Protozoan Guanylyl Cyclases - In contrast to the enzyme involved in visual transduction, guanylyl cyclase from excitable ciliary membranes of unicellular protozoans is absolutely dependent on micromolar Ca^{2+} concentrations for full activity (102). Calmodulin or calmodulin-like Ca^{2+}-binding proteins are present as endogenous enzyme subunits and are responsible for conferring calcium sensitivity to guanylyl cyclases from $Tetrahymena$ and the related genus $Paramecium$ (92). Known activators of other guanylyl cyclases like nitroso compounds or atrial natriuretic peptides are ineffective. Mg^{2+} and Mn^{2+} are equally potent as divalent metal cofactors in contrast to other guanylyl cyclases which are about 10-fold less active with Mg^{2+} (102). It is assumed that ciliary movements of these protozoans are controlled by a regenerative and graded Ca^{2+}/K^+ action potential. Ca^{2+} influx from a voltage-sensitive Ca^{2+} channel activates a ciliary reversal mechanism as well as guanylyl cyclase. Increasing cGMP initiates the recovery of the cilia forward beat (92).

Cilia of similar morphology occur also in vertebrate tissues, e.g. in bronchi epithelia, sperm and ependymal cells. Cilia-derived structures in vertebrates, besides the photoreceptor rod outer segments, are found in the hair cells of the inner ear and in olfactory epithelia. Ca^{2+}/calmodulin-regulated guanylyl cyclases have not been detected in these tissues (102). Guanylyl cyclase from olfactory and tracheal cilia is membrane-bound and resistant to solubilization by nonionic detergents. It requires however the presence of detergent and Mn^{2+} for activity. Mg^{2+} can not replace Mn^{2+} and Ca^{2+} does not support enzyme activity. Neither NO and similar compounds nor ANP affect this type of cyclase (103). It remains to be shown which enzyme form is responsible for the guanylyl cyclase activity in these tissues.

HETERODIMERIC FORMS OF GUANYLYL CYCLASE (GC-S)

Key features of the heterodimeric forms of guanylyl cyclase (GC-S) are their regulation by nitric oxide (NO) and related vasodilatory drugs, their association with a heme prosthetic group and their cytoplasmic subcellular localization (Table 1, Figure 1). Future subcellular localization studies may prove, however, that some GC-S forms actually exist on membranes.

Regulation by NO - Ca^{2+}-mobilizing hormones and neurotransmitters like acetylcholine, bradykinin, serotonin and substance P increase cGMP levels in intact cells and tissues but fail to stimulate cytosolic guanylyl cyclase activity in vitro. Originally it was thought that the receptor-mediated formation of free radicals and lipid oxidation products was responsible for the activation of soluble guanylyl cyclase (104). The activation mechanism became clear when an endothelium-derived relaxing factor (EDRF) was identified as NO (105). EDRF is generated in endothelial cells and subsequently leads to elevated cGMP concentrations in vascular smooth muscle and relaxation. Prior to the discovery of EDRF, NO, sodium nitroprusside and other nitrovasodilators had been identified as potent stimulators of guanylyl cyclase activity in the cytosolic fraction of cell and tissue homogenates (106-108). NO appears to be an important endogenous intercellular and intracellular signaling molecule and is enzymatically derived from the amino acid L-arginine by NO synthases (109). This class of enzymes was found in the endothelium, but also in a variety of other cells and tissues. Some isoenzyme forms are activated in response to an intracellular rise of Ca^{2+} evoked by the above mentioned agonists (109). Thus, Ca^{2+}-mobilizing hormones may regulate intracellular cGMP levels indirectly via a Ca^{2+}-induced activation of NO synthases and a subsequent NO-dependent stimulation of GC-S. GC-S/cGMP may therefore represent the main effector system for endogenously formed NO. However, cGMP-independent actions of NO have been described suggesting the existence of other NO-responsive enzyme systems (110).

The mechanism of GC-S activation by NO is not completely clear. Heme is associated as a prosthetic group of purified soluble guanylyl cyclase from bovine (24,25,27,111) and rat lung (23,26). This has led to a model of enzyme activation that is based on the direct interaction of NO and NO-containing compounds with the enzyme-bound heme moiety (111,112). Interestingly, heme-deficient guanylyl cyclase is not activated by NO-liberating compounds but sensitivity can be restored by the addition of heme or protoporphyrin IX (112,113). The detailed molecular mechanism of this NO-heme interaction is still unknown although heme groups are well-defined as NO binding sites in other hemoproteins (114).

<u>Primary Structures and Function</u> - Based on partial amino acid sequence data from the purified proteins, cDNAs encoding different subunits of GC-S have been cloned. The cDNA clone of the larger subunit (GC-S$_{\alpha 1}$) is predicted to encode a protein with a M_r of 77.5 kDa (39,40). In fact, α-subunits of Mr 79-82 kDa and 73 kDa, as determined by SDS-PAGE, were purified from rat and bovine lung, respectively (23,26,27). The cDNA clone of the smaller subunit has been designated as GC-S$_{\beta 1}$ and encodes a protein of M_r 70.5 kDa (37,38) corresponding to a M_r of 70-74 kDa in SDS-PAGE (23,26,27). The GC-S$_{\alpha 1}$ and GC-S$_{\beta 1}$ subunits from bovine and rat lung share 85% and 97% of the amino acids, respectively (115).

The primary structures of rat GC-S$_{\alpha 1}$ and rat GC-S$_{\beta 1}$ show about 32% identical amino acids across the entire sequence (39). The similarity is highest between the carboxy-terminal putative catalytic regions (45% amino acid sequence identity) and is somewhat lower in the amino-terminal regions (20% identity) (39,40). The carboxy-terminal region is homologous to the putative catalytic domains of plasma membrane guanylyl cyclases and to one of the two hydrophilic and predicted catalytic domains of adenylyl cyclase (30% identity; ref. 40,70). There is no such homology between the amino-terminal regions and other cyclases, other heme-containing proteins or other proteins (40). Interestingly, these GC-S subunits lack the protein kinase-like domains which are present in the membrane-bound forms. Sites responsible for subunit interaction, for heme binding and for interaction with activators have not been identified, although it is assumed that the amino-terminal regions could play a regulatory role.

Two other members of the heterodimeric GC-S family have been identified by use of degenerate oligonucleotide primers in the polymerase chain reaction. This demonstrates heterogeneity among the forms of GC-S. The first of these new sequences was obtained from rat kidney and is predicted to encode a 76.3 kDa protein. It was called GC-S$_{\beta 2}$ because of its homology to the 70 kDa GC-S$_{\beta 1}$ subunit (41). GC-S$_{\beta 2}$ mRNA is preferentially expressed in kidney and liver whereas the highest amounts of GC-S$_{\alpha 1}$ and GC-S$_{\beta 1}$ mRNA are found in lung and brain (38,40,41). The structure of GC-S$_{\beta 2}$ shows 34% amino acid identity to GC-S$_{\beta 1}$ as compared to 27% to GC-S$_{\alpha 1}$. However, GC-S$_{\beta 2}$ is unique in lacking the first 62 amino acids and having an 86 amino acid carboxy-terminal extension which shows no homology to other proteins. It also contains a consensus sequence for isoprenylation/carboxymethylation at its COOH-terminus (41). The functional consequences of this feature are currently under investigation. Isoprenylation could determine regulation and subcellular localization different from other subunits. Another cDNA, GC-S$_{\alpha 2}$ encoding a 81.7 kDa protein has been cloned and sequenced by a similar approach from human fetal brain (42). Its deduced amino acid sequence shows the highest degree of homology to the 77.5 kDa GC-S$_{\alpha 1}$ subunit with an overall identity of 48% of the amino acids as compared to 30% and 23% identical amino acids shared with GC-S$_{\beta 1}$ and GC-S$_{\beta 2}$, respectively (42).

Co-expression of both $\alpha 1$- and $\beta 1$-type subunits in the same cell was found to be necessary for GC-S enzyme activity and for sensitivity towards sodium nitroprusside (40,116). Expression in cultured mammalian cells of either subunit alone ($\alpha 1$, $\alpha 2$, $\beta 1$, $\beta 2$) showed no detectable guanylyl cyclase activity (38,40,41,116,117). The <u>in vitro</u> combination of $\alpha 1$- with $\beta 1$-type GC-S subunits expressed separately did also not reconstitute enzyme activity (117). Both α-subunits of GC-S are interchangeable, however: co-expression of GC-S($\alpha 1\beta 1$) as well as GC-S($\alpha 2\beta 1$) led to a catalytically active enzyme which was stimulated by NO-containing compounds (42). So far, co-expression experiments including GC-S$_{\beta 2}$ have not been reported. According to one model, dimerization of α- and β-type GC-S subunits is required for catalysis. Adenylyl cyclase contains two homologous hydrophilic domains in one molecule further supporting the hypothesis that at least two catalytic domains may act in concert in cyclases (70).

Soluble cyclase activity is found in most cells and tissues but the relative distribution of the known four heterodimeric guanylyl cyclases has not yet been established. GC-S$_{\alpha 1}$ and GC-S$_{\beta 1}$ mRNA are distributed similarly with the exception of cerebellum where the latter is much more abundant (40). Both GC-S$_{\beta}$ subunits are found in kidney and liver (38,41). Similarly, the physiological composition of the heterodimers as well as their specific activators and their regulatory mechanisms have to be defined further. NO is a potent activator at least of GC-S$_{(\alpha 1 \beta 1)}$ and GC-S$_{(\alpha 2 \beta 1)}$ (42). However, alternative forms of enzyme regulation are conceivable most notable points being the potential isoprenylation and membrane location of GC-S$_{\beta 2}$ and the apparent heterogeneity of GC-S due to alternative splicing (43).

CONCLUDING REMARKS

In the past years the expanding guanylyl cyclase family of receptor/enzymes has been discovered demonstrating a direct coupling of various receptors to cGMP-forming enzyme activities. With a better understanding of the structure and function at the molecular level, and knowledge of the tissue distribution of the diverse guanylyl cyclases, it should be possible to resolve the biochemistry and physiology of the cGMP signal transduction system. This should lead to the elucidation of new pharmacologic agents and to the development of new therapeutic strategies.

References

1. D.F. Ashman, R. Lipton, M.M. Melicow and T.D. Price, Biochem.Biophys.Res.Commun., 11, 330 (1963).
2. J. Tremblay, R. Gerzer and P. Hamet, Adv.Second Messenger Phosphoprotein Res., 22, 319 (1988).
3. U. Walter, Rev.Physiol.Biochem.Pharmacol., 113, 41 (1989).
4. M. Chinkers and D.L. Garbers, Annu.Rev.Biochem., 60, 553 (1991).
5. S. Schulz, P.S.T. Yuen and D.L. Garbers, Trends Pharmacol.Sci., 12, 116 (1991).
6. D.L. Garbers, Pharmacol.Ther., 50, 337 (1991).
7. D. Koesling, E. Böhme and G. Schultz, FASEB J., 5, 2785 (1991).
8. J.G. Hardman and E.W. Sutherland, J.Biol.Chem., 244, 6363 (1969).
9. G. Schultz, E. Böhme and K. Munske, Life Sci., 8, 1323 (1969).
10. A.A. White and G.D. Aurbach, Biochim.Biophys.Acta, 191, 686 (1969).
11. H. Kimura and F. Murad, J.Biol.Chem., 249, 6910 (1974).
12. D.L. Garbers, J.G. Hardman and F.B. Rudolph, Biochemistry, 13, 4166 (1974).
13. T.D. Chrisman, D.L. Garbers, M.A. Parks and J.G. Hardman, J.Biol.Chem., 250, 374 (1975).
14. D.L. Garbers, J.Biol.Chem., 251, 4071 (1976).
15. E.W. Radany, R. Gerzer and D.L. Garbers, J.Biol.Chem., 258, 8346 (1983).
16. T. Kuno, J.W. Andresen, Y. Kamisaki, S.A. Waldman, L.Y.Chang, S. Saheki, D.C. Leitman, M. Nakane and F. Murad, J.Biol.Chem., 261, 5817 (1986).
17. A.K. Paul, R.B. Marala, R.K. Jaiswal and R.K. Sharma, Science, 235, 1224 (1987).
18. R. Takayanagi, T. Inagami, R.M. Snajdar, T. Imada, M. Tamura and K.S. Misono, J.Biol.Chem., 262, 12104 (1987).
19. S. Meloche, N. McNicoll, B. Liu, H. Ong and A. DeLean, Biochemistry, 27, 8151 (1988).
20. K.-W. Koch, J.Biol.Chem., 266, 8634 (1991).
21. F. Hayashi and A. Yamazaki, Proc.Natl.Acad.Sci.USA, 88, 4746 (1991).
22. Y. Horio and F. Murad, Biochim.Biophys.Acta, 1133, 81 (1991).
23. D.L. Garbers, J.Biol.Chem., 254, 240 (1979).
24. R. Gerzer, F. Hofmann and G. Schultz, Eur.J.Biochem., 116, 479 (1981).
25. L.J. Ignarro, K.S. Wood and M.S. Wolin, Proc.Natl.Acad.Sci.USA, 79, 2870 (1982).
26. Y. Kamisaki, S. Saheki, M. Nakane, J. Palmieri, T. Kuno, B. Chang, S.A. Waldman and F. Murad, J.Biol.Chem., 261, 7236 (1986).
27. P. Humbert, F. Niroomand, G. Fischer, B. Mayer, D. Koesling, K.-D. Hinsch, H. Gausepohl, R. Frank, G. Schultz and E. Böhme, Eur.J.Biochem, 190, 273 (1990).
28. S. Singh, D.G. Lowe, D.S. Thorpe, R. Rodriguez, W.-J. Kuang, L.J. Dangott, M. Chinkers, D.V. Goeddel and D.L. Garbers, Nature, 334, 708 (1988).
29. D.S. Thorpe and D.L. Garbers, J.Biol.Chem., 264, 6545 (1989).
30. M. Chinkers, D.L. Garbers, M.-S. Chang, D.G. Lowe, H. Chin, D.V. Goeddel and S. Schulz, Nature, 338, 78 (1989).
31. D.G. Lowe, M.-S. Chang, R. Hellmiss, E. Chen, S. Singh, D.L. Garbers and D.V. Goeddel, EMBO J., 8, 1377 (1989).
32. K.N. Pandey and S. Singh, J.Biol.Chem., 265, 12342 (1990).
33. S. Schulz, S. Singh, R.A. Bellet, G. Singh, D.J. Tubb, H. Chin and D.L. Garbers, Cell, 58, 1155 (1989).
34. M.-S. Chang, D.G. Lowe, M. Lewis, R. Hellmiss, E. Chen and D.V. Goeddel, Nature, 341, 68 (1989).
35. S. Schulz, C.K. Green, P.S.T. Yuen and D.L. Garbers, Cell, 63, 941 (1990).
36. S. Singh, G. Singh, J.-M. Heim and R. Gerzer, Biochem.Biophys.Res.Commun., 179, 1455 (1991).
37. D. Koesling, J. Herz, H. Gausepohl, F. Niroomand, K.-D. Hinsch, A. Müsch, E. Böhme, G. Schultz and R. Frank, FEBS Lett., 239, 29 (1988).
38. M. Nakane, S. Saheki, T. Kuno, K. Ishii and F. Murad, Biochem.Biophys.Res.Commun., 157, 1139 (1988).

39. D.Koesling, C. Harteneck, P. Humbert, A. Bosserhoff, R. Frank, G. Schultz and E. Böhme, FEBS Lett., 266, 128 (1990).
40. M. Nakane, K. Arai, S. Saheki, T. Kuno, W. Büchler and F. Murad, J.Biol.Chem., 265, 16841 (1990).
41. P.S.T. Yuen, L.R. Potter and D.L. Garbers, Biochemistry, 29, 10872 (1990).
42. C. Harteneck, B. Wedel, D. Koesling, J. Malkewitz, E. Böhme and G. Schultz, FEBS Lett., 292, 217 (1991).
43. V. Cchajlani, P.-A. Fråndberg, J. Ahlner, K.L. Axelsson and J.E.S. Wikberg, FEBS Lett., 290, 157 (1991).
44. D.L. Garbers, Annu.Rev.Biochem., 58, 719 (1989).
45. D.L. Garbers, J.Androl., 10, 99 (1989).
46. H. Shimomura, L.J. Dangott and D.L. Garbers, J.Biol.Chem., 261, 15778 (1986).
47. S.K. Hanks, A.M. Quinn and T. Hunter, Science, 241, 42 (1988).
48. A. Ullrich and J. Schlessinger, Cell, 61, 203 (1990).
49. E.H. Fischer, H. Charbonneau and N.K. Tonks, Science, 253, 401 (1991).
50. A.J. deBold, H.B. Borenstein, A.T. Veress and H. Sonnenberg, Life Sci., 28, 89 (1981).
51. T. Inagami, J.Biol.Chem., 264, 3043 (1989).
52. P. Hamet, J. Tremblay, S.C. Pang, R. Garcia, G. Thibault, J. Gutkowska, M. Cantin and J. Genest, Biochem.Biophys.Res.Commun., 123, 515 (1984).
53. R.J. Winquist, E.P. Faison, S.A. Waldman, K. Schwartz, F. Murad and R.M. Rapoport, Proc.Natl.Acad.Sci. USA, 81, 7661 (1984).
54. S.A. Waldman, R.M. Rapoport and F. Murad, J.Biol.Chem., 259, 14332 (1984).
55. J. Tremblay, R. Gerzer, S.C. Pang, M. Cantin, J. Genest and P. Hamet, FEBS Lett., 194, 210 (1985).
56. D.B. Schenk, M.N. Phelps, J.G. Porter, R.M. Scarborough, G.A. McEnroe and J.A. Lewicki, J.Biol.Chem., 260, 14887 (1985).
57. D.C. Leitman, J.W. Andresen, T. Kuno, Y. Kamisaki, J.K. Chang and F. Murad, J.Biol.Chem., 261, 11650 (1986).
58. T. Maack, M. Suzuki, F.A. Almeida, D. Nussenzveig, R.M. Scarborough, G.A. McEnroe and J.A. Lewicki, Science, 238, 675 (1987).
59. D.B. Schenk, M.N. Phelps, J.G. Porter, F. Fuller, B. Cordell and J.A. Lewicki, Proc.Natl.Acad.Sci.USA, 84, 1521 (1987).
60. M. Shimonaka, T. Saheki, H. Hagiwara, M. Ishido, A. Nogi, T. Fujita, K. Wakita, Y. Inada, J. Kondo and S. Hirose, J.Biol.Chem., 262, 5510 (1987).
61. F. Fuller, J.G. Porter, A.E. Arfsten, J. Miller, J.W. Schilling, R.M. Scarborough, J.A. Lewicki and D.B. Schenk, J.Biol.Chem., 263, 9395 (1988).
62. M.B. Anand-Srivastava, D.J. Franks, M. Cantin, and J. Genest, Biochem.Biophys.Res.Commun, 121, 855 (1984).
63. M.B. Anand-Srivastava, M.R. Sairam and M. Cantin, J.Biol.Chem., 265, 8566 (1990).
64. M.B. Anand-Srivastava, A.K. Srivastava and M. Cantin, J.Biol.Chem., 262, 4931 (1987).
65. M. Hirata, C.H. Chang and F. Murad, Biochim.Biophys.Acta, 1010, 346 (1989).
66. S. Meloche, H. Ong and A. DeLean, J.Biol.Chem., 262, 10252 (1987).
67. T. Sudoh, N. Minamino, K. Kangawa and H. Matsuo, Biochem.Biophys.Res.Commun., 168, 863 (1990).
68. K.J. Koller, D.G. Lowe, G.L. Bennett, N. Minamino, K. Kangawa, H. Matsuo and D.V. Goeddel, Science, 252, 120 (1991).
69. J.N. Wilcox, A. Augustine, D.V. Goeddel and D.G. Lowe, Mol.Cell.Biol., 11, 3454 (1991).
70. J. Krupinski, F. Coussen, H.A. Bakalyar, W.-J. Tang, P.G. Feinstein, K. Orth, C. Slaughter, R.R. Reed and A.G. Gilman, Science, 244, 1558 (1989).
71. M. Chinkers and D.L. Garbers, Science, 245, 1392 (1989).
72. D.S. Thorpe and E. Morkin, J.Biol.Chem., 265, 14717 (1990).
73. H. Kurose, T. Inagami and M. Ui, FEBS Lett., 219, 375 (1987).
74. D.-L. Song, K.P. Kohse and F. Murad, FEBS Lett., 232, 125 (1988).
75. C.-H. Chang, B. Jiang and J.G. Douglas, Eur.J.Pharmacol., 189, 293 (1990).
76. C.-H. Chang, K.P. Kohse, B. Chang, M. Hirata, B. Jiang, J.E. Douglas, and F. Murad, Biochim. Biophys.Acta, 1052, 159 (1990).
77. A. DeLean, Life Sci., 39, 1109 (1986).
78. J. Fethiere, S. Meloche, T.T. Nguyen, H. Ong and A. DeLean, Mol.Pharmacol., 35, 584 (1989).
79. M. Chinkers, S. Singh and D.L. Garbers, J.Biol.Chem., 266, 4088 (1991).
80. L. Larose, N. McNicoll, H. Ong and A. DeLean, Biochemistry, 30, 8990 (1991).
81. L.R. Potter and D.L. Garbers, FASEB J., 6, A79 (1992).
82. D.S. Thorpe, S. Niu and E. Morkin, Biochem.Biophys.Res.Commun., 180, 538 (1991).
83. T.Iwata, K. Uchida-Mizuno, T. Katafuchi, T. Ito, H. Hagiwara and S. Hirose, J.Biochem., 110, 35 (1991).
84. M. Field, L.H. Graf, W.J. Laird and P.L. Smith, Proc.Natl.Acad.Sci.USA, 75, 2800 (1978).
85. J.M. Hughes, F. Murad, B. Chang and R.L. Guerrant, Nature, 271, 755 (1978).
86. S.A. Waldman, T. Kuno, Y. Kamisaki, L.Y. Chang, J. Gariepy, P. O'Hanley, G. Schoolnik and F. Murad, Infect.Immun., 51, 320 (1986).

87. T. Kuno, Y. Kamisaki, S.A. Waldman, J. Gariepy, G. Schoolnik and F. Murad, J.Biol.Chem., 261, 1470 (1986).
88. M.R. Thompson and R.A. Giannella, J.Recept.Res., 10, 97 (1990).
89. L.R. Forte, W.J. Krause and R.H. Freeman, Am.J.Physiol., 257, F874 (1988).
90. M.G. Currie, K.F. Fok, J. Kato, R.J. Moore, F.K. Hamra, K.L. Duffin and C.E. Smith, Proc.Natl.Acad.Sci.USA, 89, 947 (1992).
91. K.-W. Koch and L.Stryer, Nature, 334, 64 (1988).
92. J.E. Schultz and S. Klumpp in "Methods in Enzymology", Vol. 195, R.A. Johnson and J.D. Corbin, Eds., Academic Press, San Diego, CA, 1991, p. 466.
93. R.N. Lolley and R.H. Lee, FASEB J., 4, 3001 (1990).
94. L. Stryer, J.Biol.Chem., 266, 10711 (1991).
95. S. Hakki and A. Sitaramayya, Biochemistry, 29, 1088 (1990).
96. Y. Horio and F. Murad, J.Biol.Chem., 266, 3411 (1991).
97. D. Fleischman and M. Denisevich, Biochemistry, 18, 5060 (1979).
98. D. Fleischman, M. Denisevich, D. Raveed and R.G. Pannbacker, Biochim.Biophys.Acta, 630, 176 (1980).
99. S. Kawamura and M. Murakami, J.Gen.Physiol., 94, 649 (1989).
100. H.-G.Lambrecht and K.-W. Koch, EMBO J., 10, 793 (1991).
101. A.M. Dizhoor, S. Ray, S. Kumar, G. Niemi, M. Spencer, D. Brolley, K.A. Walsh, P.P. Philipov, J.B. Hurley and L. Stryer, Science, 251, 915 (1991).
102. J.E. Schultz and S. Klumpp, Adv.Cyclic Nucleotide Protein Phosphorylation Res., 17, 275 (1984).
103. S. Steinlen, S. Klumpp and J.E. Schultz, Biochim.Biophys.Acta, 1054, 69 (1990).
104. C.K. Mittal and F. Murad, J.Cyclic Nucleotide Res., 3, 381 (1977).
105. S. Moncada, R.M.J. Palmer and E.A. Higgs, Pharm.Rev., 43, 109 (1991).
106. W.P. Arnold, C.K. Mittal, S. Katsuki and F. Murad, Proc.Natl.Acad.Sci USA, 74, 3203 (1977).
107. E. Böhme, H. Graf and G. Schultz, Adv.Cyclic Nucleotide Res., 9, 131 (1978).
108. F. Murad, C.K. Mittal, W.P. Arnold, S. Katsuki and H. Kimura, Adv.Cyclic Nucleotide Res., 9, 175 (1978).
109. D.S. Bredt and S.H. Snyder, Proc.Natl.Acad.Sci.USA, 87, 682 (1991).
110. B. Brüne and E. Lapetina, J.Biol.Chem., 264, 8455 (1989).
111. R. Gerzer, E. Böhme, F. Hofmann and G. Schultz, FEBS Lett., 132, 71 (1981).
112. L.J. Ignarro, J.B. Adams, P.M. Horwitz and K.S. Wood, J.Biol.Chem., 261, 4997 (1986).
113. P.H.Craven and F.R. DeRubertis, J.Biol.Chem., 253, 8433 (1978).
114. S.L. Edwards, J. Kraut and T.L. Poulos, Biochemistry, 27, 8074 (1988).
115. D. Koesling, G. Schultz and E. Böhme, FEBS Lett., 280, 301 (1991).
116. C. Harteneck, D. Koesling, A. Sölng, G. Schultz and E. Böhme, FEBS Lett., 272, 221 (1990).
117. W.A. Buechler, M. Nakane and F. Murad, Biochem.Biophys.Res.Commun., 174, 351 (1991).

Chapter 27. HIV Vaccine Development: From Empiricism to Medicinal Chemistry

Marta J. Glass and Wayne C. Koff
Vaccine Research and Development Branch
Division of AIDS
National Institute of Allergy and Infectious Diseases
6003 Executive Blvd., Rockville MD 20892

Introduction - The history of vaccine development is replete with several examples where the combination of empiricism and enlightened serendipity led to the generation of successful vaccines which subsequently had a major beneficial impact on public health (1). Briefly, following the successes of Jenner and Pasteur, against smallpox and rabies respectively, the techniques for development of bacterial toxoids led to the generation of vaccines against tetanus and other prominent bacterial infections (2). Similarly, the development of large-scale mammalian tissue culture systems led to the generation of several attenuated virus vaccines, including polio, measles, mumps and rubella, and technologies for virus inactivation led to the generation of numerous killed vaccines including influenza and polio (3). Recent advances in recombinant DNA technologies and molecular and cellular immunology have led to the first generation of recombinant vaccines, the hallmark being hepatitis B (4). Finally, technical advances in many disciplines such as gene cloning and sequencing, antigen purification, synthetic peptide chemistry, immunochemistry, drug delivery, adjuvant development and medicinal chemistry have provided a foundation for an explosion in vaccine development which is currently in its embryonic stages (5). The present excitement in vaccine development, coupled with the urgency surrounding the global human immunodeficiency virus (HIV) epidemic has led to a formidable goal: development of safe and effective vaccines for the prevention of HIV disease/AIDS. Several recent reviews provide comprehensive updates of HIV vaccine development (6,7), and this chapter will not attempt to duplicate such efforts. Rather, the purpose of this manuscript is to discuss recent developments in adjuvant technology, antigen presentation, and vaccine delivery and indicate their impact on the development of an HIV vaccine.

The Problem. - By the year 2000, the World Health Organization estimates that approximately 40 million persons will be infected with HIV worldwide (8). Current strategies for prevention and control of AIDS, consisting of behavior modification, condom distribution, and antiviral chemotherapy have not kept pace with the international HIV pandemic (9). The morbidity, mortality, and concomitant social/economic losses due to AIDS have only recently begun to be felt. Clearly, there is a compelling need for the development of safe and effective HIV vaccines to stem the further spread of HIV infection. To meet this challenge, an international network of biomedical scientists from academia, governments, and industry has been assembled (10).

Table 1. HIV Expression Systems/Vectors

Baculovirus
Mammalian cells (e.g., Chinese Hamster Ovary)
Yeast (Saccharomyces cerevisiae)
Adenovirus
Avipox
BCG
Poliovirus
Salmonella
Vaccinia

Since the isolation and identification of HIV as the etiologic agent for AIDS (11,12), there have been several important milestones towards the development of HIV vaccines. HIV antigens have been expressed in mammalian, bacterial, and virus systems (Table 1), animal models for HIV infection and AIDS have been developed (Table 2), more than 30 experimental vaccines are currently in preclinical development (Table 3), and several of these vaccines are currently being tested in Phase 1 clinical trials (Table 4). Yet, HIV continues to be an imposing foe for vaccine developers. This is due in large part to two principal issues: HIV Variation and HIV Transmission.

Table 2. Animal Models for HIV Infection and AIDS

Chimpanzees	HIV-1
Rhesus Monkeys	SIV/macaque Rhesus, Pigtail, Cyno
Pigtail and Cyno	HIV-2
Pigtail Macaques	HIV-1
SCID Mice	(PBL and Fetal tissue)
Rabbits	HIV-1
Felines	FIV FIV/FeLV

HIV Variation.- Worldwide isolates of HIV are quite variable, particularly in immunodominant regions likely to be associated with protective immunity (13); thus, vaccine developers face a moving target of antigenic drifting, currently estimated at 1%/year in the HIV env gene which leads to major differences among international virus isolates (14). Present data suggest that a minimum of 5-7 genotypes of HIV-1 exist worldwide (14); it is quite sobering to acknowledge that experimental HIV vaccines thus far tested in Phase 1 clinical trials have for the most part failed to induce cross-reacting neutralizing antibodies (15), where cross-reactivity is defined as neutralizing other laboratory isolates within the same prominent genotype. Whether any of the vaccines will induce cross-reactive immune responses against isolates from the other 5-7 international genotypes, is the focus of intensive investigation. In addition, patterns of neutralization and tropism of clinical field isolates and laboratory isolates of HIV may differ quite substantially (16), thus complicating investigations to serotype HIV strains.

HIV Transmission - The routes of HIV transmission, predominantly sexual transmission along with perinatal and intravenous (17), provide a variety of obstacles to vaccine development similar to other sexually transmitted diseases for which vaccines have been notoriously difficult to develop (18). Among these challenges are the potential need for induction of mucosal protective immunity (19), i.e. prevention of the establishment of infection at a local mucosal site. Moreover, the capacity of HIV to infect in two forms, free virus and as virus-infected cells, provides even greater challenges than most sexually transmitted diseases.

Towards a Successful HIV Vaccine - An ideal HIV vaccine may be thought of with the following components: elicits long-term duration of protective immunity, easy to administer with few booster immunizations, elicits parenteral and mucosal immunity, induces broad spectrum cellular and humoral immune responses, capable of preventing HIV disease/AIDS caused by variable HIV isolates, and inexpensive. While first generation HIV vaccines described in Table 4 may provide some of the components of a potentially effective vaccine, none of the candidates would be considered close to an ideal HIV vaccine at this time. This leads us to conclude that greater efforts should be undertaken in developing second generation HIV vaccines with a focus on amplifying immunogenicity with concomitant optimization of dose/regimen schedules. Advances in technologies of antigen presentation, adjuvants, and vaccine vehicles may provide tools for next generation HIV vaccines, and these are summarized below.

Antigen Presentation - Design of effective HIV vaccines will rely on development of immunogens that can induce effective B and T cell immune responses. While B cells can recognize native proteins, T cells rely on antigen processing via antigen presenting cells. T cells then recognize the complex of the modified antigen in association with molecules of the major histocompatibility complex (MHC) (20). Live attenuated viral vaccines, such as measles vaccine, have in large part been effective due to their ability to mimic the induction of immune responses characteristic of natural infection. Among these responses, cytotoxic T lymphocytes (CTL) are generally considered important for clearance of virus infected cells, while neutralizing antibodies (NA) are capable of neutralizing free virus particles. For HIV, the concerns of potential reversion to wild-type from an attenuated form coupled with integration of the virus genomes into host cell genetic material has limited the enthusiasm for development of live attenuated HIV vaccines. Thus, novel approaches to antigen presentation, which might enable inactivated, subunit or peptide vaccines to efficiently induce anti-HIV CTL, has been given a high priority in vaccine development.

Table 3. Experimental Vaccines in Preclinical Development

TYPE	COMPONENT
Killed Virus	HIV-1 RF

Subunits	gp160, baculovirus
	    MN
	    THAI
	gp160, mammalian
	    MN
	    RF
	    Hybrid MN/LAI
	gp120, mammalian
	    SF-2
	p55, yeast
	Soluble CD4

Recombinant Vectors	Vaccinia-HIV env
	Attenuated Vaccinia-env
	    LAI
	    MN/LAI
	Vaccinia-HIV gag
	Vaccinia-env-pol-gag LAI
	AVIPOX - env MN/LAI
	Adenovirus-HIV env + gag
	Poliovirus-HIV env (735-752)
	Salmonella-HIV env
	BCG-HIV genes
	Retroviral vectors - HIV env

Peptides	V3 loop (gp 120) HGP-30 (p17) T1-SP10 (gp120) V3-MAPS p24/gp120 hybrid Conserved gp160 epitopes
Retrovirus-Like Particles	Pseudovirions Vaccinia expressed particles Yeast Ty gag Yeast gag-V3 virus-like particles
Virus Particles	Hepatitis B core particles
Combinations	Vaccinia-HIV-env + gp160, baculovirus Vaccinia-HIV-env + fixed autologous cells infected with vaccinia-HIV-env + gp160, mammalian

Recently, novel strategies have been developed for induction of cytotoxic T cells by modification of protein and peptide antigens. Derivatization of CTL peptide epitopes of influenza virus nucleoprotein with lipid moieties resulted in lipopetides that could prime for virus-specific CTL (21). Similarly, HIV specific CTL are induced by immunization of purified HIV-1 envelope glycoprotein presented as an immunostimulating complex (ISCOM), in which the protein antigen is incorporated into a particulate matrix of the glycoside Quil A (22). It was observed that a single immunization with a recombinant vaccinia virus-lymphocyte choriomeningitis vaccine expressing an immunodominant CTL epitope protected mice against lethal challenge (23). These strategies are now all being actively pursued in HIV vaccine development. Of particular note was the recent observation that recombinant vaccinia virus-SIV vaccine induced CTL against SIV gag epitopes in monkeys (24), and that extension of this strategy in a protocol consisting of priming with vaccinia-SIV recombinants followed by boosting with SIV envelope glycoprotein conferred protection against SIV induced AIDS in monkeys (25).

Antigen presentation strategies are also being utilized to optimize humoral immune responses against HIV and related retroviruses, leading to innovative vaccine candidates. Chemically defined synthetic antigens may be produced by linking peptides to lysine octamers in a multiple antigen peptide approach (MAPS) to elicit high titers of neutralizing antibodies against a variety of viral, bacterial, and parasitic diseases (26). These findings were extended by producing HIV-1 peptide MAPS, which induce high titers and long lasting neutralizing antibodies against HIV in animal models (27). Moreover, manipulation of MAPS with epitopes from variable HIV isolates provides an avenue for overcoming the challenge of developing HIV vaccines which can induce immunity to multiple variable isolates.

Several strategies have been developed to express HIV-1 as virus-like particles, including expression by baculovirus (28) or vaccinia infected cells (29), or through the production of noninfectious packaging mutants (30). Particles are attractive vaccine candidates, due to the expression of multiple protein components of the virus coupled with the lack of incorporation of the viral RNA genome. Preliminary immunogenicity studies have indicated that experimental particle vaccines induce HIV specific antibodies, and yeast retrotransposon-HIV gag virus-like particles are currently being evaluated in Phase 1 clinical trials. HIV-1 packaging mutants would theoretically offer all of the immunogenicity characteristics of killed vaccines, with the additional benefit of offering another methodology for virus inactivation which could complement chemical and physical inactivation in next generation vaccines.

Table 4. HIV Candidate Vaccines in Clinical Trials

IMMUNOGEN	SPONSOR
gp160: Baculovirus expression	MicroGeneSys
gp160: Mammalian expression	ImmunoAG
gp120: Yeast/deglycosylated	Biocine
gp120: Mammalian	Biocine
gp120: Mammalian	Genentech
RECOMBINANT VIRUS	
Vaccinia expressing HIV env	Bristol-Myers Squibb
Vaccinia expressing HIV env	D.Zagury/Pierr et Marie Curie Univ.
SYNTHETIC PEPTIDE	
HGP-30: Peptide of p17	Viral Technologies
VIRUS LIKE PARTICLE	
Ty-gag: Yeast retrotransposon	British Biotechnology
COMBINATIONS	
Vaccinia expressing HIV env, autologous Vac-env infected cells, gp160	D Zagury/Pierre et Marie Curie Univ.

Adjuvants - Gaston Ramon, in a series of experiments beginning in 1925 (31), noted that horses with abscesses at the site of injection developed significantly greater levels of diphtheria antitoxin, than those without localized infection. These observations led to the development of the field of vaccine adjuvants, which now is poised to reap the benefits of biotechnology in the generation of multiple avenues for immunopotentiation (32). As noted above, immune stimulating complexes (ISCOM) enable peptide and protein antigens to induce cell mediated immune responses,; however, these structures have been limited in practical use due to concerns over toxicity. Recently, scientists have identified the active adjuvant components of ISCOM, and these purified components known as Quil A derivatives have been shown to potentiate immune responses to HIV antigens in animal models (33). Similarly, the active components from Complete Freund's adjuvant have now been isolated and purified, and these muramyl dipeptide derivatives significantly increase the humoral and cellular immune responses to subunit antigens (34). Finally, incorporation of HIV antigens into liposomes has improved the immunogenicity of HIV and SIV antigens, and this strategy is currently under test in protective efficacy tests in the SIV-monkey model system. Liposomes may be formulated by varying lipid compositions, and may be targeted to antigen presenting cells in vivo (35). Adjuvants are also being utilized for induction of specific immune responses such as secretory immunity (36), which might have particular relevance for vaccines against sexually transmitted diseases including HIV. Table 5 outlines the categories of adjuvants currently being evaluated with HIV vaccines.

Table 5. Adjuvants Being Evaluated with HIV Vaccines

TYPE	EXAMPLE
Depots	Alum
	Alum + Deoxycholate
Bacterial Products	Incomplete Freunds Adjuvant
	MF-59 ± MTP-PE
	SAF
Saponin Derivatives	Quil A
	ISCOM
Lipid A Derivatives	Monophorphoryl Lipid A
Liposomes ± Cytokines	
Polylactide coglycolide polymers	
Mucosal Strategies	Cholera Toxin
	Salmonella vectors ± cytokines

Vaccine Vehicles - Three different approaches represent the recent revolution in vaccine vehicles; controlled release systems for peptides and proteins, vaccine vectors, and biolistics. Controlled release polymers offer the potential for persistent release of antigen over a period of several weeks (37). Microcapsules of SIV and HIV antigens have been developed, and are currently being evaluated in oral and systemic immunization. The capacity to maintain persistent release of HIV antigens over several days/weeks may offer an avenue to improve upon the less than ideal humoral immune responses observed with the first generation of HIV vaccines currently in clinical trials (38). Similarly, vaccine vectors offer a strategy to incorporate the genetic material of HIV into other viruses and bacteria, which then express the HIV antigens under independent promoters. This strategy has been tested with adenovirus, poliovirus, vaccinia virus, avipox, salmonella, BCG, among other vectors for HIV and SIV. These vectors incorporate the advantages of live attenuated vaccines and eliminate the safety concerns of live attenuated HIV vaccines, since only structural genes of HIV are incorporated into the vectors. Moreover, retroviral vectors are being utilized as a vehicle for gene therapy, and offer a new approach for intracellular immunization. Finally, a recent study introducing genes encoding for proteins directly into the skin of mice with a hand-held biolistic system demonstrated that DNA coated gold microprojectiles can be propelled directly into cells in vivo, providing a new strategy for genetic immunization (39).

Summary and Conclusions - Despite significant advances in HIV vaccine development, first generation candidate HIV vaccines have failed to elicit consistent broad spectrum humoral, cellular and memory immune responses which might be necessary for induction of protective immunity. As a result, there has been a tremendous push towards new strategies for antigen presentation, adjuvant development and vaccine vehicles. From low tech killed vaccines to high tech biolistic genetic immunization, multiple strategies are being utilized aimed at prevention of HIV. In addition to accelerating the development of effective HIV vaccines, it is likely that this renewed emphasis on novel approaches to potentiate immune responses of experimental vaccines will lead to the development of effective vaccines against several other important infectious diseases.

References

1. J.R. LaMontagne and G.T. Curlin. In: Vaccine Research and Developments (eds. W.C. Koff and H.R. Six); pp. 197-222. Marcel Dekker. (1991).
2. P.L. Kendrick, Am J. Public Health, 32, 615 (1942).
3. B.R. Murphy and R.M. Chanock. In: Virology (B.N. Fields, ed.), pp. 469-493 Raven Press. (1990).
4. W.J. McAleer, E.B. Bunyak, R.Z. Maigetter, D.E. Wampler. W.J. Miller, and M.R. Hilleman, Nature 307, 178 (1984).
5. G. Ada, Molecular Immunology, 28, 225 (1991).
6. W.C. Koff and D.F. Hoth, Science, 241, 426 (1988).
7. A.M. Schultz and W.C. Koff, Seminars in Immunology, 2, 351 (1990).
8. World Health Organization, Bulletin # 74, May, (1991).
9. J. Palca, Science, 252, 372 (1991).

10. W.C. Koff and A.M. Schultz, AIDS, 4 [Suppl], S179 (1990).
11. F. Barre-Sinousi, J.C. Chermann, F. Rey, M.R. Nugeyre, S. Chamaret, J, Gruest, C, Daugnet, and C. Axler-Blin, Science, 220, 868 (1983).
12. M. Popovic, M.G. Sarngadharan, E Read and R.C. Gallo, Science, 224, 497 (1984).
13. C. Cheng-Meyer, J. Homsy, L.A. Evans, and J.A. Levy, Proc Natl Acad Sci., USA 85, 2815 (1988).
14. G. Myers, HIV variation. International Conference on Advances in Aids Vaccine Development. October, 1991, Marco Island, Florida.
15. W. Koff and A. Fauci, AIDS, 3 [Suppl], S125 (1989).
16. D.C. Montefiori, J. Zhou, B. Barnes, D. Lake, E. Hersh, Y. Masuho. and L.B. Lefkowitz, Virology, 182, 635 (1991).
17. S.H. Vermund, A.R. Sheon, S.C. Ebner, R.D. Fischer, and M.A. Galbraith, AIDS Research Reviews Volume 1, 81; Marcel Dekker Inc., NY, (1991).
18. J.E. Heckels, J.E., M. Virji, and C.R. Tinsley, Vaccine, 8, 225 (1990).
19. J.R. McGhee and J. Mestecky, AIDS Research Reviews, Volume 2, 1992 (in press).
20. E.R. Unanue and P.M. Allen, Science, 236, 551 (1987).
21. K. Deres, H. Schild, K-H Weismuller, G. Jung and H-G Ramansee, Nature, 342, 561 (1989).
22. H. Takahashi, T. Takeshita, B. Morein, S. Putney, R.N. Germain, and J. Berzovsky, Nature, 344, 873 (1990).
23. L.S. Klavinskis, J.L. Whitton, and M.B.A. Oldstone. J. Virology, 63, 4311 (1989).
24. L. Shen, Z.W. Chen, M.D. Miller, V. Stallard, G.P. Mazzara, D.L. Panicali, and N.L. Letvin, Science, 252, 440 (1991).
25. S.L. Hu, K. Abrams, G.N. Barber, P. Moran, J. Zarling, A.J. Langlois, L. Kuller, W.R. MOrton, and R.E. Benveniste, Science, 255, 456 (1992).
26. J.P. Tam and Y.A. Lu, Proc Natl Acad Sci., USA, 86, 9084 (1989).
27. C.Y. Wang, D.J. Looney, M.L. Li, A.M. Walfield, J. Ye, B. Hosein, J.P. Tam, and F. Wong-Staal, Science, 254, 285 (1991).
28. D. Gheysen, E. Jacobs, F. de Foresta, C. Thiriart, M. Francotte, D. Thinnes, and M. de Wilde, Cell, 59, 103 (1989).
29. G.M. Mazzara. International Conference on Advances in AIDS Vaccine Development, Poster #27, Marco Island, Florida, October, 1991.
30. A. Aldovini and R.A. Young. In: Vaccine Research and Developments. (W.C. Koff and H.R. Six eds) Marcel Dekker, Inc. NY, pp.43-50. (1991).
31. G. Ramon, Annals de L'Institut Pasteur, 40, 1 (1926).
32. A.C. Allison and N.E. Byars, Molecular Immunology, 28(3), 279 (1991).
33. J.Y. Wu, B.H. Gardner, C.I. Murphy, J.R. Seals, C.R. Kensil, J. Recchia, G.A. Beltz, G.W. Newman and M.J. Newman, J. Immunology (in press) (1992).
34. G. Ott, G. Van Nest, and R.L. Burke. In Vaccine Research and Developments. (W.C. Koff and H.R. Six eds.) Marcel Dekker Inc., NY, pp.89-114. (1991).
35. G. Gregoriadis, Immunology Today, 11, 89 (1990).
36. S. Chattfield, R. Strugnell, and G. Dougan, Vaccine 7, 495 (1989).
37. J.H. Eldridge, J.K. Stass, J.A. Meulbrook, J.R. McGhee, T.R. Tice, and R.M. Gilley, Molecular Immunology, 28, 287 (1991).
38. R. Dolin, B.S. Graham, S.B. Greenberg, C.O. Tacket, R.B. Belshe, K. Midthun, M.L. Clements, G.J. Gorse, B.W. Horgan, R.L. Atmar, D.T. Karzon, W. Bonnez, B.F. Fernie, D.C. Montefiori, D.M. Stablein, G.E. Smith, W.C. Koff, and the NIAID AIDS Vaccine Clinical Trials Network, Ann Int Med 114, 119 (1991).
39. D. Tang, M. DeVit, and S.A. Johnson, Nature 356, 152 (1992).

Chapter 28. Inositol Trisphosphate Receptors

Suresh K. Joseph
Department of Pathology & Cell Biology
Thomas Jefferson University, Philadelphia PA 19107

Introduction - An elevation of the free calcium concentration in the cytoplasmic compartment is an integral component of the mechanism by which cells respond to certain stimuli including hormones, neurotransmitters and growth factors. Stimulation of Ca^{2+} entry across the plasma membrane and the discharge of internal Ca^{2+} stores are the key changes involved in raising the cytoplasmic free Ca^{2+} concentration. Considerable advances have been made in our understanding of the basic biochemical machinery involved in this type of signal transduction. The interactions between cell surface receptors, GTP-binding proteins and phospholipase-C results in the hydrolysis of phosphatidylinositol 4,5-bisphosphate and the release of 1,2-diacylglycerol and *myo*-inositol 1,4,5-trisphosphate (IP_3). The latter, water-soluble compound is now recognized to be the intracellular messenger responsible for the discharge of Ca^{2+} from intracellular stores. This molecule interacts with a specific intracellular membrane receptor (IP_3R) that has been purified and shown to be a ligand-gated Ca^{2+} channel. It has become apparent from cloning studies that several different forms of this protein may exist in neural and peripheral tissues. The basic aim of this chapter is to summarize recent studies on the structure, function and regulation of the IP_3R protein.

STRUCTURE

Purification Studies - IP_3 binding activity is present in high amounts in the Purkinje neuron of the cerebellum (1). This observation, and the finding that heparin potently inhibits ligand binding enabled Supattapone et al. (2) to purify the IP_3R protein to homogeneity.The receptor protein has also been purified from mouse brain (3), rat vas deferens (4) and bovine aortic smooth muscle (5). Depending on the exact gel electrophoresis conditions, the molecular weights of the brain receptor are reported to be in the range of 240-260 kDa (SDS-PAGE) or 320 kDa (agarose-PAGE). A value of 260 kDa has been estimated from target size analysis of IP_3 binding sites in brain and liver membranes (6). The native protein appears to be a tetramer of these subunits with a molecular weight >10^6 Daltons as verified by size exclusion chromatography (2,4), sucrose gradient centrifugation (7) and cross-linking studies (8). Under the electron microscope, the aortic smooth muscle protein has four-fold symmetry and has been described as resembling a 'pin-wheel with four arms radiating from a central hub' (5).

Cloning Studies - The cDNA coding for the IP_3 receptor has been cloned from mouse brain (9), rat brain (7) and Drosophila (10) with partial sequence information being available for human brain (11), human kidney (12) and mouse aortic smooth muscle (13). Recently, a second type of IP_3R has been cloned from a rat brain cDNA library (12). This type II-IP_3R has only 69% overall homology with type-I. Its identification implies that different IP_3R proteins may be present in different cells and be subject to differential regulation. However, all IP_3 receptors cloned so far share a number of common features (Figure 1) which are summarized and discussed below.

Domain Structure - The sequences of mouse and rat type I-IP_3R proteins are highly conserved and consist of 2,749 amino-acids. The cDNA encoding the protein has been transfected into a number of cell types and shown to express a protein, recognized by IP_3R antibodies, which binds IP_3 with appropriate affinity and pharmacological specificity and functions to release Ca^{2+} from intracellular stores (7,9,10,14,15). Despite initial controversies, there is now general agreement that the N- and C-termini of the protein face the cytoplasm (16). This requires an even number of transmembrane segments and hydropathy analysis indicates the presence of eight putative transmembrane helices clustered near the C-terminus of the protein (7). Cells transfected with a cDNA construct missing the 8 transmembrane helices express a soluble protein that retains ligand binding. However, analysis of this soluble protein indicates that it is monomeric suggesting that each monomer has an independent IP_3 binding site and that formation of the tetramer requires association of the transmembrane domains (10,14). Mutational analysis has also allowed the ligand-binding site to be localized to the N-terminal 650 amino-acids. All cDNA constructs with deletions in this region appear to encode proteins incapable of binding IP_3 (10,14). Evidence for the involvement of basic amino-acids in ligand binding comes from the finding that an arginine-reactive modifying agent blocks IP_3 binding to platelet membranes (17). The only other IP_3 binding protein for which sequence information is available is IP_3-3'-kinase which phosphorylates IP_3 to form IP_4. A short stretch of amino-acids in the IP_3-kinase sequence containing several basic residues is homologous to a region present in the ligand-binding domain of the IP_3R (18).

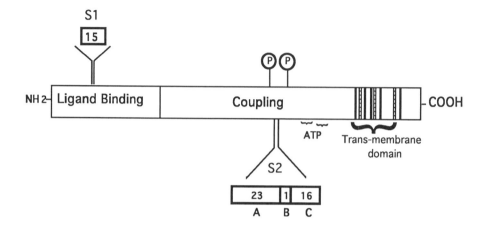

Figure 1. Domain Organization and Alternative Transcipts of the Type1-IP_3R. The schematic representation of the IP_3R depicts the proposed regions involved in ligand and nucleotide binding as well as the sites phosphorylated by A-kinase. Also shown are the location of amino-acids deleted in two alternatively spliced forms of IP_3R mRNA referred to as S1 and S2. The 40 amino-acid S2 deletion contains three sub-regions (A,B,C) that can be separately deleted.

The function of the protein requires that recognition of IP_3 transmits information within the protein to gate a Ca^{2+} channel formed from the transmembrane domains. It has been suggested that this involves a large conformational transition that has been experimentally detected as a shift in elution profile from a sizing column when IP_3 is added to a truncated, soluble form of the IP_3R expressed in COS cells (14). The N-terminal ligand binding domain is separated from the C-terminal transmembrane domain by a stretch of approximately 1600 amino acids. This region has been referred to as a 'coupling' domain (14) and contains two consensus ATP-binding sites and two serine residues that are specifically phosphorylated by cAMP-dependent protein kinase (19).

There are no extensive homologies between the IP_3R and other known ion-channels. However, the putative 6th, 7th and 8th transmembrane helices and a portion of the adjacent C-terminal tail do have homologies with the ryanodine receptor (9,10,20). This protein mediates depolarization-dependent Ca^{2+} release from skeletal muscle sarcoplasmic reticulum and is thought to be responsible for Ca^{2+}-dependent Ca^{2+} release in many experimental systems (21). The ryanodine receptor is larger than the IP_3 receptor with a subunit M_r=560 kDa (22,23). The ryanodine receptor also has a tetrameric structure with four-fold symmetry and the same orientation of C- and N-termini as the IP_3R. Some evidence that the C-terminal tail of the IP_3R is involved in Ca^{2+} channel function comes from the finding that a monoclonal antibody recognizing this region can inhibit Ca^{2+} release at sub-optimal concentrations of IP_3 (24). The homologous sequences in the ryanodine and IP_3 receptors may represent common elements present in the conductance pore of both channel proteins. Such speculations will have to await resolution of disagreements regarding the exact number of transmembrane loops in both proteins, which in the case of the ryanodine receptor is variously reported to be between 4 (22) and 12 (23).

Alternative Spliced Transcripts and Type II IP_3R - Several alternatively spliced transcripts of the Type I- IP_3R mRNA have been detected. One of these corresponds to a 15 amino-acid deletion (labelled as S1 in Fig. 1) in the ligand binding domain and contains a stretch of negatively charged amino-acids (7,25). A substantial amount (85%) of the mRNA encoding the Type I- IP_3R is estimated to be missing the S1 region in mouse cerebellum (25). Mutational analysis of the ligand binding domain would suggest that a protein lacking the S1 region would have a markedly decreased affinity for IP_3 binding (10). A second alternatively spliced transcript contains a deletion in the coupling domain corresponding to 40 amino acids, referred to as the S2 region, and located between the serine A-kinase phosphorylation sites (26). A detailed analysis of the mRNA encoding the Type I-IP_3R using the polymerase chain reaction has revealed that the S2 region is itself composed of three sub-regions that can be separately deleted (Figure 1). Region B corresponds to only one amino-acid (glutamine). Message encoding IP_3R forms S2 , S2ABC$^-$, S2BC$^-$ and S2B$^-$ have all been detected in the mouse brain with the proportion of these forms varying in different brain regions and during development (25,27). Interestingly, only the S2ABC$^-$ form is present in peripheral tissues (25,26). The functional difference between brain-specific alternatively spliced transcripts and the 'short' form (S2ABC$^-$) found in peripheral tissues has not been fully evaluated. However, purified IP_3R protein representative of each of these categories, from cerebellum and vas deferens respectively, show marked differences in their kinetics of phosphorylation by A-kinase. The vas deferens protein is phosphorylated rapidly on only one serine residue while the cerebellar protein is phosphorylated more slowly on both serine residues (26). Different combinations of the monomeric forms of these transcripts, at least in brain, offers the potential for much diversity in IP_3R function.

The Type II-IP_3R recently cloned from a rat brain cDNA library encodes a protein that is 48 amino-acids smaller than Type I (12). This form is lacking the S2 region and possesses only a single consensus A-kinase site. Both ATP binding sites are conserved in Type II-IP_3R. It has been pointed out that the sequence around the excised S2 region in Type I-IP_3R generates a third consensus ATP binding site (28). This site is absent in the Type II receptor. Overall, the least degree of homology with Type-I is in the coupling domain and in the predicted cytosolic and luminal loops connecting the transmembrane helices. Transfection with cDNA encoding the ligand and coupling domains of Type II receptor yields a protein that has a higher affinity for IP_3 than similar constructs prepared from type I (12). The mRNA for type II-IP_3R is of relatively low abundance in brain and its distribution in other tissues or its functional significance are presently unknown. Nevertheless, these cloning studies suggest that we may be only at the beginning of an expanding list of IP_3 receptors encoded by a multi-gene family. In this regard, olfactory cilia (29) and pancreatic microsomes (30) appear to contain IP_3 binding proteins specifically labelled by photoreactive IP_3 analogs that are substantially smaller (107 kDa and 37 kDa, respectively) than either Type I or Type II.

FUNCTION

Reconstitution and Electrophysiology - Ferris et al. (31) have demonstrated IP_3-stimulated $^{45}Ca^{2+}$ fluxes in liposomes reconstituted with purified cerebellar IP_3R. These studies demonstrated that the receptor is itself a Ca^{2+} channel. Direct measurements of channel activity have also been made after incorporation of purified IP_3R or IP_3-sensitive microsomal vesicles into planar lipid bilayers (8,32-34). Vesicle studies are complicated by the presence of other Ca^{2+} conductances such as those mediated by the ryanodine receptor that may also be modified by IP_3 (34). Bona fide IP_3-gated channels can be distinguished pharmacologically by their insensitivity to ryanodine, caffeine or ruthenium red. Such channels have a relatively low single channel conductance (20pS) and are voltage-independent. The transition of these channels between open and closed states is very rapid with mean open times of less than 10 msec. Maximal channel openings require the addition of ATP (or a non-hydrolyzable analog) and low concentrations of Ca^{2+} to the cis ('cytosolic') side of the bilayer. Neither of these agents opens channels by themselves. The probability of channel opening shows a bell-shaped dependence on cis Ca^{2+} concentration with peak activity at 0.3 µM (33). The steep activation and inhibition on either side of this concentration has been modeled assuming co-operative interactions between multiple stimulatory and inhibitory binding sites on an IP_3R monomer or cooperativity between monomers in the channel complex (33). A biphasic dependence on Ca^{2+} concentration has also been observed when measuring IP_3-mediated Ca^{2+} release from skinned smooth muscle fibers (35) or brain microsomal vesicles (36). This regulatory behavior may prove to be important in the generation of oscillatory Ca^{2+} transients observed in response to cell stimulation.

From a detailed analysis of the current records obtained after incorporation of cerebellum microsomal vesicles into a planar lipid bilayer, it has been concluded that channel openings involve up to 4 subconductance states with the unitary conductance step being 20pS (34). The first three levels were observed more frequently than the fourth and the openings appear to be concerted rather than sequential. On the basis of this study it was suggested that each IP_3R monomer contributes a conducting pore to the channel with one molecule of IP_3 being capable of opening one, two or three channels (34). A second molecule of IP_3 would be required to fully open the channel. Others have observed a markedly cooperative IP_3 concentration dependence for Ca^{2+} release from permeabilized basophilic leukemic cells (Hill co-efficient of 3) and suggested that all four independent binding sites must be occupied for the channel to open (37). Distinguishing between these models is a key issue remaining to be resolved.

Quantal Calcium Release - The term 'quantal' Ca^{2+} release was first used by Muallem et al. (38) to refer to the observation that sub-optimal concentrations of IP_3 failed to release all the Ca^{2+} from the IP_3-sensitive store under conditions designed to prevent re-sequestration of Ca^{2+} or metabolism of added IP_3. Subsequently, the same phenomenon has been shown in other systems using IP_3 or non-metabolizable phosphorothioate IP_3-analogues (39,40). Stopped flow analysis of the kinetics of release indicates that the process is biphasic with partial emptying of the store occurring very rapidly followed by a 30-fold slower release process (39,41). Partial emptying of the store does not reflect desensitization of the receptor because the system remains responsive to a second addition of IP_3. The ability to rapidly mobilize Ca^{2+} in a self-limiting manner in response to very small changes in IP_3 concentration makes quantal Ca^{2+} release potentially physiologically relevant. This type of response has also been called 'increment detection' by Meyer and Stryer (39). While several explanations have been proposed (42,43), the actual mechanism underlying quantal Ca^{2+} release is not known.

One hypotheses to explain this behavior proposes that quantal Ca^{2+} release reflects heterogeneity of the internal Ca^{2+} stores. Even sub-optimal concentrations of IP_3 are proposed to completely empty in an 'all-or-none' fashion those stores that are most sensitive to IP_3 (38-40). An alternative view of quantal Ca^{2+} release is that it reflects partial emptying of all of the stores (42). 'All-or-none' Ca^{2+} release has been experimentally demonstrated in permeabilized hepatocytes (44) and in response to localized photolysis of caged IP_3 in oocytes (45). A steep threshold response to IP_3 could arise if IP_3R monomer occupation was sufficient to almost fully activate the Ca^{2+} conductance as suggested by planar lipid bilayer studies (34). The 'all-or-none' hypotheses to explain quantal Ca^{2+} release visualizes the internal stores as being non-contiguous entities with a wide range of sensitivities to IP_3, determined by pre-setting the stores to respond to different threshold concentrations. How this would be accomplished is not known. One possibility is that Ca^{2+} within the intracellular store may itself be an important determinant in regulating the sensitivity of the IP_3R for IP_3 such that a fall of Ca^{2+} in the lumen would decrease the sensitivity for IP_3 and thereby limit Ca^{2+} release (42). Different levels of intraluminal Ca^{2+} and/or different receptor densities could produce stores with a range of IP_3 sensitivities. Some pieces of experimental data provide support for a regulatory role for intraluminal Ca^{2+}. In permeabilized hepatocytes, the initial rate of $^{45}Ca^{2+}$ uptake into empty stores is unaffected by a non-hydrolyzable IP_3 analogue until intraluminal Ca^{2+} reaches a critical level (46). A spontaneous release of Ca^{2+} seen when concentrated suspensions of permeabilized hepatocytes are allowed to accumulate large amounts of Ca^{2+} has been interpreted as an effect of intraluminal Ca^{2+} sensitizing the IP_3R to endogenous levels of IP_3 (47). Reduction of Ca^{2+} in the intracellular stores by 50% increased the EC_{50} for IP_3-mediated Ca^{2+} release from permeabilized hepatocytes (46). However, the same manipulation carried out in salt-gland cells did not change IP_3R sensitivity (48). A further insight into the mechanism of quantal Ca^{2+} release comes from the recent demonstration that the phenomenon can be observed in liposomes reconstituted with purified IP_3R (49). The specific experimental conditions used in this study suggest that quantal Ca^{2+} release is an intrinsic property of the IP_3R protein and may be independent of any regulation by intraluminal Ca^{2+} or heterogeneity of Ca^{2+} stores.

Table 1. Factors affecting ligand binding and/or Ca^{2+} release by the IP$_3$R

		IP$_3$ BINDING	Ca2+ RELEASE	REFS^b
external calcium	(brain)	↓	biphasic	1,56-58
	(liver)	↑	↓	59-61
cAMP mediated	(brain)	no effect	↓	62,63
phosphorylation	(liver)	no effect	↑	64,65
ATP^c		↓	biphasic	66-68
GTP		↓	no effect	68,69
Increasing pH		↑	↑	1,57,69-72
Magnesium		↓	↓	73-75
Haparin and Other Sulfated Polysaccharides		↓	↓	1,73,75,76-79
Calmodulin Antagonists		no effect	↓	8,80
Thiol agents^d		↓	↓	46,70,81-85
Cibacron blue		↓	not measured	86
ortho-Decavanadate		↓	↓	83,87,88
Potassium Channel Blockers^e		no effect	↓	89,90
Calcium Channel Blockers^f		not measured	↓	91,92

a) measured at sub-saturating concentrations of IP$_3$
b) additional references and a compendium of equilibrium binding constants can be found in Refs 55.
c) inhibition of ligand binding requires millimolar (i.e. physiological) concentrations of ATP. Direct binding of nucleotide to the purified receptor has been demonstrated (8,68).
d) ligand binding and Ca^{2+} release are inhibited by N-ethylmaleimide, p-chloromercurobenzoate and menadione. Inhibition is minimal at 0°C or if IP$_3$ is present. This suggests that free -SH groups may be masked at low temperatures and play a role in ligand binding. In hepatocytes, thiomerosal and oxidized glutathione promote Ca^{2+} mobilization mediated by sub-optimal doses of IP$_3$ (47,85)
e) Tetraethylammonium and derivatives
f) Cinnarizine and flunarizine

<u>REGULATION</u>

In many cells the application of agonists elicits Ca^{2+} transients that consist of a series of repetitive spikes (50-52). Microinjection of IP$_3$ can itself produce Ca^{2+} oscillations (53) and all models are in agreement that discharge of Ca^{2+} from the IP$_3$- sensitive store is fundamental to initiating and/or sustaining oscillatory Ca^{2+} transients. Calcium imaging of stimulated cells has also revealed that Ca^{2+} can increase in a local region of the cell and a wave of elevated Ca^{2+} can propagate across the cell (52,54). The regulation of the IP$_3$R, particularly by Ca^{2+} itself, is likely to be important in determining the precise spatial and temporal pattern of the Ca^{2+} transient seen upon agonist stimulation. A number of physiological and pharmacological effectors of the IP$_3$R are listed in Table I. Discussion is confined to regulation of the IP$_3$R by Ca^{2+} and phosphorylation and to a description of recent studies providing information on the subcellular localization of the protein.

<u>Calcium</u> - The purified IP$_3$R from cerebellum is insensitive to added Ca^{2+} but the receptor in native membranes or detergent extracts is markedly inhibited by Ca^{2+} with an EC$_{50}$ of 0.3 μM (1,2). It has been shown that the flow-through fraction on heparin agarose columns contains a factor that confers Ca^{2+} sensitivity to the purified receptor protein. This factor has been termed 'calmedin' and is attributed to an integral membrane protein which on gel-filtration columns has a native molecular

weight of 300 kDa (58). Calmedin does not appear to be present in peripheral tissues but brain calmedin will confer Ca^{2+} sensitivity to the purified vas deferens IP_3 receptor (4). The inhibitory effect of Ca^{2+} mediated by calmedin in brain is associated with a decrease in affinity of the IP_3R for its ligand (1,57). In liver, sub-micromolar concentrations of Ca^{2+} stimulate IP_3 binding and this is associated with an increased affinity of the IP_3R for its ligand (59-61). Clearly, the regulation by Ca^{2+} of ligand binding to the cerebellum and liver IP_3R are different. It has been proposed that the hepatic IP_3R exists as an equilibrium between low and high affinity forms with Ca^{2+} favoring the formation of the latter species (59). Experimental evidence suggests that the low-affinity species is active in mediating Ca^{2+} release while the high-affinity form may represent a desensitized receptor (59,60). This model may not apply to the cerebellar IP_3R because multiple affinity states of the brain IP_3R have not been detected using conventional binding assays (55). However, high and low affinity sites have been found in detergent-solubilized cerebellar extracts using a rapid binding assay (93). In the brain low concentrations of Ca^{2+} (<0.3 μM) enhance IP_3-mediated Ca^{2+} release (36) but higher concentrations are inhibitory in both brain and liver (56,57,59,60). Presumably, these tissue differences reflect the presence of different forms of IP_3R and different Ca^{2+}-binding regulatory proteins. A model to account for these differences incorporating a role for an intraluminal Ca^{2+} binding site has been proposed (94).

Ca^{2+} inhibition of IP_3-mediated Ca^{2+} release is not observed in all peripheral tissues (55). The absence of such inhibition may be due to experimental conditions (95). Exposure of cells to low levels of Ca^{2+} for even short periods during permeabilization (i.e. 1min) can remove a loosely-bound factor from intracellular stores that confers Ca^{2+} sensitivity. The nature of this factor is not known. There is some evidence that calmodulin may be involved as the purified IP_3R does adhere to calmodulin columns (8) and IP_3-mediated Ca^{2+} release can be inhibited by calmodulin antagonists (80). No consensus Ca^{2+} binding regions have been recognized in published sequences of the IP_3R although direct binding of Ca^{2+} to the purified receptor with a Kd of 50 μM has been reported (58). In summary, the effects of Ca^{2+} on the IP_3R are complex with biphasic effects on Ca^{2+} release reported in neuronal systems and inhibitory effects on Ca^{2+} release being observed in many but not all systems. The regulation of the IP_3R by Ca^{2+} has been proposed to be mediated by the integral membrane protein calmedin in brain, a loosely bound unidentified protein in peripheral tissues, and possibly by direct interaction of Ca^{2+} and calmodulin with the receptor.

Phosphorylation - Elevation of cAMP can inhibit or stimulate the action of Ca^{2+} mobilizing stimuli depending on the cell type. In neuronal tissues where the action of cAMP is generally inhibitory (96), the phosphorylation of the IP_3R in vitro has been demonstrated to decrease the sensitivity of the Ca^{2+} release system for IP_3 (62,63). In liver, where the action of cAMP is stimulatory (97), phosphorylation increases the sensitivity for IP_3 (65). Phosphorylation does not significantly affect ligand binding in either tissue, although the Ca^{2+} stimulation of ligand binding of the hepatic IP_3R may be suppressed by phosphorylation (64). It is clear that phosphorylation of the cerebellum and liver IP_3 receptors have different effects on the coupling of the ligand domain to the Ca^{2+} channel. As with Ca^{2+}, these differences may reflect the presence of different alternatively spliced variants in the two tissues. Protein kinase-C and Ca^{2+}/calmodulin-dependent protein kinase II have been found to phosphorylate the IP_3R only when reconstituted in liposomes (98). The functional consequences are not known but the sites of phosphorylation appear to be different form those used by cAMP-dependent protein kinase. Recently, it has been reported that highly purified preparations of IP_3R reconstituted into liposomes can auto-phosphorylate on serine residues and catalyze the phosphorylation of exogenously added synthetic peptides (99). This raises the novel possibility that the IP_3R protein possesses an intrinsic serine-kinase activity. This activity was not influenced by inositol phosphates and its regulatory role remains to be determined.

Where is the IP_3R in the Cell ? - Early studies established that only a proportion of the total intracellular stores were sensitive to IP_3 in most cells and it was generally believed that IP_3-sensitive stores corresponded to a specialized portion of the endoplasmic reticulum (55,100-102). However, other studies suggested that specialized organelles, referred to as 'calciosomes', may function as IP_3-sensitive stores (103). These structures were particularly evident in cell types, such as neutrophils, that are almost devoid of endoplasmic reticulum. Calciosomes were characterized as possessing IP_3 receptors, an ATP-dependent Ca^{2+} pump, and a Ca^{2+}-binding protein. In most non-muscle cells the Ca^{2+}-binding protein in calciosomes has been identified as 'calreticulin' (104,105), a protein related to 'calsequestrin', the major Ca^{2+} binding protein found in sarcoplasmic reticulum. The most recent advances in identifying the subcellular localization of the IP_3R have come from studies on the Purkinje neuron (106-111). In these neurons the IP_3R has been localized to the dendrites, cell body and to a lesser extent in the axon. High-resolution immunogold labelling studies have revealed that quantitatively the most intensely labelled structures are smooth-surfaced cisternae arranged in stacks in groups of up to a dozen. In some cases these structures appear contiguous with the rough endoplasmic reticulum (106,107). Smooth surfaced cisternae are also heavily labelled in the dendritic spine where synapses that activate inositol-lipid turnover are located. It has been difficult to study the relationship of the cisternal stacks to calciosomes because purkinje neurons of the rat do not contain immunoreactive calreticulin or calsequestrin. However, a protein cross-reacting with calsequestrin antibodies is found in calciosome-like structures in the avian purkinje neuron (109). Avian cerebellar membrane fractions analyzed on sucrose gradients show that calsequestrin and IP_3R immunoreactive proteins do not exactly overlap and under the electron microscope it is apparent that a significant proportion of calciosomes do not possess IP_3 receptors (109) and are particularly absent from dendritic spines. Dual labelling studies with antibodies to the ryanodine and IP_3 receptors indicate that both are present in the purkinje neuron and, in some cases may be co-localized to the same membranes (109,110). This has led to the suggestion that a proportion of the purkinje neuron calciosomes, lacking IP_3R but containing ryanodine receptors, may function to release Ca^{2+} in response to depolarizing stimuli that trigger Ca^{2+} influx across the plasma membrane (109,110). In the purkinje neuron the IP_3R is not found in the plasma membrane or mitochondria while some studies have noted weak labelling of the outer membrane of the nuclear envelope (107,110) and lateral tips of the golgi cisternae (108).

Similar high-resolution studies on peripheral tissues have not yet been undertaken, presumably because of the much lower expression of the IP_3R protein. However, the distribution of ligand binding and IP_3-mediated Ca^{2+} release has been monitored in many fractionation studies and the results appear to be tissue dependent (55,100-102). In the liver, a major portion of the IP_3R is associated with a plasma membrane fraction (112,113). The IP_3R does not appear to be integral to the membrane because pretreatment of liver homogenates with the microfilament disrupting drug cytochalasin-B alters the distribution of the protein (113). In some tissues a pool of IP_3R resists extraction by Triton X-100 (114). These data provide circumstantial evidence that the IP_3R can attach to cytoskeletal proteins. Such attachment may serve a structural role in helping to localize the receptor to specific regions of the cell. However, it may also play a functional role in that attachment to the sub-plasmalemmal cytoskeletal matrix could provide a means to relay information regarding the Ca^{2+} loading state of the intracellular store to Ca^{2+} transport systems in the plasma membrane. Such mechanisms are required to explain the apparent dependence of plasma membrane Ca^{2+} entry on the degree of filling of the intracellular store seen in many experimental systems (115). In T-lymphocytes, electrophysiological evidence indicates the presence of IP_3 gated Ca^{2+} channels in the plasma membrane (116) and recent studies confirm the presence of immunoreactive IP_3R in lymphocyte plasma membrane preparations (117). Lectin affinity studies on thymus IP_3R indicate the presence of sugars that are uncharacteristic of ER resident proteins (117). The IP_3R in olfactory cilia is also thought to be integral to the plasma membrane (29,117). In liver, IP_3 may bind to and release Ca^{2+} from purified nuclei (118,119). While these studies do not make clear if Ca^{2+} is being mobilized from within the nucleus or between the nuclear membranes, they do suggest that the IP_3R may be involved in mechanisms that regulate intranuclear calcium.

Summary From the studies reviewed above it is apparent that multiple IP_3R proteins may exist in cells. This diversity arises from the presence of more than one gene product as well as the presence of alternatively spliced forms of IP_3R mRNA. In addition, the receptor appears to be targeted to different membranes in different cells. The functional consequence of this diversity is not yet clear, although it is apparent that neuronal and peripheral IP_3 receptors are differentially regulated by calcium and cAMP-mediated phosphorylation. There are clues in the literature which suggest that expression of the protein may also be regulated. For example, retinoic acid induced differentiation of HL-60 cells produces a large increase in IP_3R protein (120) while chronic stimulation of muscarinic receptors decreases IP_3 binding sites in neuroblastoma cells (121). Much remains to be learnt regarding the biosynthesis, processing, targeting and turnover of this protein. Future studies will hopefully help to delineate the functional differences between IP_3R isoforms and to unravel their role in the complex spatial and temporal patterns seen in Ca^{2+} transients evoked by growth factors, neurotransmitters and hormones.

REFERENCES

1. P. F. Worley, J. M. Baraban, S. Supattapone, V. S. Wilson, S. H. Snyder, J. Biol. Chem. 262, 12132 (1987)
2. S. Supattapone, P. F. Worley, J. M. Baraban, and S. H. Snyder, J. Biol. Chem. 263, 1530 (1988).
3. N. Maeda, M. Niinobe, and K. Mikoshiba, EMBO J. 9, 6l (1990).
4. R. J. Mourey, A. Verma, S. Supattapone, and S. H. Snyder, Biochem. J. 272, 383 (1990).
5. C. C. Chadwick, A. Saito, and S. Fleischer, Proc. Natl. Acad. Sci. USA 87, 2132 (1990).
6. D. L. Nunn, B. V. L. Potter, and C. W. Taylor, Biochem. J. 265, 393 (1990).
7. G. A. Mignery, C. L. Newton, B. T. Archer III and T. C. Sudhof, J. Biol. Chem. 265, 12679 (1990).
8. N. Maeda, T. Kawasaki, S. Narade, N. Yokota, T. Takahisa, M. Kajai, and K. Mikoshiba, J. Biol. Chem. 266, 1109 (1991).
9. T. Furuichi, S. Yoshikawa, A. Miyawqaki, K. Wada, N. Maeda, and K. Mikoshiba, Nature 342, 32 (1989).
10. A. Miyawaki, T. Furuichi, Y. Ryon, S. Yoshikawa, Y. Nakagawa, T. Saito, and K. Mikoshiba, Proc. Natl. Acad.Sci.USA 88, 4911 (199l).
11. C. A. Ross, S. K. Danoff, C. D. Ferris, C. Donath, G. A. Fischer, S. Munemitsu, S. H. Snyder, and A.-Ullrich,Soc.Neurosci.Abst. 17, 18 (1991)
12. T. C. Sudhof, C. L. Newton , B. T. Archer, Y. A. Ushkaryov, and G. A. Mignery, EMBO J. 10, 3199 (1991).
13. A. R. Marks, P. Tempst, C. C. Chadwick, L. Riviere, S. Fleischer, and B. Nadal-Girard, J. Biol. Chem. 265, 20719 (1990).
14. G. A Mignery and T. C. Sudhof, EMBO J. 9, 3893 (1990).
15. A. Miyawaki, T. Furuichi, N. Maeda, and K. Mikoshiba, Neuron 5, 11 (1990).
16. P. D. Camilli, K. Takei, G. A. Mignery, and T. C. Sudhof, Nature 344, 495 (1990).
17. O'Rourke, F. and M. B. Feinstein, Biochem. J. 267, 297 (1990).
18. K. Y. Choi, H.-K. Kim, S. Y. Lee, K. H. Moon, S. S. Sim, K. W. Kim, H. K. Chung, and S. G. Rhee, Science 248, 64 (1990).
19. C. D. Ferris, A. M. Cameron, D. S. Bredt, R. L. Huganir and S. H. Snyder, Biochem. Biophys. Res. Commun. 175, 192 (1991).
20. G. A. Mignery, T. C. Sudhof, K. Takei, and P. DeCamilli, Nature 342 192 (1989).
21. W. A. Catterall, Cell 64, 871 (199l).
22. H. Takeshima, S. Nishimura, T. Matsumoto, H. Ishida, K. Kanagawa, N. Minamino, H. Matsu, M. Ueda, M. Hanaoka, T. Hirose, and S. Numa, Nature 339, 313 (1989).
23. F. Zorzato, J. Fujii, K. Otsu, M. Phillips, N. M. Green, F. A. Lai, G. Meissner, and D. H. Maclennan, J. Biol.Chem. 265, 2244 (1990).
24. S. Nakade, N. Maeda, and K. Mikoshiba, Biochem. J. 277, 125 (199l).
25. T. Nakagawa, H. Okano, T. Furuichi, J. Aruga, and K. Mikoshiba, Proc. Natl. Acad. Sci., USA 88, 6244 (199l).
26. S. K. Danoff, C. D. Ferris, C. Donath, G. A. Fischer, S. Munemitsu, A. Ullrich, S. H. Snyder, and C. A. Ross, Proc. Natl. Acad. Sci., USA 88, 2951.
27. T. Nakagawa, C. Shiota, C., H. Okano, and K. Mikoshiba, J. Neurochem. 57, 1807 (199l).
28. C.D. Ferris and S.H. Snyder, Ann. Rev. Physiol. 54, 469 (1992).
29. D. L. Kalinoski, S. B. Aldinger, A. G. Boyle, T. Huque, J. F. Maracer, G. O. Prestwich, and D. Restrepo, Biochem.J. 281 449 (1992).
30. R. Schafer, M. Nehls-Sahabandu, B. Grabowsky, M. Dehlinger-Kremer, I Schultz, and G. W. Mayr, Biochem.J. 272, 817 (1990).
31. C. D. Ferris, R. L. Huganir, S. Supattapone, and S. H. Snyder, Nature 342, 87 (1990).
32. B. E. Ehrlich and J. Watras, Nature 336, 583 (1988).
33. I. Bezprozvanny, J. Watras, and B. E. Ehrlich, Nature 357, 75l (199l).
34. J. Watras, I. Bezprozvanny, and B. E. Ehrlich, J. Neuroscience 11, 3239 (199l).
35. Iino, M., J. Gen. Physiol., 95, 1103 (1990).
36. E. A. Finch, T. J. Turner, and S. M. Goldin, Science 252, 443 (199l).
37. T. Meyer, T. Wensel, and L. Stryer, Biochemistry 29, 32 (1990).
38. S. Muallem, S. Pandol, and T. G. Beeker, J. Biol. Chem. 264, 205 (1989).
39. T. Meyer and L. Stryer, Proc. Natl. Acad. Sci. USA 87, 3841 (1990).
40. C. W. Taylor and B.V.L. Potter, Biochem. J. 266, 189 (1990).
41. P. Champeil, L. Combettes, B. Berthon, E. Doucet, S. Orlowski, M. Claret, J. Biol. Chem. 264, 17665 (1989).
42. R. F. Irvine, FEBS Lett. 263, 5 (1990).
43. S. Swillens, Mol. Pharmacol. 4l, 110 (1992).
44. K. A. Oldershaw, D. L. Nunn, and C. W. Taylor, Biochem.J. 278, 705 (199l).
45. I. Parker and I. Ivorra, Science 250, 977 (1990).
46. D. L. Nunn and C. N. Taylor, Mol. Pharmacol. 4l, 115 (1992).
47. L. Missaen, C. W. Taylor, and M. J. Berridge, Nature 352, 241 (199l).
48. T. J. Shuttleworth, J. Biol. Chem. 267, 3573 (1992).

49.   C. D. Ferris, A. M. Cameron, R. L. Huganir and S. H. Snyder, Nature 356, 350 (1992).
50.   M. J. Berridge, Cell Calcium 12, 63 (1991).
51.   T. Meyer, and L. Stryer, Ann. Rev. Biophys. Chem. 20, 153 (1991).
52.   T. A. Rooney and A. P. Thomas, Pharmac. Ther. 49, 223 (1991).
53.   M. Wakui, B. V. L. Potter, and O. H. Petersen, Nature 339, 317 (1989).
54.   J. Lechleiter, S. Girard, Clapham, D. and E. Peralta Nature 350, 505 (1991).
55.   C. W. Taylor and A. Anderson, Pharmac. Ther. 51 , 97 (1991)
56.   T. Jean and C. B. Klee, J. Biol. Chem. 261, 16414 (1986).
57.   S. K. Joseph, H. L. Rice, and J. R. Williamson, Biochem.J. 258, 261 (1989).
58.   S. K. Danoff, S. Supattopone, and S. H. Snyder, Biochem. J. 254, 701 (1988).
59.   F. Pietri, M. Hilly, and J.-P. Mauger, J. Biol. Chem. 265, 17478 (1990).
60.   F. Pietri, M. Hilly, M. Claret, and J.-P. Mauger, Cell Signalling 2, 253 (1990).
61.   D. C. Ogden, T. Capiod, J. W. Walker, and D. R. Trentham, J. Physiol. 422, 585 (1990).
62.   S. Supattapone, S. K. Danoff, A. Thiebert, S. K. Joseph, J. Steiner, and S. H. Snyder, Proc. Natl. Acad. Sci., USA 85, 8747 (1988).
63.   P. Volpe, and B. H. Alderson-Lang, Am. J. Physiol. 258, C1086 (1990).
64.   J.-P. Mauger, M. Claret, F. Pietri, and M. Hilly, J. Biol.Chem. 8821 (1989).
65.   G.M. Burgess, G.J. Bird, J. F. Obie, and J.W. Putney, J. Biol. Chem. 266, 4772 (1991).
66.   A. L. Wilcocks, A. M. Cooke, B. V. L. Potter, and S. R.Nahorski, Biochem. Biophys. Res. Commun. 146, 1071 (1987).
67.   D. L. Nunn and C. W. Taylor, Biochem. J. 270 227-232 (1990).
68.   C. D. Ferris, R. L. Huganir, and S. H. Snyder, Proc. Natl.Acad. Sci. USA 87, 2147 (1990).
69.   S. B. Hwang, Biochim. Biophys. Acta 1064, 351 (1991).
70.   G. Guillemette and J. A. Segui, Mol. Endocrinol. 2, 1249 (1988).
71.   A. M. White, M. A. Varney, S. P. Watson, S. Rigby, C. S. Liu, J. G. Ward, C. B. Reese, H. C. Graham, and R.J. A. Williams, Biochem. J. 278, 759 (1991).
72.   L. Schmitt, G. Schlewer, and B. Spiess, Biochim. Biophys.Acta 1075, 139 (1991).
73.   M. A. Varney, J. Riuera, A. L. Bernal, and S. P.Watson, Biochem. J. 269, 211 (1990).
74.   G. Guillemette, S. Lamontagne, G. Boulay, and B. Mouillac , Mol. Pharmacol 35, 339 (1989).
75.   P. Volpe, B. H. Alderson-Lang, and G. A. Nickols, Am. J. Physiol. 258 C1077 (1990).
76.   T. K. Ghosh, P. S. Eis, J. M Mullaney, C. L. Ebert, and D. L. Gill, J. Biol. Chem. 263, 11075 (1988).
77.   T. Nilsson, J. Zwiller, A. L. Boynton, and P. O. Berggren, FEBS Lett. 229, 211 (1988).
78.   M. A. Tongs, M. D. Bootman, B. F. Higgins, D. A. Lane, G. F. Pay, and J. Lindahl, FEBS Lett. 252, 105 (1989).
79.   R. A. Challiss, A. L. Willcocks, B. Mulloy, B. V. L. Potter, and S. R. Nahorski, Biochem. J. 274, 861 (1991).
80.   T. D. Hill, R. Campos-Gonzalez, H. Kindmark, and A. L. Boynton, J. Biol. Chem. 263, 16479 (1988).
81.   C. M. Yang and H. C. Lee, J. Recept. Res. 9, 159 (1989).
82.   F. B. Pruijn, J. P. Sibeijn, and A. Bast, Biochem. Pharmacol. 40, 1947 (1990).
83.   K. J. Fohr, J. Scott, G. Ahnerthilger, and M. Gratzl Biochem. J. 262, 83 (1989).
84.   S. Dimmeler, B. Brune, and V. Ullrich, Biochem. Pharmacol. 42, 1151 (1991).
85.   T. A. Rooney, D. C. Renard, E. J. Sass, and A. P. Thomas, J. Biol. Chem. 266, 12272 (1991).
86.   M. D. Bootman, G. F. Pay, C. E. Rick, and M. A. Tones, Biochem. Biophys. Res. Commun. 166, 1334 (1990).
87.   K. J. Fohr, Y. Wahl, R. Engling, T. P. Kemmer, and M. Gratzl, Cell Calcium 12, 735 (1991).
88.   J. Strupish, R. J. H. Wojcikiewicz, R. A. J. Challiss, S.T. Safrany, A. L. Willcocks, B. V. L. Potter, and S.R. Nahorski, Biochem. J. 277, 294 (1991).
89.   J. Shah and H. C. Pant, Biochem. J. 250 617, (1988).
90.   P. Palade, C. Dettbarn, P. Volpe, B. Alderson, and A. S. Otero, Mol. Pharmac. 36, 664 (1989).
91.   S. M. Seiler, A. J. Arnold, and H. C. Stanton, Biochem. Pharmacol. 36, 3331 (1987).
92.   P. Palade, C. Dettbarn, B. Alderson, and P. Volpe, Molec. Pharmacol. 36, 673 (1989).
93.   S. R. Hingorani and W. S. Agnew, Anal. Biochem. 194, 204 (1991).
94.   R. F. Irvine, Bioessays 13, 419 (1991).
95.   H. Zhao and J. Muallem, J. Biol. Chem. 265, 21419 (1990).
96.   M.D. Campbell, S. Subramanian, M.I. Kotlikoff, J.R. Williamson and S.J. Fluharty, Molec. Pharmacol. 38, 282 (1990).
97.   J. H. Exton, Hepatology 8 , 152 (1988).
98.   C. D. Ferris, R. L. Huganir, D. S. Bredt, A. M. Cameron, and S. H. Snyder, Proc. Natl. Acad. Sci. USA 88, 2232 (1991).
99.   C. D. Ferris, A. M. Cameron, D. S. Bredt, R. L. Huganir, and S. H. Snyder, J. Biol. Chem. 267 , 7036 (1992).
100.   J. Meldolesi, L. Madeddu, and T. Pozzan, Biochim. Biophys.Acta 1055, 130 (1990).
101.   K. H. Krause, FEBS Lett. 285, 225 (1991).
102.   M. F. Rossier and J. W. Putney, TINS 14, 310 (1991).
103.   P. Volpe, K. H. Krause, S. Hashimoto, F. Zorzato, T. Pozzan, J. Meldolesi, and D. P. Lew, Proc. Natl. Acad. Sci. USA 85, 1091 (1988).

104. S. Treves, M. D. Mattei, M. Lanfredi, A. Villa, N. M. Green, D. H. MacLennan, J. Meldolesi, and T. Pozzan, Biochem. J. 271, 473 (1990).

105. C. Van Delden, C. Favre, A. Spat, E. Cerny, K.-H. Krause, and D. P. Lew, Biochem. J. 281, 651 (1992).

106. T. Satoh, C. A. Ross, A. Villa, S. Supattapone, T. Pozzan, S. H. Snyder, and J. Meldolesi, J. Cell. Biol. 111, 615 (1990).

107. H. Otsu, A. Yamamoto, N. Maeda, K. Mikoshiba, and Y. Tashiro, Cell Structure and Function 15, 163 (1990).

108. A. Villa, P. Podini, D. O. Clegg, T. Pozzan, and J. Meldolesi, J. Cell. Biol. 113, 779 (1991).

109. P. Volpe, A. Villa, E. Damiani, A. H. Sharp, P. Podini, S.H. Snyder, and J. Meldolesi, EMBO J. 10, 3183 (1991).

110. P. D. Walton, J. A. Airey, J. L. Sutko, C. F. Beck, G. A. Mignery, T. C. Sudhof, T. J. Deerinck, and M. H. Ellisman, J. Cell. Biol. 113, 1145 (1991).

111. K. Takei, H. Stukenbrok, A. Metcalf, G. A. Mignery, T.C. Sudhof, P. Volpe, and P. DeCammilli, J.-Neuroscience 12, 489 (1992).

112. G. Guillemette, T. Balla, A. J. Buakal, and K. J. Catt, J. Biol. Chem. 263, 4541 (1988).

113. M. F. Rossier, G. Bird, and J. W. Putney, Biochem. J. 274, 643 (1991).

114. G. Guillemette, I. Favreau, B. Boulay, and M. Porier, Mol. Pharmacol. 38, 841 (1990).

115. J. W. Putney Jr. Adv. Pharmacol. 22, 251 (1991).

116. M. Kuno and P. Gardner, Nature 326, 301 (1987).

117. A. Khan, J. Steiner, and S. H. Snyder, Proc. Natl. Acad. Sci. 89, 2849 (1992).

118. P. Nicotera, S. Orrenius, T. Nilsson, and P. O. Berggren, Proc. Natl. Acad. Sci. ,USA 87, 6858 (1990).

119. A. N. Malviya, P. Rogue, and G. Vincendon, Proc. Natl.Acad. Sci. USA 87, 9270 (1990).

120. P. G. Bradford and M. Autier. Biochem J. 280, 205 (1991).

121. R. J. Wojcikiewicz and S. R. Nahorski, J. Biol. Chem. 266, 22234 (1991).

SECTION VI. TOPICS IN DRUG DESIGN AND DISCOVERY

Editor: Michael C. Venuti
Genentech, Inc., South San Francisco, CA 94080

SPECIAL TOPIC

Chapter 29. Macromolecular X-Ray Crystallography and NMR as Tools for Structure-based Drug Design

John W. Erickson
PRI/DynCorp, National Cancer Institute-FCRDC, Frederick, MD 21702

Stephen W. Fesik
Abbott Laboratories, Abbott Park, IL 60064

Introduction - The search for novel and patentable lead compounds is the hallmark of all drug discovery research efforts. How are these compounds discovered? Historically, most leads have been found by screening natural products and chemical libraries. However, over the past decade, advances in biotechnology have ushered in a new approach to drug discovery termed *structure-based drug design*. First envisioned by Abraham nearly 20 years ago (1) and pioneered by Goodford in the *de novo* design of antisickling agents that bind hemoglobin (2), structure-based design utilizes knowledge of the <u>structure</u> of a ligand complexed to its macromolecular target to design new lead compounds or to improve the potency and physical properties of an existing chemical series (3). In principle, all classes of macromolecules - proteins, nucleic acids, carbohydrates, and lipids - can serve as drug design targets. In practice, the experimental determination of macromolecular structure to atomic accuracy has been mainly applied to proteins or proteinaceous assemblies, such as protein-nucleic acid complexes or viral capsids. In this chapter we limit our discussion to protein targets with the understanding that many of the concepts and examples described herein may be generalized to other macromolecular systems of interest.

Despite much theoretical effort to fold protein structures based on their amino acid sequences, the determination of protein structure remains a largely experimental enterprise, and usually requires the application of macromolecular X-ray crystallography or NMR techniques. In certain cases where the three-dimensional structure of a homologue of the target enzyme or receptor is known experimentally, a plausible starting structure may be derived using homology modeling approaches (4). This strategy has been used extensively in the design of renin inhibitors (5). In any case, the key consideration for structure-based design is the generation of a <u>reliable</u> three-dimensional atomic model. Several comprehensive reviews provide earlier examples of the use of protein and small molecule crystal structures in ligand design (2,3,6). This review will discuss the uses of X-ray crystallography and NMR, either alone or in combination, in the context of recent structure-based drug design scenarios.

X-RAY CRYSTALLOGRAPHY IN DRUG DESIGN

X-ray crystallography is a well-established technique that can produce atomic resolution structures of proteins and their complexes with various ligands such as substrates, co-factors or inhibitors (7). From a purely conceptual standpoint, protein crystal structures may be used to aid drug design in several distinct manners: *a posteriori* analysis, *a priori* design, and iterative design. *A posteriori* analysis is a retrospective method whereby protein structure is used to rationalize existing SAR data and to propose design improvements. This has probably been the most frequently used mode of protein crystallography in early ligand design efforts. *A priori* design refers to the use of protein structure for the design or discovery of the initial lead compound. This mode of design holds promise for the invention of structurally novel leads, and its usefulness will undoubtedly grow in conjunction with the development of computational approaches such as GRID (8), ALADDIN (9) and DOCK (10) that utilize protein structures to design or to search for complementary small molecules that could serve as new leads. Iterative design refers to the situation where all elements of the drug design cycle are in dynamic feedback - structure determination, analysis, design, synthesis, and testing. From a practical standpoint, iterative design will entail both *a posteriori* and *a priori* design during the optimization process. In addition, iterative design requires the close coordination of an interdisciplinary team effort as well as a constant supply of target receptor for crystallization studies. For these reasons, iterative design has only recently been feasible using X-ray crystallography as an integral component.

<u>Human Rhinovirus Inhibitors - *A Posteriori* Design</u> - Human rhinoviruses cause the common cold, and over a hundred different isolates have been identified for this picornavirus. A novel class of oxazolinylphenyl isoxazole compounds (*e.g.*, **1**-**5**) were demonstrated to possess broad spectrum antiviral activity against picornaviruses (11). The mechanism of action of these compounds was shown to be due to inhibition of the viral disassembly step of the infection process. The structural basis for this activity was demonstrated with the crystallographic structure determination of a complex of WIN 52084 (**1**) with the capsid of human rhinovirus 14 (HRV14) (12). The lipophilic drug is nestled snugly within a hydrophobic core between the two β-sheets that comprise the β-barrel of the VP1 coat protein subunit (Figure 1A). There is a 25 Å deep, continuous depression or "canyon" that surrounds the icosahedral vertices of the surface of HRV14 virions. The canyon is formed by the icosahedrally-related pentamers of VP1 subunits, and has been implicated in receptor binding (13). The entrance to the drug-binding site in VP1 can be visualized as a pore at the base of the canyon floor (Figure 1B). The isoxazole moiety of **1** is buried deep within the pocket, under the floor of the canyon, and the oxazolinyl end occupies a position near the pore. These results help to explain the slow association rate (k_{on} ~ 167,000 $M^{-1}min^{-1}$) of disoxaril (**3**) as being due to restricted access to the binding pocket and to a requirement to displace water molecules occupying the pocket in order for binding to occur (11). Antiviral studies demonstrated that the (S) isomer at the 4-position of the oxazoline ring was approximately ten-fold more potent than the (R) isomer. This stereoselectivity was mirrored in the preference of the (S) over the (R) isomer binding to HRV14 in crystal soaking experiments using racemic mixtures of **1** or analogs. Energy profile analysis suggested that the 4-methyl moiety in the (S) configuration is able to make more favorable hydrophobic interactions with the protein (11). X-ray crystallographic investigation of a series of structurally-related drug/virus complexes (14) revealed the occurrence of a second mode of binding in which the inhibitor is oriented in the

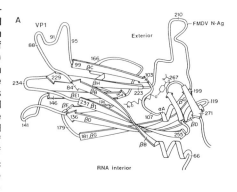

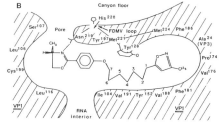

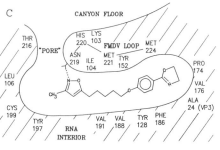

Figure 1. (A) Ribbon diagram of VP1 with WIN compound binding site. Diagrammatic representations of (B) compound **1** and (C) compound **5** bound in the WIN pocket. Reprinted with permission from the American Association for the Advancement of Science (12) and the National Academy of Sciences (14).

pocket with the opposite directionality (Figures 1C and 2) . In this case, the isoxazole moiety is closest to the opening of the canyon. Interestingly, both orientations place a methyl group in roughly the same hydrophobic region near the side chain of Leu106. This may not be coincidence, since the presence of an alkyl moiety at the 4-position on the oxazoline moiety is apparently required for the tighter mode of binding in which the oxazoline ring is positioned near the pore (**1**(S) and **2**(S)). Unsubstituted oxazolines (**3** and **5**) and compounds with shorter linkers (**4**) bind with the methylisoxazole nearer to the pore. Computational studies indicated the importance of space-filling, van der Waals interactions for the binding of this class of compounds (15).

MIC (µM)

1 (S) 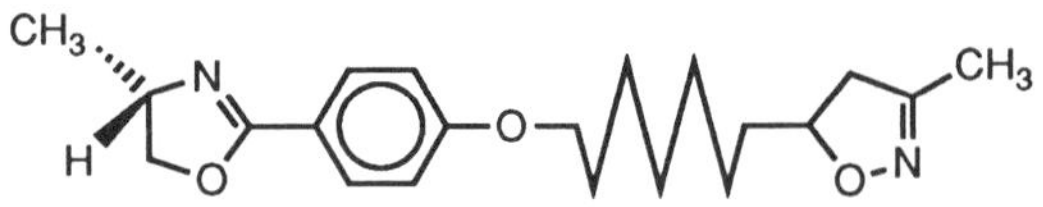   0.03

2 (S) 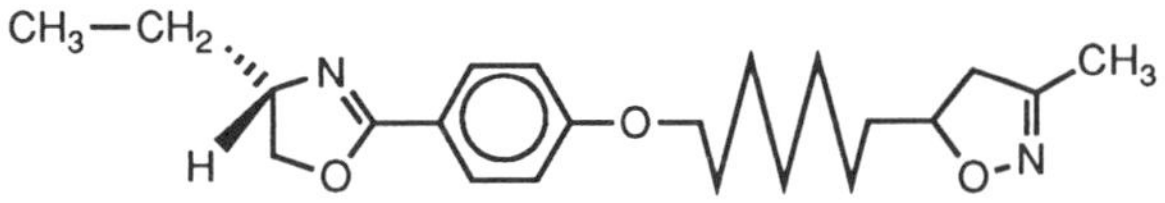   0.02

3   0.6

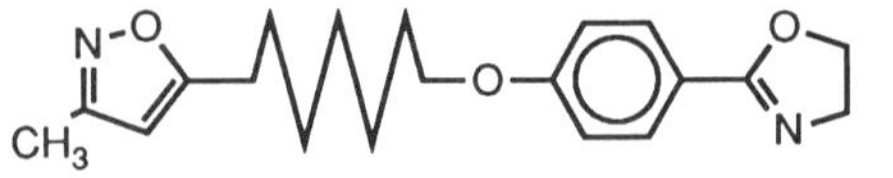

4 (S) 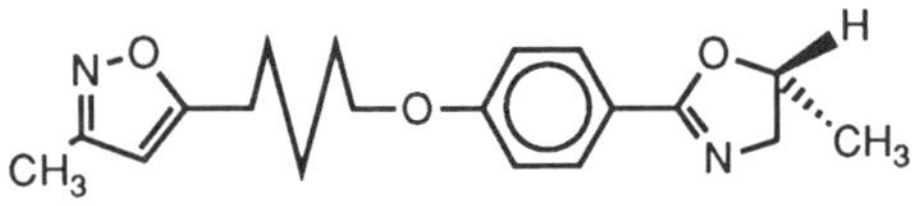   0.6

5   0.5

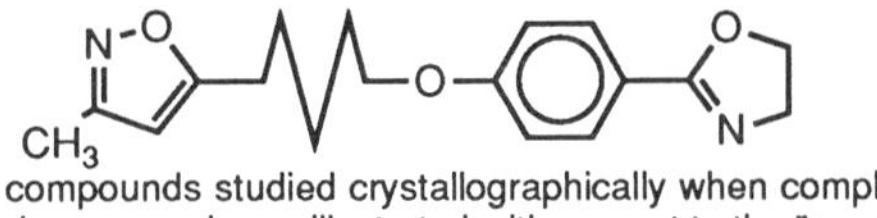

Figure 2. Some of the compounds studied crystallographically when complexed to HRV14. The relative orientations of the bound compounds are illustrated with respect to the "pore" to the canyon floor which is on the left. Adapted from (14), with permission of the National Academy of Sciences.

In addition to providing a structural basis for the antiviral effects and relative potencies of **1 - 5** , the crystallographic studies indicated that drug binding induced substantial structural changes in VP1 causing main chain atom movements up to 4.5 Å and shifts of up to 7.5 Å in the positions of side chain atoms (14). These changes near the drug binding site also caused conformational changes to the viral surface that resulted in a shallower canyon. These observations prompted speculation that drug binding might also inhibit virion attachment to the cell surface receptor. Binding studies using radiolabeled virus and cell membranes indicated that this was indeed the case (11,16). However, the cell attachment activity of the isoxazole compounds is not essential for antiviral activity since some strains of rhinovirus, such as HRV1A, that are inhibited by these compounds are unaffected in this step (17). X-ray crystallographic analysis of the HRV1A capsid structure indicated that the wild-type virus had a canyon conformation similar to that of the drug-bound form of HRV14 (17). Presumably, the antiviral compounds in this series exert their inhibitory effects through stabilization of the coat protein. The loss of flexibility accompanied by drug binding to the interior of VP1 results in the inhibition of disassembly. These structural studies have led to new insights into the mechanism of antiviral inhibition for **1 - 5** and can be used not only to assist the optimization of inhibitor potency for this class of compounds, but also to suggest new strategies for *de novo* design of antiviral agents against various strains of HRV as well as other picornaviruses. More generally, this example suggests a rationale that could be extended to the design of inhibitors of any macromolecular target whose function depends upon the flexibility of β-barrel-like structures.

HIV Protease Inhibitors - *A Priori* Design - The human immunodeficiency virus encodes a protease (HIV PR) that is essential for viral replication. HIV PR has been the subject of intensive drug design efforts and there are a number of recent excellent reviews on HIV PR structure (18) and inhibitor design (19,20). Most of the medicinal chemistry approaches employed to design HIV PR inhibitors have been based on classical substrate or transition-state analog based approaches (21). However, three recent examples emphasized structural considerations in the *a priori* design or discovery of novel lead compounds.

The first example utilized the concept of active site symmetry to design two-fold (C_2) symmetric or quasi-C_2 symmetric inhibitors (22,23). The X-ray crystal structure of the protease of Rous sarcoma virus (24) revealed that retroviral proteins are homodimeric enzymes and that they are structurally-related to the aspartic proteinases, or pepsins (25). The suggestion that the active site of HIV PR was C_2 symmetric led to the design of simple lead structures that would mimic this symmetry and also satisfy the hydrogen bonding and subsite preferences known to be important for binding based on structural studies of other aspartic proteinase/inhibitor complexes (25). For productive binding of a symmeric inhibitor to occur, it would be necessary for the C_2 axes of the enzyme and inhibitor to approximately superpose in the complex.

Modeling studies using the structures of RSV protease (24) and a complex of a reduced peptide renin inhibitor bound to rhizopuspepsin (26), a structurally related aspartic proteinase, indicated that **6** ($IC_{50} > 100$ mM) should be able to bind in the predicted manner. Elaboration of **6** by symmetrical addition of Cbz-blocked valine residues to enhance interactions with the P2 and P2' subsites in the enzyme resulted in a much more potent inhibitor, A-74704 (**7**; $IC_{50} = 3$ nM). X-ray crystallographic analysis of the structure of a complex of **7** bound to recombinant HIV PR verified that this pseudo-C_2 symmetric inhibitor bound in a highly symmetric fashion (22). This strategy has been used to design other classes of C_2 symmetric inhibitors that contain a central diol (23) or a difluoroketone (27). The C_2 or quasi-C_2 symmetric diols and difluoroketones are generally at least 10-fold more potent than their respective monohydroxy homologues. More recently, L-400,417 (**8**), a quasi-C_2 symmetric hydroxyethylene - based inhibitor of HIV protease has been reported (28). This compound differs from previous designs in that it is based on a C-terminal, rather than an N-terminal, duplication and thus contains an opposite directionality of the hydrogen bonding amides. X-ray crystallographic analysis of a complex of **8** with HIV PR revealed that the inhibitor bound in a nearly symmetric manner and formed many of the same backbone-backbone hydrogen bonds observed with **7**, with the exception of the 2-hydroxy group on the indan which was hydrogen-bonded to the carbonyl oxygen of Gly27 and the amide nitrogen of Asp 29. The high potency of **8** (Ki = 0.67 nM) may be due to the incorporation of the novel 2-hydroxy-1-aminoindan moiety in P2' which was previously shown to strongly enhance binding in a classical hydroxyethylene peptidomimetic series (29). The above examples illustrate how knowledge of enzyme structure was used for the conceptualization of novel lead compounds. The lack of sufficient quantities of recombinant enzyme combined with an unusually rapid development of sub-nanomolar inhibitors in several C_2 symmetric classes precluded an iterative drug design approach in the first case. However, the structure of the A-74704/protease inhibitor complex was used to guide the design of pharmacological improvements by focusing synthetic efforts on the peripheral N-terminal blocking groups. Substitution of a 2-pyridyl-N-methyl urea for the Cbz-carbamate in the C_2 symmetric diol series led to **9** (enzyme $IC_{50} = 0.15$ nM; viral $IC_{50} = 0.03$-0.27

μM) which was moderately soluble (197 μg/ml) and exhibited achievable antiviral blood levels in monkeys (Cmax = 0.16 μM; Tmax = 1.5 hr) and dogs (Cmax = 0.59 μM; Tmax = 0.17 hr) upon oral administration (30).

In the second example of *a priori* lead compound discovery, the HIV PR structure was used directly to search structural databases for compounds that contained shapes complementary to the enzyme active site (31). The search procedure employed the program DOCK (10) which was developed to identify accessible surfaces or cavities on macromolecules and to convert these into "negative" images that can be used to search a structural database that has been converted into a corresponding set of surface representations. Using this technique, the known compound bromperidol (10) was identified and a closely-related analogue, haloperidol (11), was shown to weakly inhibit the enzyme (Ki=100 μM) (31). Bromperidol was originally discovered and modeled into the enzyme active site using the structure of the native enzyme to construct the search image. This result was rather surprising as it was later determined that the surface structure of the active site is dramatically altered upon inhibitor binding. The twofold-related flaps undergo a topological rearrangement and movement by up to 7-10 Å in order to engage a peptide-based inhibitor (32).

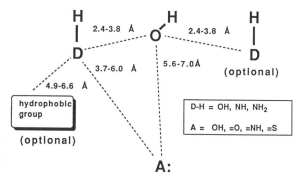

10 R = Br
11 R = Cl

In a third example of a structure-based search for new leads, the crystal structure of the A-74704/HIV PR complex was used to construct pharmacophoric targets that mimicked essential features of the inhibitor-enzyme interactions (33). Key pharmacophores were the central OH bound to the active site aspartic acids, the flanking amides bound to the main chain carbonyl oxygen atoms of Gly21 and Gly121, the buried water that bridges the flaps with the inhibitor, and the bulky phenyl group that binds in the P1 pocket. The corresponding pharmacophore search targets contained a central OH, two hydrogen bond donors, a hydrogen bond acceptor, and an aromatic moiety, respectively (Figure 3). Various combinations of these pharmacophores were used to search several chemical databases of three-dimensional structures for structurally novel compounds that might display inhibitory activity against HIV PR. A series of dibenzophenones was identified by this analysis. Several compounds were tested and shown to be inhibitory in the 10-100 μM range (Figure 4; 12-14). Structural considerations based on preliminary modeling studies indicated that the bulky R1 group on these compounds would make unfavorable van der Waals contacts with flap residues in the inhibited structure of HIV PR. Thus, either the flaps must maintain a more open conformation in complexes with these inhibitors, or the inhibitors might bind in a different orientation from that based on the model. It is interesting to note that statine-based peptidomimetics that contain an unsubstituted *p*-benzoylphenylalanyl moiety in the P1 position can be potent inhibitors of aspartic proteinases (34).

Figure 3. Schematic representation of pharmacophoric substructures used to search for non-peptidic inhibitors of HIV PR based upon the X-ray crystal structure analysis (22). D and A indicate hydrogen bond donors and acceptors, respectively. From (33), with permission from Tetrahedron Comp. Meth.

Iterative Drug Design - The convergence of rapid advances in biotechnology, particularly in the areas of recombinant gene expression and protein crystallography, has finally made it feasible to engage in the drug design cycle in an iterative and timely fashion. Two recent examples deal with the design of inhibitors for thymidylate synthase (TS) (35) and purine nucleoside phosphorylase (PNP) (36). In the TS example, the structure of the homologous enzyme from *E. coli* was used as a surrogate target for the design of human TS inhibitors. TS catalyzes the conversion of deoxyuridylate to thymidylate (dTMP) *via* an unusual, one-carbon transfer, methylation reaction that

Figure 4. Non-peptide HIV PR inhibitors identified from the ALADDIN substructure search. Shown in parenthesis are the pharmacophores used in the search. IC_{50} values are for HIV PR inhibition at pH 4.5. Adapted from (33), with permission from Tetrahedron Comp. Meth.

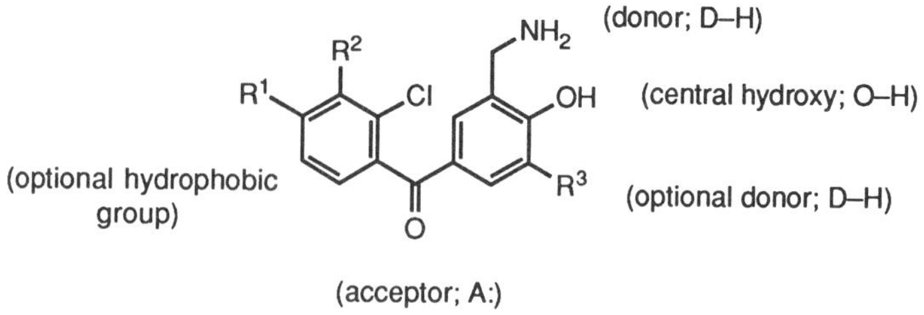

(donor; D–H)

(central hydroxy; O–H)

(optional hydrophobic group)

(optional donor; D–H)

(acceptor; A:)

	R^1	R^2	R^3	$IC_{50}, \mu M$
12	OCH_2COOEt	Cl	CH_2NH_2	11
13	OCH_2COOH	Cl	CH_2NH_2	85
14	OCH_2COOEt	H	Cl	15

utilizes 5,10-methylenetetrahydrofolate as a co-factor. TS comprises the sole and rate-limiting pathway for dTMP biosynthesis, and has been an important target for the design of antifolate antitumor agents that promise to exhibit superior activity to the clinically established dihydrofolate reductase inhibitor, methotrexate. Previous studies on the structure determination of TS from *L. casei* established the dimeric structure of the enzyme, and sequence comparison analysis suggested that the active site regions of homologous TS enzymes were highly conserved (37). A high expression recombinant system was developed for the *E. coli* enzyme, and led to X-ray crystal structure determinations for both the apo and ternary complex forms of TS (38-41). Upon analysis of these structures, it was evident that approximately 75% of the active site residues should be identical for the *E. coli* and human enzymes. Two different approaches were used for inhibitor design using the structure of *E. coli* TS: structure-based elaboration of an existing lead compound, and *de novo* design.

In the first approach, the goal was to develop a more lipophilic analog of the classical antifolate TS inhibitors, CB3717 (15) and 16 (Ki = 10 and 8 nM, respectively, for human TS) that would retain potency. Compounds related to 15 and 16 that contain the glutamate moiety are actively transported into cells and are thus subject to selective drug resistance. Removal of the *p*-CO-glutamate moiety from 16 led to a loss of 2.4 orders of magnitude in binding (17 ; Ki = 2.2 µM). Examination of the crystal structure of a complex of TS bound to 15, the 2-amino analogue of 16, indicated the presence of a small hydrophobic pocket in the enzyme proximal to the meta position of the phenyl ring. Compound 18 was modeled and synthesized and exhibited improved inhibition (Ki = 0.39 µM). The structure determination of 18 complexed with TS verified that the *m*-CF3 moiety occupied the modeled position and made van der Waals contacts with hydrophobic side chains of the enzyme. It was further predicted on electrostatic and structural grounds that a bulky hydrophobic group combined with an electron-withdrawing substituent positioned off the para position of the phenyl ring would further enhance inhibitor binding. This led to the modeling and synthesis of 19 which was nearly as potent as the lead compound 16 (Ki = 13 nM).

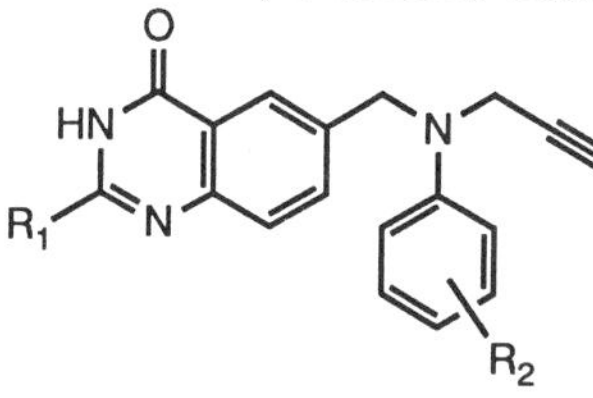

15 $R_1 = NH_2$, $R_2 = $ *p*-COGlu

16 $R_1 = CH_3$, $R_2 = $ *p*-COGlu

17 $R_1 = CH_3$, $R_2 = $ H

18 $R_1 = CH_3$, $R_2 = $ *m*-CF3

19 $R_1 = CH_3$, $R_2 = $ *p*-SO$_2$Ph

In the second study, the aim was to use the inhibited form of the enzyme structure to design novel lead compounds *de novo*. The criteria for a working lead was that the compound exhibit a Ki below 10 µM, provide a crystalline complex with TS, and have a number of positions suitable for synthetic elaboration. Two crucial considerations were pointed out that are generically applicable to *de novo* design. First, the structure of the protein active site used for modeling should be that based on an inhibitor or a cofactor-bound form. This will ensure that any structural changes in the enzyme that are naturally induced by substrate or inhibitor binding will be reflected in the structure of the model. Secondly, the design of initial leads should build in soluble, polar groups . This will help to ensure that the initial, weakly potent leads will have the necessary solubility to be able to achieve concentrations in solution required for binding to the high concentrations of protein used in crystallization experiments. Two series of novel lead compounds were developed. The first series evolved from the use of the program GRID (8) to predict the interaction energies between various functional group probes and enzyme atoms in the vicinity of the pteridine binding site. Based on this analysis and on an examination of the TS structure, compound **20** was modeled and synthesized (Ki = 3 μM). The key predictions leading to **20** were a hydrophobic pocket that could accommodate a naphthalene, a well-ordered water molecule in the crystal structure that could hydrogen bond to the carbonyl group at position 1, and the carboxylate side chain of Asp169 that would hydrogen bond to the NH at position 8. The latter two interactions could be made by bridging the 1- and 8- positions with a γ-lactam. Crystallographic analysis of a complex of **20** bound to TS revealed strategies for design improvements, and further iterative cycles of modeling, synthesis, X-ray crystallographic and computational energy analysis led to the design and synthesis of **21** with a Ki of 31 nM for human TS.

20 $R = C_2H_5$, $X = SO_2-N$⌬NH_2^+ Cl^-, $Y =$ [HN-acetyl group]

21 $R = CH_3$, $X = SO_2Ph$, $Y =$ [amidine group H_2N]

22 $X = H$, $Y = SO_2-N$⌬NH

23 $X = NH_2$, $Y = SO_2-N$⌬NH

The second class of lead compound featured an imidazole moiety based on a detailed analysis of the hydrogen-bonding requirements of residues buried deep within the active site region. Hydrophobic and previous design considerations led to the elaboration of the substituted tetrahydroquinoline **22** (Ki = 7.7 μM). Crystallographic analysis indicated that **22** bound essentially as predicted. The feasibility of additional hydrogen bonding to a protein backbone carbonyl oxygen indicated the addition of an amino group at the 2-position of the imidazole ring that led to a 120-fold enhancement in potency (**23**; Ki = 64 nM). The 2-amino groups of **23** and of the classical quinazoline series, exemplified by **15**, occupied nearly identical positions in their respective complexes, and led to substantial binding improvements in both cases. Further cycles of iterative design based on **21** and **23** have led to more potent inhibitors for both classes of TS inhibitors (42,43).

A second recent example of iterative design (36) concerns the enzyme purine nucleoside phosphorylase (PNP) which participates in the purine salvage pathway and catalyzes the reversible phosphorolysis of purine ribo- or 2'-deoxyribonucleosides to the purine and ribose- or 2'-deoxyribose-α-1-phosphate. Inhibitors of PNP tend to exhibit T-cell toxicity and may have therapeutic potential in T-cell-mediated autoimmune disorders, T-cell leukemias and lymphomas, and organ transplantation. In addition, since PNP rapidly metabolizes purine nucleosides, inhibitors of this enzyme may be beneficial when co-administered with chemotherapeutic purine nucleoside analogs, such as the anti-AIDS drug, ddI. Human PNP is specific for 6-oxypurine compounds and is a trimer of identical subunits (total molecular mass = 97 kDa). Analysis of the structural differences between the apoenzyme and a complex with guanine revealed the existence of a "swinging gate" that moves up to several Å during substrate or inhibitor binding. This gate is formed by several amino acid residues of PNP and its complex motion could not have been predicted from the apoenzyme structure alone. Hence, the most fruitful modeling attempts were done with reference to the structure of the bound form of PNP, again emphasizing the importance for drug design studies of having the atomic coordinates for the receptor complex, and not just for the native receptor target. In the PNP example, X-ray structures were determined for complexes with a number of previously synthesized PNP inhibitors with micromolar Ki's, including 8-aminoguanine, 9-benzyl-8-aminoguanine, 5'-iodo-9-deazainosine, acyclovir diphosphate, and 8-amino-9-(2-thienylmethyl)guanine (36). The X-ray studies revealed several key features of inhibitor-enzyme interactions: i) 8-amino substituents enhanced binding of guanine analogs by forming hydrogen bonds to Thr242; ii) 9-deaza guanine analogs enhanced potency through donation of a hydrogen bond to Asn243, and iii) addition of a bulky hydrophobic group at the 9-position of the purine improved binding *via* hydrophobic interactions with the ribose binding site. Based on this analysis, a number of 9-substituted 9-deazapurine analogs were modeled and synthesized (Figure 5). As expected, both 8-aminoguanine 9-substituted analogs and 9-deazaguanine analogs were potent inhibitors. Surprisingly, a combination of these features in the 8-amino-9-deazaguanine analog resulted in a relatively poor inhibitor. Analysis of 3.2 Å difference Fourier electron density maps from several of the relevant complexes revealed that two different modes of hydrogen bonding could occur with the side chains of Thr242 and Asn243 depending upon the protonation state of N-7 on the purine (Figure 5). This is a striking illustration of why it is important to experimentally determine structures of outliers during the iterative design process. The most potent compounds were those that were modeled based on a three subsite hypothesis: a purine subsite, a hydrophobic ribose subsite, and a phosphate binding site. The series of 9-deazaguanine 9-branched, benzyl-containing compounds (represented by **24**) were stated to be the most potent membrane-permeable inhibitors of PNP reported thus far.

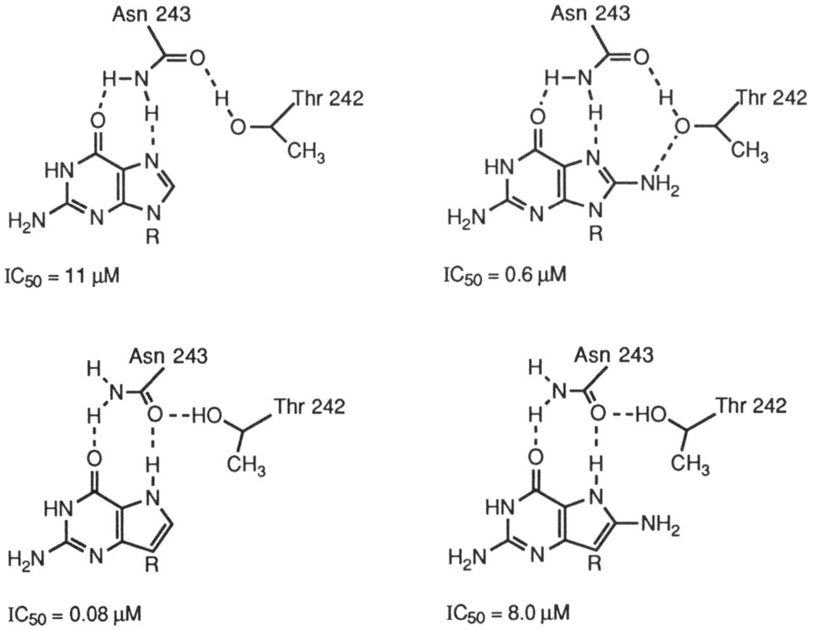

$IC_{50} = 11 \mu M$

$IC_{50} = 0.6 \mu M$

$IC_{50} = 0.08 \mu M$

$IC_{50} = 8.0 \mu M$

Figure 5. Comparison of hydrogen bonding interactions between 8-amino and 9-deaza modified purines with PNP based on X-ray crystallographic difference Fourier analysis. The R group is 2-thienylmethyl in the 8-amino purines, and 3-thienylmethyl in the 9-deazapurines. From (36), with permission from the National Academy of Sciences.

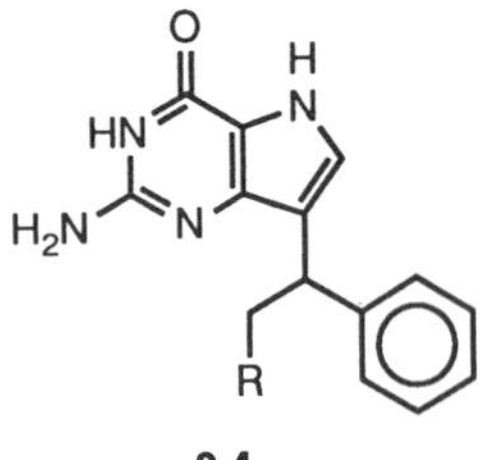

Besides the fact that structure-based design succeeded in helping to achieve nearly two orders of magnitude increase in inhibitor binding, several interesting considerations were borne out by this study. The difference Fourier maps extended to only 3.2 Å resolution - much less than the resolution that some would consider essential for drug design - and yet they proved to be exceedingly useful. The authors noted that over a 2.5 year period approximately thirty-five PNP-inhibitor complexes were evaluated using X-ray crystallographic methods, and roughly 60 active compounds were synthesized. This number of compounds is easily a tenth of the number that could be synthesized by a large medicinal chemistry team over the same period of time.

24

Thus, the iterative drug design process could be viewed in utilitarian terms as having saved a great deal of synthetic effort that was productively utilized on other therapeutic projects. Finally, the X-ray crystallographic analysis was only able to keep pace with the modeling and synthetic efforts - a key aspect of iterative design - owing to the fact that the native PNP crystals could be readily soaked with

PNP inhibitors. This was also true for the TS inhibitor iterative design example. In cases where soaking is not feasible, it is essential that co-crystallization conditions are found that permit the rapid growth (within one or two weeks) of high quality protein/inhibitor complex crystals.

Limitations and Future Developments - Of the essential steps required in a crystallographic structure determination - crystallization, X-ray diffraction data collection, data processing and reduction, phase determination, electron density map interpretation, and structure refinement - the serendipitous process of protein crystallization remains the major limitation in protein crystallography. Recent developments in crystallization robotics have improved the situation somewhat by enabling the use of large scale screening formats (44). Experience has shown that the more pure the protein is, the better are the chances of obtaining high quality crystals. The amounts of protein needed for a complete structure determination study vary from less than a mg to 100's of mg. The actual quantity required for a particular protein depends to a large extent on how rapidly one can identify reproducible conditions for crystal growth. Using microcrystallization techniques (44,45), a mg of protein can provide for numerous crystallization trials depending upon the desired protein concentration. For iterative crystallization studies, the required amount per structure drops dramatically owing to the fact that screening is no longer necessary, and in many cases individual crystals can be soaked with inhibitor. For de novo design, lead compounds should be designed to have aqueous solubilities high enough to allow for stoichiometric binding to the high protein concentrations (up to 1 mM) required for crystallization.

Phase determination for the indexed intensities, the second major hurdle in a structure determination, is done using one of several methods: molecular replacement, isomorphous (heavy atom) replacement, and anomalous dispersion. Isomorphous replacement requires measurements from one or more heavy atom-derivatized crystals and, thus, usually requires more material. The use of molecular replacement for phasing is restricted to proteins whose overall fold is suspected to be homologous to a protein of known structure. Recent advances in the direct phasing of native crystal structure amplitudes using multiple wavelength anomalous dispersion (MAD) has made it generally possible, in principle, to determine protein crystal structures without the need to resort to heavy atom substitution (46). This method avoids the unpredictability of derivatization and the lack of isomorphism that accompanies heavy atom replacement techniques. MAD uses the signal due to anomalous scatterers, like the sulfur atom of Cys or Met, that may be intrinsic to the protein, or that may be engineered into the protein by substituting selenomethionine for methionine, for example. The combination of powerful, tunable X-ray synchrotron light sources with recombinant technologies for producing anomalously-edited proteins promises an expansion of the use of MAD and accelerated phase determination for this crucial step in protein crystal structure determination.

The recent development of sensitive, high speed area detectors for rapid X-ray data collection and advances in affordable high speed computer graphics workstations are reducing the turnaround time for iterative structure determinations of drug/protein complexes to a few weeks or less. The point has been reached where structure analysis is becoming a rate-limiting step in the drug design cycle; i.e., - structures are being solved more rapidly than one can perhaps effectively utilize them. The major challenge that lies ahead is clearly computational - to improve drug binding predictions both qualitatively as well as quantitatively in the search for more potent and effective cures for disease.

NMR IN DRUG DESIGN

Besides X-ray crystallography, nuclear magnetic resonance is the only other experimental technique that can provide structural details at atomic resolution. Since high quality crystals which may be difficult to obtain are not required, data collection may begin at an early stage in the structure determination. In addition, unlike with X-ray crystallography, NMR can continuously provide useful structural information on the way to determining the complete three-dimensional structure. These features and the fact that the NMR studies are conducted under different experimental conditions (in solution vs. in the crystalline state) suggest that NMR could also be a useful tool for designing new pharmaceutical agents (47,48). Indeed, isotope-aided NMR experiments have recently been developed for determining the conformation of a drug molecule when bound to its macromolecular receptor and for identifying those portions of the ligand that interact with the target site (49-54). This information could help distinguish between those functional groups of the ligand that are important for binding to the receptor from those that could be modified without affecting binding affinity, and could be used to design active analogs with better physical properties. This information could also prove useful for designing analogs with different molecular frameworks that would position crucial functional groups of the ligand in their experimentally determined, spatial orientation. These new analogs could have the advantage of being metabolically more stable or easier to synthesize (47,48).

In addition to those NMR studies that focus on the structure of the ligand and its structural environment, NMR could also be used to determine the complete three-dimensional solution structure of a drug/receptor complex. To obtain this structural information by NMR was impossible only a few years ago. However, due to rapid advances in heteronuclear three- and four-dimensional NMR techniques (55-62) and to the availability of isotopically labeled proteins (63-65), three-dimensional structures of medium-sized proteins (15-30 KDa) and molecular complexes can now be determined by NMR in a reasonable amount of time with good precision (66,67). The structural information that can be obtained from these studies, similar to that obtained from X-ray crystal structures, is important for determining the functional groups of the receptor responsible for ligand binding and for identifying new areas of the target molecule that may interact with suitably modified analogs. In general, the same approaches described in the previous section for designing new molecules based on X-ray structures could also utilize NMR-derived structures. Since high resolution structures have only recently been determined by NMR, the application of this technique is only in its infancy and has not been extensively appied to drug design. However, drug design based on NMR-derived structures may become more common due to recent advances which markedly extend the capabilities of this technique.

<u>Advances in Multi-dimensional NMR Methods</u> - One of the most important factors leading to the dramatic improvement in both the size and quality of the structures determined by NMR is the development of heteronuclear 3D and 4D NMR techniques (55-62). In early versions of these experiments (55), a heteronuclear shift correlation and a homonuclear 2D NMR experiment (*e.g.*, COSY, NOESY) were combined in such a way that homonuclear 2D NMR spectra are edited by the heteronuclear chemical shifts. Thus, for large isotopically labeled molecules, proton NMR spectra can be resolved into many subspectra and rendered interpretable. A further increase in resolution can be achieved in a 4D NMR experiment by using another heteronuclear frequency (60-62).

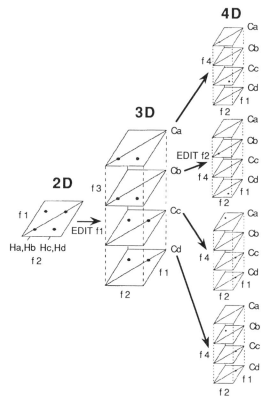

Figure 6 illustrates the utility of ^{13}C-resolved 3D and 4D NOE experiments. In a 2D NOE experiment, NOEs between overlapping protons Ha/Hb and Hc/Hd cannot be unambiguously identified. By editing in f_1 using the ^{13}C chemical shifts in a 3D NOE experiment, the NOEs can be resolved in different planes. For example, NOEs involving Ha can be distinguished from those involving Hb by their appearance on different planes corresponding to the different frequencies of their attached carbons (Ca, Cb). However, the NOE crosspeaks in f_2 are only defined by their proton frequencies. Thus, the NOE between Ha and Hc or Hd appearing on plane Ca cannot be uniquely defined. By further editing of the proton signals in f_2 in a 4D NOE experiment, the NOE between Ha and Hc is unambiguously identified, since it appears on plane Cc and not Cd.

Figure 6. Schematic illustration of 2D, 3D, and 4D NOE spectra of protons with NOEs Ha/Hc, Hb/Hd, and Hc/Hd. From (48), with permission from J. Med. Chem.

In addition to 3D and 4D NOE experiments, several important multi-dimensional NMR experiments have recently been developed to identify through-bond correlations (56-60). These experiments are extremely useful for making the ^1H, ^{13}C, and ^{15}N assignments of the protein backbone (60). They do not rely on NOEs which can be difficult to interpret, but utilize through-bond connectivities via heteronuclear couplings. In order to assign the amino acid side chains, new experiments have been developed in which through-bond connectivities are established through large one-bond ^1H-^3C and ^{13}C-^{13}C couplings (56-59), instead of the conventional manner by correlating the protons via small, conformational dependent ^1H-^1H three-bond couplings.

Isotope Labeling Techniques - In order to obtain the sensitivity required to apply many of the heteronuclear multi-dimensional NMR experiments, proteins must be isotopically labeled. Perhaps the most useful are the uniformly ^{15}N- and ^{13}C-labeled proteins. Most of the assignments and structural constraints can be obtained using double-labeled proteins. Bacterially expressed proteins can be uniformly ^{15}N- and ^{13}C-labeled using ^{13}C-/^{15}N-labeled media prepared from algal or bacterial extracts (68). Alternatively, proteins can be uniformly ^{13}C- or ^{15}N-labeled by growing bacteria in minimal media containing [U-^{13}C]glucose (or [U-^{13}C]acetate) (69) or ^{15}N-labeled ammonium chloride as the sole carbon or nitrogen source, respectively. Proteins can be selectively labeled by growing bacteria in chemically defined media containing the isotopically-labeled amino acid(s). These proteins can be used in NMR studies to unambiguously assign resonances by amino acid type using isotope-editing techniques (63). Other types of labeling have also been shown to be of value in NMR studies of proteins. Deuterated proteins have been used to simplify spectra (70), produce narrower ^1H NMR signals (64), and make stereospecific assignments (71). Non-randomly ^{13}C-labeled proteins have been employed to stereospecifically assign valine and leucine methyl groups (65).

Conformation and active site environment of bound ligands - A variety of approaches have emerged for studying large molecular complexes by NMR (48,72). One of the best approaches for studying receptor-bound ligands involves the use of isotope-editing techniques in which the proton signals of an isotopically labeled ligand are selectively observed in the presence of the many signals of its target site (49-54). Using this approach both the conformation of a bound ligand and those portions of the ligand that bind to the target site can be determined. These experiments can be conducted using two-dimensional isotope-edited NMR methods (49-52). Alternatively, for futher simplifying the spectra in cases when the proton NMR signals of the ligand overlap, heteronuclear three-dimensional NMR experiments may be employed in which the remaining proton NMR signals of the ligand are edited in a third dimension by their ^{13}C or ^{15}N frequencies (53,54). The advantages of these techniques is that important structural information for designing new analogs can be obtained on new drug/receptor complexes in as little as 2-3 weeks. Furthermore, the receptor protein does not have to be very pure, since the isotopically-labeled ligand binds selectively to the protein of interest and only the signals of the ligand and the nearby protein are detected.

An example of the type of structural information that can be obtained by NMR on an isotopically-labeled drug molecule bound to its receptor is illustrated by NMR studies on the cyclosporin A/cyclophilin complex. Cyclosporin A (CsA), **25**, a clinically useful immunosuppressant, exerts its activity through its interaction with cyclophilin (CyP), a 17.8 kD protein (165 residues) (73,74). In order to aid in the design of more potent and potentially less toxic CsA analogs, NMR studies of the CsA/CyP complex were initiated to determine the bound conformation of CsA and to identify those portions of CsA that bind to CyP. From an analysis of a ^{13}C-resolved 3D NOE spectrum of [U-^{13}C] CsA bound to CyP, three-dimensional structures of CsA (Fig. 7) were calculated using a distance geometry/dynamical simulated annealing protocol in which proton-proton distances calculated from the 3D NOE data were included as constraints (53). As shown in Figure 8, the conformation of CsA when bound to cyclophilin as determined by NMR was found (51-53) to be very different from the conformation of uncomplexed CsA determined by X-ray crystallography and NMR spectroscopy (75).

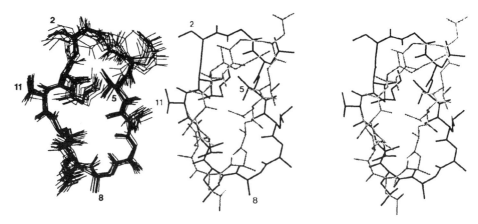

Figure 7. Superposition of the 20 CsA structures with the lowest NOE energy contribution out of 95 converged structures. From (53), with permission from Biochemistry.

Figure 8. Stereoview of CsA (restrained energy minimized average structure generated from 95 individual structures) bound to cyclophilin (black) superimposed on the X-ray crystal structure of CsA (grey) (75) determined in the absence of cyclophilin. From (53), with permission from Biochemistry.

In addition to the conformation of CsA when bound to cyclophilin, the NMR studies also revealed those portions of CsA that interact with cyclophilin from CsA/CyP NOEs (53,76). As shown in Figure 9, the protons attached to the filled and checkered carbon atoms were found to be in close proximity to cyclophilin as evidenced by NOEs between these CsA protons and the protein. These data suggest that CsA residues 1, 2, 9, 10, and 11 are involved in binding to cyclophilin and are consistent with the structure/activity relationships (77,78) that indicate the importance of these CsA residues for cyclophilin binding and immunosuppressant activity. Those portions of CsA that bind to CyP and those that are exposed to solvent were also distinguished by measuring the proton T_1 values of ^{13}C-labeled CsA in the presence and absence of a paramagnetic relaxation reagent, 4-hydroxy-2,2,6,6-tetramethylpiperidinyl-1-oxy (HyTEMPO) (79). Large effects of the spin label were observed for the protons attached to the jagged and checkered filled carbon atoms (Fig. 9), indicating that CsA residues 4, 6,

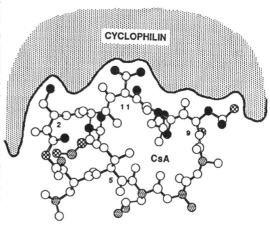

Figure 9. Three-dimensional structure of CsA bound to cyclophilin. CsA/cyclophilin NOEs were observed from CsA protons attached to the filled and checkered carbon atoms (53). The protons attached to the jagged and checkered carbon atoms exhibited the largest change in relaxation rate upon the addition of HyTEMPO. From (79), with permission from J. Am. Chem. Soc.

7,8 are exposed to solvent. These results are consistent with the structure/activity relationships of CsA analogs in which cyclophilin binding was found to be relatively insensitive to modifications at these sites (77,78).

The NMR studies described thus far were performed with cyclophilin A. Additional cyclophilins have been identified which could also play an important role in the immune response. One of these, cyclophilin B (CyPB), contains a hydrophobic N-terminal signal sequence and is 64% identical to CyPA (80). From the similar ^1H and ^{13}C chemical shifts and NOE data of [U-^{13}C]CsA bound to CyPB and CyPA, it was concluded that the conformation and active site environment of CsA in the two complexes are nearly identical, except in the vicinity of the MeVal^{11} CsA residue

(81). This approach was extremely useful for identifying subtle structural differences between two drug/receptor complexes in only a few days and could prove useful for designing analogs that bind selectively to only one of the proteins. For example, on the basis of the NMR studies of the CsA/CyPB complex, selectivity might be achieved with a CsA analog modified at the 11-position.

<u>Macromolecular Structure Determination</u> - The general approach for determining complete three-dimensional structures by NMR involves the following steps: i) assigning the NMR signals, ii) obtaining ^{1}H-^{1}H distance constraints from NOE data, iii) obtaining dihedral angle constraints from coupling constants, and iv) calculating structures using distance geometry and restrained molecular dynamics (67,82). Steps (i) and (ii) are greatly facilitated using 3D and 4D NMR experiments. Not only do these experiments resolve the data, thereby simplifying the analysis, but they also allow assignments of the ^{15}N and ^{13}C chemical shifts. In addition to these experiments, new methods have also been devised (83-86) for measuring ^{1}H-^{1}H, ^{1}H-^{15}N, and ^{1}H-^{13}C three-bond J-couplings from 2D and 3D NMR spectra. From these J-values, the stereospecific assignments of $C\beta$ methylene protons can be determined and dihedral angle constraints can be obtained which improve the quality of the calculated structures (87,88).

Currently, the largest protein structure determined by NMR is interleukin-1β (IL-1β) (66). Conventional 2D NMR methods were largely unsuccessful for determining the structure of this 153 residue protein due to its relatively large size. However, a variety of heteronuclear double- and triple-resonance 3D NMR experiments resulted in the assignment of a complete set of ^{1}H, ^{13}C, and ^{15}N resonances, the determination of the secondary structure, the location of bound water molecules, and the investigation of backbone dynamics (66 and references therein). Initial low-resolution structures of IL-1β were obtained using only 446 NOEs obtained from ^{15}N-resolved 3D NOE experiments. The accuracy and precision of these structures was dramatically improved using a much

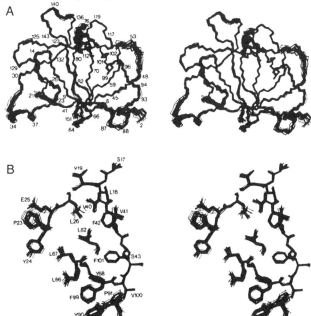

Figure 10. Stereoviews showing the best fit superposition of the (A) backbone atoms of IL-1β and (B) all heavy atoms of a selected region of IL-1β. Adapted from (66), with permission from Biochemistry.

larger number of distance constraints derived from heteronuclear 4D NOE experiments (66). The final structure calculations were done using 2780 distances and 366 torsion angles for a total of 3,146 experimental constraints. With this many constraints, the 3D structure of IL-1β was well-defined (Fig. 10). The atomic rms distribution about the average coordinate positions of the calculated structures was 0.41 +/- 0.04 Å for the backbone atoms and 0.82 +/- 0.04 Å for all atoms except those belonging to residues 1, 152, and 153. From the high resolution structure of IL-1β determined by NMR, the biological activity and receptor binding displayed by chemically modified and mutant proteins could be interpreted. Three distinct binding sites for the IL-1 receptor involving surface residues of IL-1β were postulated (66).

<u>Limitations and future developments</u> - Currently, NMR can be practically applied to the study of macromolecules with molecular weights less than about 40 kDa. This limitation is mainly due to the broad NMR signals for molecules of this size and the corresponding loss in signal intensity observed

in multi-dimensional NMR experiments. A second limitation is the requirement that the protein or molecular complex be soluble to at least 1 mM. For certain studies, a third limitation is the requirement for isotopically-labeled ligands, particularly for those studies aimed at determining the complete high resolution structure of a protein larger than 100 residues. This requires that the protein can be expressed in active form at relatively high levels (1 mg/ml) in a bacterial expression system. Ideally, both labeled proteins and labeled receptors would be available for NMR studies. Initial studies should involve labeled ligands bound to unlabeled receptors to determine the bound conformation of the ligand and to identify those portions of the ligand that interact with the receptor. Alternatively, unlabeled ligands could be studied when bound to labeled proteins using recently developed NMR techniques (89) with the goal of determining the bound structure of the ligand and its active-site environment. The next step would involve the complete three-dimensional structure determination of the drug/receptor complex using isotopically labeled receptor proteins and multi-dimensional NMR techniques. Once an initial structure is determined, additional three-dimensional structures of the same receptor bound to different ligands could be rapidly obtained, facilitating the design of new analogs in an interactive manner.

Future developments are expected to improve the speed and accuracy in which three-dimensional structures can be determined by NMR. The accuracy of the structures, important for designing analogs with a precision required for tight binding to the receptor, may be improved by more accurately interpreting the NOE data and by obtaining more constraints. Three-dimensional structures may be obtained more rapidly using new heteronuclear multi-dimensional NMR experiments and improved software tools, reducing the time required both for signal assignment and extraction of the relevant structural parameters used in calculating the structures.

COMPARISON OF X-RAY CRYSTALLOGRAPHY AND NMR

NMR Solution and X-Ray Crystal Structures: Similarities and Differences - Comparisons of protein structures determined using both NMR and X-ray crystallography have shown that the solution and crystal structures of a protein are in general agreement overall (90-92). This is not surprising since protein crystals are highly hydrated (27-65% solvent content) and can be treated as well ordered, highly concentrated protein solutions. Well-determined NMR and X-ray crystal structures usually do not deviate by more than about 1 Å rms for backbone atoms (90-92). The degree of structural correspondence between X-ray and NMR structures is greatest for residues in well-ordered secondary structural elements and in the interior, or core, of a protein, and is least for surface residues. It should be noted that different crystal forms of the same protein may deviate by 0.4-0.5 Å rms for backbone atoms. These differences, which are also most pronounced at the protein surface, are commonly attributed to the effects of different crystal packing arrangements as well as experimental errors.

Protein crystals are formed by a regular, three-dimensional network of protein-protein interactions stabilized by the high protein concentrations (typically on the order of 10 mM) within crystalline environments. These environments may induce order-disorder transitions for localized regions of a structure, stabilize flexible side chains or polypeptide segments that may assume various conformations in dilute solutions, as well as influence quaternary structure. Thus, crystal packing forces may influence a segment of a protein to assume a conformation not ordinarily found in dilute solution. In contrast, NMR structures are unbiased by crystal packing forces and can give a more accurate picture of the relative mobilities of side chains located at a protein's surface. Moreover, the extent of crystal packing influences on protein structures can be revealed by NMR studies. These influences may be localized to individual side chains, or may extend over secondary structural elements. For example, in the crystal structure of C3a (93), a 77 residue complement protein, both the N- and C-terminal regions make intermolecular crystal contacts. The crystal structure of the protein contains a long C-terminal helix and a disordered N-terminus. NMR studies of C3a (94) clearly indicated that the C-terminal hexapeptide was in a random coil conformation, while the N-terminus exhibited a helix in a region that was disordered in the X-ray crystal structure. The C-terminus of C3a contains the receptor recognition site, and an accurate conformation for this region is critical for the structure-based drug design of C3a antagonists.

In an example of possible crystal packing effects on subunit-subunit interactions, comparison of the X-ray (92) and NMR (95) structures of interleukin-8 (IL-8) revealed that this dimeric immunoregulatory protein can undergo a rearrangement of quaternary structure that controls the size of the cleft formed by the separation of two, C_2-related α-helices from individual monomers. In addition, the crystal and solution structures exhibited strikingly different hydrogen-bonding patterns for His33, which accepts a hydrogen bond from a backbone amide of Gln8 in the solution structure, but donates a hydrogen bond to the backbone carbonyl of Glu29 in the crystal structure.

The conformation of a small molecule ligand when bound to its receptor in solution may differ dramatically from the conformation exhibited in the uncomplexed form either in solution or in the crystalline state. Thus, small molecule X-ray crystal or NMR structures must be regarded cautiously, and one should not assume that these structures are representative of the bioactive conformation. As an illustration of this point, NMR solution studies demonstrated that the conformation of CsA when complexed to CyP differed substantially from the crystal structure of CsA (51-53). Similarly, X-ray crystallographic studies of the structure of the immunosuppressant, FK-506, complexed with its receptor, FK-506 binding protein (FKBP), indicated that FK-506 exhibited dramatically different conformations in the bound and free crystalline states (96). These results were confirmed by NMR studies on the solution structure of an FK-506 analogue, ascomycin, complexed with FKBP (54). The solution structure of FK-506 in chloroform was also determined by NMR (97) and found to be conformationally dissimilar to both the uncomplexed and bound crystal forms.

X-Ray Crystallography and NMR as Complementary Techniques - Structural information obtained by X-ray crystallography can aid in the structure determination of the same or of a homologous system by NMR spectroscopy and vice versa. For example, once the structure of a receptor is known from X-ray studies, this information can be used to aid the interpretation of NOE signals from isotopically-labeled ligand/receptor complexes, and thereby facilitate the structural analysis of ligand/receptor interactions in solution. This strategy was employed in the determination of the bound conformation of a renin inhibitor complexed with pepsin (49). Knowledge of the crystal structure of a related inhibitor/pepsin complex aided in assigning enzyme-ligand cross peaks, and resulted in a model for the docked inhibitor. This strategy is particularly important for cases where crystals of the relevant complex do not exist. In a more recent example (98), the NMR solution structure of the bound conformation of CsA (52) was docked with the X-ray crystal structure of native CyP (99). The complex was refined by molecular dynamics using the crystal structure of CyP and the intermolecular NOEs as constraints. Comparison of the resultant model structure with a crystal structure of a tetrapeptide/CyP complex provided experimental evidence to support the hypothesis that the MeBmt^1 carbinol group of CsA mimics the transition state or intermediate structure in the PPIase reaction (100).

NMR can also provide valuable information for the X-ray crystallographic structure determination of proteins. The secondary structure of a protein in solution can be accurately and quickly determined by NMR studies and used to help fit a polypeptide backbone into an electron density map. NMR structural data were used in this manner to assist in the electron density map interpretations for the recent three-dimensional X-ray crystal structure determinations of CyP (101) and the FK-506/FKBP complex (96). In addition, three-dimensional protein structures determined by NMR could be used as starting models for molecular replacement phasing. This was shown in test experiments using the known crystal structures of crambin (102) and tendamistat (103). Recently, the NMR solution structure of IL-8 was used in a molecular replacement study to solve the unknown crystal structure of IL-8 at 1.8 Å resolution (92). Previous attempts to solve IL-8 using the crystal structure of a closely related protein, platelet factor 4, gave unsatisfactory results, and attempts to prepare suitable heavy atom derivatives were unsuccessful. Thus, the availability of the NMR structure of IL-8 was critical to the successful crystal structure determination for this immunoregulatory factor. This experience clearly demonstrates the high degree of structural accuracy that NMR is capable of providing, and it is expected that there will be a growing demand for NMR-derived structures to assist in X-ray crystallographic studies.

FUTURE PROSPECTS

The promise of using three dimensional atomic structures of receptors to aid medicinal chemists in their drug design efforts is coming of age. Concomitant advances in numerous fields are converging to make the interdisciplinary approach of structure-based drug design applicable to an ever-growing number of important targets. As a recent example, the AIDS epidemic has resulted in the most intensive search for an antiviral agent ever witnessed in spite of the remarkable fact that no viral disease has ever been completely cured or controlled by chemotherapy. The reassessment of our approach to antivirals that is behind this effort is largely due to targeted and structure-based strategies to anti-HIV drug design that have become possible owing to a number of rapid scientific advances. In the past four years, the structures of HIV-1 protease (104-106), HIV-1 RNase H (107) and the HIV gp120-binding domain of the CD4 receptor (109,109) have been determined using X-ray crystallographic methods, and the solution structure of the HIV-1 p7 nucleocapsid protein has been determined by NMR (110). This short litany of HIV-related structural studies illustrates the power that modern molecular and structural biology approaches can bring to drug discovery programs in almost any area.

Although structural protein or enzymes are primary targets for drug design, protein/nucleic acid complexes and hormone/receptor complexes comprise two other classes of therapeutically important targets. No structures of medically relevant protein/nucleic acid complexes have been reported thus far, but the growing number of structures of regulatory proteins complexed with oligodeoxyribonucleotides (111) indicates that they are certainly feasible targets for structure-based design. Recently, the crystal structure of human growth hormone (hGH) complexed to the recombinant soluble extracellular domain of its receptor was determined by X-ray crystallography (112). Besides providing a detailed picture of the interactions between the polypeptide ligand and its receptor, this structure provides new insights into the structural basis of signal transduction which may be mediated through hormone-induced dimerization of receptor molecules. This study was possible owing to the soluble nature of the hormone-binding domain of the hGH receptor. Many receptors are relatively insoluble, integral membrane-bound proteins whose structure elucidation by either X-ray crystallography or NMR is much more difficult owing to the technical problems of achieving large, well-ordered three dimensional crystals and the broad linewidths of the NMR signals expected for these molecules in solution. However, if these obstacles can be surmounted, the structure determination of this important and abundant class of targets would become accessible. Advances in crystallization of membrane proteins (113) has led to the X-ray structure determination of a bacterial photosynthetic reaction center (114) and a trimeric pore-forming protein, porin, from a bacterial outer membrane (115). Two dimensional crystals, or well-ordered sheets, of membrane proteins are often easier to obtain than three-dimensional crystals. In these cases, high resolution electron microscopy can provide valuable structural information. Recently, this technique was employed to determine the 3.0 Å structure of bacteriorhodopsin from a purple membrane (116). High resolution electron microscopy promises to become a more widely used method particularly for determining structural data on well-ordered aggregates of membrane proteins and other macromolecular assemblies.

The limits of the structure-based drug design approach have not yet been realized, and many of the difficulties in successfully applying this method are largely theoretical. Major conceptual challenges lie ahead in the areas of structure and binding affinity prediction. While methods are being developed in these areas, experimental structure determination efforts using X-ray crystallography and NMR will continue to provide critical information that can be used to design more effective drugs. The ultimate success of the structure-based drug design approach will depend, in any case, on the wisdom of target selection and on the ability to marry together teams of scientists that contain the requisite scientific and collaborative skills.

<u>References</u>

1. D.J. Abraham, Intra-Science Chem. Rept., <u>8</u>, 1 (1974).
2. P.J. Goodford, J. Med. Chem., <u>27</u>, 557 (1984).
3. W.G.J. Hol, Tibtech, <u>5</u>, 137 (1987).
4. J. Greer, Ann. N.Y. Acad. Sci. <u>439</u>, 44 (1985).
5. C. Hutchins and J. Greer, Crit. Rev. Biochem. Mol. Biol., 26, 77 (1991).
6. J.J. Stezowski and K. Chandrasekhar, Ann.Rep.Med.Chem., <u>26</u>, 293 (1986).
7. T. L. Blundell and L.N. Johnson, "Protein Crystallography," Academic Press, New York, N.Y., 1976.
8. P.J. Goodford, J. Med. Chem., <u>28</u>, 849 (1985).
9. J. H. Van Drie, D. Weininger and Y. C. Martin, J. Comp.-Aided Mol. Des . <u>3</u>, 225 (1989).
10. I.D. Kuntz, J.M. Blaney, S.J. Oatley, R. Langridge and T.E. Ferrin, J.Mol.Biol., <u>161</u>, 269 (1982).
11. F.J. Dutko, G.D. Diana, D.C. Pevear, M.P. Fox and M.A. McKinlay in "Use of X-Ray Crystallography in the Design of Antiviral Agents," W.G. Laver and G.M. Air, Eds., Academic Press, New York, N.Y., 1990, p. 187.
12. T.J. Smith, M.J. Kremer, M. Luo, G. Vriend, E. Arnold, G. Kamer, M.G. Rossman, M.A. McKinlay, G.D. Diana and M.J. Otto, Science, <u>233</u>, 1286 (1986).
13. M..G. Rossmann, E. Arnold, J.W. Erickson, E.A. Frankenberger, J.P. Griffith, H.J. Hecht, J.E. Johnson, G. Kamer, M. Luo, A.G. Mosser, R.R. Rueckert, B. Sherry and G. Vriend, Nature, <u>317</u>, 145 (1985).
14. J. Badger, I. Minor, M.J. Kremer, M.A. Oliveira, T.J. Smith, J.P. Griffith, D.M.A. Guerin, S. Krishnaswamy, M. Luo, M.G. Rossmann, M.A. McKinlay, G.D. Diana, F.J. Dutko, M. Francher, R.P. Rueckert and B.A. Heinz, Proc.Natl.Acad.Sci.USA, <u>85</u>, 3304 (1988).

15. G.D. Diana, A.M. Treasurywala, T.R. Bailey, R.C. Oglesby, D.C. Pevear and F.J. Dutko, J. Med. Chem., <u>33</u>, 1306 (1990).
16. D.C. Pevear, M.J. Fancher, P.J. Felock, M.G. Rossman, M.S. Miller, G. Diana, A.M. Treasurywala, M.A. McKinlay and F.J. Dutko, J. Virol., <u>63</u>, 2002 (1989).
17. S. Kim, T.J. Smith, M.S. Chapman, M.G. Rossmann, D.C. Pevear, F.J. Dutko, P.J. Felock, G.D. Diana and M.A. McKinlay, J. Mol. Biol., <u>210</u>, 91 (1989).
18. A.G. Tomasselli, J.W. Howe, T.K. Sawyer, A. Wlodawer and R. L. Heinrikson, Chimica Oggi , <u>9</u>, 6 (1991).
19. D.W. Norbeck and D.J. Kempf, Ann.Rep.Med.Chem. <u>26</u>, 141 (1991).
20. J.R. Huff, J.Med.Chem. <u>34</u>, 2305 (1991).
21. M. Szelke in "Aspartic Proteinases and Their Inhibitors," V. Kostka, Ed., de Gruyter, New York, N.Y., 1985, p. 421.
22. J. Erickson, D.J. Neidhart, J. VanDrie, D.J. Kempf, X.C. Wang, D.W. Norbeck, J.J. Plattner, J.W. Rittenbouse, M. Turon, N. Wideburg, W.E. Kohlbrenner, R. Simmer, R. Helfrich, D.A. Paul and M. Knigge, Science <u>249</u>, 527 (1990).
23. D.J. Kempf, L. Codacovi, X.C. Wang, D.W. Norbeck, W.E. Kohlbrenner, N.E. Wideburg, D.A. Paul, M.F. Knigge, S.Vasavanonda, A. Craig-Kennard, A. Saldivar, W. Rosenbrook, Jr., J.J. Clement, J.J. Plattner and J. Erickson, J. Med.Chem., <u>33</u>, 2687 (1990).
24. M. Miller, M. Jaskolski, J.K.M. Rao, J. Leis and A. Wlodawer, Nature <u>337</u>, 576 (1989).
25. D. Davies, Ann. Rev. Biophys. Biophys. Chem., <u>19</u>, 189 (1990).
26. K. Suguna, E.A. Padlan, C.W. Smith, W.D. Carlson and D.R. Davies, Proc.Natl. Acad.Sci. USA <u>84</u>, 7009 (1987).
27. H.L. Sham, N.E. Wideburg, S.G. Spanton, W.E. Kohlbrenner, D.A. Betebenner, D.J. Kempf, D.W. Norbeck, J.J. Plattner and J. Erickson, J.Chem.Soc., Chem. Comm. 110-112, (1990).
28. R. Bone, J.P. Vacca, P.S. Anderson and M.K. Holloway, J.Med.Chem. <u>113</u>, 9382 (1991).
29. T.A. Lyle, C.M. Wiscount, J.P. Guare, W.J. Thompson, P.S. Anderson, P.L. Darke, J.A. Zugay, E.A. Emini, W. Schleif, R.A.F. Dixon, I.S. Sigal and J.R. Huff, J.Med.Chem., <u>34</u>, 1230 (1991).
30. D.J. Kempf, K. Marsh, D.H. Paul, M.F. Knigge, D.W. Norbeck, W.E. Kohlbrenner, L. Codacovi, S. Vasavanonda, P. Bryant, X.C. Wang, N.E. Wideburg, J.J. Clement, J.J. Plattner and J. Erickson, Antimicrob.Agents Chemother., <u>35</u>, 2209, (1991).
31. R. L. DesJarlais, G.L. Seibel, I.D. Kuntz, P.S. Furth, J.C. Alvarez, P.R. Ortiz De Montellano, D.L. DeCamp, L.M. Babe and C.S. Craik, Proc.Natl.Acad.Sci. USA, <u>87</u>, 6644 (1990).
32. M. Miller, J. Schneider, B.K. Sathyanarayana, M.V. Toth, G.R. Marshall, L. Clawson, L. Selk, S.B.H. Kent and A. Wlodawer, Science, <u>246</u>, 1149 (1989).
33. M.G. Bures, C.W. Hutchins, M. Maus, W. Kohlbrenner, S. Kadam, J.W. Erickson, Tetrahedron Comp.Meth., in press (1992).
34. P. Kuzmic, C.-Q. Sun and D.H. Rich, in "Proc. 11th Ann. Pep. Symp.," 1991, p. 129.
35. K. Appelt, R.J. Bacquet, C.A. Bartlett, C.L.J. Booth, S.T. Freer, M.A.M. Fuhry, M.R. Gehring, S.M. Herrmann, E.F. Howland, C.A. Janson, T.R. Jones, C. Kan, V. Kathardekar, K.K. Lewis, G.P. Marzoni, D.A. Matthews, C. Mohr, E.W. Moomaw, C.A. Morse, S.J. Oatley, R.C. Ogden, M.R. Reddy, S.H. Reich, W.S. Schoettlin, W.W. Smith, M.D. Varney, J.E. Villafranca, R.W. Ward, S. Webber, S.E. Webber, K.M. Welsh, and J. White, J.Med.Chem., <u>34</u>, 1925 (1991).
36. S.E. Ealick, Y.S. Babu, C.E. Bugg, M.D. Erion, W.C. Guida, J.A. Montgomery and J.A. Secrist, III, Proc.Natl.Acad. Sci. USA, <u>88</u>, 11540 (1991).
37. L.W. Hardy, J.S. Finer-Moore, W.R. Montfort, M.O. Jones, D.V. Santi and R.M. Stroud, Science, <u>235</u>, 448 (1982).
38. D.A. Matthews, K. Appelt, S.J. Oatley, and Ng.H. Xuong, J.Mol.Biol., <u>214</u>, 923 (1990).
39. D.A. Matthews, J.E. Villafranca, C.A. Janson, W.W. Smith, K. Welsh and S. Freer, J.Mol.Biol., <u>214</u>, 937 (1990).
40. W.R. Montfort, K.M. Perry, E.B. Fauman, J.S. Finer-Moore, G.F. Maley, L. Hardy, F. Maley and R.M. Stroud, Biochemistry, <u>29</u>, 6964 (1990).
41. J.S. Finer-Moore, W.R. Montfort and R.M. Stroud, Biochemistry, <u>29</u>, 6977 (1990).
42. M.D. Varney, G.P. Marzoni, C.L. Palmer, J.D. Deal, S. Webber, K.M. Welsh, R.J. Bacquet, C.A. Bartlett, C.A. Morse, C.L.J. Booth, S.M. Herrmann, E.F. Howland, R.W. Ward and J. White, J. Med. Chem., in press (1992).
43. S.H. Reich, M.M. Fuhry, D. Nguyen, M.J. Pino, K.M. Welsh, S. Webber, C.A. Janson, S.R. Jordan, D.A. Matthews, W.W. Smith, C.A. Bartlett, C.L.J. Booth, S.M. Herrmann, E.F. Howland, C.A. Morse, R.W. Ward and J. White, J. Med. Chem., in press (1992).
44. P.C. Weber, Adv. Prot. Chem., <u>41</u>, 1 (1991).
45. A. McPherson, "Preparation and Analysis of Protein Crystals", Wiley, New York, (1982).
46. W. A. Hendrickson, Science <u>254</u>, 51 (1991).
47. S.W. Fesik. In "Computer-Aided Drug Design. Methods and Applications"; T.J. Perun and C.L. Propst, Eds., Marcel Dekker, New York, 1989, p. 133 .
48. S.W. Fesik, J. Med. Chem., <u>34</u>, 2937 (1991).
49. S.W. Fesik, J.R. Luly, J.W. Erickson and C. Abad-Zapatero, Biochemistry, <u>27</u>, 8297 (1988).
50. G. Wider, C. Weber, R. Traber, H. Widmer and K. Wüthrich, J. Am. Chem. Soc., <u>112</u>, 9015 (1990).
51. C. Weber, G. Wider, B. von Freyberg, R. Traber, W. Braun, H. Widmer and K. Wüthrich, Biochemistry, <u>30</u>, 6563 (1991).

52. S.W. Fesik, R.T. Gampe, Jr., T.F. Holzman, D.A. Egan, R. Edalji, J.R. Luly, R. Simmer, R. Helfrich, V. Kishore and D.H. Rich, Science, 250, 1406 (1990).
53. S.W. Fesik, R.T. Gampe, Jr., H.L. Eaton, G. Gemmecker, E.T. Olejniczak, P. Neri, T.F. Holzman, D.A. Egan, R. Edalji, R. Simmer, R. Helfrich, J. Hochlowski and M. Jackson, Biochemistry, 30, 6574 (1991).
54. A.M. Petros, R.T. Gampe, Jr., G. Gemmecker, P. Neri, T.F. Holzman, R. Edalji, J. Hochlowski, M. Jackson, J. McAlpine, J. R. Luly, T. Pilot-Matias, S. Pratt and S.W. Fesik, J. Med. Chem., 34, 2925 (1991).
55. S.W. Fesik and E.R.P. Zuiderweg, Q. Rev. Biophys., 23, 97 (1990).
56. S.W. Fesik, H.L. Eaton, E.T. Olejniczak, E.R.P. Zuiderweg, L.P. McIntosh and F.W. Dahlquist, J. Am. Chem. Soc., 112, 886 (1990).
57. L.E. Kay, M. Ikura and A. Bax, J. Am. Chem. Soc., 112, 888 (1990).
58. L.E. Kay, M. Ikura, R. Tschudin and A. Bax, J. Magn. Reson., 89, 496 (1990).
59. A. Bax, G.M. Clore and A.M. Gronenborn, J. Magn. Reson., 88, 425 (1990).
60. L.E. Kay, G.M. Clore, A. Bax and A.M. Gronenborn, Science, 249, 411 (1990).
61. E.R.P. Zuiderweg, A.M. Petros, S.W. Fesik and E.T. Olejniczak, J. Am. Chem. Soc., 113, 370 (1991).
62. G.M. Clore, L.E. Kay, A. Bax and A.M. Gronenborn, Biochemistry, 30, 12 (1991).
63. D.C. Muchmore, L.P. McIntosh, C.B. Russell, D.E. Anderson and F.W. Dahlquist, Meth. Enzymol., 177, 44 (1989).
64. D.M. LeMaster and F.M. Richards, Biochemistry, 27, 142 (1988).
65. D. Neri, T. Szyperski, G. Otting, H. Senn and K. Wüthrich, Biochemistry, 28, 7510 (1989).
66. G.M. Clore, P.T. Wingfield and A.M. Gronenborn, Biochemistry, 30, 2315 (1991).
67. G.M. Clore and A.M. Gronenborn, Science, 252, 1390 (1991).
68. D.M. LeMaster and F.M. Richards, Anal. Biochem., 122, 238 (1982).
69. R.A. Venters, T.L. Calderone, L.D. Spicer and C.A. Fierke, Biochemistry, 30, 4491 (1991).
70. J. Reisman, I. Jariel-Encontre, V.L. Hsu, J. Parello, E.P. Geiduschek and D.R. Kearns, J. Am. Chem. Soc., 113, 2787 (1991).
71. D.M. LeMaster, Q. Rev. Biophys., 23, 133 (1990).
72. S.W. Fesik, E.R.P. Zuiderweg, E.T. Olejniczak and R.T. Gampe, Jr., Biochem. Pharmacol., 40, 161 (1990).
73. R.E. Handschumacher, M.W. Harding, J. Rice, R.J. Drugge and D. Speicher, Science 226, 544 (1984).
74. J. Liu, J.D. Farmer, Jr., W.S. Lane, J. Friedman, I. Weissman and S.L. Schreiber, Cell, 66, 807 (1991).
75. H.R. Loosli, H. Kessler, H. Oschkinat, H.P. Weber, T.J. Petcher and A. Widmer, Helv. Chim. Acta, 68, 682 (1985).
76. P. Neri, R. Meadows, G. Gemmecker, E. Olejniczak, D. Nettesheim, T. Logan, R. Simmer, R. Helfrich, T. Holzman, J. Severin and S. Fesik, FEBS Lett., 294, 81 (1991).
77. V.F.J. Quesniaux, M.H. Schreier, R.M. Wenger, P.C. Hiestand, M.W. Harding and M.H.V. VanRegenmortel, Eur. J. Immunol., 17, 1359 (1987).
78. P.L. Durette, J. Boger, F. Dumont, R. Firestone, R.A. Frankshun, S.L. Koprak, C.S. Lin, M.R. Melino, A.A. Pessolano, J. Pisano, J.A. Schmidt, N.H. Sigal, M.J. Staruch and B.E. Witzel, Transplant. Proc., 20, 51 (1988).
79. S.W. Fesik, G. Gemmecker, E.T. Olejniczak and A.M. Petros, J. Am. Chem. Soc., 113, 7080 (1991).
80. E.R. Price, L.D. Zydowsky, M. Jin, C.H. Baker, F.D. McKeon and C.T. Walsh, Proc. Natl. Acad. Sci. USA, 88, 1903 (1991).
81. P. Neri, G. Gemmecker, L.D. Zydowsky, C.T. Walsh and S.W. Fesik, FEBS Lett., 290, 195 (1991).
82. K. Wüthrich, NMR of Proteins and Nucleic Acids, John Wiley and Sons, New York (1986).
83. G.T. Montelione and G. Wagner, J. Am. Chem. Soc., 111, 5474 (1989).
84. G.T. Montelione, M.E. Winkler, P. Rauenbuehler and G. Wagner, J. Magn. Reson., 82, 198 (1989).
85. G. Gemmecker and S.W. Fesik, J. Magn. Reson., 95, 208 (1991).
86. L.E. Kay and A. Bax, J. Magn. Reson., 86, 110 (1990).
87. P. Guntert, W. Braun, M. Billeter and K. Wüthrich, J. Am. Chem. Soc., 111, 3997 (1989).
88. P.C. Driscoll, A.M. Gronenborn and G.M. Clore, FEBS lett., 243, 223 (1989).
89. G. Gemmecker, E.T. Olejniczak and S.W. Fesik, J. Magn. Reson., 96, 199 (1992).
90. M. Billeter, A.D. Kline, W. Braun, R. Huber and K. Wuthrich, J. Mol. Biol., 206, 677 (1989).
91. T.A. Holak, W. Bode, R. Huber, J. Otlewski and T. Wilusz, J. Mol. Biol., 210, 649 (1989).
92. E.T. Baldwin, I.T. Weber, R. St. Charles, J.-C. Xuan, E. Appella, E. Yamada, K. Matsushima, B.F.P. Edwards, G.M. Clore, A.M. Gronenborn and A. Wlodawer, Proc. Natl. Acad. Sci. USA, 88, 502 (1991).
93. R. Huber, H. Scholze, E.P. Paques and J. Deisenhofer, Hoppe-Seyler's Z. Physiol. Chem., 361, 1389 (1980).
94. D.G. Nettesheim, R.P. Edalji, J.W. Malison, J. Greer and E.R.P. Zuiderweg, Proc. Natl. Acad. Sci, USA, 85, 5036 (1988).
95. G.M. Clore, E. Appella, E. Yamada, K. Matsushima, and A.M. Gronenborn, Biochemistry, 29, 1689 (1990).
96. G.D. Van Duyne, R.F. Standaert, P.A. Karplus, S.L. Schreiber and J. Clardy, Science, 252, 839 (1991).
97. P. Karuso, H. Kessler and D.F. Mierke, J. Am. Chem. Soc., 112, 9434 (1990).

98. S.W. Fesik, P. Neri, R. Meadows, E.T. Olejniczak and G. Gemmecker, J. Am. Chem. Soc. in press (1992).
99. H. Ke, L.D. Zydowsky, J. Liu and C.T. Walsh, Proc. Natl. Acad. Sci. USA, 88,9483 (1991).
100. S.L. Schreiber, Science, 251, 283 (1991).
101. J. Kallen, C. Spitzfaden, M.G. M. Zurini, G. Wider, H. Widmer, K. Wuthrich and M.D. Walkinshaw, Nature, 353, 276 (1991).
102. A.T. Brunger, R.L. Campbell, G.M. Clore, A.M. Gronenborn, M Karplus, G.A. Petsko and M.M. Teeter, Science, 235, 1049 (1987).
103. P.L. Weber, O. Epp, K. Wuthrich and R. Huber, J. Mol. Biol., 206, 669 (1989).
104. M.A. Navia, P.M.D. Fitzgerald, B.M. McKeever, C.-T. Leu, J.C. Heimbach, W.K. Herber, I.S. Sigal, P.L. Drake and J.P. Springer, Nature, 337, 615 (1989).
105. A. Wlodawer, M. Miller, M. Jaskollski, B.K.Sathyanarayana, E. Baldwin, I.T. Weber, L.M. Selk, L. Clawson, J. Schneider and S.B.H. Kent, Science, 245, 616 (1989).
106. R. Lapatto, T. Blundell, A. Hemmings,, J. Overington, A. Wilderspin,S. Wood,J.R. Merson,P.J. Whittle, D.E. Danley,K.F. Geoghegan,S.J. Hawrylik,S.E. Lee,K.G. Scheld, and P.M. Hobart, Nature, 342, 299 (1989).
107. J.F. Davies, Z. Hostomska, Z. Hostomsky, S.R. Jordan and D.A. Matthews, Science, 252, 88 (1991).
108. J. Wang, Y. Yan, T.P.J. Garrett, J. Liu, D.W. Rodgers, R.L. Garick, G.E. Tarr, Y. Husain, E.L. Reinherz and S.C. Harrison, Nature, 348, 411 (1990).
109. S.-E. Ryu, P.D. Kwong, A. Truneh, T.G. Porter, J. Arthos, M. Rosenberg, X. Dai, N.-H. Xuong, R. Axel, R.W. Sweet and W.A. Hendrickson, Nature, 348, 419 (1990).
110. M.F. Summers, T.L. South, B. Kim, D. Hare, Biochemistry, 29, 329 (1990).
111. S. Harrison, Nature, 353, 715 (1991).
112. A.M. De Vos, M. Ultsch, A.A. Kossiakoff, Science, 255, 306 (1992).
113. H. Michel, Trends Biochem. Sci., 8, 56 (1983).
114. J. Diesenhofer, O. Epp, K. Miki, R. Huber and H. Michel, Nature, 318, 618 (1985).
115. M. Weiss, U. Abele, J. Weckesser, W. Welte, E. Schiltz and G.E. Schultz, Science, 254, 1627 (1991).
116. R. Henderson, J.M. Baldwin and T.A. Ceska, J. Mol. Biol., 213, 899 (1990).

Chapter 30. Three-dimensional Models of G-Protein Coupled Receptors

Christine Humblet and Taraneh Mirzadegan
Parke-Davis Pharmaceutical Research Division
Warner-Lambert Company, Ann Arbor, MI 48105

Introduction - An increasingly detailed understanding of the structural biology of G-protein coupled receptors (GPCRs) now offers the opportunity to construct 3-dimensional (3D) functional models to explain their mechanisms of action (1). Beyond the distinct chemical nature of their ligands (*i.e.*, cationic neurotransmitters, hormonal peptides), the molecular details of the GPCR family (*i.e.*, amino acid composition and secondary structures) continue to reveal similarities and differences that enlighten our understanding of their various functions and selectivities.

In the absence of crystallographic data, experimental results combined with predictive techniques based on the protein primary sequences can be used to elaborate 3D models. Predictive algorithms have been mostly applied to determine the location of hydrophobic domains. Other technological breakthroughs (*e.g.*, electron cryo-microscopy, EPR studies of spin labelled-mutants) have dramatically improved the ability to describe the membrane protein topology beyond the limited primary sequence, toward the construction of hypothetical 3D molecular models (2,3). The current 3D molecular models of the GPCR membrane protein remain speculative and are mostly limited to the transmembrane helices. However, their use in combination with comparative sequence analyses, biophysical experiments, site-directed mutagenesis, and structure-activity relationships provides a potentially invaluable molecular template for innovative drug design.

STRUCTURAL BASES FOR THE SEVEN TRANSMEMBRANE PROTEINS

The cloning and sequencing of the GPCRs has revealed a superfamily of GPCR proteins which activate various guanine-nucleotide binding *(G) proteins* (Table 1) (1,4-6). Throughout this chapter, a distinction is made between the GPCRs on the basis of the chemical nature of their ligands. The non-peptide GPCRs include the opsins, which bind retinal analogs, and the neurotransmitter receptors, which bind cationic amines. The peptide GPCRs include a variety of hormonal receptors. The odorant receptors are not included in this classification since the nature of their ligands remains unknown.

The Seven Transmembrane Helical Proteins - Bacteriorhodopsin (BR), the major light-sensitive protein of the purple membrane of *Halobacterium halobium* was the first membrane bound protein found to be organized into seven transmembrane spanning domains (7). More recently, reconstruction of BR by electron microscopy has been accomplished at high resolution (3Å), showing the description of the seven transmembrane domain helical packing, the interhelical loop connection pattern, and the location of the bound all-*trans* retinal chromophore. This structural topology is shown in Figure 1(2). Although BR is not a GPCR, the sensory receptors (visual opsins, rhodopsin and blue, green and red cone pigments) do require an intermediary G-protein called transducin to activate the second messenger system. In common with BR, rhodopsin (RH) binds a retinal chromophore through Schiff base formation. The chromophore is a distinct retinal isomer (11-*cis* retinal) which also covalently binds a lysine in the seventh transmembrane helix. RH has been the subject of multiple chemical and biophysical studies (8,9), which have determined many aspects of its mechanism of action at the molecular level, and have also provided indirect indications for its structural topology which is reminiscent of the seven membrane-spanning helices observed in BR. Subsequently, the whole field of GPCRs changed significantly when it was recognized that the opsins, the adrenergic receptor, and other neurotransmitter receptors are encoded by genes with

Table 1: The GPCR Families and Recent Sequences

Non-Peptide Ligand		Ref.	Peptide Ligand			Ref.
α- and β-adrenergic		(10)				
adenosine	A1	(11,12)	angiotensin II	A1	bovine	(16)
	A2	(13)			rat	(17)
cAMP			bombesin		human	(10)
cannabinoid			bradykinin	B2	rat	(18)
dopamine	D1 to D5		c5a		human	(19)
histamine			endothelin	ET_a	human	(20)
muscarinic	M1 to M5		endothelin	ET_b	human	(21)
serotonin	5HT1,5HT2	(14,15)	fMLP		human	(22)
tyramine			FSH		human	(23)
			interleukin-8		human	(1)
			LH-CG			
			neurotensin			
Sensory			neuropeptide	Y1	rat	(24)
			neuropeptide	Y3	bovine	(25)
light (Opsin)			tachykinin	Nmk		
rod	rhodopsin		tachykinin	Sk		
cone	red		tachykinin	Sp	human	(26)
	green		TRH			
	blue		TSH		human	(27)
			thrombin		human	(28)
Odorant			thromboxane	A2		
			ste2/a-mating factor			
			VIP		human	(29)

Figure 1. Structural topology of the GPCRs and highly conserved residues. Residues studied by site-directed mutagenesis are indicated in bold and intracellular regions involved in signal transduction are highlighted with parentheses.

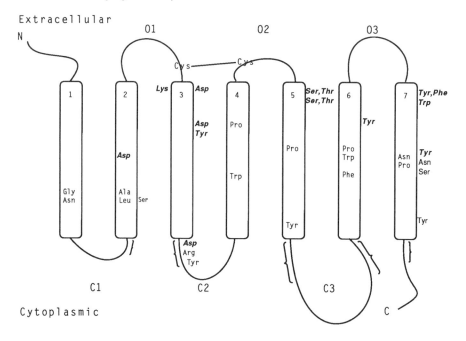

similar features, including high sequence homologies, the presence of key residues involved in folding, binding or activation and the canonical seven-membrane spanning organization.

Hydrophobic Transmembrane Regions - To identify the hydrophobic segments in the GPCR primary sequences, various theoretical methods, such as hydrophobicity profiles, hydrophobic moments, helix amphiphilicity and secondary structure prediction can be applied. A large variety of multi-sequence alignments have appeared in the recent literature (30-36). They present varying degrees of similarity, the details of which are well beyond the scope of this chapter. One proposed multi-sequence alignment is presented in Table 2 (37). For clarity, the results are limited to the hydrophobic segments of a selection of GPCR sequences (ETB hum: human endothelin ETb; AT1 rat: rat angiotensin II A1; M1 hum: human muscarinic M1; D1 rat: rat dopamine D1; Opsin hum: human opsin). A Kyte-Doolittle scale (window size of 19 residues) has been applied to extract the hydropathy plots of the selected primary sequences.

Conserved Residues - Highly conserved residues, either totally conserved (inside the helix) or partially conserved (outside the helix) have been incorporated in Figure 1. Among the conserved residues, four prolines are buried in helices 4,5,6 and 7. Due to their cyclic nature, Pro residues restrict the psi torsional properties as compared to other natural amino acids (38,39) and are thus likely to kink the helices away from their usual linear axes. The transmembrane helices contain various polar amino acid residues. Multiple reports describing the respective roles of the aspartic residues have appeared in the recent literature (1,16,40). The most conserved non-charged residues generally appear in the lower part of the transmembrane helices, close to the cytoplasmic region. Many of them are speculated to be involved in the transduction cascade. The tryptophan residue in transmembrane helix 6 is one of the most conserved in all GPCRs. This suggests a critical structure-function role. From its location directly under the chromophore in BR and the opsins, it is speculated to be involved in a gating mechanism responsible for the transduction triggering (41). The major difference between non-peptide and peptide GPCR subsets is observed in the upper part of helix 3 where Asp is substituted with a Lys or Arg, thus leading to a switch in ionic character.

Table 2: Multi-sequence Alignment of the Helical Regions for Selected GPCRs (plain text, hydrophilic; *italic*; hydrophobic; underline; acidic; and shadow; basic residues).

```
Label      :.........HELI.........................HELII.............HELIII.
ETB hum    :YINTVVSCLVFVLGIIGNSTLLR---GPNILIASLALGDLLHIVIDIPI---CKLVPFIQKA
AT1 rat    :VMIPTLYSIIFVVGIFGNSLVVI---VASVFLLNLALADLCFLLTLPLW---CKIASASVSF
M1 hum     :ASIGITTGLLSLATVTGNLLVLI---VNNYFLLSLACADLIIGTFSMNL---CDLWLALDYV
D1 rat     :ILTACFLSLLILSTLLGNTLVCA---VTNFFVISLAVSDLLVAVLVMPW---CNIWVAFDIM
OPSIN hum  :SMLAAYMFLLIVLGFPINFLTLY---PLNYILLNLAVADLFMVLGGFTS---CNLEGFFATL

Label      :...HELIII.................HELIV.................HELV.....
ETB hum    :SVGITVLSLCALSIDRY---EIVLIWVVSVVLAV-PEAIGFD---WWLFSFYFCLPLAIT
AT1 rat    :NLYASVFLLTCLSIDRY---TCIIIWLMAGLASL-PAVIHRN---LTKNILGFLFPFLII
M1 hum     :ASNASVMNLLLISFDRY---MIGLAWLVSFVLWA-PAILFWQ---FGTAMAAFYLPVTVM
D1 rat     :CSTASILNLCVISVDRY---LISVAWTLSVLISFIPVQLSWH---ISSSLISFYIPVAIM
OPSIN hum  :GGEIALWSLVVLAIERY---GVAFTWVMALA-CAAPPLAGWS---IYMFVVHFTIPMIII

Label      :HELV...........HELVI...................HELVII........
ETB hum    :AFFYT---TVFCLVLVFALCWLPLHLSRIL---IGINMASLNSCINPIALYLVS
AT1 rat    :LTSYT---IIMAIVLFFFFSWVP-HQIFTF---ITICIAYFNNCLNPLFYGFLG
M1 hum     :CTLYW---TLSAILLAFILTWTPYNIMVLV---LGYWLCYVNSTINPMCYALCN
D1 rat     :IVTYT---TLSVIMGVFVCCWLPFFISNCM---VFVWFGWANSSLNPIITYAFNA
OPSIN hum  :FFCYG---MVIIMVIAFLICWVPYASVAFY---IPAFFAKSAAIYNPVIYIMMN
```

Connecting Pattern - A wide variety of experimental results (*e.g.*, antibody mapping (42), molecular biology, probe insertion,) have provided ample support for a straightforward interhelical connection scheme, at least for selected non-peptide GPCRs (*e.g.*, adrenergic, serotonin, muscarinic, rhodopsin). The extracellular stretches connect helices 2-3, 4-5 and 6-7 while the intracellular loops link helices 1-2, 3-4, and 5-6 as described in Figure 1.

Extracellular Regions - The N-terminal segment length varies to extreme across the GPCR families. Protein chemistry experiments have identified various N-glycosylation sites (43). While this segment has not been shown to interact with the non-peptide ligands, enhanced ligand binding has been observed in particular peptide receptors (44). The first extracellular loop (O1), where lengths vary from 18 to 25 residues, tends to be slightly more basic than acidic, and presents the lowest polar character when compared to the other loops. The highest degree of residue conservation occurs in the eight residue stretch that precedes the membrane span of helix 3. Until recently, this stretch was characterized by the tryptophan-proline-x-glycine-x-x-x-cysteine (WPxGxxxC) sequence in which the last cysteine residue is totally conserved and has been shown to form a disulfide bridge with a cysteine in the next, proximal extracellular loop. In the peptide receptors, the Trp residue is conserved throughout. In the non-peptide GPCRs, it is totally conserved except for the A1 and A2 adenosine and the opsin receptors. The second extracellular loop (O2) spans lengths ranging from 20 to 40 residues with a large proportion of charged or hydrogen-bonding residues. This often leads to a net negative charge. O2 includes, at varying positions, the conserved Cys residue known to form a disulfide bridge with loop O1 (45). The third extracellular loop (O3) has very short spans (11 to 21 residues) in the non-peptide receptors with slightly longer ranges (18 to 25 residues) in the peptide GPCRs. Interestingly, the loop incorporates two cysteines in the former subset and none or only one in the latter.

Intracellular Regions - The intracellular loops and C-terminal fragments have been shown to affect signal transduction through coupling to the G-protein (1,34,46). Throughout all GPCRs, they are characterized by a heavy content of basic amino acid residues. The first intracellular loop (C1) has a very short range (8 to 13 residues) containing up to five basic residues in all GPCRs. Interestingly, the occurrence of 1 or 2 cysteine residues is only observed in some of the peptide GPCRs. The second intracellular loop (C2) contains 22 to 25 residues. Until recently, this loop was invariably characterized by the presence of the totally conserved Asp-Arg-Tyr (DRY) triad believed to interact very specifically with the G protein. It now appears that the tyrosine residue can be replaced by Phe, Cys or His (22,24,28). Five residues downstream from the DRY triad, all non-peptide GPCRs except the α_2-adrenergic subtype, present a highly conserved proline residue. In the peptide GPCRs, this proline consensus is not as clear. A number of peptide receptors present 1 or 2 proline residues much closer to helix 4, but the interleukin-8 and bradykinin sequences do not contain proline. The third intracellular loop (C3), characterized by the richest overall polarity, varies considerably and can incorporate more than 100 residues. Numerous experimental results have clearly demonstrated the importance of loop C3 in G-protein binding. The C-terminal intracellular sequence varies in sizes and contains multiple phosphorylation sites (1,34) and a cysteine residue that is rather conserved across non-peptide GPCRs. In the peptide GPCR, the cysteine position is not constant. It has been shown that loop C3 and the amino-terminus segment of the carboxyl terminal have a cooperative role in anchoring the receptor in the membrane (4).

Mutation, Chimeric Studies, Antibody Mapping and Site-specific Synthetic Peptides - Results from the multiple site-directed mutations in the major GPCRs are summarized in Table 3. Mutated residues have also been highlighted in Figure 1 (**bold characters**). Various chimeric studies applied to the α_2-adrenergic and muscarinic GPCRs have been recently reviewed (1,47). These studies confirm earlier reports indicating that C2, C3 as well as a segment of the C-terminus, are implicated in the G-protein coupling as highlighted in Figure 1. Site-specific synthetic peptides based on sequence of the connecting loops in avian β-adrenergic receptor were recently described (46). C1 is thought to impair structural parameters while C2, the N-terminal region of C3 and N-terminal region of the C-terminal intracellular segment have been observed to block the effect on adenylate activation at micromolar doses.

Table 3: Particular Residue Functions Determined by Mutagenesis

GPCR	Residue	Affect Binding	Affect Activation	Ref.
Rhodopsin	E113-Q	counterion	yes/indirectly	(16)
	E134,Q	no	yes/GTP	(16)
	Q135-Q	no	yes/GTP	
	D83-N	no	yes/indirectly	(16)
	K248-A	no	yes/GTP	(48)
	E122-Q		yes/GTP	(16)
β_2-adrenegic	C106,C184,	*		(45)
	C190,C191	*		
	D130-N	agonist, not antagonist	yes	(49)
	D113-N	agonist/antagonist, binds N+		(50)
	D113-E	antagonist-partial agonist		(51)
	S204-A	binds m-OH		(52)
	S207-A	binds p-OH		(52)
	D79-N	no/indirectly	yes (effector)	(50)
	D130-N	no	yes (GP)	(49)
α_2-adrenergic	D113-N	agonist/antagonist, binds N+		(40)
	S204-A	binds p-OH		(40)
	S200-A	no direct H-bond		(40)
	F412-N	increase β-antagonism decrease α-antagonism		(53)
	D79-N	no/indirectly	yes/indirectly	(54)
	D130-N	no	yes	(40)
muscarinic M1	D71-N	no/indirectly	GTP binding	(55)
	D122-N	no	GTP binding	(55)
	D99-N	agonist/antagonist		(55)
	D105-N	agonist/antagonist, binds N+		(55)
muscarinic M3	T231-A	agonist	reduction	(56)
	T234-A	agonist	reduction	(56)
	T148-F	agonist	reduction	(56)
	Y506-F	agonist	reduction	(56)
	Y533-F	agonist	no	(56)
	Y529-F	agonist	no	(56)
	S120-A	antagonist	no	(56)
	T502-A	no	no	(56)
	T537-A	no	no	(56)
dopamine D2	D80-A,E	no/indirectly	impairs	(57)
	D114-N	agonist/antagonist, binds N+	no/indirectly	(58)
	S194-A	binds catechol p-OH	?	(58)
	S197-A	binds catechol m-OH	?	(58)
lutropin	D383-N	yes	yes	(44)
	E429-N	no	no	(44)
	D556-N	no	no	(44)
endothelin ET_b	K181-D	agonist	no	(59)

* affect folding

THREE-DIMENSIONAL STRUCTURAL MODELS

A number of recently proposed structural models for the non-peptide GPCRs suggest schematic 2-dimensional (2D) representations (60-63), 3D models limited to selected helices (64) or highly refined 3D models which sometimes incorporate the connecting loops (8,35,65-71). The molecular details of the 3D models are very difficult to describe and compare in depth due to differences in the procedures used in model building and, more importantly, the unavailability of final atomic coordinates.

Sequences and Hydrophobic Segments - The identification of the hydrophobic bundles remains challenging and relies on the use of various hydrophobicity scales such as those of Kyte-Doolittle (72), Engelman et al. (73), and the Hopp and Woods matrix (74) to derive hydropathy plots and extract the presumed transmembrane segments. The presence of spare hydrophilic or charged residues buried in the hydrophobic core and the increasing degree of hydrophilicity observed toward the exposed extracellular and intracellular regions further hamper the precise localization of the helical end residues. This leads to major variation in the delineation of the hydrophobic segments, the residue window size varying from 19 to 27 residues. The selection of the primary sequences and their alignments will not only be a necessary starting point before any structural model can be assembled, they will also greatly impact the details of the final 3D models. Although BR presents only a remote primary sequence homology to RH and other GPCRs, it remains the most frequently used template for initial sequence alignments. The incorporation of RH, which would be expected to be more closely related to BR than the other GPCRs themselves, has only appeared in few reports (8,9,66). In general, linear sequential alignment is undertaken between the primary sequences. Considering that exon shuffling could have taken place during gene evolution a non-sequential alignment of the hydrophobic segments led to improved homology scores (69).

Helical Packing - Relative helical positioning (i.e., helix packing, individual depth of each helix and orientation and rotation of the vertical axes) is the next step toward 3D model assembly. In attempting to pack seven transmembrane helices, most approaches are based on the simple and convenient assumption that, despite the rather distant sequence homology, the helical packing observed in the crystal structure of BR can be used as the initial 3D template. Controversial alternatives have also been proposed such as the non-sequential alignment (69) and the de novo modeling approach (65). In the former, the resulting helical packing is drastically rearranged, while the latter proposes a "mirror image" arrangement that is also distinct from BR. Once the helix packing has been selected, molecular modeling methods can be applied to orient the individual helices along their axes (height, rotation and tilting). Various criteria have been considered including the antiparallel arrangement of adjacent helices, the hydrophobic character of the interhelical interactions, the proline-induced kinks in the helices, the side chain orientation as to present hydrophobic residues toward the membrane bilayer and conserved or charged residues toward the internal cavity (39,70,75). Computational methods (e.g., molecular mechanics, molecular dynamics, electrostatic potentials) are then applied to optimize details of the structural model.

Loop Connectivities and Construction - Very few modeling reports have been expanded to include the challenge of loop modeling. The methods used have mostly relied upon molecular dynamics simulations, distance geometry approach (DGEOM) and a method based on structural knowledge using protein crystal structures deposited in the Brookhaven Databases (35,65,66,68).

Binding Site and Ligand Docking - The internal proline residues impose a curving effect in the helices, molding the transmembrane domains in an overall oblong shape (65,66). The key charged residues are trapped in an elongated cavity approximately in a plane horizontal to the helical axes, at the depth of the proline occurrence. In most studies, the molecular description of the binding site has been clarified with ligand docking experiments using the guidelines derived from available mutation studies (see Table 3) (65,68). The cationic amine is positioned to interact electrostatically with the second, lower aspartic residue present in the third transmembrane helix (e.g., Asp114 in dopamine D2). An exception to this has appeared in a D2 dopamine model which suggests that Asp80 of the second helix interacts with the cationic head of dopamine (68) . Such an interaction is not corroborated by the mutation results. In rhodopsin, the 11-cis-retinal chromophore was docked to bind Lys296 to form the protonated Schiff base (66). Beyond the general acceptance that the core aspartic residue interacts with the cationic amines, the 3D models concur in their description of two hydrophobic regions including various Phe or Trp residues. One is located in the vicinity of the cationic heads of the ligand, while the other surrounds the aromatic rings in the ligands. Serine residues are shown to undertake hydrogen-bonding interactions with the ortho and para-hydroxy phenyl substituents. One D2 dopamine model, however, implied the involvement of the Asp80 residue instead of the serines, in contradiction with the mutation results presented in Table 2 (76). The binding cavities of representative GPCRs are schematized in Figure 3. The approximate side chain orientations of the key residues are represented in the cross-sectional plane of the transmembrane helices. With the exception of the RH model (66) which incorporated extensive structure-activity data, only a few substrates or ligands have been docked in the GPCR 3D models. The docking studies have assisted the elucidation and elaboration of the molecular details of the binding pockets. For example, two models of the β_2-adrenergic receptor were able to justify the R configuration of the β-hydroxyl moiety of epinephrine (35,65). Structural differences have also been suggested to differentiate agonist from antagonist binding site requirements. Thus, residues belonging to the helices 3,6,5 and 7 seem to impact the agonist specificity while those belonging to the helices 2,7 and 3 are related to antagonistic profiles (see Figure 2). Further work is needed to conclusively understand ligand function and more importantly, their selectivity.

G-protein Coupling and Transduction Triggering - It has long been speculated that a receptor conformational change, triggered by agonist binding, would be responsible for initiating the resulting transduction mechanism. Recent theoretical, experimental, and modeling data suggest an active proline involvement toward this effect (65,69). In one study, two amphiphilic helical segments predicted in the third intracellular loop have been shown to align parallel to each other, in close proximity to the helical bundles 5 and 6 (65). It is suggested that a conformational change in the helical bundles could alter the orientation of these cytoplasmic helices leading to the interaction with the G_s coupling.

Peptide GPCRs - Variations on the familiar non-peptide GPCR structural theme have been reported in very few peptide receptors (e.g., thrombin (28) and lutropin hormone (LH) (44) receptors). The thrombin receptor sequence presents a putative cleavage site for the serine protease thrombin in the N-terminal extracellular domain of the receptor. The new terminus created by cleavage at this site apparently constitutes the physiological "tethered" ligand for the receptor. In the follicle stimulating hormone (FSH/LH) family, additional extracellular N-terminal domains have been incorporated into the basic structure. It has been shown that the 335 residue N-terminal domain binds the lutropin hormone. These two peptide receptors thus provide the first evidence for differences in the ligand binding mode and/or the subsequent cAMP activation mechanism. In contrast, results obtained for the endothelin and LH receptors indicate that the N-terminal domain is not necessary for ligand binding or activation of the G-protein (34,59). Although it seems premature to describe a peptide GPCR binding mode, preliminary evidence indicates that there might be multiple binding sites, including an attachment to the extracellular N-terminal fragment. Simultaneously, the presence of the highly conserved basic (Arg or Lys) residue in helix 3 for the peptide receptors tends to indicate the involvement of this residue in an electrostatic interaction within the transmembrane domain (59).

Conclusions - The structural biology of the GPCRs decisively progressed when the first seven transmembrane protein structure was recently solved at high crystallographic resolution. Simultaneously, synergistic application of site-directed mutagenesis, protein chemistry, immunology and computer-assisted molecular modeling have generated a vast amount of experimental results that have enabled the assembly of 3D models. Clearly, their structural validation awaits the crystal structure determination for one or more GPCRs. This is unfortunately not likely to happen until cloning and expression methods successfully overcome current limitations in producing appropriate expression and purity levels.

The 3D models present differences that arise from the hydropathy analyses, the primary sequence alignments, the assignment for helical boundaries and the helix packing. In light of these differences, the consistent localization of particular residues, i.e. key pharmacophoric moieties, in some proposed binding sites is rather surprising. Various binding site models have been shown to corroborate structure-activity data known for the ligands. Although the docking experiments will necessitate an expanded set of ligand structures to further validate and define the molecular details of the binding sites, further studies hold promises for rational drug design. On the contrary, the majority of the studies reported have poorly addressed the dynamic aspects of the connecting loops as well as the impact of conformational changes upon ligand binding. A better understanding of the subtle molecular events involved in the G-protein activation remains the challenge of forthcoming studies.

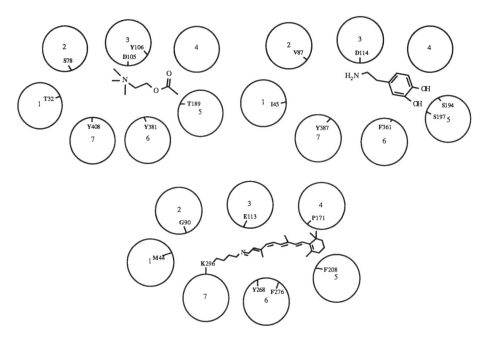

Figure 3: Muscarinic M1, Dopamine D2 , and Rhodopsin binding pockets.

References

1. T. Jackson, Pharmacol.Ther., 50, 425 (1991).
2. R. Henderson, J.M. Baldwin, T.A. Ceska, F. Zemlin, E. Beckmann and K.H. Downing, J.Mol.Biol., 231, 899 (1990).
3. T.A. Nakayama and H.G. Khorana, J.Biol.Chem., 265, 15762 (1990).
4. E.M. Ross, Neuron, 3, 141 (1989).
5. C.W. Taylor, Biochem. J., 272, 1 (1990).
6. R.A.F. Dixon, C.D. Strader and I.S. Sigal, Annu.Rep.Med.Chem., 23, 221 (1988).
7. R. Henderson and P.N.T. Unwin, Nature, 24, 1487 (1975).
8. J.B.C. Findlay and P.F. Zagalsky, Proc. Yamada Conf. XXI, (1988).
9. P.A. Hargrave, J.H. McDowell, R.J. Feldmann, P.H. Atkinson, R.K.M. Rao and P. Argos, Vision Res. 24, 1487 (1984).
10. J.F. Battey, J.M. Way, M.H. Corjay, H. Shapira, K. Kusano, R. Harkins, J.M. Wu, T. Slattery, E. Mann and R.I. Feldman, Proc. Natl. Acad. Sci.USA, 88, 395 (1991).
11. L.C. Mahan, L.D. Mcvittie, R.E.M. Smyk, H. Nakata, F.J.J. Monsma, C.R. Gerfen and D.R. Sibley, Mol.Pharmacol., 40, 1 (1991).
12. S.M. Reppert, D.R. Weaver, J.H. Stehle and S.A. Rivkees, Mol. Endocrinol, 5, 1037 (1991).
13. F. Libert, E. Passage, M. Parmentier, M.J. Simons and M.G. Mattei, Genomics, 11, 225 (1991).
14. Y. Fujiwara, D.L. Nelson, K. Kashihara, E. Varga, W.R. Roeske and H.I. Yamamura, Life Sci., 47, 127 (1990).
15. M.W. Hamblin and M.A. Metcalf, Mol.Pharmacol., 40, 143 (1991).
16. T.P. Sakmar, R.R. Franke and H.G. Khorana, Proc.Natl.Acad.Sci.USA, 86, 8309 (1989).
17. N. Iwai, Y. Yamano, S. Chaki, F. Konishi, S. Bardhan, C. Tibbetts, K. Sasaki, M. Hasegawa, Y. Matsuda and T. Inagami, Biochem.Biophys.Res.Commun., 177, 299 (1991).
18. A.E. McEachern, E.R. Shelton, S. Bhakta, R. Obernolte, C. Bach, P. Zuppan, J. Fujisaki, R.W. Aldrich and K. Jarnagin, Proc.Natl.Acad.Sci.USA, 88, 7724 (1991).
19. N.P. Gerard and C. Gerard, Nature, 349, 614 (1991).
20. K. Hosoda, K. Nakao, A. Hiroshi, S.I. Suga, Y. Ogawa, M. Mukoyama, G. Shirakami, Y. Saito, S. Nakanishi and H. Imura, FEBS Lett., 287, 23 (1991).
21. A. Sakamoto, M. Yanagisawa, T. Sakura, Y. Takuwa, H. Yanagisawa and T. Masaki, Biochem.Biophys.Res.Commun., 178, 656 (1991).
22. P.M. Murphy and D. McDermott, J.Biol.Chem., 19, 12560 (1991).
23. T. Minegish, K. Nakamura, Y. Takakura, Y. Ibuki and M. Igarashi, Biochem.Biophys.Res. Commun., 175, 1125 (1991).

24.   C. Eva, E.J. Kinanen, H. Monyer, P. Seeburg and R. Sprengel, FEBS Lett., 271, 81 (1990).
25.   J. Rimland, W. Xin, P. Sweetnam, K. Saijol, J. Nestler and R. Duman, Mol.Pharmacol., 40, 869 (1991).
26.   Y. Takeda, K.B. Chou, J. Takeda, B.S. Sachais and J.E. Krause, Biochem.Biophys.Res. Commun., 179, 1232 (1991).
27.   G.C. Huang, M.J. Page, A.J. Roberts, A.N. Malik, H. Spence, A.M. McGregor and J.P. Banga, FEBS Lett., 264, 193 (1990).
28.   T.K.H. Vu, D.T. Hung, V.I. Wheaton and S.R. Coughlin, Cell, 64, 1057 (1991).
29.   S.P. Sreedharan, A. Robichon, K.E. Peterson and E.J. Goetzl, Proc.Natl.Acad.Sci.USA, 88, 4986 (1 (1991).
30.   H. Moereels, L. De Bie and J.P. Tollenaere, J.Comput.Aided Mol.Des., 4, 131 (1990).
31.   T.K. Attwood, E.E. Eliopoulos and J.B.C. Findlay, Gene, 98, 153 (1991).
32.   H.G. Dohlman, J. Thorner, M.G. Caron and R.J. Lefkowitz, Annu.Rev.Biochem., 60, 653 (1991).
33.   R. Henderson, F.R.S. Schertler and G.F.X. Schertler, Phil.Trans.R.Soc. (Lond.), 326, 379 (1990).
34.   W.C. Probst, L. Snyder A., D.I. Schuster, J. Brosius and S.C. Sealfon, DNA and Cell Biol., 11, 1 (1992).
35.   M.F. Hibert, S. Trumpp-Kallmeyer, A. Bruinvels and J. Hoflack, Mol.Pharmacol., 40, 8 (1991).
36.   E.C. Hulme, C.A.M. Curtis, M. Wheatley, A. Aitken and A.C. Harris, Fourth International Symposium on Sub-types of Muscarinic Receptors, Wiesbaden, West Germany, 22 (1989).
37.   T. Mirzadegan, R.S.H. Liu and L.U. Colmenares, U.S. Japan Joint Seminar on the Biophysical Chemistry of Rehinal Proteins, V. 1, (1992).
38.   D.J. Barlow and J.M. Thornton, J.Mol.Biol., 201, 601 (1988).
39.   G. Von Heijne, J.Mol.Biol., 218, 499 (1991).
40.   S.K.F. Wong, E.M. Parker and E.M. Ross, J.Biol.Chem., 265, 6219 (1990).
41.   T. Nakayama and H.G. Khorana, J.Biol.Chem., 266, 4269 (1991).
42.   H.Y. Wang, L. Lipfert, C.C. Malbon and S. Bohouth, J.Biol.Chem., 264, 14424 (1989).
43.   E. Rands, M.R. Candelore, A.H. Cheung, W.S. Hill, C.D. Strader and R.A.F. Dixon, J.Biol.Chem., 265, 10759 (1990).
44.   I. Ji and T.H. Ji, J.Biol.Chem., 266, 14953 (1991).
45.   M.R. Candelore, S.L. Gould, W.S. Hill, A.H. Cheung, E. Rands, B.A. Zimcik, I.S. Sigal, R.A.F. Dixon and C.D. Strader in "Biology of Cellular Transducing Signals," J.Y. Vanderhoek, Ed., Plenum, New York, N.Y., 1989, p.190
46.   G. Munch, C. Dees, M. Hekman and D. Palm, Eur.J.Biochem., 198, 357 (1991).
47.   J. Lameh, R.I. Cone, S. Maeda, M. Philip, M. Corbani, L. Nadasdi, J. Ramachandran, G.M. Smith and W. Sadee, Pharm.Res., 7, 1213 (1990).
48.   R.R. Franke, T.P. Sakmar, D.D. Oprian and H.G. Khorana, J.Biol.Chem., 263, 2119 (1988).
49.   C.M. Fraser, F.Z. Chung, C.D. Wang and J.C. Venter, Proc.Natl.Acad.Sci.USA , 85, 5478 (1988).
50.   C.D. Strader, I.S. Sigal, M.R. Candelore, E. Rands, W.S. Hill and R.A.F. Dixon, J.Biol.Chem, 263, 10267 (1988).
51.   C.D. Strader, R.A.F. Dixon, A.H. Cheung, C.M. Rios, A.D. Blake and I.S. Sigal, J. Biol. Chem., 262, 16439 (1987).
52.   R.A.F. Dixon, I.S. Sigal and C.D. Strader, Cold Spring Harbor Symp. Quant. Biol., 53, 487 (1988).
53.   S. Suryanarayana, D.A. Daunt, Z.M. Von and B.K. Kobilka, J.Biol.Chem., 266, 15488 (1991).
54.   D.A. Horstman, S. Brandon, A.L. Wilson, C.A. Guyer, E.J.J. Cragoe and L.E. Limbird, J.Biol.Chem., 265, 21590 (1990).
55.   C.M. Fraser, C.D. Wang, D.A. Robinson, J.D. Gocayne and J.C. Venter, Mol.Pharmacol., 36, 840 (1989) 56.   J. Wess, D. Gdula and M.R. Brann, EMBO J., 10, 3729 (1991).
57.   K.A. Neve, B.A. Cox, R.A. Henningsen, A. Spanoyannis and R.L. Neve, Mol.Pharmacol., 39, 733 (1991).
58.   F. Meng, A. Mansour, J. Meador-Woodruff, L.P. Taylor, and H. Akil, 21st Annual Neuroscience Soc. Mt., 17, 599 (1991).
59.   F.Z. Chung, L.H. Wu, G. Zhu, C. Mauzy, A.M. Egloff, M. Vartanian, B. Wu, W. Cody, T. Mirzadegan, S. Rapundalo and D. Oxender, J.Cell. Biochem., in press (1992).
60.   H.G. Dohlman, M.G. Caron and R.J. Lefkowitz, 26, 2657 (1987).
61.   J.C. Venter and C.M. Fraser in "Drug Discovery Technology," C.R. Clark, W.H. Moos, Eds., Ellis Horwood Ltd., Chichester, UK., 1990, p. 231.
62.   G. Berstein and T. Haga in "Current Aspects of the Neurosciences," Vol. 1, N.N. Osborne, Ed., Macmillan Press, UK., 1990, p. 245.
63.   P.G. Strange, Trends Neurosci., 13 , 373 (1990).
64.   H.-D. Holtje, A. Batzenschlager, H. Briem and J. Bruggmann, Pharm.Unserer.Zeit., 20, 59 (1991).
65.   K. MaloneyHuss and T.P. Lybrand, J.Mol.Biol., 40, 8 (1992).
66.   T. Mirzadegan and R.S.H. Liu, Prog. Retinal Res., 11, 57 (1992).
67.   N.J. Vogelaar, Diss.Abst.Int., 50, 231 (1989).
68.   S.G. Dahl, O. Edvardsen and I. Sylte, Proc.Natl.Acad.Sci.USA, 88, 8111 (1991).
69.   L. Pardo, J.A. Ballesteros, R. Osman and H. Weinstein, Proc.Natl.Acad.Sci.USA, in press (1992).
70.   D. Garvey, C. Hutchins, J. Chung, Y.K. Shue and M. McKinney, Soc. Neuroscience Absts., 16, (1990).
71.   V.B. Cockcroft, B. Brewster and G.G. Lunt ,19th Annual. Neuroscience Soc. Mt. Abst. (1989).

72.   J. Kyle and R.F. Doolittle, J.Mol.Biol., 157,105 (1982).
73.   A. Goldman, D. Engelman and T. Steitz, Ann.Rev.Biophys.Chem., 15, 321 (1986).
74.   T.P. Hopp and K.P. Woods, Proc.Natl.Acad.Sci.USA , 78, 3824 (1981).
75.   D. Donnelly, M.S. Johnson, T.L. Blundell and J. Saunders, FEBS Lett., 251, 109 (1989).
76.   C. Hutchins, ISQBP International Society of Quantum Biology and Pharmacology, Stanford University, Abst. 1.3, (1991).

Chapter 31. Carbohydrates as Drug Discovery Leads

John H. Musser
Glycomed Inc., Alameda, CA 94501

<u>Introduction</u> - Carbohydrates are currently underutilized as leads in drug discovery. Indeed, this Chapter is the first to review saccharide medicinal chemistry in this forum (1). The neglect by the pharmaceutical and academic communities is due to the fact that the biological impact of this class of biopolymer is underestimated. Pharmacologists have long considered these molecules as merely energy stores or integral parts of intracellular matrices. Their apparent random distribution and the lack of adequate research tools have made the localization and functional characterization of carbohydrates difficult. Furthermore, medicinal chemists have considered carbohydrate synthesis, in theory, as involving only carbon-oxygen bond formation. In practice, however, purification and characterization of carbohydrates were found to be tedious and often intractable.

Research on carbohydrates is now undergoing considerable growth and promises to be a major focus as drug discovery leads (2). Carbohydrates are critical in the operation of two fundamental but opposing cellular processes: the proper maintenance of the body's immune defense, and the initiation of bacterial and viral infections. Also, when cellular regulatory mechanisms go astray, as in autoimmune disease or cancer, cell surface carbohydrates change in structure and composition. Finally, the initiation of life itself in egg-sperm recognition may be mediated by carbohydrate-protein interaction. In parallel, chemists are recognizing that carbohydrates are innately exquisite forms of concise informational packages. In comparison to polypeptides or oligonucleotides on a unit basis, carbohydrates have the potential for greater complexity. Two identical amino acids or nucleotides when joined together yield only one dipeptide or one dinucleotide; however, when two identical saccharides are linked, 11 different disaccharides are possible. Thus, of all the biopolymeric structural types, carbohydrates have the greatest theoretical potential for specificity and new lead generation.

<u>Physical Methods</u> - Evolutional advances in analytical instrumentation, including both structural analysis and computational methodologies, are greatly facilitating new lead discovery based on oligosaccharides and polysaccharides. X-ray crystallographic data are providing insight into the atomic features of protein-carbohydrate interactions (3). The importance of polar and non-polar residues for carbohydrate recognition and affinity is becoming more clear with respect to hydrogen bonds, van der Waals contacts per sugar unit and the involvement of water molecules. Apparently, most protein-carbohydrate complexes show stacking of aromatic residues of the protein against both sides

of the sugar ring. Also, the new technique of electrospray is making high quality mass spectra of charged polysaccharides a routine task. Finally, multidimensional NMR techniques allow for the first time the determination of solution conformations of oligosaccharide and ligand/protein interactions.

Computational chemistry is a new area of multidisciplinary research in carbohydrates and involves *ab initio* molecular orbital calculations, semiempirical MO methods, molecular mechanics, and molecular dynamics in the design of new drug candidates. Computer-aided modeling of oligosaccharides employing molecular dynamics simulations is providing useful tools for the study of carbohydrate structure (4). Although it is difficult to obtain reliable parameterization of anomeric effects, molecular mechanics force fields have been calculated which may provide reliable data for the generation of useful oligosaccharide structural information (5).

Progress is being made in the synthesis of biologically active glycopeptides and glycolipids as a result of developments in glycosylation methods and protective group strategies (6). With respect to coupling methodology, the Koenigs-Knorr heavy metal glycosylation is being supplemented with more efficient reactions, such as those using trichloroacetimidate, fluoride ion, or thioether activation. Indeed, the total synthesis of the cell adhesion receptor ligand, Sialyl Lewis X (*vide infra*), is facilitated with stereochemical control using the using a thiophenyl auxiliary (7).

A general criticism of oligosaccharides as leads for drug discovery is their low receptor binding affinities, with K_i's often in the millimolar range. One way of making synthetic oligosaccharides more effective is to prepare multivalent presentation constructs with identical carbohydrate groups attached. These constructs might mimic what is thought to occur in nature, in that each individual carbohydrate interaction contributes to the strength of the whole. For example, the binding between trivalent glycopeptide structures with defined geometry to the asialoglycoprotein receptor of rat heptocytes has been reported (8).

GLYCOCONJUGATES

Many carbohydrates of current interest are expressed glycoconjugates on the cell surface. They can be classified as glycolipids, glycoproteins or proteoglycans and may either be anchored in the cell surface membrane, secreted as semisoluble substances or deposited as insoluble substances (9). A brief description of each glycoconjugate follows.

Glycolipids - Glycolipids are structurally diverse oligosaccharides which are covalently attached to lipids embedded in the plasma membrane on cell surfaces (10). A class of glycolipids which contain ceramide (glycosphingolipids) and and at least one sialic acid is called gangliosides. Gangliosides may be of value for nerve growth and neural repair, although this is not universally accepted since their mechanisms of action remain elusive (11). Apparently, gangliosides act in a multifactorial fashion modulating cAMP levels, protein kinases and nerve growth factors. Clinical trials are underway with GM1 **1** which could determine whether this agent can play a significant role in repairing damaged nervous tissue due to stroke (12).

$\underline{1}$

Gangliosides may modulate other types of cell proliferation by interacting with a number of transmembrane proteins, including the insulin receptor and epidermal growth factor. GM_3 regulates cell growth by inhibition of EGF-dependent receptor autophosphorylation, as indicated in a freeze-thaw protocol using intact A431 cells (13). In contrast, the ganglioside, 2,3-sialosyl-paragloboside, inhibits insulin-dependent cell proliferation via effects on the tyrosine kinase activity of the receptor (14). Apparently, GM_3 is able to inhibit EGF-stimulated phosphorylation of the EGF receptor without affecting ligand/receptor binding.

Glycosaminoglycans - Glycosaminoglycans (GAGS) are composed of long chains of repeating disaccharides and are classified by their carbohydrate constituents, such as iduronic acid, glucuronic acid and glucosamine and sulfation patterns (15). One of the most well known GAGs is heparin.

Heparin is an old drug which has many pharmacological and clinical activities including the well known properties of anticoagulation and anticlotting as well as the less widely studied effects on angiogenesis, cell proliferation and interactions with lipases and growth factors (16). It is interesting to note that GAGs can displace anticoagulant-active heparin from plasma protein binding sites (17). Although heparin has potential as a drug for the treatment of thrombosis, its potency as an anticoagulant and its low bioavailability preclude its development. Low molecular weight heparins (LMWH) are being studied to circumvent these problems (18). The mechanism of actions of LMWHs is understood with respect to antithrombic therapy (19) (Figure I). Although LMWHs are uniform with respect to the inability to inactivate thrombin, an analysis of LMWHs in the prophylaxis of venous thrombosis revealed that antifactor Xa activity can vary significantly among different preparations with similar MWs (20). There are several LMWHs currently undergoing clinical trials for prophylaxis of thromboembolism and perioperative deep vein thrombosis in general surgery. These LMWHs include Fragmin (21), Logiparin (22), Lomoparin (23), Enoxaparin (24), Fraxiparin (25), and RD Heparin (26).

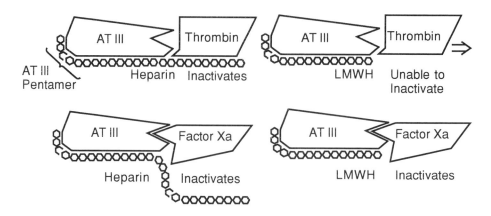

Figure 1   Mechanism of Clotting Factor Inactivation for Heparin and LMWH

Theoretically, millions of sequences consisting of distinct di- to decasaccharides may exist along heparin chains, each with the unique potential to interact with specific protein-binding sites. Indeed, heparin binds to a number of plasma proteins including fibronectin (27) and von Willebrand factor (28). Therefore, heparin may be considered to contain wealth of carbohydrate structures with the potential to act as drug discovery leads. As a specific illustration, the exact structural sequence responsible for antithrombin III binding is known. It is the high affinity AT III synthetic pentasaccharide, **2**, which inhibits clotting factors X and V (29).

$$\underline{\mathbf{2}}\ R = H\ \underline{\mathbf{3}}\ R = OSO_3^-\ (ORG\ 31550)$$

A model to explain many of the aspects of the binding of the ATIII pentasaccharide sequence to is under development (30). SAR studies of **2** indicate that regiosulfation and carboxylate stereochemistry are critical for biological activity (31). For example, an extra 3-O-sulfate group on the reducing sugar [ORG 31550 **3**] enhances the interaction with the protein, whereas, an extra 3-O-sulfate on the glucuronate impedes binding to ATIII. Epimerization of the nonreducing end anomeric center also produces an inactive analog (32). Taken together it appears that these highly negatively charged oligosaccharides have specific requirements with respect to charge distribution and spacial orientation. The existence of other heparin oligosaccharide sequences specific for binding and modulation of biologically important proteins is, thus, possible.

In contrast to direct binding, a novel way to modulate biologically important proteins is to induce the endogenous synthesis of heparin oligosaccharide sequences. A series of lipophilic β-D xylosides was synthesized to examine the structural requirements of the aglycone for GAG priming in CHO cells (33). Of particular interest is estradiol β-D-xyloside <u>**4**</u> which primes heparin sulfate efficiently in a number of cell types including CHO, BHK, BAE, and Balb/cs.

<u>Glycoproteins</u> - Carbohydrates are attached to mammalian proteins primarily to hydroxyl groups of serines or threonines as O-glycosides or to amido groups of asparagines as N-glycosides (34). The carbohydrate groups on natural glycoproteins confer important physical properties such as conformational stability, protease resistance, charge and water-binding capacity. Although some common mannose branching schemes are found in nature, diversity, including heterogeneous protein glycosylation, is the rule. The glycosylation pattern of a protein or glycoform can affect a protein-based drug's potency and determine its target tissue. Indeed, lack of proper glycosylation by microorganisms has mandated the use of mammalian cells in production, instead of simpler bacterial fermentations. Thus, carbohydrate research may affect second-generation protein products, improving their efficacy and possibly reducing dosage levels and side effects. Redesigned glycosylation patterns may also provide glycoproteins that will target specific tissues, have longer half-lives, or be more stable or soluble. Many protein-based therapeutics either fail to function or cause immune reactions when not properly glycosylated. Erythropoetin (EPO), for example, loses all biological activity if deglycosylated. Researchers are learning from nature with respect to transport systems in biological systems in addressing absorption, distribution and metabolism of drugs. For example, significant improvement is observed with peptide renin inhibitors when they are synthetically converted into glycopeptides (35).

With respect to regulatory and proprietary issues, protein glycoforms are becoming part of the strategy for patent applications and drug development. Both regulatory agencies and industry are concerned with carbohydrate characterization because it affects protein drug efficacy with respect to pharmacokinetics and tissue distribution, safety as it relates to immune reactions and innovation as it pertains to patent strategy (36).

<u>Glycosyltransferases and Glycosidases</u> - Glycosyltransferases and glycosidases modulate the biosynthesis of cell-surface complex carbohydrates and have a considerable effect on both the biophysical and biochemical actions of glycoconjugates. Glycosyltransferases have the ability to specifically transfer

glycosyl residues from sugar nucleotides to hydroxyl groups of acceptor oligosaccharides whereas glycosidases process or trim N-linked oligosaccharides. These enzymes represent prime targets for the inhibition.

Glycosidase inhibitors, such as, 1-deoxynorjiramycin **5** and its corresponding N-substituted derivatives continue to generate interest as anti-diabetic and anti-HIV agents (37). A number of deoxygenated oligosaccharide acceptor analogs were evaluated as specific inhibitors of glycosyltransferases (38). For example, **6** inhibited alpha (1,2) fucosyltransferase isolated from pig submaxillary with a K_i of 0.8 mM and a related analog inhibited beta (1,6) glucosyl-N-acetyltransferase V isolated from hamster kidney with a K_i of 0.063 mM. Both inhibitors are competitive.

5   **6**

CARBOHYDRATE-MEDIATED CELL-CELL INTERACTIONS

Cell-surface glycoconjugates may act as cell-cell recognition molecules via specific binding between carbohydrates on one cell and protein receptors on an opposing cell or specific interactions directly between carbohydrates on opposing cells (39) Indeed, blood groups A, B, and O are genetically-regulated carbohydrate antigens found on cell surfaces (40). Adhesion of cells in the immune system is clinically important when considering such indications as cancer, microbial infection, and inflammatory, allergic and autoimmune diseases.

Lectins are simply carbohydrate binding proteins. The structure of the calcium-dependent (C-type) lectin domain from a rat mannose-binding protein was determined by X-ray crystallography employing multiwavelength anomalous dispersion phasing (41). These results are significant since the structure may be used to gain insight into other carbohydrate binding proteins, such as the selectins.

Selectin-Mediated Adhesion - Selectins are a family of glycoproteins that are implicated in the adhesion of leukocytes to platelets or vascular endothelium (42). Adhesion is an early step in leukocyte extravasation which sequela includes thrombosis, recirculation, and inflammation. An excellent review on leukocyte adhesion in inflammation and the immune system was recently published. (43). Three protein receptors, E-, L-, and P-selectins are assigned to the selectin family based on their cDNA sequences. Each contains a domain similar to calcium-dependent lectins or C-lectins, an epidermal growth factor-like domain, and several complement binding protein-like domains (44). An intense effort is ongoing to define the native carbohydrate ligands for each selectin receptor. In the case of E-selectin (ELAM-1, LECAM-2) several laboratories have described the ligands from myeloid cells as the sialyl Lewis X epitope $\underline{7}$ (45). Current information suggests that sialyl Lewis X is also a carbohydrate ligand for P-selectin (GMP-140, LECAM-3) (46) and L-selectin (LECAM-1) (47). Other ligands for the selectins have been partially identified. For example, a sulfated 50 kD glycoprotein which contains both sialic acid and fucose is reported to be a ligand for the L-selectin receptor (48). Finally, there is data which suggests that the heterogeneous 3-sulfated galacosyl ceramides $\underline{8}$, referred to as sulfatides, bind to both the L-Selectin receptor (49) and P-Selectin (50). However, this binding does not appear to calcium dependent.

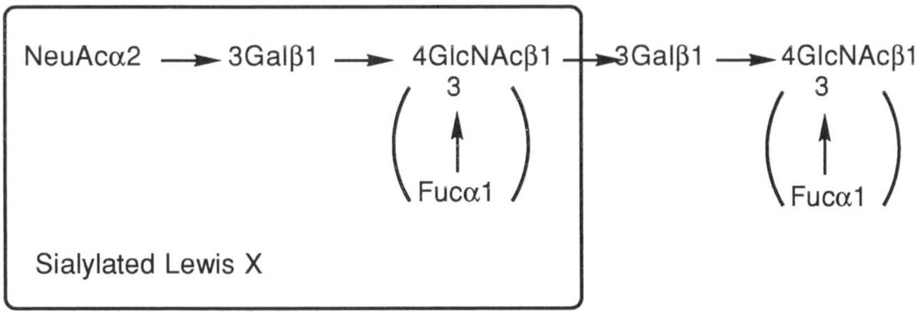

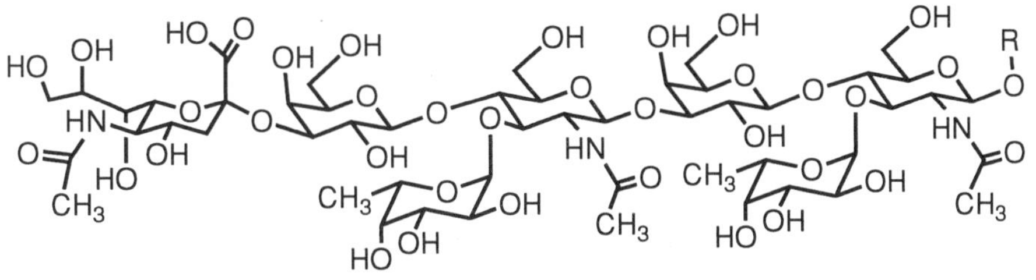

NeuAcα2 ⟶ 3Galβ1 ⟶ 4GlcNAcβ1 ⟶ 3Galβ1 ⟶ 4GlcNAcβ1

Sialylated Lewis X

$\underline{7}$

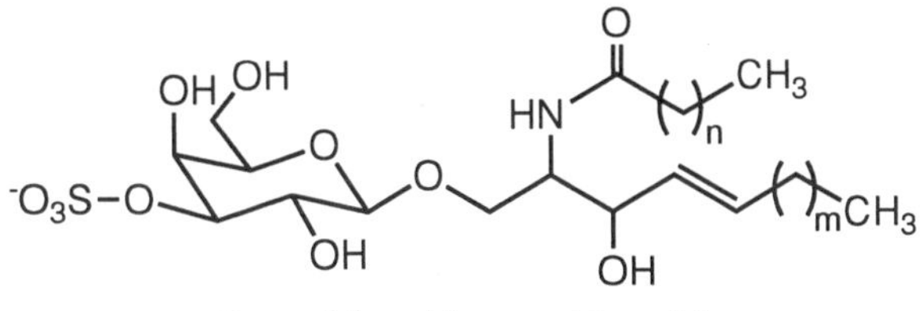

$\underline{8}$ n = 14 or 16, m = 12 or 14

A synthetic peptide, CQNRYTDLVAIQNKNE, inhibits neutrophil binding to P-Selectin (51). This protein was designed based on an epitope spanning residues 19-34 of P-Selectin domain recognized by a blocking monoclonal antibody. Although neutrophil recognition is thought to require a calcium-induced conformational change in the lectin domain of GMP-140, this result indicates that a small polypeptide may be used to block this interaction.

<u>Sperm-Egg Adhesion</u> - Evidence is accumulating that mammalian sperm cells in general express proteins that interact specifically with saccharide receptors

on eggs. For example, a 56 kDa mouse sperm cell surface protein (ZP3) that specifically binds to the oligosaccharide part of a glycoprotein on unfertilized mouse eggs is similar in size to lectin-like receptors on sperm of other animal species (52). Indeed, attachment of mouse blastocysts to endometrial cells is inhibited by a distinct pentasaccharide. Thus, new possibilities for contraception may be possible by modulating fertilization processes that are based on protein-carbohydrate interactions.

Microorganism-Cell Interaction - Carbohydrates are critical in the the initiation of bacterial and viral infections. It is well established that bacterial lectins play a significant role in host-cell recognition, adherence and initiation of infection (53). For example the α-mannosides **9** inhibits *E. coli* adherence to epithelial cells and is 10-fold more potent than naturally occuring oligosaccharides and 1000-fold more potent than α-methyl mannoside. This demonstrates that mannoside derivatives can specifically inhibit bacterial infections at least in rodent experimental models. Both glycoproteins and glycolipids contain terminal sialic acids which bind influenza virus *in vitro* (54). Another factor is multivalent binding via the viral hemagglutinins which are trimeric in the viral envelope and bind most avidly to multiple optimally spaced carbohydrate determinants. In addition to influenza, several other viruses have been reported to bind to glycoconjugates, including rabies virus, Sendai virus, Newcastle disease virus and retroviruses (55).

9

Cancer Metastasis - There is strong evidence that specific cell-surface carbohydrates expressed on colonic, pancreatic and other tumors are involved in metastasis (56). Colonic and pancreatic tumors exhibit high levels of sialyl Lewis X as they progress through metastatic stages of malignancy; human urinary bladder carcinoma metastatic potential is correlated with expression of sialyl Lewis X; and a large proportion of stomach intramucosal carcinomas, gastric adenomas and goblet cells of intestinal metaplasia exhibit sialyl Lewis X (57). Indeed, the primary adhesion event for circulating metastatic cells may use the same identical adhesive system as that for lymphocytes. Thus, a carbohydrate lead may be of value in prophylactic therapy after primary tumors are discovered and during and after surgery where metastatic-potentiated cells may be dislodged into the circulation.

Summary - Carbohydrate-containing biomolecules are found on all cell surfaces and because of their inherent structural diversity, many oligosaccharides are information carriers and recognition molecules. Carbohydrate groups provides signals for protein targeting and cell-cell interactions and serve as receptors for binding toxins, viruses, and hormones.

They control vital events in fertilization and early development, regulate many critical immune system recognition events and target aging cells for destruction. Carbohydrates can alter drug pharmacokinetics and efficacy. Taken together, knowledge of carbohydrate synthesis, structure, and function can serve the medicinal chemist well in the design of new drug candidates (58).

References

1.   Cumulative Chapter Titles Keyword Index in "Ann.Rep.Med.Chem." Vol. 26, J.A. Bristol, Ed., Academic Press, New York, N.Y., 1991, p329.
2..   J.H. Musser, in IUPAC Monograph "Medicinal Chemistry for the 21st Century" C.G. Wermuth, Ed., Blackwell Scientific, Oxford, UK, in press.
3.   N.C. Vyas, Curr.Opin.Struct.Biol., 1 732 (1991).
4.   J.W. Brady, Curr.Opin.Struct.Biol. 1 711 (1991).
5.   S.W. Homans, Biochemistry, 29, 9110, (1990).
6.   K.H. Jung and R.R. Schmidt, Curr.Opin.Struct.Biol., 1 721 (1991).
7.   K.C. Nicolaou, C.W. Hummel, N.J. Bockovich, and C.H. Wong, J.Chem.Soc., Chem. Commun., 871, (1991).
8.   K.G. Rice, O.A. Weisz, T. Barthel, R.T. Lee, and Y.C. Lee, J.Biol.Chem., 265, 18429 (1990).
9.   K.A. Karlsson, Trends Pharm.Sci., 121, 265 (1991).
10.   C.L.M. Stults, C.C. Sweeley, and B.A. Macher, in "Methods of Enzymology", Acadmic Press, San Diego, CA, 1989, p167.
11.   A.C. Cuello, in "Advances in Pharmacology." Vol. 21, Academic Press, New York, N.Y., 1990, p1.
12.   D.B. Jack, Drug News Prosp. 3, 292 (1990).
13.   W. Song, M.F. Vacca, R. Welti and D.A. Rintoul, J.Biol.Chem., 266, 10174 (1991).
14.   H. Nojiri, J.Biol.Chem., 266, 4531 (1991).
15.   H.E. Conrad, Ann.N.Y.Acad.Sci. 556, 18 (1989).
16.   R.J. Linhardt, Chem.Ind. 21, 45 (1991).
17.   E. Young, P. Petrowski and J. Hirsh, Thromb.Haeomost., 65, 933, (1991).
18.   J. Harenberg, Semin. Throm.Hemosat., 16, 12 (1990).
19.   F.A. Ofosu and T.W. Barrowcliffe in "Antithrombic Therapy, Bailliere's Clinical Haematology," Vol. 3, J. Hirsh, Ed., Bailliere Tindall, London, U.K. 1991, p505.
20.   F.R. Rosendaal, M.T. Nurmohamed, H.R. Buller, E. Dekker, J.P. Vandenbrouche and E. Briet, Thromb.Haemost. 65, 927, (1991).
21.   P. Hartle, P. Brucke, E. Dienstl and H. Vinazzer, Thromb.Res. 57, 577 (1990).
22.   A. Leizorovicz, H. Picolet, J.C. Peyrieux and J. P. Borssel, Br.J.Surg., 78, 412 (1991).
23.   J. Leclerc, L. Desjardins, W. Geerts, F. Jobin, F. Delorme and J. Bourgouin, Thromb.Haemost., 65, 753, (1991).
24.   M.N. Levine, J. Hirsh, M. Gent, A.G.G. Turpie, J. Leclerc, P.J. Powers, and R. M. Jay, Ann.Intern.Med., 114, 545, (1991).
25.   P.F. Leyvraz, F. Bachman, J. Hoek, H.R. Bueller, M. Postel, M. Samama, M.D. Vandenbroek, Brit.Med.J., 303, 543 (1991).
26.   J. Heit, C. Kessler, E. Mammen, H. Kwaan, J. Neemah, V. Cabanas, A. Trowbridge, and B. Davidson, Blood, 79, (1992).
27.   J. Pawes and N. Pavak, Thromb.Haemost. 65, 829, (1991).
28.   M. Sobel, P.M. McNeil, P.L. Carlson, J.C. Kermode, B. Adelman, R. Conroy and D. Marques, J.Clin.Invest., 87, 1787 (1991).
29.   F.A. Ofosu, J. Choay, N. Anvari, L.M. Smith and M.A. Blajchman, Eur.J.Biochem., 193, 485 (1990).
30.   P.D.J. Grootenhuis and C.AA.A. van Boekel, J.Am.Chem.Soc., 113, 2743 (1991).
31.   C.A.A. van Boeckel, P.D.J. Grootenhuis and C.A.G. Haasnoot, Trends Pharm.Sci., 12, 241 (1991).
32.   H. Lucas, Tetrahedron, 46, 8207 (1990).
33.   F.L. Lugemwa and J.D. Esko, J.Biol.Chem., 266, 6674, (1991).

34. J.C. Paulson, Trends Biol.Sci., 272 (1989).
35 J.F. Fisher, A.W. Harrison, G.L. Bundy, K.F. Wilkinson, B.D. Rusht and M.J. Ruwart, J.Am.Chem.Soc., 34, 3140 (1991).
36. P. Knight, Biotechnology, 7, 35 (1989).
37. Y. Yoshikuni, Trends Glycosci.Glycotech., 3, 184 (1991).
38. O. Hindsgaul, K.J. Kaur, G. Srivastava, M. Blaszczyk-Thurin, S.C. Crawley, L.D. Heerze and M. Palcic, J.Biol.Chem., 266, 17858 (1991).
39. N. Kojima and S. Hakomori, J.Biol.Chem., 264, 20159 (1989).
40. T. Feizi, Trends Biochem.Sci., 15, 330 (1990).
41. W.I. Weis, R. Kahn, R. Fourme, K. Drickamer, and W.A. Hendrickson, Science, 254, 1608 (1991).
42. L. Osborn, Cell, 62, 3 (1990).
43. T.A. Springer, Nature, 346, 425 (1990).
44. L. Lasky, Cell, 56, 1045 (1989).
45. B.K. Brandley, S.J. Sweidler, P.W. Robbins, Cell, 63 (1990).
46. M.J. Pauley, M.L. Phillips, E. Warner, E. Nudleman, A. K. Singhal, S.I. Hakomori and J.C. Paulsen, Proc.Natl.Acad.Sci.USA, 88, 6224 (1991).
47. C. Foxall, S.R. Watson, D. Dowbenko, C. Fennie, L.A. Lasky, M. Kiso, A. Hasegawa, D. Asa and B.K. Brandley, J. Biol. Chem. in press.
48. Y.Imai, M.S. Singer, C. Fennie, L.A. Lasky and S.D. Rosen, J.Cell Biol., 113, 1213, (1991).
49. Y. Imai, D.D. True, M.S. Singer and S.D. Rosen, J. Cell Biol., 111, 1225 (1990).
50. A. Aruffo, W. Kolanus, G. Walz, P. Fredman and B. Seed, Cell, 67, 35 (1991).
51. J.G. Geng, K.L. Moore, A.E. Johnson, R.P. McEver, J.Biol.Chem., 266, 33 (1991).
52. J. D. Bleil and P.M. Wassarman, Proc.Natl.Acad.Sci.USA, 87, 5563 (1990).
53. I. Ofek and N. Sharon, Curr.Top.Microbiol.Immunol., 151, 91 (1990).
54. T.J. Pritchett and J.C. Paulson, J.Biol.Chem., 264, 9850, (1989).
55. R.E. Willoughby, R.H. Yolken, and R.L. Schnaar, J.Virol., 64, 4830, (1990).
56. T. Imura, Y. Matsushita, S.D. Hoff, T. Yamori, S. Nakamori, M.L. Frazier, G.G. Giacco, K.R. Cleary, and D.M. Ota, Semin.CancerBiol., 2, 129 (1991).
57. T. Matsusako, H. Muramatsu, T. Shirahama, T. Muramatsu and Y. Ohi, Biochem.Biophys.Res.Commun., 181, 1218, (1991).
58. R.L. Schnaar, in "Advances in Pharmacology" Vol. 23, Academic Press, New York, N.Y., 1992, p35.

Chapter 32. Sequence-Specific DNA Binding and the Regulation of Eukaryotic Gene Transcription

Regan G. Shea and John F. Milligan
Gilead Sciences, Foster City, California 94404

<u>Introduction</u> - In this chapter, we will review some recent data which explain how eukaryotic gene expression is regulated at the level of protein-DNA interactions. Additional information is contained in more specialized reviews (1-4). Sequence-specific interactions between DNA and proteins (5), transcription factor domain taxonomy (6,7), and transcription factor binding sites have been reviewed (8). We will conclude with an examination of current efforts to produce sequence-specific DNA binding molecules (9,10). Such molecular design has already yielded artificial nucleases which are useful as molecular biology tools (11) and holds promise for the development of therapeutic strategies using small molecules (12) or oligonucleotides (13) which specifically inhibit transcription.

<u>Promoters and Transcription Factors</u>- The rate at which RNA polymerase II (Pol II) transcribes eukaryotic genes is dictated by the interaction of trans-acting transcription factors with cis-acting DNA promoter elements. Traditionally, the elements proximal to the transcriptional start site were considered promoter elements and those distal were considered upstream activating sequences (UAS) or enhancer elements (14). As the mechanism of transcriptional inititiation has become more clear, these distinctions have changed. Promoters contain two classes of cis-acting elements; the core (basal) elements and regulatory elements. The basal promoter elements consist of the TATA box at -30 and often an initiator region at +1 (15). The basal elements bind a conserved set of transcription factors, the TFII family, and Pol II, which are common to all promoters. All other cis-acting elements are generally considered to be regulatory whether they act proximally or distally (or both) from the site of transcription initiation (16). These regulatory elements bind factors which act to enhance or repress basal level transcription. The number and identity of the factors which bind to regulatory elements are different for each promoter and it is the combined activity of these factors which determines the rate of transcription for each promoter. Some regulatory factors such as Sp1, Oct-1, and the AP-1 family (which include the oncogenes *fos* and *jun*) are fairly ubiquitous and are found over a broad distribution of tissue types (2). Other factors are tissue specific such as NF-AT in T cells, the HNF family of factors in hepatocytes, and PU.1 in macrophages and B cells (2,17). The cis elements which bind these regulatory factors are usually within a few hundred basepairs (bp) of the transcriptional start site but can be over 1 kilobase (kb) away. The mechanism by which these regulatory factors stimulate transcription from such a linear distance is still not well understood. It is fairly clear that transcription is stimulated by a physical interaction between the regulatory transcription factors and the basal transcriptional apparatus, either directly or through coactivator proteins.

BASAL TRANSCRIPTION

In eukaryotic cells, transcription of protein-coding genes operates through the core promoter elements, a TATA-box at -30 and and often an initiator region at +1, and a set of general (basal) transcription factors which appear to be required for the transcription of all promoters (Table I). In addition to Pol II, at least five well characterized protein factors have been implicated as necessary for transcription initiation *in vitro*: TFIIA, -IIB, -IID, -IIE, -IIF (14,15,18). Four other factors have been recently described and have also been implicated in basal transcription: TFIIG, -IIH, -II-I and -IIJ (19-22). Although the mechanisms of transcription initiation are still poorly understood, it is clear that these factors bind DNA and Pol II in a temporal and spatial order to form a complex capable of basal (unregulated) levels of transcription. It is this basal transcriptional complex (pre-initiation complex) which is then acted upon by regulatory factors to increase or decrease the level of transcription (23-25).

<u>TFIID</u> - Of the core factors implicated in basal transcription, only TFIID has been shown to have sequence specific DNA binding properties, binding to the promoter element TATAAA or related sequences (the TATA box) (25). *In vitro* the binding of TFIID to the DNA is a prerequisite for the binding of the other factors and Pol II to form an active pre-initiation complex. TFIID has been purified

as a 120-140 kDa complex which contains a core TATA-binding protein (TBP) and six tightly associated polypeptides known as TAFs (TBP associated factors) (25). The 38 kDa TBP has been cloned from a number of organisms and is highly conserved in the 180 amino acid carboxyl terminus, which has been implicated in DNA binding (26-28). TBP alone has been shown by footprint and bandshift analysis to bind to DNAs containing a TATA box (26). Unlike most DNA binding proteins, TBP from both humans and yeast recognizes and binds in the minor groove of the TATA box (29,30). The binding of TBP to DNA causes anomolous migration of the complexes in polyacrylamide gels suggesting that TBP may cause DNA bending (31). In partially purified extracts, TBP can replace TFIID in supporting basal transcription in $vitro$. However, TBP cannot functionally replace TFIID in responding to transcriptional activators such as Sp1, CTF, and GAL4-VP16 (32,33). Since TFIID does respond to these activators in $vitro$, it appears that the TAFs, alone or with other coactivator proteins, may mediate the interaction of transcriptional activators with the basal transcriptional complex (34).

Formation of the Pre-initiation Complex - In $vitro$ studies have revealed an ordered pathway for the formation of the pre-initiation complex (see Figure 1) (35). First, TFIID binds to the TATA box of a promoter in a slow step (25). Next TFIIA binds to form the TFIID/A complex (36,37). TFIIA is not required for the in $vitro$ binding of TFIID to DNA although it appears to stabilize TFIID on the promoter, perhaps anchoring it to the TATA box through multiple rounds of transcription (38). The stable interaction of TFIID and TFIIA on the promoter creates a template-committed complex which is resistant to challenges by free DNA templates (35). TFIIB binds to the template-committed complex to form a TFIID/A/B complex (39). TFIIB has been postulated to have a structural role in forming the pre-initiation complex and may be responsible for selecting the site of transcriptional initiation (40). It is after TFIIB binds that Pol II in association with TFIIF is recruited, forming the TFIID/A/B/Pol/F pre-initiation complex (41,42). TFIIE has been shown to bind to Pol II after it is bound to the complex, forming the TFIID/A/B/Pol/F/E final pre-initiation complex (shown in Figure 2) (20,43).

Transcription From TATA-less Promoters - For Pol II transcription the TATA box is generally considered to be ubiquitous, although there are many cellular and viral promoters which contain no obvious TATA-like sequences. Transcription from TATA-less promoters has been shown to be dependent upon TFIIA, B, D, E, F and Pol II, the same factors known to be essential for transcription from TATA-containing promoters (32). Most confusing is the absolute requirement for TFIID, whose requisite role in complex formation is binding to the TATA box (44). One common property of many TATA-less promoters is the presence of multiple binding sites for the transcriptional activator Sp1 (45). In $vitro$ studies have revealed that basal level transcription from TATA-less promoters does not occur in the absence of Sp1 (32). It appears then, that in TATA-less promoters, Sp1 recruits TFIID, tethering it to the promoter so that complex formation can occur in the absence of direct DNA binding by TFIID. The interaction between Sp1 and TFIID does not occur at the level of TBP as the cloned TBP alone is not sufficient for activity, implying that the TAFs or some other co-activator protein may be involved in tethering TFIID to Sp1.

TFII-I has also been implicated as a DNA binding protein which assists in the formation of the pre-initiation complex on TATA-less promoters. TFII-I has been demonstrated to bind to the consensus sequence YAYTCYYY, which is often found near the initiation region (Inr) for Pol II transcribed genes (21,46). Within the adenovirus major late (AML) core promoter, basal transcription can be reconstituted by TFIIB, D, E, F, and Pol II if either TFIIA or TFII-I is added (21). The AML core promoter contains both a TATA box and a consensus TFII-I binding site at the initiation region. The behavior of the basal transcription machinery with these two factors implies that two distinct mechanisms of transcriptional activation may be occurring. In one mechanism, TFIIA stabilizes the binding of TFIID to the TATA box to promote an active pre-initiation complex. In the absence of TFIIA, TFII-I binds to the Inr, stabilizes TFIID binding, and promotes formation of the pre-initiation complex. Other experiments have demonstrated that in TATA-less promoters, Sp1 binding sites alone, weakly promote transcription (46). When a consensus TFII-I binding site is inserted into the promoter the transcription is strongly stimulated indicating that transcriptional activators such as Sp1 can act concertedly with TFII-I to stimulate transcription of TATA-less promoters. Regardless of the promoter type, the basal transcription factors are always required, although the auxiliary factors and the mechanism of recognition of the promoter may change (47).

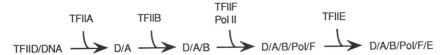

Figure 1: Ordered assembly of the basal factors to form a pre-initiation complex.

Table I: Summary of Basal Transcription Factors

Factor	MW(s)	Proposed Role on Complex Formation	Refs.
TFIIA	14, 19, 34 kDa	Stabilizes TFIID binding to DNA to form TFIID/A complex	36,39, 48,49
TFIIB	33 kDa	Promotes Pol II binding to the bound TFIID/A complex; interacts with acidic activators	20,43, 50
TFIID	120-140 KDa; 37 kDa (TBP) + 6 proteins (TAFs)	Binds to TATA box at the first step of initiation; possible target of activators which stimulate transcription	25-34, 51
TFIIE	34 (b), 57 (a) KDa $(\alpha_2\beta_2)$ tetramer	Binds to Pol II after formation of the TFIID/A/B/Pol/F complex; ATPase?	20, 43
TFIIF	30, 74 KDa (RAP30/74)	Required for Pol II binding to the pre-initiation complex; binds after TFIID/A/B complex formation; helicase?	41,42, 52-54
TFIIG	?	Functionally related to TFIIA	19
TFIIH	35, 90 kDa	?	20
TFII-I	120 kDa	DNA binding activity, Inr recognition; promotes TFIID binding	21,46, 47
TFIIJ	?	TAF of TFIID?	22

REGULATED TRANSCRIPTION

Trancription factors bound to DNA operate jointly and spatially to direct the transcription of each gene. The large number of possible combinations allows complex patterns of regulation with relatively few factors. Regulatory transcription factors fall into two broad categories corresponding to those which are more ubiquitous and those which are tissue or cell type specific. A tabular presentation of vertebrate transcription factors along with their consensus binding sequences, binding motifs, family relationships, and tissue specificities has been prepared (17). Ubiquitous regulators include Sp1 and Oct-1. Tissue specific regulation is directed by factors such as C/EBP, NF-AT, and HNF-1 among others. In general these factors have distinct DNA binding and transcriptional activation domains, a property uncovered in domain swap experiments and the subject of recent reviews (1,2). The following is a brief overview of several important factors.

Transcription Factor Sp1 - This ubiquitous factor activates transcription from many eukaryotic promoters and is important in modulating promoter activity. The Sp1 protein uses three zinc fingers to bind at the GC box GGGGCGGGGC and similar sequences. Available evidence places critical sequence-specific contacts in the major groove at cytosine N4 and guanosine O6 at the center of the consensus site (55). Equilibrium footprinting as well as gel mobility shift data yield dissociation constants of 410-530 pM, indicating binding somewhat weaker than for many other mammalian factors. Transcriptional activation by Sp1 also requires at least one of two glutamine-rich motifs.

Activation has been demonstrated from either proximal or distal binding sites (56). The presence of both kinds of sites leads to synergistic activation by Sp1, apparently through protein-protein interactions and DNA looping. Recently, electron microscopy has been applied to this question and the results show that Sp1 is capable of self-association when bound to the SV40 early promoter or the thymidine kinase GC6 promoter (57). The interaction requires the presence of at least one of the Sp1 glutamine-rich segments. There is evidence that self-association proceeds beyond the dimer level to include tetramer, octamer, and dodecamer formation (58). Sp1 also mediates synergistic activation by interacting with other factors, such as bovine papillomavirus type 1 protein E2 (59).

POU-Domain Transcription Factors - The highly conserved POU-domain family, which includes Pit-1, Oct-1, Oct-2, Pdm-1, Pdm-2, Cf1a, Brn-1, Brn-2, Brn-3, Tst-1, Ceh-6, unc-86, i-POU, and Oct-3/4, has been the subject of recent reviews (60,61). These proteins appear to play a key role in the regulation of cell development and differentiation. This family of proteins has generally been characterized as a subset of the homeodomain proteins (3), but they should perhaps be considered as a separate class of regulatory proteins since they contact DNA in a bipartite manner in contrast to the monovalent binding of homeodomains. POU binding in all cases is through a POU-specific domain as well as a POU-homeodomain. The POU-homeodomain forms a helix-turn-helix in common with all other homeodomains. Details of the structure and DNA interactions of the POU-specific domain remain unclear in the absence of NMR or X-ray crystallographic data. A key POU-domain protein is Oct-1, a ubiquitous mammalian transcriptional activator. Although Oct-1 expression is widespread, it can participate in tissue-specific regulation by interacting with more specialized factors such as Pit-1, an anterior pituitary gland protein (62), and the herpes virus activator protein, VP16 (63).

One of the most intriguing results of homeodomain research has been the discovery of important sequence-specific minor groove contacts which greatly enhance the affinity of homeodomains for DNA. Alkylation interference experiments using dimethylsulfate and diethylpyrocarbonate have shown that Oct-1 contacts a 14 bp region, including every base of the core octamer element ATGCAAAT (64). The size of the binding site is relatively large for a monomeric factor, and contrasts with the 9 bp contact region for homeodomains. The POU-specific domain of Oct-1 is binding incompetent in the absence of the POU-homeodomain and its main contribution to binding is additional sequence-specificity with only modest impact on binding affinity. The independent POU-homeodomain is sufficient for binding, but displays micromolar nonspecific DNA binding. POU-specific domain binding has been localized to the left end of the octamer sequence while the POU-homeodomain occupies the right end. The strongest Oct-1 binding is not obtained by marrying together sequences best for each half. Changing the 8 bp sequence above to ATGATAAT improves POU-homeodomain kd from 54 to 2.7 nM but weakens the binding of intact Oct-1 from 90 to 370 pM.

CCAAT Box/Enhancer Binding Protein (C/EBP) - C/EBP, a bZIP factor which dimerizes via a leucine zipper, binds CCAAT boxes, cAMP response elements, and viral enhancer core sites through major groove contacts covering 12-14 bp (65-67). These binding sites appear to be critical for the control of mRNA synthesis. Consequently C/EBP is emerging as an important regulator of genes involved in growth arrest and terminal cell differentiation in liver and adipose tissue (68). A single gene is responsible for the production of a family of related transcription factors. Circular dichroism and protease digestion measurements have revealed that the basic portion of the protein becomes fully helical only when bound to DNA (69,70). Domain swap experiments showed that this basic domain is sufficient to impart sequence-specificity (71).

Still unresolved is the molecular basis for the promiscuous binding referred to in the name of the protein. Both the highly conserved CCAAT sequence (plus related sequences) and the enhancer consensus TGTGG(A/T)(A/T)(A/T)G are avidly bound by C/EBP. The unrelated protein CTF/NFI also binds the highly conserved CCAAT box. Despite sharing specificity for the same DNA sequence as C/EBP, CTF/NFI is not a bZIP protein (nor does it contain a helix-turn-helix or zinc finger motif). A summary of similarities and differences between CCAAT-binding proteins has been published (4).

The AP-1 Family Of Factors - This family controls both the expression of oncogenes (e.g. *ras* and *src*) and growth factor stimulation of cellular genes. Important members of this family include the proto-oncogene products Fos and Jun. This pair of bZIP proteins form heterodimers via leucine zippers and Jun is also capable of homodimerization (72,73). The observation that the length of the contact region between Fos-Jun and DNA was incompatible with idealized alpha-helical proteins bound to B-DNA combined with the fact that the very similar heterodimer and Jun homodimer can have opposing activities has led to an investigation into the nature of the protein/DNA interaction (74-76). As with C/EBP and other bZIP proteins, the basic region of the dimers becomes more helical upon binding to DNA. Circular permutation and phasing analyses also demonstrated bending of the DNA by the dimers. Fos-Jun bends DNA toward the major groove, but Jun homodimers bend DNA toward the minor groove. The molecular basis for the size and direction of DNA bending has not been elucidated yet. Another clue to the binding of Jun and Fos is that it is modulated by reduction and oxidation of of a conserved cysteine and is stimulated by a nuclear protein that reduces the two proteins (77).

Nuclear Factor of Activated T Cells (NF-AT) - This factor is critical for the regulation of genes involved in activation of immune response in T cells, particularly through binding to an antigen receptor response element (ARRE-2) of the IL-2 gene (78). The NF-AT gene appears to be induced early in the T cell activation and differentiation pathway. By using inhibitors of protein synthesis it was shown that binding of NF-AT at the IL-2 regulatory element precedes and is required for IL-2 mRNA production (79). FACS analysis of NF-AT-driven transcription of β-galactosidase has shown a bimodal distribution between high expression and no expression in monoclonal Jurkat T cells (80). This indicates that there may be a threshold NF-AT concentration which must be surpassed before transcription is initiated. Two immunosuppressant drugs, cyclosporin A (CsA) and FK-506 (81) owe at least some of their activity to their ability to inhibit the binding activity of NF-AT (82). It has been shown that this inhibition results from blocking nuclear localization of a cytoplasmic subunit of NF-AT subsequent to an antigen receptor signal. CsA also inhibits the activity and/or synthesis of other T lymphocyte-specific factors (83). A detailed treatment of transcriptional aspects of T lymphocyte activation has been published (84).

MECHANISMS OF TRANSCRIPTIONAL ACTIVATION

One of the most striking features of transcriptional activator proteins is their ability to activate transcription at a distance. DNA binding sites for most activator proteins fall within a few hundred bp of the transcription initiation site, but some have been reported to be as far as 100 kb away. Activator proteins in general have two domains, a DNA binding domain and an activation domain. These two domains may reside within the same polypeptide or may be formed as part of a multi-protein complex. Activation domains are less well defined than DNA binding domains, but in general comprise a region which can interact, either directly or indirectly (see Figure 2), with the basal transcriptional complex to stimulate transcription. The activation domains tend to fall into one of three classes: those rich in acidic amino acids (acid blobs), rich in glutamine, or rich in proline. For example, GAL4 is a known yeast transcriptional activator protein which contains both a DNA binding domain and an acidic domain. GAL4 has been shown to activate transcription in a wide variety of cell types when a GAL4 binding site is introduced into the promoters of reporter genes. Since GAL4 can activate transcription in a species independent manner, it seems likely that the acidic activation domain stimulates transcription by interacting with a highly conserved component of the basal transcriptional machinery. The herpes virus VP16 protein also has a highly acidic transcriptional activation domain, but no intrinsic activation properties and very low affinity for DNA (63). Oct-1 is a DNA binding protein which has no known activation domain (64). To become a functional transcriptional activator VP16 forms a complex with Oct-1 (85,86). As further evidence of the mechanism of activation by VP16, a fusion protein in which the DNA binding region of GAL4 is attached to to the acidic domain of VP16 (GAL4-VP16) is a potent activator of transcription from promoters which contain a GAL4 DNA binding site (87). The acidic activation domain of VP16 stimulates transcription of specific genes in conjunction with a DNA binding activity which gives it promoter specificity. This DNA binding activity can be intrinsic, as in GAL4-VP16, or extrinsic, as in Oct-1/VP16, and provides a model for the general mechanism of transcriptional activation. Despite their dissimilarities, GAL4, Oct-1 and VP16 are all considered transcriptional activators.

Acidic Activators - The highly acidic domains of activator proteins interact either directly or indirectly with some component of the basal transcriptional machinery. Affinity chromatography using the acidic domain of VP16 as a ligand revealed that TFIID was selectively and specifically retained from a HeLa cell extract, implying that TFIID is a direct target for acidic activation domains (88). However, other experiments have shown that GAL4-VP16 interacts with a factor not known to be part of the basal transcriptional apparatus, suggesting that an adaptor molecule(s) mediates the signal from the acidic domain to the basal transcriptional machinery (33,89). The addition of excess GAL4-VP16 has been shown to inhibit stimulated, but not basal, transcription from promoters which contain a dA:dT UAS (upstream activating sequence). This phenomenon is known as "squelching" and occurs when the large excess of the activator soaks up all the adaptor molecules, effectively inhibiting the stimulatory action of the dA:dT UAS binding protein. Extracts purified for TFIID and Pol II were added but did not relieve the inhibition, indicating that neither of these components are the direct targets of the acidic domains (89).

Mechanistic studies of transcriptional activation using the activator GAL4-AH indicate that TFIIB is the direct target of acidic activation domains (50). *In vitro* it has been shown that GAL4-AH stimulates transcription by facilitating the formation of the pre-initiation complex, and not by stimulating pre-existing complexes. This step is apparently ATP-independent and occurs prior to formation of the first phosphodiester bond in transcription. It appears that the acidic activation domains can stimulate transcription by enhancing the recruitment of TFIIB to the promoter bound TFIID/A complex. The interaction of an acidic domain with TFIIB is in contrast to earlier reports of VP16 interacting with TFIID (88). The original experiments with the VP16 affinity column detected TFIID after a 100 mM KCl wash; however, upon further analysis it was revealed that both TFIID and TFIIB are retained under these conditions (50,88). Under more stringent washing conditions of 200 mM KCl, only TFIIB is retained, with the TFIID eluted in the flow through fractions (50). It appears that TFIIB may be a major target of the acidic activating domains of transcriptional activator proteins although it is possible (indeed likely) that activators will have more that one target.

Glutamine-Rich Activators - The mechanism of transcriptional activation from factors which do not contain acidic domains is less clear. The transcriptional activator Sp1 has two glutamine-rich domains which are known to be essential for activation of transcription. *In vitro*, addition of an Sp1 site induces transcription about five-fold over the basal level when drosophila TFIID is added to the reaction (32,90). The addition of the cloned TBP is not sufficient to stimulate transcription above the basal level. However, when drosophila extract is added in addition to the cloned TBP, transcriptional activation returns, even at concentrations in which binding of TBP saturates the promoters. A mutant of drosophila TFIID which has a deletion in the amino terminus is competent for basal transcription but cannot be stimulated by the addition of Sp1 and drosophila extract. Together, these results indicate that coactivator proteins are required to mediate the signal between Sp1 and the basal transcriptional complex and that the ultimate target of this signal is TFIID.

Synergistic Activation - Multiple interactions from activator proteins have a synergistic effect upon transcription. Models suggest that multiple activators work by simultaneously contacting factors of the basal transcriptional apparatus (91,92). By placing multiple binding sites upstream of a promoter it has been determined that acidic activators work synergistically under conditions in which the DNA binding sites are saturated. When GAL4-VP16 (strong activator) or GAL4-AH (weaker activator) are used separately, on promoters with increasing numbers of GAL4 binding sites, a cooperative increase in the transcriptional activation was observed (91). This cooperative increase was not at the level of enhanced protein binding to DNA, but rather at the level of transcriptional activation. Interestingly, the differences in the activation by the stronger GAL4-VP16 and the weaker GAL4-AH diminish as the number of activator binding sites increase, until saturation of activation is observed. It takes a greater number of binding sites to saturate activation with GAL4-AH than it does with GAL4-VP16 (91). Similar experiments were conducted using the yeast derived GAL4-AH and the mammalian activator ATF simultaneously. At a saturating concentration of GAL4-AH, increasing amounts of ATF were still able to stimulate transcription further. Similarly, at saturating amounts of ATF, increasing concentrations of GAL4-AH were also still able to stimulate transcription (90). It is unlikely that a yeast and a mammalian activator protein have evolved to interact with each other; more likely is that the activator proteins are all acting with the same target to activate transcription. The target of the activators may be TFIIB and/or TFIID, both identified previously as interacting with acidic activation domains (50,88) or some unidentified factor(s). Regardless, this factor(s) must have the incredible ability to interact with multiple activator proteins to stimulate transcription as a reflection of both their number and strength. Perhaps there are indeed multiple targets for acidic activator domains which might include factors in the basal transcriptional complex, coactivator proteins, or regions of Pol II itself.

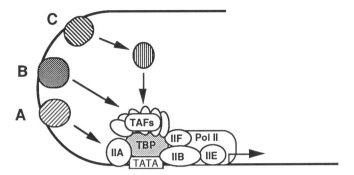

Figure 2: Models of regulatory factors interaction with the basal transcriptional complex. A. Direct activation through one of the basal factors. B. Activation of TFIID through the TBP-associated factors (TAFs). C. Indirect activation through a coactivator.

APPROACHES TO THERAPEUTIC AGENTS

Improved understanding of molecular interactions of transcriptional machinery makes rational design of drugs which act at the level of transcription an increasingly realistic goal. In principle both DNA and DNA binding proteins are viable targets for specific drug action. One approach to the inhibition of transcription factors is by use of duplex DNA analogs which can compete with the native DNA element (93,94). Double stranded phosphorothioate oligonucleotides containing either octamer or NF-κB consensus sequences were reported to inhibit activation of both a reporter plasmid and the human immunodeficiency virus enhancer in transiently transfected B cells, as well as decreasing IL-2 secretion in Jurkat T leukemia cells (93). But several fundamental problems remain, including poor permeation of DNA into cells, the high concentration of transcription factors (relative to gene concentrations), and the possibility that a targeted protein factor may bind only weakly to its consensus sequence in the absence of long-range enhancer mediated interactions. Thus, molecules capable of sequence-specific binding to DNA represent more attractive drug design targets for now.

Peptide DNA Binders - It has been possible to effectively bind specific DNA sequences with peptide portions of DNA binding proteins. In one case a 34 amino acid dimer in which the leucine zipper of GCN4 was replaced by a disulfide linkage exhibited nanomolar affinity for DNA as determined by gel shift assay, and the binding was shown to be sequence-specific by DNase I footprinting (95). Research on the DNA binding domain of Hin recombinase has shed light on minor groove protein-DNA contacts. The presence or absence of an arginine residue determined the sequence-specificity of the peptide through the minor groove recognition of an AAA sequence (96). Attachment of a metal-binding site to the amino terminus of this Hin fragment afforded a sequence-specific DNA-cleaving metalloprotein to which copper and nickel impart differing site selectivities for DNA strand cleavage (97). A more general approach involves design of short structure-forming peptides through incorporation of non-natural residues which, for example, promote helix initiation (98).

Small Molecule DNA Binders - Interruption of transcription has been demonstrated *in vitro* for a variety of intercalators, groove binders, and hybrid molecules capable of both kinds of interaction (99). The sequence-specificity of these ligand-DNA interactions is generally insufficient to limit binding to a single site (9). It is now recognized that minor groove binding by small molecules can influence major groove interactions between proteins and DNA. In a study of the distamycin-induced inhibition of homeodomain-DNA complexes, a mutant homeodomain peptide lacking the N-terminal arm responsible for minor groove contacts showed a distamycin-dependent decreased affinity for DNA (100). Distamycin, **1**, and longer oligo(N-methylpyrrolecarboxamide)s have been shown to accelerate open complex formation at a promoter *in vitro*, through minor groove binding in a spacer region (101). In both cases, a small molecule-induced conformational change in DNA was postulated to cause the observed effects. Rational changes to known small molecule DNA binders can provide altered sequence-specificity (9). By replacing selected base-contacting atoms of molecules such as distamycin and Hoechst 33258, **2**, with different atoms it has been possible to engineer G/C binding preferences into these A/T recognizing molecules. Recently, such altered molecules have been covalently linked to intercalators to improve binding (10,102,103).

Another class of sequence-specific small DNA binders, the calicheamicins (3), is generating considerable interest (104). These enediyne antitumor antibiotics recognize tetrameric sequences such as TCCT and CTCT and initiate double-stranded cleavage. Much of the interest comes from the potent anti-cancer activity of members of the family, which in some cases is much greater than for currently used therapies. Synthetic efforts have yielded simplified molecules which mimic the natural products (105). In another example of structural simplification, mitomycin C (4) analogs were prepared which retained the ability of the parent molecule to recognize 5'-CG in preference to 5'-GC containing sites (106). Greater understanding of the features of DNA recognition should allow the design of molecules with higher sequence-specificity and may lead to less toxic anti-cancer agents.

Oligonucleotides - Binding to the major groove of duplex DNA by oligonucleotides to form triple helices has proven sufficient to block binding of transcription factors such as Sp1 (107). Further studies in artificially constructed promoters show that oligonucleotides can inhibit transcription, not only by inhibiting transcription factor binding, but also through their proximity to the basal transcriptional apparatus (108). Oligos with chemically reactive moieties which covalently alkylate DNA downstream from a promoter can block the progress of an actively transcribing polymerase, effectively derailing transcription (109). Triple helix forming oligonucleotides have also been reported to inhibit transcription in cells (110). High specificity of binding makes this class of molecules extremely attractive for the promoter-specific inhibition of gene expression. Improvements in the binding affinity, permeation characteristics, and stability of such oligos will dramatically influence future therapeutic strategies (13).

References

1. A.D. Frankel and P.S. Kim, Cell, 65, 717 (1991).
2. P.J. Mitchell and R. Tjian, Science, 245, 371 (1989).
3. A. Laughon, Biochemistry, 30, 11357 (1991).
4. P.F. Johnson and S.L. McKnight, Annu.Rev.Biochem., 58, 799 (1989).
5. M.E.A. Churchill and A.A. Travers, Trends Biochem.Sci., 16, 92 (1991).
6. K.Struhl, Trends Biochem.Sci., 14, 137 (1989).
7. S.C. Harrison, Nature, 353, 715 (1991).
8. J.Locker and G. Buzard, DNA Sequence-J.DNA Sequencing Mapping, 1, 3 (1990).
9. P.E. Nielsen, Bioconjugate Chem., 2, 1 (1991).
10. C. Bailly and J.-P. Henichart, Bioconjugate Chem., 2, 379 (1991).
11. D.S. Sigman, Biochemistry, 29, 9097 (1990).
12. L.H. Hurley, J.Med.Chem., 32, 2027 (1989).
13. M.D. Matteucci and N. Bischofberger, Annu.Rep.Med.Chem., 26, 287 (1991).
14. A.G. Saltzman and R. Weinman, FASEB J., 3, 1723 (1989).
15. M. Sawadogo and A. Senrenac, Annu.Rev.Biochem., 59, 711 (1990).
16. W. Wand and J.D. Gralla, Mol.Cell.Biol., 11, 4561 (1991).
17. S. Faisst and S. Meyer, Nucleic Acids Res., 20, 3 (1992).
18. R.G. Roeder, Trends Biochem.Sci., 16, 402 (1991).
19. H. Sumimot, Y, Ohkuma, T. Yamamoto, M. Horikoshi and R.G. Roeder, Proc.Natl.Acad.Sci.USA, 87, 9158 (1990).
20. J. Inostroza, O. Flores and D. Reinberg, J.Biol.Chem., 266, 9304 (1990).
21. A.L. Roy, M. Meisterernst, P. Pognec and R.G. Roeder, Nature, 354, 245 (1991).
22. P. Cortes, O. Flores and D. Reinberg, Mol.Cell.Biol., 12, 413 (1992).
23. M. Ptashne, Nature, 335, 683 (1988).
24. M. Ptashne and A.A.F. Gann, Nature, 346, 329 (1990).
25. F.B. Pugh and R. Tjian, J.Biol.Chem., 267, 679 (1992).
26. M.G. Peterson, N. Tanese, B.F. Pugh and R. Tjian, Science, 248, 1625 (1990).
27. M. Meisterernst, M. Horikoshi and R.G. Roeder, Proc.Natl.Acad.Sci.USA, 87, 9153 (1990).
28. M.L. Muhich, C.T. Iida, M. Horikoshi, R.G. Roeder and C.S. Parker, Proc.Natl.Acad.Sci.USA, 87, 9148 (1990).
29. D.B. Starr and D.K. Hawley, Cell, 67, 1231 (1991).
30. D.K. Lee, M. Horikoshi and R.G. Roeder, Cell, 67, 1241 (1991).
31. M. Horikoshi, C. Bertccioli, R. Takada, J. Wang, T. Yamamoto, and R.G. Roeder, Pro.Natl.Acad.Sci. USA, 89, 1060 (1992).
32. B.F. Pugh and R. Tjian, Cell, 61, 1187 (1990).
33. S.L. Berger, W.D. Cress, A. Cress, S.J. Triezenberg and L. Guarente, Cell, 61, 1199 (1990).
34. B.D. Dynlacht, T. Hoet and R. Tjian, Cell, 66, 563 (1991).
35. S. Buratowski, S. Hahn, L. Guarente and P. Sharp, Cell, 56, 549 (1989).
36. S. Hahn, S. Buratowski, P. Sharp and L. Guarente, EMBO J., 8, 3379 (1989).
37. S. Buratowski and H. Zhou, Science, 255, 1130 (1992).
38. D. Reinberg, M. Horikoshi and R.G. Roeder, J.Biol.Chem., 262, 3322 (1987).
39. E. Maldonado, I. Ha, P. Cortes, L. Weis and D. Reinberg, Mol.Cell.Biol., 10, 6335 (1990).
40. I. Ha, W.S. Lane and D. Reinberg, Nature, 352, 689 (1991).
41. O. Flores, I. Ha and D. Reinberg, J.Biol.Chem., 265, 5629 (1990).
42. O. Flores, H. Lu. M. Killeen, J. Greenblatt, Z.F. Burton and D. Reinberg, Proc.Natl.Acad.Sci.USA, 84, 9999 (1991).
43. Y. Ohkuma, H. Sumimoto, M. Horikoshi and R.G. Roeder, Proc.Natl.Acad.Sci.USA, 87, 9163 (1990).
44. B.F. Pugh and R. Tjian, Genes Dev., 5, 1935 (1991).
45. W.S. Dynan, Trends Genet., 2, 196 (1986).
46. E. Seta, Y. Shi and T. Shenk, Nature, 354, 241 (1991).
47. B. Zenzie-Gregory, A. O'Shea-Greenfield, and S.T. Smale, J.Biol.Chem., 267, 2823 (1992).
48. Y. Usuda, A. Kubota, A.J. Berk and H. Hanada, EMBO J., 10, 2305 (1991).
49. J.A. Ranish, W.S. Lane, and S. Hahn, Science, 255, 1127 (1992).
50. Y.S. Lin and M.R. Green, Cell, 63, 971 (1991).
51. M.R. Van Dyke, M. Sawadogo and R.G. Roeder, Mol.Cell.Biol., 10, 342 (1989).
52. M.T. Killeen and J.F. Greenblatt, Mol.Cell.Biol., 12, 30 (1992).
53. A. Finkelstein, C.F. Kostrub, J. Li, D.P. Chavez, B.Q. Wang, S.M. Fang, J. Greenblatt and Z.F. Burton, Nature, 355, 464 (1992).
54. T. Aso, H.A. Vasavada, T. Kawaguchi, J.F. Germino, S. Ganguly, S. Kitajima, S.M. Weissman and Y. Yasukochi, Nature, 355, 461 (1992).
55. J. Letovsky and W.S. Dynan, Nucleic Acids Res., 17, 2639 (1989).
56. A.J. Courey, D.A. Holtzman, S.P. Jackson and R. Tjian, Cell, 59, 827 (1989).

57. W. Su, S. Jackson, R. Tjian and H. Echols, Genes Dev., 5, 820 (1991).
58. I.A. Mastrangelo, A.J. Courey, J.S. Wall, S.P. Jackson and P.V.C. Hough, Proc.Natl.Acad.Sci.USA, 88, 5670 (1991).
59. R. Li, J.D. Knight, S.P. Jackson, R. Tjian and M.R. Botchan, Cell, 65, 493 (1991).
60. M.G. Rosenfeld, Genes Dev., 5, 897 (1991).
61. G. Ruvkun and M. Finney, Cell, 64, 475 (1991).
62. J.W. Voss, L. Wilson and M.G. Rosenfeld, Genes Dev., 5, 1309 (1991).
63. R.F. Greaves and P. O'Hare, J. Virology, 64, 2716 (1990).
64. C.P. Verrijzer, A.J. Kal and P.C. van der Vliet, Genes Dev., 4, 1964 (1990).
65. W.H. Landschulz, P.F. Johnson and S.L. McKnight, Science, 243, 1681 (1989).
66. C.R. Vinson, P.B. Sigler and S.L. McKnight, Science, 246, 911 (1989).
67. J.A. Nye and B.J. Graves, Proc.Natl.Acad.Sci.USA, 87, 3992 (1990).
68. R.M. Umek, A.D. Friedman and S.L. McKnight, Science, 251, 288 (1991).
69. K.T. O'Neil, J.D. Shuman, C. Ampe and W.F. DeGrado, Biochemistry, 30, 9030 (1991).
70. J.D. Shuman, C.R. Vinson and S.L. McKnight, Science, 249, 771 (1990).
71. P. Agre, P.F. Johnson and S.L. McKnight, Science, 246, 922 (1989).
72. R. Turner and R. Tjian, Science, 243, 1689 (1989).
73. R. Gentz, F.J. Rauscher III, C. Abate and T. Curran, Science, 243, 1695 (1989).
74. T.K. Kerppola and T. Curran, Cell, 66, 317 (1991).
75. T.K. Kerppola and T. Curran, Science, 254, 1210 (1991).
76. L. Patel, C. Abate and T. Curran, Nature, 347, 572 (1990).
77. C. Abate, L. Patel, F.J. Rauscher III and T. Curran, 249, 1157 (1990).
78. E.A. Emmel, C.L. Verweij, D.B. Durand, K.M. Higgins, E. Lacy and G.R. Crabtree, Science, 246, 1617 (1989).
79. J.-P. Shaw, P.J. Utz, D.B. Durand, J.J. Toole, E.A. Emmel and G.R. Crabtree, Science, 241, 202 (1988).
80. S. Fiering, J.P. Northrop, G.P. Nolan, P.S. Mattila, G.R. Crabtree and L.A. Herzenberg, Genes Dev., 4, 1823 (1990).
81. J.R. Luly, Annu.Rep.Med.Chem., 26, 211 (1991).
82. W.M. Flanagan, B. Corthesy, R.J. Bram and G.R. Crabtree, Nature, 352, 803 (1991).
83. C. Randak, T. Brabletz, M. Hergenrother, I. Sobotta and E. Serfling, EMBO J., 9, 2529 (1990).
84. G.R. Crabtree, Science, 243, 355 (1989).
85. C.R. Goding and P. O'Hare, Virology, 173, 363 (1989).
86. J.R. Nevins, Trends Biochem.Sci., 16, 435 (1991).
87. I. Sadowski, J. Ma, S. Triezenberg and M. Ptashne, Nature, 335, 563 (1988).
88. K.F. Stringer, J.C. Ingeles and J. Greenblatt, Nature, 345, 783 (1990).
89. R.J. Kelleher, D.M. Flanagan and R.D. Kornberg, Cell, 61, 1209 (1990).
90. B. Lewin, Cell, 61, 1161 (1990).
91. Y.S. Lin, M. Carey, M. Ptashne and M.R. Green, Nature, 345, 359 (1990).
92. M. Carey, Y.S. Lin, M.R. Green and M. Ptashne, Nature, 345, 361 (1990).
93. A. Bielinska, R.A. Shivdasani, L. Zhang and G.J. Nabel, Science, 250, 997 (1990).
94. H. Wu, J.S. Holcenberg, J. Tomich, J. Chen, P.A. Jones, S.-H. Huang and K.L. Calame, Gene, 89, 203 (1990).
95. R.V. Talanian, C.J. McKnight and P.S. Kim, Science, 249, 769 (1990).
96. J.P. Sluka, S.J. Horvath, A.C. Glasgow, M.I. Simon and P.B. Dervan, Biochemistry, 29, 6551 (1990).
97. D.P. Mack and P.B. Dervan, J.Am.Chem.Soc., 112, 4604 (1990).
98. F. Ruan, Y. Chen and P.B. Hopkins, J.Am.Chem.Soc., 112, 9403 (1990).
99. D.R. Phillips, R.J. White, H. Trist, C. Cullinane, D. Dean and D.M. Crothers, Anti-Cancer Drug Des., 5, 21 (1990).
100. A. Dorn, M. Affolter, M. Muller, W.J. Gehring and W. Leupin, EMBO J., 11, 279 (1992).
101. P.A. Martello, J.P. Bruzik, P. deHaseth, R.S. Youngquist and P.B. Dervan, Biochemistry, 28, 4455 (1989).
102. C. Bailly, N. Helbecque, J.-P. Henichart, P. Colson, C. Houssier, K.E. Rao, R.G. Shea and J.W. Lown, J.Mol.Recognit., 3, 26 (1990).
103. F. Subra, S. Carteau, J. Pager, J. Paoletti, C. Paoletti, C. Auclair, D. Mrani, G. Gosselin and J.L. Imbach, Biochemistry, 30, 1642 (1991).
104. M.D. Lee, G.A. Ellestad and D.B. Borders, Acc.Chem.Res., 24, 235 (1991).
105. K.C. Nicolaou, P. Maligres, J. Shin, E. deLeon and D. Rideout, J.Am.Chem.Soc., 112, 7825 (1990).
106. M.F. Weidner, S.T. Sigurdsson and P.B. Hopkins, Biochemistry, 29, 9225 (1990).
107. L.J. Maher, B. Wold and P.B. Dervan, Science, 245, 725 (1989).
108. L.J. Maher, P.B. Dervan and B. Wold, Biochemistry, 31, 70 (1992).
109. S.L. Young, S.H. Krawczyk, M.D. Matteucci and J.J. Toole, Proc.Natl.Acad.Sci.USA, 88, 10023 (1991).
110. E.H. Postel, S.J. Flint, D.J. Kessler and M.E. Hogan, Proc.Natl.Acad.Sci.USA, 88, 8227 (1991).

SECTION VII. TRENDS AND PERSPECTIVES

Editor: James A. Bristol
Parke-Davis Pharmaceutical Research Division
Warner-Lambert Co., Ann Arbor, MI 48105

Chapter 33. To Market, To Market - 1991

June D. Strupczewski and Daniel B. Ellis
Hoechst-Roussel Pharmaceuticals Inc., Somerville, NJ 08876

New chemical entities (NCEs), including several genetically engineered molecular entities, introduced for human therapeutic use into the world marketplace for the first time during 1991 totaled 36. This was comparable to the 37 introduced in 1990 (1), up slightly from the 33 in 1989 (2) but still well below the 52 NCEs introduced in 1988 (3).

The United States, for the first time in many years, held the lead with 11 NCE introductions, 2 more than Japan the perennial leader in recent years, followed by the United Kingdom with 4. Over 50% of the new NCEs launched in 1991 originated in two countries; 14 in the United States and 8 in Japan while the United Kingdom and Switzerland had 4 and 3, respectively. Cardiovascular drugs (including two new calcium channel blockers and one new ACE inhibitor) and antiinfectives were the leading therapeutic categories introduced in 1991, followed by cancer-related agents and psychotropics. Novel NCEs included alglucerase, a modified glucocerebrosidase enzyme for the treatment of Gaucher's disease. Other NCEs of special interest include the antifungal agent amorolfine hydrochloride, which inhibits two stages of ergosterol biosynthesis; the CSFs, filgrastim and sargramostim for the treatment of neutropenia associated with cancer chemotherapy and bone marrow transplants; sumatriptan succinate, a highly selective $5HT_1$-like receptor agonist as a new treatment for migraine and the new antihypertensive moxonidine, which acts as an agonist at CNS imidazole receptors.

During 1991, 30 NCE therapeutic agents were approved in the United States (4,5) tieing the record set in 1985 for the most NCEs approved by the FDA in one year. Included in these were five designated by the FDA as having a significant therapeutic gain with alglucerase, didanosine and fludarabine phosphate being the first worldwide introductions. Didanosine is the second NCE after AZT to be approved for the treatment of AIDS in the United States.

Alglucerase (enzyme) (6, 7)

Country of Origin:	**U.S.A.**
Originator:	**Genzyme**
First Introduction:	**U.S.A.**
Introduced by:	**Genzyme**
Trade Name:	**Ceredase**
CAS Registry No.:	**37228-64-1**

Alglucerase, a modified version of glucocerebrosidase, is an orphan drug for the treatment of Type I Gaucher's disease. It acts as a replacement for the natural enzyme. In human studies, alglucerase increased hemoglobin levels and reduced the size of the liver and spleen.

Alpidem (anxiolytic) (8-10)

Country of Origin:	**France**
Originator:	**Synthelabo**
First Introduction:	**France**
Introduced by:	**Synthelabo**
Trade Name:	**Ananxyl**
CAS Registry No.:	**82626-01-5**

Alpidem is an imidazopyridine antianxiety agent with anticonvulsant properties. It is claimed to be the first non-benzodiazepine anxiolytic to show selectivity for the omega-1 modulatory site of the $GABA_A$ receptors. Alpidem is reported to be superior to buspirone and similar to the benzodiazepines, but is significantly better tolerated with lower abuse potential.

Amorolfine Hydrochloride (topical antifungal) (11-13)

Country of Origin:	**Switzerland**	Introduced by:	**Roche**
Originator:	**Roche**	Trade Name:	**Loceryl**
First Introduction:	**United Kingdom**	CAS Registry No.:	**78613-38-4**

Amorolfine hydrochloride is the first morpholine antifungal agent useful in the topical treatment of onychomycosis. Mechanistically, it appears to inhibit two stages of the ergosterol biosynthetic pathway, which is necessary for fungal cell viability. Active against a broad spectrum of fungi, amorolfine hydrochloride is particularly effective against yeasts such as Histoplasma capsulatum, dermatophytes and dematiaceous fungi.

Benidipine Hydrochloride (antihypertensive) (14, 15)

Country of Origin:	**Japan**
Originator:	**Kyowa Hakko**
First Introduction:	**Japan**
Introduced by:	**Kyowa Hakko**
Trade Name:	**Coniel**
CAS Registry No.:	**91599-74-5**

Benidipine hydrochloride is a new, long acting dihydropyridine calcium antagonist useful in the treatment of hypertension and angina pectoris. In experimental animals, benidipine hydrochloride dose dependently induced hypotension with a potency of 6 and 12 times more than nifedipine and nicardipine, respectively. It also exhibited a slow onset of action.

Calcipotriol (antipsoriatic) (16-18)

Country of Origin:	**Denmark**
Originator:	**Leo Denmark**
First Introduction:	**Denmark**
Introduced by:	**Leo Denmark**
Trade Name:	**Daivonex**
CAS Registry No.:	**112965-21-6**

Calcipotriol is a topical vitamin D_3 derivative effective in the treatment of psoriasis vulgaris. The drug acts by binding to vitamin D_3 receptors in the skin keratinocytes, producing an elevation in cell differentiation and a reduction in cell proliferation. Although its efficacy is comparable to calcitriol, calcipotriol exhibits at least 100 times less effect on calcium metabolism in rats.

Carvedilol (antihypertensive) (19-21)

Country of Origin:	**Germany**	Introduced by:	**Boehringer Mannheim**
Originator:	**Boehringer Mannheim**	Trade Name:	**Dilatrend**
First Introduction:	**Germany**	CAS Registry No.:	**72956-09-3**

Carvedilol is a vasodilating beta-blocker useful in the treatment of hypertension and angina pectoris. In addition to lowering blood pressure, carvedilol decreases total vascular resistance without the reflex tachycardia usually occurring with vasodilators. It is reported to be well tolerated with renal sparing effects.

Cefdinir (antibiotic) (22, 23)

Country of Origin:	**Japan**
Originator:	**Fujisawa**
First Introduction:	**Japan**
Introduced by:	**Fujisawa**
Trade Name:	**Cefzon**
CAS Registry No.:	**91832-40-5**

Cefdinir is an orally active, beta-lactamase stable cephalosporin with a broad spectrum of activity. Compared to other oral cephalosporins, cefdinir is more potent against Gram-positive bacteria, especially Staphylococci. Its activity against Gram-negative bacteria such as E. coli, K. pneumoniae and P. mirabilis is similar to cefixime, but superior to cefaclor and cephalexin.

Cefotiam Hexetil Hydrochloride (antibiotic) (24-26)

Country of Origin:	**Japan**	Introduced by:	**Takeda**
Originator:	**Takeda**	Trade Name:	**Pansporin T**
First Introduction:	**Japan**	CAS Registry No.:	**95789-30-3**

Cefotiam hexetil hydrochloride is an orally active, broad-spectrum cephalosporin useful in the treatment of bacterial infections. It is a prodrug hydrolyzed in the GI tract and absorbed as cefotiam. Cefotiam hexetil hydrochloride is reported to be especially potent against Streptococci strains.

Centchroman (antiestrogen) (27-29)

Country of Origin:	**India**
Originator:	**Central Drug Research Institute**
First Introduction:	**India**
Introduced by:	**Hindustan Latex**
Trade Name:	**Choice-7**
CAS Registry No.:	**31477-60-8**

Centchroman is a new, weekly, orally active contraceptive with weak estrogenic and potent antiestrogenic properties. It is reported to be the first non-hormonal, non-steroidal oral contraceptive. Centchroman acts by inhibiting the implantation of fertilized eggs in the uterus. With the exception of longer menstrual cycles, no side effects have been reported. Another indication under study is the treatment of hormone dependent breast cancer.

Centoxin (immunomodulator) (30-32)

Country of Origin:	**U.S.A.**
Originator:	**Centocor**
First Introduction:	**Netherlands**
Introduced by:	**Centocor**
Trade Name:	**HA-1A**

Centoxin is an anti-endotoxin human monoclonal antibody indicated for the treatment of Gram-negative sepsis and septic shock. The drug was reported to resolve organ failure within one week of treatment. The results also suggest that centoxin blocks the release of tumor necrosis factor in septic patients. It is also under investigation for the treatment of burn patients and meningococcal septicemia.

Cloricromen (antithrombotic) (33-35)

$$H_5C_2OCCH_2O-\text{(coumarin ring with Cl, O, =O, CH}_2CH_2N(C_2H_5)_2, CH_3\text{)}$$

Country of Origin:	**Italy**	Introduced by:	**Sinax (Fidia)**
Originator:	**Sinax (Fidia)**	Trade Name:	**Proendotel**
First Introduction:	**Italy**	CAS Registry No.:	**68206-94-0**

Cloricromen is a new coumarin derivative useful as a coronary vasodilator and antithrombotic. It differs from existing antithrombotic agents in that it does not act chiefly on platelets. Cloricromen is reportedly well tolerated, improves exercise tolerance and decreases usage of nitrates by angina patients.

Clospipramine Hydrochloride (neuroleptic) (36-38)

Country of Origin:	**Japan**
Originator:	**Yoshitomi**
First Introduction:	**Japan**
Introduced by:	**Yoshitomi**
Trade Name:	**Cremin**
CAS Registry No.:	**98043-60-8**

• 2HCl

Clospipramine hydrochloride is a metabolite of clocapramine reportedly useful in the treatment of schizophrenia. In schizophrenic patients, its clinical efficacy and overall safety rating were not significantly different from clocapramine. However, at a lower dosage, clospipramine hydrochloride was more effective than clocapramine.

Dezocine (analgesic) (39-41)

Country of Origin:	**U.S.A.**
Originator:	**Wyeth-Ayerst (American Home Products)**
First Introduction:	**U.S.A.**
Introduced by:	**Astra**
Trade Name:	**Dalgan**
CAS Registry No.:	**53648-55-8**

Dezocine is an injectable agonist/antagonist analgesic indicated when an opioid analgesic is suitable for the management of pain. The drug differs from other opioid agonist/antagonists by having a high affinity for both the mu and delta receptors and a low effect on kappa receptors. Dezocine is reported to have a rapid onset of action, short half-life and low abuse potential with insignificant side effects.

Didanosine (antiviral) (42-44)

Country of Origin:	**U.S.A.**
Originator:	**National Cancer Institute (NIH)**
First Introduction:	**U.S.A.**
Introduced by:	**Bristol-Myers Squibb**
Trade Name:	**Videx**
CAS Registry No.:	**69655-05-6**

Didanosine is an orally active purine dideoxynucleoside analog indicated for adult and pediatric patients with advanced HIV infection who are either intolerant or significantly deteriorated on zidovudine. It appears to increase CD4 cell counts and decrease p24 antigen levels. Major adverse effects are pancreatitis, peripheral neuropathy and diarrhea. Unlike zidovudine, didanosine exhibits insignificant bone marrow suppression.

Doxacurium Chloride (muscle relaxant) (45-47)

Country of Origin:	**United Kingdom**	Introduced by:	**Wellcome**
Originator:	**Wellcome**	Trade Name:	**Nuromax**
First Introduction:	**U.S.A.**	CAS Registry No.:	**106819-53-8**

Doxacurium chloride is an injectable, noncumulative, nondepolarizing neuromuscular blocking agent which exhibits no significant cardiovascular effects. It is a mixture of (1R, 1'S, 2S, 2'R), (1R, 1'R, 2S, 2'S) and (1S, 1'S, 2R, 2'R) isomers of a bis-benzylisoquinolinium diester. Doxacurium chloride provides satisfactory intubation with duration of action similar to d-tubocurarine or pancuronium. However, it is reported to be 2 to 3 times as potent as pancuronium.

Filgrastim, Recombinant (immunostimulant) (48-50)

Country of Origin:	**U.S.A.**
Originator:	**Amgen**
First Introduction:	**U.S.A.**
Introduced by:	**Amgen**
Trade Name:	**Neupogen**
CAS Registry No.:	**121181-53-1**

Filgrastim is a recombinant human granulocyte colony stimulating factor (G-CSF) indicated as an adjunct to cancer chemotherapy for patients with non-myeloid malignancies. It reduces the duration of chemotherapy-induced neutropenia by forcing immature leucocytes to differentiate into neutrophils, thereby decreasing the incidence and length of infections. Bone pain is the most significant side effect. Chugai launched the second G-CSF, lenograstim, in December 1991. Other potential uses include bone marrow transplants, myelodysplastic syndromes and febrile/chronic neutropenia.

Fludarabine Phosphate (antineoplastic) (51-53)

Country of Origin:	**U.S.A.**
Originator:	**Southern Research Institute**
First Introduction:	**U.S.A.**
Introduced by:	**Berlex (Schering AG)**
Trade Name:	**Fludara**
CAS Registry No.:	**75607-67-9**

Fludarabine phosphate is an antimetabolite indicated for the treatment of B cell lymphocytic leukemia. It is reportedly effective in patients refractory to other therapies. Fludarabine phosphate acts by inhibiting primer RNA synthesis. Its side effects include bone marrow suppression, anemia, thrombocytopenia and neutropenia.

Fosinopril Sodium (antihypertensive) (54-56)

Country of Origin:	**U.S.A.**	Introduced by:	**Bristol-Myers Squibb**
Originator:	**Bristol-Myers Squibb**	Trade Name:	**Staril**
First Introduction:	**United Kingdom**	CAS Registry No.:	**88889-14-9**

Fosinopril sodium, an ester prodrug of fosinoprilat, is the first of a new generation of phosphinic acid ACE inhibitors indicated for the once-daily treatment of hypertension. Unlike captopril and lisinopril, it is reportedly effective in increasing the left ventricular peak filling rates and peak ejection rates in hypertensive patients at rest. Another advantage is that fosinopril sodium is metabolized equally by renal and hepatic routes, thereby avoiding the requirement to modify dosage in patients with renal insufficiency.

Gallium Nitrate (calcium regulator) (57-59)

Country of Origin:	**U.S.A.**
Originator:	**Memorial Sloan-Kettering Cancer Center**
First Introduction:	**U.S.A.**
Introduced by:	**Fujisawa**
Trade Name:	**Ganite**
CAS Registry No.:	**13494-90-1**

Gallium nitrate, initially developed as an anticancer agent, was introduced by Fujisawa as an orphan drug for the treatment of cancer-related hypercalcemia and bone metastases that do not respond to adequate hydration. The compound acts specifically on bone by inhibiting calcium resorption and also possibly by stimulating bone formation. Compared with calcitonin and etidronate, gallium nitrate is more potent and substantially longer acting. Other potential uses could be in the treatment of osteoporosis and Paget's disease.

Granisetron Hydrochloride (antiemetic) (60-63)

Country of Origin:	**United Kingdom**	Introduced by:	**SmithKline Beecham**
Originator:	**SmithKline Beecham**	Trade Name:	**Kytril**
First Introduction:	**South Africa**	CAS Registry No.:	**107007-99-8**

Granisetron hydrochloride is the second selective $5HT_3$-antagonist approved for the management of nausea and vomiting induced by cancer chemotherapy. Like ondansetron, granisetron hydrochloride is superior to metoclopramide in both efficacy and side effect profile. The compound appears to be effective both as a prophylactic agent and in blocking vomiting once started.

Halobetasol Propionate (topical antiinflammatory) (64-66)

Country of Origin:	**Switzerland**
Originator:	**Ciba-Geigy**
First Introduction:	**U.S.A.**
Introduced by:	**Bristol-Myers Squibb**
Trade Name:	**Ultravate**
CAS Registry No.:	**66852-54-8**

Halobetasol propionate is a new ultra-potent topical corticosteroid exhibiting both antiinflammatory and antihyperplasia properties. The compound is recommended for the treatment of moderate to severe corticosteroid-responsive dermatoses and appears to be particularly effective in patients with plaque psoriasis.

Interferon Gamma-1b, Recombinant (immunostimulant) (67, 68)

Country of Origin:	**U.S.A.**
Originator:	**Genentech**
First Introduction:	**U.S.A.**
Introduced by:	**Genentech**
Trade Name:	**Actimmune**
CAS Registry No.:	**98059-61-1**

Interferon gamma-1b, a recombinant DNA product, has orphan drug status for the treatment of chronic granulomatous disease. It is also under investigation as a treatment for trauma-related infections and atopic dermatitis.

Lacidipine (antihypertensive) (69-71)

Country of Origin:	**United Kingdom**
Originator:	**Glaxo**
First Introduction:	**Italy**
Introduced by:	**Glaxo; Guidotti; Zambon**
Trade Name:	**Lacipil; Lacirex; Viapres**
CAS Registry No.:	**103890-78-4**

Lacidipine is a new second-generation dihydropyridine calcium antagonist introduced as a once a day treatment for mild to moderate hypertension. It is reported to have high selectivity for vascular smooth muscle and also a long duration of action. The use of lacidipine as an antiatherosclerotic agent is currently under investigation.

Levocabastine Hydrochloride (antihistamine) (72-74)

Country of Origin:	**U.S.A.**
Originator:	**Janssen (Johnson & Johnson)**
First Introduction:	**Denmark**
Introduced by:	**Janssen (Johnson & Johnson)**
Trade Name:	**Livostin**
CAS Registry No.:	**79547-78-7**

Levocabastine hydrochloride is a new, highly potent and specific histamine H_1-receptor antagonist without anticholinergic or antiserotonergic side effects. It is reported to have both a fast onset and long duration of action. Indications are the treatment of allergic conjunctivitis and allergic rhinitis.

Moxonidine (antihypertensive) (75-77)

Country of Origin:	**Germany**
Originator:	**Beiersdorf**
First Introduction:	**Germany**
Introduced by:	**Beiersdorf; Giulini (Solvay)**
Trade Name:	**Cynt; Physiotens**
CAS Registry No.:	**75438-57-2**

Moxonidine, which is structurally related to clonidine, is a new centrally acting antihypertensive that acts as a stronger agonist at imidazole receptors and a weaker agonist at α_2-adrenergic receptors than clonidine. It is also reported to have less side effects and a much reduced potential to produce a rebound in blood pressure on withdrawal. Clinically, moxonidine appears to have comparable antihypertensive efficacy with the ACE inhibitors and calcium antagonists.

Nemonapride (neuroleptic) (78-80)

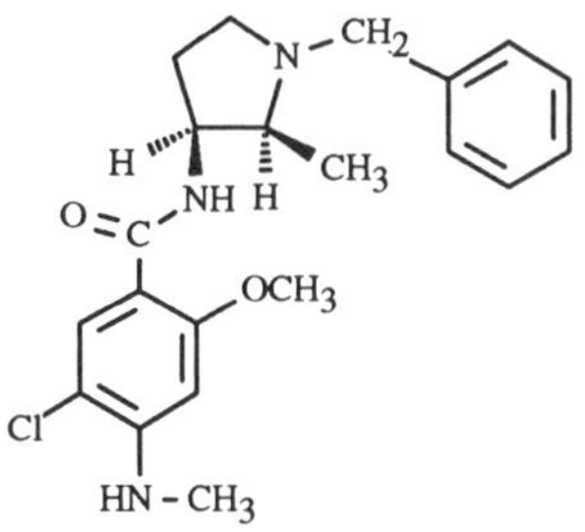

Country of Origin: **Japan**
Originator: **Yamanouchi**
First Introduction: **Japan**
Introduced by: **Yamanouchi**
Trade Name: **Emirace**
CAS Registry No.: **75272-39-8**

Nemonapride is a potent dopamine D_2 receptor antagonist antipsychotic reported to have superior efficacy against hallucinations and delusions and also to be effective in the treatment of negative symptoms. The compound is a benzamide diastereomer and, although more potent than haloperidol, side effects are reported to be few primarily akinesia and some muscle rigidity with extrapyramidal symptoms being rare.

Paroxetine (antidepressant) (81, 82)

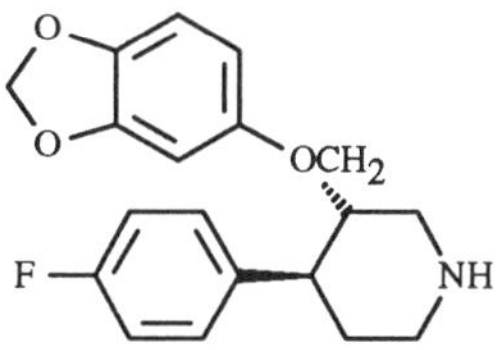

Country of Origin: **Denmark**
Originator: **As Ferrosan (Novo-Nordisk)**
First Introduction: **United Kingdom**
Introduced by: **SmithKline Beecham**
Trade Name: **Seroxat**
CAS Registry No.: **61869-08-7**

Paroxetine is a new highly selective serotonin reuptake inhibitor, mechanistically similar to fluoxetine, fluvoxamine and sertraline, introduced for the treatment of all types of depressive illnesses including depression associated with anxiety. It is reportedly non-sedating and non-stimulatory and compared to fluoxetine has a shorter duration of action (half-life of 24 hours versus 2 to 3 days). Paroxetine is also being investigated as a treatment for obesity, alcoholism and obsessive-compulsive disorders.

Pemirolast Potassium (antiasthmatic) (83, 84)

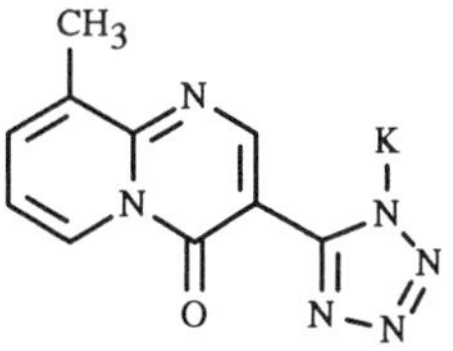

Country of Origin: **U.S.A.**
Originator: **Bristol-Myers Squibb**
First Introduction: **Japan**
Introduced by: **Bristol-Myers Squibb; Tokyo Tanabe**
Trade Name: **Pemilaston; Alegysal**
CAS Registry No.: **100299-08-9**

Pemirolast potassium is a new potent antiallergic mast cell stabilizer with a similar mechanism to sodium cromoglycate. Its main indication is bronchial asthma. Another indication under investigation is as a gastroprotective agent.

Pilsicainide Hydrochloride (antiarrhythmic) (85-87)

Country of Origin:	**Japan**
Originator:	**Suntory**
First Introduction:	**Japan**
Introduced by:	**Suntory; Daiichi**
Trade Name:	**Sunrhythm**
CAS Registry No.:	**88069-49-2**

Pilsicainide hydrochloride is a new pyrrolizidine lidocaine derivative with antiarrhythmic activity. It is the first pharmaceutical product introduced by Suntory and was solely developed in Japan. The compound is reported to be highly effective in the treatment of premature ventricular contraction. It has no anticholinergic or CNS activity and is not as likely as other class 1C antiarrhythmics to have restricted indications.

Romurtide (immunostimulant) (88, 89)

Country of Origin:	**Japan**		Introduced by:	**Daiichi**
Originator:	**Daiichi**		Trade Name:	**Nopia**
First Introduction:	**Japan**		CAS Registry No.:	**78113-36-7**

Romurtide is a muramyl dipeptide which augments the production of G-CSF. The compound is indicated as an immunomodulator for the treatment of leukopenia caused by anticancer chemotherapy/radiotherapy and also thrombopenia.

Sargramostim, Recombinant (immunostimulant) (90, 91)

Country of Origin:	**U.S.A.**
Originator:	**Immunex**
First Introduction:	**U.S.A.**
Introduced by:	**Immunex; Hoechst-Roussel**
Trade Name:	**Leukine; Prokine**

Sargramostim, a recombinant DNA product, is a granulocyte macrophage colony stimulating factor (GM-CSF) indicated for the treatment of cancer patients after autologous bone marrow transplants. It stimulates immature cells to develop into granulocytes and macrophages and can also switch on mature granulocytes and macrophages. Other indications include improved bone marrow recovery in non-Hodgkins lymphoma and acute lymphoblastic leukemia.

Succimer (chelator) (92, 93)

Country of Origin:	**U.S.A.**
Originator:	**Johnson and Johnson (J&J)**
First Introduction:	**U.S.A.**
Introduced by:	**McNeil Consumer Products (J&J)**
Trade Name:	**Chemet**
CAS Registry No.:	**304-55-2**

$$\begin{array}{c} COOH \\ | \\ H-C-SH \\ | \\ H-C-SH \\ | \\ COOH \end{array}$$

Succimer is a newly available oral chelator of lead and other heavy metals. The compound is indicated for the treatment of lead poisoning in children. Succimer does not bind iron, calcium or magnesium and preferentially binds lead, mercury and arsenic over copper or zinc. The metal-bound succimer is excreted in the urine. The drug may also be effective in adults.

Sumatriptan Succinate (antimigraine) (94, 95)

Country of Origin:	**United Kingdom**	Introduced by:	**Glaxo**
Originator:	**Glaxo**	Trade Name:	**Imigran**
First Introduction:	**Netherlands**	CAS Registry No.:	**103628-48-4**

Sumatriptan succinate is a highly selective $5HT_1$-like receptor agonist introduced as a new treatment for migraine. It is indicated for the acute relief of migraine and cluster headache. Oral administration is reported to be free of substantial side effects. The compound appears to be a significant advance over the use of ergotamine and other agents in the treatment of migraine.

Temafloxacin Hydrochloride (antibacterial) (96, 97)

Country of Origin:	**U.S.A.**
Originator:	**Abbott**
First Introduction:	**Sweden**
Introduced by:	**Abbott; ICI**
Trade Name:	**Temac**
CAS Registry No.:	**105784-61-0**

Temafloxacin hydrochloride is a new fluoroquinolone antibacterial structurally and mechanistically related to tosufloxacin, another microbial DNA topoisomerase inhibitor introduced in 1990. It is particularly useful against beta-lactamase producing organisms and is indicated for the treatment of respiratory and urinary tract infections as well as sexually transmitted diseases, skin and soft tissue infections. The compound has high tissue penetration and is reported to have a low propensity to cause adverse effects and drug interactions.

Terbinafine Hydrochloride (antifungal) (98, 99)

Country of Origin:	**Switzerland**
Originator:	**Sandoz**
First Introduction:	**United Kingdom**
Introduced by:	**Sandoz**
Trade Name:	**Lamisil**
CAS Registry No.:	**78628-80-5**

Terbinafine hydrochloride is the first orally active allylamine antifungal with 30-fold greater antifungal activity than naftifine. The compound is indicated for the treatment of ringworm and fungal nail infections. Terbinafine hydrochloride acts on a single fungal enzyme, squalene epoxidase, interfering with the biosynthesis of ergosterols in cell membranes. Unlike other antifungal agents, it does not inhibit cytochrome P450 enzymes.

References

1.   J. D. Strupczewski, D. B. Ellis and R. C. Allen, Annu. Rep. Med. Chem., 26, 297 (1991).

2.   H. H. Ong and R. C. Allen, Annu. Rep. Med. Chem., 25, 309 (1990).

3.   H. H. Ong and R. C. Allen, Annu. Rep. Med. Chem., 24, 295 (1989).

4.   J. Rosenberg and C. Starr, Drug Topics, 48 (February 3, 1992).

5.   F-D-C Reports, 13, (January 6, 1992).

6.   N. W. Barton, R. O. Brady, J. M. Dambrosia, A. M. Di Bisceglie, S. H. Doppelt, S. C. Hill, H. J. Mankin, G. J. Murray, R. I. Parker, C. E. Argoff, R. P. Grewal, K.-T. Yu, O. C. Graham, C. A. Holder, K. D. Howard, C. R. Kaneski, K. L. Oliver, S. Riesz, C. L. Verderse and G. C. Zirzow, New Engl. J. Med., 324, 1464 (1991).

7.   E. Beutler, A. Kay, A. Saven, P. Garver, D. Thurston, A. Dawson and B. Rosenbloom, Blood, 78, 1183 (1991).

8.   S. Z. Langer, S. Arbilla, J. Benavides and B. Scatton, Adv. Biochem. Psychopharmacol., 46, 61 (1990).

9.   S. Z. Langer, S. Arbilla, S. Tan, K. G. Lloyd, P. George, J. Allen and A. E. Wick, Pharmacopsychiatry, 23 (Suppl. 3), 103 (1990), and following papers.

10.   B. I. Diamond, H. Nguyen, E. O'Neal, R. Ochs, M. Kaffeman and R. L. Borison, Psychopharmacol. Bull., 27, 67 (1991).

11.   A. del Palacio-Hernanz, S. Lopez-Gomez, P. Moreno-Palancar and F. Gonzalez-Lastra, Clin. Exp. Dermatol., 14, 141 (1989).

12.   W. Melchinger, A. Polak and J. Muller, Mycoses, 33, 393 (1990).

13.   A. del Palacio, F. Sanz, M. Garcia-Bravo, C. Gimeno, S. Cuetara, P. Miranda and A. R. Noriega, Mycoses, 34, 85 (1991).

14.   K. Muto, T. Kuroda, H. Kawato, A. Karasawa, K. Kubo and N. Nakamizo, Arzneimittelforschung, 38, 1662 (1988), and following papers.

15.   J. R. Prous, ed., Drugs Future, 15, 1116 (1990).

16.   K. Kragballe, Dermatologica, 181, 211 (1990).

17.   K. Kragballe and I. L. Wildfang, Arch. Dermatol. Res., 282, 164 (1990).

18.   A. M. Kissmeyer and L. Binderup, Biochem. Pharmacol., 41, 1601 (1991).

19.   R. R. Ruffolo Jr., M. Gellai, J. P. Hieble, R. N. Willette and A. J. Nichols, Eur. J. Clin. Pharmacol., 38 (Suppl. 2), S82 (1990).

20.   K. Strein and G. Sponer, Z. Kardiol., 79 (Suppl. 3), 89 (1990).

21.   P. Omvik and P. Lund-Johansen, Eur. Heart J., 12, 736 (1991).

22.   J. R. Prous, ed., Drugs Future, 13, 224 (1988).

23.   N. Fukushima, Y. Wagatsuma, A. Takase, A. Ishikawa and S. Takahashi, Jpn. J. Antibiot., 43, 1783 (1990), and following papers.

24.   J. R. Prous, ed., Drugs Future, 13, 231 (1988).

25.   S. C. Chang, W. C. Hsieh, K. T. Luh and S. W. Ho, Taiwan I Hsueh Hui Tsa Chih, 89, 661 (1990).

26.   H. C. Korting, M. Schafer-Korting, F. Kees, A. Lukacs and H. Grobecker, Eur. J. Clin. Pharmacol., 39, 33 (1990).

27.   N. C. Misra, P. K. Nigam, R. Gupta, A. K. Agarwal and V. P. Kamboj, Int. J. Cancer, 43, 781 (1989).

28.   J. K. Paliwal, R. C. Gupta, P. K. Grover, O. P. Asthana, J. S. Srivastava and S. NityaNand, Pharm. Res., 6, 1048 (1989).

29.   N. Majumdar and J. K. Datta, Indian J. Exp. Biol., 28, 717 (1990).

30.   R. C. Bone, JAMA, 266, 1686 (1991).

31.   K. A. Schulman, H. A. Glick, H. Rubin and J. M. Eisenberg, JAMA, 266, 3466 (1991).

32.   B. J. Zarowitz, DICP, 25, 778 (1991).

33.   R. A. Travagli, A. Zatta, N. Banzatto, M. Finesso, R. Mariot, F. Tessari and M. Prosdocimi, Thromb. Res., 54, 327 (1989).

34.   A. Del Maschio, G. Bazzoni, A. Zatta, Z. M. Chen, E. Dejana and M. Prosdocimi, Eur. J. Pharmacol., 187, 541 (1990).

35.   A. Zatta, M. Prosdocimi, G. Bazzoni, E. Dejana and A. Del Maschio, Eur. J. Pharmacol., 198, 97 (1991).

36.   M. Mikuni, I. Yamashita, S. Matsubara, Y. Odagaki, M. Setoguchi and T. Fukuda, Clin. Neuropharmacol., 9 (Suppl. 4), 319 (1986).

37.   J. Ishigooka, M. Murasaki, H. Wakatabe, S. Miura, K. Hikida, M. Shibata and H. Nobunaga, Psychopharmacology, 97, 303 (1989).

38.   J. R. Prous, ed., Drugs Future, 15, 434 (1990).

39.   J. E. Stambaugh Jr. and J. McAdams, Clin. Pharmacol. Ther., 42, 210 (1987).

40.   J. J. O'Brien and P. Benfield, Drugs, 38, 226 (1989).

41.   P. J. Hoskin and G. W. Hanks, Drugs, 41, 326 (1991).

42.   K. J. Connolly, J. D. Allan, H. Fitch, L. Jackson-Pope, C. McLaren, R. Canetta and J. E. Groopman, Am. J. Med., 91, 471 (1991).

43.   N. R. Hartman, R. Yarchoan, J. M. Pluda, R. V. Thomas, K. M. Wyvill, K. P. Flora, S. Broder and D. G. Johns, Clin. Pharmacol. Ther., 50, 278 (1991).

44.   V. A. Johnson, D. P. Merrill, J. A. Videler, T. C. Chou, R. E. Byington, J. J. Eron, R. T. D'Aquila and M. S. Hirsch, J. Infect. Dis., 164, 646 (1991).

45.   R. S. Emmott, B. J. Bracey, D. R. Goldhill, P. M. Yate and P. J. Flynn, Br. J. Anaesth., 65 480 (1990).

46.   D. R. Cook, J. A. Freeman, A. A. Lai, K. A. Robertson, Y. Kang, R. L. Stiller, S. Aggarwal, M. M. Abou-Donia and R. M. Welch, Anesth. Analg., 72, 145 (1991).

47.   E. Ornstein, R. S. Matteo, J. A. Weinstein, J. D. Halevy, W. L. Young and M. M. Abou-Donia, J. Clin. Anesth., 3, 108 (1991).

48.   K. Eguchi, S. Sasaki, T. Tamura, Y. Sasaki, T. Shinkai, K. Yamada, Y. Soejima, M. Fukuda, Y. Fujihara, H. Kunitou, K. Tobinai, T. Ohtsu, K. Suemasu, F. Takaku and N. Saijo, Cancer Res., 49, 5221 (1989).

49.   K. Eguchi, T. Shinkai, Y. Sasaki, T. Tamura, Y. Ohe, K. Nakagawa, M. Fukuda, K. Yamada, A. Kojima, F. Oshita, M. Morita, K. Suemasu and N. Saijo, Jpn. J. Cancer Res., 81 1168 (1990).

50.   H. Tanaka and T. Tokiwa, Cancer Res., 50, 6615 (1990).

51.   W. Hiddemann, R. Rottmann, B. Wormann, A. Thiel, M. Essink, C. Ottensmeier, M. Freund, T. Buchner and J. van de Loo, Ann. Hematol., 63, 1 (1991).

52.   C. A. Puccio, A. Mittelman, S. M. Lichtman, R. T. Silver, D. R. Budman and T. Ahmed, J. Clin. Oncol., 9, 1562 (1991).

53.   S. A. Taylor, J. Crowley, F. S. Vogel, J. J. Townsend, H. J. Eyre, K. A. Jaeckle, H. E. Hynes and J. T. Guy, Invest. New Drugs, 9, 195 (1991).

54.   K. L. Duchin, A. P. Waclawski, J. I. Tu, J. Manning, M. Frantz and D. A. Willard, J. Clin. Pharmacol., 31, 58 (1991).

55.   T. Forslund, P. Franzen and R. Backman, J. Intern. Med., 230, 511 (1991).

56.   D. A. Sica, R. E. Cutler, R. J. Parmer and N. F. Ford, Clin. Pharmacokinet., 20, 420 (1991).

57.   I. H. Krakoff, Semin. Oncol., 18 (4 Suppl. 5), 3 (1991), and following papers.

58.   P. A. Todd and A. Fitton, Drugs, 42, 261 (1991).

59.   R. P. Warrell Jr., W. K. Murphy, P. Schulman, P. J. O'Dwyer and G. Heller, J. Clin. Oncol., 9, 1467 (1991).

60.   J. Carmichael, B. M. J. Cantwell, C. M. Edwards, B. D. Zussman, S. Thompson, W. G. Rapeport and A. L. Harris, Cancer Chemother. Pharmacol., 24, 45 (1989).

61.   D. R. Nelson and D. R. Thomas, Biochem. Pharmacol., 38, 1693 (1989).

62.   P. J. Hesketh and D. R. Gandara, J. Natl. Cancer Inst., 83, 613 (1991).

63.   J. P. Logue, B. Magee, R. D. Hunter and R. D. Murdoch, Clin. Oncol., 3, 247 (1991).

64.   W. A. Watson, R. E. Kalb, S. B. Siskin, J. P. Freer and L. Krochmal, Pharmacotherapy, 10, 107 (1990).

65.   C. P. Robinson, Drugs Today, 27, 304 (1991).

66.   S. Yawalkar, I. Wiesenberg-Boettcher, J. R. Gibson, S. B. Siskin and W. P. Basel, J. Am. Acad. Dermatol., 25 (6 Pt. 2), 1137 (1991), and following papers.

67.   K. Bartnes and K. Hannestad, Eur. J. Immunol., 21, 2365 (1991).

68.   J. I. Gallin, Rev. Infect. Dis., 13, 973 (1991).

69.   D. Micheli, A. Collodel, C. Semeraro, G. Gaviraghi and C. Carpi, J. Cardiovasc. Pharmacol., 15, 666 (1990).

70.   M. E. Heber, P. A. Broadhurst, G. S. Brigden and E. B. Raftery, Am. J. Cardiol., 66, 1228 (1990).

71.   S. Soro and L. A. Ferrara, Eur. J. Clin. Pharmacol., 41, 105 (1991).

72.   K. L. Dechant and K. L. Goa, Drugs, 41, 202 (1991).

73.   M. Schata, W. Jorde and U. Richarz-Barthauer, J. Allergy Clin. Immunol., 87, 873 (1991).

74.   P. H. Van de Heyning, J. Claes, J. Van Haesendonck and M. Rosseel, Clin. Exp. Allergy, 21 (Suppl. 2), 21 (1991).

75.   V. Planitz, J. Clin. Pharmacol., 27, 46 (1987).

76.   M. C. Michel, O. E. Brodde, B. Schnepel, J. Behrendt, R. Tschada, H. J. Motulsky and P. A. Insel, Mol. Pharmacol., 35, 324 (1989).

77.   W. Kirch, H. J. Hutt and V. Planitz, J. Clin. Pharmacol., 30, 1088 (1990).

78.   M. Terai, K. Hidaka and Y. Nakamura, Eur. J. Pharmacol., 173, 177 (1989).

79.   Y. Kudo, G. Ikawa, Y. Kawakita, M. Saito, T. Sakai, T. Nakajima, T. Nishimura, Y. Higashi and K. Hitomi, Eur. J. Pharmacol., 183, 591 (1990).

80.   A. S. Unis, J. G. Vincent and B. Dillon, Life Sci., 47, 151 (1990).

81.   K. L. Dechant and S. P. Clissold, Drugs, 41, 225 (1991).

82. G. C. Dunbar, J. B. Cohn, L. F. Fabre, J. P. Feighner, R. R. Fieve, J. Mendels and R. K. Shrivastava, Br. J. Psychiatry, 159, 394 (1991).

83. Y. Yanagihara, H. Kasai, T. Kawashima and K. Ninomiya, Jpn. J. Pharmacol., 51, 93 (1989).

84. D. G. Tinkelman and R. B. Berkowitz, Ann. Allergy, 66, 162 (1991).

85. N. Inomata and T. Ishihara, Eur. J. Pharmacol., 145, 313 (1988).

86. N. Inomata, T. Ishihara and N. Akaike, Br. J. Pharmacol., 98, 149 (1989).

87. T. Terazawa, M. Suzuki, T. Goto, R. Kato, H. Hayashi, A. Ito, S. Isikawa and I. Sotobata, Am. Heart J., 121, 1437 (1991).

88. E. Tsubura, I. Azuma and T. Une, Arzneimittelforschung, 38, 951 (1988), and following papers.

89. K. Shimoda, S. Okamura, C. Kawasaki, F. Omori, T. Matsuguchi and Y. Niho, Int. J. Immunopharmacol., 12, 729 (1990).

90. J. M. Burr and B. B. McCall, Clin. Pharm., 10, 947 (1991).

91. M. Gibaldi, Perspect. Clin. Pharm., 9, 57 (1991).

92. J. J. Chisolm Jr., Environ. Health Perspect., 89, 67 (1990).

93. P. Grandjean, I. A. Jacobsen and P. J. Jorgensen, Pharmacol. Toxicol., 68, 266 (1991).

94. P. J. Goadsby, A. S. Zagami, G. A. Donnan, G. Symington, M. Anthony, P. F. Bladin and J. W. Lance, Lancet, 338, 782 (1991).

95. P. A. Fowler, L. F. Lacey, M. Thomas, O. N. Keene, R. J. Tanner and N. S. Baber, Eur. Neurol., 31, 291 (1991), and following papers.

96. B. I. Davies, F. P. Maesen, H. L. Gubbelmans and H. M. Cremers, J. Antimicrob. Chemother., 26, 237 (1990).

97. A. Iravani, Antimicrob. Agents Chemother., 35, 1777 (1991).

98. E. B. Smith, N. Noppakun and R. C. Newton, J. Am. Acad. Dermatol., 23 (4 Pt. 2), 790 (1990).

99. J. E. White, P. J. Perkins and E. G. Evans, Br. J. Dermatol., 125, 260 (1991).

GENERIC NAME	INDICATION	YEAR INTRODUCED	ARMC VOL., PAGE	
acarbose	antidiabetic	1990	26,	297
acetohydroxamic acid	hypoammonuric	1983	19,	313
acipimox	hypolipidemic	1985	21,	323
acitretin	antipsoriatic	1989	25,	309
acrivastine	antihistamine	1988	24,	295
adamantanium bromide	antiseptic	1984	20,	315
adrafinil	psychostimulant	1986	22,	315
AF-2259	antiinflammatory	1987	23,	325
afloqualone	muscle relaxant	1983	19,	313
alacepril	antihypertensive	1988	24,	296
alclometasone dipropionate	topical antiinflammatory	1985	21,	323
alfentanil HCl	analgesic	1983	19,	314
alfuzosin HCl	antihypertensive	1988	24,	296
alglucerase	enzyme	1991	27,	321
alminoprofen	analgesic	1983	19,	314
alpha-1 antitrypsin	protease inhibitor	1988	24,	297
alpidem	anxiolytic	1991	27,	322
alpiropride	antimigraine	1988	24,	296
alteplase	thrombolytic	1987	23,	326
amfenac sodium	antiinflammatory	1986	22,	315
aminoprofen	topical antiinflammatory	1990	26,	298
amisulpride	antipsychotic	1986	22,	316
amlexanox	antiasthmatic	1987	23,	327
amlodipine besylate	antihypertensive	1990	26,	298
amorolfine hydrochloride	topical antifungal	1991	27,	322
amosulalol	antihypertensive	1988	24,	297
amrinone	cardiotonic	1983	19,	314
amsacrine	antineoplastic	1987	23,	327
APD	calcium regulator	1987	23,	326
apraclonidine HCl	antiglaucoma	1988	24,	297
APSAC	thrombolytic	1987	23,	326
arbekacin	antibiotic	1990	26,	298
argatroban	antithromobotic	1990	26,	299
arotinolol HCl	antihypertensive	1986	22,	316
artemisinin	antimalarial	1987	23,	327
aspoxicillin	antibiotic	1987	23,	328
astemizole	antihistamine	1983	19,	314
astromycin sulfate	antibiotic	1985	21,	324
auranofin	chrysotherapeutic	1983	19,	314
azelaic acid	antiacne	1989	25,	310
azelastine HCl	antihistamine	1986	22,	316
azithromycin	antibiotic	1988	24,	298
azosemide	diuretic	1986	22,	316
aztreonam	antibiotic	1984	20,	315
bambuterol	bronchodilator	1990	26,	299
beclobrate	hypolipidemic	1986	22,	317
befunolol HCl	antiglaucoma	1983	19,	315
benazepril hydrochloride	antihypertensive	1990	26,	299
benexate HCl	antiulcer	1987	23,	328
benidipine hydrochloride	antihypertensive	1991	27,	322
betaxolol HCl	antihypertensive	1983	19,	315
bevantolol HCl	antihypertensive	1987	23,	328
bifemelane HCl	nootropic	1987	23,	329
binfonazole	hypnotic	1983	19,	315
binifibrate	hypolipidemic	1986	22,	317
bisantrene hydrochloride	antineoplastic	1990	26,	300

GENERIC NAME	INDICATION	YEAR INTRODUCED	ARMC VOL., PAGE	
bisoprolol fumarate	antihypertensive	1986	22,	317
bopindolol	antihypertensive	1985	21,	324
brotizolam	hypnotic	1983	19,	315
brovincamine fumarate	cerebral vasodilator	1986	22,	317
bucillamine	immunomodulator	1987	23,	329
bucladesine sodium	cardiostimulant	1984	20,	316
budralazine	antihypertensive	1983	19,	315
bunazosin HCl	antihypertensive	1985	21,	324
bupropion HCl	antidepressant	1989	25,	310
buserelin acetate	hormone	1984	20,	316
buspirone HCl	anxiolytic	1985	21,	324
butoconazole	topical antifungal	1986	22,	318
butoctamide	hypnotic	1984	20,	316
butyl flufenamate	topical antiinflammatory	1983	19,	316
cadexomer iodine	wound healing agent	1983	19,	316
cadralazine	hypertensive	1988	24,	298
calcipotriol	antipsoriatic	1991	27,	323
camostat mesylate	antineoplastic	1985	21,	325
carboplatin	antibiotic	1986	22,	318
carumonam	antibiotic	1988	24,	298
carvedilol	antihypertensive	1991	27,	323
cefbuperazone sodium	antibiotic	1985	21,	325
cefdinir	antibiotic	1991	27,	323
cefixime	antibiotic	1987	23,	329
cefmenoxime HCl	antibiotic	1983	19,	316
cefminox sodium	antibiotic	1987	23,	330
cefodizime sodium	antibiotic	1990	26,	300
cefonicid sodium	antibiotic	1984	20,	316
ceforanide	antibiotic	1984	20,	317
cefotetan disodium	antibiotic	1984	20,	317
cefotiam hexetil hydrochloride	antibiotic	1991	27,	324
cefpimizole	antibiotic	1987	23,	330
cefpiramide sodium	antibiotic	1985	21,	325
cefpodoxime proxetil	antibiotic	1989	25,	310
ceftazidime	antibiotic	1983	19,	316
cefteram pivoxil	antibiotic	1987	23,	330
cefuroxime axetil	antibiotic	1987	23,	331
cefuzonam sodium	antibiotic	1987	23,	331
celiprolol HCl	antihypertensive	1983	19,	317
centchroman	antiestrogen	1991	27,	324
centoxin	immunomodulator	1991	27,	325
cetirizine HCl	antihistamine	1987	23,	331
chenodiol	anticholelithogenic	1983	19,	317
choline alfoscerate	nootropic	1990	26,	300
cibenzoline	antiarrhythmic	1985	21,	325
cicletanine	antihypertensive	1988	24,	299
cilostazol	antithrombotic	1988	24,	299
cimetropium bromide	antispasmodic	1985	21,	326
cilazapril	antihypertensive	1990	26,	301
cinitapride	gastroprokinetic	1990	26,	301
ciprofibrate	hypolipidemic	1985	21,	326
ciprofloxacin	antibacterial	1986	22,	318
cisapride	gastroprokinetic	1988	24,	299
citalopram	antidepressant	1989	25,	311
clarithromycin	antibiotic	1990	26,	302

GENERIC NAME	INDICATION	YEAR INTRODUCED	ARMC VOL., PAGE	
clobenoside	vasoprotective	1988	24,	300
cloconazole HCl	topical antifungal	1986	22,	318
clodronate disodium	calcium regulator	1986	22,	319
cloricromen	antithrombotic	1991	27,	325
clospipramine hydrochloride	neuroleptic	1991	27,	325
cyclosporine	immunosuppressant	1983	19,	317
dapiprazole HCl	antiglaucoma	1987	23,	332
defibrotide	antithrombotic	1986	22,	319
deflazacort	antiinflammatory	1986	22,	319
delapril	antihypertensive	1989	25,	311
denopamine	cardiostimulant	1988	24,	300
dezocine	analgesic	1991	27,	326
diacerein	antirheumatic	1985	21,	326
didanosine	antiviral	1991	27,	326
dilevalol	antihypertensive	1989	25,	311
disodium pamidronate	calcium regulator	1989	25,	312
divistyramine	hypocholesterolemic	1984	20,	317
dopexamine	cardiostimulant	1989	25,	312
doxacurium chloride	muscle relaxant	1991	27,	326
doxazosin mesylate	antihypertensive	1988	24,	300
doxefazepam	hypnotic	1985	21,	326
doxifluridine	antineoplastic	1987	23,	332
doxofylline	bronchodilator	1985	21,	327
dronabinol	antinauseant	1986	22,	319
droxicam	antiinflammatory	1990	26,	302
droxidopa	antiparkinsonian	1989	25,	312
ebastine	antihistamine	1990	26	302
emorfazone	analgesic	1984	20,	317
enalapril maleate	antihypertensive	1984	20,	317
enalaprilat	antihypertensive	1987	23,	332
encainide HCl	antiarrhythmic	1987	23,	333
enocitabine	antineoplastic	1983	19,	318
enoxacin	antibacterial	1986	22,	320
enoxaparin	antithrombotic	1987	23,	333
enoximone	cardiostimulant	1988	24,	301
enprostil	antiulcer	1985	21,	327
eperisone HCl	muscle relaxant	1983	19,	318
epidermal growth factor	wound healing agent	1987	23,	333
epirubicin HCl	antineoplastic	1984	20,	318
epoprostenol sodium	platelet aggreg. inhib.	1983	19,	318
eptazocine HBr	analgesic	1987	23,	334
erythromycin acistrate	antibiotic	1988	24,	301
erythropoietin	hematopoetic	1988	24,	301
esmolol HCl	antiarrhythmic	1987	23,	334
ethyl icosapentate	antithrombotic	1990	26,	303
etizolam	anxiolytic	1984	20,	318
etodolac	antiinflammatory	1985	21,	327
exifone	nootropic	1988	24,	302
famotidine	antiulcer	1985	21,	327
felbinac	topical antiinflammatory	1986	22,	320
felodipine	antihypertensive	1988	24,	302
fenbuprol	choleretic	1983	19,	318
fenticonazole nitrate	antifungal	1987	23,	334
filgrastim	immunostimulant	1991	27,	327
fisalamine	intestinal antiinflammatory	1984	20,	318
flomoxef sodium	antibiotic	1988	24,	302

GENERIC NAME	INDICATION	YEAR INTRODUCED	ARMC VOL., PAGE	
fluconazole	antifungal	1988	24,	303
fludarabine phosphate	antineoplastic	1991	27,	327
flumazenil	benzodiazepine antag.	1987	23,	335
flunoxaprofen	antiinflammatory	1987	23,	335
fluoxetine HCl	antidepressant	1986	22,	320
flupirtine maleate	analgesic	1985	21,	328
flutamide	antineoplastic	1983	19,	318
flutazolam	anxiolytic	1984	20,	318
fluticasone propionate	antiinflammatory	1990	26,	303
flutoprazepam	anxiolytic	1986	22,	320
flutropium bromide	antitussive	1988	24,	303
fluvoxamine maleate	antidepressant	1983	19,	319
formoterol fumarate	bronchodilator	1986	22,	321
foscarnet sodium	antiviral	1989	25,	313
fosfosal	analgesic	1984	20,	319
fosinopril sodium	antihypertensive	1991	27,	328
fotemustine	antineoplastic	1989	25,	313
gallium nitrate	calcium regulator	1991	27,	328
gallopamil HCl	antianginal	1983	19,	319
ganciclovir	antiviral	1988	24,	303
gemeprost	abortifacient	1983	19,	319
gestodene	progestogen	1987	23,	335
gestrinone	antiprogestogen	1986	22,	321
goserelin	hormone	1987	23,	336
granisetron hydrochloride	antiemetic	1991	27,	329
guanadrel sulfate	antihypertensive	1983	19,	319
halobetasol propionate	topical antiinflammatory	1991	27,	329
halofantrine	antimalarial	1988	24,	304
halometasone	topical antiinflammatory	1983	19,	320
hydrocortisone aceponate	topical antiinflammatory	1988	24,	304
hydrocortisone butyrate	topical antiinflammatory	1983	19,	320
ibopamine HCl	cardiostimulant	1984	20,	319
ibudilast	antiasthmatic	1989	25,	313
idarubicin hydrochloride	antineoplastic	1990	26,	303
idebenone	nootropic	1986	22,	321
imipenem/cilastatin	antibiotic	1985	21,	328
indalpine	antidepressant	1983	19,	320
indeloxazine HCl	nootropic	1988	24,	304
indobufen	antithrombotic	1984	20,	319
interferon, gamma	antiinflammatory	1989	25,	314
interferon gamma-1b	immunostimulant	1991	27,	329
interleukin-2	antineoplastic	1989	25,	314
ipriflavone	calcium regulator	1989	25,	314
irsogladine	antiulcer	1989	25,	315
isepamicin	antibiotic	1988	24,	305
isofezolac	antiinflammatory	1984	20,	319
isoxicam	antiinflammatory	1983	19,	320
isradipine	antihypertensive	1989	25,	315
itraconazole	antifungal	1988	24,	305
ivermectin	antiparasitic	1987	23,	336
ketanserin	antihypertensive	1985	21,	328
ketorolac tromethamine	analgesic	1990	26,	304
lacidipine	antihypertensive	1991	27,	330
lamotrigine	anticonvulsant	1990	26,	304
lenampicillin HCl	antibiotic	1987	23,	336
lentinan	immunostimulant	1986	22,	322

GENERIC NAME	INDICATION	YEAR INTRODUCED	ARMC VOL., PAGE	
leuprolide acetate	hormone	1984	20,	319
levacecarnine HCl	nootropic	1986	22,	322
levobunolol HCl	antiglaucoma	1985	21,	328
levocabastine hydrochloride	antihistamine	1991	27,	330
levodropropizine	antitussive	1988	24,	305
lidamidine HCl	antiperistaltic	1984	20,	320
limaprost	antithrombotic	1988	24,	306
lisinopril	antihypertensive	1987	23,	337
lobenzarit sodium	antiinflammatory	1986	22,	322
lomefloxacin	antibiotic	1989	25,	315
lonidamine	antineoplastic	1987	23,	337
loprazolam mesylate	hypnotic	1983	19,	321
loratadine	antihistamine	1988	24,	306
lovastatin	hypocholesterolemic	1987	23,	337
loxoprofen sodium	antiinflammatory	1986	22,	322
mabuterol HCl	bronchodilator	1986	22,	323
malotilate	hepatroprotective	1985	21,	329
manidipine hydrochloride	antihypertensive	1990	26,	304
medifoxamine fumarate	antidepressant	1986	22,	323
mefloquine HCl	antimalarial	1985	21,	329
meglutol	hypolipidemic	1983	19,	321
melinamide	hypocholesterolemic	1984	20,	320
mepixanox	analeptic	1984	20,	320
meptazinol HCl	analgesic	1983	19,	321
metaclazepam	anxiolytic	1987	23,	338
metapramine	antidepressant	1984	20,	320
mexazolam	anxiolytic	1984	20,	321
mifepristone	abortifacient	1988	24,	306
milrinone	cardiostimulant	1989	25,	316
miokamycin	antibiotic	1985	21,	329
misoprostol	antiulcer	1985	21,	329
mitoxantrone HCl	antineoplastic	1984	20,	321
mizoribine	immunosuppressant	1984	20,	321
moclobemide	antidepressant	1990	26,	305
mometasone furoate	topical antiinflammatory	1987	23,	338
moricizine hydrochloride	antiarrhythmic	1990	26,	305
moxonidine	antihypertensive	1991	27,	330
mupirocin	topical antibiotic	1985	21,	330
muromonab-CD3	immunosuppressant	1986	22,	323
muzolimine	diuretic	1983	19,	321
nabumetone	antiinflammatory	1985	21,	330
nafamostat mesylate	protease inhibitor	1986	22,	323
nafarelin acetate	hormone	1990	26,	306
naftifine HCl	antifungal	1984	20,	321
naltrexone HCl	narcotic antagonist	1984	20,	322
nedocromil sodium	antiallergic	1986	22,	324
nemonapride	neuroleptic	1991	27,	331
nicorandil	coronary vasodilator	1984	20,	322
nilutamide	antineoplastic	1987	23,	338
nilvadipine	antihypertensive	1989	25,	316
nimesulide	antiinflammatory	1985	21,	330
nimodipine	cerebral vasodilator	1985	21,	330
nipradilol	antihypertensive	1988	24,	307
nisoldipine	antihypertensive	1990	26,	306
nitrefazole	alcohol deterrent	1983	19,	322
nitrendipine	hypertensive	1985	21,	331

GENERIC NAME	INDICATION	YEAR INTRODUCED	ARMC VOL., PAGE	
nizatidine	antiulcer	1987	23,	339
nizofenzone fumarate	nootropic	1988	24,	307
nomegestrol acetate	progestogen	1986	22,	324
norfloxacin	antibacterial	1983	19,	322
norgestimate	progestogen	1986	22,	324
octreotide	antisecretory	1988	24,	307
ofloxacin	antibacterial	1985	21,	331
omeprazole	antiulcer	1988	24,	308
ondansetron hydrochloride	antiemetic	1990	26,	306
ornoprostil	antiulcer	1987	23,	339
osalazine sodium	intestinal antinflamm.	1986	22,	324
oxaprozin	antiinflammatory	1983	19,	322
oxcarbazepine	anticonvulsant	1990	26,	307
oxiconazole nitrate	antifungal	1983	19,	322
oxiracetam	nootropic	1987	23,	339
oxitropium bromide	bronchodilator	1983	19,	323
ozagrel sodium	antithrombotic	1988	24,	308
paroxetine	antidepressant	1991	27,	331
pefloxacin mesylate	antibacterial	1985	21,	331
pegademase bovine	immunostimulant	1990	26,	307
pemirolast potassium	antiasthmatic	1991	27,	331
pergolide mesylate	antiparkinsonian	1988	24,	308
perindopril	antihypertensive	1988	24,	309
picotamide	antithrombotic	1987	23,	340
piketoprofen	topical antiinflammatory	1984	20,	322
pilsicainide hydrochloride	antiarrhythmic	1991	27,	332
pimaprofen	topical antiinflammatory	1984	20,	322
pinacidil	antihypertensive	1987	23,	340
pirarubicin	antineoplastic	1988	24,	309
piroxicam cinnamate	antiinflammatory	1988	24,	309
plaunotol	antiulcer	1987	23,	340
pravastatin	antilipidemic	1989	25,	316
prednicarbate	topical antiinflammatory	1986	22,	325
progabide	anticonvulsant	1985	21,	331
promegestrone	progestogen	1983	19,	323
propacetamol HCl	analgesic	1986	22,	325
propentofylline propionate	cerebral vasodilator	1988	24,	310
propofol	anesthetic	1986	22,	325
quazepam	hypnotic	1985	21,	332
quinapril	antihypertensive	1989	25,	317
quinfamide	amebicide	1984	20,	322
ramipril	antihypertensive	1989	25,	317
ranimustine	antineoplastic	1987	23,	341
rebamipide	antiulcer	1990	26,	308
remoxipride hydrochloride	antipsychotic	1990	26,	308
repirinast	antiallergic	1987	23,	341
rifapentine	antibacterial	1988	24,	310
rifaximin	antibiotic	1985	21,	332
rifaximin	antibiotic	1987	23,	341
rilmazafone	hypnotic	1989	25,	317
rilmenidine	antihypertensive	1988	24,	310
rimantadine HCl	antiviral	1987	23,	342
rokitamycin	antibiotic	1986	22,	325
romurtide	immunostimulant	1991	27,	332
ronafibrate	hypolipidemic	1986	22,	326
rosaprostol	antiulcer	1985	21,	332

GENERIC NAME	INDICATION	YEAR INTRODUCED	ARMC VOL., PAGE	
roxatidine acetate HCl	antiulcer	1986	22,	326
roxithromycin	antiulcer	1987	23,	342
RV-11	antibiotic	1989	25,	318
salmeterol hydroxynaphthoate	bronchodilator	1990	26,	308
sargramostim	immunostimulant	1991	27,	332
schizophyllan	immunostimulant	1985	22,	326
setastine HCl	antihistamine	1987	23,	342
setiptiline	antidepressant	1989	25,	318
setraline hydrochloride	antidepressant	1990	26,	309
sevoflurane	anesthetic	1990	26,	309
simvastatin	hypocholesterolemic	1988	24,	311
sodium cellulose PO4	hypocalciuric	1983	19,	323
sofalcone	antiulcer	1984	20,	323
somatropin	hormone	1987	23,	343
spizofurone	antiulcer	1987	23,	343
succimer	chelator	1991	27,	333
sufentanil	analgesic	1983	19,	323
sulbactam sodium	B-lactamase inhibitor	1986	22,	326
sulconizole nitrate	topical antifungal	1985	21,	332
sultamycillin tosylate	antibiotic	1987	23,	343
sumatriptan succinate	antimigraine	1991	27,	333
suprofen	analgesic	1983	19,	324
surfactant TA	respiratory surfactant	1987	23,	344
tazanolast	antiallergic	1990	26,	309
teicoplanin	antibacterial	1988	24,	311
temafloxacin hydrochloride	antibacterial	1991	27,	334
temocillin disodium	antibiotic	1984	20,	323
tenoxicam	antiinflammatory	1987	23,	344
teprenone	antiulcer	1984	20,	323
terazosin HCl	antihypertensive	1984	20,	323
terbinafine hydrochloride	antifungal	1991	27,	334
terconazole	antifungal	1983	19,	324
tertatolol HCl	antihypertensive	1987	23,	344
thymopentin	immunomodulator	1985	21,	333
tiamenidine HCl	antihypertensive	1988	24,	311
tianeptine sodium	antidepressant	1983	19,	324
tibolone	anabolic	1988	24,	312
timiperone	neuroleptic	1984	20,	323
tinazoline	nasal decongestant	1988	24,	312
tioconazole	antifungal	1983	19,	324
tiopronin	urolithiasis	1989	25,	318
tiquizium bromide	antispasmodic	1984	20,	324
tiracizine hydrochloride	antiarrhythmic	1990	26,	310
tiropramide HCl	antispasmodic	1983	19,	324
tizanidine	muscle relaxant	1984	20,	324
toloxatone	antidepressant	1984	20,	324
tolrestat	antidiabetic	1989	25,	319
toremifene	antineoplastic	1989	25,	319
tosufloxacin tosylate	antibacterial	1990	26,	310
trientine HCl	chelator	1986	22,	327
trimazosin HCl	antihypertensive	1985	21,	333
troxipide	antiulcer	1986	22,	327
ubenimex	immunostimulant	1987	23,	345
vesnarinone	cardiostimulant	1990	26,	310
vigabatrin	anticonvulsant	1989	25,	319

GENERIC NAME	INDICATION	YEAR INTRODUCED	ARMC VOL., PAGE	
vinorelbine	antineoplastic	1989	25,	320
xamoterol fumarate	cardiotonic	1988	24,	312
zidovudine	antiviral	1987	23,	345
zolpidem hemitartrate	hypnotic	1988	24,	313
zonisamide	anticonvulsant	1989	25,	320
zopiclone	hypnotic	1986	22,	327
zuclopenthixol acetate	antipsychotic	1987	23,	345

GENERIC NAME	INDICATION	YEAR INTRODUCED	ARMC VOL., PAGE	
gemeprost	ABORTIFACIENT	1983	19,	319
mifepristone		1988	24,	306
nitrefazole	ALCOHOL DETERRENT	1983	19,	322
quinfamide	AMEBICIDE	1984	20,	322
tibolone	ANABOLIC	1988	24,	312
mepixanox	ANALEPTIC	1984	20,	320
alfentanil HCl	ANALGESIC	1983	19,	314
alminoprofen		1983	19,	314
dezocine		1991	27,	326
emorfazone		1984	20,	317
eptazocine HBr		1987	23,	334
flupirtine maleate		1985	21,	328
fosfosal		1984	20,	319
ketorolac tromethamine		1990	26,	304
meptazinol HCl		1983	19,	321
propacetamol HCl		1986	22,	325
sufentanil		1983	19,	323
suprofen		1983	19,	324
propofol	ANESTHETIC	1986	22,	325
sevoflurane		1990	26,	309
azelaic acid	ANTIACNE	1989	25,	310
nedocromil sodium	ANTIALLERGIC	1986	22,	324
repirinast		1987	23,	341
tazanolast		1990	26,	309
gallopamil HCl	ANTIANGINAL	1983	19,	319
cibenzoline	ANTIARRHYTHMIC	1985	21,	325
encainide HCl		1987	23,	333
esmolol HCl		1987	23,	334
moricizine hydrochloride		1990	26,	305
pilsicainide hydrochloride		1991	27,	332
tiracizine hydrochloride		1990	26,	310
amlexanox	ANTIASTHMATIC	1987	23,	327
ibudilast		1989	25,	313
pemirolast potassium		1991	27,	331
ciprofloxacin	ANTIBACTERIAL	1986	22,	318
enoxacin		1986	22,	320
norfloxacin		1983	19,	322
ofloxacin		1985	21,	331
pefloxacin mesylate		1985	21,	331
rifapentine		1988	24,	310
teicoplanin		1988	24,	311
temafloxacin hydrochloride		1991	27,	334
tosufloxacin tosylate		1990	26,	310

GENERIC NAME	INDICATION	YEAR INTRODUCED	ARMC VOL., PAGE	
arbekacin	ANTIBIOTIC	1990	26,	298
aspoxicillin		1987	23,	328
astromycin sulfate		1985	21,	324
azithromycin		1988	24,	298
aztreonam		1984	20,	315
carboplatin		1986	22,	318
carumonam		1988	24,	298
cefbuperazone sodium		1985	21,	325
cefdinir		1991	27,	323
cefixime		1987	23,	329
cefmenoxime HCl		1983	19,	316
cefminox sodium		1987	23,	330
cefodizime sodium		1990	26,	300
cefonicid sodium		1984	20,	316
ceforanide		1984	20,	317
cefotetan disodium		1984	20,	317
cefotiam hexetil hydrochloride		1991	27,	324
cefpimizole		1987	23,	330
cefpiramide sodium		1985	21,	325
cefpodoxime proxetil		1989	25,	310
ceftazidime		1983	19,	316
cefteram pivoxil		1987	23,	330
cefuroxime axetil		1987	23,	331
cefuzonam sodium		1987	23,	331
clarithromycin		1990	26,	302
erythromycin acistrate		1988	24,	301
flomoxef sodium		1988	24,	302
imipenem/cilastatin		1985	21,	328
isepamicin		1988	24,	305
lenampicillin HCl		1987	23,	336
lomefloxacin		1989	25,	315
miokamycin		1985	21,	329
rifaximin		1985	21,	332
rifaximin		1987	23,	341
rokitamycin		1986	22,	325
RV-11		1989	25,	318
sultamycillin tosylate		1987	23,	343
temocillin disodium		1984	20,	323
mupirocin	ANTIBIOTIC, TOPICAL	1985	21,	330
chenodiol	ANTICHOLELITHOGENIC	1983	19,	317
lamotrigine	ANTICONVULSANT	1990	26,	304
oxcarbazepine		1990	26,	307
progabide		1985	21,	331
vigabatrin		1989	25,	319
zonisamide		1989	25,	320
bupropion HCl	ANTIDEPRESSANT	1989	25,	310
citalopram		1989	25,	311
fluoxetine HCl		1986	22,	320
fluvoxamine maleate		1983	19,	319
indalpine		1983	19,	320
medifoxamine fumarate		1986	22,	323

GENERIC NAME	INDICATION	YEAR INTRODUCED	ARMC VOL., PAGE	
metapramine		1984	20,	320
moclobemide		1990	26,	305
paroxetine		1991	27,	331
setiptiline		1989	25,	318
sertraline hydrochloride		1990	26,	309
tianeptine sodium		1983	19,	324
toloxatone		1984	20,	324
acarbose	ANTIDIABETIC	1990	26,	297
tolrestat		1989	25,	319
granisetron hydrochloride	ANTIEMETIC	1991	27,	329
ondansetron hydrochloride		1990	26,	306
centchroman	ANTIESTROGEN	1991	27,	324
fenticonazole nitrate	ANTIFUNGAL	1987	23,	334
fluconazole		1988	24,	303
itraconazole		1988	24,	305
naftifine HCl		1984	20,	321
oxiconazole nitrate		1983	19,	322
terbinafine hydrochloride		1991	27,	334
terconazole		1983	19,	324
tioconazole		1983	19,	324
amorolfine hydrochloride	ANTIFUNGAL, TOPICAL	1991	27,	322
butoconazole		1986	22,	318
cloconazole HCl		1986	22,	318
sulconizole nitrate		1985	21,	332
apraclonidine HCl	ANTIGLAUCOMA	1988	24,	297
befunolol HCl		1983	19,	315
dapiprazole HCl		1987	23,	332
levobunolol HCl		1985	21,	328
acrivastine	ANTIHISTAMINE	1988	24,	295
astemizole		1983	19,	314
azelastine HCl		1986	22,	316
ebastine		1990	26,	302
cetirizine HCl		1987	23,	331
levocabastine hydrochloride		1991	27,	330
loratadine		1988	24,	306
setastine HCl		1987	23,	342
alacepril	ANTIHYPERTENSIVE	1988	24,	296
alfuzosin HCl		1988	24,	296
amlodipine besylate		1990	26,	298
amosulalol		1988	24,	297
arotinolol HCl		1986	22,	316
benazepril hydrochloride		1990	26,	299
benidipine hydrochloride		1991	27,	322
betaxolol HCl		1983	19,	315
bevantolol HCl		1987	23,	328
bisoprolol fumarate		1986	22,	317
bopindolol		1985	21,	324

GENERIC NAME	INDICATION	YEAR INTRODUCED	ARMC VOL., PAGE	
budralazine		1983	19,	315
bunazosin HCl		1985	21,	324
carvedilol		1991	27,	323
celiprolol HCl		1983	19,	317
cicletanine		1988	24,	299
cilazapril		1990	26,	301
delapril		1989	25,	311
dilevalol		1989	25,	311
doxazosin mesylate		1988	24,	300
enalapril maleate		1984	20,	317
enalaprilat		1987	23,	332
felodipine		1988	24,	302
fosinopril sodium		1991	27,	328
guanadrel sulfate		1983	19,	319
isradipine		1989	25,	315
ketanserin		1985	21,	328
lacidipine		1991	27,	330
lisinopril		1987	23,	337
manidipine hydrochloride		1990	26,	304
moxonidine		1991	27,	330
nilvadipine		1989	25,	316
nipradilol		1988	24,	307
nisoldipine		1990	26,	306
perindopril		1988	24,	309
pinacidil		1987	23,	340
quinapril		1989	25,	317
ramipril		1989	25,	317
rilmenidine		1988	24,	310
terazosin HCl		1984	20,	323
tertatolol HCl		1987	23,	344
tiamenidine HCl		1988	24,	311
trimazosin HCl		1985	21,	333
AF-2259	ANTIINFLAMMATORY	1987	23,	325
amfenac sodium		1986	22,	315
deflazacort		1986	22,	319
droxicam		1990	26,	302
etodolac		1985	21,	327
flunoxaprofen		1987	23,	335
fluticasone propionate		1990	26,	303
interferon, gamma		1989	25,	314
isofezolac		1984	20,	319
isoxicam		1983	19,	320
lobenzarit sodium		1986	22,	322
loxoprofen sodium		1986	22,	322
nabumetone		1985	21,	330
nimesulide		1985	21,	330
oxaprozin		1983	19,	322
piroxicam cinnamate		1988	24,	309
tenoxicam		1987	23,	344
fisalamine	ANTIINFLAMMATORY,	1984	20,	318
osalazine sodium	INTESTINAL	1986	22,	324

GENERIC NAME	INDICATION	YEAR INTRODUCED	ARMC VOL., PAGE	
alclometasone dipropionate	ANTIINFLAMMATORY,	1985	21,	323
aminoprofen	TOPICAL	1990	26,	298
butyl flufenamate		1983	19,	316
felbinac		1986	22,	320
halobetasol propionate		1991	27,	329
halometasone		1983	19,	320
hydrocortisone aceponate		1988	24,	304
hydrocortisone butyrate propionate		1983	19,	320
mometasone furoate		1987	23,	338
piketoprofen		1984	20,	322
pimaprofen		1984	20,	322
prednicarbate		1986	22,	325
pravastatin	ANTILIPIDEMIC	1989	25,	316
artemisinin	ANTIMALARIAL	1987	23,	327
halofantrine		1988	24,	304
mefloquine HCl		1985	21,	329
alpiropride	ANTIMIGRAINE	1988	24,	296
sumatriptan succinate		1991	27,	333
dronabinol	ANTINAUSEANT	1986	22,	319
amsacrine	ANTINEOPLASTIC	1987	23,	327
bisantrene hydrochloride		1990	26,	300
camostat mesylate		1985	21,	325
doxifluridine		1987	23,	332
enocitabine		1983	19,	318
epirubicin HCl		1984	20,	318
fludarabine phosphate		1991	27,	327
flutamide		1983	19,	318
fotemustine		1989	25,	313
idarubicin hydrochloride		1990	26,	303
interleukin-2		1989	25,	314
lonidamine		1987	23,	337
mitoxantrone HCl		1984	20,	321
nilutamide		1987	23,	338
pirarubicin		1988	24,	309
ranimustine		1987	23,	341
toremifene		1989	25,	319
vinorelbine		1989	25,	320
ivermectin	ANTIPARASITIC	1987	23,	336
droxidopa	ANTIPARKINSONIAN	1989	25,	312
pergolide mesylate		1988	24,	308
lidamidine HCl	ANTIPERISTALTIC	1984	20,	320
gestrinone	ANTIPROGESTOGEN	1986	22,	321
acitretin	ANTIPSORIATIC	1989	25,	309
calcipotriol		1991	27,	323

GENERIC NAME	INDICATION	YEAR INTRODUCED	ARMC VOL., PAGE	
amisulpride	ANTIPSYCHOTIC	1986	22,	316
remoxipride hydrochloride		1990	26,	308
zuclopenthixol acetate		1987	23,	345
diacerein	ANTIRHEUMATIC	1985	21,	326
octreotide	ANTISECRETORY	1988	24,	307
adamantanium bromide	ANTISEPTIC	1984	20,	315
cimetropium bromide	ANTISPASMODIC	1985	21,	326
tiquizium bromide		1984	20,	324
tiropramide HCl		1983	19,	324
argatroban	ANTITHROMBOTIC	1990	26,	299
defibrotide		1986	22,	319
cilostazol		1988	24,	299
cloricromen		1991	27,	325
enoxaparin		1987	23,	333
ethyl icosapentate		1990	26,	303
ozagrel sodium		1988	24,	308
indobufen		1984	20,	319
picotamide		1987	23,	340
limaprost		1988	24,	306
flutropium bromide	ANTITUSSIVE	1988	24,	303
levodropropizine		1988	24,	305
benexate HCl	ANTIULCER	1987	23,	328
enprostil		1985	21,	327
famotidine		1985	21,	327
irsogladine		1989	25,	315
misoprostol		1985	21,	329
nizatidine		1987	23,	339
omeprazole		1988	24,	308
ornoprostil		1987	23,	339
plaunotol		1987	23,	340
rebamipide		1990	26,	308
rosaprostol		1985	21,	332
roxatidine acetate HCl		1986	22,	326
roxithromycin		1987	23,	342
sofalcone		1984	20,	323
spizofurone		1987	23,	343
teprenone		1984	20,	323
troxipide		1986	22,	327
didanosine	ANTIVIRAL	1991	27,	326
foscarnet sodium		1989	25,	313
ganciclovir		1988	24,	303
rimantadine HCl		1987	23,	342
zidovudine		1987	23,	345